International Conference on Civil Engineering and Building Services (CIBv 2019)

AA007760

IOP Conference Series: Materials Science and Engineering
Volume 789

Brasov, Romania
1 - 2 November 2019

Editors:

Ioan Tuns
Florin-Lucian Tamas
Radu Muntean
Teofil-Florin Galatanu
Dorin Radu

ISBN: 978-1-7138-3504-2
ISSN: 1757-8981

Printed from e-media with permission by:

Curran Associates, Inc.
57 Morehouse Lane
Red Hook, NY 12571

Some format issues inherent in the e-media version may also appear in this print version.

Institute of Physics
Dirac House, Temple Back
Bristol BS1 6BE UK

Phone: 44 1 17 929 7481
Fax: 44 1 17 920 0979

techtracking@iop.org

Additional copies of this publication are available from:

Curran Associates, Inc.
57 Morehouse Lane
Red Hook, NY 12571 USA
Phone: 845-758-0400
Fax: 845-758-2633
Email: curran@proceedings.com
Web: www.proceedings.com

TABLE OF CONTENTS

IOP Conf. Series: Materials Science and Engineering **789** (2020) 011001 doi:10.1088/1757-899X/789/1/011001

International Conference CIBv2019
Civil Engineering and Building Services

1st - 2nd November 2019, Brasov, Romania

Selected, peer reviewed papers from the International Conference Civil Engineering and Building Services CIBv2019

Edited by:

Ioan TUNS
Transilvania University of Brasov, Romania

Florin-Lucian TĂMAȘ
Transilvania University of Brasov, Romania

Radu MUNTEAN
Transilvania University of Brasov, Romania

Teofil-Florin GĂLĂȚANU
Transilvania University of Brasov, Romania

Dorin RADU
Transilvania University of Brasov, Romania

IOP Conf. Series: Materials Science and Engineering **789** (2020) 011001 doi:10.1088/1757-899X/789/1/011001

FOREWORD

The International Scientific Conference CIBv represents one of the most important and traditionally scientific events of Civil Engineering and Building Services in Romania, enjoying increasingly higher visibility at the international level.

The aim of the CIBv conference is to provide a platform for researchers, engineers, academicians as well as professionals from all over the world to present their research results and development activities. It also provides opportunities for the participants to exchange new ideas and application experiences, to establish business or research relations and to find other partners and even friends for future collaboration.

The Conference has been organized each year at Transilvania University of Brașov, since 2000. In November 2019, Faculty of Civil Engineering was honoured to organize, in partnership with Shenyang Jianzhu University from China and to host the 14th edition of "International Scientific Conference – Civil Engineering and Building Services CIBv2019" in Brasov. More than 100 participants from China and Romania, as well as from Serbia, Croatia, Slovakia, Japan, Germany, Russia, Finland and Norway submitted over 80 papers to the scientific board which were presented during the conference. After a rigorous peer-review process (which included for each article at least two reviewers), 71 papers were finally accepted for publication in the CIBv2019 proceedings volume.

We would like to address special thanks to all those who were involved in the success achieved by the CIBv2019 International Conference, as well as to our distinguished publisher.

Prof. Eng. Ioan TUNS, PhD

Dean of Faculty of Civil Engineering,
Transilvania University of Brasov, Romania

IOP Conf. Series: Materials Science and Engineering **789** (2020) 011001 doi:10.1088/1757-899X/789/1/011001

International Conference CIBv2019
Civil Engineering and Building Services

Faculty of Civil Engineering, Transilvania University of Brasov, Romania in collaboration with Shenyang Jianzhu University, China

HONORARY COMMITTEE:

Ioan Vasile ABRUDAN, Rector of Transilvania University of Brasov

Yupeng LI, Vice rector of Shenyang Jianzhu University

Dan STEMATIU, President of Civil Engineering and Urbanism section of the Romanian Academy of Technical Sciences, Professor of Technical University of Civil Engineering of Bucharest

ORGANIZING COMMITTEE

Chairman:

Ioan TUNS, Dean of Civil Engineering Faculty of Brasov

Members:

Ioan BOIAN, Transilvania University of Brasov, Romania
Lucia BOERIU, Transilvania University of Brasov, Romania
Sorin BOLOCAN, Transilvania University of Brasov, Romania
Marius Florin BOTIŞ, Transilvania University of Brasov, Romania
Christiana CAZACU, Transilvania University of Brasov, Romania
Cristina CHIŢONU, Transilvania University of Brasov, Romania
Vasile CIOFOAIA, Transilvania University of Brasov, Romania
Lucian CÎRSTOLOVEAN, Transilvania University of Brasov, Romania
Mircea CONŢIU, Transilvania University of Brasov, Romania
Ovidiu DEACONU, Transilvania University of Brasov, Romania
Mihai DOBROVOLSCHI, Transilvania University of Brasov, Romania
Adam DOŞA, Transilvania University of Brasov, Romania
George DRAGOMIR, Transilvania University of Brasov, Romania
Liting FAN, Shenyang Jianzhu University, China
Mariana FRATU, Transilvania University of Brasov, Romania
Teofil-Florin GĂLĂŢANU, Transilvania University of Brasov, Romania
Mircea HORNEŢ, Transilvania University of Brasov, Romania
Nicolae IORDAN, Transilvania University of Brasov, Romania
Marius MĂNTULESCU, Transilvania University of Brasov, Romania
Paraschiva MIZGAN, Transilvania University of Brasov, Romania
Gavrilă MUNTEAN, Transilvania University of Brasov, Romania
Radu MUNTEAN, Transilvania University of Brasov, Romania
Gabriel NĂSTASE, Transilvania University of Brasov, Romania
Costel PLEŞCAN, Transilvania University of Brasov, Romania
Dorin RADU, Transilvania University of Brasov, Romania
Yuanyuan SUN, Confucius Institute from Brasov
Daniel TAUS, Transilvania University of Brasov, Romania
Florin-Lucian TĂMAŞ, Transilvania University of Brasov, Romania
Xing WANG, Shenyang Jianzhu University, China

IOP Conf. Series: Materials Science and Engineering **789** (2020) 011001 doi:10.1088/1757-899X/789/1/011001

INTERNATIONAL ACADEMIC SCIENTIFIC COMMITTEE

Peer review statement

All papers published in this volume of *IOP Conference Series: Materials Science and Engineering* have been peer reviewed through processes administered by the proceedings Editors. Reviews were conducted by expert referees to the professional and scientific standards expected of a proceedings journal published by IOP Publishing.

IOP Conf. Series: Materials Science and Engineering **789** (2020) 012001 doi:10.1088/1757-899X/789/1/012001

Green roofs – modern solutions for greening buildings

I R Baciu[1*], N Taranu[1], D N Isopescu[1], M L Lupu[1], T C Dragan[2] and S G Maxineasa[1]

[1] "Gheorghe Asachi" Technical University of Iasi, Faculty of Civil Engineering and Building Services, Department of Civil and Industrial Engineering, Iasi, Romania
[2] SC Next Studio SRL, Iasi, Romania

[*] ioanaroxana.baciu@yahoo.com

Abstract. In a society intensively urbanized, nature and green spaces have an extremely important and determined role. Solutions that are integrating different modern techniques related to a new or an existing building, with a tremendously positive effect over the natural building and horticultural sectors will represent an important factor in reducing the negative effect over the environment on a global scale. These smart solutions could extend the garden effect from the ground to the elevated levels of the building. Even if at the present moment there are different classic solutions to realize a green roof, in the last decades, have been developed different new and modern techniques for the greening environment. Taking into account all the above mentioned, the aim of the paper is to present the modern green roof systems considering all the changes that have occurred over time, and also to analyze a series of examples of how this concept has been adopted in different parts of the world and for varies structural systems. The authors consider that the study is very important in order to completely understand the characteristics of these systems and their main components, in order to properly design and implement them according to the present environmental needs. It is believed that the knowledge of this concept and its evolution and tendencies will help solve a series of ecological and social problems.

1. Introduction

In the last decades, all over the world, it was observed a tendency of living in cities, densifying the urban population, and expanding the urban areas quickly at the expense of the green spaces [1]. The process of urbanization in its development is connected with a series of negative environmental issues, such as urban heat island effects, increased air pollution, stormwater runoff, sound pollution, and last but not least esthetical urban problems, caused by reduced green areas and chaotic building mode [2].

These unfavorable impacts are preceded by different effects, like health problems, increased psychic discomfort and anxious, social and also ecological problems, and high demand for cooling and heating of buildings, which lead to increased energy consumption [3].

Taking into account the problems previously listed, there is a tendency of searching and created a series of sustainable solutions for the new and existing buildings, to diminish the negative effects of this continuous process of urbanization [4].

The main concepts integrated into urban areas were green roof and living greenery, both important parts of the greening infrastructure used and developed in the last years [5]. In this way, the built sector became the support of these systems and the principal pilot in the actual current in terms of sustainability.

Their primary goal is not only the amelioration of the urban enlargement but the re-location of the urban areas in zones like urban forestry or dedicated green spaces.

In this context, the smart solutions of green roofs have come in the attention of society and became one of the most popular inventions used for their technical, ecological, economic, and aesthetical benefits. The expansion of the construction of the green roof has experienced an immediate spreading, now becoming part of the residential, public, commercial and government buildings. With all the positive characteristics and remarkable benefits, the researches considered that the green roof systems are the most innovative solutions of greening the buildings, in our days [6]. Researchers in Japan [7] have studied and found that the green roofs helped in the reduction of the heat flux on the order of 50%, meanwhile in Ottawa [8], the percentage is much encouraging, around 95%. Studies in Madrid [9] showed a reduction of the cooling load on an eight-story residential building with 6% during the summertime.

Taking into account all the above mentioned, the present paper has the principal aim to present the modern green roof systems considering all the changes that have occurred over time, with greater attention granted to the new technologies developed based on the oldest systems and the improvement of these systems and all its components. Another part of the paper is concentrated on analyzing a series of examples of how this concept has been adopted in different parts of the world and for various structural systems. This study has an important role in order to completely understand the characteristics of these systems and the changes appeared in the main components, in order to properly design and implement them according to the present environmental needs.

2. The modern green roof technologies

In the literature, there is a knowledge gap related to the past evolution of the green roof systems considering the materials, the technology, and the design, fact which ends with an insure connection with modern development [10].

It is well known that the roots of the green roof systems are in Antiquity, based on the development of the vernacular green roof which represents the way people tried to resists harsh environments with extreme climate [11]. On the other hand, sod-roof innovation technology continued to be honed and used from ancient to modern eras. Even if the beginning of these systems was demonstrated to be in the Ancient period, and their main benefit was to protect the people of the extreme climatic conditions, in the Middle Ages, they became an elegant symbol of the aristocratic families and religious bodies, being used as terrace garden with a particular landscape design [12]. This period was the intermediary period between the elaborated ground-level garden from the Renaissance to the artificial structures.

In the 1850s, the modern reinforced concrete technology was an important step in the evolution of the green roofs, based on the multi-story buildings with flat roofs which became the support of these systems. In 1867, the Paris World expo represented the ideal framework for promoting the technique of installing gardens on flat roofs [13].

After this incredible start, Germany integrates the spontaneous vegetation on a flat roof which is the demonstration of nature's capacity of fixing the green roof systems on the houses [14]. This was the decisive moment for the scientific interest and which inspired and insight to make a big step from the spontaneous green roofs incidental established to the analyzed and researched green roofs systems on modern building with flat roofs.

In conclusion, a revived ancient system, green roof system was at first used for the restoration of the environment and for the protection of the roof membranes, once with the passage of time the scientific interest developed in the direction of finding different flexibility benefits, such as versatility and simplicity in design, low weight of the roof components, improved capacity of retention of the stormwater, careful selection of the plants species and also of the growing medium.

In this context, one of the best solutions developed in this direction is represented by the modular green roof system, which can be defined as modern efficient construction technology [5]. This greened innovation presents some advantages and defined characteristics, such as simple installation, simple and modern design, accessibility for the maintenance works and also repairing interventions using the

IOP Conf. Series: Materials Science and Engineering **789** (2020) 012001 doi:10.1088/1757-899X/789/1/012001

clicking features of modules. The modular green roof represents an advanced technology that includes innovative solutions and optional elements which makes it suitable for the extensive and intensive green roof [15].

By definition, a green roof module represents a technology for covering with a layer composed of vegetation a growing medium, a roof surface [16]. On the other hand, the modular green roof (figure 1) is the system that includes these modules, such as trays, blocks, containers, boxes and similar. As a result, these modules are the structural support for the growing medium and the plants, and also present an advantage regarding execution time, maintenance, and money. Green roof modules are a self-contained portable roof greening product that presents the same characteristics as a roof paver.

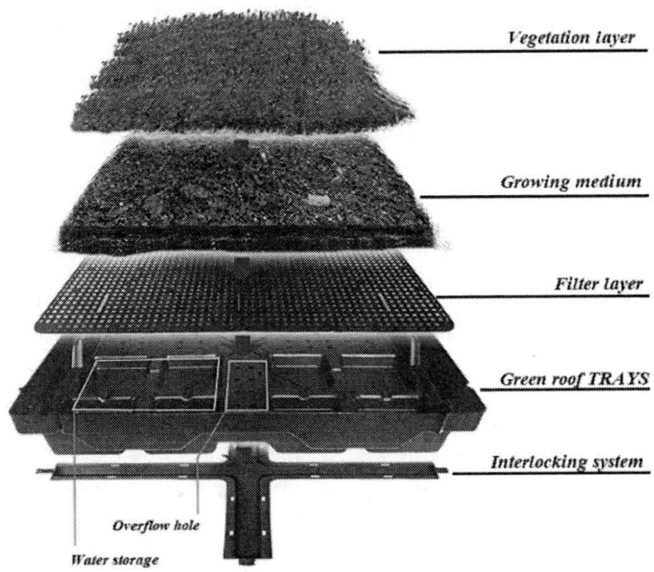

Figure 1. The components of a modular green roof (source: https://sedumgreenroof.co.uk).

In the specialty literature, modular green roof technology is presented as an energy-saving alternative and more attractive than a conventional green roof system. In order to establish the impact of these structures, it will be presented a series of advantages [17]:

- simplicity and flexibility in design and execution;
- lightweight structural components which are prefabricated and assembled on-site;
- accessibility reflected in the maintenance of the green roof;
- the possibility of integration of different devices or accessories, like solar panels, automatic irrigation system, lights, and decorative elements;
- the usage of recycling and ecological materials for the modules.

In conclusion, it can be advocated that the modular green roofs offer ecological, technological, aesthetical and well-being benefits, thus they represent a sustainable practice, by helping to fulfill the ecological dimension in the construction field [18]. The main characteristic making into account in promoting these systems is the economic benefit, sustained by the low execution and maintenance costs.

Another trend regarding green roof systems remarked in recent years, is the Hybrid Photovoltaic (PV) – green roof. This system presents an interest in matters of new alternative applications and findings, due to its combined characteristics both the green roof benefits and improvement of PV electrical yield [19]. The technology of these systems depends on the temperature of the PV modules and the surrounding ambient air temperature [20], meaning that cooler temperature improves the PV performance.

The success of this hybrid green roof is sustained by the evapotranspiration potential of a green roof which cools the surface and the temperature air, a fact that improves the performance of PV cells [21]. On the other hand, the PV panels help the impact of the green roof by reducing the sun's exposure over the surface and by improving the evaporation rates. In a study realized in Spain, the results showed the PV – green roof efficiency compared with a PV – gravel roof, and the percentage was 1.29% in the case of PV – gazania green roof and 3.33% in the case of PV – sedum green roof [22]. Another conclusion of these experiments pointed out the importance of the plants' selection according to their ability to adapt to different environmental conditions.

The case study conducted in Hong Kong highlighted the increased percentage of electricity produced by the PV – green roof compared with the stand-alone PV cell [23]. In this context, it can be concluded that PV technology even if it is a mature one and widely used in different parts of the world, can be improved by a green roof system that acts as an important element in a sustainable direction.

Research in the field has shown certain concerns about irrigation source of the green roofs. A solution in this direction is presented by the greywater [24]. It is well known that the greywater totaling (65-90)% of the domestic wastewater production, coming from laundry, kitchen activities, and bathroom [25]. The utilization of greywater presents other benefits like covering the water requirement for green roof maintenance and the possibility of choosing a wider variety of plants, not only succulents.

Several studies showed that greywater collected from the kitchen usually contains nutrients fact that minimizes the fertilization requirement of the plants' layer. On another hand, there are concerns in terms of quantity and quality of these nutrients, which can be guided by the growth substrate according to some experiments [26]. Therefore, a green roof can become a filter for the greywater, taking into account the selected plants and their needs.

Taking into consideration all above mentioned, it can be concluded that green roof system represents a topic of major interest, based on the studies carried out in recent years, in terms of technology, structure, component elements, benefits and last but not least the best management practice to combat urbanization.

3. Notable modern green roof exemplars

In order to understand and to exemplify the way green roof systems developed over time, it is of overwhelming importance to analyze a series of examples of the notable modern green roof, adopted in different parts of the world, taking into account the various structural systems. The analysis was divided into three categories thus being evidenced by some of the significant modern green roofs from Europe, America and Asia.

From European countries, it is well known that France is remarkable for its elaborate and sophisticated landscape design. In the post-industrial period, many old infrastructures and installation from all France have been abandoned and left unused for a long period of time. In this context, in 1859, an obsolete railway viaduct stretched 4.7 km and 9 m above the street, was designed as a pleasant and impressive green roof system, presented in figure 3. It is known as Promenade Plantée (French for the planted promenade) or Coulée Verte (French for the green course). The entire design is composed of a diverse assemblage of plants, passing from the herbs to climbers, trees and shrubs, different species of flowers and decorative herbs, vegetation completed by a fountain, ponds and diverse water elements with unique aquatic creations. The well-vegetated surface has provided different habitats for spontaneous plants, which successfully completes the whole image of the place. This project has an important impact in the old neighborhood by adding 18500 m^2 of offices and 7000 m^2 of shops. Another advantage of this project is that it creates noise barriers and protect the surroundings from air pollutants.

Figure 2. The Derry and Toms Department roof garden, High Street Kensington, London (source [10]).

Figure 3. The Promanede Plantée – railway viaduct green roof, France (source [10]).

Passing to another country with history in the landscape, the United Kingdom is one of the countries that have a history of pioneering green-roof designs in some cities. A formidable example is the pace-setting green roof from the heart of London, at High Street Kensington, installed in 1938. With an amazing surface around 6000 m^2, this structure is part of the Derry and Toms Department Store [27]. On the top of a retail structure, was built for the first time a sizeable roof garden (figure 2). Due to its attractive and varied landscape design, it remained Europe's most elegant and elaborate roof garden. After different modifications, the green roof gains a new picture made up of three distinct genres gardens, as follows Morish Spanish, English Woodland, and Tudor Walled. With an elaborate and refreshed design completed with diversities of hard and soft landscape features, with fine crossings from formal-geometric, designed by stone arches, cloister walk, towers, clusters, flower beds, tree rows to informal-naturistic, pointed out by the woodlands, herbaceous covers, pools, and waterfalls. The structural elements of the entire green roof system [28, 29] were the following:

- bitumen waterproofing base;
- heavy drainage layer composed of loosely laid brick and ruble fragments;
- fan-shaped pipe-drain system;
- soil layer at the beginning with a width of 60 cm, increased to 90 cm for the accommodation of some species of trees;
- the vegetation layer comprises a large variety of different species starting from floricultural species, decorative herbs, shrubs up to trees of the most different.

This green roof represents a symbol for horticultural England, as it is mentioned by Bridge, in 2014 [30]. It is also known as one of the wonders of England due to the fact that it is an eminent historical garden and unique rooftop.

In another part of the world, in America, the Union Square, from San Francisco, opened in 1850 [31] was one of the few unnamed city-center public open spaces in this country. At first, the roof garden was composed of barren coastal sand dunes which some local plants, but it was later restored. In 1870, the Union Square become a public park, re-designed in 1903. The new project includes shrubs, trees, and peripheral tree rows. After the Great Earthquake of 1906, an innovative idea to build the world's first underground car park at the park was implemented in 1910. This is the moment when the park helps to increase commercial business. This idea was better outlined in 1939 when a private company re-build in with private funds. The new park was the world's first urban park on the top of a reinforced multi-leveled underground structure. The entire structure has four-story with 1700 parking spaces, and 160 massive columns, which represented the support of the roof garden.

Another notable American green roof system is the Casino Theatre at Broadway in New York, which is the first building with a roof garden with flowers covers, a cafe place, seats and stage [32]. It was the

IOP Conf. Series: Materials Science and Engineering **789** (2020) 012001 doi:10.1088/1757-899X/789/1/012001

first record of a green roof system in America, being completely built-in 1882. Even if it is emblematic in the history of these systems, some lessons can be learned from the project of the Casino Theatre. The main objective of this project was performances and commercial activities, which led to a largely hard-pavement with sporadic vegetation such as flower beds, herbs, and climbers in small regions. With large spaces covered by pavement and some vegetation randomly disposed of, it could be described as a rooftop entertainment space designed with some plants. Even if the project didn't complete all the needs for which was built, its principal contribution was the rapid spread of its main idea reflected in more roof gardens built with more vegetation and better implementation.

In 2008, at California Academy of Sciences in the Golden Gate Park, San Francisco, was opened a modern museum whose main aim was sustainable design. In its top, was created a signature green roof designed with hills and domes imitates the relief of the city. This impressive green garden has an approximate surface of 17652 m^2, covered with 50 native wildflowers and other species that bring back urban biodiversity. The extensive green roof is composed of different indigenous ecosystems of a natural hill, a fact that attracts local wildlife and shelter. The 50000 porous trays made of coconut coir, allow roots to penetrate into adjoining trays thus forming a stable and unified substrate [33]. In this growing medium were planted different types of vegetation, especially chosen for creating a noise and air barrier, and for the thermal insulation and their evapotranspiration characteristics [34]. This exemplar of green roof excels by creating a completely natural ecosystem that offers both a new site image and various ecological, economic and sustainable benefits.

The last and not least, countries studied are Asian ones, because of culture and history in terms of landscape, architecture, engineering, and innovation. The first example is the green roof installed on the top of the Mitsukoshi Department Store in Tokyo. This roof garden was designed based on the traditional Japanese style and after the Great Kanto earthquake in 1923, the building was destroyed. The next period was marked by the construction of a relatively large number of green roofs, so this concept spreads rapidly on the territory of Japan. In this context, will be presented some exemplary cases which denoted the provision of natural space despite the intensive urban development.

Fukuoka in Kyushu, southern Japan, the Asian Crossroads over the Sea Institution, presents a special interest in this subject, due to the outstanding innovative design, which successfully combines the concepts of living wall and green roof. Due to the extensive surface of this institution, it was desirable to compensate for the loss of green spaces by creating impressive green facades that almost fully dress the walls of the building. The 15 levels of terraces were designed with a collection of flowers, herbs, shrubs, and trees. The growing medium on these terraces had a thickness of 30-60 cm, a fact that supported the planted of 50000 plants around 120 species [35]. With a height of 60 m and an area of 8300 m^2 of contiguous greenery, the institution's building viewed from the park resembles a charming green hill designed with natural woodland vegetation (figure 4).

Figure 4. Green roof Fukuoka, ACROS (source: https://commons.wikimedia.org).

It is the largest green cover on one building in the world, a fact that provides the special design and the impressive overview [36].

The Municipal Central Gymnasium in Yahatya Park, Osake completed built-in 1996, represents another world green-roof record. The entire stadium was concealed underground by the new, innovative and impressive structure. The main arena of the stadium, which can accommodate around 10000 persons was covered by a dome with the widest span of 110 m and the greatest height of 26 m. The second stadium, the smaller one, has a 52 m wide dome [37]. These domes are the structural support of a layer of soil on which were planted different types of vegetation, thus were formed two circular green roofs. The most representative garden and the largest one measure around 14500 m^2 and the smaller one 2500 m^2. This green garden represents the first one established on the top of a sports structure and the largest green roof implemented on a domed roof.

These notable modern green roof exemplars previously presented, are just a small part of the largest specialty literature studied and represents the example of the green roof built in different parts of the world, and in a difficult period of the world's history. They are the living proof that nature can always find its place in the crowded urban environment and in continuous motion.

In Romania, there are some examples of natural green roofs where the plants have grown spontaneously, [38], but only in the last years, the development of green roofs have been analyzed, especially for the identification of the species of plants that can adapt to the specific meteorological conditions of the geographical area. The concept of green roofs began to be implemented in isolation, without feasible studies to establishing the vegetation solutions that can successfully withstand a gap in temperature from (-30)°C in winter to +(40-60)°C in summer, the temperature is measured in the support layer for plants. The design and execution of these types of systems in the Romanian urban environment have known development in recent years, due to the appearance of a relatively large number of specialized companies. It can be noted the fact that the implementation of green roofs is mostly in private dwellings, especially houses, but can be also noticed the desire for implementation at the level of collective dwellings. Also, private companies or various chains of stores bring to the attention of the inhabitants of Romania these systems, being an example by using these.

4. Implications and conclusion

Research work regarding green roofs has been challenging and provides an open gate to opportunities and provocations for future studies and researches. As a conclusion of this study, the green roof subject is a topical topic, even if it is very much analyzed, discussed and studied one. Even if at the present moment different classic solutions of a green roof are often used in different parts of the world, in the last decades, they have been developing new, innovative, and modern techniques for greening a new or an existing building.

At the moment the attention was directed to the environmental problems, which resulted in drastic changes in the structure of the green roof and also the emergence of the latest technologies or the combining of several technologies to achieve the ecological dimension of the construction sector. Taking into account all the above mentioned, the aim of the paper was to define and analyze the modern green roof systems considering all the changes that have occurred over time. As a conclusion of this part of the study, it can be said that the modern solutions of the green roof as modular green roof or PV – green roof systems, are both innovative and special technologies that solve parts of the problems in terms of adapting the concept of green roof to the current conditions surrounding the environment. The modular green roof brings a solution to the issue of execution, in terms of time and costs, while the PV – green roof systems complement the classic green roof technology with photovoltaic panels technology that balances the benefits of the two technologies.

On the other hand, were analyzed a series of examples of how this concept has been adopted in different parts of the world as in Europe, America, and Asia. As the study highlighted, this concept has been developed in all parts of the world with greater or lower speed. Most of the time, earthquakes or tense moments in the history of a decisive country the emergence of a new solution that brings ecological, social, economic and mental benefits.

The characteristics of the different species of plants, the requested growing conditions, as well as the specificity of the location environment, are important aspects for identifying the optimal solutions of green roofs, regarding the declared purpose of reducing pollution. Plant species have different abilities to eliminate atmospheric pollutants and reduce emissions, therefore they need to be selected to maximize the improvement of air quality. The percentages of reduction for dust, ozone, and harmful gases are maximum during the growth periods of the plants, and the leaves play a major role so that the plants with permanent leaves (conifers) can give better results in the intention of reducing the polluting factors for the air.

On the other hand, Romania is a particular example that provides in recent years the need for implementation of these types of systems. Taking into account the climatic and geographic characteristics, the vegetal specifics and urban modifications, it is noted that the implementation of green roofs at the national level can raise certain problems, but also the basic concept can be adopted ease due to the flexibility of adapting the plant species.

The authors consider that the study is important in order to completely understand the characteristics of these systems and their main components, in order to properly design and implement them according to the present environmental needs. It is believed that the knowledge of this concept and its evolution and tendencies will help solve a series of ecological and social problems.

Acknowledgments

This paper was elaborated with the support of the "Ecoinnovative Products and Technologies for Energy Efficiency in Constructions – EFECON" research grant, project ID P_40_295/105524, Program co-financed by the European Regional Development Fund through Operational Program Competitiveness 2014-2020.

References

[1] Antrop M 2004 Landscape change and the urbanization process in Europe *Landsc. Urban. Plan.* **67** 9-26

[2] Dunnett N and Kingsbury N 2004 *Planting green roofs and living walls* Portland OR: Timber Press

[3] Gagliano A, Detommaso M, Nocera F and Berardi U 2016 The adoption of green roofs for the retrofitting of existing buildings in the Mediterranean climate *Int. J. Sustain. Build. Techno. Urban. Dev.* **7** 116-29

[4] Berardi U 2016 Benefits of green roof retrofits as a local intervention for mitigation of the urban heat island effect in Toronto *Regenerative and Resilient Urban Environments*

[5] Korol E and Shushunova N 2016 Benefits of a modular green roof Technology *Procedia Engineering* **161** 1820-26

[6] The Multifunctionality of Green Infrastructure *Directorate-General for the Environment European Commission* 2012

[7] Onmura S, Matsumoto M and Hokoi S 2001 Study on evaporative cooling effect of roof lawn gardens *Energy and Buildings* **33** 653-66

[8] Liu K 2004 Engineering performance on rooftop gardens through field evaluation *Journal of Roog Consultants Institute* **22** 4-12

[9] Saiz S, Kennedy C, Bass B and Pressnail K 2006 Comparative life cycle assessment of standard and green roofs *Environ. Sci. Technol.* **40** 4312-16

[10] Jim C Y 2017 Green roof evolution through exemplars: Germinal prototypes to modern variants *Sustain. Cities Soc.* **35** 69-82

[11] Guisepi R A 2004 Ancient voices: The birth of a civilization: A history of the Mesopotamian world *Civilization Series Press*

[12] Elborombaly H and Molina-Prieto L F 2014 Sustainable development and eco-roof *International Journal of Science and Research* 2319-64

[13] Dunnet N and Kingsbury N 2008 *Planting green roofs and living walls* (revised edition) Portland

OR: Timber Press

[14] Köhler M 2006 Long-term vegetation research on two extensive green roofs in Berlin *Urban Habitats* **4** 3-26

[15] Bau-Show Lin, Chin-Chung Yu, Ai-Tsen Su and Yann-Jou Lin 2013 Impact of climatic conditions on the thermal effectiveness of an extensive green roof *Build. Environ.* 26-33

[16] Lameraa C, Becciua G and Rullia M C 2014 Green roofs effects on the urban water cycle components *Procedia Engineering* **70** 988-97

[17] Sfakianaki A, Pagalou E, Pavlou K, Santamouris M and Assimakopoulos M N 2009 Theoretical and experimental analysis of the thermal behavior of a green roof system installed in two residential buildings in Athens *Int. J. Energ. Res.* **33** 1059-69

[18] Takao U 2016 Evaluation of Sustainable Roof from Various Aspects and Benefits of Agriculture Roofing in Urban Core *Procedia-Social and Behavioral Sciences* **216** 850-60

[19] Lamnatou C and Chemisana D 2015 A critical analysis of factors affecting photovoltaic-green roof performance *Renew. Sustain. Energy Rev.* **43** 264-80

[20] Kaiser A S, Zamora B, Mazón R, Garcia J R and Vera F 2014 Experimental study of cooling BIPV modules by forced convection in the air channel *Appl. Energy* **135** 88-97

[21] Chemisan D and Lamnatou Chr 2014 Photovoltaic-green roofs: An experimental evaluation of system performance *Appl. Energy* **119** 246-56

[22] Hui S C M and Chan S C 2011 Integration of Green Roof and Solar Photovoltaic System *Joint Symposium 2011: Integrated Building Design in the New Era of Sustainability* Hong Kong

[23] Van Mechelen C, Dutoit T and Hermy M 2015 Adapting green roof irrigation practices for a sustainable future: a review *Sustain. Cities Soc.* **19** 74-90

[24] Katukia A Y, Ronteltap M, Niwagabla C B, Kansiime F and Lens P N L 2014 A two-step crushed lava rock filter unit for grey water treatment at household level in an urban slum *J. Environ. Manag.* **133** 258-67

[25] Cameron R and Berghage R 2009 Integrated biofilters converting wastewater to a water resource *Greening rooftops for sustainable cities conference, Atlanta, GA.* In: Proceedings of 7th North American Greenroof Conference: Greening Rooftops for Sustainable Communities, MD

[26] Peel D W 1960 *A garden in the sky: The story of Barkers of Kensington, 1870-1957* London: W H Allen

[27] Landscape Institute 1994 *A visitor's guide to 20th-century British landscape design* London: Landscape Institute

[28] Historic England 2016 *The Roof Garden, 99 High, St Kensington* https://historicengland.org.uk/listing/the-list/list-entry/1001406

[29] Bridge L 2014 *The roof gardens, Kensington, London.* In Ed M Harney (Gardens and landscapes in historic buildings conservation) New York: Wiley

[30] Dalzell R 2014 *The gilded age origins of New York city's roof gardens* https://ny.curbed.com/2014/7/16/10076106/the-gilded-age-origins-of-new-york-citys-rooftop-gardens

[31] Greenroof.com 2016 *Rockefeller Center roof gardens* https://www.greenroofs.com/projects/rockefeller-center-roof-gardens/

[32] Nuno G J 1993 A history of Union Square: Historical essay *Argonaut* 4 http://www.foundsf.org/index.php?title=A_HISTORY_OF_UNION_SQUARE

[33] American Society of Landscape Architects 2017 *ASLA 2009 professional awards, general design category, Honor Award, California Academy of Sciences, San Francisco, CA* https://www.asla.org/2009awards/111.html

[34] Greenroof.com 2016 *ACROS Fukuoka Prefectural International Hall* https://www.greenroofs.com/projects/acros-fukuoka-prefectural-international-hall/

[35] Velazquez L 2011 *ACROS Fukuoka Prefectural International Hall*Sky Gardens Blog https://www.greenroofs.com/2015/12/03/video-project-of-the-week-for-november-30-2015-acros-fukuoka-prefectural-international-hall/

[36] Osaka City 2016 *Osaka Municipal Central Gymnasium* (https://osaka-info.jp/en/page/osaka-municipal-central-gymnasium)

[37] Osaka City 2016 *How to build the OMC gymnasium* (https://www.mystays.com/en-us/attractions-maruzen-intec-arena-osakaosaka-municipal-central-gymnasium-osaka-japan/)

[38] Vasiliu-Oromulu L, Bărbuceanu D and Vierbergen G 2012 *Thysanoptera in a green roof micro-ecosystem in Bucharest*, Romania, Acta Phytopathologica et Entomologica Hungarica **47** (1) 161-6

IOP Conf. Series: Materials Science and Engineering **789** (2020) 012002 doi:10.1088/1757-899X/789/1/012002

Current tendencies in welding of steel bridges; choice of material and use of thick plates

R Băncilă[1], E Petzek[1], A Feier[2] and D Radu[3*]

[1] Faculty of Civil Engineering, Politehnica University of Timisoara, Romania
[2] Department of Materials and Manufacturing Engineering, Politehnica University of Timisoara, Romania
[3] Faculty of Civil Engineering, Transilvania University of Braşov, Romania

[*] dorin.radu@unitbv.ro

Abstract. The emergence of new efficient materials, the high demands on the quality of the welded joints, while ensuring their safety in operation, outstanding aesthetic forms, the complexity of the new European welding standards, the pressure of the short execution times, had as result the adoption of modern solutions for steel and composite bridges. Durability is also one of the demands of the European Standards. The use of thick heavy plates can lead to reduction of the constructive depth, possibility for improved fatigue details, short execution time and easy maintenance. In this direction the fabrication efficiency is important, the need of a correct technology is vital. The role of the welding specialist is determinant. The paper presents some examples in the realization of modern steel and composite bridges with thick plates for the Romanian highways network.

1. Introduction

The steel making industry is continuously trying to improve the steel quality by developing more and more sophisticated steel grades [1]. The application of modem high-strength steels offers to constructors, the opportunity to realize economical and environment-friendly structures. A new and efficient tendency is the use of thick plates in structures with large spans, for example in European Infrastructure projects (figure 1). Using heavy plates, the total weight of the structure can be reduced as well as the structural safety, can be increased.

Today, heavy plates can be delivered in a nearly unlimited range of dimensions facilitating the designing problems. Plates in thicknesses up to 120 mm have been widely used in the construction of bridges and multi-story buildings in all over Europe. Steel is due to its unlimited recyclability the ideal choice for a sustainable material in construction [2].

The main advantages of thick plates are:

- reduction of the constructive depth;
- possibility for improved fatigue details;
- elimination of the cover plates (difficult welding procedure);
- multiple prefabrication possibilities;
- reduced transport energy, as well as transport costs;

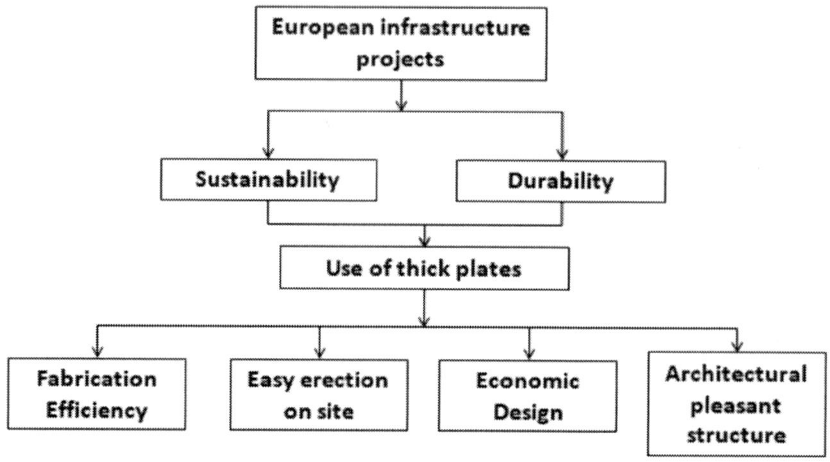

Figure 1. Thick plates used in the infrastructure projects – characteristics.

- short execution time (easy erection on site); short interruption of the traffic;
- easy maintenance (large flat surfaces);
- low acoustic emissions due to plate thickness;
- fabrication costs can be saved due to the outstanding strength, toughness and fabrication properties;
- furthermore, thermomechanical rolled steels (S355M/L, S420M/L and S460M/L) have an excellent weldability despite their high strength and an excellent brittle crack-arrest-toughness;
- structural safety and reliability;
- reduced costs by using higher strength steels and savings in welding as shown in figure 2 [3].

The rational exploitation of these possibilities opens new advantageous constructive solutions for steel and composite structures.

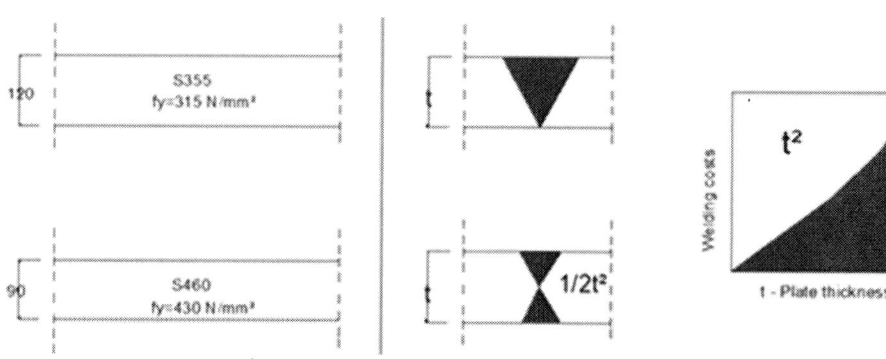

Figure 2. a) Reduction of plate thickness; b) Welding cost savings in function of plate thickness.

However, a major disadvantage of heavy plates is crack-induced through-thickness failures and lamellar tearing; these failures have been discovered during the processes of welding and construction of steel structures with thick plates. A special attention must be paid in this direction.

These modem steels represent the right answer to the increasing requirements on modern constructional materials [4].

2. Choice of the material

According to Eurocode [5], the choice of the steel grade had to follow a special methodology (figure 3).

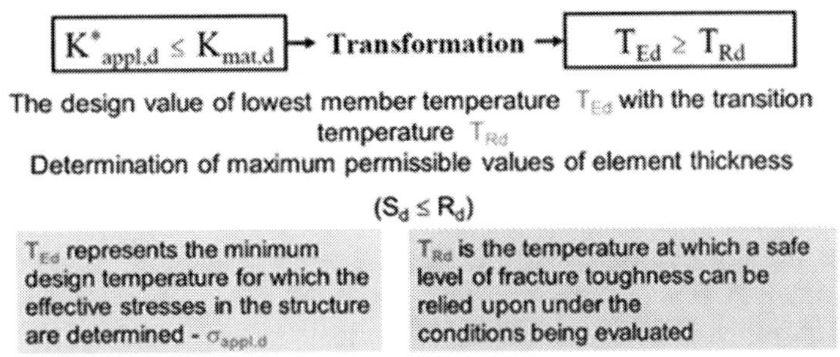

Figure 3. Choice of material in welded steel constructions.

Some initial recommendations can be done [6, 7]:

- Gradually replacement of the classical S235 with S355, or a higher steel grade.
- Recommendation for using weathering steels (like S355J0 W – EN 10025-5).
- Choice from the designing stage of the correct execution category, according to EN 1090.
- At these large thicknesses the use of thermomechanical laminated steels is recommended. They have a very good weldability, together with a high ductility. For the verification of weldability, the equivalent carbon content is determined [3]:

$$CEV\ (IIW) = C + Mn/6 + (Cr + Mo + V)/5 + (Ni + Cu)/15 \quad (1)$$

- Above the CEV % values for three S460 steels are presented in table 1.

Table 1. Different types of steel.

Steel	S460NL	S460QL	S460ML
CEV%	0.50	0.42	0.41

- Due to the fine granularity and a slender chemical composition, thermomechanical laminated steels should not be preheated.

For the choice of material [5], the following steps should be taken:

- calculus of the maximum applied stress σ_{Ed}, according to the design combination corresponding to an accidental combination (unfavourable occurrence of lowest temperature, flaw size, location of flaw and material property).

$$E_d = E\{A[T_{Ed}]\ "+" \sum G_K\ "+" \psi_1 Q_{K1}\ "+" \sum \psi_{2,i} Q_{Ki}\} \quad (2)$$

where the leading action A is the reference temperature T_{Ed} that influences the toughness of material of the element considered and might also lead to stress from restraint of movement. $\sum G_K$ are the permanent actions, $\psi_1 Q_{K1}$ is the frequent value of the variable load and $\psi_{2i} Q_{Ki}$, are the quasi-permanent values of the accompanying variable loads, that govern the level of stresses on the material.

- Calculation of the maximum applied stress σ_{Ed}, which should be the nominal stress at the location of the potential crack initiation, determined as for the serviceability limit state considering all combinations of permanent and variable actions and will not exceed 75% of the yield strength. The reference stresses σ_{Ed} should be determined using an elastic analysis.
- Calculus of the reference temperature T_{Ed} at the potential fracture location

$$T_{Ed} = T_{md} + \Delta T_r + \Delta T_\sigma + \Delta T_R + \Delta T_\varepsilon + T_{\varepsilon cf} \quad (3)$$

where:

- T_{md} is the lowest air temperature defined at the stress level corresponding to $\sigma_{appl.d}$ (for example for middle Europe is defined as -25°C); where $\sigma_{appl.d}$ is the stress produced by the permanent and variable actions at the minimum operating temperature (T_{Ad}),
- ΔT_r is an adjustment for radiation loss (can be considered -5°C),
- ΔT_σ is the temperature range determined by the crack size ad and the effective value of stress $\sigma_{appl.d}$, an adjustment for stress and yield strength of material, crack imperfection and member shape and dimensions (resulted from the Failure Assessment Diagram FAD (R6 Method - Assessment of the Integrity of Structures Containing Defects - CEGB), Wallin-Sanz-correlation between the stress intensity factor K and the temperature T (Master Wallin Curve which include the relation between the stress intensity factor K_{Jc} and temperature and Sanz correlation T100 = T27J – 18°C),
- ΔT_R is a safety allowance, if required, to reflect different reliability levels for different applications (in relation with the reliability index β and the failure probability P),
- ΔT_ε is the adjustment for a strain rate other than the reference strain rate ε_0,
- $\Delta T_{\varepsilon cf}$ is the adjustment for the degree of cold forming. Normally it is considered a material without cold deformation so that $\varepsilon cf = 0\% \rightarrow \Delta T\varepsilon cf = 0$. Currently there are practical executed examples of steel plate cold forming facilitating the development of composite bridge sections (figure 4).

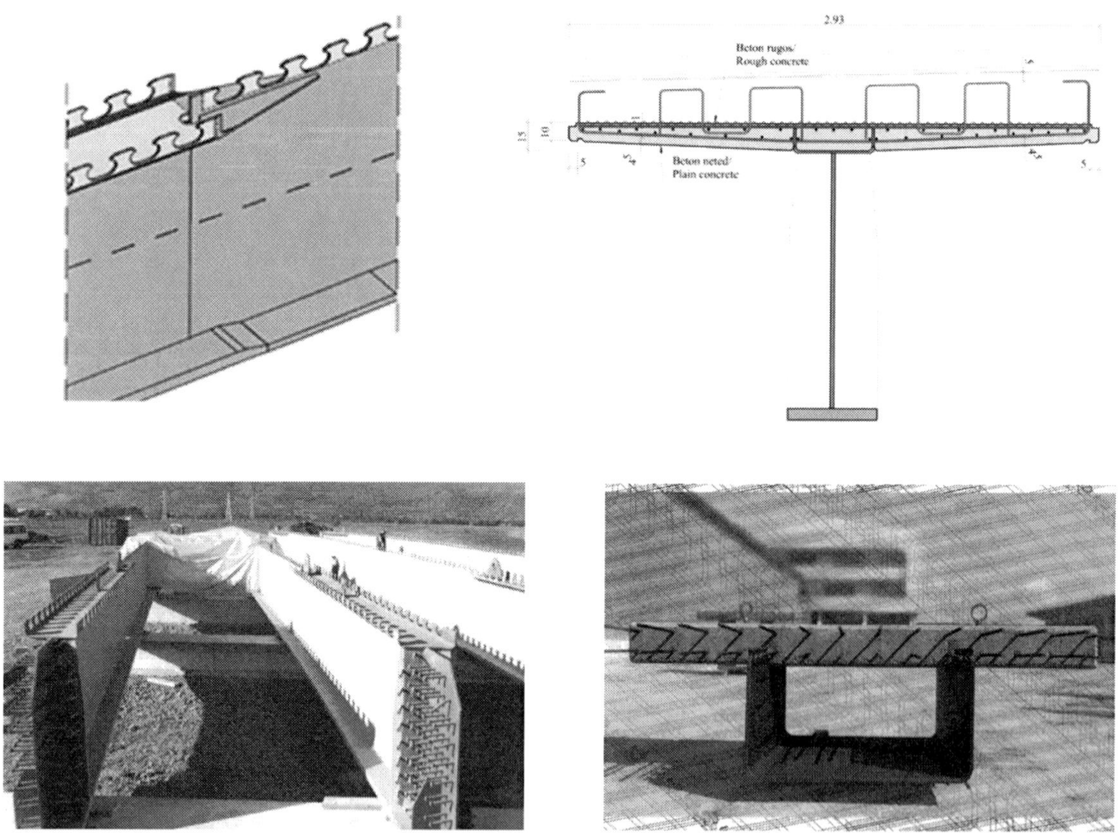

Figure 4. Composite bridge sections with cold bend steel plates.

The importance of applying the principles of fracture mechanics results also from the fact that the European Standards EUROCODE provides as a criterion of material choice, respective as a fundamental element to preventing brittle fracture, the relation (4).

$$K_{mat} \geq K_I = Y \cdot \sigma \cdot (\pi \cdot a)^{1/2} \tag{4}$$

Usually T_{Ed} is equal to T_{mdr}; according to the DASt Richtlinie 009 [6], it can be taken – for Central Europe - table 2.

Table 2. Different types of steel.

No.	Member	Reference Temperature T_{mdr} (°C)
1.	*Steel and composite bridges*	-30
2.	*Buildings*	
	- Members exposed to external climate	-30
	- Members protected from external climate	0
3.	*Crane runways*	
	- Members exposed to external climate	-30
	- Members protected from external climate	0
4.	*Hydraulic structures*	
	- Members almost fully emerged from water	-30
	- Members with one sided contact with water	-15
	- Members partially submerged in water	-15
	- Members fully submerged in water	-5

The maximum allowable values of element thickness is given in Eurocode 1993 Annex 3 table 2, in function of:

- the maximum permissible values of element thickness appropriate to a steel grade,
- its toughness quality in terms of KV-value,
- the reference stress value σ_{Ed} and the reference temperature T_{ed} (°C), in terms of three stress levels expressed as proportions of the nominal yield strength:

$$\sigma_{Ed} = 0.75\, f_y(t) \text{ (N/mm}^2\text{)}$$
$$\sigma_{Ed} = 0.50\, f_y(t) \text{ (N/mm}^2\text{)}$$
$$\sigma_{Ed} = 0.25\, f_y(t) \text{ (N/mm}^2\text{)}$$

where $f_y(t) = f_y - 0.25 \cdot t/t_0$ (t – plate thickness in mm and t_0 = 1 mm).

The values are presented in terms of a choice of seven reference temperatures: +10, 0, -10, -20, -30, -40 and -50 °C.

3. Case study - composite highway bridge

The highway bridge is situated at km 26 + 350 over three electrified, main railway lines, the CF300 and the double CF201. Because the VTR® solution is well suited for skew bridges, the middle span could be reduced from the initial 80 m length, to 48 m length. The main advantage in this case is that the main girders can be lifted and mounted at once, without the use of temporary supporting towers. This is especially important when working near electrified railway lines. The bridge has a length of 217.00 m (figure 5), with 6 spans (28.00 m + 32.00 m + 48.00 m + 40.00 m + 28.00 m + 28 m). The substructure follows the skew of 31° between the highway axis and the railways. The bridge is made of two parallel structures, each 13.60 m wide (figure 6). The structure is an integral bridge, with rigid connection between the sub- and superstructure. Most of the longitudinal deformation is taken over by pendulum-like columns in the abutment area. No bearings are used, and expansion joints are needed only at the abutments.

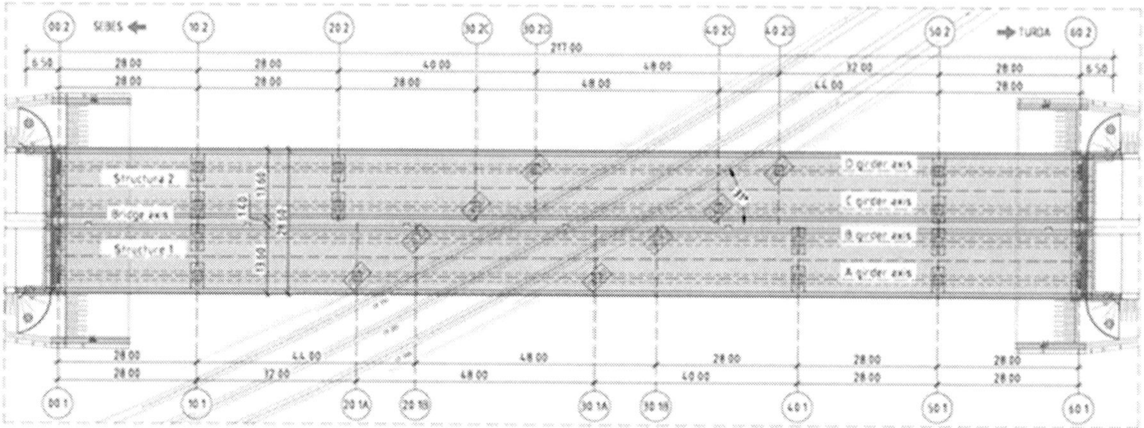

Figure 5. View of the bridge – plan.

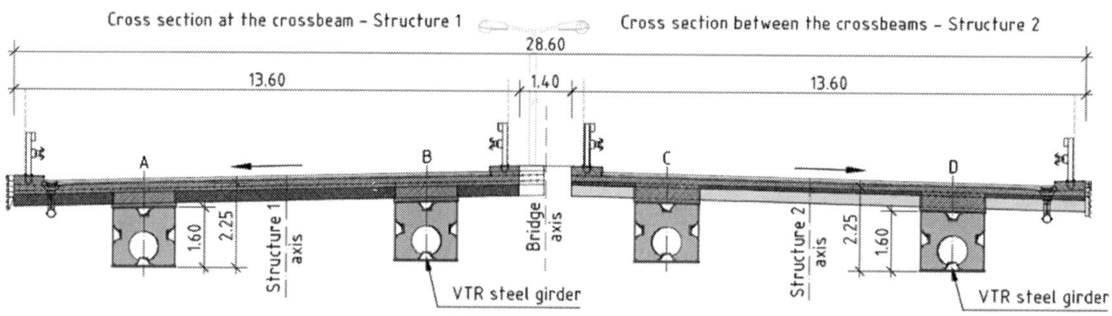

Figure 6. View of the bridge – cross section.

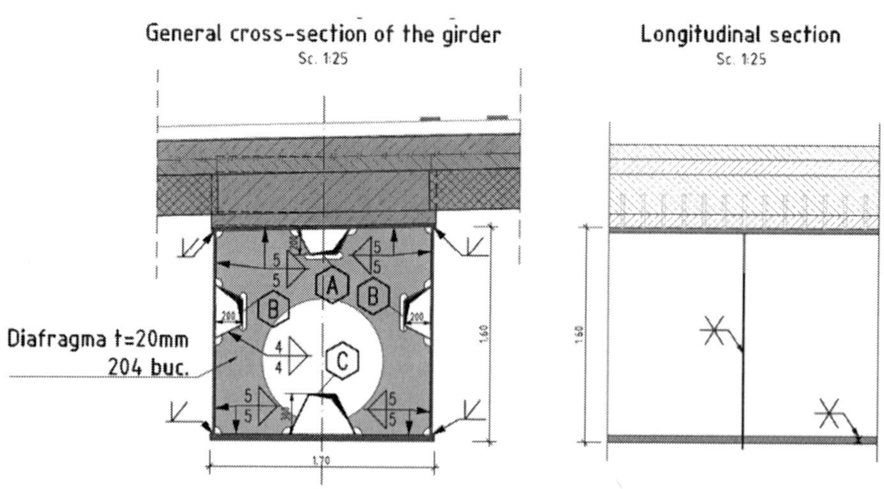

Figure 7. Cross section of the bridge and types of welding joints.

The maximum stress value in the steel structure – lower flange, is presented in figure 8.

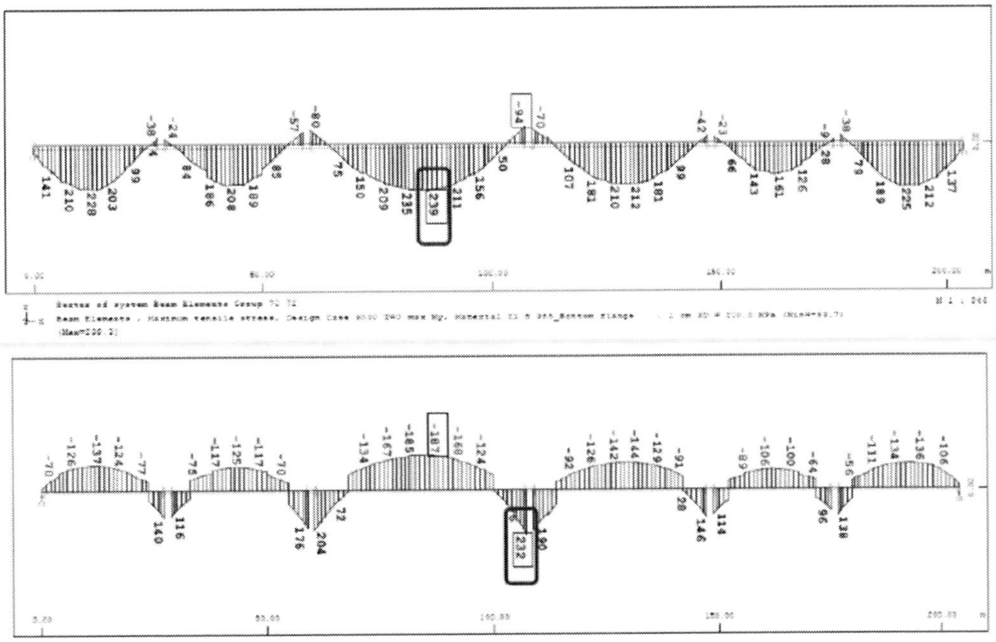

Figure 8. Maximum stress values in the lower flange of the main girder.

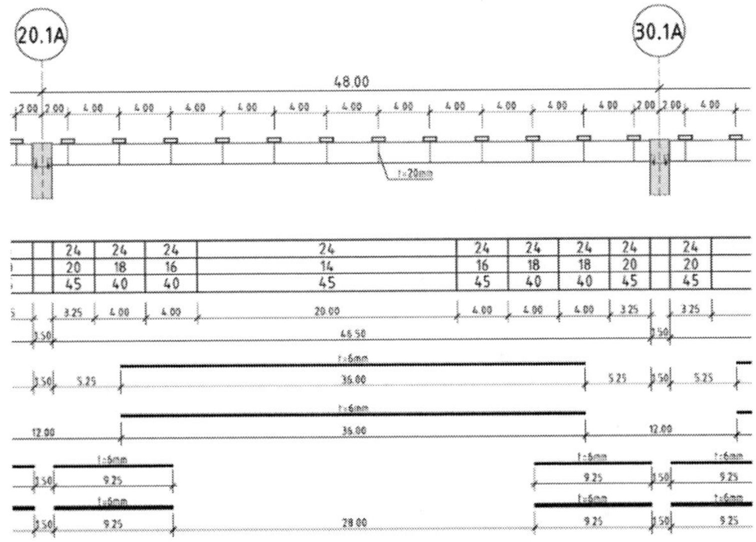

Figure 9. The thickness of the material in the section with maximum stress.

Entrance parameters [5]: $f_y(t) = f_y - 0{,}25 \times \frac{t}{t_0};\ cu\ t_0 = 1{,}0$

$$f_y(t) = 355 - 0{,}25 \times \frac{45}{1} = 343{,}75\ N/mm^2$$

$$\frac{\sigma_{Ed}}{f_y(t)} = \frac{23{,}9}{34{,}375} = 0{,}695$$

$$\Rightarrow \sigma_{Ed} \cong 0{,}695 \times f_y(t)$$

Marcă	Calitate	Energia Charpy K_V la T [°C]	J_{min}	Temperatura de referință T_{Ed} [°C]																				
				σ_{Ed} = 0,75 $f_y(t)$							σ_{Ed} = 0,50 $f_y(t)$							σ_{Ed} = 0,25 $f_y(t)$						
				10	0	-10	-20	-30	-40	-50	10	0	-10	-20	-30	-40	-50	10	0	-10	-20	-30	-40	-50
S235	JR	20	27	60	50	40	35	[illegible]	25	20	90	75	65	55	[illegible]	40	35	135	115	100	85	75	65	60
	J0	0	27	90	75	60	50	[illegible]	35	30	125	105	90	75	[illegible]	55	45	175	155	135	115	100	85	75
	J2	-20	27	125	105	90	75	[illegible]	50	40	170	145	125	105	[illegible]	75	65	200	200	175	155	135	115	100
S275	JR	20	27	55	45	35	30	[illegible]	20	15	80	70	55	50	[illegible]	35	30	125	110	95	80	70	60	55
	J0	0	27	75	65	55	45	[illegible]	30	25	115	95	80	70	[illegible]	50	40	165	145	125	110	95	80	70
	J2	-20	27	110	95	75	65	[illegible]	45	35	155	130	115	95	[illegible]	70	55	200	190	165	145	125	110	95
	M,N	-20	40	135	110	95	75	[illegible]	55	45	180	155	130	115	[illegible]	80	70	200	200	190	165	145	125	110
	ML,NL	-50	27	185	160	135	110	[illegible]	75	65	200	200	180	155	[illegible]	115	95	230	200	200	200	190	165	145
S355	JR	20	27	40	35	25	20	[illegible]	15	10	65	55	45	40	[illegible]	25	25	110	95	80	70	60	55	45
	J0	0	27	60	50	40	35	[illegible]	20	15	90	80	60	50	[illegible]	40	30	150	130	110	95	80	70	60
	J2	[illegible]	[illegible]	[illegible]	[illegible]	[illegible]	[illegible]	40	35	[illegible]	[illegible]	[illegible]	[illegible]	[illegible]	65	55	45	200	175	150	130	110	95	80
	K2,M,N	-20	40	110	90	75	60	50	40	35	155	135	110	95	80	65	55	200	200	175	150	130	110	95
	ML,NL	-50	27	155	130	110	90	75	60	50	200	180	155	135	110	95	80	210	200	200	200	175	150	130
S420	M,N	-20	40	95	80	65	55	45	35	30	140	120	100	85	70	60	50	200	185	160	140	120	100	85
	ML,NL	-50	27	135	115	95	80	65	55	45	190	165	140	120	100	85	70	200	200	200	185	160	140	120
S460	Q	-20	30	70	60	50	40	30	25	20	110	95	75	65	55	45	35	175	155	130	115	95	80	70
	M,N	-20	40	90	70	60	50	40	30	25	130	110	95	75	65	55	45	200	175	155	130	115	95	80
	QL	-40	30	105	90	70	60	50	40	30	155	130	110	95	75	65	55	200	200	175	155	130	115	95
	ML,NL	-50	27	125	105	90	70	60	50	40	180	155	130	110	95	75	65	200	200	200	175	155	130	115
	QL1	-60	30	150	125	105	90	70	60	50	200	180	155	130	110	95	75	215	200	200	200	175	155	130
S690	Q	0	40	40	30	25	20	15	10	10	65	55	45	35	30	20	20	120	100	85	75	60	50	45
	Q	-20	30	50	40	30	25	20	15	10	80	65	55	45	35	30	20	140	120	100	85	75	60	50
	QL	-20	40	60	50	40	30	25	20	15	95	80	65	55	45	35	30	165	140	120	100	85	75	60
	QL	-40	30	75	60	50	40	30	25	20	115	95	80	65	55	45	35	190	165	140	120	100	85	75
	QL1	-40	40	90	75	60	50	40	30	25	135	115	95	80	65	55	45	200	190	165	140	120	100	85
	QL1	-60	30	110	90	75	60	50	40	30	160	135	115	95	80	65	55	200	200	190	165	140	120	100

Figure 10. Eurocode table for the choice of the thickness

Determination of maximum permissible values of element thickness is done according with Eurocode norm. Considering that the base material is S355 J2 with t_{max} = 45 mm, it results the admissible thickness t_{adm} = 47.5 mm (interpolation).

Welding details and edge preparation are presented in figure 11.

(a) (b) (c)

Figure 11. (a) Welding detail on the site; (b) Edge preparation of the thick plate; (c) Steel assembly in the workshop.

4. Conclusions

In the present context of the important European Infrastructure Projects, the Eurocodes Standards, the necessity of safe and sustainable steel constructions, the role of the Welding Specialist is decisive. A special attention must be paid to the education in the welding field Education in welding IWE (International welding Engineer) and IWSD (International Welding Designer). During the Round Table Discussion "Developments in welded steel constructions" held in Timisoara in September 2018, a final Statement regarding the importance and necessity of the education in the welding field, was adopted [8, 9].

References

[1] Lehnert T and Schroeter F 2014 How modern steel developments can help cost and sustainability of bridge constructions, Dillinger, Germany

[2] Băncilă R, Petzek E, Feier A and Radu D 2019 The place and role of the welding specialist in the design and execution of welded steel constructions, *Proceedings of the 7th International Conference Contemporary Achievement in Civil Engineering*, 23-24 April 2019 Subotica, Serbia

[3] Lehnert T, Kühn B and Krieglstein T 2019 Einsatz dicker Grobbleche im Brückenbau, *Stahlbau* **88** – Heft 2

[4] * * * Proceedings of the 10th Danube Bridges Conference 2019 Vienna

[5] * * * Eurocode 3, Annex 10 - EN 1993-1-10/2006

[6] * * * Stahlsortenauswahl für geschweisste Stahlbauten 2009 *Deutscher Ausschuss für Stahlbau*, DASt Richtlinie 009

[7] Băncilă R, Petzek E and Feier A 2019 *Course notes EWE & IWSD (European Welding Engineer & International Structural Welding Engineer)*, ASR & ISIM Timişoara

[8] Băncilă R, Petzek E and Feier A 2019 The place and role of the welding expert in the design and execution of welded steel constructions; interdisciplinary connection between the architect, the design engineer and the welding expert (in Romanian), SUDURA ISSN 1453-0384

[9] Round Table Developments in welded steel constructions held in Timişoara on 6 September 2018, organized by the Romanian Alliance of Technical Universities (ARUT), Politehnica University of Timişoara (UPT), and the Romanian Welding Society (ASR).

Testing methods for the assessment of material properties in historical masonry structures: a review

M Barnaure[1*] and M Cincu[1]

[1] Faculty of Civil, Industrial and Agricultural Buildings, Technical University of Civil Engineering Bucharest, Bucharest, Romania

* mircea.barnaure@utcb.ro

Abstract. The current state of testing methods for assessing the mechanical properties of old masonry buildings is reviewed, with a focus on the methods adapted for heritage structures. For these buildings, an accurate assessment is needed in order to minimise the interventions extent while ensuring an adequate structural safety level. As old masonries are non-homogenous and exhibit a highly anisotropic behaviour, numerous parameters are required for correctly simulating the behaviour under a given load scenario. In earthquake-prone areas, numerous failure patterns, in-plane and out-of-plane, can occur for the structural walls, each of them depending on different characteristics of the material. If for new masonries the testing procedures are standardised, for old buildings less standardised tests exist, while investigations are often based on modified or empiric methods. The testing methods can be classified as destructive, minor destructive and non-destructive. The destructive methods are the most precise but are extremely intrusive and may not be acceptable for heritage listed buildings. The non-destructive methods mostly provide qualitative results. The minor-destructive tests are often the most adapted for the study of historical buildings, as they can provide relevant information while affecting only in a limited extent the existing elements.

1. Introduction

For the evaluation of the vulnerability of the existing buildings, an important aspect is the correct assessment of the mechanical properties of the materials. Masonry is a highly anisotropic material. Moreover, in old buildings, multiple changes often occurred through time (local degradations, reparations or modifications using different materials etc.). As the material characteristics are not uniform as would be for new constructions, there are numerous uncertainties regarding these properties. It is difficult to define, in a unique way, the procedures and techniques to be used in an experimental program. For a particular case, some types of tests might not be acceptable at all, if they were to alter the architectural or cultural value of the building, such as wall paintings.

Also, if for the modern materials the codes contain some correlations between properties measured separately on brick and mortar and the global properties of masonry, these cannot always be used for historical structures.

In highly seismic areas, the various in-plane and out-of-plane failure patterns need to be considered. These are governed by various characteristics of the masonry, such as the compression strength and elastic moduli of the blocks and mortar, the friction coefficient of the interface, the bond strength in tension or shear, the past actions that might have partially broken this bond in various areas etc.

In this context, the present study, aimed at reviewing the current methods in assessing material properties for old masonry buildings, is of particular interest. Only by understanding the advantages and shortcomings of various methods, practicing engineers can correctly establish relevant survey and testing programs for a given project, that respect the specific architectural and cultural values. For the researchers, this could be a starting point in improving the existing methods or establishing new testing procedures, more adapted towards assessing the vulnerability of historical monumental buildings.

2. Testing methods for assessing the mechanical properties of masonry

The testing methods can be classified as destructive, minor destructive and non-destructive. Many methods are standardised, while some, in particular the non-destructive ones, are not.

2.1. Destructive methods

The destructive methods are the most precise but are extremely intrusive and may not be acceptable for heritage listed buildings. They are based either in extracting large specimens from the existing buildings and testing these using standard procedures used for new buildings, or in performing tests directly in-situ.

2.1.1. Laboratory testing on large samples consists in using the standardized methods for new buildings (such as [1-8]). But, in place of constructing the samples in the laboratory, as would be done for new buildings, they are extracted from the existing building and transported before being tested. In order for the results to be relevant, several tests should be performed for each material characteristic (compression, bending, shear). Some of the required samples, are very large, such as 1.2x1.2 m for [7]. The number of extracted samples must also account for the risk of damaging them during extraction, transport or handling, in particular for masonry with weak mortar. Therefore, this approach is highly intrusive and rarely acceptable for heritage buildings.

2.1.2. Shear-compression test in-situ. The shear-compression test consists of the application of an increasing shear force on a masonry wall subjected to constant compression stress, figure 1. This in-situ test is not standardized. The test is performed by separating the panel from the masonry (side and above) and applying a certain level of axial compression loading or separating the panel from the masonry (left-right) using the real value of the compression effort due to the gravitational loads that can be measured by performing single-jack test. The typical dimension of the tested masonry panel is 90x180 cm [9, 10].

Figure 1. In situ shear tests done by UTCB for a 20th century building in Constanta.

2.1.3. Diagonal compression test in-situ. This type of test is performed in-situ and is based on the standardized method that is performed in the laboratory to determine the diagonal compression strength [7]. The tested masonry panel must have dimensions greater than or equal to 1.2 m x 1.2 m, and the width between 25cm and 70 cm. The panel is separated by the adjacent masonry on three sides. The test results are influenced by the quality of the wall and especially by the local defects [9].

2.2. Minor destructive methods

Minor-destructive tests can be performed relatively easily and have a small impact on the construction, but they provide only local information and the results obtained are quite scattered.

2.2.1. Tests using flat jacks consist in performing a split, perpendicular to the direction of the stresses to be measured, followed by the introduction into the split of a flat press, in order to generate a pressure inside the split, until the initial state of deformations is restored, figure 2 [11, 12]. The initial state of deformations is determined by extensometer measurements, performed before splitting. Based on this test several parameters can be established: the existing compressive stress, the compression strength, the σ-ε law and the modulus of elasticity, the shear strength against sliding in the horizontal bed joint [13].

Figure 2. Flat jack tests performed on a 19th century masonry building in Bucharest.

2.2.2. Shove tests are used in order to establish the shear strength against sliding in the horizontal bed joint for a given vertical compression level, figure 3. The method consists in imposing the sliding of a brick between two horizontal mortar layers [14]. The bricks adjacent to the tested one are removed and a hydraulic press, acting horizontally on the brick, is installed. The force corresponding to obtaining movement in the testing brick gives the measure of the shear strength against sliding in the bed joint. As the shear force depends on the adherence between bricks and mortar as well as on the friction force under vertical loads, the tests should be performed both at the ground floor level (where the compressive stress from gravitational loads is maximum) as well as under window openings (where the value is nearly zero). Even if the shear strength of a single mortar joint does not fully represent the shear resistance of a horizontal wall, this test provides the most accurate and direct estimate of shear strength in the horizontal joints that can be made in-situ.

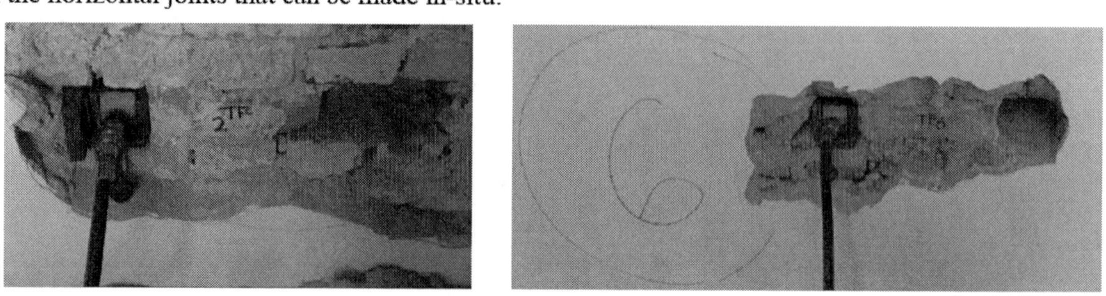

Figure 3. Shove tests done by UTCB for a 20th century building in Constanta.

2.2.3. Splitting tests on masonry cores with inclined mortar joint are a recently proposed (unstandardized) method for establishing the shear strength in mortar joints [10], [15, 16]. For these tests, cores with a diameter between 70 mm and 110 mm are extracted from the existing walls, figure 4. Cores are extracted in such way so that the mortar layer is near the centre of the section. The cylinders are then placed horizontally in a vertical press and tested under axial force. The inclination of the mortar joint in relation to the direction of application of the axial force is varied between 45° and 60°. Based on the Mohr-Coulomb criteria, the shear strength is computed. A special problem in the case of the poor quality of the mortar is the extraction and transport of intact specimens.

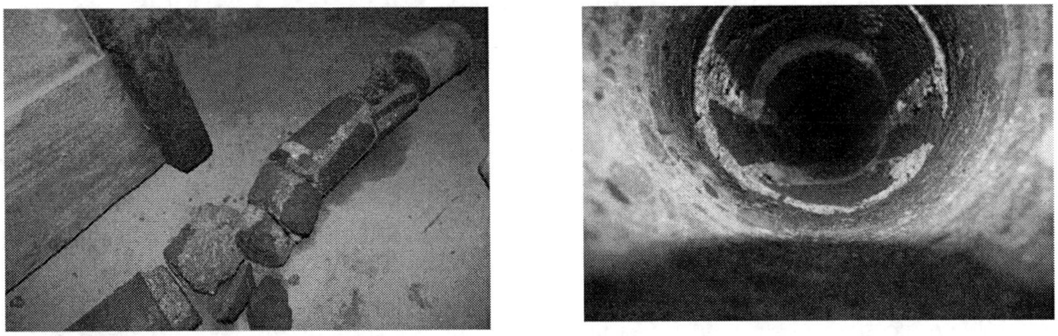

Figure 4. Extraction of cores from a 19th century masonry building in Bucharest.

2.2.4. Bond wrench tests are standardised tests performed based on [17] or [18]. The method consists in fixing a standardized wrench on the top brick of an assembly then twisting until failure. Based on this test, the adherence strength in the mortar layer under tensile stresses in bending can be assessed. The test can be performed in-situ at the top of the walls or below window openings. Otherwise, they can be performed in laboratory on reconstituted walls using bricks extracted from the building and replicated similar mortar (discussed further below). It is to be noted that for low-strength mortar it is recommended that samples of only 2 bricks and 1 mortar layer are used.

2.2.5. Pull-out tests consist in inserting a screw (typically a 60 mm one), by drilling, into the mortar bed joint, for a depth of 35 mm. A correlation was observed between the shear strength developed by the small cylinder around the screw and the compressive strength of the mortar. Therefore, the force required to pull out the screw gives a rough estimate of the strength of the mortar. The advantage of the test is that it can be used for standard 10 mm mortar joints and the damage done while testing is minor and easily repaired.

2.2.6. Extracting bricks and mortar samples from the wall is one method that is less intrusive than extracting large panels. For instance, if a given wall is deemed to be jacketed with reinforced concrete, the extracted bricks create useful connection points between the jacket and the existing wall (figure 5). Bricks and mortar can be tested in the laboratory using standardized methods for new buildings. Also, based on the analysis of the mortar, a similar formulated mortar can be created such that wall panels using the extracted bricks and the similar mortar can be constructed and tested. The usual tests are discussed further below. It is important to mention that it is very difficult to reconstitute a mortar if the original formula is unknown. The chemical analysis of existing mortars [19] can lead to inaccurate results if the original ingredients are not available and no check can be made for soluble lime or silica parts in the aggregates. Even if all of the ingredients are known, the reconstituted mortar might exhibit different properties from the one in the building, since the original air content, water/binder ratio and sand grading cannot be fully determined. Finally, even if the reconstituted mortar were to have the same compressive strength and elastic moduli as the original, many failure patterns of masonry structures are governed by the bond strength between mortar and bricks. Tests on reconstituted samples should always be compared with tests on intact elements extracted from the existing building.

Figure 5. Extraction of individual bricks from a 1930's building in Bucharest and connecting detail between RC jacket and existing wall during the retrofitting works

2.2.7. Tests on blocks can be performed as for new buildings. Based on the standard tests [20-23], the compressive strength of blocks made either from masonry or natural stone can be measured. For masonry units, the splitting tensile strength can be obtained. A special attention must be given to the preparation of the blocks extracted from the work prior to testing [24].

2.2.8. Tests on mortar can be performed as for new buildings using standardized methods [25-27]. For lime-based mortars, a particular attention must be given to the curing time and conditions. The minimum time should be 90 days (as compared with 28 days for cement-based mortars). The curing conditions should be, if possible, adapted so that they reproduce the climate from the location of the existing building, as curing conditions highly influence the properties of the mortar at early stages.

2.2.9. Tests on reconstituted samples from bricks and mortar can be performed using the standardized methods for new buildings [1-8]. As for the reconstituted mortar, a long curing time is needed for the lime-based mortars, of at least 90 days, before tests can be performed.

2.3. Non-destructive methods

The non-destructive methods do not compromise the existing masonry constructions and can be carried out quickly and with relatively low costs. The use of non-destructive methods allows both detection and localization of defects. As there are numerous methods, we chose to focus on those mentioned in [28] as reliable for assessing the existing buildings. Unfortunately, the non-destructive methods do not provide any exact values of the parameters of the masonry necessary to establish the vulnerability, so they must always be accompanied by minor destructive or destructive methods. It is recommended to combine different non-destructive methods in order to obtain information as relevant as possible.

2.3.1. Acoustic emission (AE) is a non-destructive method that is based on recording the elastic waves released when sudden redistribution of stresses occurs in the material. AE sensors are very sensitive and can record extremely low motions that are transmitted to the surface of the masonry when cracks occur or grow and can provide useful information regarding the strength and risk of failure of a component. This method has been successfully used for historical monumental masonry constructions that have been subjected to numerous seismic actions [29]. The method allows to identify cracks and deteriorations in the walls and to evaluate the stability of the structure based on the evolution of the degradation phenomena.

2.3.2. Ultrasonic pulse velocity tests are used to check the quality of existing masonry based on the velocity of ultrasonic pulses passing through the material. The equipment includes a pulse generator, a transducer that transforms electronic pulses into mechanical pulses and a reception circuit. Generally, 40 to 50 kHz frequency devices are used.

This method allows to detect the density variations and the modulus of elasticity of the masonry, as well as the presence of cracks and discontinuities. Higher velocities indicate a good and continuous material, while slower velocities may indicate voids or cracks. This method cannot be applied for poor quality masonry or with many defects and cracks, as the results are sensitive to the condition of the masonry surface.

2.3.3. Mechanical Pulse Velocity method consists in hitting the masonry wall with a special hammer and measuring the time needed for the wave to travel over a certain distance. This method is not standardized but can be used in place of the ultrasonic pulse velocity method when determining the properties of masonry on larger surfaces.

2.3.4. Impact – echo is an acoustic method is based on generating waves on the surface of the structural element (often done by using a special hammer). The waves are reflected by internal flaws and external surfaces.

The speed of the wave depends on the properties of the material but is also influenced by gaps, cracks or unfulfilled joints. So, by using this method, the presence of voids or embedded steel elements inside the structural wall can be identified.

2.3.5. Radiography tests are similar to those used in medicine. A beam of high energy radiation is passed through the structure and the signal is then detected on the remote face by transducers.

This method permits identifying the inhomogeneities and discontinuities in the masonry (e.g. reinforcements or other embedded elements). Due to special precautions regarding the safety for the people when using this technique and to possible interferences with equipment, special permits are nowadays required, and this technique is often replaced by radar scans.

2.3.6. Infrared thermography is a method of surface testing. A thermal energy flux is applied to the surface of the material or generated by the tensioning of the material. The flux is affected by the insulating properties of the material and by the degree to which its surface radiates energy. The surface temperature differences are recorded using infrared cameras or special thermal coatings.

For the masonry, data can be obtained regarding the damaged areas, or those with different moisture content in the brick and porous stone.

The method is capable of reporting only surface anomalies and their location. However, it cannot determine their depth and thickness. The results are influenced by external environmental factors (temperature and humidity).

2.3.7. Surface hardness using a Schmidt hammer. The testing method is described in [30, 31]. A certain correlation was observed between the rebound value and the uniaxial compressive strength. By performing this test, a rough estimate of the compressive strength can be obtained.

The method is sensitive enough so that differences in the strength on masonry elements can be identified. It is not, however, sensitive to the level of cracks in the wall. The roughness of the *analysed* surface can also influence the results.

2.3.8. Radar scan (microwaves). The penetration power of microwaves allows in-depth analysis of a material for locating defects, discontinuities or foreign bodies. In diagnosing the old buildings, especially the historical monuments, the method can be used for detecting the presence of several layers in the masonry, detecting the important voids, the presence of embedded steel elements, the cracked or damaged areas, or the moisture content of the masonry.

3. Discussion and conclusions

For assessing the structural safety level for old masonry buildings, it is fundamental to correctly determine the characteristics of the materials as well as the pre-existing damage. For buildings with high architectural or cultural value it is extremely important to minimise the interventions extent to a minimum. The same goes for the experimental campaigns, where extensive destructive investigations should be avoided.

In the same time, destructive methods are the most reliable in providing the necessary information for the structural assessment. The standardised testing methods for new buildings require numerous tests on large samples. As this approach is generally unacceptable for historically listed buildings, as a first step in the experimental campaign, non-destructive methods should be used. These have been proved to provide useful information regarding the geometry and damage levels of the building. But the interpretation of the results is more qualitative then quantitative and by themselves the non-destructive tests are not sufficient. They must always be accompanied by destructive or minor-destructive tests.

The minor-destructive evaluation methods are often preferred over the destructive techniques. The problems concerning the reliability of their results are related to the lack of sufficient experimental data on the effect of specimen size on the measured property so that tests performed on small-scale tests could be correlated with standardised tests on large specimens. Another problem arises from the difficulty of extracting and transporting the intact samples to be tested in the laboratory.

In the specific case of masonry structures, extracting brick and mortar samples and building similar specimens that reproduce the original characteristics of the masonry is highly unreliable. This is mostly related to the impossibility of correctly assessing the formulation of the mortar and the curing conditions, which impact on the mechanical properties. The characteristics of reconstructed samples must be compared to results performed on samples collected form the building. As the curing time is long in the case of lime-base mortars, this impacts on the total duration of the evaluation campaign.

Two directions should be pursued in order to increase the reliability of minor-destructive tests campaigns in the scope of correctly assessing the structural safety of old buildings. The first one consists in performing more investigations regarding the correlations between the measurements performed on small-scale specimens and the corresponding values obtained through large-scale tests. The second direction consists in creating extensive databases containing information regarding masonry types, material characteristics and constructive details, mechanical properties, construction period and region. A large database might serve for a reliable statistical analysis of experimental data with the possibility of identifying correlations between various measured properties and this way reducing the extent of mechanical tests required for each individual building. Such database could be developed in Romania by a consortium of Technical Universities, with financial support from the government. This type of database has, for example, been developed in Italy and its acronym is MADA (Masonry Database) [32]. In Italy, MADA was developed by ReLUIS (The Laboratories University Network of Seismic Engineering), over more than a decade (2005-2016), and has been funded through 3 research programs.

References

[1] EN 1052-1:1999, *Methods of test for masonry. Determination of compressive strength*
[2] EN 1052-2:2016, *Methods of test for masonry. Determination of flexural strength*
[3] EN 1052-3:2002, *Methods of test for masonry. Determination of initial shear strength*
[4] EN 1052-4:2000, *Methods of test for masonry. Determination of shear strength including damp proof course*
[5] ASTM C1314-18, *Standard Test Method for Compressive Strength of Masonry Prisms*
[6] ASTM C1717-19, *Standard Test Methods for Conducting Strength Tests of Masonry Wall Panels*
[7] ASTM E519-15, *Standard Test Method for Diagonal Tension (Shear) in Masonry Assemblages*
[8] ASTM E2126-19, *Standard Test Methods for Cyclic (Reversed) Load Test for Shear Resistance of Vertical Elements of the Lateral Force Resisting Systems for Buildings*
[9] Ferretti F, Ferracuti B, Mazzotti C and Savoia M 2019 Destructive and minor destructive tests on masonry buildings: Experimental results and comparison between shear failure criteria

Construction and Building Materials **199** 12-29

[10] Corradi M, Borri A and Vignoli A 2003 Experimental study on the determination of strength of masonry walls *Construction and building materials* **17** (5) 325-337

[11] ASTM C1196-14, *Standard Test Method for In Situ Compressive Stress Within Solid Unit Masonry Estimated Using Flatjack Measurements*

[12] ASTM C1197-14, *Standard Test Method for In Situ Measurement of Masonry Deformability Properties Using the Flatjack Method*

[13] Alecci V, Stipo G, La Brusco A, De Stefano M and Rovero L 2019 Estimating elastic modulus of tuff and brick masonry: A comparison between on-site and laboratory tests *Construction and Building Materials* **204** 828-838

[14] ASTM C1531-16, *Standard test methods for in situ measurement of masonry mortar joint shear strength*

[15] Pelà L, Roca P and Benedetti A 2016 Mechanical characterization of historical masonry by core drilling and testing of cylindrical samples *International Journal of Architectural Heritage* **10** (2-3) 360-374

[16] Pelà L, Kasioumi K and Roca P 2017 Experimental evaluation of the shear strength of aerial lime mortar brickwork by standard tests on triplets and non-standard tests on core samples *Engineering Structures* **136** 441-453

[17] EN 1052-5:2005, *Methods of test for masonry. Determination of bond strength by the bond wrench method*

[18] ASTM C1072-19, *Standard Test Methods for Measurement of Masonry Flexural Bond Strength*

[19] ASTM C1324-15, *Standard Test Method for Examination and Analysis of Hardened Masonry Mortar*

[20] EN 772-1:2011, *Methods of test for masonry units. Determination of compressive strength*

[21] EN 1926:2006, *Natural stone test methods. Determination of uniaxial compressive strength*

[22] ASTM C67-19, *Standard Test Methods for Sampling and Testing Brick and Structural Clay Tile*

[23] ASTM C1006-19, *Standard Test Method for Splitting Tensile Strength of Masonry Units*

[24] ASTM C1587-15, *Standard Practice for Preparation of Field Removed Manufactured Masonry Units and Masonry Specimens for Testing*

[25] EN 1015-2:1999, *Methods of test for mortar for masonry. Bulk sampling of mortars and preparation of test mortars*

[26] EN 1015-11:2019, *Methods of test for mortar for masonry. Determination of flexural and compressive strength of hardened mortar*

[27] EN 12390-6:2009 *Testing hardened concrete. Tensile splitting strength of test specimens*

[28] ASCE/SEI 41-13 *Seismic evaluation and retrofit of existing buildings*

[29] Ohtsu, M (Ed.) 2008 *Innovative AE and NDT Techniques for On-Site Measurement of Concrete and Masonry Structures* RILEM State-of-the-Art Reports vol. 20, Springer

[30] ASTM C805-18 *Standard Test Method for Rebound Number of Hardened Concrete*

[31] ASTM D5873-14 *Standard Test Method for Determination of Rock Hardness by Rebound Hammer Method*

[32] Augenti N, Parisi F and Acconcia E 2012 *MADA: online experimental database for mechanical modelling of existing masonry assemblages proc.* 15th World Conference on Earthquake Engineering, Lisbon, Portugal

Structural design of residential buildings according to Eurocode

M Bešević[1*], M V Purcar[1], L Kozarić[1] and S Zivković[1]

[1] Faculty of Civil Engineering Subotica, Subotica, Serbia

[*] miroslav.besevic@gmail.com

Abstract. In this paper, the dimensioning of the load-bearing structures A, B and C of building complex in the surroundings of Salzburg, Austria is given. The floors of object A are underground floor, ground floor and two floors, and for objects B and C it is underground floor, ground floor, first floor and the attic. These buildings have an underground shared garage. The foundation is 30 cm thick reinforced concrete slab, with capitals 200/200 cm, thickness d_{pp} = 20 cm that are constructed in the middle part of the garage under the columns. Other underground parts of the buildings are the basements of each individual building. In addition to the common garage, the buildings have a common part of the ground floor, which connects the separate buildings. The construction of the garage at ground level is in the form of reinforced concrete slab with thickness d_p = 35 cm with reinforcement-capitals under the columns with thickness $d_{kpl.}$ = 20 cm. The foundation and the ground floor slab are separated by an expansion joint in the axis 7. The waterproofing of the ground floor slab is achieved uniquely for the whole slab of the complex. Special attention is paid to the thermal calculation.

1. Introduction

The paper presents an example of design project of structure according to EC standards. The residential building consists of three independent units - building A, B and C, which are shown in figure 1 (situation with buildings). The complex of three buildings is located in the Strasbourg district of Austria, and consists of a common underground garage, basement rooms for each building and floors. The floors of object A are underground floor, ground floor and two floors, and for objects B and C it is underground floor, ground floor, first floor and the attic. Due to the extremely low temperatures at the place of construction of the complex of objects of approx. 3000.00 m^2, walls of greater thickness (30 cm) were applied with thermal insulation outer layers of thickness d_{iz} = 20 cm.

The windows and doors of the balconies are designed with a thermal break bridge to provide savings in heating during the winter. In addition to the thermal insulation of all A, B and C buildings, sound insulation was also taken into account between the apartments and the external noise. Vertical communication in the buildings is achieved by stairs and an elevator. The construction works on buildings are carried out in phases. Access to the complex is made possible from the access road.

Special attention was paid to the thermal disconnection of the floor and balcony slabs, using the certified element "Isokorb" [1], which allows the acceptance of influence on the balcony slab by applying the tightened reinforcement in the upper zone of the balcony slab, as well as receiving the pressure force on the lower part of the floor from high quality concrete reinforced with steel microfibers, which are performed on construction site as strips of width greater than 100 mm.

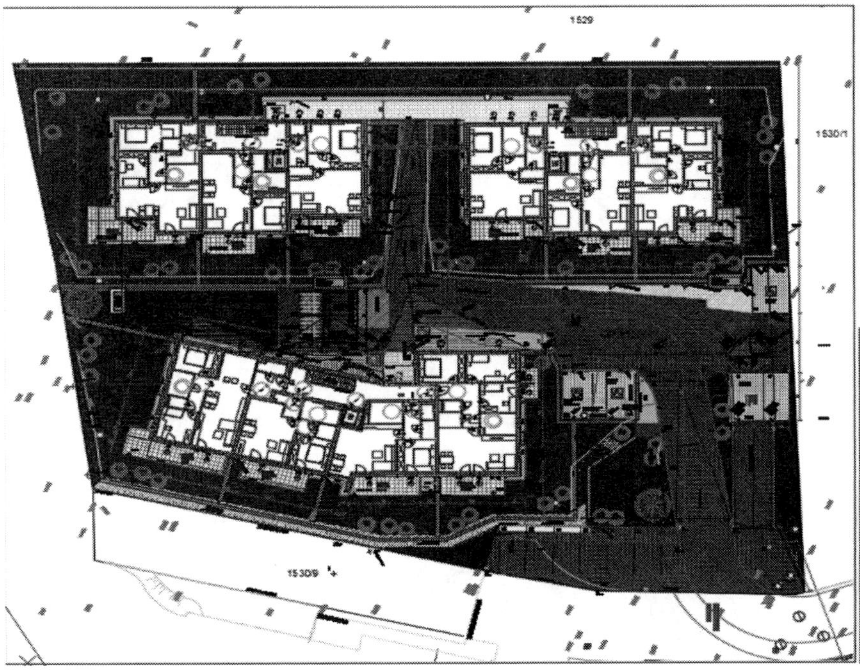

Figure1. Situation – buildings A, B and C.

The layout of the ground floor and view of building A, B and C and cutaway view of the building C are shown in figures 2, 3, 4 and 5 [2].

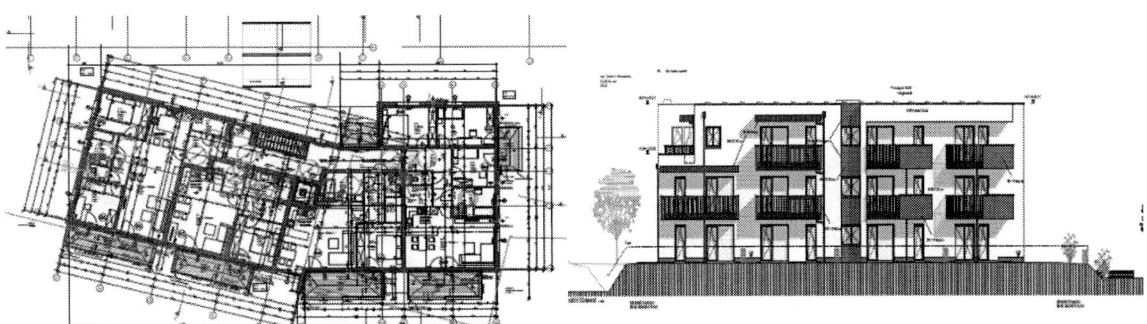

Figure 2. Layout of the ground floor and view of building A.

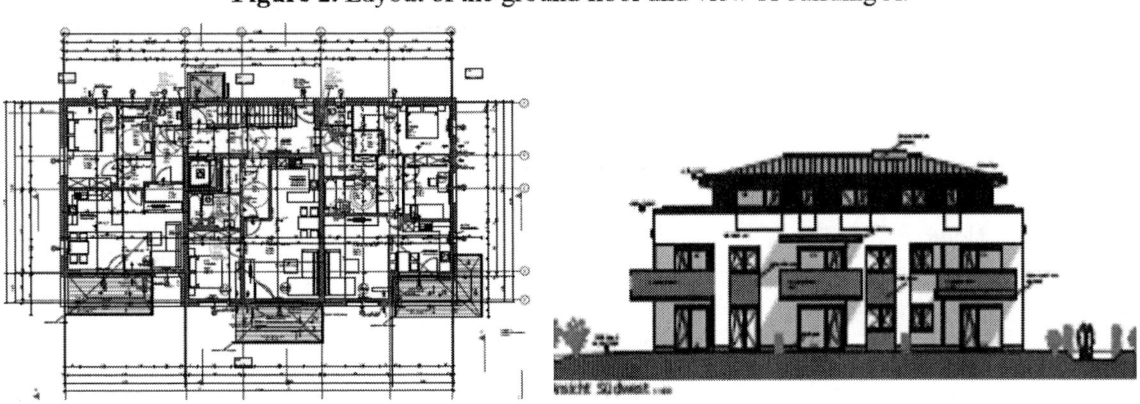

Figure 3. Layout of the ground floor and view of building B.

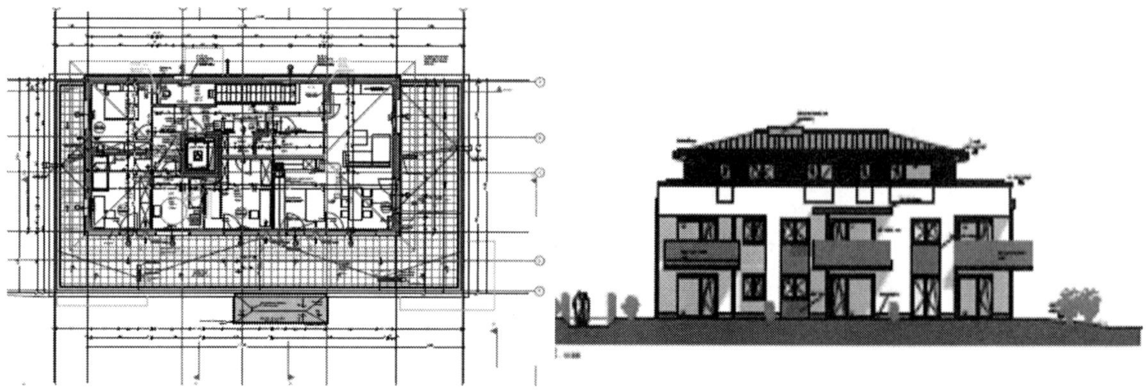

Figure 4. Layout of the ground floor and view of building C.

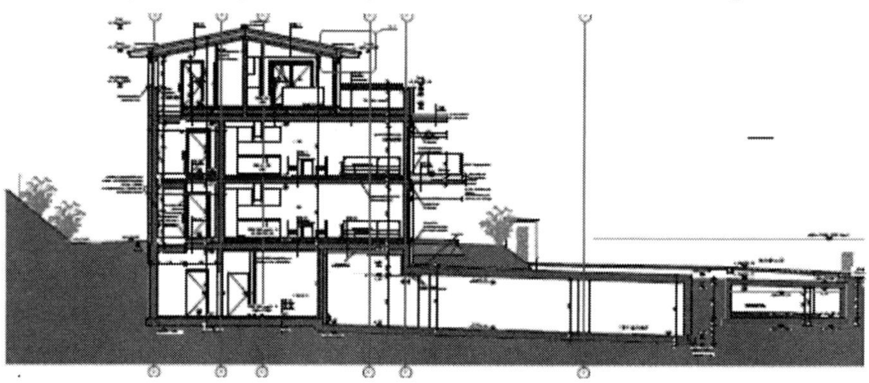

Figure 5. Cutaway view of the building C.

2. Structure of the building

The main bearing structure is designed in the form of load-bearing reinforced concrete solid slabs, supported by Porotherm load-bearing walls from the ground floor to the roof slabs, vertical columns and elevator reinforced concrete wall panels. Reinforced concrete floor slabs are constructed in-situ and their thickness is: d = 22.30 cm and 35 cm. At ground level these slabs are supported by a system of longitudinal and transverse beams for buildings B and C, where beams are supported by basement walls and garage columns, while at building A they are supported on internal walls, columns and beams. The ground floor slab also has capitals in the part of the columns of the garage, measuring 200/200 cm, the thickness of the capitals d_{kp} = 30 cm. The foundations is in the form of reinforced concrete slab d = 30 cm thick with capitals below the columns d_{kpt} = 20 cm. The foundation slab is divided into two parts - by expansion in axis 7. The connection of the parts of the foundation slab is achieved by specially built horizontal anchors in the middle of the thickness of the slab. Underneath the slab and walls of the building complex, a unique classic waterproofing is envisaged with the help of the waterproofing of the concrete foundation slab.

The following loads were analyzed in the structure calculation of the building:

- imposed load = 2.0 kN/m^2,
- snow load = 3.93 kN/m^2,
- partition walls = 1.20 kN/m^2,
- floor layers = 2.50 kN/m^2,
- wind load = 0.80 kN/m^2,
- balconies = 4.00 kN/m^2,
- seismic loading is irrelevant,
- the self-weight of the structure is taken into account by the program.

3. Structural analysis

The calculation of the structure included the calculation of the foundation slab of the garage and complete buildings A, B and C. Figure 6 shows a 3D model of a structure.

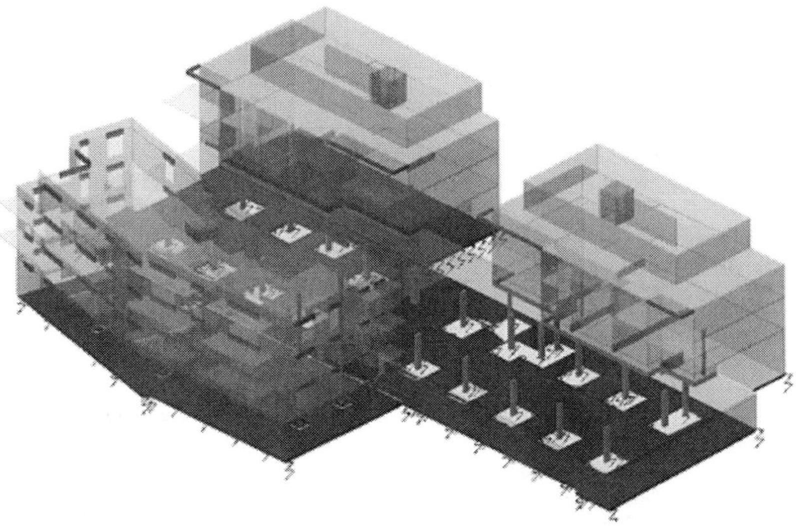

Figure 6. 3D model.

Dimensioning of structural bearing elements was carried out for the relevant influences according to load combinations, in all as shown in the static calculation. For simplicity of construction, the reinforcements in lower zone were Ø 12/15, in both directions and additional reinforcement is Ø 10/15 at an angle of 90° and in the upper zone Ø 7/10 and additional reinforcement above the columns Ø 20/12.5, and in middle of the slabs Ø 12/15. Reinforcements was shown in figure 7 and figure 8. The concrete class of all load-bearing structural elements of the structure, cast on site, is C25/30, and reinforcement B500.

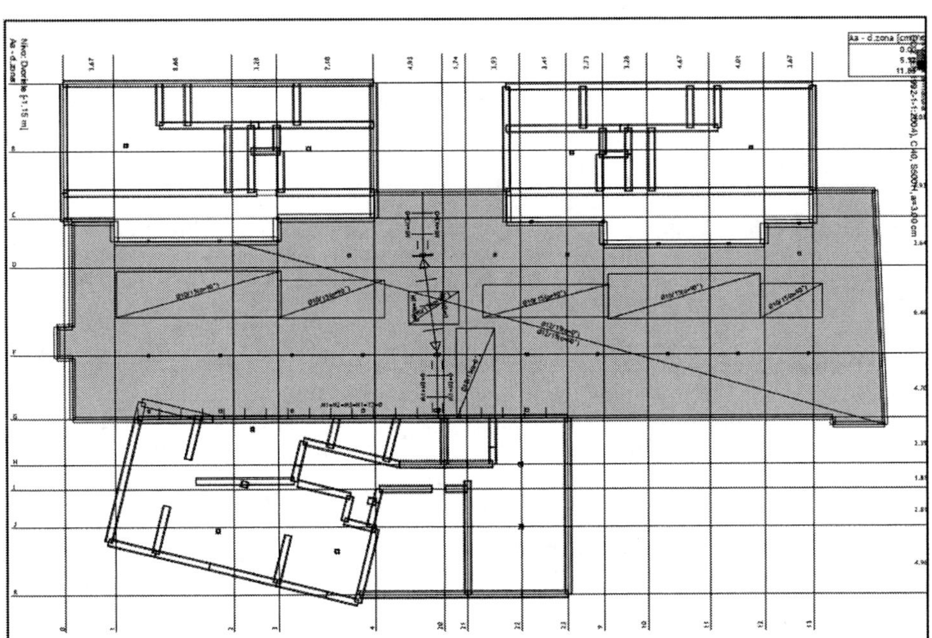

Figure 7. Reinforcement of the slab – lower zone.

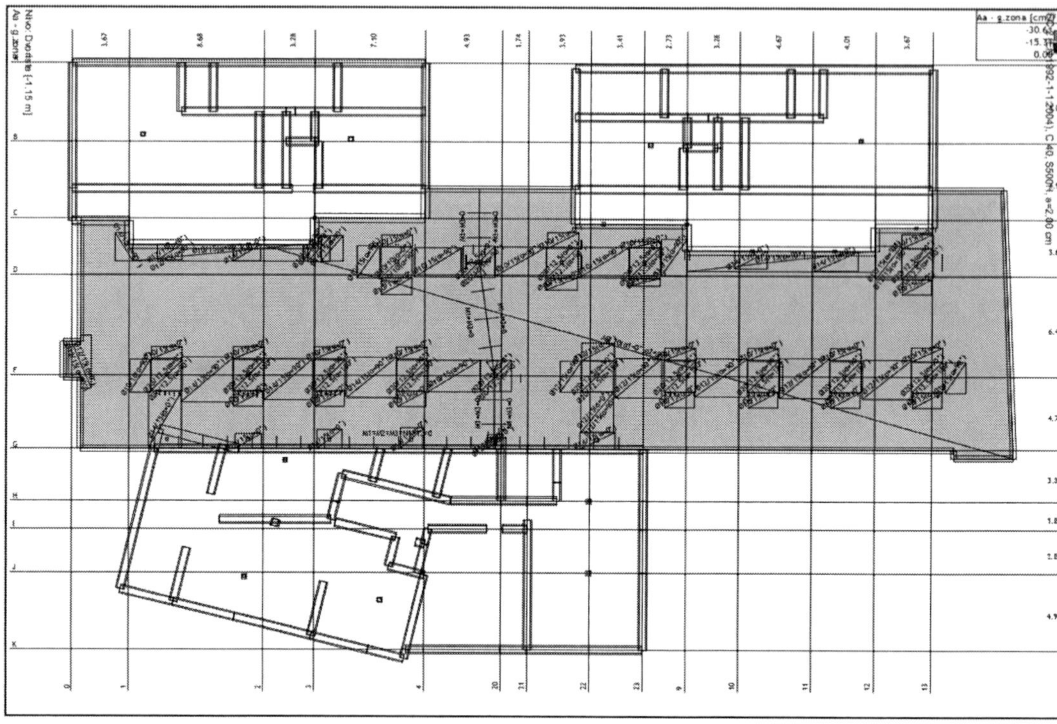

Figure 8. Reinforcement of the slab – upper zone.

Special attention was paid for checking punching shear of the slab. The certified computer program the Schöck BOLE system was used. The following text shows the calculation [3]:

Force

Punching shear load	V_{Ed}= 930 kN
Dynamic part	$V_{Ed,dyn}$= 0 kN
Load increasing factor	β=1.15
Dimensions – Column	
Width of the column	a=250 mm
Thickness of the column	b=200 mm
Thickness of the slab	h=300 mm
Statical height	d=270 mm
Concrete protection layer	c_o; c_u=20; 20 mm
Material	
Concrete	C35/45 (f_{ck}=35 N/mm^2)
Steel	B500 (f_{yk}=500 N/mm^2)
Reinforcement	ρ=0.78%

A_{sx}=20.9 cm^2/m (Ø20/150 mm); A_{sy}=20.9 cm^2/m (Ø20/150 mm)

Punching shear calculation DS EN EC2:2015 + NA:2017 + ETA(v3=0.95)

Factor k	$k=\min\{1+(200/d)^{1/2};2\}=1.86$
Influence of the slab width	$\eta=1+(d-200)/1000$ {min 1.0; max 1.6}=1.07
Factor $C_{RD,c}$	$C_{Rd,c}=0.18/\gamma_c=0.12$
Minimum load-bearing capacity of concrete	$v_{min}=(0.0525/\gamma_c)\,k^{3/2}\,f_{ck}^{1/2}=525.5$ kN/m^2
Load-bearing capacity of concrete	$V_{Rd,c}=\max\{C_{Rd,c} \times (\rho \times f_{ck})^{1/3};v_{min}\}=671.1$ kN/m^2
Edge of the column u_o	
Perimeter of cross section	u_o=0.900 m
Load-bearing capacity of concrete	$V_{Rd,c,max,uo}=0.4\ v \times f_{cd}=4816.0$ kN/m^2
Load-bearing capacity of concrete	$V_{Rd,c,max,uo}= V_{Rd,c,max,uo}\ d \times u_o=1170.3$ kN

Critical circular cross section u_{crit}

Critical distance	a_{crit}=2d=540 mm
Range of circular cross section	u_{crit}=4.293 m
Transverse force	$V_{Ed,\beta}=\beta \times V_{Ed}$=1069.5 kN
Load-bearing capacity of concrete	$V_{Rd,c,crit}= V_{Rd,c} \times d \times u_{crit}$=777.8 kN
Maximum load-bearing capacity	$V_{Rd,max,crit}= V_{Rd,c,crit,(CRdc=0.11)} \times 1.96$=1422.9 kN
min{ $V_{Rd,c,crit}$; $V_{Rd,c,max,uo}$ }=	777.8 kN $\leq V_{Ed,\beta}$= 1069.5 kN $\leq V_{Rd,max,crit}$=1422.9 kN

Punching shear reinforcement required!

Chosen:	8xSchock BOLE U 20/260-3/A570-CV20
Proof of load-bearing capacity of steel	$V_{Ed,\beta}$= 1069.5 kN$\leq V_{Rd,sy,crit}=m_c \times n_c \times A_{s,I} \times f_{yd}/\eta$=2042
External circular cross section	$u_{out(ls+1.5d)}$
Length of reinforced area	s=475 mm
Range of circular cross section	u_{out}=6.429 m
Load increasing factor	$\beta_{red}=\beta$=1.15
Transverse force	$V_{Ed,out}=\beta_{red} \times V_{Ed}$=1069.5 kN
Load-bearing capacity of concrete	$V_{Rd,c,out}=\max\{C_{Rd,c,out} \times k(\rho \times f_{ck})^{1/3};v_{min}\}$=671.1 kN/m^2
Load-bearing capacity of concrete	$V_{Rd,c,out}= V_{Rd,c,out} \times d \times u_{out}$=1164.9 kN
	$V_{Ed,out}$=1069.5 kN $\leq V_{Rd,c,out}$= 1164.9 kN

Punching shear reinforcement sufficient!

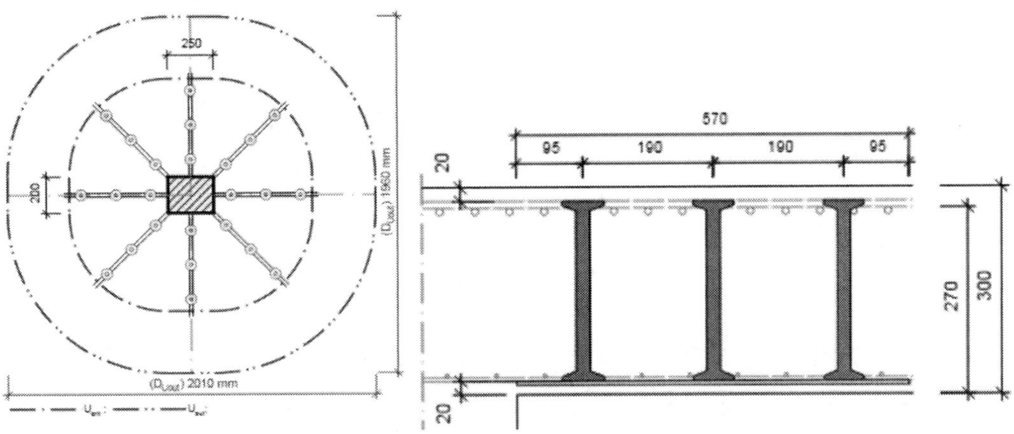

Figure 9. Additional reinforcement for punching shear strength of the slab.

Figure 9 illustrates a method of securing a slab from punching shear by the use of round steel dowel. The protective layer to the reinforcement is 20 mm.

4. Thermal bridges in buildings

At low outdoor temperatures, such as is the case for the location of the above buildings, in the area of insufficient insulated connections the surface temperature in the interior of the building decreases more than in other areas. Thermal bridges are created in these places, with two types being distinguished:

- material-induced thermal bridges that occur when building elements of significantly different thermal conductivity are found next to one another;
- geometric thermal bridges, which arise when the surface that gives the heat is much larger than the one that receives the heat.

Particularly critical areas are balconies-terraces, because they are conditioned by the material and geometry conditioned thermal bridges. Condensation is created in the area of the thermal bridge, which leads to permanent damage to the ceilings and walls of the apartments. The consequence of thermal bridges is higher energy consumption, which in turn has the effect of increasing the heating costs of the building but also increasing the environmental impact. This structural design uses the certified innovative element Schöck Isokorb. Schöck refers to solutions to problems in the field of structural physics, statics and structures. Also, this innovative element is used in the areas of thermal insulation and sound protection in new buildings. The following figures 10 and 11 show the effects of heat with and without a thermal bridge.

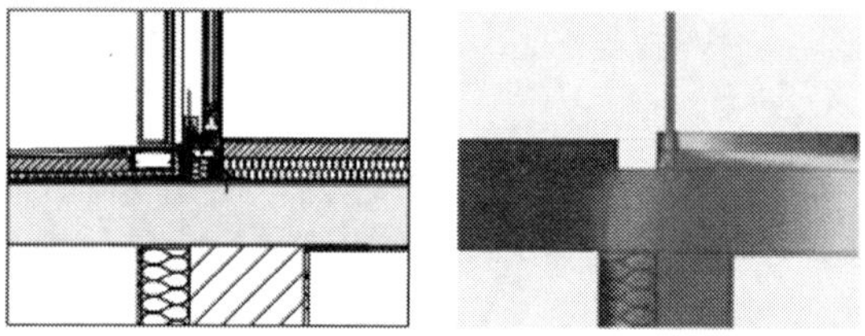

Figure 10. Without thermal bridge interruption.

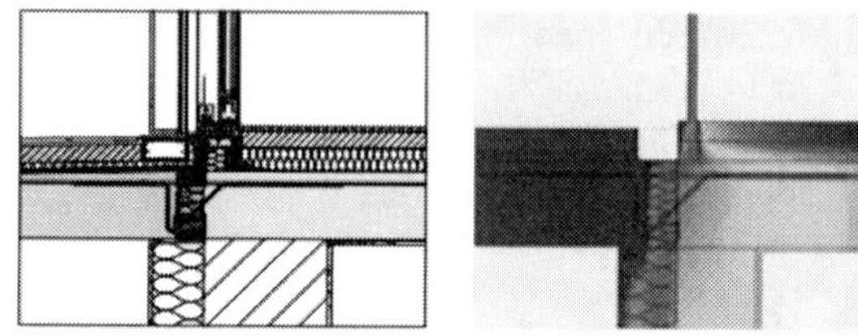

Figure 11. With thermal bridge interruption.

The Isokorb, Novomur and Tronsole Schöck product groups represent outstanding solutions, the latest technology. Figures 12, 13 and 14 show the static influences in the balcony-terrace by applying this system to one terrace of object A and also, there is the case of slab-slabs, balcony length l = 7.00 m and cantilever a = 1.94 m.

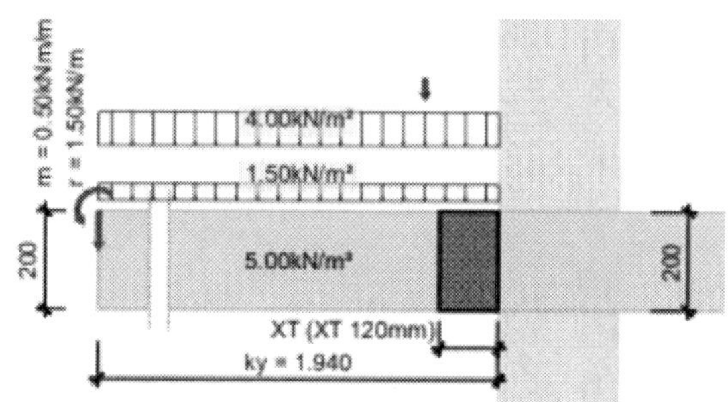

Figure 12. Loading of the balcony.

Figure 13. Reinforcement in the slab.

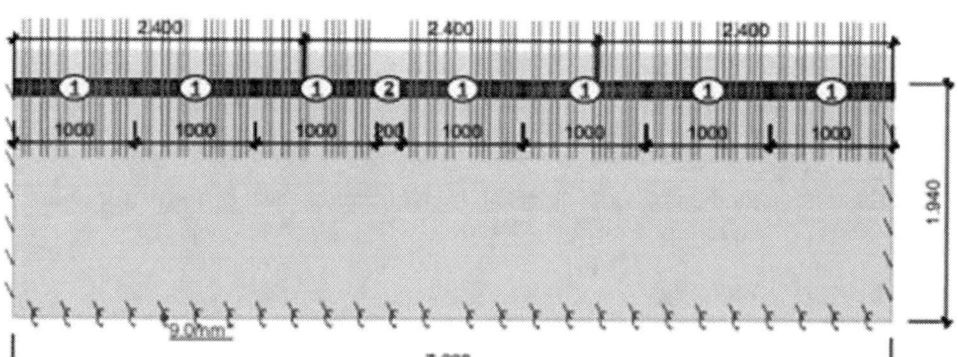

Figure 14. Balcony with thermal bridge interruption.

The bending moment of the cantilever (balcony) is accepted by the coupling of internal forces, tension and pressure.

In the upper zone, the tensile force receives the reinforcement and the pressure force is transmitted directly through the lower part of the innovative element of the Shock Bole. The balcony slab is made of high grade steel fiber. Figure 15 gives a detail of the construction of the balcony with the application of the innovative solution.

Figure 15. Balcony with reinforcement for connection with the slab (with innovative element).

The installation of Schöck Isokorb in the balcony connection area interrupts the thermal conductivity of the reinforced concrete slab. The thermal bridge was constructively broken and the energy flow at the critical location was minimized.

5. Conclusion

This paper analyses in detail the static analysis and optimized dimensioning of reinforced concrete structural elements. According to the EC code, the punching shear strength of the slab was also analysed and one of the methods of punching protection was adopted. Section 4 deals with thermal bridges and methods of reducing the influence heat transfer from the structural elements of the balcony-terraces with the interior reinforced-concrete slabs of the floors of the buildings.

Acknowledgement

The present work has been supported by The Ministry of Education and Science of the Republic of Serbia (Project No. ON174027).

References

[1] Sertifikat Bauteile GmbH, Austrija in Schöck Bauteile GmbHBaden-Baden Nemacka Isokorb XT bzw. T Typen H
[2] Besevic M and Gajic M *Structural design of buildings A, B and C at Salzburg* Austria
[3] Eurocode 2 *Design of concrete structures - Part 1-1: General rules and rules for buildings* European Committee for Standardization (CEN) Brussels Belgium

IOP Conf. Series: Materials Science and Engineering **789** (2020) 012005 doi:10.1088/1757-899X/789/1/012005

Reducing heat losses – first step toward nZEB

I Boian[1] and I Tuns[2*]

[1] Department of Building Services, Transilvania University of Brasov, Romania
[2] Department of Civil Engineering, Transilvania University of Brasov, Romania

[*] ioan.tuns@unitbv.ro

Abstract. Energy consumption for space heating and air conditioning holds a significant share in the total energy consumption, among them residential and tertiary buildings being on the first places. This strong energetic dissipative character of different elements of the building envelope offers multiple possibilities of intervention aiming to reduce the heat losses. The Romanian requirements concerning the elements evaluating conformity with the Standard C107-2010 and the Order no.2641/2017 are given with a focus on thermal bridges influence aside with a short description of different ways to estimate this influence in EU countries. The paper presents in general, concepts to be used in order to achieve the nZEB standard developed by the Energy Performance of Building Directive having in mind a mitigation of greenhouse gases, GHG emission. In this context, is presented the role of balcony elements as undesirable "heat exchanger" and this thermal bridge is depicted together with some ways to avoid this inconvenience.

1. Introduction

Energy consumption in the residential sector holds a relative significant share being responsible for the Greenhouse Gases, GHG emission, specifically the carbon dioxide, CO_2 contributing to the climate change. Reducing the climate change effects implies some measures and technical issues necessary to be implemented by professionals to improve the energy efficiency of buildings. In this context, the energy auditor holds a role to play not only when certifying the energy performance of buildings but also in case of consultancy given to architects during the design phase.

Improving the energy efficiency needs a high level of thermal insulation and airtightness as part of design process of new buildings, and an effort to renovate the existing ones based on clear principles, too. Traditional design and practice have been concentrated on the envelope insulation, i.e. on exterior walls, the floor and the roof of buildings aiming to reduce their thermal transmittance, U. Thermal bridging is an important consideration in case on nearly Zero Energy Building design.

Until recently little attention was paid to heat losses involved by joining elements and by building openings, resulting in uncontrolled air infiltration. Without measures limiting thermal bridges from joining elements, and a better envelope sealing, the share of heat losses caused by them tends to increase, as the insulation standards have evolved, in the context of the renovation of the building envelope aiming to a better efficiency. To highlight the significance of thermal bridges it must be pointed out that the effect of the thermal bridges on the energy requirement can equal, or even exceed, the energy provided by the solar collectors for domestic hot water, which is beyond the current concerns of the general public.

As a result, it is not sufficient to estimate the U-value for main parts of walls, roofs and floors, being necessary to consider the non-repeated thermal bridges.

IOP Conf. Series: Materials Science and Engineering **789** (2020) 012005 doi:10.1088/1757-899X/789/1/012005

Researches have shown that about 30% of the total heat loss of building envelopes result from thermal bridges, and annual CO_2 emissions can be reduced by up to 10% through the use of the enhanced construction details and an improved airtightness. Therefore, it is necessary as far as possible the design and execution of works to be based on the principles that lead to improved performance by over 85% in the case of lintels, the joints between walls and floor, ceiling and wall and of the gable respectively [1].

2. Compliance of the building thermal performance with the Romanian requirements

The Romanian Standard C107-2010 [2] defines a limit for the annual energy demand for heating, Q measured in kWh/(m^2 year)

$$Q < QN \tag{1}$$

In case of residential buildings this limitation depends on the surface area to volume (A/V) ratio and it makes a distinction that refers to the design contracting date: before, or after January 1, 2011, as depicted in figure 1.

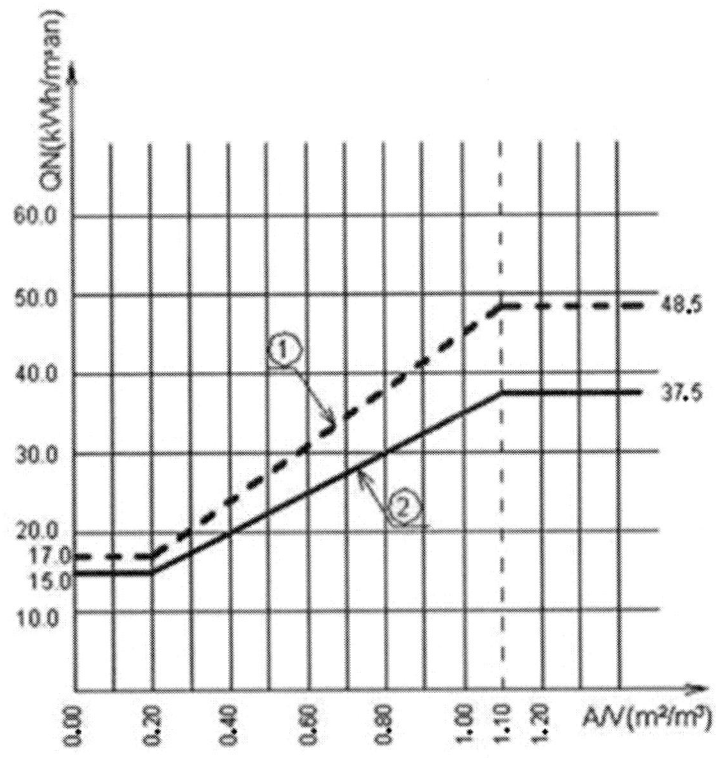

Figure 1. The Annual normative energy demand for heating QN, in case of residential buildings correlated with the Surface area to volume (A/V) ratio: 1 – before January 1, 2011; 2 – after January 1, 2011.

Later, in 2017 the Order no. 2641 specified the maximal annual energy consumption from non-renewable sources $q_{an, max}$ as 153 kWh/(m^2 year) for low-rise residential buildings (ground floor and no more than 3 stories), and 117 kWh/(m^2 year) for buildings higher than ground floor and 3 stories, respectively.

Considering the thermal insulation of a building envelope, the Standard C107-2010 and the Order no. 2641/2017 [3] both require that thermal insulation overall coefficient, G, related to the building-as-a-whole, to be limited by the normative overall coefficient of thermal insulation of the building, GN:

$$G < GN \tag{2}$$

The thermal insulation overall coefficient, G, expressed in $W/(m^3K)$, depends on the surface area to volume (A/V) ratio, and on the mean thermal resistance, R_m, and on the number of air changes per hour, ACH:

$$G = \frac{A/V}{R_m} + 0.34 \cdot ACH \tag{3}$$

The normative values of the overall coefficient GN are tabulated in the Standard C107-2010 and in the Order no. 2641/2017being a function of the surface area to volume (A/V) ratio and having the number of floors, N as parameter. Figure 2 presents in a graphical form these tabulated values, GN and the calculated values G must be under the curve corresponding to the number of floors, N.

As can be noted, the surface area to volume (A/V) ratio, has higher values for single-family houses than in case of multifamily buildings. As a result, to keep G < GN – the requirement of the standard – makes necessary a better insulation for single-family houses, i.e. higher values for the mean thermal resistance, R_m is necessary.

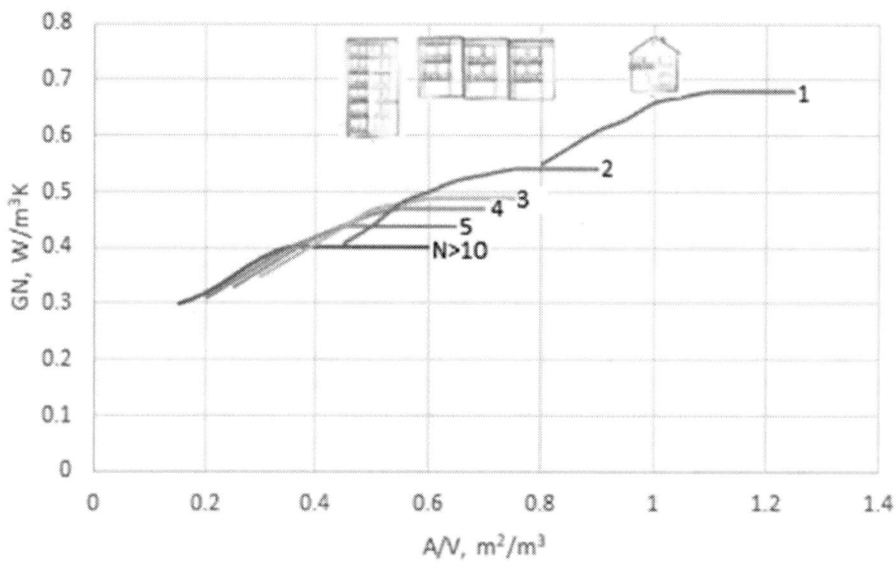

Figure 2. The normative overall coefficient GN correlated with the surface area to volume (A/V) ratio and having the number of floors, N as parameter.

Besides the above requirement, G < GN, the thermal resistance, R' of different elements of the building envelope must respect the condition

$$R' > R'_{min} \tag{4}$$

where R' refers to the corrected value that takes into consideration the influence of thermal bridges and the normative values for R'_{min} in case of different building elements are specified by C107-2010 and by the Order no. 2641/2017, being presented in figure 3.

Thermal bridges result in an overall reduction in thermal resistance of the envelope element with a higher heat transfer / heat leaks associated with increased carbon emissions, and a thermal discomfort.

Thermal bridges can occur in different locations of an envelope, usually at junction between elements of the envelope, like balcony-wall, slab-grade, floor-wall, roof or ceiling-wall, window, or door-wall etc.

Thermal bridges are characterized by two- or three-dimensional heat transfer, with fluctuations of temperatures, meaning a non-steady-state process.

To avoid calculation difficulties of such a complex model the concept of thermal bridge was introduced. Thermal bridging is an important consideration in Low energy Building design. It has been estimated that up to 30% of the heat loss in a well-insulated building is through these non-repeating thermal bridges.

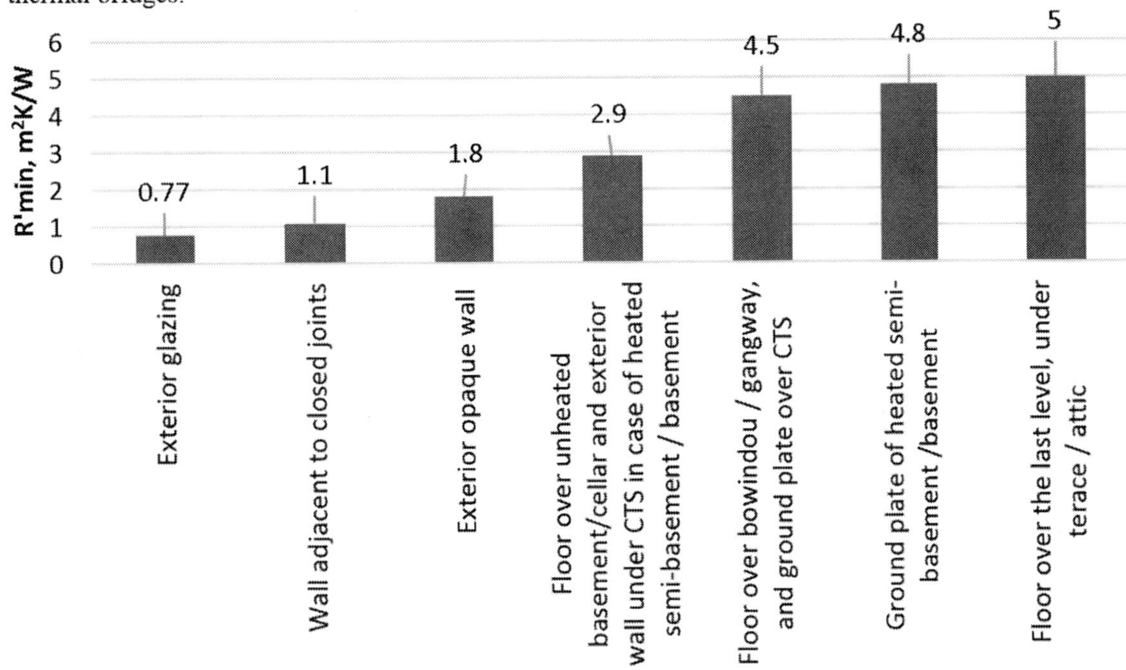

Figure 3. The normative values for corrected thermal resistance, R'_{min} specified by the Standard C107-2010 and by the Order no. 2641/2017.

3. Thermal bridges and insulation

The corrected thermal resistance, R' as a reverse value of the transmittance, U' can be written as

$$U' = \frac{1}{R'} = \frac{1}{R} + \frac{1}{R_\psi} + \frac{1}{R_\chi} \tag{5}$$

where, besides the unidirectional (or repeating) thermal resistance, R some other two elements are considered: Thermal resistance of linear (or non-repeating) thermal bridges, R_ψ, and Thermal resistance of point thermal bridges, R_χ, measured in m^2K/W and defined as

$$R_\psi = \frac{A}{\Sigma(\psi l)} \tag{6}$$

$$R_\chi = \frac{A}{\Sigma \chi} \tag{7}$$

A linear thermal bridge adds an extra heat flow to the unidirectional one being specific for building elements junction like corners (floor-to-wall junction, balcony-to-wall or wall-lintels /jambs), or in case of the openings (windows, doors) installed in walls.

As mentioned before equation (6) the Thermal resistance of a linear thermal bridge, R_ψ, depends on the linear thermal transmittance, ψ expressed in W/(m K), and on its length, l.

Dividing the surface area A of the envelope by the sum of ψ_l calculated for all thermal bridges existing on this surface results in a value representing the resistance of linear thermal bridges, R_ψ. Cantilevered balconies and exposed slab edges are the most critical thermal bridges in a building envelope.

In case of a point thermal bridge, occurring in one spot, energy losses are caused by elements that penetrate the insulating layer, like anchor bolts, or curtain wall supports, fastening elements, steel balconies etc. The point thermal transmittance, χ specific for every situation is measured in W/K.

Considering that thermal bridges can result in additional heat losses, that impact the amount of energy required to heat a space in cold climates, it is obvious the necessity to prevent thermal bridges by creating thermal breaks.

From figure 4 it can be noticed that a thermal resistance of a linear thermal bridge, R_ψ having the same magnitude as the unidirectional thermal resistance, R will reduce to one half the corrected thermal resistance, R' when compared to R (factor r).

In case of a cumulated R_ψ being only 50% of R the corrected thermal resistance, R' will be lowered to about 35% of the unidirectional thermal resistance, R (point thermal bridges was neglected in this evaluation). As a result, heat will follow this low-resistance path resulting in higher heat losses.

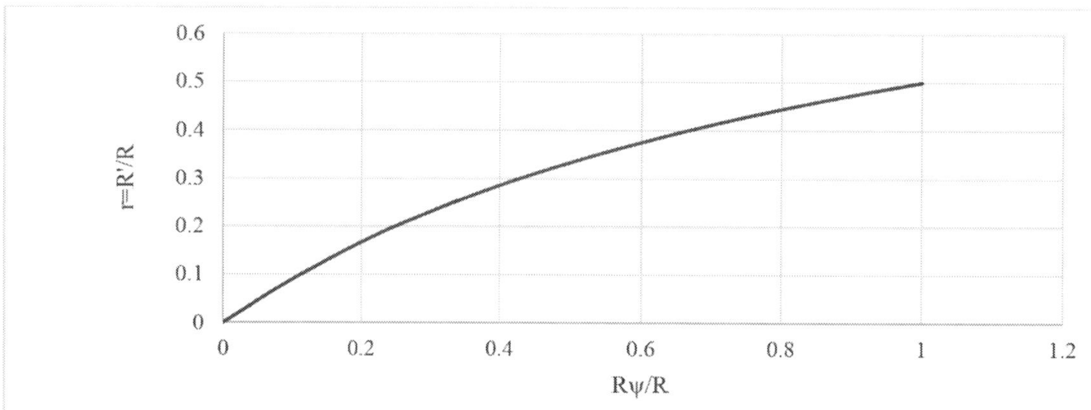

Figure 4. The influence of thermal resistance of a linear thermal bridge, R_ψ on the corrected thermal resistance, R'.

Heat loss is becoming significant when thermal bridges contribute to a lower thermal resistance of the envelope, as figure 5 shows.

Different linear thermal bridges have various influence on heat loss, depending on psi-value and on the corresponding length, but usually floor-to-wall and balcony-to-junction are on poll position followed by slab-on-grade, roof, or ceiling-to-wall junctions. Window-to-wall and door-to-wall junction can have a significant influence because of their number.

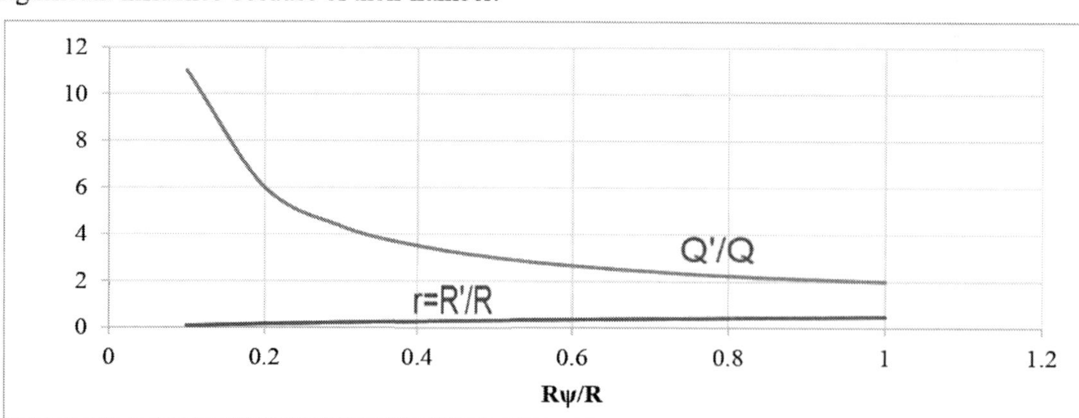

Figure 5. The thermal resistance of a linear thermal bridge, R_ψ is affecting the heat loss: the lower the R_ψ, the higher the heat loss Q'.

Evaluating the additional heat losses due to thermal bridges through the explicit method is the most accurate way but considering its tedious character it represents an important disadvantage. It is not only the evaluation of the lengths and the number of every type of thermal bridges, but the access to an atlas offering the psi-values is necessary too.

Reducing and avoiding thermal bridges is not only recommended but necessary. A continuous thermal insulation layer consisting of mineral wool, rigid foam, or polystyrene board is the first issue, which will be accompanied with double/triple pane windows with gas filler and low-emissivity coating for installed in thermally broken frames resulting from low conductivity material. Once the envelope is insulated the relative contribution of thermal bridges is amplified in the frame of the heat balance: ψ-values increase after the thermal insulation as a result of thicker layers near the thermal bridges [4]. Higher heat losses and lower temperatures on the adjacent interior surfaces will result with local condensation and a reduced comfort.

Lower temperatures inside the structure will be followed by supplementary stress in material elements with possible interstitial condensation resulting in damage due to humidity. Critical building elements are masonry lintels, floor-to-wall junctions, window-to-wall and door-to-wall junctions, steel studs and others.

Small buildings like single family houses having an unfavorable surface area to volume (A/V) ratio, need to be thermal bridge-free as much as possible. Cantilevered balconies are examples of critical thermal bridges involving expensive compensation measures and preventing high standards as the passive house.

3.1. Cantilevered balconies before and after wall insulation

Balconies as an extended surface through which occurs a heat exchange increases heat losses of the building: it works like a heat exchanger an undesirable one having higher temperatures in comparison with the near surfaces, as shown in figure 6.

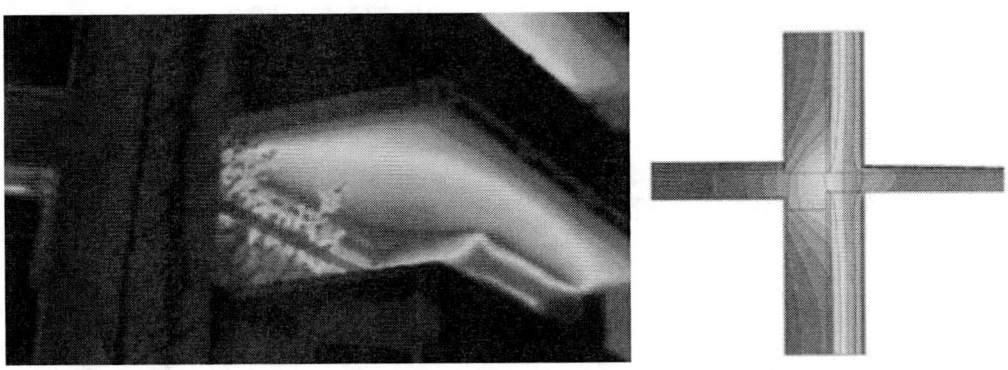

(a) (b)

Figure 6. Thermal bridge of a balcony slab: a) infrared scan; b) numerical simulation.

This numerical example [5] is intended to show what changes occur in the balcony slab after insulating the façade wall on which it is connected. The slab having the dimensions shown in figure 7 (a) and being of the same length as the junction wall considered as of 10 m^2 was investigated through the finite difference method.

Temperatures and heat fluxes calculated for every element resulted after the discretization of the slab are shown in figure 7 (b) and 7 (c). The heat loss of the wall, before insulation is $Q_{w\ nonins}$= 3.85 kWh/day for a unidirectional thermal resistance, $R_{w\ nonins}$= 2.56 m^{2}K/W (a 10 cm layer of mineral wool exists between the two concrete structure elements) and being exposed to an outside temperature of -21°C (Romania, Climatic zone IV) and the interior temperature of +20°C (q=16.05 W/m^2). In the same time the slab "extracts" from the building an amount of heat $Q_{s\ nonin}$= 4.3 kWh/day.

As can be noted from figure 7 (b), almost the entire length of the slab is at the same temperature as the ambient one: only the temperature of the first three elements of the slab being in the vicinity of the wall differ with less than one degree centigrade from the outside temperature. This means that the contribution of elements no. 5, 6...11 is negligible, only the first 3-4 elements being active in transferring heat.

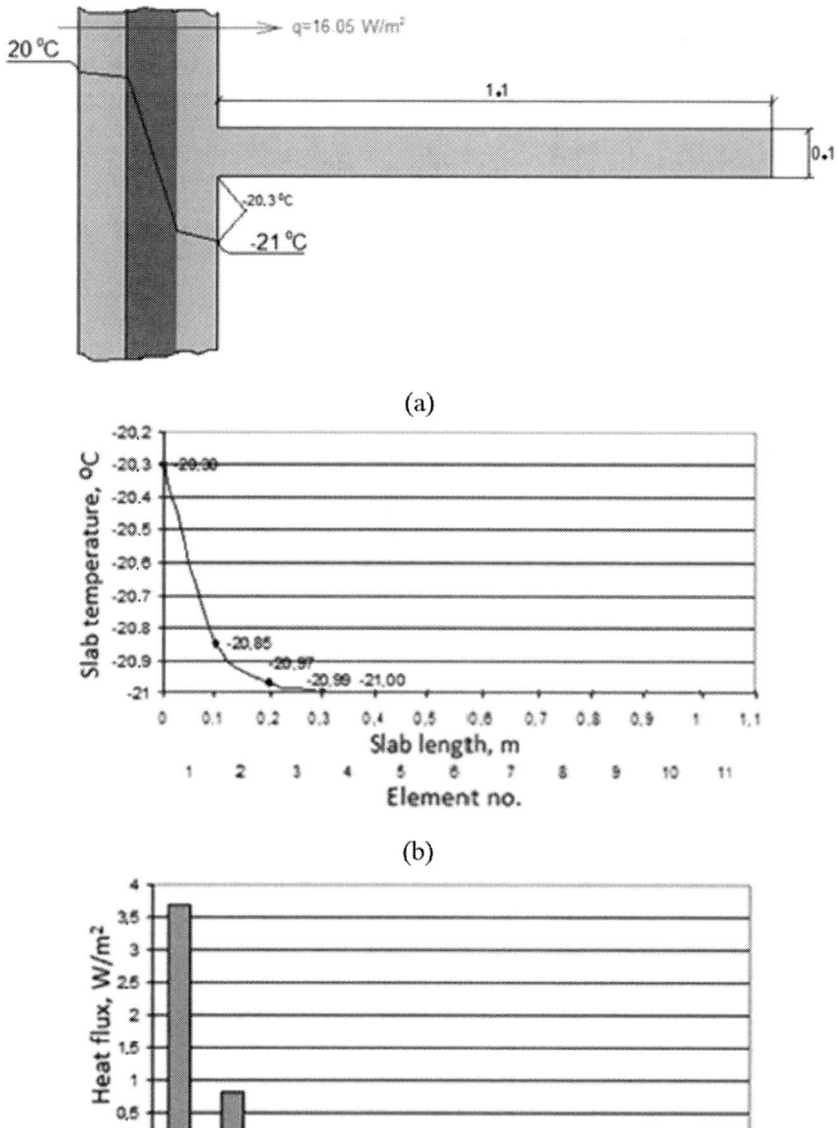

Figure 7. The balcony slab (a) and its temperature distribution (b), with the element corresponding heat flux (c) for the insulated wall.

After the insulation of the wall with 10 cm of polystyrene or mineral wool its thermal resistance is improved to $R_{w\ ins}$=4.83 m^2K/W resulting in lower heat losses $Q_{w\ ins}$=2.0 kWh/day that is almost the half of the previous case. Considering that wall insulation keeps the first element of the slab separated from the cold air, (i.e. no heat loss for this element) the temperature distribution in the slab is modified as depicted in figure 8 (b): higher temperatures compared to the previous situation when no insulation was added (-1.4°C instead of (-20.3)°C for the first element, (-17)°C instead of 20.85°C for the second element). The heat flux through the insulated wall has been lowered to 8.49 W/m^2 instead of 16.05 W/m^2 in case of uninsulated wall. In the meantime, heat losses from the slab to the surrounding air increased, as a result of higher temperatures in the slab after insulating the junction wall. Figure 8 (c) shows.

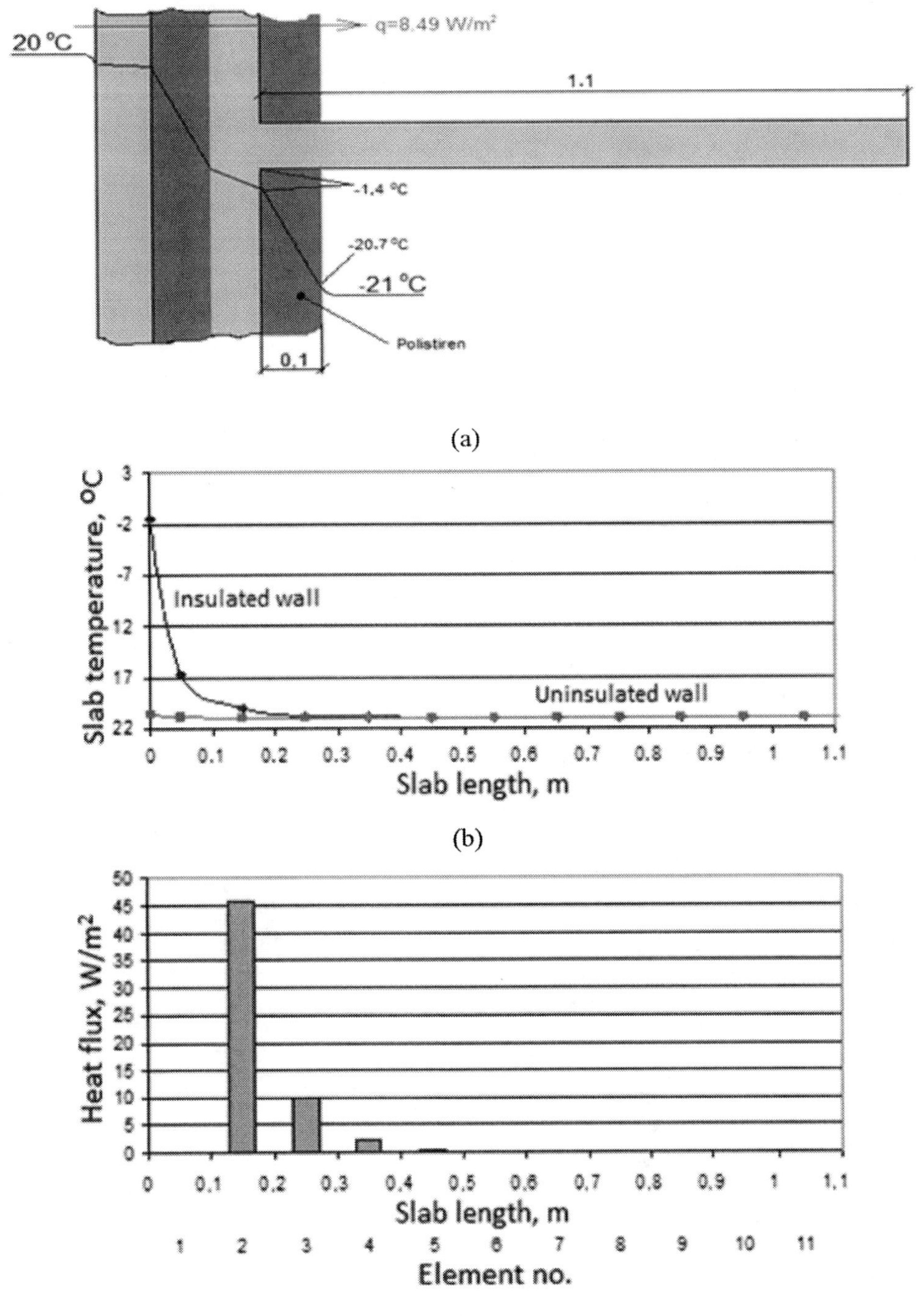

Figure 8. The balcony slab (a) and its temperature distribution (b), with the element corresponding heat flux (c) for the insulated wall.

A comparison of heat fluxes per every element of the balcony slab is given in figure 9: if the first element has a heat flux of 3.70 W/m^2 before insulation of the wall, and after the insulation it became zero having no contact with the exterior cold air; but as a result of the higher temperature the second element has a heat flux of 45.05 W/m^2 instead of 0.82 W/m^2.

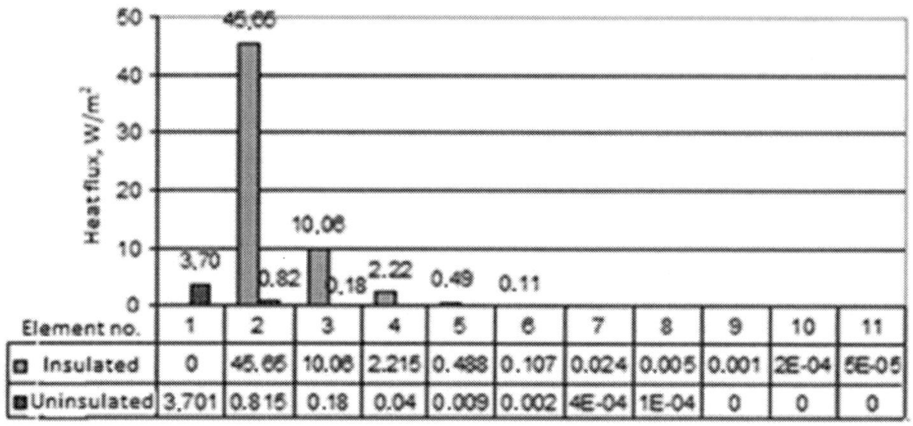

Figure 9. Heat flux of slab elements before and after the wall insulation.

Finally, insulating the wall its heat loss is lowered from 3.85 to 2.0 kWh/day, but this effort is not effective in case of the adjacent balcony slab, because of higher temperature inside the slab: the 4.30 kWh/day heat loss of the slab before insulation became 5.30 kWh/day, as shown in figure 10. In total only 8.15-7.3=0.85 kWh/day is saved, that is about 10%, when only the wall is insulated.

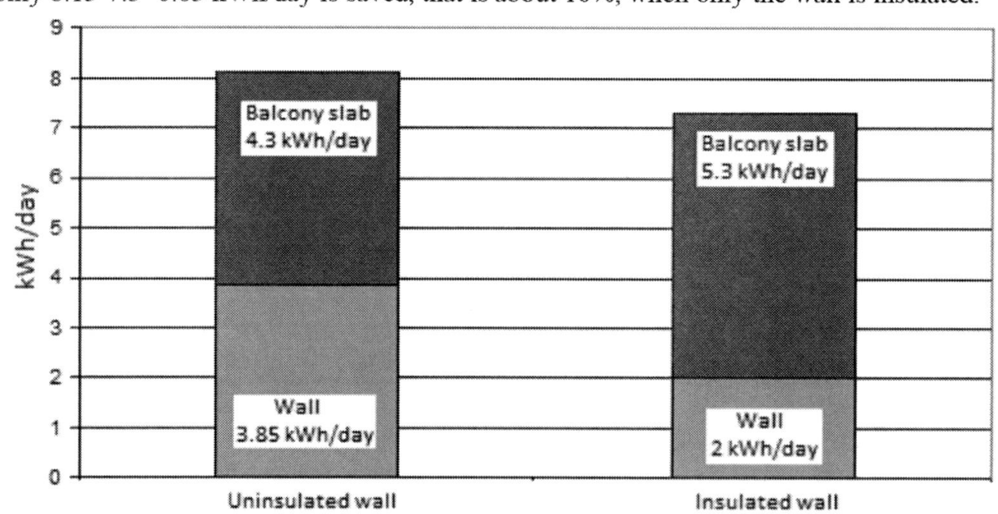

Figure 10. Heat losses of the balcony slab and of the wall before and after the insulation of the wall.

Adding insulation on the opaque surface of the walls is not effective to improve the thermal performance of a building, as shown before, if not inserted a thermal break between the exterior structure and the interior one.

The easy heat flow path around the insulation prevent minimizing the energy use and thermal bridges must be identified in order to be mitigated. In case of balconies, the exterior structure must be separated from the interior one by means of a thermal break like an expanded polystyrene block having a minimum thickness of 8 cm. Reinforcement bars connecting the exterior elements with the interior structure help to transfer loads and reduce the heat transfer because of the low conductivity of stainless steel (k=15 W/mK), figure 11.

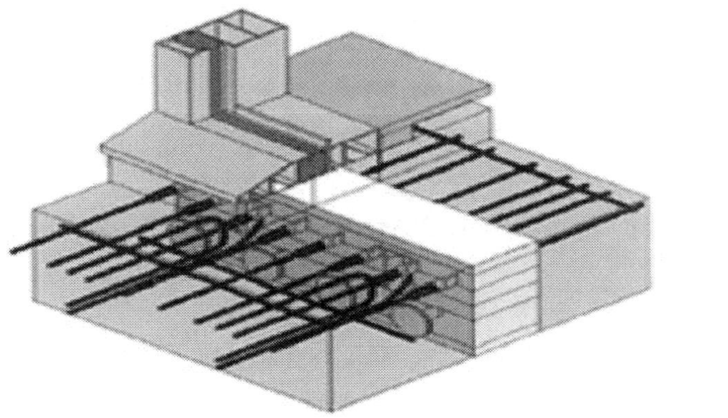

Figure 11. A thermal break (expanded polystyrene block) is inserted to separate the balcony slab from the interior structure.

4. U-values, Psi-values and y-values

Thermal bridge heat loss generally represents around 20% of the whole primary energy consumption of a low energy building. Reducing heat losses of a building requires the mitigation of the transmittance U' meaning to improve the thermal resistance of the envelope, not only the repeating one, R, but the linear thermal resistance, R_ψ too as shown before, equation (6). Considering the building as a free - point thermal bridges one, an identification of each linear thermal bridge followed by the evaluation of their thermal resistance R_ψ is necessary. Each linear thermal transmittance, ψ (numerical calculated values of psi with an accuracy of ±5%, or extracted from thermal bridge catalogues, 20%) multiplied by every measured length are then summed up to get the transmission heat loss transfer coefficient:

$$H_{TB} = \Sigma(\psi \cdot l) \quad (8)$$

Instead of this cumbersome procedure some EU countries like Germany, The Netherlands, Poland and Italy use a more practical method to evaluate this transmission heat loss transfer coefficient:

$$H_{TB} = y \cdot \Sigma A_{ext} \quad (9)$$

The y-value is referred to what we call as an increment to the U-value denoted ΔU and A_{ext} meaning the total external heat loss area of the building:

$$y = U' - U = \frac{1}{R'} - \frac{1}{R} \quad (10)$$

The default y-value is 0.15 but can be reduced to 0.08 and by thermally modelling the junction details, conforming to at least Accepted Construction Details this figure can be get down to less than 0.04 to determine the total heat losses. Using the value of y=0.04 result in ranges of U-values for individual building elements as:

- floors 0.18 – 0.12 W/m^{2}K;
- walls 0.25 – 0.16 W/m^{2}K;
- roofs 0.13 – 0.09 W/m^{2}K.

The condition to have as a user-defined value for $y = 0.04$ is the aggregate percentage change from the respective target U-values should not exceed 20%. For example, if the wall U-value was increased 10% above the wall target U-value (0.22 instead of 0.20), then the roof U-value could be at most 10% below the roof target U-value (0.10 instead of 0.11), so the aggregate change would then be 20%.To keep the U-values for the building elements in the above mentioned range means to design the elements and the entire building in conformity with specific prescriptions established in those countries that adopted such heat losses reduction.

5. Conclusions

Energy Performance of Buildings Directive recently recast sets as a target for 2050 to limit the global warming by a GHG emissions mitigation of about 90% in comparison with those existing in 1990. The built environment responsible for approx. 40% of the total GHG emissions will support this target through the nearly Zero Energy Building concept, the first step being the reduction of the existing heat losses. Envelope insulation without a thermal bridge reduction is only a half-step, therefore the identification and the evaluation of their magnitude being necessary.

The usual way to evaluate the thermal resistance of a linear thermal bridge with accuracy of about 20% involving an atlas containing psi-values will have as a final point a "good practice" construction detail. This is a time-consuming issue that has been replaced by some EU countries through the y-value method meaning that the final target can be achieved by the designer only by using a set of prescribed construction details giving a "Thermal Bridge Free" building, which is a thermal bridge coefficient lower than 0.01 W/mK. Junction performance can be improved by: isolating the thermal bridge with insulation, changing the thermal bridge geometry, increasing the thermal bridge heat path, and changing the thermal bridge material.

References

[1] Enhanced Construction Details 2009 *Guidance on energy efficient construction* Energy Saving Trust
[2] C107-2005, Annex 2 *Normative regarding the thermotechnical calculation of the building elements*
[3] Order no. 2641/2017 on the modification and completion of the technical regulation *Methodology for calculating the energy performance of buildings*
[4] BUILD-UP > 2010 *P188_Thermal bridge guidance principles ASIEPI-WP4.doc Good practice guidance on thermal bridges & construction details, Part I: Principles*
[5] Boian I 2003 *Heat and mass transfer with numerical applications* Editing House of Transilvania University of Brasov

IOP Conf. Series: Materials Science and Engineering **789** (2020) 012006 doi:10.1088/1757-899X/789/1/012006

Comparative study statically determined trusses with trapezoidal and parabolic shape with large span

M F Botiş[1*] and A Doşa[1]

[1] Department of Civil Engineering, Faculty of Civil Engineering, Transilvania University of Brasov, Romania

[*] mbotis@unitbv.ro

Abstract: In this work the authors propose a parametric study for beams with the opening of 30 m and 60 m. The analysed structures are statically determined and the optimal height and the related weight were established based on strength and stability conditions of structure elements. The structure analysis was performed using a program developed in the Matlab environment based on the displacement method. The analysed beams are manufactured of S355 steel bars with annular section. The beams were analysed for loads applied both the upper and lower nodes of structures. Based on the obtained results a comparison was made concerning the efficient use of material used for the trusses in respect to their weight.

1. Introduction

The statically determined truss type structures have a minimal number of connections required for geometrical invariability and for fixed structure in plan. To determine the sectional efforts only the static equilibrium condition is necessary, but for displacements determination both the static equilibrium condition and the condition of displacements compatibility with the connections are used. The trusses are used to cover large surfaces (sports halls, hangars), for antenna pillars, cranes [1]. The main simplifying hypotheses for truss calculation are:

- the bar axes are concurrent in joints;
- the joints are considered perfect hinged;
- the concentrated forces are applied in joints only.

The consequence of these hypotheses is that only axial effort of tensile or compression appear in the bars composing these structures. The experimental tests and calculations made for beams highlighted that if the largest dimension of cross section is lower than 1/10 of the bar length, the effect of the rigid joint could be considered as a secondary effect in respect to the efforts calculated in the hypothesis of perfect articulated joints [2].

The trusses could have a geometric invariability independently provided by the support base or together with this base. The simplest geometrical form with geometric invariability is the triangle formed by three articulated bars, figure 1. To two triangle joints another joint can be attached using two articulated bars. Therefore, for each joint, except the first three ones, two bars are necessary. Based on this consideration one can determine the relation establishing the condition of the own geometrical invariability of trusses [3].

$$b + 3 - 2n = 0; \tag{1}$$

IOP Conf. Series: Materials Science and Engineering **789** (2020) 012006 doi:10.1088/1757-899X/789/1/012006

where b is the number of bars; n is the number of joints; 3 represents the three cinematic degrees of freedom which must be blocked to prevent the planar displacement of a rigid body.

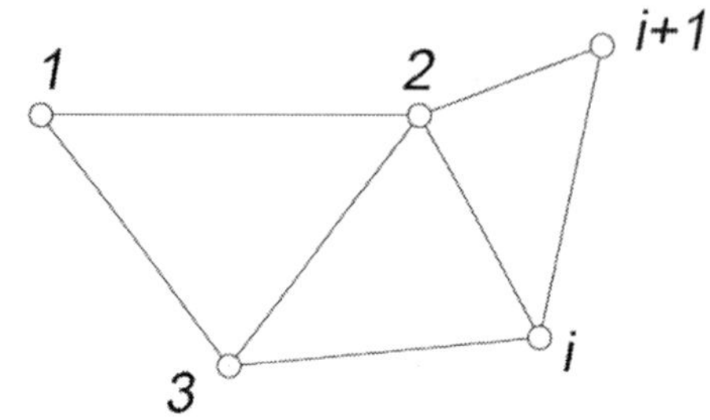

Figure 1. Generation articulated bars.

To attach the structure of trusses to the base, 3 external connections are additionally needed. Therefore, the condition of geometrical invariability together with the connections to the support base can be defined as:

$$b + r - 2n = 0; \tag{2}$$

where r is the number of the simple support;

The condition number 2 is the condition of static determination, which can be defined as:

$$N = b + r - 2n. \tag{3}$$

There are three possible cases depending on the N value:

- $N = 0$ statically determined trusses;
- $N > 0$ statically undetermined trusses;
- $N < 0$ mechanisms.

Because the number of unknowns is equal to the number of the efforts in the b bars and of the reactions in the r simple connections to the support base and, on the other hand, the number of equations of static equilibrium is twice the number of joints (two equilibrium equations for each joint), the result is that the equilibrium equation system is a determined compatible system in which the number of equations is equal to the number of unknowns.

Taking in account their composition, the trusses can be classified as follows:

- simple trusses including a succession of triangles; this configuration starts with a basic triangle to which one adds successively segments of one joint and two bars; the optimization of such trusses is discussed in this paper;
- composed trusses consisting of superposition of simple systems in such a way that the assembly has the geometrical invariability ensured; such structures are used when a reduced buckling length is required for the compressed elements;
- complex trusses where at least three bars are connected to each joint and the intersection points of the bars are not considered joints.

To calculate the efforts in the truss bars can use the following methods: the method of joint isolation, the method of simple sections and the method of finite elements. The first two methods are used for simple structures. The main advantage of the statically determined bars is that they are not sensitive to the loads caused by: manufacturing imperfections of the structure bars, structure mounting imperfections and thermal variations applied to the structures. Because all these loads do not introduce additional reactions or efforts in the structure and the sectional efforts in bars do not depend on the relation between the bar stiffnesses, each bar can be dimensioned and tested to the axial tensile and compression independently from the other bars. The main disadvantage of the statically determined structures resides in the fact that there is no structural redundancy because the undetermined degree is zero. Therefore, if the structure has displacements in the post-elastic field when the first plastic articulation is formed, the structure becomes a mechanism.

2. Calculation model

The trusses analyzed in this paper were calculated using a Matlab program which allows the determination of reactions and sectional efforts for the planar truss type structures. Based on the analysis with calculation program one establishes the sectional efforts in the compressed and stretched bars. To optimize the annular bar sections, the calculation is performed for each bar, depending on the sectional efforts developed in each bar.

2.1 Calculation of the stretched bar elements

The structure being a statically determined one, the section area for i element of a stretched bar is calculated with the formula:

$$f_{yd} = \frac{N_i}{A_i} \rightarrow A_i = \frac{N_i}{f_y}\; i = 1 \ldots number\ of\ bars, \tag{4}$$

where f_{yd} is the allowable strength and N_i is the stretching sectional effort in the i bar.

For an annular section one supposes that the inner diameter d is a function of outer diameter D according to the formula:

$$d = kD;\; k = 0.7. \tag{5}$$

The diameters are calculated according to the obtained area with the formulas:

$$A = \frac{\pi \cdot D^2}{4}(1 - k^2); D = \left(\frac{4 \cdot A}{\pi \cdot (1-k^2)}\right)^{1/2}; d = kD. \tag{6}$$

2.2 Calculation of the compressed bar elements

For the compressed bars one should take into account the buckling phenomenon and therefore one calculates the critical force of stability loss using the Euler's formula [4]:

$$F_{cr} = \frac{\pi^2 \cdot E \cdot I_{min}}{l_f^2}; \tag{7}$$

where E is the transversal elastic modulus, I_{min} is the minimal axial moment of inertia of the cross section, and l_f is buckling length.

The buckling safety factor for the compressed i bar is:

$$c_{sf} = \frac{F_{cr}}{N_i}. \tag{8}$$

Based on the condition to not reach the critical force of stability loss and taking into account a buckling safety factor of $c_{sf} = 2$, one can find the minimum axial moment of inertia I_{min1} of the cross section:

$$\frac{F_{cr}}{N_i} = c_{sf} \rightarrow \frac{\pi^2 \cdot E \cdot I_{min1}}{l_f^2 \cdot N_i} = c_{sf}\; ; I_{min1} = \frac{l_f^2 \cdot N_i \cdot c_{sf}}{\pi^2 \cdot E}. \tag{9}$$

Using the condition that the buckling should be in the elastic range, one can find the minimal axial moment of inertia I_{min2} :

$$f_y = \frac{F_{cr}}{A_i} = \frac{\pi^2 \cdot E \cdot I_{min2}}{l_f^2 \cdot A_1} = \frac{\pi^2 \cdot E}{\lambda_0^2}; \tag{10}$$

$$\lambda_0 = \left(\frac{\pi^2 \cdot E}{f_y}\right)^{1/2}; \lambda_0 = \left(\frac{3{,}14^2 \cdot 210000}{355}\right)^{1/2} \approx 77(S355); \tag{11}$$

$$\lambda = \lambda_0; \frac{l_f}{i} = l_f\left(\frac{A_i}{I_{min2}}\right)^{1/2}; \lambda_0^2 = l_f^2\frac{A_i}{I_{min2}}; I_{min2} = \frac{A_i \cdot l_f^2}{\lambda_0^2}. \tag{12}$$

The moment of inertia is calculated in such a way that the bar remains in the elastic buckling range, using the relation:

$$I_{min} = min(I_{min1}, I_{min2}) \tag{13}$$

Taking into account an annular section with the outer diameter D and the inner diameter d and observing the condition $d = kD$, the following formulas result for the annular section diameters:

$$A = \frac{\pi \cdot D^2}{4}(1 - k^2); I = \frac{\pi \cdot D^4}{64}(1 - k^4); \tag{14}$$

$$k^2 = \left(1 - \frac{4 \cdot A}{\pi \cdot D^2}\right); k^4 = \left(1 - \frac{64 \cdot I}{\pi \cdot D^4}\right); 1 - \frac{4 \cdot A}{\pi \cdot D^2} = \left(1 - \frac{64 \cdot I}{\pi \cdot D^4}\right)^2; \tag{15}$$

$$A = \frac{\pi \cdot d^2}{4}\left(\frac{1}{k^2} - 1\right) = \frac{\pi \cdot d^2}{4}\left(\frac{1-k^2}{k^2}\right); I = \frac{\pi \cdot d^4}{64}\left(\frac{1-k^4}{k^4}\right); \tag{16}$$

$$k^2 = \left(\frac{4 \cdot A}{\pi \cdot D^2} + 1\right); k^4 = \left(\frac{64 \cdot I}{\pi \cdot d^4} + 1\right); \frac{4 \cdot A}{\pi \cdot D^2} + 1 = \left(\frac{64 \cdot I}{\pi \cdot d^4} + 1\right)^2; \tag{17}$$

$$D = \left(\frac{8 \cdot I_{min}}{A} + \frac{2 \cdot A}{\pi}\right)^{1/2}; d = \left(\frac{8 \cdot I_{min}}{A} - \frac{2 \cdot A}{\pi}\right)^{1/2} \tag{18}$$

The critical force of stability loss and the safety factor for the compressed i bar become:

$$F_{cr} = \frac{\pi^2 \cdot E \cdot I_{min}}{l_f^2} = \pi^2 \cdot E \cdot \frac{\pi}{64}(D^4 - d^4) \cdot \left(l_f^2\right)^{-1}; \tag{19}$$

$$c_{sf} = \frac{F_{cr}}{N_i} = \pi^2 \cdot E \cdot \frac{\pi}{64}(D^4 - d^4) \cdot \left(l_f^2 \cdot N_i\right)^{-1}. \tag{20}$$

3. Studied cases

One assumes that the structure is manufactured of S355 steel according to SR EN 1993-1-1:2006 [5] having the following physical and mechanical characteristics:

- yield strength fy=355 MPa;
- ultimate strength fu=510 MPa;
- longitudinal elastic modulus E=210000 MPa;
- transversal elastic modulus G=81000 MPa;
- Poisson coefficient for transversal contraction ν=0.3.

The cross section of the bars included in the structure has an annular form. The analysed beams are articulated and supported at the ends.

For the parametric study of trusses, one has used trapezoidal and parabolic variations of the elevation along the beam.

For the parabolic variation, one considers the following elevation variation equation of the beam:

$$y(x) = A \cdot x^2 + B \cdot x + C; \tag{21}$$

If the beam span is d and the elevation in the beam middle is f, the following equation system results based on the condition that the parable passes through the $A(0,0), B(\frac{d}{2}, f)$ and $C(d, 0)$ points:

$$\begin{cases} y(0) = 0 = A \cdot 0^2 + B \cdot 0 + C \\ y\left(\frac{d}{2}\right) = f = A \cdot \left(\frac{d}{2}\right)^2 + B \cdot \left(\frac{d}{2}\right) + C \\ y(d) = A \cdot d^2 + B \cdot d + C \end{cases} \tag{22}$$

After solving the equation system, the following parable equation results:

$$A = -\frac{4 \cdot f}{d^2}; B = \frac{4 \cdot f}{d}; C = 0. \tag{23}$$

$$y(x) = -\frac{4 \cdot f}{d^2} \cdot x^2 + \frac{4 \cdot f}{d} \cdot x = \frac{4 \cdot f \cdot x}{d}\left(1 - \left(\frac{x}{d}\right)\right); \quad (24)$$

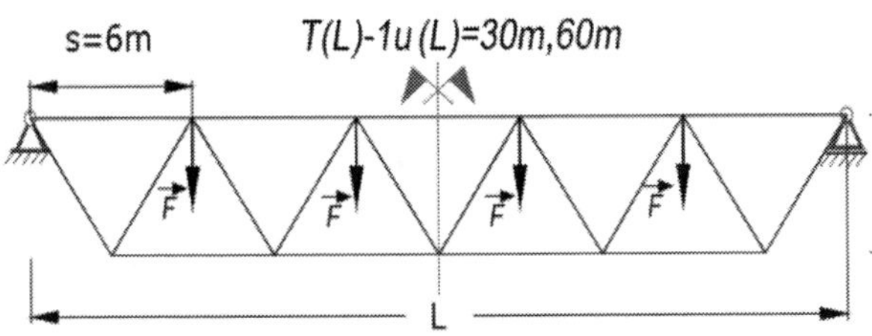

Figure 2. Beams loaded up.

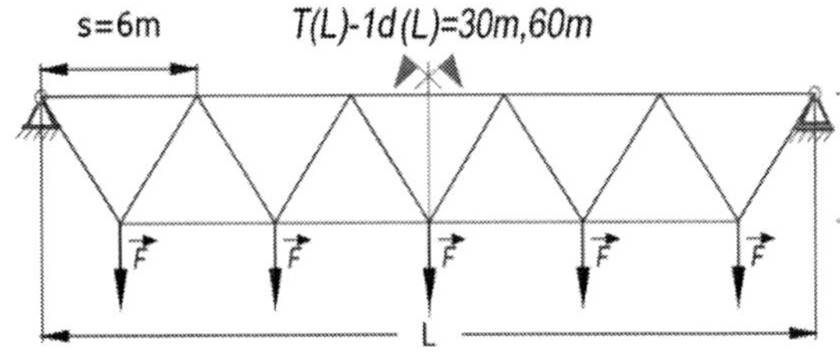

Figure 3. Beams loaded down.

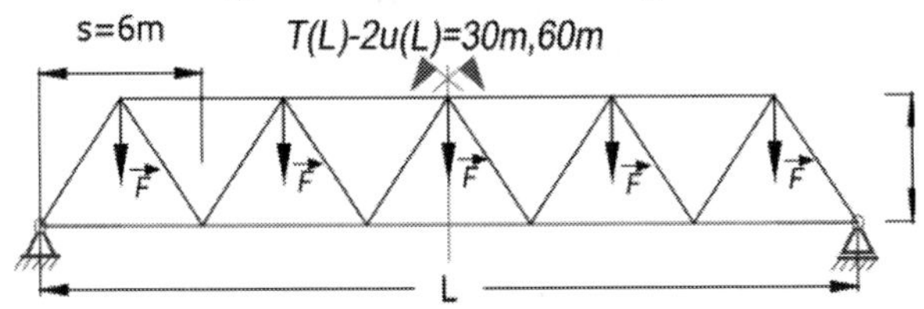

Figure 4. Beams loaded up.

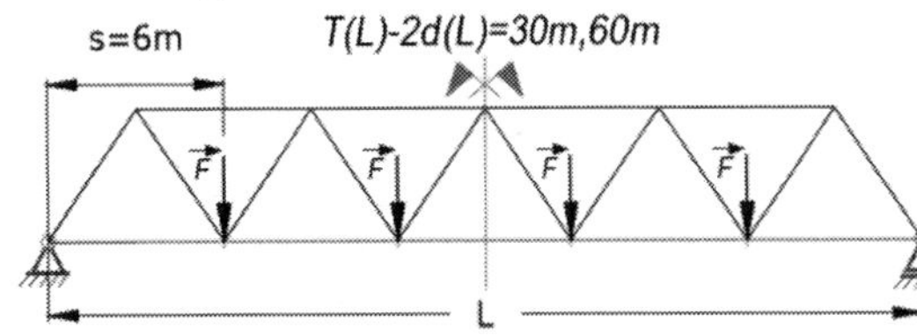

Figure 5. Beams loaded down.

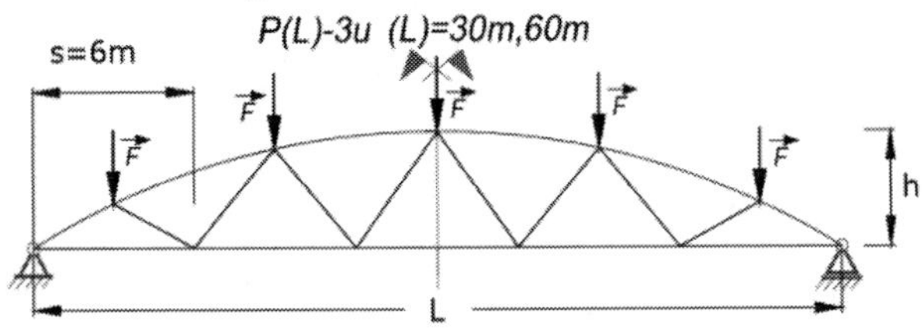

Figure 6. Beams loaded up.

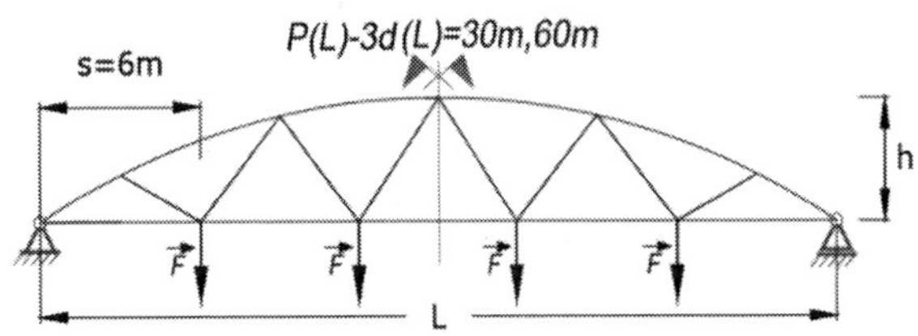

Figure 7. Beams loaded down.

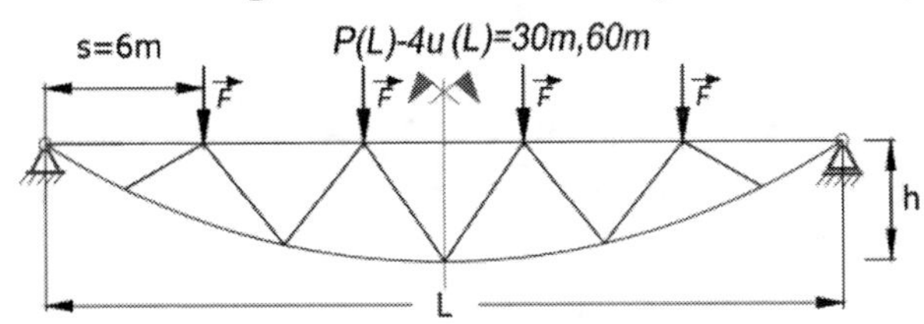

Figure 8. Beams loaded up.

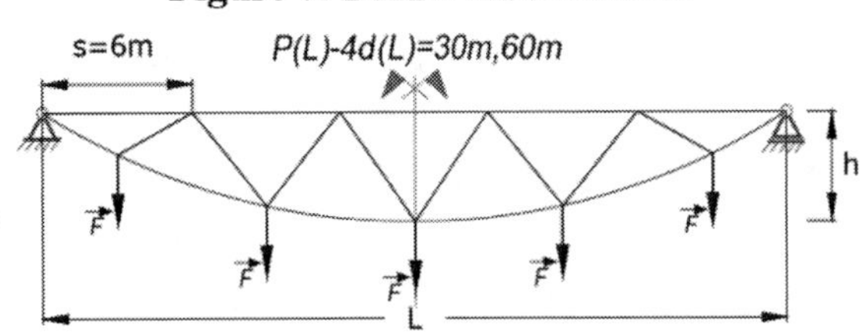

Figure 9. Beams loaded down.

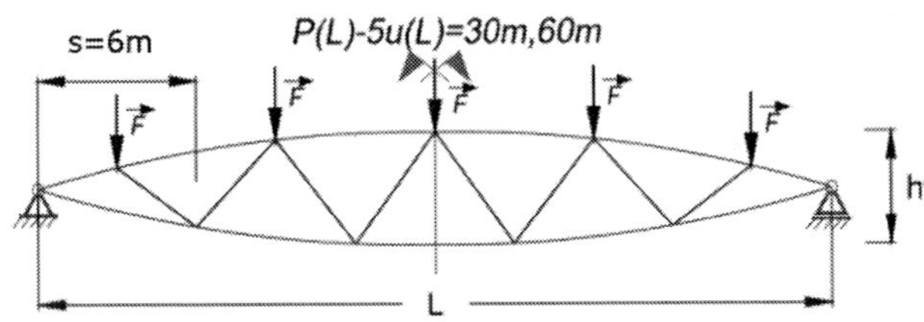

Figure 10. Beams loaded up.

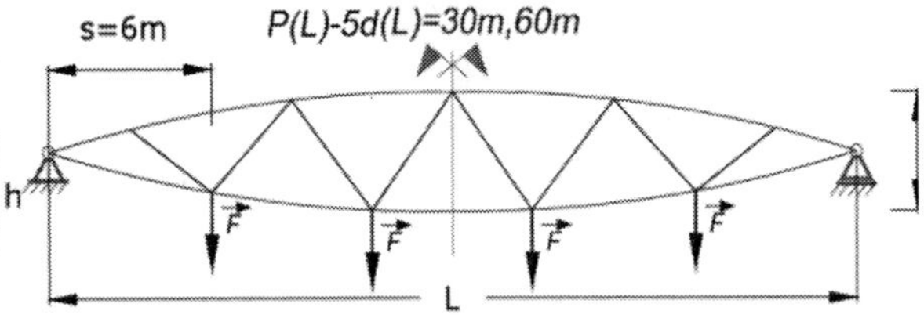

Figure 11. Beams loaded down.

The parametric study was performed on the following beam types:

- beams with trapezoidal variation on the elevation with span of 30 m - T30-1u and span of 60 m - T60-1u, loaded at the upper side - figure 2;
- beams with trapezoidal variation on the elevation with span of 30 m - T30-1d and span of 60 m - T60-1d, loaded at the lower side - figure 3;

- beams with trapezoidal variation on the elevation with span of 30 m - T30-2u and span of 60 m - T60-2u, loaded at the upper side - figure 4;
- beams with trapezoidal variation on the elevation with span of 30 m - T30-2d and span of 60 m - T60-2d, loaded at the lower side - figure 5;
- beams with parabolic variation on the elevation with span of 30 m - P30-3u and span of 60 m - P60-3u, loaded at the upper side - figure 6;
- beams with parabolic variation on the elevation with span of 30 m - P30-3d and span of 60 m - P60-3d, loaded at the lower side - figure 7;
- beams with parabolic variation on the elevation with span of 30 m - P30-4u and span of 60 m - P60-4u, loaded at the upper side - figure 6;
- beams with parabolic variation on the elevation with span of 30 m - P30-4d and span of 60 m - P60-4d, loaded at the lower side - figure 7;
- beams with parabolic variation on the elevation with span of 30 m - P30-5u and span of 60 m - P60-5u, loaded at the upper side - figure 6;
- beams with parabolic variation on the elevation with span of 30 m - P30-6d and span of 60 m - P60-6d, loaded at the lower side - figure 7.

The load applied to the trusses with the span of 30 m was determined considering a gravitational afferent of 4 m x 30 m x 3.5 kN/m = 420 KN. The load was applied using concentrated forces in upper or lower side joints according to figure 2 to figure 11.

The load applied to the trusses with the span of 60 m was determined considering a gravitational afferent of 4 m x 60 m x 3.5 kN/m = 840 KN. The load was applied using concentrated forces in upper or lower side joints according to figure 2 to figure 11. The structure analysis was performed with a Matlab program [6] which allows the determination of sectional efforts in the structure bars. To establish the optimal height for the beams having the spans of 30 m and 60 m, the beam height h was varied till the strength f_{yd} was reached and the safety factor of buckling in the compressed bars was $c_{sf} > 2$. The changing process of the height at the beam middle is represented in figure 12 for the beams with span of 30 m and in figure 16 for the span of 60 m. The values for the optimal height are represented in figure 15 for the beams with the span of 30 m and in figure 19 for the beams with the span of 60 m. The values of structure weights for the optimal h are represented in figure 13 for the span of 30 m and in figure 17 for the span of 60 m. The buckling calculation for the compressed bars was performed in the elastic range because the slenderness $\lambda > \lambda_0$ was provided for all the compressed bars. The buckling safety factors obtained for the optimal height at the beam are represented in figure 14 for the beams with the span of 30 m and in figure 18 for the beams with the span of 60 m. Comparative analysis of results was performed in Excel program [7].

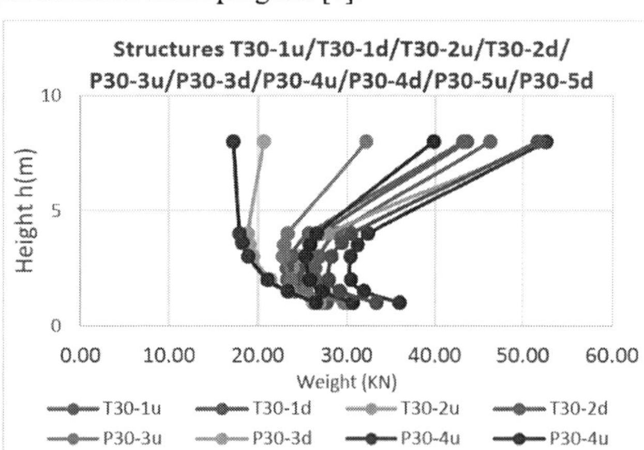

Figure 12. Variation of height-span 30 m.

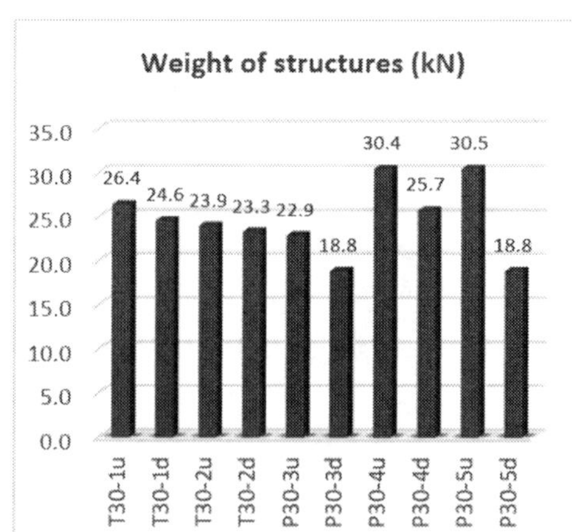

Figure 13. Weight of structure-span 30 m.

Figure 14. Safety coefficient-span 30 m.

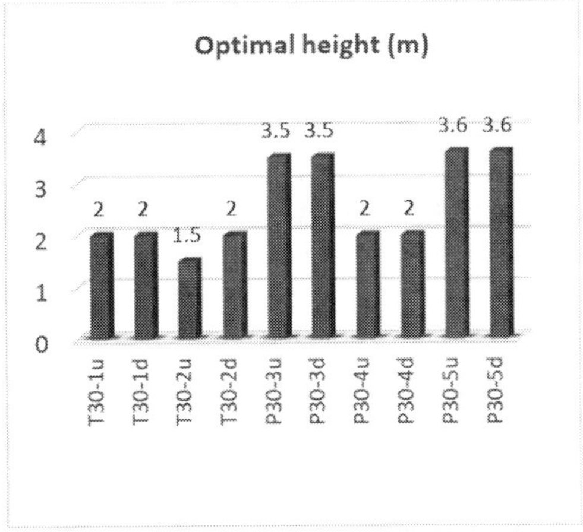

Figure 15. Optimal height-span 30 m.

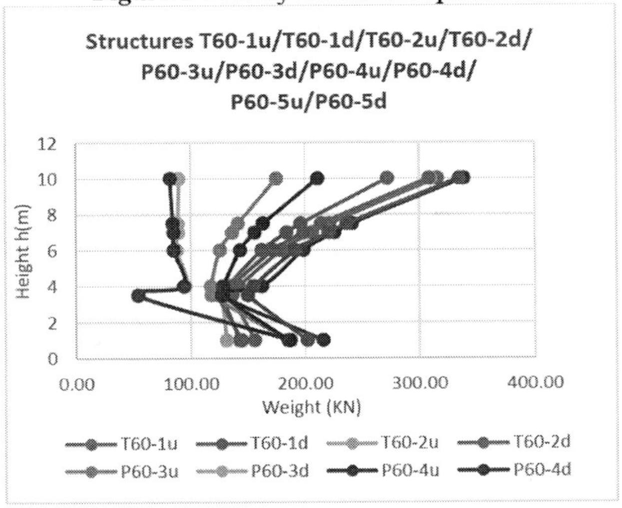

Figure 16. Variation of height-span 60 m.

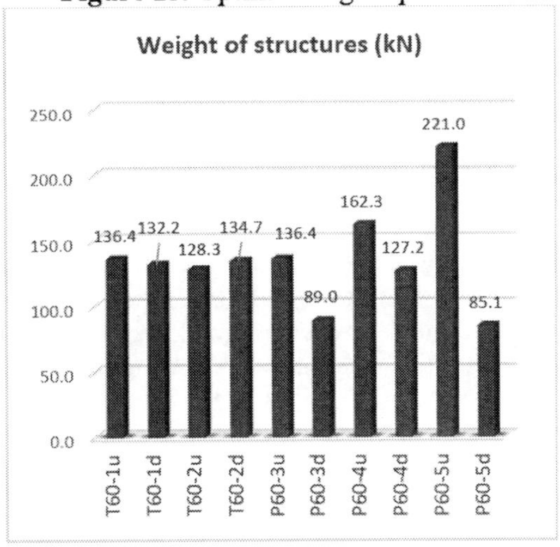

Figure 17. Weight of structure-span 60 m.

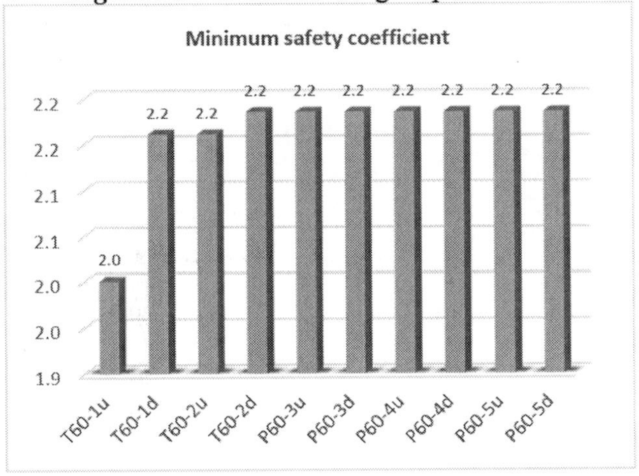

Figure 18. Safety coefficient-span 60 m.

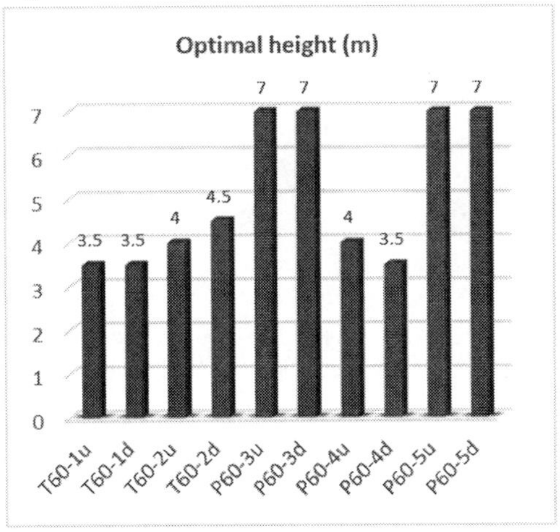

Figure 19. Optimal height-span 60m.

IOP Conf. Series: Materials Science and Engineering **789** (2020) 012006 doi:10.1088/1757-899X/789/1/012006

4. Results and conclusions

From figure 13 and figure 17 one can deduct that the structures having the smallest weight are P30-3d, P30-5d, P60-3d and P60-5d. The reduced weight obtained for these structures is due to the largest number of stretched bars which are dimensioned based on the condition of stretching strength only.

In the process of optimal height determination one can notice that, by increasing the height, at a certain moment one reaches the necessary safety factor of buckling in the compressed bar in the middle of the structure (upper or lower side, depending on the structure and load) and by further increasing of the height one reaches the necessary safety factor of buckling for compressed diagonals from the end of beam.

For trusses with large spans one recommends a greater s step (longer bars) and steels of higher class having a smaller slenderness λ_0 to remain in the range of the elastic buckling.

The advantages resulting from keeping the compressed bars in the range of the elastic buckling are:

- the ratio between the length of a bar element and the largest of the cross-section dimensions is greater than 10, so the rigid joints can be considered articulated joints;
- the truss type structures being statically determined, it results that they have no structural redundancy, therefore the occurrence of the first plastic deformation converts the structure into a mechanism.

References

[1] Ambrose J 1994 *Design of Building Trusses* (New York:Wiley)
[2] Feodosiev V I 2005 *Advanced Stress and Stability Analysis* (Berlin: Springer Verlag)
[3] Teodorescu M E 2001 *Statics of Constructions* (Bucharest: MatrixRom)
[4] Hibbeler R C 2017 *Statics and Mechanics of Materials* (Boston: Pearson)
[5] SR EN 1993-1-1 2006 *Design of steel structure* (Romania: Romanian standard)
[6] https://www.mathworks.com/
[7] https://products.office.com/ro-ro/excel

Coordination of construction manipulation robotic system using UAV

A G Bulgakov[1*] and E G Pakhomova[1]

[1] Southwest State University, Kursk, Russia

[*] agi.bulgakov@gmail.com

Abstract. Drone technology has demonstrated to be of great contribution to the construction field. Tasks such as photogrammetry, plot surveillance and assets inspections are successfully automated with the aerial robotic technology with some level of contribution of a human operator. In this paper we analyse the possibility to achieve fully automated construction task using dispersed robotic system: a drone and mobile robotic crane. The role of the drone is to study the surroundings and generate a desired trajectory for the crane boom arm in order to lay-in pergola blades. The paper focuses on the layers of the dispersed cyber-physical system allowing to achieve such complex task and optimize the trajectory of the crane boom arm by resolving two-dimensional Chebyshev-Gauss collocation method in order to have minimum-jerk path.

1. Introduction

The practical implementation of the cyber-physical systems in our day to day life is well established. We recognize the development in such solutions leading to Internet of Things (IoT), augmented and virtual reality contributing to a new level of end-user's satisfactions and extended useful experience. In contrast, the paradigm of applying cyber-physical systems to the construction field is yet to explore and set its own potential.

This is majorly due to the late and slow development of the autonomous and robotized platforms which has a projected application in the construction field.

A techno-commercial solution of the autonomous systems in construction is rarely considered as it is always cheaper and less risky to deploy man-power (carpenters, steel fixers) rather than purchasing and customizing a robotic solution.

In one hand, the implementation of sophisticated solution is costly and require specialized staff present at site for monitoring, and on the other hand, the solution is not widely tested and its results are barely registered and recognized.

In this paper, we tackle the first challenge aiming to reduce the specialized and idle man power by studying the automation level required to lay-in pergola blades on top of a building using dispersed cyber-physical system composed of a drone and a mobile lifting crane. By cyber-physical system, we understand the concept of reliable fusion between computational and physical platform embedded within the same body [1-5], however, enabling better interaction capabilities with the surrounding (world and human operator) though many modalities and processes. In light of that, the terms of interaction, integration and enhancement are key areas where the research of cyber physical systems is occurring.

Traditionally, the physical system is a consolidated system and not dispersed over selected area. Recently, with the introduction of Internet of things, the probability of designing distributed physical systems became achievable, as it overcame the disadvantages of radio systems and the huge cost related to GPS and GLONASS integration in navigation platforms, which made this solution not optimal techno-commercially. In this paper, we introduce the cyber-physical system consisting of the following levels:

- aerial: a UAV scanning the roof and generating the positioning tasks for the subsequent component of the pergola;
- cybernetic: analyzing the video-input of the UAV and generating trajectory for the manipulator using visual odometry algorithm;
- ground: a Mobile robotic crane, receiving the coordinates from the cyber system and performing the laying-out of the pergola blades.

A concept of the solution is illustrated in figure 1 according to which, the UAV will dictate the following parameters as a potential control task to the cybernetic system: the swing angle ϕ, the luff angle θ, the hoist elevation h and the mobility across axis OX and OY, which can be replaced by the angle formed by the hoist and the luff axis [6-10].

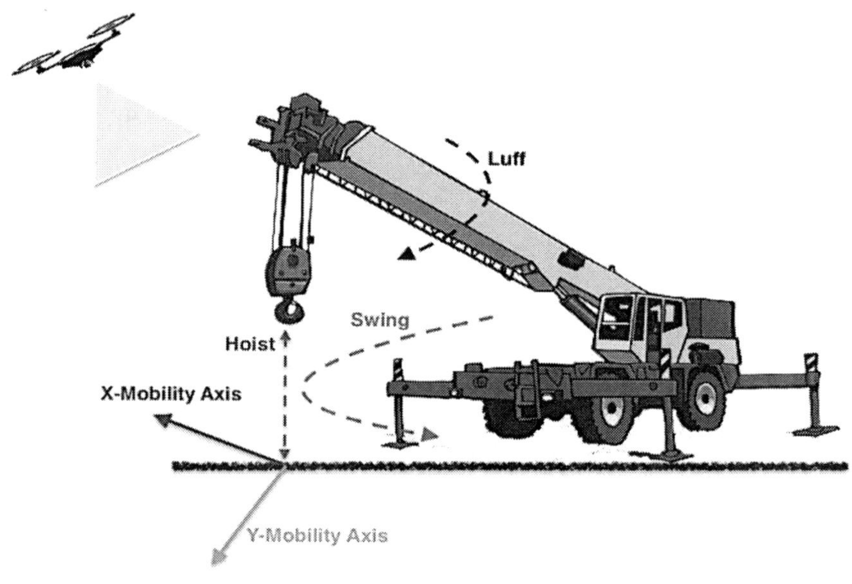

Figure 1. Concept of dispersed cyber-physical system.

2. Paradigm of proposed cyber physical system

By paradigm, we underline a worldview underlying theories and methodologies of a particular scientific subject. To some extent, the notion of the cyber-physical system is well defined in the IT fields. However, its application in other engineering subjects is yet to be defined.

Our target in this paper is to make notions more tangible and hence allowing better evaluation and optimization. In line of that, we will be studying the generic system architecture, the enabling knowledge and technology assets, and distinguishing system characteristics allowing for expansion and optimization.

Cyber physical systems (CPS) have emerged from the synergetic merging of information, sensory and communication system with a physical system. This is a comparable notion to mechatronic system, from where we will begin integrating CPS architecture to the electromechanical word of construction.

In addition, we have to consider the human-machine interface as the CPS will not be deployed at site in isolation with the other construction activities. Hence, an additional layer in the CPS will be figuring out and influencing on the overall performance of the system. Figure 2 and 3 illustrate generalized CPS layers and compared to the applied proposed CPS [11, 12].

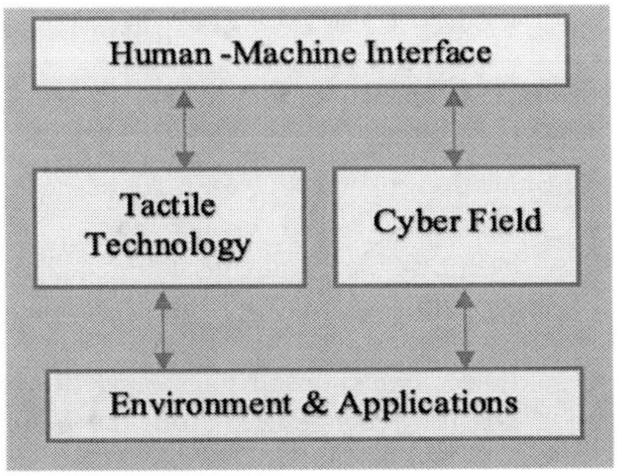

Figure 2. Traditional CPS.

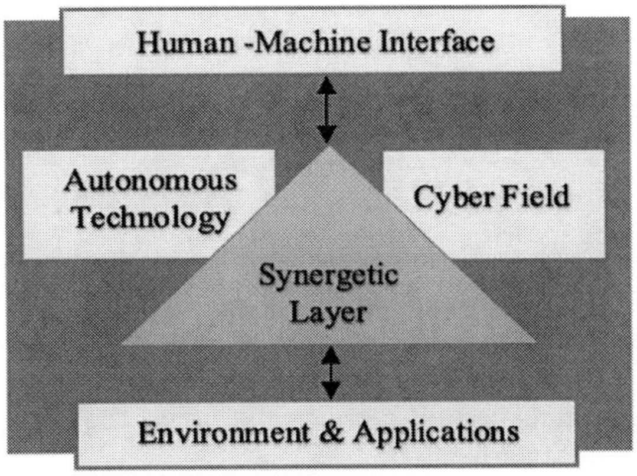

Figure 3. Proposed CPS.

As one can see from figures 2 and 3, the difference between a traditional and the proposed CPS can be highlighted in several points: firstly, the human-machine interface is encompassed inside the CPS in the traditional system, however, in the proposed layout, the layer is peripheral and has no direct interaction with the autonomous technology and cyber field.

Secondly, the tactile technology is substituted with an autonomous system as we are aiming to robotize construction task and lastly, the direct interaction with the environment and application is done by a synergetic "mechatronic layer" responsible of communication and execution of the task. In addition, the environment layer is neither connected directly to the autonomous and cyber system. In short, the difference lays in omitting the human factor influencing the process and replacing it with actuating mechanism.

The proposed model complies with the worldwide notions of CPS especially point C5 stating the system should be articulated and heterogeneous. By articulated, we define the connectivity of the system with its embedded components and environment; and by heterogeneous - we mean the different physical origin of each systems: having a cyber part, a synergy and actuators. The same is import as the CPS can suffer possible downtime leading to task failure. In this case, the CPS analysed by its Cyber part has to plan for a functional (graceful) degradation [13].

3. System architecture

In this paragraph, we will discuss the core parts of the CPS, its parameters, requirement and functions. We will study the cyber system component, set the algorithm of trajectory generation for the crane boom arms using UAV and identify the transformation equations between the pixel and metric references.

The last part of the paragraph will address the optimization of the generated trajectory and transformation of the boom arm movement to a two-dimensional Chebyshev-Gauss collocation method in order to achieve minimum jerk movement considering zero initial conditions.

3.1. Cyber system

Firstly, regarding the cyber field, an embedded software command the generation of the trajectory of the crane boom arms and requirement of the geometric coordinates explained in paragraph 1. In

IOP Conf. Series: Materials Science and Engineering **789** (2020) 012007 doi:10.1088/1757-899X/789/1/012007

addition, the cyber system is responsible of intra and inter-objects transmissions. As per the current standards of CPS the following range of transmission for a multi-scale CPS need to be identified:

Table 1. Transmission range for intra and inter-Objects communication.

Intra-Object Transmission		Inter -Objects Transmission	
Not Available		Global	10^6 m
Nano Scale	from 10^{-9} m	Long range	10^4 m
Micro Scale	up to 10^{-6} m	Medium Range	10^2 m
Macro Scale	up to 10^{-3} m	Short Range	10^0 m

From table 1, we need to pay attention to the following: the intra-object transmission needs to be respected and achieved by all means, since it is a mandatory requirement for possible functional degradation in case of system failure.

Pertaining to the inter-objects' transmission, a least of medium range should be achieved allowing the communication between the drone and the crane with possible engagement from the human-machine interface.

The other part of the cyber system is the trajectory generating software. We will be deploying the drone, equipped with a camera, to generate desired trajectory for the boom arm, study possible collision and assign subsequent step to lay the pergola blade. Using machine vision and odometry, CPS can autonomously define the trajectory of the boom arm. Figure 4 and equations (1-3) illustrate the transition between pixel to metric coordinates.

The distance from the camera to a specific coordinate (Earth-axis) is computed using the following equations:

$$d(x) = H_{\Gamma} = H \cdot \tan(\phi + \alpha); \quad (1)$$

$$d(y) = H_{B} = H \cdot \tan(\theta + \beta); \quad (2)$$

Using equations (1) and (2), we can identify the transformation formulas between different reference systems.

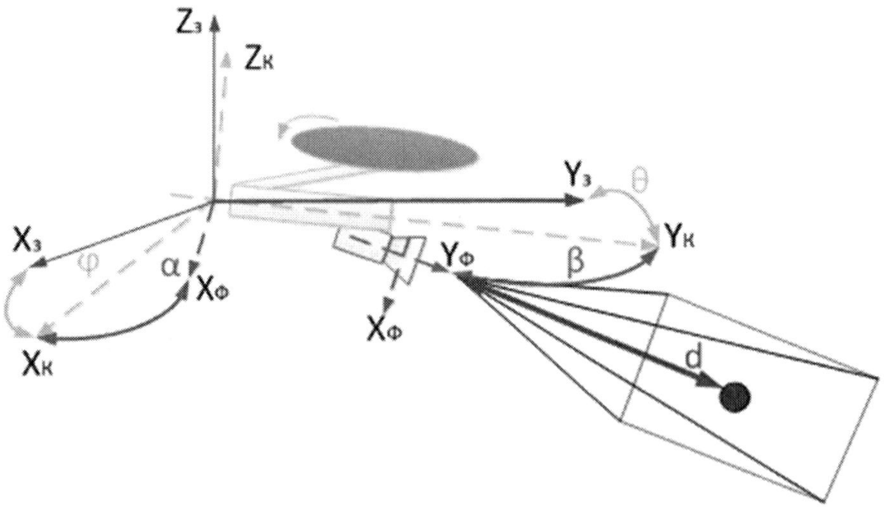

Figure 4. Illustration of different reference systems and coordinates.

$$\begin{cases} X_{п} = \frac{x}{Пx} \cdot \cos\left(\arctan\left(\frac{H_B}{H_г}\right)\right) \\ Y_{п} = \frac{y}{Пy} \sin\left(\arctan\left(\frac{H_B}{H_г}\right)\right) \\ \rho_x = \sqrt{{X_{п}}^2 + {Y_{п}}^2} \cdot cos(\arctan\left(\frac{H_B}{H_г}\right)) \\ \rho_y = \sqrt{{X_{п}}^2 + {Y_{п}}^2} \cdot sin(\arctan\left(\frac{H_B}{H_г}\right)) \end{cases} \quad (3)$$

where $X_{п}$ and $Y_{п}$ – are the pixel coordinates, x and y – are the body axis coordinate and ρ_x and ρ_y are the angular pixel coordinates.

3.2. Actuating system

The actuators are responsible of fulfilling the generated trajectory, to lift and lay the pergola blades according to the defined coordinates. In addition, the process has to be jerk-free movement in order to avoid collisions and inverted pendulum effect, especially that wind might affect the positioning of the blade at a certain height. To achieve an optimal trajectory to the manipulator, the movement will be posed and solved as quadratic programming using multi-segment Chebyshev orthogonal collocation for transcription. Using the two-dimensional Chebyshev-Gauss collocation method to obtain possible numerical solution of differential equation by partially is differentiating it in time and discretizing it using finite difference method [14]. Taking into consideration that the trial or candidate trajectories $f_k(x)$ will be set on homogenous Dirichlet boundary conditions [13], the approximate solution can be presented in the following equation:

$$U^N(x,t) = \sum_{k=0}^{N} a_k(t)\, f_k(x) \quad (4)$$

In the collocation approach, the differential equation has to be satisfied by the approximate solution at the collocation points in the assigned domain [15-18]. The derivative at each node can be found by multiplying the value at any Chebyshev point by a differentiation matrix.

4. Results

By implementing the proposed model, and running the quadratic programming listing on Matlab, we obtained the result of multi-segment trajectory simulation as depicted in figure 5.

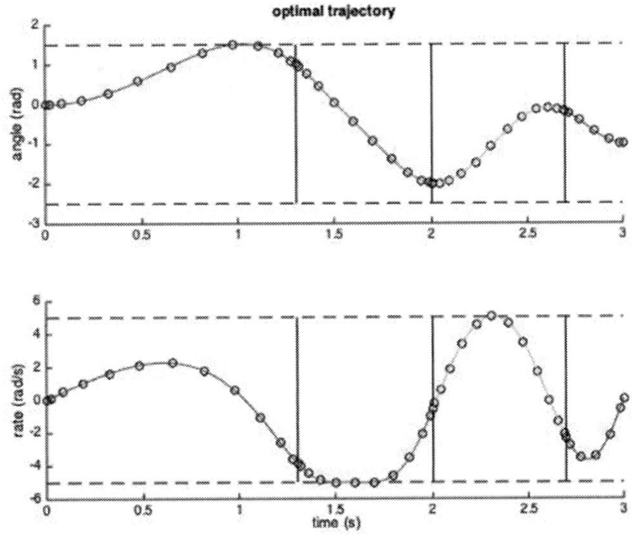

Figure 5. Results of multi-segment trajectory.

In figure 5 can be seen that the trajectory is passing by all the Chebyshev node points, where the initial a final velocity rated in rad/s is zero and the angle of the final trajectory is almost negligible.

By applying the same concept to the both sections of the manipulator, with specific boundaries and input trajectory, we can see that the arm end-effector holding the bald is travelling smoothly, on the doted lines, shown in figure 5, taking into consideration for joint 1 and 2, the initial velocity is zero (starting point of lifting task), and the resultant final velocity of the end-effector is also zero rad/s, which corresponds to discharge of the manipulator after laying task.

5. Conclusion

This paper studied the architecture of multi layered cyber-physical system consisting of UAV and robotic crane. The role of the UAV is to generate desired path for the crane boom arm in order to lift and lay pergola blades on top of a building. The paper addressed different layers of the cyber physical systems and proposed an additional synergetic part interacting solely with human machine interface and acquire information for the environment. In such scenario the cyber physical system will be bordered and allow for better functional (graceful) degradation if required. Later, the paper focused on the trajectory planning and generation of transportation coordinates of the pergola blades. The problem is then transformed as two-dimensional Chebyshev-Gauss orthogonal collocation method aiming to achieve jerk-free transportation and positioning of the blades. The obtained simulation results illustrated both the joints of the lifting crane have followed the optimal trajectory, passing by all Chebyshev points with minimum jerk considering that the initial velocity of the joints is zero and the final velocity should be obtained zero rad/s with minimum to zero overshooting angles.

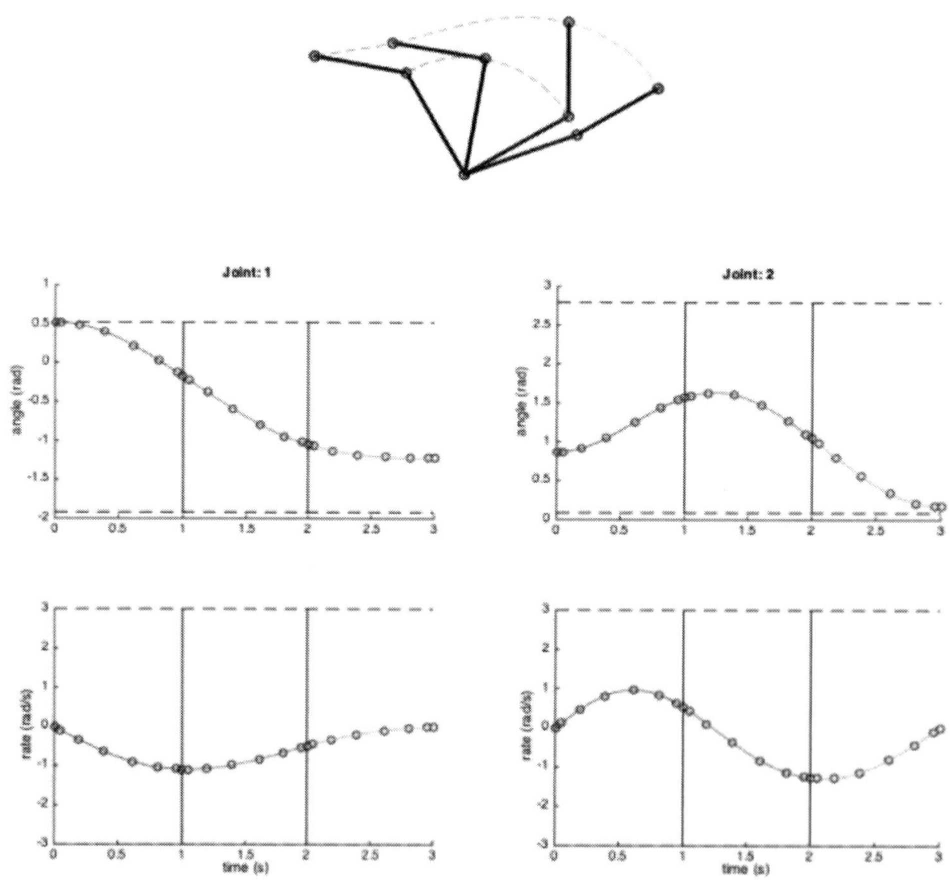

Figure 6. Minimum-Jerk trajectory of the robotic manipulator corresponding to Chebyshev nodes.

An analysis of the values of the volume modulus K in poly disperse composites with spherical inclusions defined by formula (1) showed that for studies of an infinite number of values of this module it is advisable to proceed to the consideration of dimensionless functions k and φ - functions of three dimensionless parameters. A numerical study of formulas (3) revealed the possibility of a graphical representation of the functions k and φ in a flat space (k - w) (see figure 1). By values, this space is divided into internal (lower) and external (upper).

For values of θ=0, formulas (3) are reduced to two degenerate formulas (8) represented in the same k - w space by figures k_0 and φ_0 (see figure 6). The contours of these figures limit the region of location of points that display the values of the modules k_0 for two ordered subsets of degenerate composites (with an undeformable filler). In this case, the upper figure displays (for given values of the modules of the matrix) the highest values of the module k (η_M, θ, w).

The space between the figures k_0 and φ_0 (see figure 2, 3) is intended to place the figured regions of the values of the modules k (η_M, θ, w) of the composites in which the value of θ is nonzero. The figures of the regions for such composites (with elastic filler) continuously fill the named intermediate space. Examples of such figures are presented in figure 4 and 5.

The graphical visibility of the solutions (see figures 2, 3, 4 and 5) given by equations (1), (3) and (8) significantly expands the scope of the applied use of these formulas. At the same time, there is a real possibility of conducting targeted experimental studies (mechanical compression tests of standard sized composite samples) to confirm the predicted properties of the composites. In this case, there is no need for physical analogies that were used by Christensen [19-29].

References

[1] Sayfeddine D, Bulgakov A and Kruglova T 2017 Intelligent passively stabilized quadrotor *IOP Conf. Series: Earth and Environmental Science* DOI:10.1088/1755-1315/87/3/032036

[2] Bulgakov A, Kruglova T and Bock T 2019 Formulation of the Optimization Problem of the Cyber-physical Diagnosis System Configuration Level for Construction Mobile Robots *Proceedings of the 36th International Symposium on Automation and Robotics in Construction (ISARC 2019)* Alberta p 704-708

[3] Cho Y, Haas C T, Liapi K, and Sreenivasan S V 2002 A framework for rapid local area modelling for construction automation, *Autom. Constr.* **11** no. 6 p 629–641

[4] Lin J J, Han K K and Golparvar-Fard M A 2015 Framework for Model-Driven Acquisition and Analytics of Visual Data Using UAVs for Automated Construction Progress Monitoring, *Computing in Civil Engineering*

[5] Wang F, Cui J and Chen B M 2013 Comprehensive UAV Indoor Navigation System Based on Vision Optical Flow and Laser FastSLAM. *Acta Autom Sin.*

[6] Upadhay A 2018 UAV-Robot relationship for coordination of robots on a collision free path, *International conference on robotics and smart manufacturing* (RoSMa) p 424-31

[7] Ham Y, Han K K, Lin J J and Golparvar-Fard M 2016 Visual monitoring of civil infrastructure systems via camera-equipped Unmanned Aerial Vehicles (UAVs): a review of related works. *Vis Eng.*

[8] Ellenberg A, Branco L, Krick A, Bartoli I and Kontsos A 2015 Use of Unmanned Aerial Vehicle for Quantitative Infrastructure Evaluation *J Infrastruct Syst.*

[9] Gheisari M, Irizarry J and Walker B N 2014 UAS4SAFETY: The Potential of Unmanned Aerial Systems for Construction Safety Applications. In, *Construction Research Congress*

[10] Liu T, Yang L, Lin Q, Guo Y and Liu Y 2014 Anchor-free backscatter positioning for RFID tags with high accuracy *Proceedings - IEEE INFOCOM*

[11] Bulgakov A, Evgenov A and Weller C 2015 Automation of high-rise structures inspection using quadrotor vol. **123** *Procedia Engineering* p 101 – 109

[12] Horvath I and Gerritsen B 2012 Cyber-physical systems: concepts, technologies and implementation principles *Proceedings of TMCE* May 7–11 p 19-36

[13] Cheng A and Cheng D T 2005 Heritage and early history of the boundary element method vol. **29**, *Engineering Analysis with Boundary Elements* p 268–302
[14] Sameeh M 2016 Chebyshev Collocation Method for Parabolic Partial Integrodifferential Equations *Advances in Mathematical Physics* p 7
[15] https://arxiv.org/pdf/gr-qc/0609020.pdf
[16] Parthasarathy S A 1997 Tight Analysis of the Greedy Algorithm for Set Cover, *J. Algorithms*, vol. **25** 2 p 237–254,
[17] Bouabdallah S 2007 Design and control of quadrotors with application to autonomous flying. *Master's thesis*, Swiss Federal Institute of Technology
[18] Rajhans A, Cheng S W, Schmerl B, Garlan D, Krogh B H, Agbi C and Bhave A 2009 An architectural approach to the design and analysis of cyber-physical systems vol. **21** *Proceedings of the 3rd International Workshop on Multi-Paradigm Modeling-MPM EASST* pp 1-10
[19] Sayfeddine D 2019 Functional degradation of cyber-physical aerial system for trajectory planning and agent tracking. *The International Conference on Information Technologies in Business and Industry*
[20] Kuo C-H, Kuo C-M, Leber A and Boller C 2013 Vector thrust multi-rotor copter and its application for building inspection. *International Micro Air Vehicle Conference and Flight Competition (IMAV2013),* Toulouse France.
[21] Bulgakov A, Emelianov S, Bock T and Sayfeddine D. 2014 Control of hovering altitude of a quadrotor with shifted center of gravity for inspection of high-rise structures. *Proceedings of the 31st International Symposium on Automation and Robotics in Construction (ISARC)* Sydney Australia
[22] Bertram T, Bock T, Bulgakov A and Evgenov A. 2014 Generation the 3D Model Building by Using the Quadcopter, *31st International Symposium on Automation and Robotics in Construction and Mining,* Sydney Australia University of Technolog p 778-783
[23] Pascal W H, Wegner J D and Schindler K 2015 Globally consistent registration of terrestrial lase scans via graph optimization, *ISPRS Journal of Photogrammetry and Remote Sensing* Vol. **109** p 126-138
[24] Kusovkov N 1976 Modal control and observer devices *Machine industry proceedings*, Moscow p.184
[25] Eschmann C, Kuo C M, Kuo C-H and Boller C 2016 Unmanned Aircraft Systems for Remote Building Inspection and Monitoring. *6th European Work-shop on Structural Health Monitoring*
[26] Tanaka S, Nakamura K, Imai R, Kubota S and Sakurai J 2016 Design Data Generation System for Progress Control of Working Form *Proceedings of Civil Engineering Conference in the Asian Region* Hawaii USA, p 1-12
[27] Kubota S, Kawai, Y and Kadotani R 2017 Accuracy Validation of Point Clouds of UAV Photogrammetry and its Application for River Management *The International Archives of the Photogrammetry, Remote Sensing and Spatial Information Sciences* p 195-199
[28] Sayfeddine D, Nonlinear autoregressive neural network with exogenous inputs based solution for local minimum problem of agent tracking using quadrotor. *Electronic scientific journal. Engineering Journal of Don.* Vol. **2** 2014
[29] Akinci B, Boukamp F, Gordon C, Huber D, Lyons C and Park K 2006 A formalism for utilization of sensor systems and integrated project models for active construction quality control *Autom. Constr.* vol. **15** p 124–138

IOP Conf. Series: Materials Science and Engineering **789** (2020) 012008 doi:10.1088/1757-899X/789/1/012008

GIS-based landslide susceptibility for environment protection of Black Sea Romanian coast

C Buta[1], C Maftei[2,3*], G Dobrica[3,4] and G Draghici[4]

[1] Ovidius University of Constanta, Civil Engineering Faculty, Constanta, Romania
[2] Transilvania University of Brasov, Faculty of Civil Engineering, Brasov, Romania
[3] Ovidius University of Constanta, Doctoral School of Applied Sciences, Constanta, Romania
[4] Romanian Water Agency, Dobrogea-Littoral branch, Constanta, Romania

[*] cemaftei@gmail.com

Abstract. In the last decades the landslide assessment plays a crucial role in land use development in order to minimize the negative effect of economic live and people's health. This paper present a GIS based landslide susceptibility assessment. This model is applied on the Littoral Coast line from Constanta to 2 Mai village, which is about 50 km along the Black Sea coast, crossing the Danube – Black Sea Channel at Agigea and passing through several resorts – Eforie Nord, Eforie Sud, Techirghiol, Costinesti and Mangalia. The data used are: geological map (lithology per geologic group), topographic maps of relevant scale to define slope angle (β), and some geotechnical parameters per geological group. For each of them a GIS map was produced. Based on the research results, we could conclude that Landslide Susceptibility Area (LSA) maps represent a great tool used in the identification of areas predisposed to landslides and could be used as support in the decision-maker process in order to propose environmental work protection or county planning.

1. Introduction

Landslide is a part of natural hazard which could produce negative effect to natural and social environment. According the International Disaster Database of the Centre for Research on the Epidemiology of Disasters (CRED) (EM-DAT), landslide represents 5.2% of natural hazard in terms of number of disasters for the 1997-2017 periods [1].

In Romania, the most affected area is the Sub Carpathian region, especially the Curvature Sector, Transylvanian Plateau and Moldavian Plateau [2, 3] where 50% of territory is affected by gravitational movement mass [4]. In Dobrogea region the most affected areas are Danube valley especially due to loess Quaternary deposits.

In this context, landslide mapping is an important tool for the risk management and improve the resilience of society against landslide disaster and protect the environment.

Earl E. Brabb [5], the pioneer in landslide mapping, in his publication entitled: "The World Landslide Problem" (1991) wrote: "[...] Landsliding is a worldwide problem that probably results in thousands of deaths and tens of billions of dollars of damage each year. Much of this loss would be avoidable if the problems were recognized early, but less than one percent of the world has landslide-inventory maps that show where landslides have been a problem in the past, and even smaller areas have landslide

IOP Conf. Series: Materials Science and Engineering **789** (2020) 012008 doi:10.1088/1757-899X/789/1/012008

susceptibility maps that show the severity of landslide problems in terms decision makers understand. Landslides are generally more manageable and predictable than earthquakes, volcanic eruptions, and some storms, but only a few countries have taken advantage of this knowledge to reduce landslide hazards."

Despite all the efforts, the situation concerning the landslide cartography has not changed significantly. There are several methodological approaches to model the landslide hazard or/and susceptibility. Firstly, a controversy exists between the terms landslide "susceptibility" and landslide "hazard". Landslide susceptibility is the likelihood of a landslide occurring in an area on the basis of local terrain conditions [6]. Landslide hazard maps indicate the probability that a landslide of a given magnitude will occur in a given period and in a given area.

The goal of this paper is to assess the landslide susceptibility area maps (LSA) based on FEMA (Federal Emergency Management Agency) method on the Black Sea Littoral using GIS technique.

2. Materials and methods

FEMA (Federal Emergency Management Agency) method is used to assess the landslide susceptibility maps. According to FEMA methodology [7], landslide susceptibility of geologic groups under static conditions is described as a function of the site condition that refers to the topography, geology and groundwater level (table 1). Landslide susceptibility is measured on a scale of I to X where I is less susceptible and X most susceptible. Groundwater condition is divided in dry and wet condition.

Table 1. Landslide susceptibility of geologic groups (scale I-less susceptible, scale X-most susceptible).

	Geological Group	Slope Angle, degrees					
		0-10	10-15	15-20	20-30	30-40	>40
	(a) DRY (groundwater below level of sliding)						
A	Strongly Cemented Rocks (crystalline rocks and well-cemented sandstone, c'=300 psf, Φ'=35°)	None	None	I	II	IV	VI
B	Weakly Cemented Rocks and Soils (sandy soils and poorly cemented sandstone, c'=0, Φ'=35°)	None	III	IV	V	VI	VII
C	Argillaceous Rocks (shales, clayey soil, existing landslides, poorly compacted fills, c'=0, Φ'=20°)	V	VI	VII	IX	IX	IX
	(b) WET (groundwater level at ground surface)						
A	Strongly Cemented Rocks (crystalline rocks and well-cemented sandstone, c'=300 psf, Φ'=35°)	None	III	VI	VII	VIII	VIII
B	Weakly Cemented Rocks and Soils (sandy soils and poorly cemented sandstone, c'=0, Φ'=35°)	V	VIII	IX	IX	IX	X
C	Argillaceous Rocks (shales, clayey soil, existing landslides, poorly compacted fills, c'=0, Φ'=20°)	VII	IX	X	X	X	X

To apply these models at a regional scale, the following data are necessary: (i) geological maps of relevant scale (lithology per geologic group); (ii) topographic maps of relevant scale to define slope angle (β); (iii) some geotechnical parameters per geological formation must be estimated / calculated (Φ' - effective angle of friction of geomaterial (°), c'- effective cohesion of geomaterial (kPa), γ- specific weight KN/m^3); (iv) groundwater level.

This methodology is applied on the Littoral Coast line from Constanta to 2 Mai village, which is about 50 km along the Black Sea coast, passing through several resorts – Eforie Nord, Eforie Sud, Techirghiol, Costinesti and Mangalia, shown in figure 1.

From a geological point of view, this area includes three tectonic units – Northern, Central and Southern Dobrogea, presented in figure 2. The tectonic units are separated by two major crustal faults, approximately oriented NW-SE: Peceneaga-Camena (between North and Central Dobrogea) and Capidava-Ovidiu (between Central and the Southern units). The common feature of the three units of

Dobrogea is the vast Quaternary cover, having various thicknesses of loess layers, as in figure 3. There are in small percentage: green schist, limestone and reddish clay.

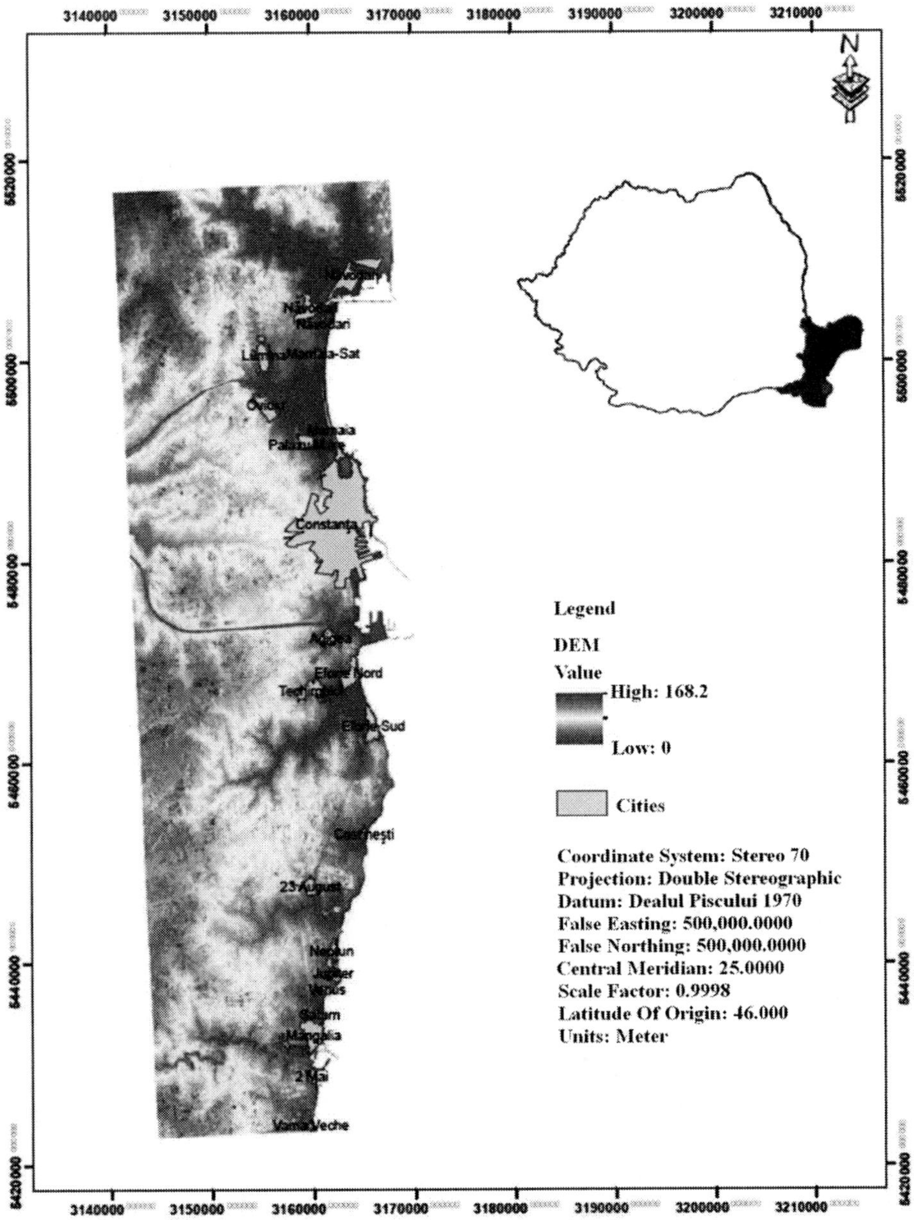

Figure 1. Location of investigated area and DEM (Digital Elevation Model).

An ASTER DEM (resolution 30x30 m) was used for generating the Digital Elevation Model, as in figure 1, which has been used for the processing that followed. Maximum elevation in the region is 168.2 m. The major part of the investigated area is flat (0-10 degree). The steeper slopes are on the coast line, on the Danube-Black Sea Canal and on the valley. The geological group was identified based on geological map, presented in figure 3; for the geological materials identified (figure 3) the effective angle of friction of geomaterial (Φ') varies between 19° and 35°, effective cohesion of geo-material varies between 5-23 kPa and the specific weight is situated in the range 14.50-20.37 KN/m^3. Based on the groundwater level map we establish the condition: dry or wet.

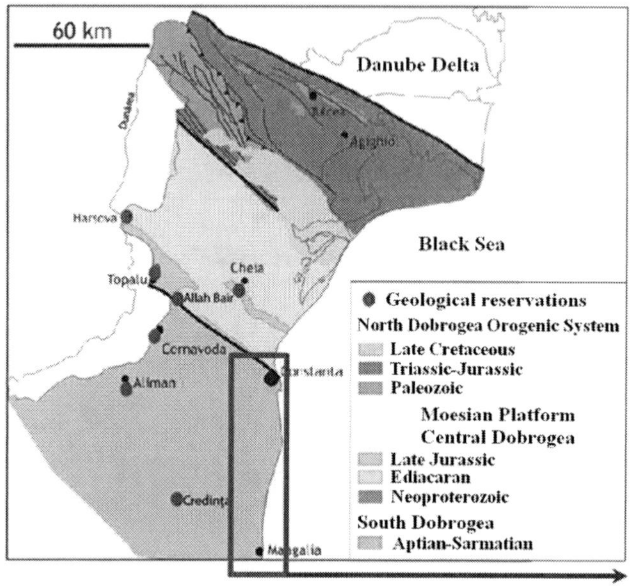

Figure 2. The three tectonic units from Dobrogea.

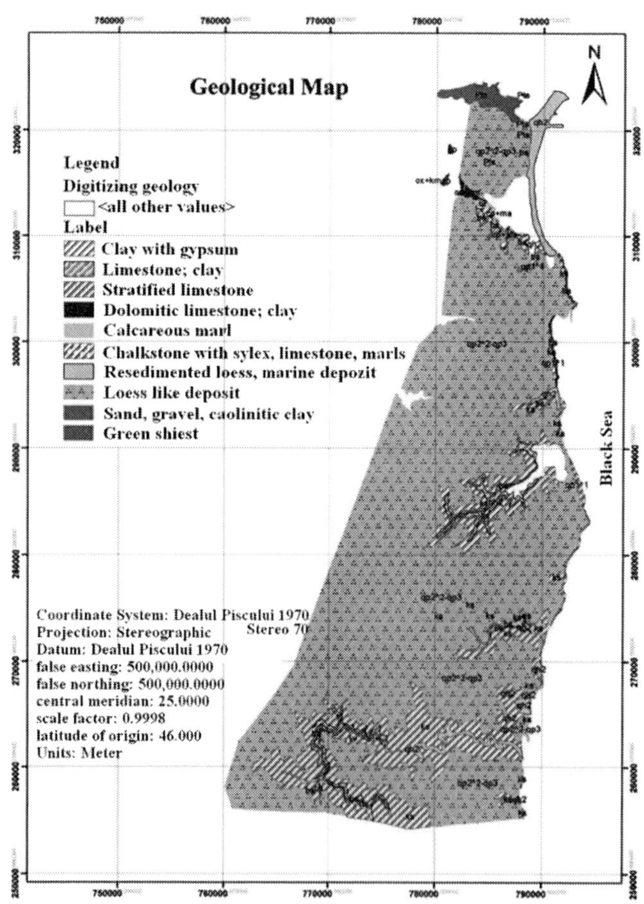

Figure 3. Common feature of the three units of Dobrogea - Quaternary cover.

3. Results and discussion

In order to assess the landslide susceptibility model under ArcView GIS, the methodology proposed is described in the figure 4:

- derived slope from DEM, figure 1, and reclassify the slope according FEMA method, as in table 1 and figure 4;
- developed geology map and a raster for the geological group described in table 1;
- developed water table map and divided the groundwater level in the two categories: dry or wet;
- intersection of all raster maps in order to obtain LSA map;
- reclassify the LSA raster map in the classes proposed by FEMA given the scales. All maps are presented in Stereo 70 projection.

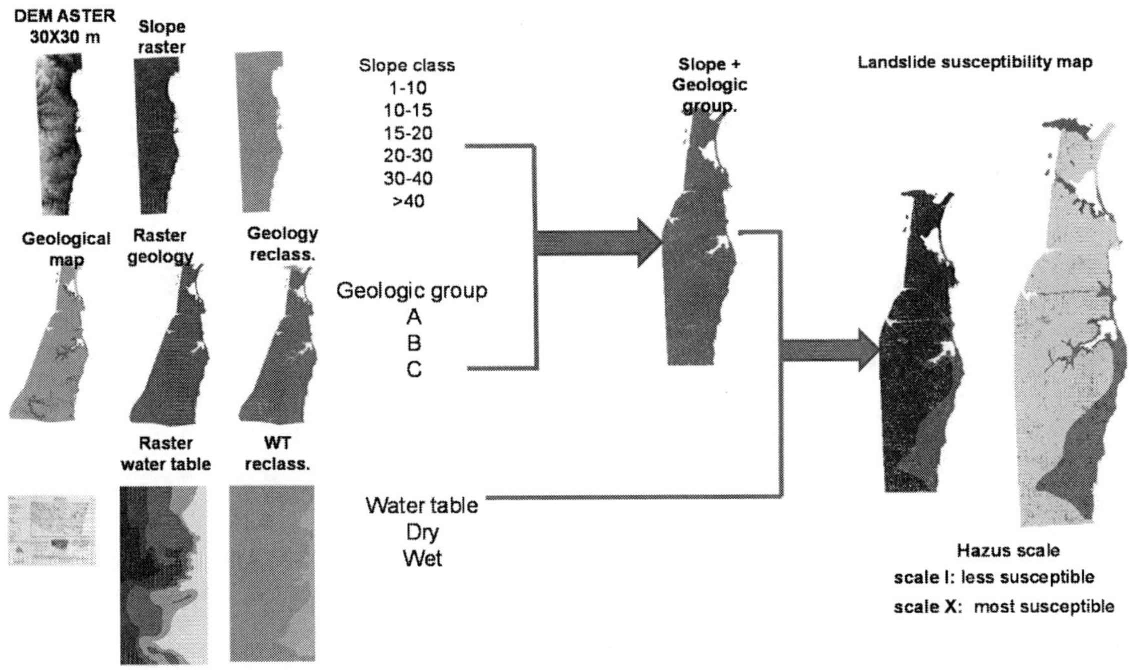

Figure 4. Methodology used for determining Landslide Susceptibility map.

Based by the methodology proposed, we obtained the landslide susceptibility map, in figure 5b and 5c (a zoom of Eforie area). This map shows that the area fall within V and VI class without the cliffs where the territory falls within VII class. The cliffs in Eforie Sud, 2 May and Vama Veche are the subject to landslide susceptibility (VII and IX classes). For evaluating the effectiveness of the LSA map this is compared with the landslide prone area identified by Constanta County Inspectorate for Emergency Situations (CCIES), as in figure 5a, using a method establish by Governmental Decision (447 from April 2003).

According this method, an average hazard coefficient (AHC) is calculated based on a large variety of data. The comparison of two maps (figure 5a and 5b) shows that there is a remarkable convergence in the location of landslide prone areas at regional scale especially for Eforie area (in rectangle – figure 5a and figure 5c), despite the fact that the two models have very different input data requirements and backgrounds.

Currently, the Eforie area is strongly affected by landslides and collapses of the cliff due to local geological conditions: the clay layer at the base of loess deposits. Some consolidation works and sand nourishment project are in developing in the Eforie area.

Taking into consideration that the FEMA models have minimal data requirements (ASTER DEMs, topographical data, geological map, groundwater map), this model can be used to reliably delineate landslide prone areas on a regional (1:25000-1:100000) scale with minimal data requirements in order to proceed with risk assessment in that scale.

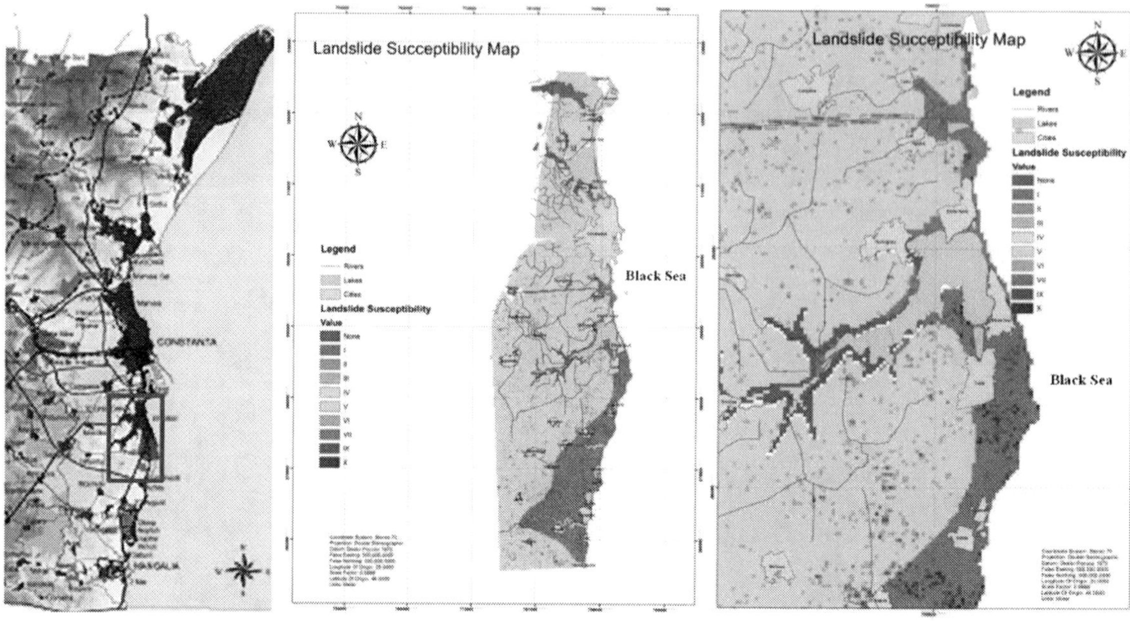

(a) (b) (c)

Figure 5. Landslide susceptibility map: (a) landslide prone areas identified by AHC method; (b) using FEMA method for the Black Sea coast line. (c) zoom for Eforie area.

4. Conclusions

Landslides are a local phenomenon which depends on a large variety of factors (internal and external). The choice of the most appropriate method depends on data availability and investigation scale. We could conclude that the LSA maps at regional scale based on FEMA method, represent a useful tool in the identification of areas predisposed to landslides and could be used as support in the decision-maker process in order to propose environmental work protection or county planning. At a next stage, deterministic model based on calculating the stability factor of safety factor which can be used especially on site-specific (local) scales in order to determine the landslide hazard map, thus helping make decisions about designing effective preventive measures. Data requirements in this case, especially for highly detailed topographic data of the area, are very demanding.

References

[1] Centre for Research on the Epidemiology of Disasters – CRED 2019 *Natural Disasters 2018* CRED Belgium
[2] General Inspectorate for Emergency Situations 2016 *National Risk Assessment – RO RISK Country report* Bucharest
[3] Marinescu M, Mafteiu M, Andra-Toparceanu A, Marunteanu C, Gheorghe R, Neata Ghe, Popa I, Andra-Toparceanu M, Mihailescu D and Bugiu S 2015 The need of interdisciplinarity approach on landslides research in Romania *European Scientific Journal* p 479
[4] Posea Gr, Popescu N and Ielenicz M 1974 *The Relief of Romania* Scientific Publishing House, Bucharest (in Romanian)
[5] Brabb E E 1991 Th*e world landslide problem. Episodes* **14** (i) p 52
[6] Brabb EE 1984 Innovative approaches to landslide hazard mapping *Proc 4thInternational Symposium on Landslides (Toronto)* vol. **1** (Canadian Geotechnical Society. BiTech Publishers, Vancouver, B.C) p 307
[7] Federal Emergency Management Agency 2013 *Hazus – MH Multi-hazard Loss Estimation Methodology Washington* D.C. 718

Influence of link length on re-centring capability of dual eccentrically braced frames

A Chesoan[1*], A Stratan[1] and D Dubină[1,2]

[1] Politehnica University of Timisoara, Department of Steel Structures and Structural Mechanics, Timisoara, Romania
[2] Romanian Academy, Fundamental and Advanced Technical Research Centre (CCTFA), Timisoara, Romania

[*] adriana.chesoan@upt.ro

Abstract. Repair costs and downtime of a structure hit by an earthquake can be reduced, by implementing the concepts of removable dissipative members (bolted links which are intended to provide the energy dissipation capacity and to be easily replaceable) and re-centring capability (which is provided by the more flexible moment resisting frames) in a dual structure, obtained by combining steel eccentrically braced frames (EBFs) with removable bolted links, with moment resisting frames (MRFs). The seismic performance, re-centring capability and link replacement feasibility of a dual re-centring EBF with removable flush end-plate bolted links were already studied and experimentally validated. In this case, very short shear links were used in order to achieve the flush end-plate connection over-strength, so the connection can be kept elastic in order to make the link removal possible. Re-centring capability of EBFs with other types of removable links has not been yet approached. One of the main objectives of an ongoing research project is to extend the validation of re-centring capability and link replacement feasibility on extended end-plate bolted links. This type of connection geometry provides a larger connection capacity, thus shear links can be designed longer. The present paper numerically investigates the influence of using different link lengths (and connection detailing) on the re-centring capability of two height levels structures.

1. Introduction

Mankind is more and more concerned lately about the vast amount of material and life loss due to earthquakes. Therefore, structural engineers strive to find solutions that ensure safety without incurring high costs. Eccentrically braced frames are such a solution that ensures a balance between structural stiffness and ductility. In order to reduce costs and downtime of a structure hit by earthquake, the concepts of removable dissipative elements (which ensure energy dissipation capacity and are easily replaceable) and re-centring capability (ensured by the more flexible moment resisting frames) are implemented in a dual structure, obtained by combining eccentrically braced frames (EBFs) with removable links with moment resisting frames (MRFs) [1]. The seismic performance, re-centring capability and link replacement feasibility of a dual re-centring EBF with removable flush end-plate bolted links were experimentally validated within the DUAREM project [2]. In this case, very short links were used in order to achieve the flush end-plate connection over-strength, so it can be kept elastic in order to be able to remove the link. Re-centring capability of EBFs with other types of removable links has not been yet approached. One of the main objectives of an ongoing research project is to extend the validation of re-centring capability and link replacement feasibility on extended end-plate bolted links. This type of connection geometry was previously investigated [3] being proved more rigid than the flush end-plate one and providing a larger connection capacity, thus longer links can be used.

2. Investigated structures

The present paper numerically investigates the influence of using different link lengths (and connection detailing) on the re-centring capability of two height levels structures: low-rise frames with 3 stories (P+2E) and mid-rise frames with 6 stories (P+5E) (figure 1).

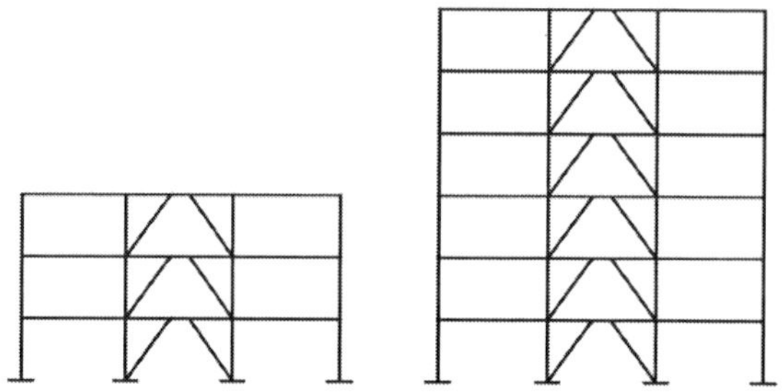

Figure 1. Elevation of low-rise (a) and mid-rise (b) frames.

Each structure was designed and numerically investigated using two link connection layouts: flush end-plate (F) and extended end-plate (E) (figure 2).

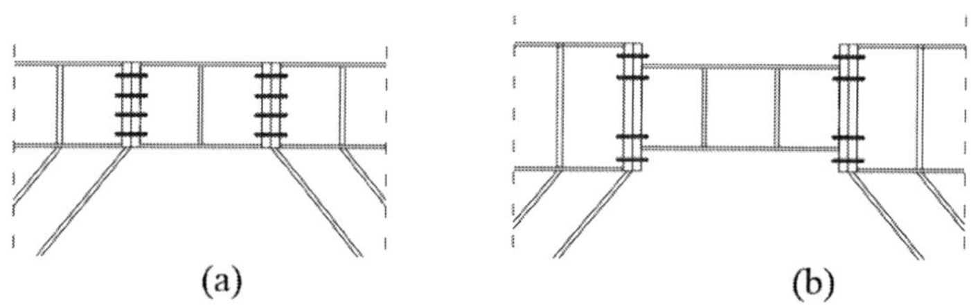

Figure 2. Flush end-plate (a) and extended end-plate (b) link connections.

The plan layout of the investigated structures is presented in figure 3, having three spans of 6 m on one direction and five spans of 6 m on the other direction. The height of each story is of 3.5 m.

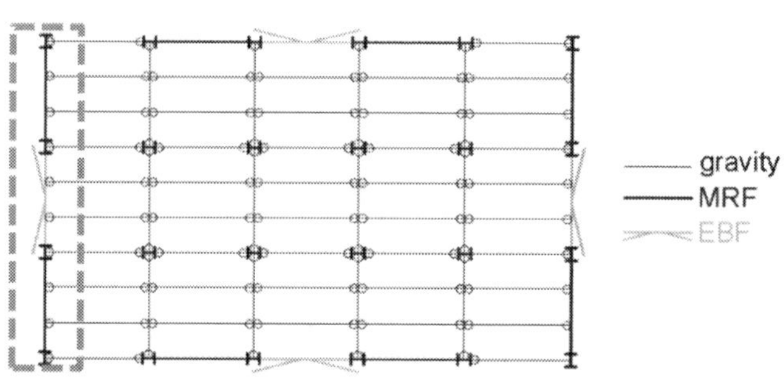

Figure 3. Plan layout of investigated structures.

The main lateral load resisting system is composed of four MRFs and two EBFs on each direction. The marginal frames are dual steel frames that combine two MRFs (ensure the necessary re-centring capability of the structure, providing the restoring forces after an earthquake) with one central EBF with replaceable bolted links (aimed to ensure the energy dissipation and to be easily replaceable) (figure 1).

These are the plane frames that were designed and investigated (figure 3). All the other inner frames are only gravitational loads resisting frames.

The main beams, columns and braces are made from European I profiles (IPE, HEA, HEB and HEM), and the dissipative bolted links are made from welded I profiles. S355 steel is used for structural elements, as well as S690 high-strength steel in MRFs (in order to ensure the re-centring capability in some cases).

Gravitational loads were considered as uniformly distributed loads on secondary beams and applied as concentrated forces on the main frames. Permanent loads of 5 kN/m^2 were assigned on current floors and 5.5 kN/m^2 on the roof. The live load was considered according to building destination (offices - class B) and partition walls, amounting to 3.8 kN/m^2 for current floors and 3 kN/m^2 on the roof. All gravitational loads assigned to investigated, frames correspond to half the span (3m).

High ductility class (DCH) was considered. Type 1 design response spectrum (spectral acceleration versus period), for soil type C (Figure), determined according to EC8 [4], was used in design. A peak ground acceleration of 0.3g and a behaviour factor q=4 were adopted [5].

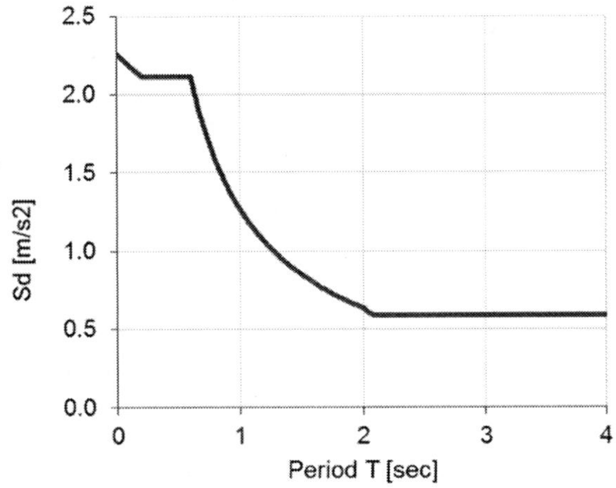

Figure 4. Type 1 Design response spectrum for soil type C.

3. Frames analyses

3.1. Structural analysis

For all frames, the first two modes of vibration have activated 90% of the total mass. Spectral analysis was used to study the structural behaviour.

Replaceable links and MRF beams are the dissipative elements of the frames.

Links must be removable and replaceable. This objective may be attained using a bolted end-plate connection that must be kept elastic. In order to achieve the connection over-strength, short and very short links were adopted (table 1).

Table 1. Shear links parameters.

Frame	Limit	Physical length e (m)	Type	Considered over-strength
P+2E_F	$e \approx 0.8\, M_{p,link}/\, V_{p,link}$	0.5	Very short	1.8
P+2E_E	$e \approx 1.6\, M_{p,link}/\, V_{p,link}$	1.0	Short	1.5
P+5E_F	$e \approx 0.8\, M_{p,link}/\, V_{p,link}$	0.8	Very short	1.8
P+5E_E	$e \approx 1.6\, M_{p,link}/\, V_{p,link}$	1.6	Short	1.5

Links were modelled in the global analysis with a reduced equivalent stiffness, to account for the flexibility of the bolted connections.

Frames were designed as dual structures [6]. A structural over-strength of 2.38 (P+2E_F), 2.24 (P+2E_E), 2.57 (P+5E_F) and 2.97 (P+5E_E) was used in the load combination for non-dissipative elements design: columns, braces and EBF beams.

Inter-story drifts were limited to 0.0075h. In order to avoid soft-story mechanisms, "strong column-weak beams" condition was satisfied.

The rigid diaphragm effect was assigned at each level to account for the presence of concrete slabs.

Structural masses were computed from gravitational loads corresponding to half the structure (Figure) and assigned to analysed frames nodes, because only the two marginal frames represent the lateral load resisting system.

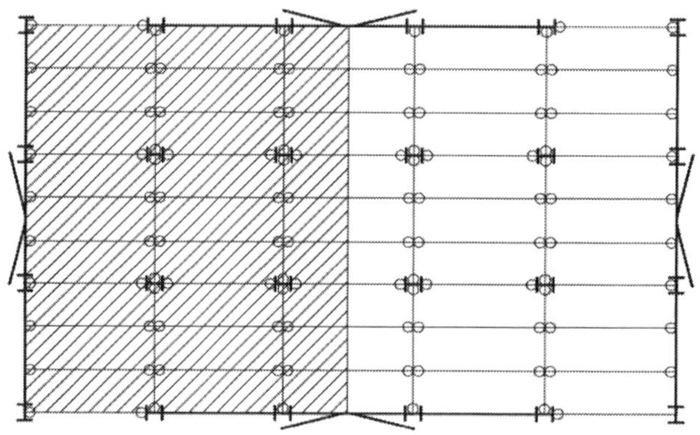

Figure 5. Plan layout of investigated structures.

A leaning (P-Δ) column was used in order to account for the effect of gravitational loads that correspond to the inner frames missing from the 2D analysis (figure 6), thus considering second order effects.

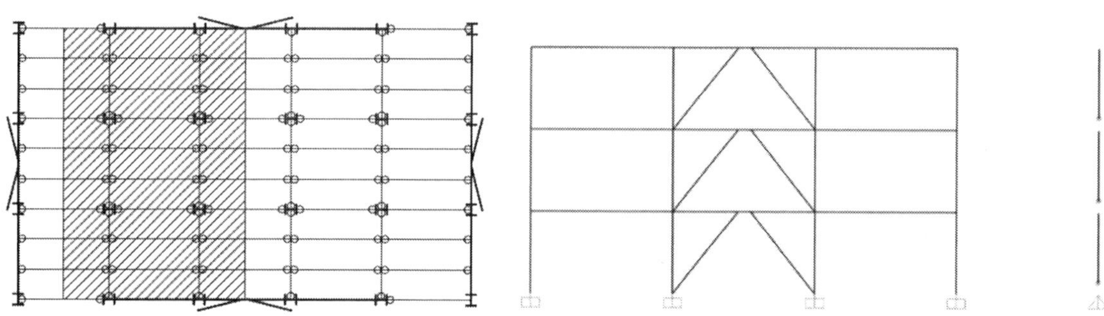

Figure 6. Leaning (P-Δ) column model.

3.2. Re-centring capability

A static nonlinear (pushover) analysis was performed in order to check the re-centring capability.

In case of the low-rise building with flush end-plate connections (P+2E_F) and very short links (e= 500 mm≈$0.8M_{p,link}/V_{p,link}$), yielding was observed outside the links only immediately after reaching 0.14 rad plastic rotation (ultimate deformation capacity at Ultimate Limit State) in these removable elements, the basic design requirement for dual frames with removable links being accomplished. When the peak rotation of a link reaches 0.14 rad (second floor), complete plastic mechanism is attained, with rotations of approximately 0.1 rad in the other links.

Table 2. P+2E_F elements sections.

Links (S355)	Central columns (S355)	Corner columns (S355)	Braces (S355)	EBF beams (S355)	MRF beams (S355)
330x200x16x9	HE300B	HE240B	HE260B	HE340A	IPE330
290x190x15x8	HE300B	HE240B	HE240B	HE300A	IPE330
250x180x14x7	HE300B	HE240B	HE200B	HE260A	IPE330

For the other frames (P+2E_E, P+5E_F, P+5E_E), yielding was observed in elements outside links before reaching 0.14 rad in the removable dissipative bars.

In case of the low-rise building with extended end-plate connections (P+2E_E) and short links (e= 1000 mm≈1.6$M_{p,link}/V_{p,link}$), S690 high-strength steel was used for the re-design of MRFs, thus ensuring frame re-centring (up to the ultimate plastic deformation of 0.14 rad), after eliminating one by one the damaged links. When the peak rotation of a link reaches 0.14 rad (second floor), complete plastic mechanism is attained, with rotations reaching between 0.09 rad and 0.13 rad in the other links.

Table 3. P+2E_E elements sections.

Links (S355)	Central columns (S690)	Corner columns (S690)	Braces (S355)	EBF beams (S355)	MRF beams (S690)
330x200x16x9	HE340B	HE240B	HE260B	HE340A	IPE270
290x190x15x8	HE340B	HE240B	HE240B	HE300A	IPE270
250x180x14x7	HE340B	HE240B	HE200B	HE260A	IPE270

In case of the mid-rise building with flush end-plate connections (P+5E_F) and very short links (e= 500 mm≈0.8$M_{p,link}/V_{p,link}$), S690 high-strength steel was used for the re-design of MRFs, thus ensuring frame re-centring (up to the ultimate plastic deformation of 0.14 rad), after eliminating one by one the damaged links. When the peak rotation of a link reaches 0.14 rad (third floor), complete plastic mechanism is attained, with rotations reaching between 0.08 rad and 0.13 rad in the other links.

Table 4. P+5E_F elements sections.

Links (S355)	Central columns (S690)	Corner columns (S690)	Braces (S355)	EBF beams (S355)	MRF beams (S690)
490x240x20x8	HE340M	HE300B	HE320B	HE450A	IPE300
490x240x20x8	HE340M	HE300B	HE320B	HE450A	IPE300
490x240x20x8	HE340M	HE300B	HE320B	HE450A	IPE300
390x240x20x8	HE340B	HE300B	HE280B	HE400A	IPE300
290x240x20x8	HE340B	HE300B	HE260B	HE300A	IPE300
250x190x16x5	HE340B	HE300B	HE200B	HE260A	IPE300

But in case of the mid-rise building with extended end-plate connections (P+5E_E) and short links (e= 1000 mm≈1.6 $M_{p,link}/V_{p,link}$), using S690 steel to re-design the MRFs, ensures frame re-centring only up to reaching 0.07 rad (half the ultimate deformation capacity at ULS) in links, the basic design requirement for dual frames with removable links not being accomplished in this last case.

Table 5. P+5E_E elements sections.

Links (S355)	Central columns (S690)	Corner columns (S690)	Braces (S355)	EBF beams (S355)	MRF beams (S690)
490x240x20x8	HE340M	HE300B	HE320B	HE450A	IPE300
490x240x20x8	HE340M	HE300B	HE320B	HE450A	IPE300
490x240x20x8	HE340M	HE300B	HE320B	HE450A	IPE300

390x240x20x8	HE300M	HE300B	HE280B	HE400A	IPE300
290x240x20x8	HE300M	HE300B	HE260B	HE300A	IPE300
250x190x16x5	HE300M	HE300B	HE200B	HE260A	IPE300

Following the conventional elastic design of the four frames, it was observed that adopting extended (E) end-plate connections for the shear removable links, allows the increase of their lengths with respect to the flush (F) end-plate configuration. This leads to increased frame ductility for extended end-plate configurations, as illustrated in the capacity curves (obtained from the pushover analyses), for both low-rise and mid-rise frames (figure 7). Also, larger lateral drifts are obtained for the extended end-plate configurations, indicating the possibility of adopting a behaviour factor larger than 4, but on the other hand, experiencing extended reinforced concrete slab degradation.

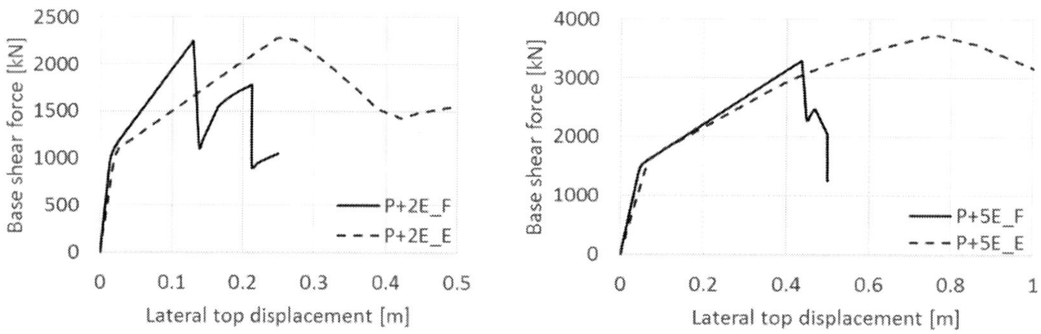

(a) (b)

Figure 7. Capacity curves for investigated low-rise (P+2E) (a) and mid-rise (P+5E) (b) frames, with both flush (F) (very short links) and extended (E) (short links) end-plate configurations

Using extended end-plate removable shear links (e= 1000mm≈1.6$M_{p,link}/V_{p,link}$) leads to increased connection design efforts, needing replacement of HEA profiles used for EBF beams, with taller IPE profiles, providing the necessary geometry to achieve the bolted connection.

4. Link removal

For studying the link removal procedure and frame re-centring, a static non-linear "staged construction" analysis was performed on the P+5E_F structure (figure 8), the procedure being similar also for the other frames.

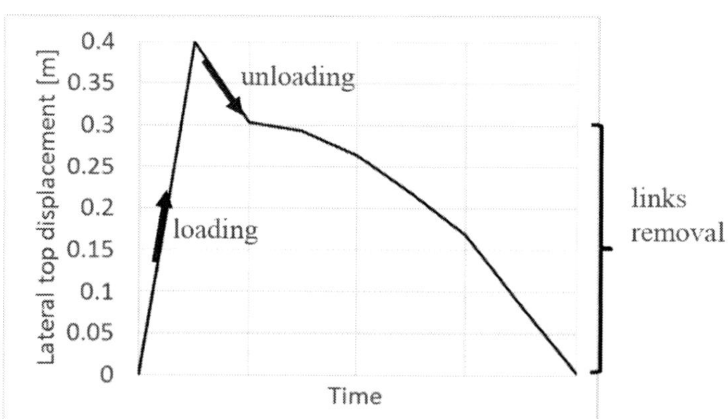

Figure 8. Top displacement during link removal

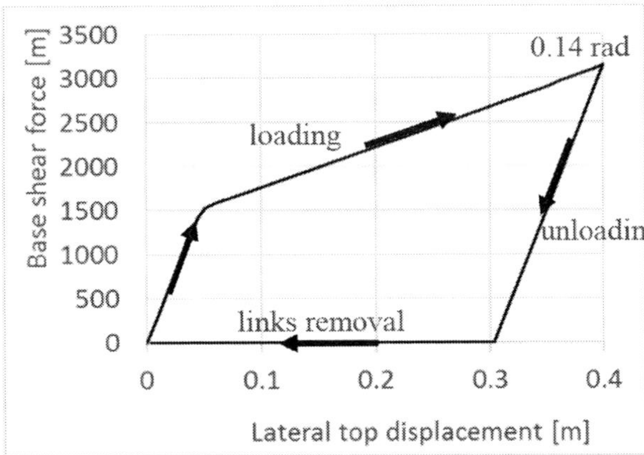

Figure 9. Force-displacement relationship during link removal and frame re-centring.

The analysis stages are the following: firstly, gravitational loads and lateral forces (until reaching the ultimate plastic deformation of 0.14 rad in links) are applied (figure 10 a), simulating the loading of the structure, and then the frame is unloaded (figure 10 b) (remaining deformed due to plastic deformations constrained only in links), secondly, links are removed story by story, from the bottom one towards the top [7] (figure 10 c-h), the lateral drifts being reduced gradually. After removing the last link, the structure almost recovers its initial position (see figures 8, 9 and 10 h), having a permanent top displacement of 3 mm.

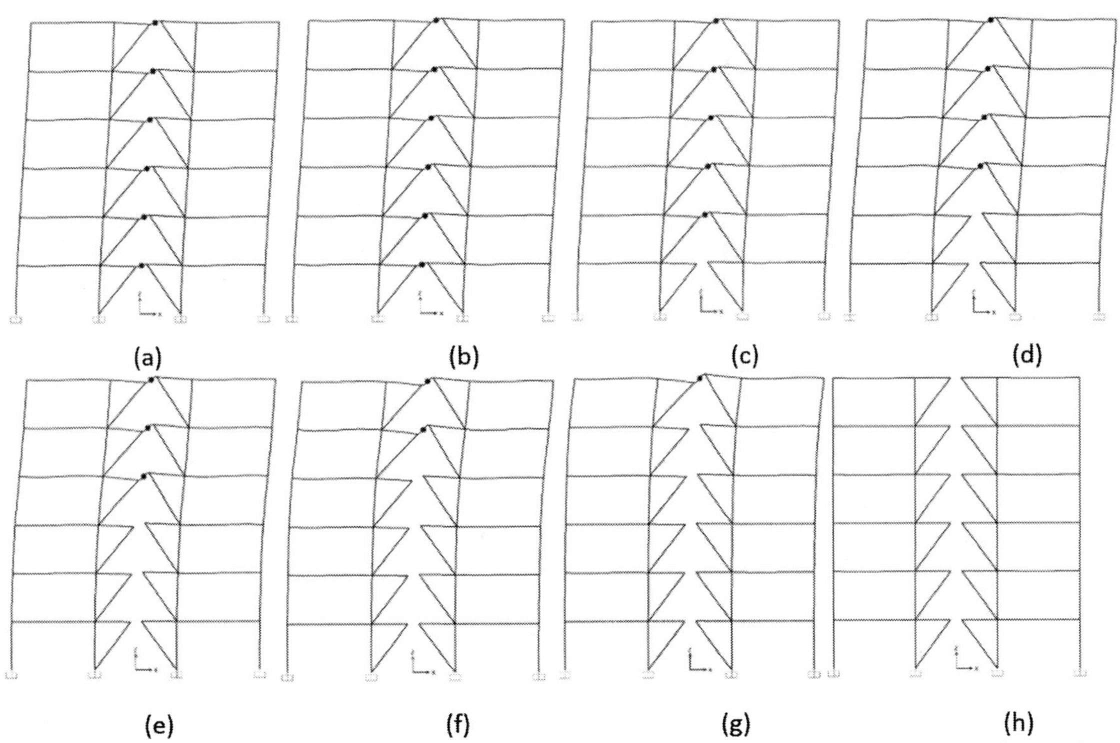

Figure 10. Link removal stages: frame loading (a), frame unloading (b) and link removal story by story (c-h).

From technical point of view, the easiest way to remove links is by flame cutting the web and flanges of the bar [8] if large permanent lateral drifts are recorded or by unbolting otherwise, story by story [7], as already proven within the procedure adopted for the experimental specimen of DUAREM project [2].

5. Conclusions

When using flush-end plate connections for the bolted links in dual EBFs with re-centring capability, very short lengths (around half the range for shear links but not exceeding it, e $\leq$0.8 $M_{p,link}$/ $V_{p,link}$) are necessary in order to provide the connection capacity within the given geometry. This leads to less ductile frames and larger links over-strength, but ensures the re-centring capacity up to the ultimate plastic rotation in links at ULS (MRFs providing the necessary restoring forces by remaining elastic, plastic deformations being constrained in links only).

The removable shear links could also be designed longer (around the upper limit of the range for shear links but not exceeding it e $\leq$1.6 $M_{p,link}$/ $V_{p,link,}$), but by adopting extended end-plate connections in order to provide the connection capacity within the given geometry. This leads to a larger frame ductility (indicating the possibility of using larger behaviour factors) and smaller links over-strength, but ensures the re-centring capacity only for low-rise frames (and by using high-strength steel in MRFs). In case of mid-rise (and taller) frames, the MRFs are not able to provide the necessary restoring forces (even if high-strength steel is adopted), as they yield before ultimate plastic rotation is attained in links.

Acknowledgement

This work was supported by a grant of Ministry of Research and Innovation (Romania), CNCS - UEFISCDI, project number PN-III-P1-1.1-PD-2016-1655, within PNCDI III.

References

[1] Dubina D, Stratan A and Dinu F 2008 Dual high-strength steel eccentrically braced frames with removable links *Earthq Eng Struct D* Vol. **37** p 1703-1720

[2] Ioan A, Stratan A, Dubina D, Poljansek M, Molina F J, Taucer F, Pegon P and Sabau G 2016 Experimental validation of re-centring capability of eccentrically braced frames with removable links *Eng. Struc.t* **113** p 335-346

[3] Mansour N, Christopoulos C and Tremblay R 2011 Experimental validation of replaceable shear links for eccentrically braced steel frames *J Struct Eng* **137** p 1141-1152.

[4] * * * EN 1998-1 2004 Eurocode 8: *Design of Structures for Earthquake Resistance* 1st ed Brussels BSI

[5] Dubina D, Stratan A and Chesoan A 2017 Design recommendations for dual moment – eccentric braced frames with replaceable links *ce/papers (Wiley)* Vol. **1** Iss. 2-3 Collections: *Proceedings of Eurosteel*

[6] Stratan A, Dinu F and Dubina D 2010 Replacement of bolted links in dual eccentrically braced frames *14th European Conference on Earthquake Engineering* Republic of Macedonia

[7] Ioan A, Stratan A and Dubina D 2012 Evaluation of restoring capacity of dual steel EBFs with removable links *8th International PhD & DLA Symposium* Hungary

[8] Stratan A, Ioan A and Dubina D 2012 Re-centring capability of dual eccentrically braced frames with removable bolted links *Proc STESSA 2012* Chile p 723-728

IOP Conf. Series: Materials Science and Engineering **789** (2020) 012010 doi:10.1088/1757-899X/789/1/012010

Accessibility to Transylvania´s cultural heritage through BIM-Heritage and community involvement

G C Chițonu[1*], C N Rolfsen[2] and O Deaconu[1]

[1] Department of Civil Engineering, Faculty of Civil Engineering, Transilvania University of Brasov, Brasov, Romania
[2] Department of Civil Engineering and Energy Technology, Faculty of Technology, Art and Design, Oslo Metropolitan University, Oslo, Norway

[*] gabriela.chitonu@unitbv.ro

Abstract. The rural areas of Transylvania offer important resources for developing research in diversity of the cultural heritage. Many of these communities have seen great changes in demographics over the last 30-50 years, and important cultural heritage from previous generations is being lost in lack of understanding, human resources and willingness to preserve the value of our joint cultural heritage. The research seeks a digitisation approach to preserving and promoting the local cultural heritage, as well as a social science approach, which seeks to find answers to how one can be able to preserve, maintain and promote this cultural heritage, something which, according to the study hypothesis, can only be done if the communities see the benefits of doing so. The digital shift has the ability to create another level of understanding, opening a vast horizon for accessibility and progress linked with BIM technology. As an answer to dealing with the climate changes, degradation of the buildings due to natural disasters or more from the indifference of the young generations in regard of history inheritance, the study could be a valuable instrument in enhancing BIM-Heritage methodologies.

1. Introduction

The purpose of this study is to generate a research basis in some of Transylvania's rural areas, villages and towns, capable of promoting a sustainable management of their cultural heritage, with a main focus on the traditional cultural area and to contribute to the development of the multicultural society, while preserving the knowledge of the local history.

The presented study is an early phase of a sizable research project in Transylvanian rural areas. The project examines areas rich in cultural heritage, where Romanians, Hungarians, Saxons and the Roma population are living together in communities, with their own various traditions.

The European Commission has set as an important objective to all Member States to involve and develop strategies and policies on digitisation of cultural heritage. "The EU's strategy for digitisation and preservation builds on the work done over the last few years in the digital libraries initiative." Digitisation is an important means for ensuring greater access to and use of cultural material" [1].

In line with this objective, the research group has developed a strategy for documenting and preserving the local cultural heritage, while promoting its value for economic, cultural and sustainable development in the future.

The tangible cultural heritage includes traditional buildings, structures or landscapes. This includes locations, the environmental relationship, and the building materials. Moreover, the intangible heritage includes practices, representations, expressions, memories, knowledge, crafts and the abilities which are used and transmitted by these communities and groups from generation to generation [2].

Romania is permanently improving the tourism sector, developing sustainable tourism, cultural-tourism, eco-tourism, trying to help in making European strategies to safeguard the natural environment, natural heritage and cultural heritage. Transylvania's cultural heritage (not exploited and preserved enough) is very much appreciated by foreign tourists who are interested in the mixed multiethnic heritage because is easier to integrate as part of Europe [3].

2. Background, approaches and preliminary investigations

Many countries in Europe are digitising on a larger scale immovable cultural heritage and more than 1/3 of Member States reported funding programs for digitisation of monuments, historical buildings and archaeological sites [1].

Romania, an EU Member, has on Digital Day 2019 signed up to the Declaration of cooperation on advancing digitisation of cultural heritage. "The European Year of Cultural Heritage 2018 also created a momentum by bringing all Member States and stakeholders together around Europe's rich cultural heritage and contributed to enhance its visibility" [4].

Created in the frame of developing international cooperation between higher education Institutions, three research teams from Transilvania University of Brasov, Oslo Metropolitan University and University for Architecture and Urbanism "Ion Mincu", Bucharest, combined their knowledge to create a basis for a common research field with a focus on studying the existing Transylvanian cultural heritage. The goal is bringing together advanced expertise in building engineering, architecture, innovation and knowledge in BIM and HBIM technologies by using the opportunity of the field investigations in rural areas. The main research objective is an inventory of the most representative traditional elements from the tangible and intangible heritage in rural areas, from Transylvania's Central Region.

The research focuses on old Saxon-villages from Brasov county like Ticus, Racos, Soars villages and other un-exploited rural mixed population communities. Apold and Dealu Frumos, villages from Mures and Sibiu counties, are amongst other villages in the region which are studied with a view to finding ways of preserving the cultural history while enabling sustainable development, and tackling complicated societal problems which effect development in the region.

The academic teams involved in this study have since autumn 2018 been working on development of the academic platform required for in depth studies of the communities and successful dissemination of the acquired knowledge.

A key goal of the research is to directly use the knowledge of the past to help maintain the societal systems, infrastructure and management strategies which have been working well, and using this in education and aid of the communities that are currently residing there. Additionally, the research aims to use the knowledge acquired about the landscape, building mass, natural resources, culture etc. to help the communities develop long term sustainable strategies, infrastructure and systems for improving their lives and the society.

The research strategy involves investigating and documenting the region, and presenting the findings in easily understandable and accessible formats. The researchers are using advanced digitisation techniques and presenting many of the findings through 3D modelling, CAD, BIM, HBIM, image, video and virtual techniques.

All the participants in the study from the teams from Romania and Norway have different approaches and profiles offering diversity to the research project. Transilvania University of Brasov with the Faculty of Civil engineering brings expertise in field research, topographic measurements, urban planning, construction and rehabilitation. Oslo Metropolitan University has strength in advanced BIM/HBIM technologies, GIS and virtual presentations, and the "Ion Mincu" University for Architecture and Urbanism in vernacular architecture, cultural heritage and landscape preservation.

The cooperation supports creativity, and the discipline interference will sharpen their identity. The scientific experience from the engineering field focuses on technical BIM and HBIM, the architecture field focuses on historical data and 3D modeling, spatial and volumetric approach, while the urban-planning field focuses on territorial developments and cultural landscape. The research strategy is structured in 3 main phases:

- the initial phase includes in the first part exploration and selection of the areas with the highest cultural potential, where the feed-back from the communities is the most positive and which has most relevant perspectives in developing a sustainable development;
- the second phase is a digitisation approach to preserving and promoting the local cultural heritage and also a social science approach. It seeks to find answers in relation to the rural communities and how one can be able to preserve, maintain and promote this cultural heritage. This can only be done successfully if the communities become aware of the benefits of such activities.
- The two initial fields of research will lead to a third phase and field, which seeks to highlight the red line, and the interconnectivity, between the previous two fields and the key topics of the research field.
- The third field entails using state of the art technological solutions like 3D laser-scanning, advanced measuring techniques, and BIM modulation, and turning it into an interactive solution with focus on documenting and preserving the cultural heritage. The HBIM approach will provide an inventory of documentation in construction and conservation of the cultural heritage studied [5].

The computational approach will preserve knowledge and scientific information but will also be a bridge for providing communities with an overview of their civilization background and support continuous development. Using the latest methods of presenting the study results (know-how, media, video dissemination, interviews), will provide a basis for cultural education activities, especially in disadvantaged communities.

BIM – Building Information Modeling has been defined as "a digital representation of physical and functional characteristics of a facility, a reliable basis for decisions during its life-cycle; defined as existing from earliest concept to demolition" (The American National Building Information Model Standard Committee) and the last advance technologies confirm that BIM has been classified into several maturity levels:

- level 0D – refers to the zero baseline,
- level 1D – refers to the (the initial) point,
- level 2D – refers to traditional 2D line drawings,
- level 3D – refers to the three primary spatial dimensions (width, height and depth),
- level 4D – refers to the fourth dimension of time and sequence,
- level 5D – refers to 3D + time (schedule) + costs,
- level 6D – refers to the Lifecycle Management and Facility Management in what is delivered [6].

The transition to the HBIM – Historical Building Information Modeling, has been gradually done over the last 10-12 years due to the necessity of considering part of the Lifecycle Management additionally to the existing buildings and also the historical monuments.

So, the actual BIM approach, which stats from 0 to 6, 7D can't be applied for something which has already been proven over the decades or centuries and where the Lifecycle Assessment do not have a data basis containing 2D, 3D/CAD information. The goal in HBIM is focusing on Conservation, Repair and Maintenance (CRM). The approach is clearly broader then including CRM as a definition, also integrating digital technologies support, laser scanning, survey options, thermal scanning etc, the results become complex with multitude use and users in specific communities [6].

Current maps and historic maps will be compared to determine the nature and the landscape changes over time. GIS (Computerized Geographic Information Systems) and topographic measurements will determinate the rural landscape characteristics and spatial settlements by using typical operations such as map overlays, distance calculations, and interpolation.

These operations can also be used to define the boundaries of properties and to assess the visual impact of land use changes [7].

3. Feed-back and answers from Transylvania's communities

Argumentations in selecting areas:

- the database of Brasov's county strategy contains little information about the selected rural areas;
- the sustainable development principal focus points are: equity and equality;
- the objectives selected for the study are complementing the existing touristic maps.

The selected areas are not included in the main touristic routes of Transylvania Center region, and the local communities need support in promoting their own values. By doing research in these territories, the future development will bring about economic, social, cultural integration while diminishing the migration of local population to the big cities.

Transylvania features many historic cultural resources, the Saxon villages which have been confronted in the last 3 decades with massive depopulation, due to the migration of Saxons when their German-speaking inhabitants emigrated to Germany in the 1990s. Similarly, part of Romanian and Hungarian populations migrated in search for better life quality to the bigger cities of the country or abroad.

In Brasov county there are numerous depopulated or abandoned Saxon villages, with only a small number of inhabitants, Romanians, Hungarians or Roma population in need of assistance to revitalize their rural habitat. Beautiful medieval fortresses, and fortified churches in poor condition, ruined structures are specific for the local rural areas.

Villages where in the last 10-20 years, the birth rate dropped rapidly could still be very interesting areas with valuable cultural heritage for tourists, also due to their pristine natural and wildlife surroundings.

The research team of Transilvania University of Brasov (UTBv), being located in Brasov city, has conducted preliminary investigations in the rural communities, which are willing to enhance their cultural heritage. During the initial phase, the researchers have been developing working relationships with a few local foundations/NGOs, cooperating with local and regional authorities to establish partnerships and collect important public data, including links with rural planning and local strategies.

Descriptions of some of the communities that the research project is studying:

- Ticus is first mentioned in 1373 as an old medieval Saxon village, as in figure 1, having mixed population (Romanian 65.08%, Hungarian 18.94%, Roma 12.11% and others), a Fortified Evangelical Church (XVth- XIXth century), monument in Ticusul Vechi (figure 1) and Reformed Church (XVth- XIXth century) in Cobor, as in figure 2 and figure 3.

Figure 1. Ticuş village – the main street (photo from author´s collection).

Figure 2. Cobor village – Fortified Reformed church tower (photo from author's collection).

Figure 3. Cobor village – Fortified Reformed church (photo from author's collection).

- Racos a medieval Saxon village with mixed population (Romanian 22.27%, Hungarian 52.18%, Roma 20.27% and others), having an old monument Sukosd-Bethlen Castle built in 1625 in the Transylvanian Renaissance style, as showed in figure 4 and figure 5, surrounded by beautiful landscapes, with an old volcano, a basaltic scoria quarry (The Brazi Quarry) and the Emerald Lake, as great touristic attractions [8].

Figure 4. Racoş village - Sukosd-Bethlen Castle (photo from author's collection).

Figure 5. Racoş village - Sukosd-Bethlen Castle (photo from author's collection).

- Apold village, Mures county, has been founded before the Saxons arrived and first mentioned in 1309 (Romanian 22.27%, Hungarian 52.18%, Roma 20.27% and others). The 13th century Church (figures 6 and 7) erected in a strategic position on top of a hill surrounded by the river Şaeş, evolved later into the famous fortified church of Apold. The ensemble survived turbulent times and stays even today almost entirely preserved, with fortification walls, defense towers and storage rooms [9].

Figure 6. Apold village - The Evangelic Fortified Church (photo from author's collection).

Figure 7. Apold village - The Evangelic Fortified Church (photo from author's collection).

IOP Conf. Series: Materials Science and Engineering **789** (2020) 012010 doi:10.1088/1757-899X/789/1/012010

The enhancement on multiple levels of the rural heritage implies identifying and analyzing the natural-risk factors, with a view to determining the conditions wherein regeneration, restoration activities, as well as other types of human interventions, may be initiated and conducted.

The present study starts from the method of zoning on geographical sections, which allows identifying several habitats, as well as creating zoning categories.

Approaching the issue of the traditional habitats in the light of the cross- cut zoning is achieved in the context of the old traditional urban planning, historically inherited in each village with the general frame for conducting the aforementioned activities.

As regards the Transylvania region, and especially rural villages, reaching the proposed objectives is conditioned by the following aspects:

- inclusion in the analysis of all components and activities afferent to the rural habitat (buildings, landscaping elements, restoration and real-estate activities etc.);
- inventory of the natural and historic-patrimony elements specific to the studied habitat. An investigation in situ was made upon all the habitat elements which convey overall the specificity of the studied region;
- rehabilitation of the landscaping component of the habitat, against the background of the conservation and development of the community's traditions within the natural environment, as strategic stimulating element on a rational basis, of the rural tourism and agro-tourism;
- identification of solutions with a view to maintaining, if possible, the plotting of the area and the existing structure of the households, keeping withal the neighbors areas, given the significant influence of these administrative aspects on the natural landscape and the architectonic traditional elements [10].

4. HBIM contribution in Transylvania's areas

The BIM-sub-group of OsloMET University is doing research with a broad spectrum: processes related to utilization of BIM-related technology, implementation of BIM in the industry and education, BIM supported sustainability assessment, transformation of processes for professional work and collaboration.

The research is focusing on the information flow through the entire life cycle of the built environments. HBIM will be an important tool in the research project and the way of presenting findings, but it is only one of the tools needed to reach the researcher's goals of delivering an end product generating benefits for the local communities, tourists, historians and our globally shared cultural heritage [11].

Linking the technologies from measuring, imaging, CAD to HBIM for the future phases of the research is a key technical challenge that has been integral to this study.

Much of the "state of the art" aspect lies in the combination of technologies and in how they are applied. 3D laser scanning, advanced measuring, high detail photography, non- to minimum invasive investigation tools (tiny inspection cameras for example, or samples) of buildings have been done before. This project delivers amongst other outputs, very advanced "passports" or libraries, with extensive information for selected cultural heritage objects. For buildings this could be everything from volumes, materials, history, functions, state of maintenance, energy needs, clues for improvements, risk analysis, and much more.

The process of studying in great detail the cultural objects, including buildings and monuments will present several technological challenges:

- individual passports for each of the studied cultural heritage objects, and next,
- creating an extensive topographic analysis, and a 3D map of the village and surroundings, and a high detail model of selected parts of the village;
- taking the individual passports and implementing them in a wider presentation of a research project, and through a map function linking the individual passports to each other, to the geographical location and allowing the users to explore these objects in detailed ways.

5. Conclusions

This study which was conducted as a preliminary investigation and pathfinder for a greater research project "Re-discovering the Traditional and Cultural Heritage through nonintrusive techniques and HBIM based digitisation for sustainable Transylvanian rural communities" has identified important parameters in the continuous development of the research platform.

The research will build knowledge and competences in the following fields: the cultural heritage approach, innovative BIM technologies, advanced graphic presentations with high intuitive and educational focus, and rural communities and environment curricula. Due to inter-disciplinarity, inter-institutional, and international cooperation this approach will allow the new generations of students to participate in future research; based on multicultural links and diversity in a networked environment.

HBIM concepts in addition to providing advanced data about the existing historical buildings, which includes architectural and structural parametric rules, will connect the objects with the areas generating a mapping system with specific libraries. The local communities will benefit directly by acquiring knowledge about the building types they live in, and about methods to maintain and improve them, as well as from being known and placed on the map of cultural heritage, to several levels like educational, preservation, cultural, international visibility and socio-economic development. Very careful considerations must be taken to balance the requirements and constraints of HBIM tools, ease of sharing through close compliance with suggested standards, with delivering a sophisticated graphical presentation of the totality of the data and the cultural heritage one aims to share. The goal of the researchers is that the complex research and IT-platforms developed within the project, be used by the local communities for enhancing and preserving on its basis their cultural life and traditions. It is essential that the platform development operates with a focus on limiting the financial expenses, while being capable to deliver a high quality graphical output, with less power consumption. The need of practicing optimized environmental sustainability in the operations of the graphical end product should be a guideline for further development and innovation of the digitised solution.

References

[1] Official Journal of the European Union, Commission Recommendation of 27 October 2011, on the digitisation and online accessibility of cultural material and digital preservation (2011/711/EU)
[2] UNESCO 2003 Convention for the safeguarding of the intangible cultural heritage, Paris
[3] Dulău A V and Coros M M 2009 Is cultural tourism attractive in Transylvania? A focus on Cluj and Sibiu Counties, Romania WSES *Transaction on Business and Economics* p 13
[4] https://ec.europa.eu/digital-single-market/en/news/eu-member-states-sign-cooperate-digitising-cultural-heritage
[5] Murphy M 2012 *Historic Building Information Modelling (HBIM) for Recording and Documenting Classical Architecture in Dublin 1700 to 1830* PhD Thesis p 19-50
[6] Maxwell I 2014 *Integrating digital technologies in support of historic building information modelling: BIM4Conservation (HBIM)* A COTAC Report p 7-13
[7] McClelland L F, Keller J T, Keller G P and Melnick R Z 2012 Guidelines for evaluating and documenting rural historic landscapes *National Register Bulletin* U.S.A. p 12
[8] http://www.romania-travel-guide.com/en/attractions/spiritual/homorod-racos-hoghiz.html
[9] https://kirchenburgen.org/en/fortified-churches/
[10] Chiţonu G C 2014 *Habitatul traditional branean. Perspective socio-economice si culturale in dezvoltarea durabila in peisajul rural (The traditional habitat of Bran. Socio-economic and cultural perspectives of sustainable development in rural landscape)* Ph.D thesis "Ion Mincu" University of Architecture and Urban Planning, Bucuresti
[11] https://cotac.global/resources/COTAC-HBIM-Report-Final-A-21-April-2014-2-small.pdf

Applying the mathematical optimization model in water distribution management

V Cojanu[1*] **and E Helerea**[1]

[1] Electrical Engineering and Applied Physics Department, Transilvania University of Brasov, Romania

[*] vlad.cojanu@unitbv.ro

Abstract. This paper aims to analyse the concept of water distribution system management and the modalities of including the optimization tools for increasing the process efficiency. Water distribution systems are often unable to respond quickly and efficiently to consumers demands. Water is a non-renewable resource and should therefore be treated with special interest. Sustainable development and optimization are the steps to take. A mathematical optimization model is presented in the second part of the paper with explanations regarding the steps followed. The result demonstrates the efficiency, speed and accessibility of using graphical-analytical optimization methods in the field of water distribution system.

1. Introduction

Water is an essential source for life. Human society has developed a number of systems in an attempt to facilitate access to water, improve water quality and use it responsibly. There is a wide range of water that is used in urban agglomeration of the large cities: rainwater, bodies of water, drinking water, wastewater. The urban water system is a complex consisting of several subsystems: the water treatment station, the water storage, the water distribution system, the consumers (industrial and household), the waste water collection system [1]. These systems are generally equipped with intelligent SCADA elements. The first attempts of automation were made in the 1970s on water treatment systems in order to reduce operating costs. The systems were oversized and often their operation was not correlated with the load fluctuations. Nowadays, more emphasis is placed on efficiency in order to take over the peaks of loads that have increased with the population and the standard of living [2]. The same problem is encountered in water distribution systems. In this scientific work, it is desired to applying the mathematical optimization model in a water distribution system to determine the best parameters for optimal management. The mathematical model is based on the graphical-analytical optimization methods.

2. Water distribution management

Water distribution management is a challenging issue that is closely linked to global warming, quality of life, the water circuit in nature. Managing water distribution in big cities means managing all the functions of the water distribution system.

The water distribution system consists of pipes, valves, meters, pumps, storage tanks, etc. and has the role of transporting water from the source to the consumer.

At the consumer level the water must have the microbiological quality, the required flow and pressure. Water distribution is the "urban metabolism" [3]. All over the world, efforts are being made to reduce losses, adapt to the demands of people and ensure the reliability of water infrastructure. The concept of water sensitive urban design aims to incorporate all the water systems that serve a city, focusing on stormwater management and domestic wastewater [3].

Water losses can have different causes: system leaks system damage, unauthorized consumers (consumers that are not counted) or apparent losses. Leakage volumetric losses represent over 70% of total water losses [4]. Some of the main causes that can lead to the appearance of volumetric losses are: excessive pressure in certain points of the network, mechanical factors that may occur (road traffic loads, accidental hit), the irresponsible operation of the network or the existence of an old network.

The most efficient way to control the losses in the system is with the help of pressure monitoring indifferent points of the network [5]. In such a system the pressure anomalies are analysed.

According to the severity of the problems losses in the water system are classified as follows [6]:

- background losses (low flow, they are very hard to detect using classic methods like acoustic detection),
- unreported losses (moderate flow and long duration, their identification requires time),
- reported losses (high flow losses and short duration, this are visible problems).

Long-term constructing transitions of water distribution management require a correlation between society, urban development of the region and the structure of the water system. These three factors are interdependent, the modification of one leading to implicit modification of the others.

In [7] an overall approach is proposed based on three modules:

- Social Transitions Module is related to the stormwater management and its role is to explore the possible outcomes and to generate from a large portfolio a set of typical pathways,
- the Urban Development Module is an agent-based modelling solution that aims to simulate the population density evolution on certain urban areas,
- biophysical module is related to design and performance of the urban water system.

This approach based on three basic modules has a high degree of accuracy in optimizing long-term urban water systems. The algorithm underlying these models is a mathematical optimization algorithm. Optimization is the act of obtaining the best result under given circumstances. In practice, any system or installation must be controlled and management decisions are required in several stages. The ultimate goal of all these decisions is to minimize the effort required and maximize the desired benefit. Since the required effort or desired benefit in any practical situation can be expressed as a function of certain decision variables, optimization can be defined as the process of finding the conditions that give the maximum or minimum value of a function. Optimization can also be defined as choosing, from the multitude of possible solutions to a problem, the one that is best in relation to a predefined criterion.

This definition implies the existence of the following components [8-9]:

1. a technical problem consisting of the mathematical calculation of a solution;
2. existence of several solutions for the same problem;
3. a criterion for selecting the optimal solution.
4. establishing the optimal solution.

There is no single method available for efficiently solving all optimization problems. Therefore, a series of optimization methods have been developed to solve different types of optimization problems. Optimal search methods are also known as mathematical programming techniques and are generally studied as part of operational research. Linear programming takes data from real-world problems and transposes them into a series of mathematical formulas. Such a problem is characterized by a linear objective function of the form:

$$f(x_1, x_2, \dots, x_n) = c_1x_1 + c_2x_2 + \dots + c_3x_3 \quad (1)$$

and a number of constraints represented by a system of linear equations of equality type. The process of transposing constraints from the real world into linear functions is an essential step in solving linear programming problems [10-11]. A constraint is restrictive if the right and the left member are equal when the decision variables are replaced by the optimal values. If the two members of the constraint are not equal after the decision variable is replaced, constraint is not restrictive.

In order to solve such a problem, the mathematical function must have a certain form, known as the standard form.

This form has the positive variables and constraints are equal which have the positive terms on the right side. The general mathematical form of a linear programming problem must comply with the restrictions:

$$\sum_{j=1}^{n} a_{ij}x_j = d_i, i = 1, \dots, m \tag{2}$$

$$x_j \geq 0, j = 1, \dots, n \tag{3}$$

If the constraint (2) is not an equality, a variable will be added or subtracted (it is known as a programmable element) so that the mathematical relation is brought in the form of an equality. The importance of the optimization problem lies in the fact that it corresponds to the purpose general optimization of the used resources in order to meet certain objectives [12]. The numerical solution of linear programming problems is based on the fact that the solution of the problem is on the border of the admitted solutions.

The simplest method of solving is the geometric or graphical-analytical method. This method is limited to solve problems that have less than or equal to three decision variables. On the other hand, the simplex method is considered one of the oldest solving methods and it is based on George Dantzig's algorithm [13], whose complexity increases directly in proportion to the number of variables associated with the problem.

3. Case study

An application of the linear programming is proposed as case study considering the problem of optimizing the benefit in the case of an urban water system.

Problem description: drinking water in the urban system considered is provided from two sources: raw water and used water. The quality assurance of the drinking water is obtained throughby a complex treatment process consisting of a series of treatment steps:

- mechanical treatment step, with the role of removing the floating solids in water, here is the disintegration, separation of fats and decanting of water;
- chemical treatment step, with the role of removing some of the impurities found in water, most often using the coagulation-flocculation process, in which the basic reactant is the chlorinated ferrous sulphate $FeSO_4$;
- biological treatment step, which uses cultures of microorganisms to transform hazardous substances into degradation products and sludge that does not affect the environment;
- a tertiary treatment has the role of removing excess substances and disinfecting the water.

In this case study, the chemical step of water treatment for an urban area is considered. The treatment consists of introducing of the chemical substances as follows:

- Raw water (C1) treatment:

S1- 150 g/m^3;S2-250 g/m^3;S3-125 g/m^3;S4-150 g/m^3.

- Used water (C2) treatment:

S1-300 g/m^3; S2-500 g/m^3; S3-400 g/m^3; S4-150 g/m^3.

From the infrastructure point of view, the water treatment capacities for each substance are: S_{m1} = 90 kg, S_{m2} =180 kg, S_{m3} = 150 kg, S_{m4} = 60 kg. The cost of producing the drinking water is: 4 lei / m^3 for water providing from used water, and 2 lei /m^3 for water obtained from raw water. The optimization problem is defined: how to get the maximum benefit for obtaining drinking water from the two categories of water sources, at a charge in the given circumstances.

Problem solving includes the following solving steps are applied:

3.1. Initial data centralization

The problem data are centralized in table 1.

Table 1. Quantities and capacities of chemical treatment substances needed for obtaining drinking water from raw and used water for a cycle of treatment

Treatment restrictions	Row water (C1) (g/m^3)	Used water (C2) (g/m^3)	Capacity (kg)
S1	150	300	90
S2	250	500	180
S3	125	400	150
S4	150	150	60
Water selling price[lei/m^3]	2	4	

3.2. Setting variables
Variable are: x for raw water treatment and y for used water treatment.

3.3. Establishing the mathematical model of the restrictions and the objective function
The mathematical model which expresses all restrictions described in table 1 consists of the following system:

$$150x + 300y \leq 90\,000 \quad (4)$$

$$250x + 500y \leq 180\,000 \quad (5)$$

$$125x + 400y \leq 150\,000 \quad (6)$$

$$150x + 150y \leq 60\,000 \quad (7)$$

$$x \geq 0 \quad (8)$$

$$y \geq 0 \quad (9)$$

Based on the given data, the objective function is written as:

$$f = 2x + 4y \quad (10)$$

3.4. Solving by graphical method
The graphical representation was done using software from the Autodesk package. The package contains software services for architecture, engineering, manufacturing and education industries. In figure 1 the graphical representation of lines D1, D2, D3, D4, corresponding to the relations (4)-(7) is shown.

3.5. Establishing the admissible solution area
The set of admissible solutions is determined taking into account the conditions of inequalities in (4)-(9). Shaded area of figure 2 represents the admissible solution.

3.6. Obtaining the graphical solution
The mathematical function

$$(d_0) = -\frac{2}{4}x = -0.5\,x \quad (11)$$

is represented graphically with the help of the extreme points Also, the values of the intersection points A, B, C, in figure 3, are determined using the graphical method.

3.7. Establishing the values of objective function for ABCD points
The following values for the objective function (4) are obtained, corresponding to the intersection points A, B, C:

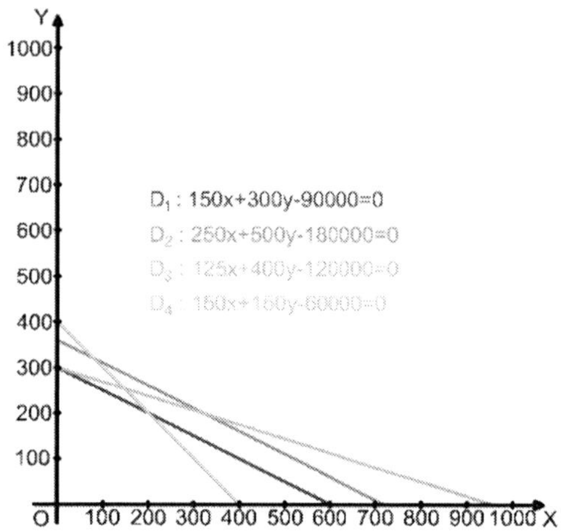

Figure 1. Graphical representation of lines D1, D2, D3, D4, corresponding to relations (4)-(7).

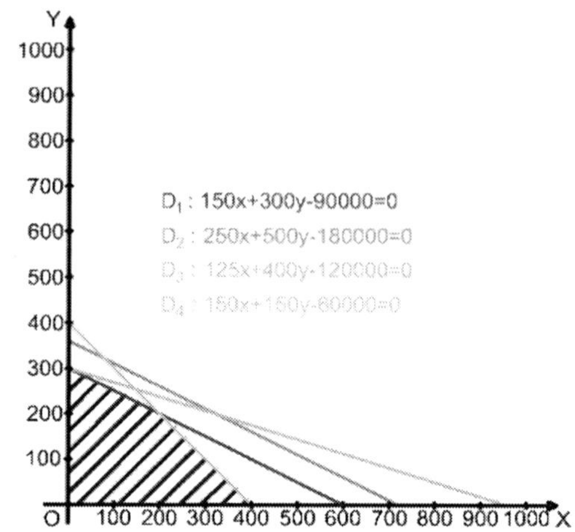

Figure 2. Admissible solution is shaded area.

For point A:

$$f(0, 300) = 2 \times 0 + 4 \times 300 = 1200 \quad (12)$$

For point B;

$$f(200, 200) = 2 \times 200 + 4 \times 200 = 1200 \quad (13)$$

For point C:

$$f(400, 0) = 2 \times 400 + 4 \times 0 = 800 \quad (14)$$

3.8. Solving the system:

$$150x + 300y = 90000 \quad (15)$$
$$150x + 150y = 60000 \quad (16)$$

it is possible to determinate the solution: x=200 and y=200 which means the best optimization plan is C1=200 m^3 and for C2=200 m^3.

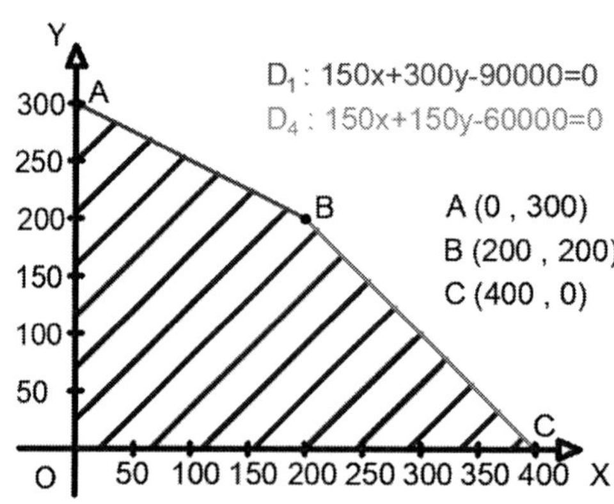

Figure 3. Representation of d_0 and establishing the coordinates of the points ABCD.

3.9. Establishing the benefit and the optimum planning

The benefit is determined:

$$f = 2x + 4y = 2 \times 200 + 4 \times 200 = 1200 \quad (17)$$

and the optimum planning is obtained:

$$S1: 150 \times 200 + 300 \times 200 = 90\ 000 \text{ g/m}^3$$

$$S2: 250 \times 200 + 500 \times 200 = 150\ 000 \text{ g/m}^3$$

$$S3: 125 \times 200 + 400 \times 200 = 105\ 000 \text{ g/m}^3$$

$$S4: 150 \times 200 + 150 \times 200 = 60\ 000 \text{ g/m}^3$$

4. Conclusions

For an efficient management of the water treatment system presented in the studied case, the parameters obtained from the calculations must be respected. This example demonstrates the simple and efficient way by which short and very short-term optimizations can be made without the use of complex computing systems. The graphical methods of solving optimization problems can be adapted to a number of problems. A disadvantage of this method is that the mathematical formulation of the problem must be defined as a system of linear functions.

Although this method has limitations, it can be used in operation without investing a considerable amount of money in different software and hardware solutions. The graphical method for solving linear problems is efficient and simple to be used but it is only a first step that forms the basis of an efficient management system of water resources in urban agglomerations.

To be able to maintain a sustainable evolution of water usage we must integrate modules like Societal Transitions, Urban Development, and Biophysical which will help to predict the future challenges.

References

[1] Yuan Z Olsson G Cardell-Oliver R van Schagen K and Jiang G 2019 Sweating the assets – the role of instrumentation, control and automation in urban water systems *Water Research* **155** 381-402

[2] Olsson G Carlsson B Comas J Copp J Gernaey K V Ingildsen P Jeppsson U Kim C Rieger L Rodríguez-Roda I Steyer J P Takács I Vanrolleghem P A Vargas A Yuan and Z Åmand L 2014 Instrumentation, control and automation in wastewater – from London 1973 to Narbonne 2013 *Water Sci. Technol.* **69** 1373–85

[3] Winker M Gehrmann S Schramm E Zimmermann and Rudolph-Cleff M 2019 Greening and cooling the city using novel urban water systems: a European perspective *Approaches to Sensitive Urban Design* 431-54

[4] Farley M 2001 Leakage management and control *World Health Organization Geneva* 1-157

[5] Samir N Kansoh R Elbarki W and Fleifle A 2017 pressure control for minimizing leakage in water distribution systems *Alexandria Engineering* **58** 601-12

[6] Mutikanga H 2012 *Water loss management: tools and methods for developing countries* (Delft: Netherlands Institute for Water Education) 1-170

[7] Rauch W Urich C Bach PM Rogers and BC Deletic A 2017 modeling transitions in urban water systems *Water Research* **126** 501-14

[8] Pătrăşcoiu C 2005 *Numerical techniques for optimization* (in Romanian) (Bucureşti MatrixRom) p 9-253

[9] Singiresu S R 2009 *Engineering optimization: theory and practice* (New Jersey John Wiley & Sons, Inc) p 5-788

[10] Necoară I 2013 *Numerical optimisation methods* (in Romanian) (Bucureşti Politehnica University) p 6-274

[11] Precup R E 2002 *Computer added mathematics: applications* (in Romanian) (Timişoara Polytechnics) p 113-150

[12] Bebea N 1978 *Methods for solving optimization problems* (in Romanian) (Bucureşti Editura didactică şi pedagogică) p 5-170

[13] Grötschel M and Nemhauser G L 2008 George Dantzig's contributions to integer programming *Discrete Optimization* **5** 168-73

IOP Conf. Series: Materials Science and Engineering **789** (2020) 012012 doi:10.1088/1757-899X/789/1/012012

Masonry strength influence on the buildings' performance

S Constantinescu[1*]

[1] Department of Civil Engineering, Technical University of Construction, Bucharest, Romania

[*] sorina.constantinescu@yahoo.com

Abstract. This paper studied the behaviour of a medium height, confined masonry walls building, in a medium seismic area (0.25*g*, where *g* is the gravity acceleration), in Iasi, Romania. The study focuses on different masonry strengths impact on the building's elastic and plastic behaviour. It is important to establish if the structure can bare the gravitational and seismic loads if weaker bricks are used and also how the plastic mechanism is formed if the walls are made of stronger and stiffer masonry. It is also highlighted how the beams stiffness affects the structure ductility and how it can be improved. The building does not have the same stiffness on both *X* and *Y* directions, so this will also be taken into account.

1. Introduction

The paper compares the behaviour of medium height buildings, with different masonry strengths, in medium seismic areas. Load bearing walls can be made of other materials than clay bricks. Rammed earth walls are traditional and easy to make in some countries [1]. Bricks made of mud, water and straws are also used for masonry walls. There is no accurate design code for these materials to follow. The experimental results show that the adding of high cement ratios, can lead to optimum compressive strengths [2]. Concrete blocks can be used for masonry walls just like clay bricks. The compressive strengths of concrete and clay bricks are similar. The concrete bricks require more mortar, and they have a more brittle failing mechanism [3]. Clay brick walls have higher shear strength than concrete blocks walls, but lower deformation capacity [4]. Both clay and concrete of masonry walls subjected to eccentric loads crack at 40% to 60% of the maximum load at the time of the collapse [5]. According to tests on masonry prisms made using different bricks and mortars, the bricks elasticity modulus and mortar strength have an important influence on the masonry walls [6]. Confined masonry behaves well in moderate earthquakes. Masonry walls are thought to fail due to diagonal shear or sliding shear [7]. Well defined continuous load paths help the masonry buildings bare the earthquake loads. They can be achieved with regular structural layout with symmetry and uniformity both in plan and elevation [8]. Masonry walls reinforcement may be useful in medium seismic areas. The flexural strengths of reinforced masonry walls are 5-16 times higher than for the non-reinforced walls [9]. Horizontal walls reinforcement or connecting bars improve the seismic performance of confined masonry walls, by enhancing the ductile behavior to some extent and lateral load carrying capacity [10]. Masonry walls continuous horizontal reinforcement helps achieve a ductile behavior [11].

IOP Conf. Series: Materials Science and Engineering **789** (2020) 012012 doi:10.1088/1757-899X/789/1/012012

2. Building description

The study is used for a new building that will be situated in Iasi, Romania. The building 3D image can be seen in figure 1. The codes in force used to design the building are according with [12–18].

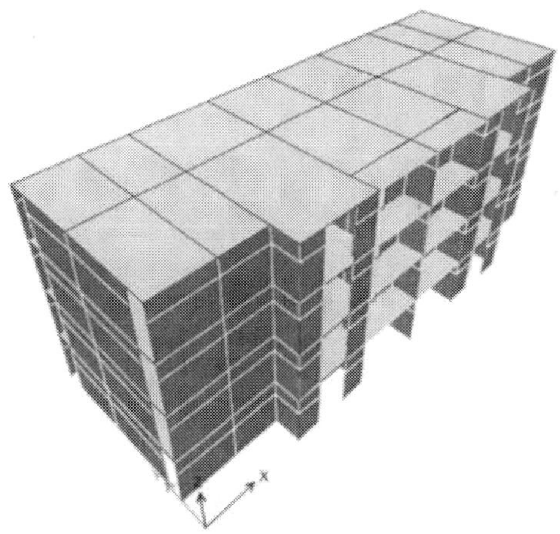

Figure 1. 3D building image.

This is considered a medium seismic area as the seismic acceleration is 0.25g (where g is the gravity acceleration). The building contains a ground floor and 3 stories above it. Story height is 3m. The structure has masonry walls confined with tie beams and reinforced concrete tie-columns designed according to [18]. Hollowed bricks are used for the masonry walls. According to the masonry code, it is possible to use bricks of higher or lower strengths. The paper will highlight the buildings behavior according to the bricks strength. Two cases (A and B) of bricks strengths are considered here. In both cases some beams are slender and some are stiff. Case A uses stronger bricks. Case C highlights the buildings plastic stage behavior with the weaker bricks from case B and slender beams.

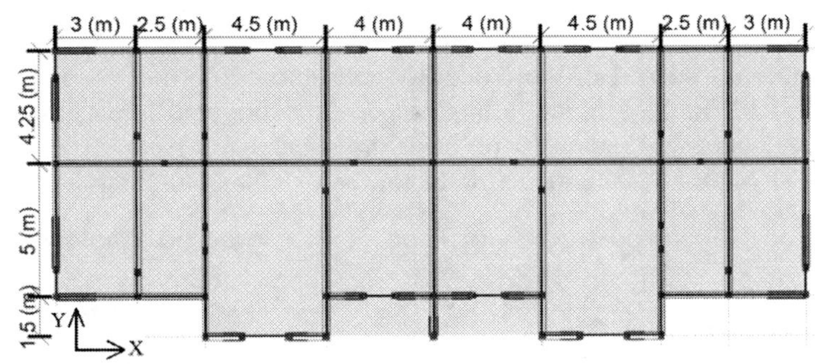

Figure 2. Bays dimensions (8 bays on X and 3 on Y).

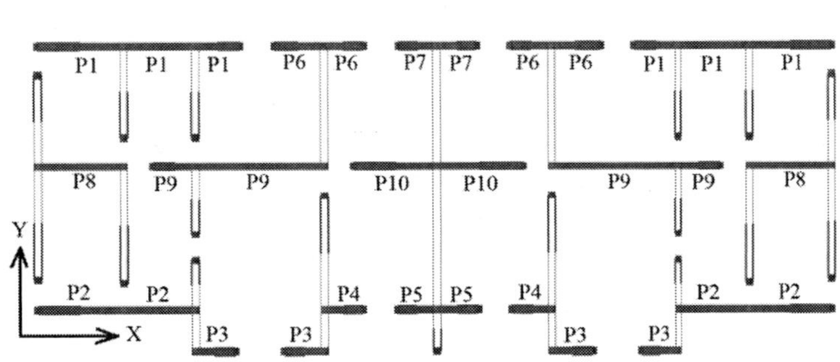

Figure 3. Bearing masonry walls on direction X.

The floor plan with spans dimensions is in figure 2. The walls names on directions X and Y are in figures 3 and 4. Walls are red, beams and tie beams are blue, slabs are grey and reinforced concrete tie- columns are green. The program used to design is ETABS 2017.

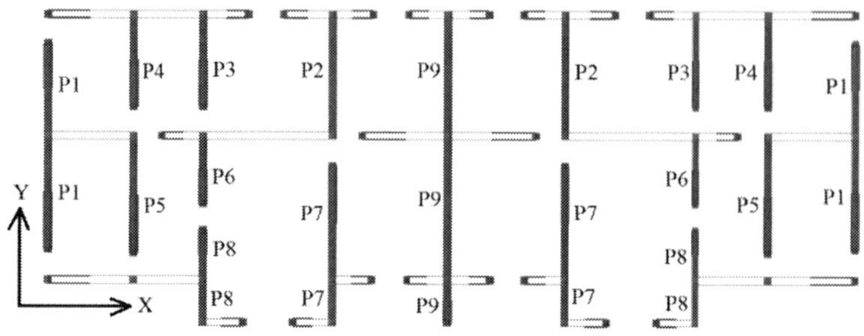

Figure 4. Bearing masonry walls on direction Y.

2.1. *Materials used*

Hollowed bricks 250x115x88 mm ($Lxlxh$) are used for the masonry walls. For the first case studied, case A, the bricks standard strength is f_b =15 N/mm^2, mortar M10 and elasticity modulus E_M = 6000 N/mm^2. For the second case B the bricks standard strength is f_b = 7.5 N/mm^2, mortar M7.5 and elasticity modulus E_M = 3350 N/mm^2 [12]. For the reinforced concrete tie beams and reinforced concrete tie-columns, the concrete class is C16/20 [16], with elasticity modulus E_C = 29000 N/mm^2. Tie beams are beams placed in the masonry wall at each floor. Beams are placed between the masonry walls at each floor. Both beams and tie beams are attached to the floor slabs. They may have the same transversal sections and reinforcements.

Sometimes beams may have greater sections and reinforcement percentages. Reinforcement bars are S345 with elasticity modulus E_S = 210000 N/mm^2 [16].

3. Theory elements used in the paper

3.1. *Seismic force design*

The seismic action is introduced in the model by coefficient c_s. The base force F_b (1) is calculated using [17, 18]. $\gamma_{I,e}$ = 1.2 is the building's importance-exposure coefficient, β_0 = 2.5 is the maximum value in the elastic spectrum and q is the structure's behavior factor, q = 2.25·1.25 [18], m is building's mass [18]. η = 0.88 is the reduction factor, λ = 0.85, a_g = 0.25g [18].

$$F_b = \gamma_{I,e} \cdot \beta_0 \cdot a_g/q \cdot m \cdot \eta \cdot \lambda = c_s \cdot m{\cdot}g = 0.17{\cdot}m{\cdot}g \text{ [kN]} \quad (1)$$

3.2. *Materials strengths*

The stresses analyzed for masonry walls are σ_x, σ_z and τ_{xz}. They are compared to the masonry design strengths [12]: horizontal compression f_{dh}, vertical compression f_d, and shear strength for horizontal direction $f_{vd,0}$. The design strengths (2) to (4) are determined from their characteristic values: f_{kh}, f_k and $f_{vk,0}$. The masonry strengths insurance factor is γ_M=1.9 [12]. The concrete compression, tension and shear design strengths (f_{cd}, f_{ctd} and f_{cvd}) (7) to (10) are determined using the characteristic strengths (f_{ck} and f_{ctk}) with (9) if $\sigma_{cp} \leq \sigma_{c,lim}$ and with (10) if $\sigma_{cp} > \sigma_{c,lim}$. For steel, f_{yd} is the design strength and f_{yk} is the characteristic value. γ_M is 1.5, for concrete and 1.15 for steel [16]. N_{Ed} is the axial force and Ac is the section area of the reinforced concrete tie-columns for which f_{cvd} is calculated. The masonry design bending strengths perpendicular to the wall f_{xd1} and f_{xd2} (5) and (6) are calculated using their characteristic values f_{xk1} and f_{xk2} [12]. (A) and (B) mean cases A and B respectively.

$$f_{dh} = f_{kh}/\gamma_M = 1.18/1.9 = 0.62 \text{ N/mm}^2 \text{ (A)}; 0.86/1.9 = 0.45 \text{ N/mm}^2 \text{ (B)} \quad (2)$$

$$f_d = f_k/\gamma_M = 6.0/1.9 = 3.15 \text{ N/mm}^2 \text{ (A)}; 3.35/1.9 = 1.76 \text{ N/mm}^2 \text{ (B)} \quad (3)$$

$$f_{vd,0} = f_{vk,0}/\gamma_M = 0.3/1.9 = 0.158 \text{ N/mm}^2 \text{ (A)}; 0.25/1.9 = 0.13 \text{ (N/mm}^2\text{) (B)} \quad (4)$$

$$f_{xd1} = f_{xk1}/\gamma_M = 0.24/1.9 = 0.126 \text{ N/mm}^2 \quad (5)$$

$$f_{xd2} = f_{xk2}/\gamma_M = 0.48/1.9 = 0.25 \text{ N/mm}^2 \quad (6)$$

$$f_{cd} = f_{ck}/\gamma_M = 16/1.5 = 10.67 \text{ N/mm}^2 \quad (7)$$

$$f_{ctd} = f_{ctk}/\gamma_M = 1.3/1.5 = 0.87 \text{ N/mm}^2 \quad (8)$$

$$f_{cvd} = (f_{ctd}^{\,2} + \sigma_{cp} \cdot f_{ctd})^{0.5} \text{ (N/mm}^2\text{)} \quad (9)$$

$$f_{cvd} = [f_{ctd}^{\,2} + \sigma_{cp} \cdot f_{ctd} - (\sigma_{cp} - \sigma_{c,lim})^2/4]^{0.5} \text{ (N/mm}^2\text{)} \quad (10)$$

$$\sigma_{cp} = N_{Ed}/Ac \text{ (N/mm}^2\text{)} \quad (11)$$

$$\sigma_{c,lim} = f_{ctd} - 2 \cdot [f_{ctd} \cdot (f_{ctd} + f_{cd})]^{0.5} \text{ (N/mm}^2\text{)} \quad (12)$$

$$f_{yd} = f_{yk}/\gamma_M = 345/1.15 = 300 \text{ N/mm}^2 \quad (13)$$

Here, the values for σ_{cp} are 1.28 N/mm^2, 0.88 N/mm^2, 0.48 N/mm^2 and 0.112 N/mm^2 from stories 1 to story 4. f_{cvd} values have resulted as 1.37 N/mm^2, 1.23 N/mm^2, 1.08 N/mm^2 and 0.924 N/mm^2 from stories 1 to story 4. Reinforcement for tie beams and reinforced concrete tie-columns is designed using the minimum reinforcement percentage 0.8% for a_g = 0.25g seismic areas. The load combinations ECX and ECY used to design the structure is 1.0·permanent loads+0.4·variable loads+1.0·earthquake loads [14] on directions X and Y.

3.3. Masonry walls bearing efforts parallel to the walls plane

M_{Rd} is the wall's bearing bending moment [12]. $M_{Rd(M)}$ and $M_{Rd(As)}$ are the bearing bending moments from the masonry area and from the reinforced concrete tie-columns reinforcements respectively [12]. Ac is the wall's compressed area [12].

$$M_{Rd} = M_{Rd(M)} + M_{Rd(As)} \text{ (kNm)} \quad (14)$$

$$Ac = N_{Ed}/(0.85 \cdot f_d) \text{ (mm}^2\text{)} \quad (15)$$

$$M_{Rd(M)} = N_{Ed} \cdot y_c \text{ (kNm)} \quad (16)$$

In figure 5, y_c is the distance between the compressed masonry area weight center CG and the wall's weight center WG [12]. The wall in figure 5 is used to explain theory elements used in the paper. It is not one of the piers P1 to P10. l_s is the distance between the reinforced concrete tie-columns at the wall's edges. A_s is the reinforced concrete tie-columns horizontal reinforcement area. b_w = 25 cm is the walls width and E is the earthquake direction. V_{Rd} is the masonry wall bearing shear force and V_{Ed} is the horizontal shear force from the seismic loads combination [12].

$$M_{Rd(As)} = l_s \cdot A_s \cdot f_{yd} \text{ (kNm)} \quad (17)$$

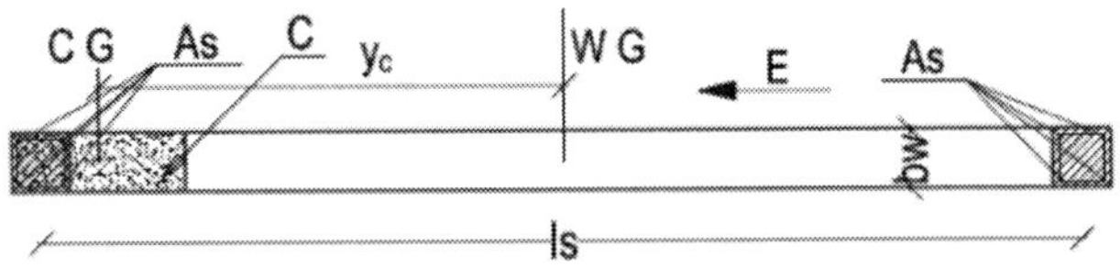

Figure 5. Confined masonry wall section.

$$V_{Rd} = V_{Rd1}^{\,*} + V_{Rd2} + V_{RSC} \text{ (kN)} \quad (18)$$

$$V_{Rd1}^{\,*} = 0.4 \cdot (N_{Ed} + 0.8 \cdot V_{Ed} \cdot h_{pan}/l_{pan}) \text{ (kN)} \quad (19)$$

$$V_{Ed} \leq l_{pan} \cdot t \cdot f_{vd,0} \text{ (kN)} \quad (20)$$

where h_{pan} = story height – 30 cm (tie beam height) and $l_{pan}=l_s-b_w$ are the height and length of the masonry area panel. V_{Rd2} is the bearing horizontal shear force from the reinforcement in the reinforced concrete tie-columns at the walls compressed edge [12]. A_s is the reinforcement area in the reinforced concrete tie-columns at the walls compressed edge. λ_c is the reinforcement participation factor. Here, λ_c=0.20, for longitudinal reinforcement Φ16. V_{RSC} is the shearing capacity for reinforced concrete tie-columns at the walls compressed area.

$$V_{Rd2} = \lambda_c \cdot A_s \cdot f_{yd} \text{ (kN)} \qquad (21)$$

$$V_{RSC}=A_c \cdot f_{cvd} \text{ (kN)} \qquad (22)$$

3.4. *Masonry walls bearing efforts perpendicular to the walls plane*

The bearing bending moments perpendicular to the masonry walls M_{Rxd1} (horizontal) and M_{Rxd2} (vertical) values will be compared to the design bending moments values, M_{Exd1} and M_{Exd2} [12] calculated from the model respectively.

$$M_{Rxd1} = W_w \cdot (f_{xd1} + \sigma_{dw}) \text{ (kNm/m)} \qquad (23)$$

$$M_{Rxd2} = W_w \cdot f_{xd2} \text{ (kNm/m)} \qquad (24)$$

$$\sigma_{dw} = \gamma_{mas} \cdot H_S \text{ (N/mm}^2\text{)} \qquad (25)$$

$$W_w = 1000 \cdot t^2/6 = 1000 \cdot 250^2/6 = 10416666 \text{ mm}^3\text{/m} \qquad (26)$$

where W_w= 1000 · t^2/6 is the wall resistance modulus (in mm^3/m), t is the wall thickness, σ_{dw} is the compression stress at the wall's middle height section [12], γ_{mas} =18 kN/m^3 is masonry weight per cubic meter and H_S is masonry story walls height from the building bottom to the level where σ_{dw} is calculated. The values for σ_{dw} are 0.189 N/mm^2, 0.135 N/mm^2, 0.081 N/mm^2, 0.027 N/mm^2 for stories 1, 2, 3 and 4 respectively.

M_{Rxd1} is 3.28 kNm/m, 2.72 kNm/m, 2.15 kNm/m and 1.59 kNm/m for stories 1, 2, 3 and 4 respectively and M_{Rxd2}= 2.60 kNm/m for all 4 stories.

4. Elastic analysis results

Dimensions and longitudinal reinforcement in beams, tie beams and reinforced concrete tie-columns can be found in table 1. As is the longitudinal reinforcement area. The bars are blue discs and the diameter (Φ) of bars (in mm) is written for each element.

Table 1. Dimensions and longitudinal reinforcement in beams, tie beams and reinforced concrete tie-columns.

Tie beam (beam) and lintel 25x30 As → 4Φ14	Reinforced concrete tie-columns 25x25 As → 4Φ16
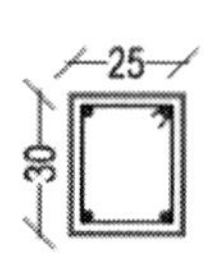	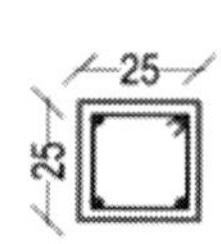

4.1. *Efforts and stresses in masonry walls parallel to the walls plane*

Efforts parallel to the walls planes are presented as graphics to compare the values easier. The M_{Ed} values are split to 100 and the N_{Ed} and V_{Ed} values are split to 10 to fit the numbers in the tables.

4.1.1. Masonry walls efforts on direction X.

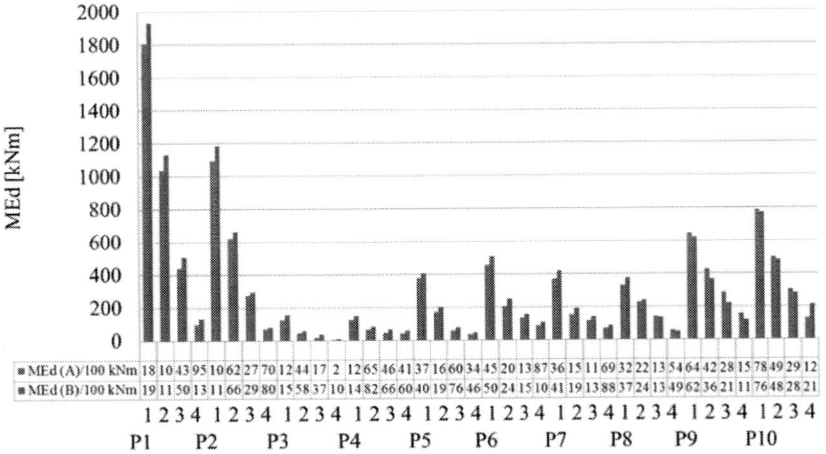

Figure 6. M_{Ed} for cases A and B.

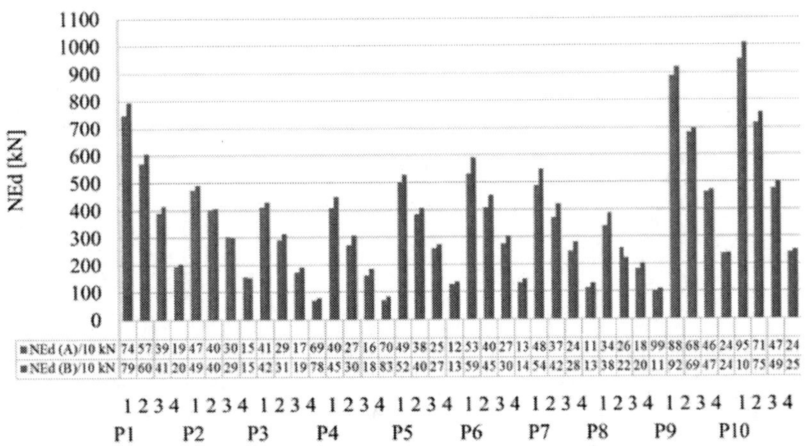

Figure 7. N_{Ed} for cases A and B.

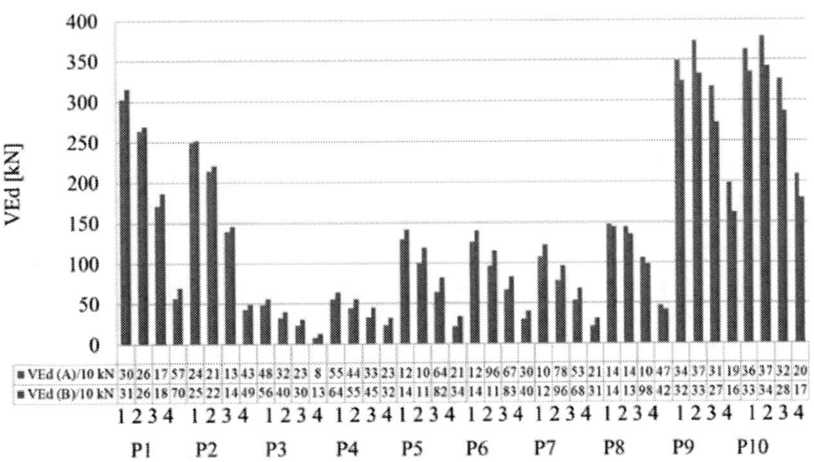

Figure 8. V_{Ed} for cases A and B.

4.1.2. Masonry walls efforts on direction Y.

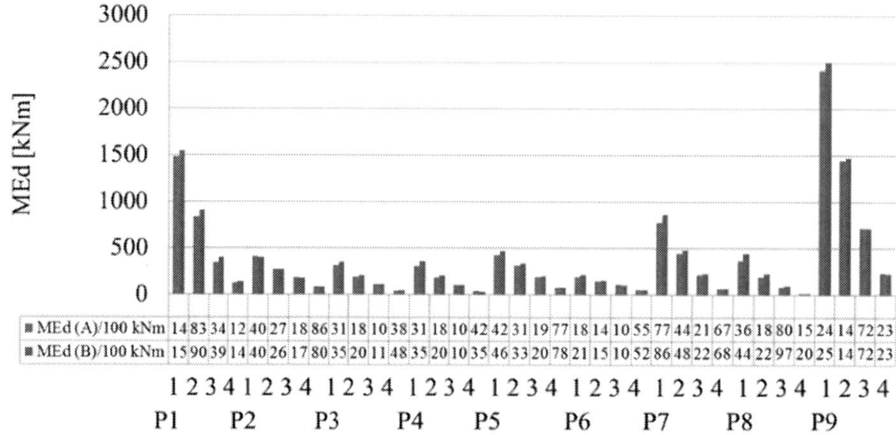

Figure 9. M_{Ed} for cases A and B.

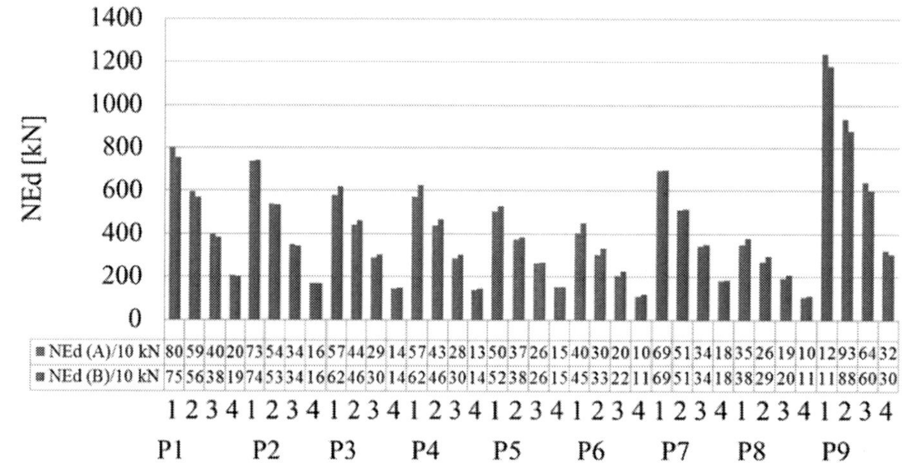

Figure 10. N_{Ed} for cases A and B.

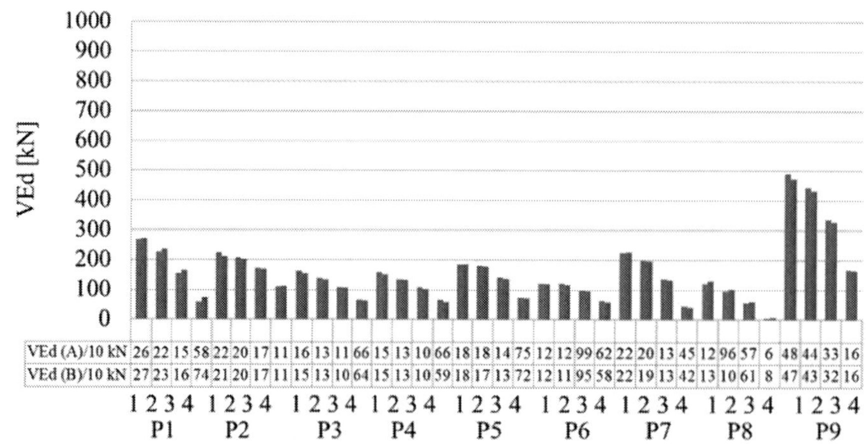

Figure 11. V_{Ed} for cases A and B.

Figures 8 to 11 show efforts in the masonry walls: M_{Ed} (bending moment), N_{Ed} (axial force) and V_{Ed} (shear force) and the bearing efforts respectively M_{Rd}, N_{Rd} and V_{Rd}. In the elastic stage, the effective efforts N_{Ed}, M_{Ed} and V_{Ed} in cases A and B are comparable for both directions *X* and *Y*.

4.2. Masonry walls efforts perpendicular to the walls plane

M_{Exd1} values reach from -0.5 kNm/m to 1 kNm/m, below M_{Rxd1} and M_{Exd2} values are from -2.1 to 2.4 kNm/m below M_{Rxd2}, for both cases A and B, on both directions *X* and *Y* (figures 12 and 13).

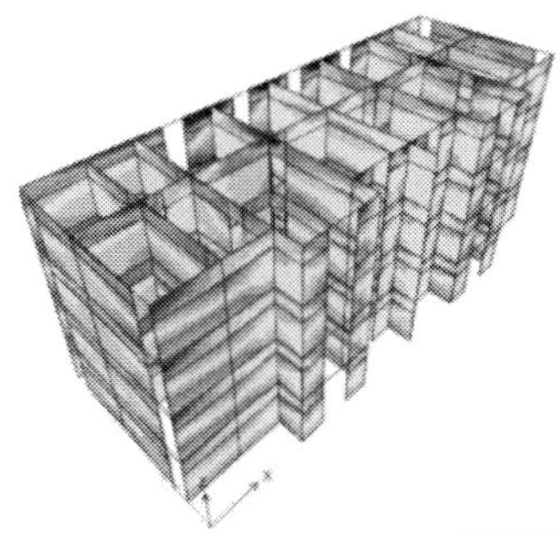

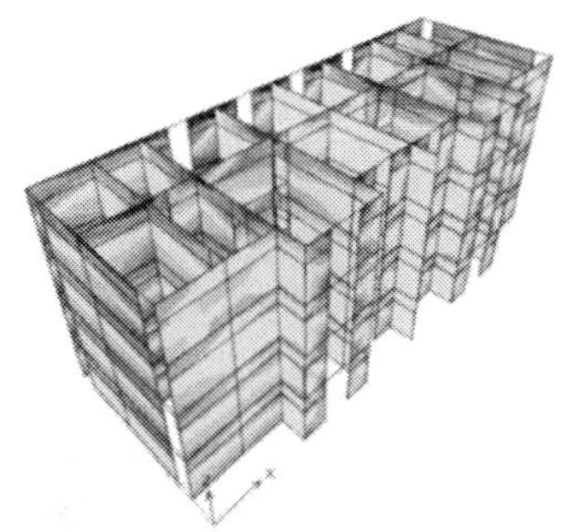

Figure 12. M_{Exd2} distribution A (kNm/m). **Figure 13.** M_{Exd2} distribution B (kNm/m).

The highest positive values are blue, yellow means 0 and the lowest negative values are purple.

5. Analysis results in the plastic stage

The structure's behavior in the plastic stage is analyzed using the pushover diagrams PX1 and PY1 for case A and PX2 and PY2 for case B for directions *X* and *Y* respectively.

5.1. Plastic hinges development and pushover diagrams

The plastic hinges colors are: green when the hinge is formed, blue when the hinge reaches its limit, pink when the load is redistributed and red at collapse.

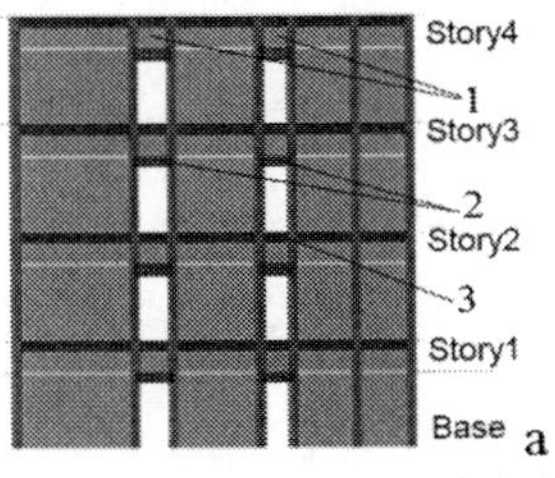

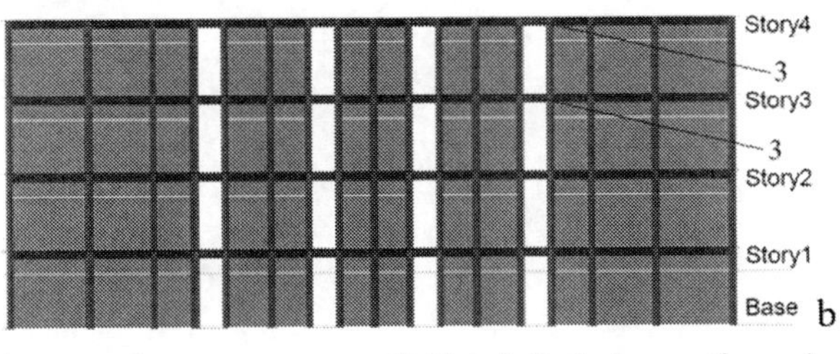

Figure 14. a) thick beams, b) thin beams: 1-masonry area, 2-lintel, 3-tie beam (beam).

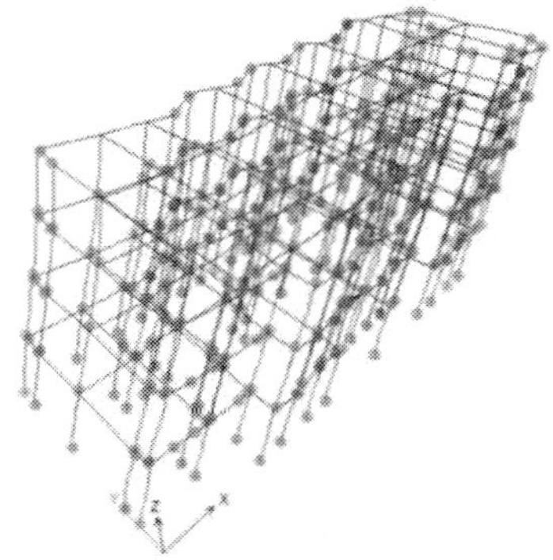

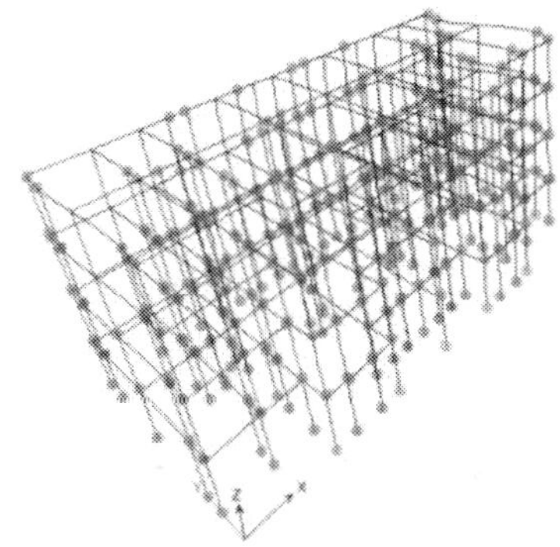

Figure 15. PX1 stage 46. **Figure 16.** PY1 stage 10.

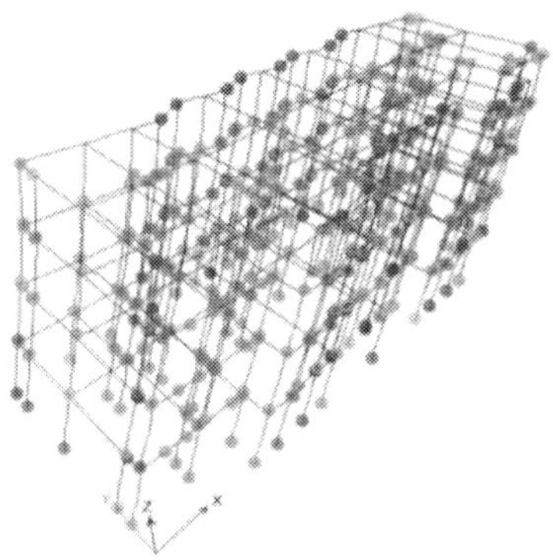

Figure 17. PX2 stage 85.

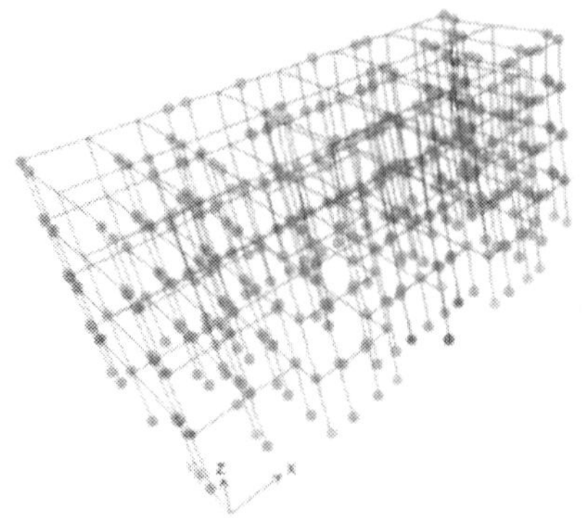

Figure 18. PY2 stage 25.

In case A, (figures 15 and 16) on direction *Y*, most beams connecting walls are thick.

They are placed above doors and are accompanied by wall panels (figure 14 a), so they do not develop high stages in plastic hinges. In case B (figures 17 and 18) the structure reaches the displacement 700 mm on direction *X* as many plastic hinges reach the collapse stage. For direction *Y*, the plastic hinges in advanced stages are seen in the thin beams (figure 14 b) at the edges and at the reinforced concrete tie-columns bottoms. There are both thick and thin beams for both directions. There are more thin beams on direction *X* (the longitudinal side).

5.2. Pushover diagrams

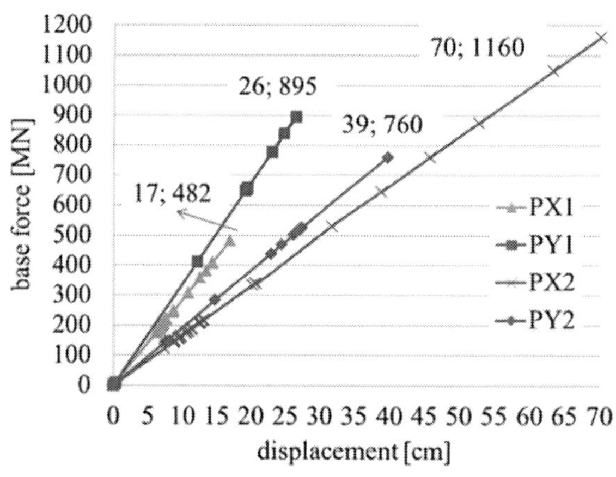

Figure 19. Pushover diagrams: PX1, PY1, PX2, PY2.

The pushover diagrams are shown in figure 25. In case A, the maximum base force reached for direction *Y* is twice greater than for *X*. The top displacement is also greater for *Y*. This is caused by the plastic hinges in the short beams connecting walls on *X*, giving in early. These beams, placed above windows, are thin. This is not valid for the internal walls. Thin beams are not accompanied by wall panels (figure 14). In case B, the maximum top displacement is smaller on direction *Y* because plastic hinges reach high stages at the reinforced concrete tie-columns bottoms.

The structure's ability to reach such a high displacement on direction *X* is explained by the weaker masonry walls in this case. The beams work better with walls of lower bricks strength. This is also seen on direction *Y*, but it is less evident because the structure has a higher rigidity on *Y*. Rigidity is regarded as base force/displacement. For all 4 pushover cases the building's rigidity is maintained the same throughout the analysis.

IOP Conf. Series: Materials Science and Engineering **789** (2020) 012012 doi:10.1088/1757-899X/789/1/012012

5.3. Walls stresses

Stresses σ_x and σ_z reach the highest values in the lower stories. They surpass the masonry strengths f_{dh} and f_d respectively from the second step of the analysis. At that moment, σ_x values range from -0.7 N/mm^2 to 0.7 N/mm^2 for case A (PX1 and PY1) and from -0.5 N/mm^2 to 0.5 N/mm^2 for case B (PX2 and PY2). The stresses distribution is seen in figures 20 to 25. Stresses σ_z values reach from -3.15 N/mm^2 to 3.15 N/mm^2 for case A and from -1.8 N/mm^2 to 1.8 N/mm^2 for case B.

The highest positive values are blue, yellow is for 0 and the lowest negative values are purple. Stress τ_{xz} values range from -0.16 N/mm^2 to 0.16 N/mm^2 for case A and from -0.13 N/mm^2 to 0.13 N/mm^2 for case B. These values are greater than the strength $f_{vd,0}$ from the first step of the analysis. τ_{xz} values are greater in walls parallel to the seismic action. In some walls, τ_{xz} values are greater than the masonry strength for all stories, so the all the wall area is cracked there. All stress values decrease from the building bottom to the top. For cases PX1 and PY1 all stresses reach higher values. The masonry is cracked before the plastic mechanism is formed.

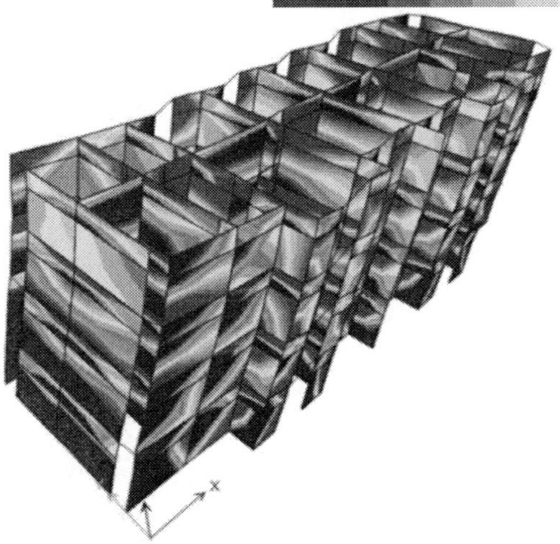

Figure 20. σ_x distribution.

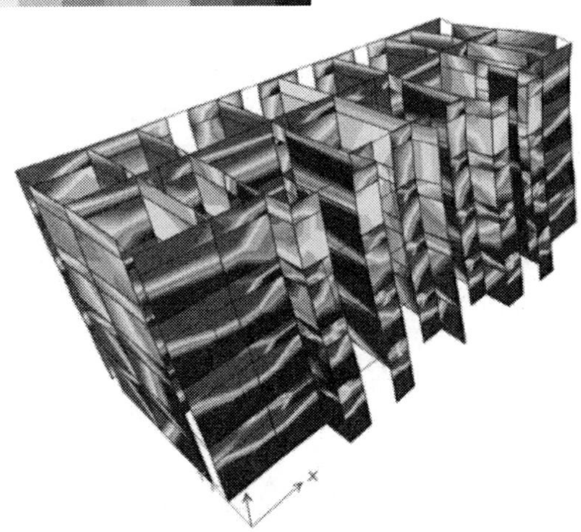

Figure 21. σ_x distribution.

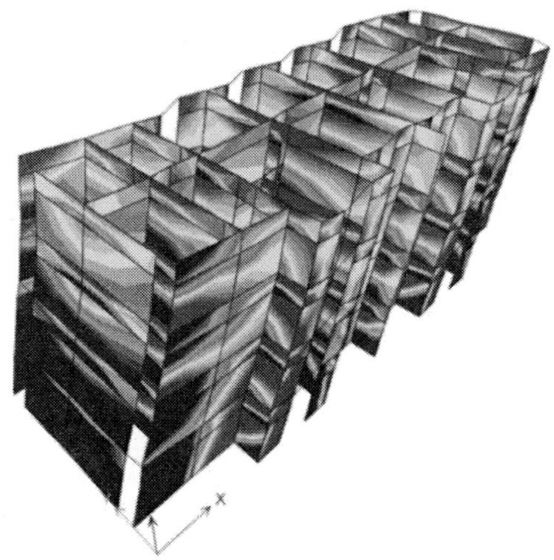

Figure 22. σ_z distribution.

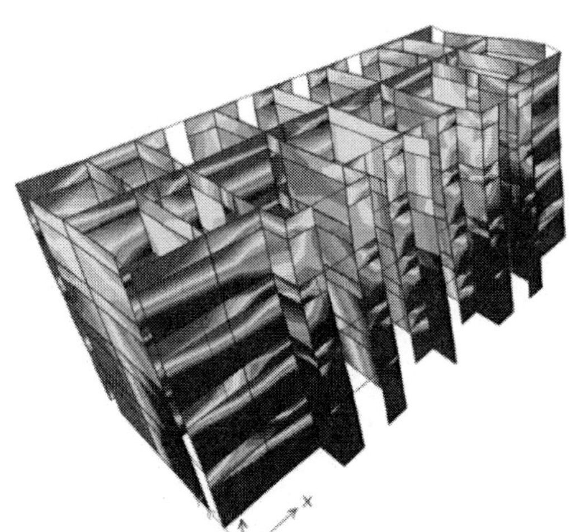

Figure 23. σ_z distribution.

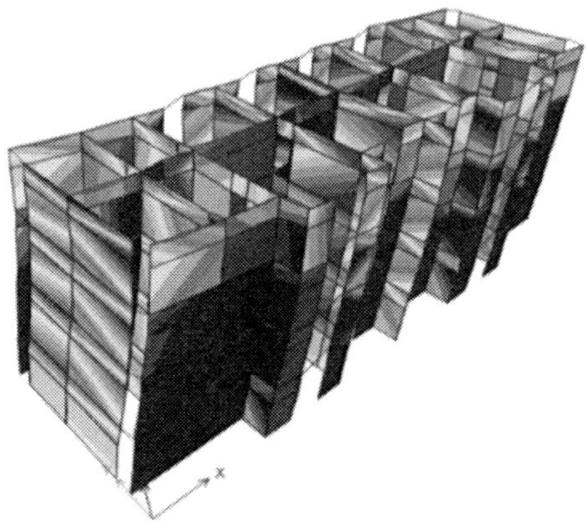

Figure 24. τ_{xz} distribution.

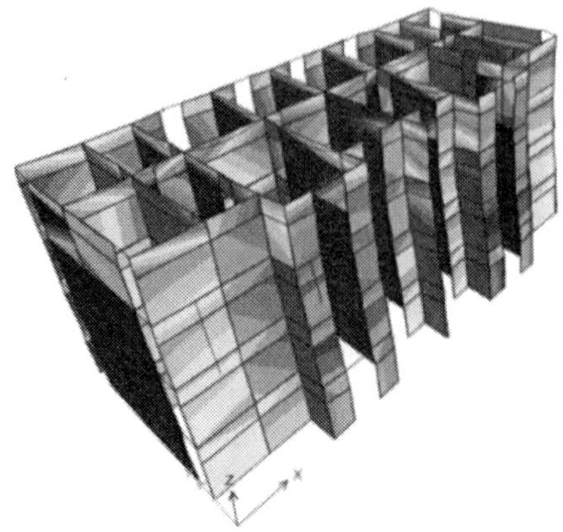

Figure 25. τ_{xz} distribution.

6. Structure enhancement

It is interesting to further study the influence the thick beams (figure 14.a) have on the structure's plastic behavior. If all beams are thin (figure 14.b) as in case C, then the structure becomes a mechanism at smaller top displacements and the rigidity value on direction *Y* decreases. The plastic hinges development for case C is in figures 26 and 27 and the pushover diagrams for cases A, B and C are in figure 28. The structure's rigidities are in table 2. For case C, the plastic mechanism is better than for cases A and B because there are less plastic hinges at the reinforced concrete tie-columns bottoms.

6.1. Plastic hinges development

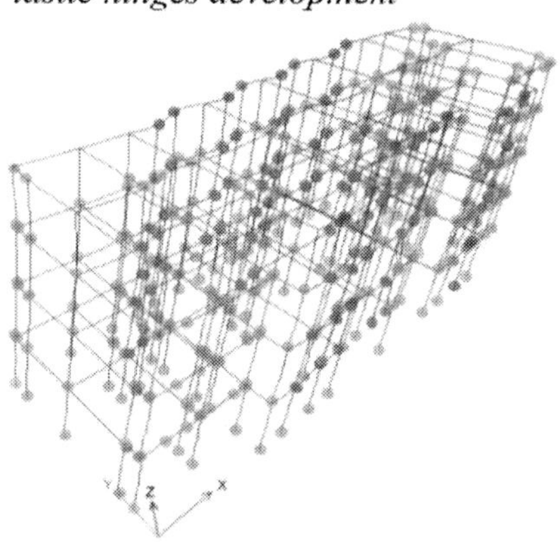

Figure 26. PX3 stage 5.

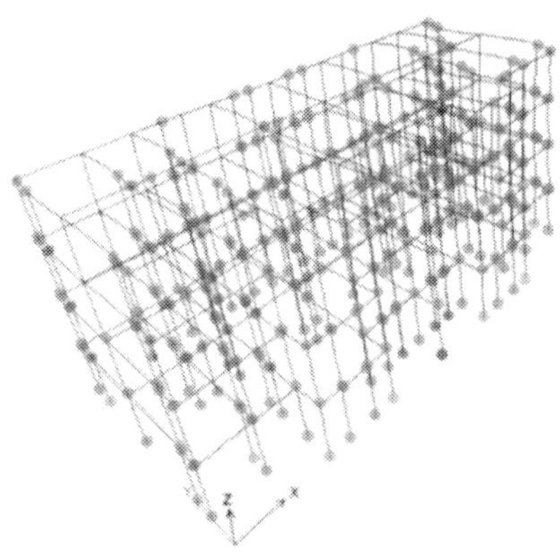

Figure 27. PY3 stage 9.

6.2. Pushover diagrams for cases A, B and C

Table 2. Structure rigidities

Case	Rigidity	Direction
A	28.3	X
	34.4	Y
B	16.6	X
	19.5	Y
C	15.6	X
	17.0	Y

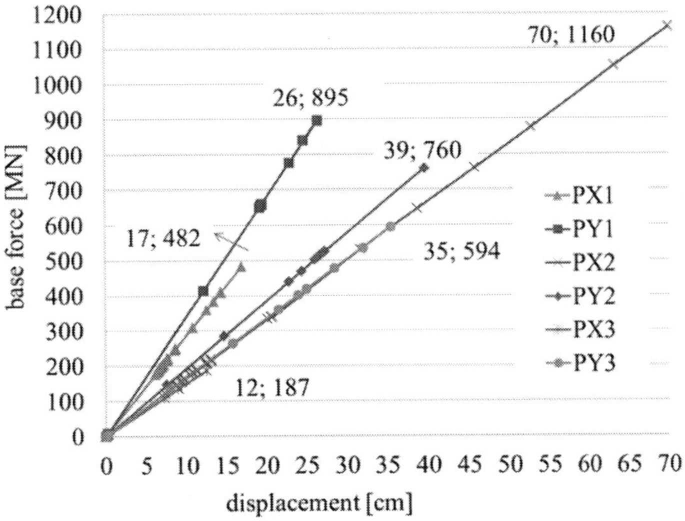

Figure 28. Pushover diagrams: PX1, PY1, PX2, PY2, PX3, PY3.

7. Conclusions

- The structure can bare the efforts it is subjected to, for both cases A and B.
- In the plastic stage it behaves better in case B, so here lower strength bricks are better.
- Thick beams above doors assure larger displacements on direction X in the plastic stage in case B.
- If only thin beams are used, as in case C, rigidities decrease for both directions X and Y.
- The plastic mechanism is better in case C than in A and B, as there are less plastic hinges at the reinforced concrete tie-columns bottoms.

References

[1] Ahimbisibwe A and Ndibwami A 2016 Demystifying Fired Clay Brick: Comparative analysis of different materials for walls, with fired clay brick *PLEA Los Angeles - 32nd International Conference on Passive and Low Energy Architecture Cities, Buildings, People: Towards Regenerative Environments*

[2] Al-Ajmi F Abdalla H, Abdelghaffar M and Almatawah J 2016 Strength Behavior of Mud Brick *Building Construction Open Journal of Civil Engineering* DOI 10.4236/ojce.2016.63041

[3] San Bartolome A, Angles P and Quinn D 2012 Seismic behavior comparison of confined masonry walls of clay and concrete bricks *15th Interantional Brick and Block Masonry Conference Florianopolis-Brazil*

[4] Claderon S, Vargas L, Sandoval C and Araya-Letelier G 2018 Behavior of partially grouted reinforced masonry shear walls built with concrete blocks and clay bricks: a comparative study *10th International Masonry Conference, Milano Italy*

[5] Cornado C, Rosell J R, Leiva J and Diaz C 2016 Experimental study of brick masonry walls subjected to eccentric and axial load *RILEM International Conference on Materials, Systems and Structures in Civil Engineering Conference segment on Historical Masonry* Technical University of Denmark, Lyngby Denmark p 33–40 www.rilem.net/publications/proceedings-500218

[6] Jaber A 2010 Effect of Masonry Units Type and Concrete Grouting on Compressive Strength of Prisms *Eng. & Tech Journal* Vol. **28** No.132010

[7] Dhanasekar M 2011 Shear in reinforced and unreinforced masonry: response, design and construction *The Twelfth East Asia Pacific Conference on Structural Engineering and Construction, Elsevier Procedia Engineering* Vol **14** DOI 10.1016/j.proceng.2011.07. 260

[8] Hancilat U, Durukal E, and Erdik M 2008 Seismic assessment and rehabilitation of a historical unreinforced masonry (URM) building in Istambul *The 14th world conference on earthquake engineering* Beijing, China

[9] Triwiyono A, Nugroho A S B, Firstyadi A D and Ottama F 2015 Flexural strength and ductility of concrete brick masonry wall strengthened using steel reinforcement *The 5th International Conference of Euro Asia Civil Engineering Forum* (EACEF-5) Elsevier Procedia Engineering **125** p 940

[10] Yoshimura K, Kikuchi K, Kuroki M, Nokana H, Tae Kim K, Wangdi R, and Oshikata A 2004 Experimental study for developing higher seismic performance of brick masonry walls *13th World Conference on Earthquake Engineering Vancouver, B.C.* Canada, paper No. 1597

[11] Wijaya W, Kusumastuti D, Suarjana M, Rildova R and Pribadi K 2011 Experimental Study on Wall-Frame Connection of Confined Masonry Wall *The Twelfth East Asia –Pacific Conference on Structural Engineering and Construction* Elsevier Procedia Engineering 14 p 2094-2102 DOI 10.1016/j.proceng.2011.07.263

[12] CEN EN 1996-1-1 2006 Eurocode 6: Design of masonry structures - Part 1-1: General rules for reinforced and unreinforced masonry structures

[13] CEN EN 1991-1-1 2004 Eurocode 1: Actions on structures - Part 1-1: General actions- Densities, self-weight, imposed loads for buildings

[14] CEN EN 1990-2004 Eurocode 0: Basics of structural design

[15] CEN EN 1991-1-3 2005 Eurocode 1: Actions on structures - Part 1-3: General actions- Snow loads

[16] CEN EN 1992-1-1 2004 Eurocode 2: Design of concrete structures - Part 1-1: General rules and rules for buildings

[17] CEN EN 1998-1 2004 Eurocode 8: Design of structures for earthquake resistance. Part 1: General rules, seismic actions and rules for buildings

[18] P100-1 2013 Seismic design code – Part 1- General rules for buildings

IOP Conf. Series: Materials Science and Engineering **789** (2020) 012013 doi:10.1088/1757-899X/789/1/012013

Adding a masonry floor on top of a reinforced concrete high building in a seismic area

S Constantinescu[1*]

[1] Department of civil engineering, Technical University of Construction, Bucharest, Romania

[*] sorina.constantinescu@yahoo.com

Abstract. The paper studies the influence of adding an aerated autoclaved concrete (AAC) masonry walls story above a reinforced concrete structure of a high building. This structure is situated in a high seismic area (0.30*g*, where *g* is the gravity acceleration) in Bucharest, Romania. The new story will be lightweight because of the AAC. It is necessary to establish if the existing structure can bare the new story gravity loads and to determine the new building's seismic behaviour. It is also determined if the new AAC masonry walls can bare the seismic loads they are to be subjected to. Both elastic and plastic analysis are present in the study.

1. Introduction

The paper presents the effect of adding a confined masonry (CM) walls story on top of an existing building. The existing structure is a dwelling building with reinforced concrete (RC) walls. The added story contains aerated autoclaved concrete (AAC) confined masonry walls. This is only a study. It is not about an already erected building. According to literature, masonry-partitioning walls contribute significantly to the buildings seismic behaviour [1]. Masonry can also be used in bearing walls. In confined masonry walls, the number of confining columns has an important influence on the energy dissipation capacity, ductility, cracking pattern and strength of the walls. It is preferred to use more confining columns [2]. Confined masonry is a suitable solution for building low to medium rise buildings. Both earthquakes and tests show the effectiveness of confined masonry systems. This performance is also influenced by the masonry properties and structure configuration, reinforcement detailing in tie beams, beams and slender columns [3]. Masonry walls failure is consisted mostly of diagonal cracking and out of plane displacements according to finite element analysis [4]. An AAC story on top of a RC building becomes a soft story. A soft story presents lower stiffness than the others. It may be called flexible story [5]. Aerated autoclaved concrete walls crack trough the aerated concrete blocks, showing that the full strength capacity of the blocks was reached. The effect of the glue mortar is expected to be most significant in enhancing the flexural strength of the AAC walls under loading [6]. In multi-story residential building, the AAC load bearing blocks can reduce the buildings weight and improve the seismic behavior, as the AAC blocks have smaller volume density and are homogenous [7]. The results of AAC building seismic analysis point out that for 1-2 story buildings and for low to medium seismic areas, the damage limit may not be reached. For high seismic areas and multi-story buildings, significant damage can be expected. It is recommended to use geometrical regularity in plan to improve AAC structures performance [8]. According to linear and nonlinear analysis, autoclaved concrete walls can be used as bearing elements for low buildings in seismic areas [9]. According to shaking table tests, AAC masonry walls have a great energy dissipation capacity [10]. Walls slenderness is also important to take into account as it may lower the masonry walls lateral load bearing capacity [11].

2. Building description

This is a study used for a building that will have a penthouse built over the top story. The structure is situated in Bucharest, Romania. The codes in force used to design the building are: [12–18]. This is considered a high seismic area as the seismic acceleration is 0.30 g (g is the gravity acceleration) [18]. The building contains a ground floor and 7 stories above it. Story height is 3 m. It is a reinforced concrete walls dwelling building. The penthouse will have confined masonry bearing walls made of AAC. The analysis is done for two stages. Stage 1 is without the penthouse (figure 1) and stage 2 is with the penthouse (figure 2). Reinforced concrete walls plan is in figure 3 and confined masonry walls plan is in figure 4. Walls are red, beams and tie beams are blue, slender columns are green and slabs are grey. The program used to design is ETABS 2017.

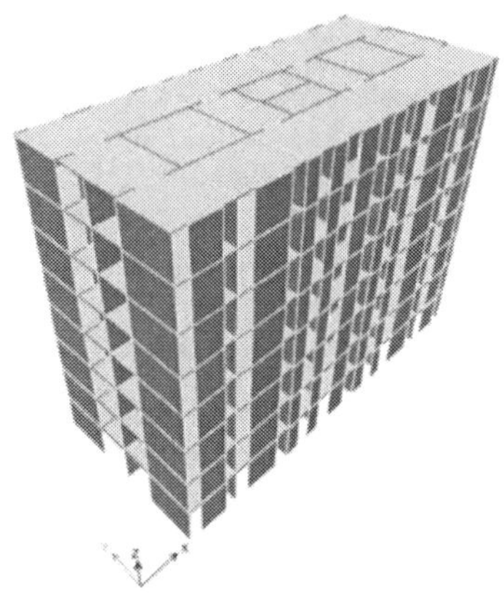

Figure 1. 3D image for stage 1.

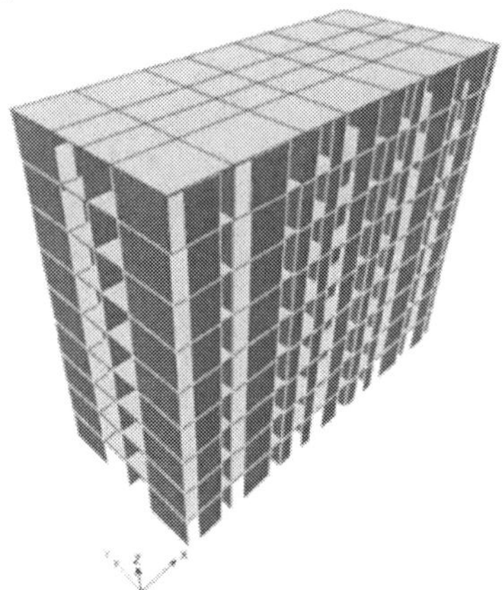

Figure 2. 3D image for stage 2.

The masonry walls will be made of autoclaved concrete bricks, because this material is light weight. These walls will be placed on top of the reinforced concrete walls. The confining elements are concrete slender columns and tie beams. According to the masonry design code in force autoclaved concrete blocks can be used for bearing walls [12].

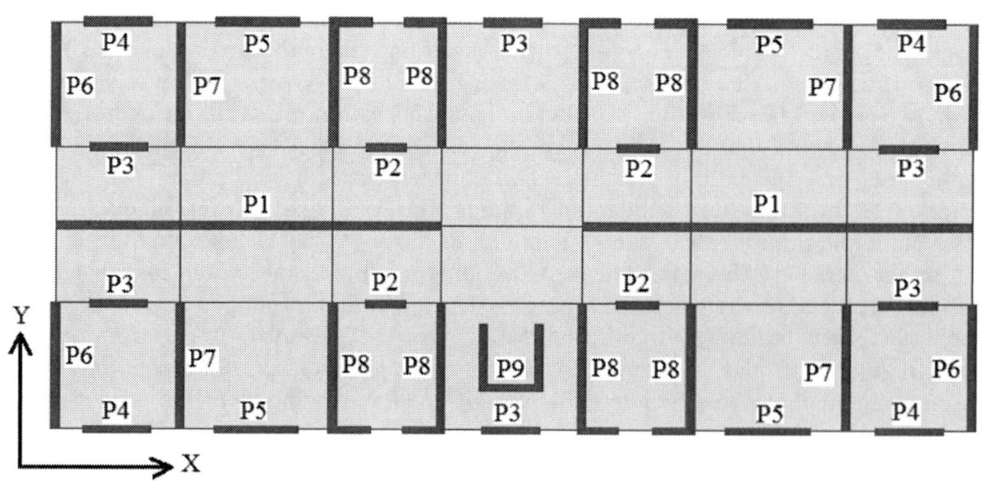

Figure 3. Reinforced concrete walls names.

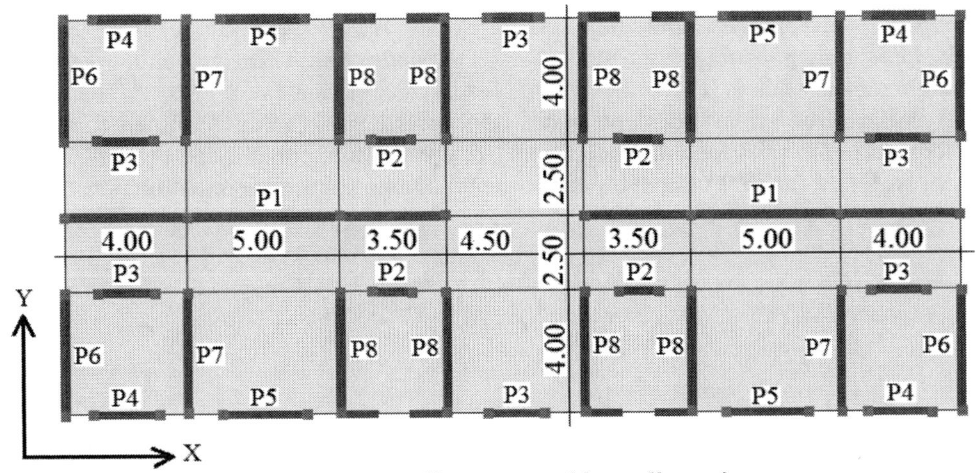

Figure 4. AAC walls names and bays dimensions.

For the reinforced concrete walls, the class is C25/30 [16], with elasticity modulus E_C=31000 N/mm^2. For the masonry aerated autoclaved concrete walls, the blocks are 600·100·240 (mm) with standard strength f_b =8 N/mm^2, mortar M10 and elasticity modulus E_M= 3760 N/mm^2 [12]. Concrete C16/20 is used for the confining elements at the top story, Ec=29000 N/mm^2. Reinforcement bars are S345 with elasticity modulus E_S= 210000 N/mm^2 [16].

3. Theory elements used

3.1. Seismic load

The seismic load is introduced by coefficient c_s. The base force F_b in (1) is calculated using [17-18]. $\gamma_{I,e}$= 1.2 is the building's importance-exposure coefficient, β_0= 2.5 is the maximum value in the elastic spectrum and q is the structure's behavior factor, q=3.5·1.15 [18], m = building's mass [18]. η = 1 is the reduction factor, λ = 0.85, a_g = 0.30g [18].

$$F_b = \gamma_{I,e} \cdot \beta_0 \cdot a_g/q \cdot m \cdot \eta \cdot \lambda \; = c_s \cdot m{\cdot}g = 0.22{\cdot}m{\cdot}g \text{ (kN)} \tag{1}$$

3.2. Materials design strengths

The stresses analyzed for masonry penthouse walls are σ_x, σ_z, τ_{xz}, τ_{xy} and τ_{yz}. They are compared to the masonry design strengths in (2) to (6) [12]: horizontal compression f_{dh}, vertical compression f_d, shear strength for horizontal direction $f_{vd,0}$, horizontal and vertical strengths perpendicular to the wall f_{xd1} and f_{xd2}. The design strengths are determined from their characteristic values: f_{kh}, f_k, $f_{vk,0}$, f_{xk1} and f_{xk2}. The masonry strengths insurance factor is γ_M=1.9 [12]. The concrete compression design strength f_{cd} in (7) is determined using the characteristic strength f_{ck}. For steel, f_{yd} in (8) is the design strength and f_{yk} is the characteristic value. γ_C=1.5, for concrete and γ_S=1.15 for steel [16].

$$f_{dh} = f_{kh}/\gamma_M = 2.09/1.9 = 1.1 \text{ N/mm}^2 \tag{2}$$

$$f_d = f_k/\gamma_M = 4.7/1.9 = 2.47 \text{ N/mm}^2 \tag{3}$$

$$f_{vd,0} = f_{vk,0}/\gamma_M = 0.25/1.9 = 0.13 \text{ N/mm}^2 \tag{4}$$

$$f_{xd1} = f_{xk1}/\gamma_M = 0.1/1.9 = 0.05 \text{ N/mm}^2 \tag{5}$$

$$f_{xd2} = f_{xk2}/\gamma_M = 0.2/1.9 = 0.10 \text{ N/mm}^2 \tag{6}$$

$$f_{cd} = f_{ck}/\gamma_C = 25/1.5 = 16.67 \text{ N/mm}^2 \tag{7}$$

$$f_{yd} = f_{yk}/\gamma_S = 345/1.15 = 300 \text{ N/mm}^2 \tag{8}$$

3.3. Concrete walls design theory

N_{Ed} is the axial force from the seismic loads combinations. $M_{Ed,0}$ (9) and $M_{Ed,s}$ (10) are the ground floor and upper floors design bending moments and M_{Rd} (15) is the wall's bearing bending moment. M'_{Ed} is the bending moment value given by the seismic loads combination for any story [16]. l_w is the walls length. As is the vertical reinforcement area at the walls edges. It is determined form (13) for the ground floor and from (14) for the other floors. Ac is the walls compressed area. b=250 mm is the walls width. k_M=1.15 for DCM (medium ductility buildings). S is the seismic direction.

$$M_{Ed,0} = M'_{Ed} \quad (9)$$

$$M_{Ed,s} = k_M \cdot \Omega \cdot M_{Ed}' \leq \Omega \cdot M'_{Ed} \quad (10)$$

$$\Omega = M_{Rd,0}/M'_{Ed,0} \quad (11)$$

$$Ac = N_{Ed}/f_{cd} \quad (12)$$

$$As = (M_{Ed,0} - N_{Ed} \cdot 0.9 \cdot l_w/2)/(f_{yd} \cdot 0.9 \cdot l_w) \quad (13)$$

$$As = (M_{Ed,S} - N_{Ed} \cdot 0.9 \cdot l_w/2)/(f_{yd} \cdot 0.9 \cdot l_w) \quad (14)$$

$$M_{Rd} = As \cdot f_{yd} \cdot 0.9 \cdot l_w + N_{Ed} \cdot 0.9 \cdot l_w/2 \quad (15)$$

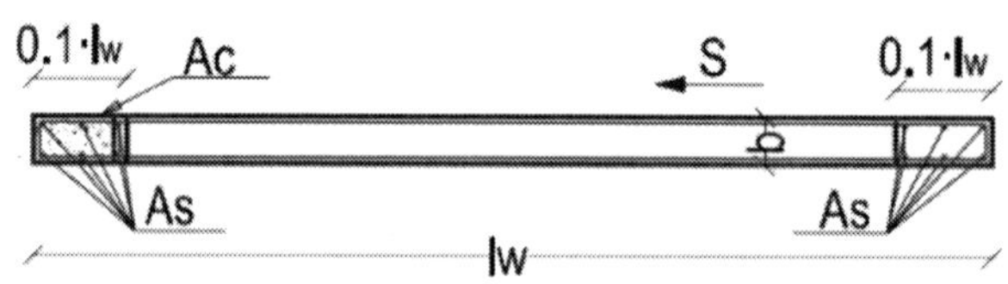

Figure 5. Reinforced concrete wall section.

3.4. Concrete beams design theory

Reinforcement in beams, detailed in figure 6, is designed according to M_{Ed} (16) [16]. λ_x is the beam section compressed area height [16]. $A_{s,min}$ is the minimum horizontal reinforcement area for beams. f_{ctm} = 2.6 N/mm^2 is the medium value of the concrete tensile strength. Reinforcement for tie beams and slender columns at the top story is designed using the minimum reinforcement percentage 1% for a_g=0.30g seismic areas.

$$M_{Ed} = b \cdot \lambda x \cdot f_{cd} \cdot (d - \lambda x/2) = A_s f_{yd} z \text{ (kNm)} \quad (16)$$

$$m = M_{Ed}/(b \cdot d^2 \cdot f_{cd}) \quad (17)$$

$$z = d - \lambda x/2 = d - d \cdot (1-(1-2m)^{0.5})/2 \text{ (mm)} \quad (18)$$

$$A_{s,min} = min\{0.26 \cdot f_{ctm}/f_{yk} \cdot b \cdot d;\ 0.0013 \cdot b \cdot d\} \quad (19)$$

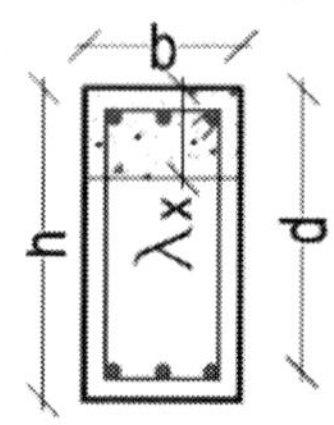

Figure 6. Reinforced concrete beam section.

3.5. Confined masonry walls design theory

M_{Rd} (20) is the wall's bearing bending moment [12]. $M_{Rd(M)}$ (22) and $M_{Rd(As)}$ (23) are the bearing bending moments from the masonry area and from the slender columns reinforcements respectively [12]. Ac is the wall's compressed area [12].

$$M_{Rd} = M_{Rd(M)} + M_{Rd(As)}\ \text{(kNm)} \tag{20}$$

$$Ac = N_{Ed}/(0.85 \cdot f_d)\ (\text{mm}^2) \tag{21}$$

$$M_{Rd(M)} = N_{Ed} \cdot y_c\ \text{(kNm)} \tag{22}$$

In figure 7 y_c is the distance between the compressed masonry area weight center and the wall's area weight center [12]. l_s is the distance between the slender columns at the wall's edges. A_s is the slender columns horizontal reinforcement area.

$$M_{Rd(As)} = l_s \cdot A_s \cdot f_{yd}\ \text{(kNm)} \tag{23}$$

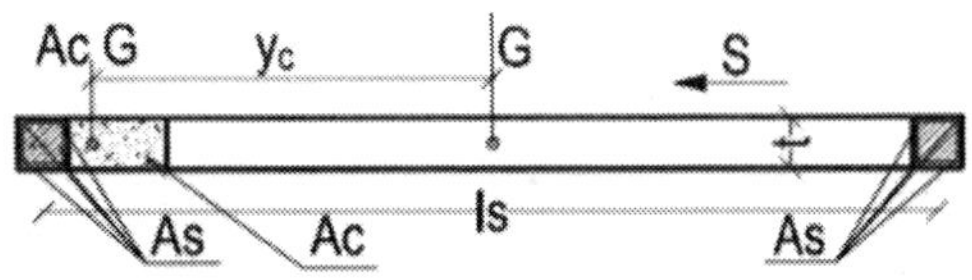

Figure 7. Confined masonry wall section.

A_C G is the compressed area gravity center. G is the wall section gravity center. S is the earthquake action. t = 250mm is the wall's width. [12]. V_{Rd} is the masonry wall bearing shear force and V_{Ed} is the horizontal shear force from the seismic loads combination.

$$V_{Rd} = V_{Rd1}^{*} + V_{Rd2}\ \text{(kN)} \tag{24}$$

$$V_{Rd1}^{*} = 0.4 \cdot (N_{Ed} + 0.8 \cdot V_{Ed} \cdot h_{pan}/l_{pan})\ \text{(kN)} \tag{25}$$

$$V_{Ed} \leq l_{pan} \cdot t \cdot f_{vd,0} \tag{26}$$

$$V_{Rd2} = \lambda_c \cdot A_s \cdot f_{yd}\ \text{(kN)} \tag{27}$$

where h_{pan} is the story height – 30 cm (tie beam height) and $l_{pan}=ls- t$ are the height and length of the masonry area panel. V_{Rd2} (27) is the bearing horizontal shear force from the reinforcement in the slender column at walls compressed edge [12]. A_s is the reinforcement area in the slender column at the walls compressed edge. λ_c is the reinforcement participation factor. Here, λ_c=0.25, for longitudinal reinforcement Φ14 mm. The load combination used to design the structure is 1.0·permanent loads + 0.4·variable loads + 1.0·seismic loads [14].

4. Elastic analysis results

The beams, tie beams and slender columns dimensions and longitudinal reinforcement are seen in table 1. As is the longitudinal reinforcement area [16]. The bars are seen as red discs and the diameter (Φ) of bars (in mm) is written for each element.

Table 1. Beams, tie beams and slender columns dimensions and reinforcements.

Beam 25x50 As →2Φ22 and 1 Φ20 up and down	Tie beam 25x30 As →6Φ12	Slender column 25x25 As →4Φ14
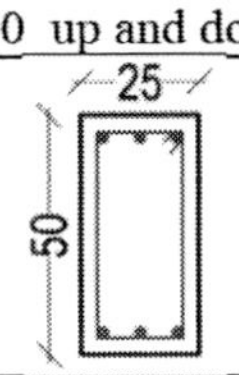	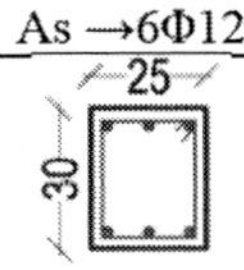	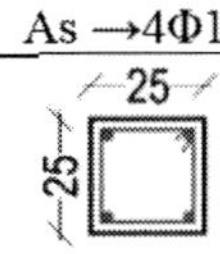

4.1. Concrete walls efforts

The autoclaved walls effect to the existing structure will increase the efforts in the concrete walls. It is important to establish if the concrete walls can withstand the new efforts. In figures 8 to 15 N_{Ed1} and N_{Ed2} (in kN) are the axial forces in the concrete walls in the initial stage 1 and the final stage, after the autoclaved walls story has been built, stage 2. M_{Ed1}, M_{Ed2} and M_{Rd} (in kNm) are the bending moments in the concrete walls in stages 1 and 2. The axial forces and bending moments analyzed are from the seismic loads combinations. No AAC wall will be built on top of P9.

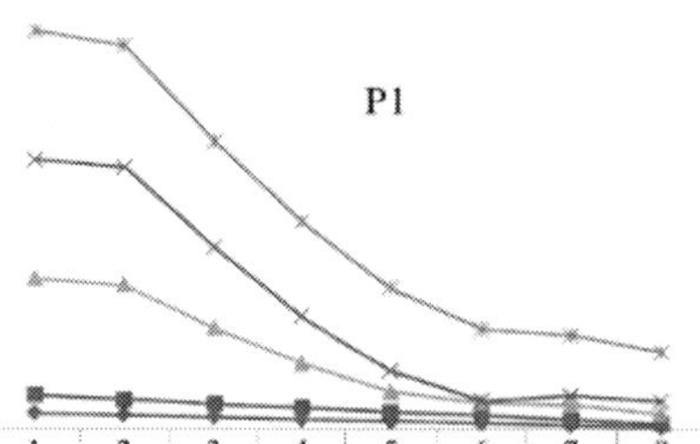

	1	2	3	4	5	6	7	8
MRd	31873	30053	26096	23289	20516	17760	15009	12265
MEd2	29188	28954	19803	11563	5104	425	2369	3024
MEd1	28948	28329	18723	10982	4926	3212	3555	2405
NEd2	4358	3837	3333	2842	2358	1877	1397	914
NEd1	3978	3459	2951	2452	1959	1469	980	492

Figure 8. P1 efforts.

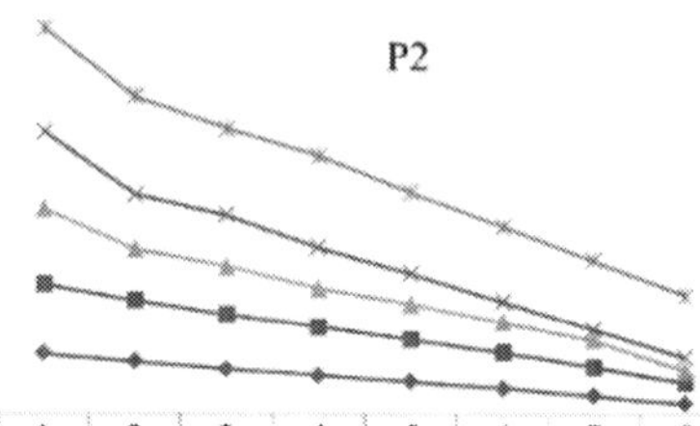

	1	2	3	4	5	6	7	8
MRd	582	544	482	511	452	421	387	349
MEd2	428	305	285	228	173	113	58	74
MEd1	421	286	269	215	194	169	149	64
NEd2	389	344	305	270	236	200	161	119
NEd1	336	290	250	215	179	141	100	54

Figure 9. P2 efforts.

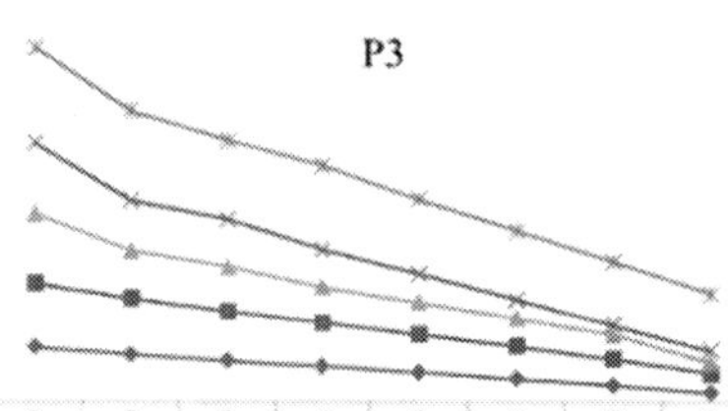

	1	2	3	4	5	6	7	8
MRd	582	544	482	511	452	421	387	349
MEd2	428	305	285	228	173	113	58	74
MEd1	421	286	269	215	194	169	149	64
NEd2	389	344	305	270	236	200	161	119
NEd1	336	290	250	215	179	141	100	54

Figure 10. P3 efforts.

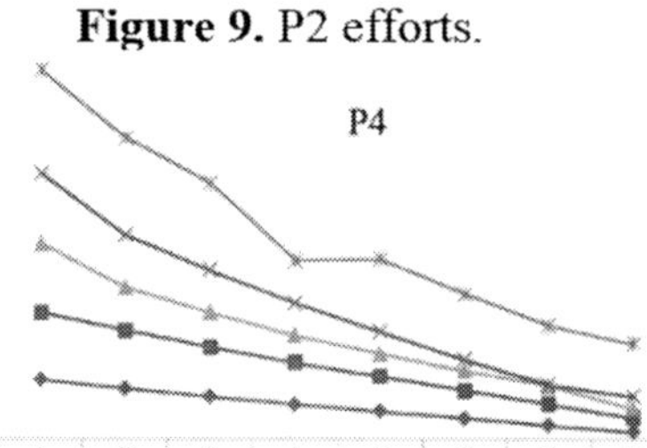

	1	2	3	4	5	6	7	8
MRd	818	769	687	338	573	520	467	417
MEd2	550	413	332	253	170	86	10	95
MEd1	543	339	274	210	183	157	143	72
NEd2	529	459	394	334	277	221	166	111
NEd1	470	400	335	275	219	164	110	58

Figure 11. P4 efforts.

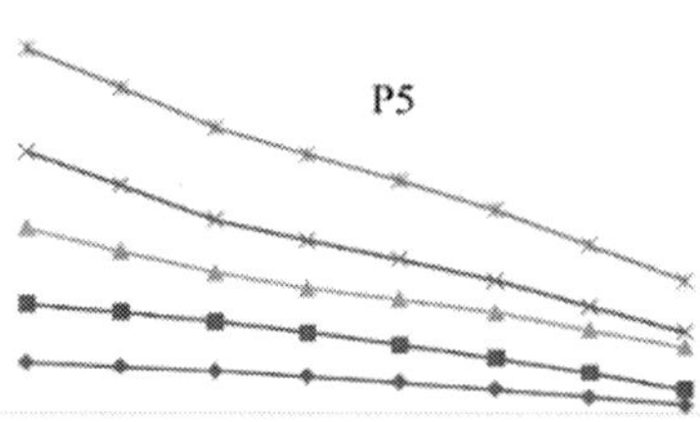

	1	2	3	4	5	6	7	8
MRd	1276	1212	1140	1059	968	867	754	628
MEd2	940	809	645	582	489	389	286	188
MEd1	928	752	602	544	558	559	529	525
NEd2	722	671	613	547	474	392	299	195
NEd1	635	583	524	458	384	301	209	106

Figure 12. P5 efforts.

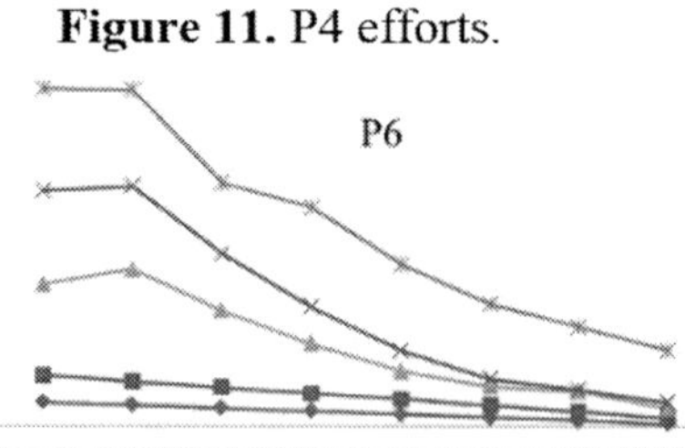

	1	2	3	4	5	6	7	8
MRd	3561	3370	2451	3439	2994	2578	2182	1830
MEd2	3229	2878	1975	1282	717	267	53	207
MEd1	3176	3892	2668	1730	970	675	717	335
NEd2	952	846	742	638	533	425	314	197
NEd1	848	742	638	535	431	326	219	109

Figure 13. P6 efforts.

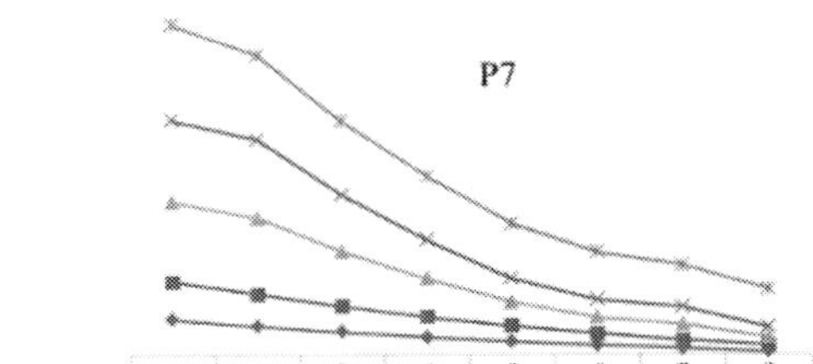

	1	2	3	4	5	6	7	8
MRd	3874	3413	2978	2585	2247	1957	1718	1531
MEd2	3300	3166	2291	1579	965	715	713	415
MEd1	3254	3099	2244	1551	957	656	637	318
NEd2	1546	1289	1045	826	637	475	342	239
NEd1	1429	1173	931	713	525	365	231	128

Figure 14. P7 efforts.

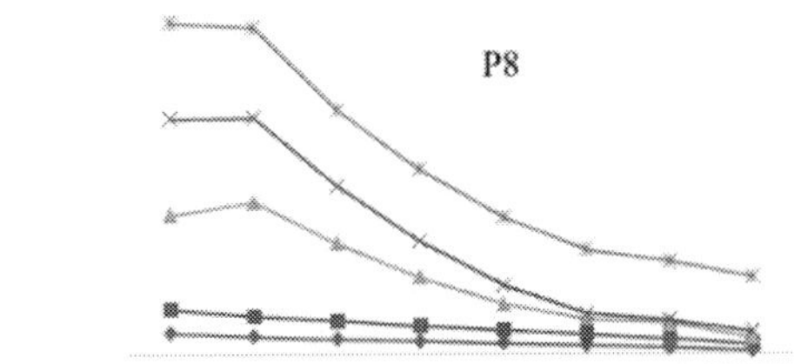

	1	2	3	4	5	6	7	8
MRd	5199	4877	4079	3855	3642	3423	3188	2947
MEd2	5183	4529	3094	1943	1024	314	167	330
MEd1	5089	6168	4210	2643	1398	823	860	353
NEd2	1295	1112	972	849	730	607	475	337
NEd1	1147	963	821	696	574	449	315	177

Figure 15. P8 efforts.

Figures 8 to 15 help compare the piers efforts values. Both axial forces and bending moments values decrease from story 1 to 8 for all piers. The axial forces decrease is faster for shorter walls (P2, P3, P4 and P5), slower for longer walls (P6, P7 and P8) and very slow for the longest wall P1. This can be caused by the walls weights. The bending moments decrease in a linear pattern and M_{Ed1} remains lower than M_{Ed2} for short walls (P2, P3, P4 and P5). The bending moments decrease more rapidly for the other piers. For P6, P7, P8 and P1, the M_{Ed1} values are significantly lower than M_{Ed2} for lower stories. In some cases the bending moments values slightly increase form story 1 to story 2. This is caused by the design formula, more specifically by the great difference between M_{Rd} and M_{Ed} for story 1.

The forces transmitted to the foundation in stage 1 are: base axial force N_1=45642 kN, base shear force V_1=9751 kN for stage 1 and N_2=50398 kN, V_2=9928 kN for stage 2. The base bending moments for the load combination with seism on direction *X* are M_{1X}= 292709 kNm and M_{2X}= 323517 kNm. On direction Y the values are M_{1Y}= 1017925 kNm and M_{2Y} = 1112075 kNm. For the load combination with seism on direction Y, M_{1X}= 454846 kNm and M_{2X}= 490625 kNm. On direction *Y* the values are M_{1Y}= 855788 kNm and M_{2Y}= 949676 kNm. The axial loads and bending moments transmitted to the foundations increase by approximately 10%. The shear forces increase by about 2%.

4.2. Concrete walls compression lengths, stresses and strength

The compression lengths (l_c) and axial stresses (σ_{cp}) reached at the bottom story (*S1*) increase from stage 1 to stage 2 [16-18]. The l_c appears due to the seismic loads combinations and σ_{cp} due to the axial loads for the ultimate limit state combination. l_c should not surpass the maximum compression length x_u, thus the walls will have a ductile behavior and σ_{cp} values need to remain smaller than f_{cd} thus the concrete will not be crushed at the walls bottoms. Table 2 contains these values for stages 1 and 2 [16].

Table 2. RC walls compression lengths, stresses and strength.

		P1	P2	P3	P4	P5	P6	P7	P8
l_c	1	954	119	80	113	152	203	343	275
(mm)	2	1046	125	93	127	173	228	371	311
x_u	1	3876	577	626	754	928	1248	1276	1208
(mm)	2	3956	582	640	772	954	1264	1281	1221
σ_{cp}	1	1.85	1.71	1.73	1.66	1.73	1.	1.	2.43
(N/mm^2)	2	2.07	1.93	1.91	1.83	1.91	1.68	2.1	2.68
f_{cd} (N/mm^2)		16.67	16.67	16.67	16.67	16.67	16.67	16.67	16.67

4.3. Masonry walls efforts, stresses and strength

The autoclaved concrete walls need to bear the efforts they are subjected to, as seen in table 3. The axial stresses at the walls base σ_d need to be lower than the vertical masonry strength f_d [12].

Table 3. AAC walls efforts, stress and strength.

	N_{Ed}	M_{Ed}	M_{Rd}	V_{Ed}	V_{Rd}	σ_d	f_d
P1	402	776	3831	130	218	0.22	2.47
P2	58	9	239	2	71	0.19	2.47
P3	65	26	352	6	76	0.21	2.47
P4	57	32	409	12	75	0.16	2.47
P5	85	37	573	12	85	0.16	2.47
P6	113	71	831	20	80	0.22	2.47
P7	66	133	945	27	106	0.20	2.47
P8	138	86	953	38	159	0.26	2.47

4.4. Natural periods of vibration

Natural periods of vibration for the first 3 vibration modes: *T1, T2* and *T3* are seen in table 4. They increase by up to 16% from stage 1 to stage 2.

Table 4. Natural vibration periods for stages 1 and 2.

Stage 1	*T1*=0.285s	*T2*=0.220s	*T3*=0.187s
Stage 2	*T1*=0.335s	*T2*=0.258s	*T3*=0.215s

5. Plastic analysis results

The building's behavior in the plastic stage is analyzed by using the pushover diagrams for stages 1 and 2 for both directions *X* and *Y*. Pushover cases used are PX1, PY1, PX2 and PY2. The pushover diagrams are shown in figure 16. The maximum displacements considered are 700 mm for stage 1 and 800 mm for stage 2.

5.1. Pushover diagrams

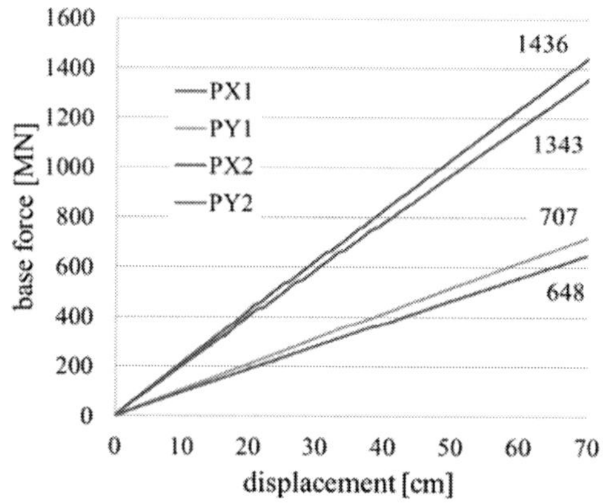

Figure 16. Pushover diagrams.

For direction *X*, the base force is twice the one reached for direction *Y* for both stages 1 and 2. The structure is stiffer on *X*. This may be caused by the 2 long walls seen on that direction. The maximum base forces reached in stage 2 are lower than these in stage 1 by 100 MN for direction X and 50 MN for *Y*. This means the added masonry walls affect the existing concrete structure more on direction *X*.

In all cases, the structures mostly maintain the same rigidity, as the diagrams are all straight lines. Rigidity is regarded as base force/displacement. The diagrams for direction *X* do show some waves, thus there are some slight stiffness decreases.

5.2. Plastic hinges development

The final stages analyzed are seen in figures 17 to 20. The plastic hinges in beams and slender columns need to be shown clearly thus only these appear in the figures. The plastic hinges color code is: green means the plastic hinge is formed, blue is for the plastic hinge reaching the limit, pink represents the load being redistributed and red means collapse.

Figures 17 to 20 show more plastic hinges reach collapse on *X* than on *Y* when the same top displacement is reached. The top displacement is greater for stage 2, as the building is higher, but the plastic hinges are less developed in this stage. This is because in stage 2, the masonry walls at the top story help to dissipate the seismic energy, as they are flexible compared to the concrete walls. This is in accordance with the pushover curves that show a reduced rigidity for stage 2. It is clear that the plastic hinges at the masonry walls story are less developed. This is explained as the AAC walls are more flexible than concrete ones, thus they can bend easier and not get thus much damage.

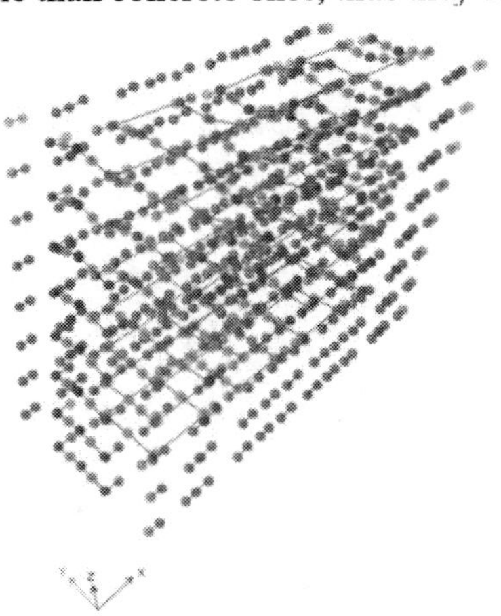

Figure 17. PX1 step 182 d=70 cm.

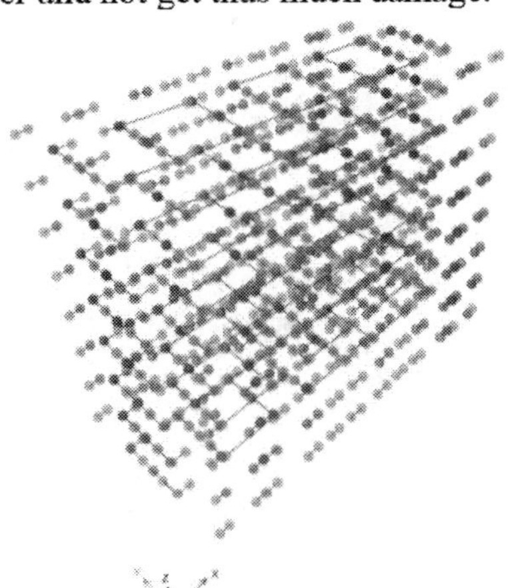

Figure 18. PY1 step 233 d=70 cm.

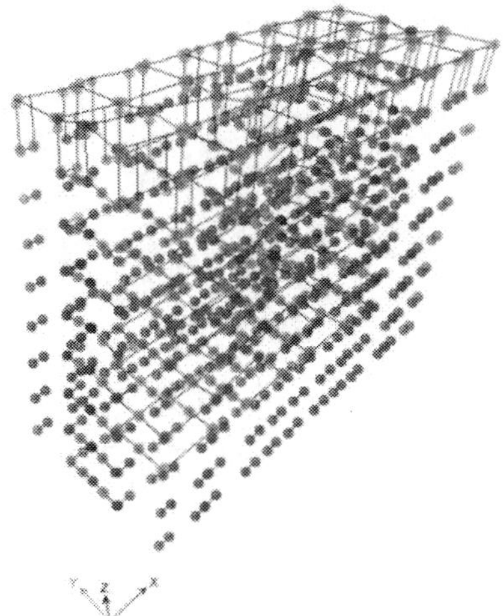

Figure 19. PX2 step 195 d=80 cm.

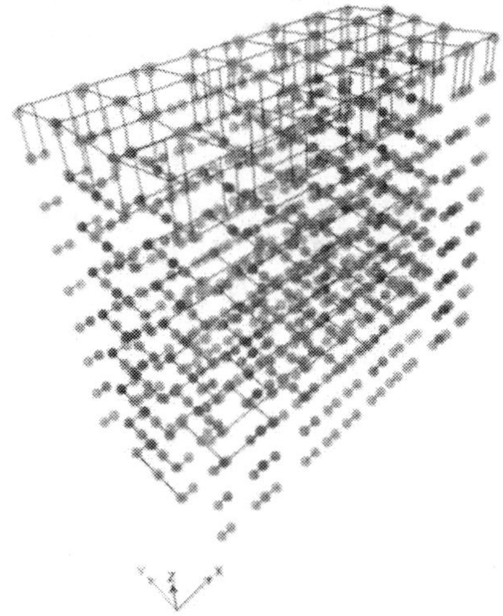

Figure 20. PY2 step 198 d=80 cm.

5.3. Walls stresses in the plastic stage

It is important to study the masonry walls stresses in the plastic stage. Figures 21 to 30 show the stress values at stages when they surpass the masonry strengths. Stresses σ_x and σ_z from PX2, reach high values at the walls corners. The masonry gives out by being crushed there. The high stresses are distributed more evenly among walls for PX2. Higher stress values are seen in walls on direction *Y*.

Stresses σ_x and σ_z from PY2 are focused mostly on the 2 long walls on direction X. They reach high values at the walls tops and bottoms. This means the walls are subjected more to bending perpendicular to their plane.

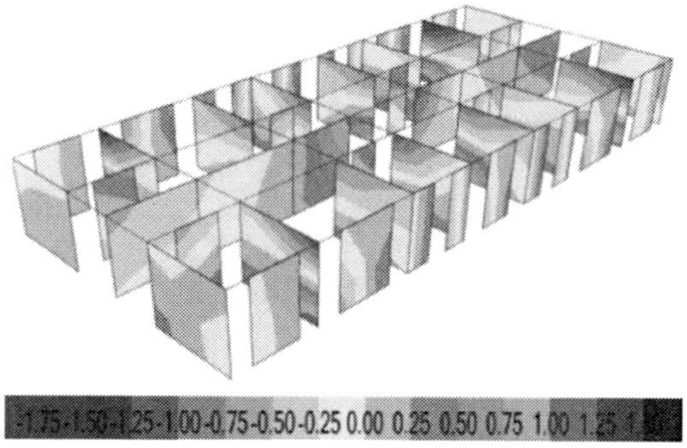

Figure 21. PX2 σ_x step 9.

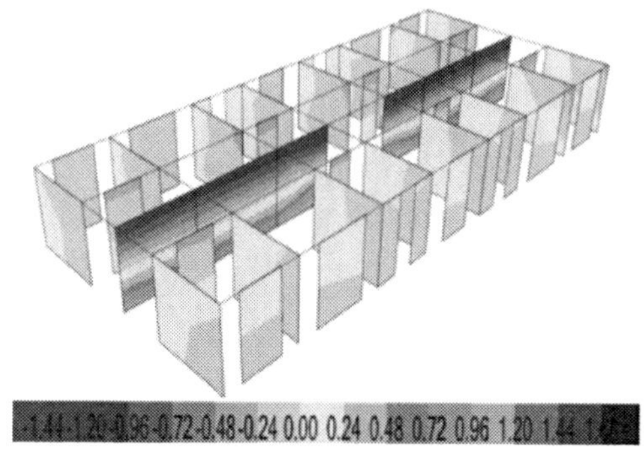

Figure 22. PY2 σ_x step 7.

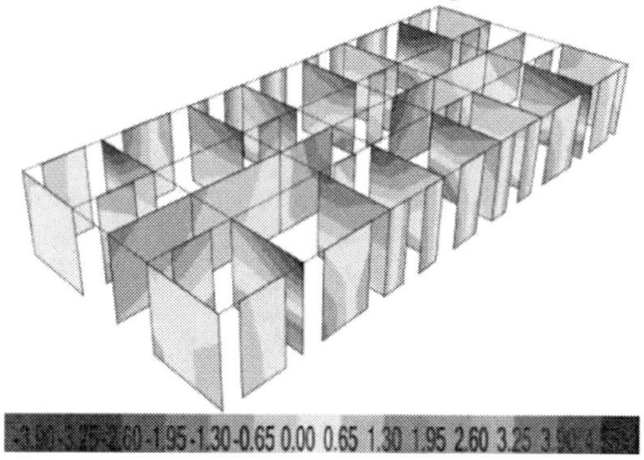

Figure 23. PX2 σ_z step 6.

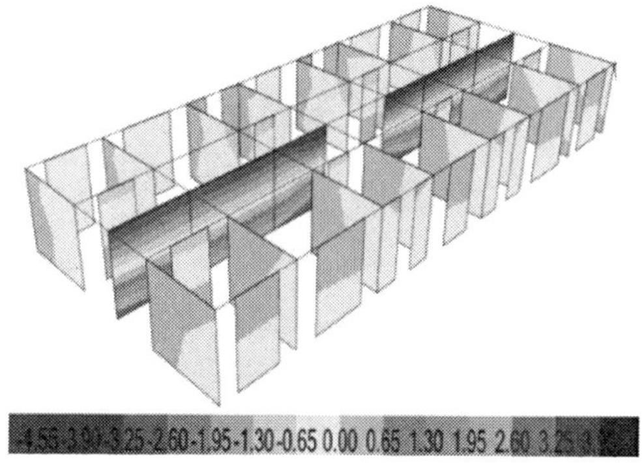

Figure 24. PY2 σ_z step 7.

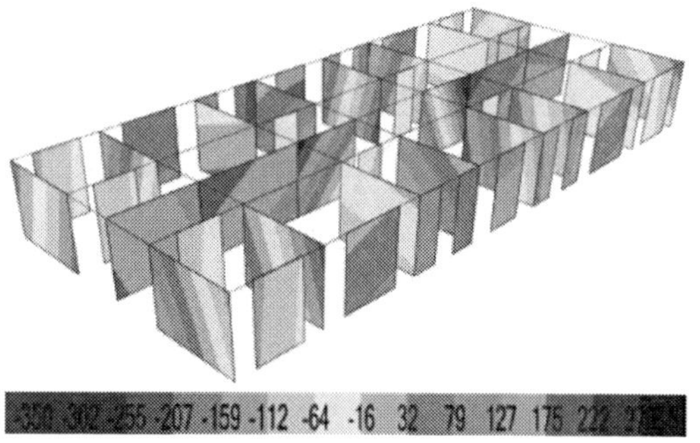

Figure 25. PX2 τ_{xz} step 2.

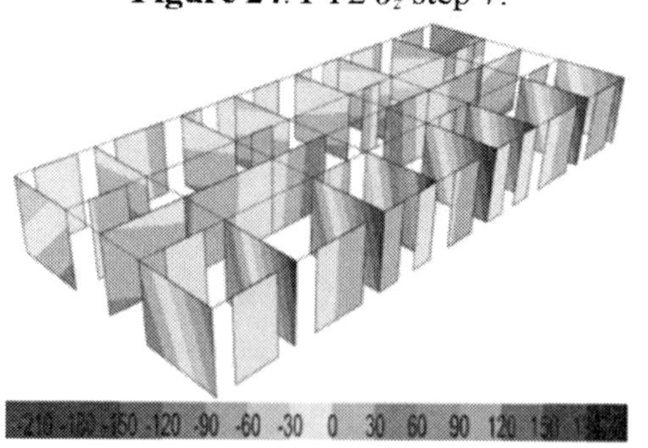

Figure 26. PY2 τ_{xz} step 2.

Stresses τ_{xz} reach high values in walls placed on the same direction as the pushover case. The highest stresses can be found at the walls corners.

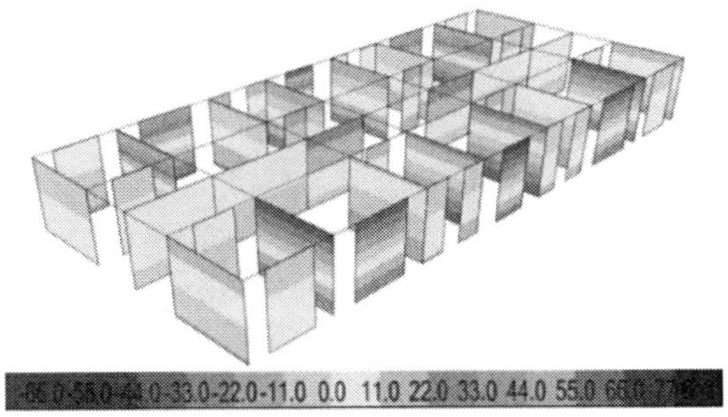

Figure 27. PX2 τ_{xy} step 145.

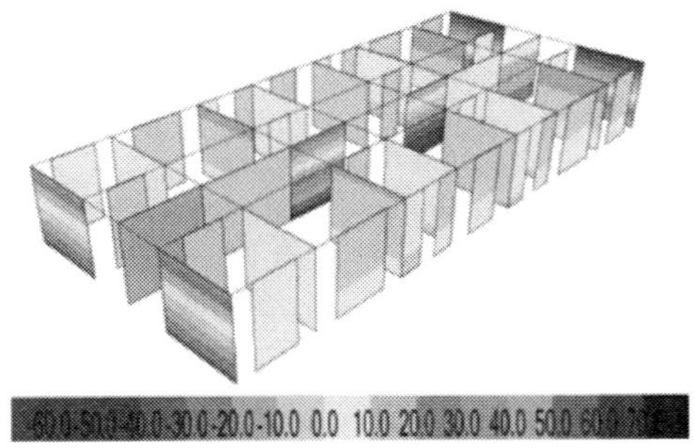

Figure 28. PY2 τ_{xy} step 165.

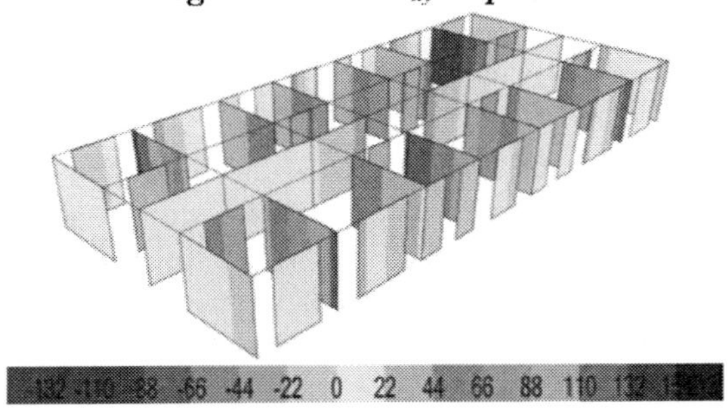

Figure 29. PX2 τ_{yz} step 9.

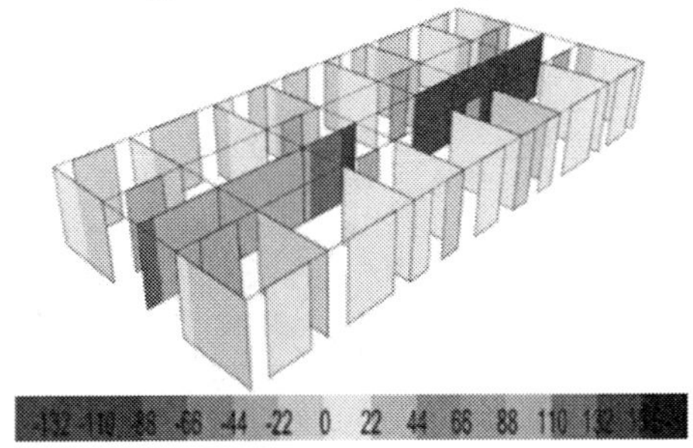

Figure 30. PY2 τ_{yz} step 7.

Stresses τ_{xy} clearly surpass the masonry strength at steps 145 and 165. The walls would give in to this stress long after many plastic hinges have been formed. An explanation may be that the walls can bend as the plastic mechanism is formed. Stresses τ_{yz} reach high values in walls perpendicular to the pushover case direction, because this is a perpendicular stress.

Both stresses reach the highest values in slender walls with high axial loads. This can happen because walls with lower axial forces can bend more perpendicular to their plane and not get that much damage.

6. Conclusions

- The additional story can be built above the existing structure as the old structure can bare it. The design efforts remain below the bearing efforts after the additional story has been built. The compression lengths and axial stresses at the walls bottoms stay below the maximum compression length and the vertical compression strength respectively for both stages 1 and 2. The AAC walls can bare the efforts they are subjected to.
- The new story changes the structures plastic behavior. The new story helps dissipate the seismic energy. The top displacement is greater in stage 2. The building is higher, but the plastic hinges are less developed in this stage. In stage 2, the masonry walls at the top story help dissipate the seismic energy, because the AAC story can bend easier and not get as damaged as the concrete stories. The pushover diagrams show a reduced rigidity for stage 2. The plastic hinges at the masonry walls story are less developed.
- In the plastic stage, stresses σ_x and σ_z from PX2, reach high values at the walls corners and the masonry is crushed there. High stresses are distributed more evenly among walls for PX2. Stresses σ_x and σ_z from PY2 are focused mostly on P1. P1 walls are long and situated on direction X. They are subjected mostly to bending perpendicular to their plane. Stresses τ_{xz} reach high values in piers situated on their direction. Both stresses τ_{xy} and τ_{yz} reach the highest stress values in slender walls with high axial loads because walls with lower axial forces can bend perpendicular to their plane without giving in.

References

[1] Teguh M 2017 Experimental Evaluation of Masonry Infill Walls of RC Frame Buildings Subjected to Cyclic Loading *Sustainable Civil Engineering Structures and Construction Materials, SCESCM 2016, Elsevier Procedia Engineering* **171** p 191–200 DOI 10.1016/j.proceng.2017.01.326

[2] Marinilli A and Castillia E 2004 Experimental Evaluation of Confined Masonry Walls with Several Confinig Columns *13th World Conference on Earthquake Engineering Vancouver,B.C.,Canada*, paper No.1597

[3] Strength Behavior of Mud Brick in Building Construction. *Open Journal of Civil Engineering* **6**, 482–94 DOI 10.4236/ojce.2016.63041

[4] Ahimbisibwe A and Ndibwami A 2016 Demystifying Fired Clay Brick: Comparative analysis of different materials for walls, with fired clay brick *PLEA Los Angeles -32nd International Conference on Passive and Low Energy Architecture Cities, Buildings, People: Towards Regenerative Environments*

[5] Chourasia A, Bhattacharyya S K, Bharagava P K and Bhandari N M 2013 Influential aspects on seismic performance of confined masonry conctruction, *Natual Science* **5** No.8A1 56–62 DOI 10.4236/ns.2013.58A1007

[6] Alexandris A, Protopapa E and Psycharis I 2004 Collapse mechanisms of masonry buildings derived by the distinct element method *13th World Conference on Earthquake Engineering Vancouver,B.C.,Canada*, paper No.1597

[7] Guevara-Perez T 2012 Soft Story and Weak Story in Earthquake Resistant Design: A multidisciplinary Approach *15 World Conferences on Earthquake Engineering (WCEE)* Lisboa

[8] Al-Mudhaf H A and Attiogbe E K 1996 Performance of autoclaved aerated-concrete masonry walls in Kuwait *Materials and Structures 29,* August /September p 448–52 DOI 10.1007/BF02485996

[9] Jinghai Y, Jianfeng C and Tianhui F 2013 Experimental study on improving seismic behavior of load-bearing masonry wall made of autoclaved aerated concrete *Transactions Tianjin Univ.* **19** Issue 6 pp 419-24 DOI 10.1007/s12209-013-2121-6

[10] Costa A A, Penna A, Magenes G and Galasco A 2008 Seismic performance assessment of autoclaved aerated concrete (AAC) masonry buildings *Proc. The 14th World Conference on Earthquake Engineering* , October 12-17 Beijing

[11] Costa A A, Penna A and Magenes G 2011 Seismic Performance of Autoclaved Aerated Concrete (AAC) Masonry: From Experimental Testing of the in-Plane Capacity of Walls to Building Response Simulation *Journal of Earthquake Engineering*, **15**:1–31 ISSN: 1363-2469 DOI 10.1080/13632461003642413

[12] Tomaževič M, and Gams M 2010 Seismic Behaviour of Confined Autoclaved Aerated Concrete Masonry Buildings: a Shaking Table Study, *Mauerwerk European Journal of Masonry*, **14**(3) p 153–60 DOI 10.1002/dama.201002047

[13] Vermeltfoort A T and Martens D R W 2017 Preliminary tests on the lateral load-bearing capacity of slender masonry walls. *Proceedings 13th Canadian Masonry Symposium*, 1–12

[14] CEN EN 1996-1-1 2006 Eurocode 6: Design of masonry structures - Part 1-1: General rules for reinforced and unreinforced masonry structures

[15] CEN EN 1991-1-1 2004 EC 1 Actions on structures - Part 1-1: General actions- Densities, self-weight, imposed loads for buildings

[16] CEN EN 1990 2004 Eurocode 0: Basics of structural design

[17] CEN EN 1991-1-3 2005 EC 1 Actions on structures 1-3 General actions Snow loads

[18] CEN EN 1992-1-1 2004 EC 2 Design of concrete structures. General rules for buildings

[19] CEN EN 1998-1 2004 Eurocode 8: Design of structures for earthquake resistance. Part 1: General rules, seismic actions and rules for buildings

[20] P100-1 2013 Seismic design code – Part 1- General rules for buildings

The impact of thermal rehabilitation of the reinforced concrete water tanks ROMANIAN built fund over its lifecycle

D Covatariu[1*], A Filip[1] and G Covatariu[1]

[1] "Gheorghe Asachi" Technical University of Iasi, Faculty of Civil Engineering and Building Services, Romania

[*] daniel.covatariu@tuiasi.ro

Abstract. Liquid storage tanks are special structures and strategically very important. In particular, water tanks are important to assure a continued operation of water distribution system in the event of unpredicted events. In order to ensure the operating conditions in normal parameters, the temperature of the stored liquid (drinking water) have to not exceed certain limits. The aim of this article is to investigate the most suitable thermal insulation system for a reinforced concrete overground water tank. In this regard, a comparative multicriteria analysis has been performed presenting the main advantages / disadvantages of each material (extruded polystyrene, basaltic mineral wool, organic ecological foam) taking into account several criteria. The scores obtained have been compared in order to highlight the most effective material in the case of the reinforced concrete water tanks.

1. Introduction

In Romania, the concept of quality in constructions is defined as the sum of all the requirements for the exploitation of an engineering structure, imposed by the current norms and by the conditions of their use [1].

Therefore, the defining factors (evaluated in order to assure the quality requirements) are defined by the structure's durability and the sustainability of the process ensured by it.

In this sense, the life cycle analysis of a structure becomes mandatory for a special engineering structure (reinforced concrete drinking water tank), being a structure with a considerable life span (more than 100 years) and with high requirements of exploitation.

Life cycle analysis (LCA) is an environmental management technique for evaluating the environmental aspects of the product systems and their associated potential impacts (according to SR EN ISO 14040:2002 – Environmental management. Life cycle assessment. Principles and framework), which is led by going through four phases:

- defining the purpose and scope;
- inventory analysis for relevant input and output elements of a product system;
- assessing potential environmental impacts;
- interpretation of the results of the inventory analysis and the impact assessment phase.

Because, this life cycle is sometimes referred to as "from cradle-to-grave" it is necessary that any relevant environmental impact be analysed and all the processes for reducing its effects are taken into account.

Regarding the water tanks, for the design, construction and operation of a compliant structure, the following fundamental requirements are required to be maintained, during the entire exploitation period of the constructions:

- mechanical strength and stability;
- fire safety;
- hygiene, health and the environment;
- safety and accessibility in operation;
- sustainable use of natural resources
- energy saving and thermal insulation [2].

This last requirement imposes the analysis of a stage (less applied in Romania), namely that of thermal insulation (for those in the conception / execution stage) or of thermal rehabilitation (for those in operation).

This stage becomes compulsory following an LCA, because the absence of such a protective layer of the envelope decreases the life of the water tank, leading to a negative impact on the environment.

Thermal insulation in the final form, called thermal insulation system, includes:

- the thermal insulation layer;
- fasteners;
- protective cover;
- radiant barriers (optional);
- waterproofing / soundproofing and / or anti-corrosion coating (optional).

The major advantage of the thermal insulation layer is not only the reduction of mass transfer between the indoor and outdoor environment, but also the protection of the envelope (vertical circular wall and horizontal floor - roof - figure 1) against the aggression of the environmental agents, which cancels the possibility of degradation / significant structural damage.

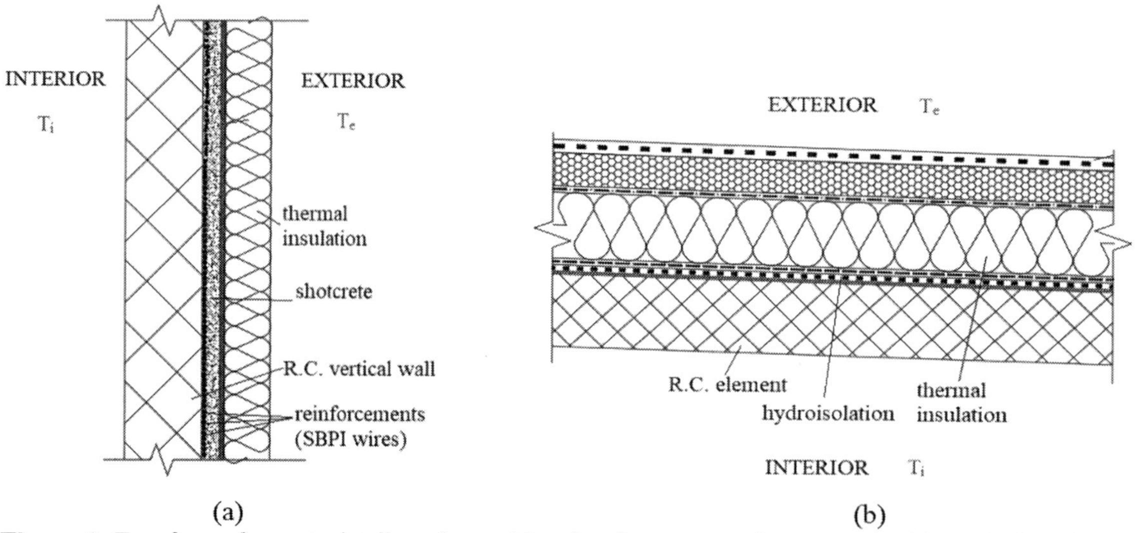

Figure 1. Envelope elements details – thermal insulated overground water tank: (a) vertical perimeter wall, (b) horizontal floor made of prefabricated reinforced concrete elements – roof.

Moreover, operating conditions under normal conditions of the water tank require that the temperature of the stored liquid (drinking water) does not exceed certain limits. It is obvious that the thermal impact on the stored liquid will be high in the absence of the thermal insulation layer applied on the tank envelope [3].

The presence of the thermal insulation layer from the execution phase will significantly extend the time (T_1 – figure 2) at which the intervention on the structure (structural rehabilitation) is necessary, because the moment of the degradation of the envelope elements will be delayed, which will reduce the negative impact on the environment, it will have a positive economic effect and may even extend the operating period under normal conditions. The process of degradation of water tanks can have many causes (improper exploitation of the construction, cracking of the concrete, corrosion of the reinforcement etc.), which leads to the necessity of carrying out the rehabilitation works [4, 5].

It can be seen from figure 2, that the delay of the moment of beginning the rehabilitation process (T_1) under the conditions of the occurrence and evolution of the degradation, prolongs the execution period of the rehabilitation works (T_2-T_1) which has as negative effects the temporary removal from the exploitation of the tank but also an increase of the financial resources involved [6].

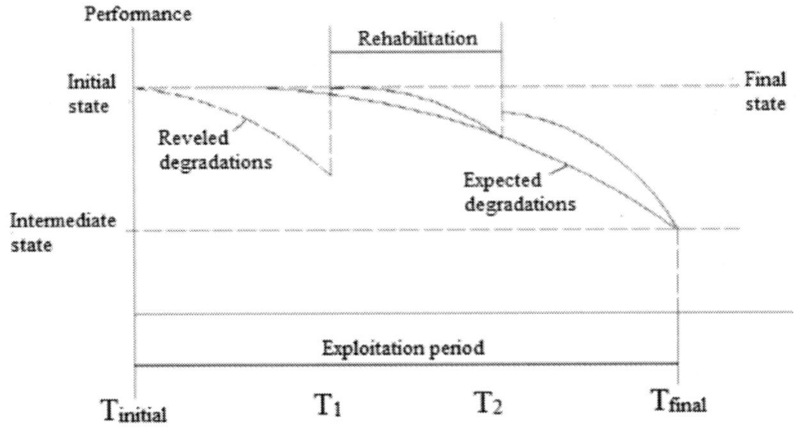

Figure 2. The relationship between the operating life of the structure and the level of degradation

The thermal protection solutions used in Romania are multiple, but in the case of water tanks, their choice and application require special attention due to the curved surfaces of the vertical walls of the tanks (to be mounted).

The factors that influence the choice of thermal protection / rehabilitation method are:

- the energy efficiency of the insulation system (calculation parameters: the thermal conductivity of the material/s used and the thickness of the applied layer);
- the temperature difference between the external and the internal environment (calculation parameters: the extreme temperature of the external environment and the standard temperature of the stored liquid - water);
- mounting technology;
- the impact of using the system on the environment.

Are evaluated:

- the sustainability characteristics of thermal insulation materials,
- ecological origin: natural materials (organic / inorganic or synthetic)
- the price of the thermal insulation system;
- its effect (during the exploitation period) on the inhabited environment and on its inhabitants.

The ability of the material to be transformed from "industrial waste" or material from disassembly / demolition into "insulating material" (through reintegration) is also appreciated.

Of particular importance is in this regard the aggregation state of the insulating material used (the form of the insulating layer); such materials could be:

- rigid (plates / panels, blocks / bricks, segments);
- pulverous (foam / foam);
- fibrous (mattresses, strips, fibres);
- granular / powdery;
- pellicular (paint).

It is obvious that from a technological point of view, the pulverizing, fibrous and pellicular materials will be preferred.

2. Thermal insulation materials used in Romania

In Romania, the most used thermal insulation for thermal rehabilitation works for water tanks are made of extruded polystyrene, table 1 [7] and basaltic mineral wool, table 2.

Table 1. Graphite extruded polystyrene.

Appearance: rigid plates	Advantages	Disadvantages
	- high thermal efficiency; - presents the best thermal resistance / cost ratio; - reduced load on the structural elements.	- difficult to mount on curved surfaces (such as perimeter walls on cylindrical tanks) - slightly flammable - high impermeability (water vapor); - relatively low durability due to the action of environmental factors (solar radiation, humidity etc.)

Table 2. Basaltic mineral wool.

Appearance: rolled material	Advantages	Disadvantages
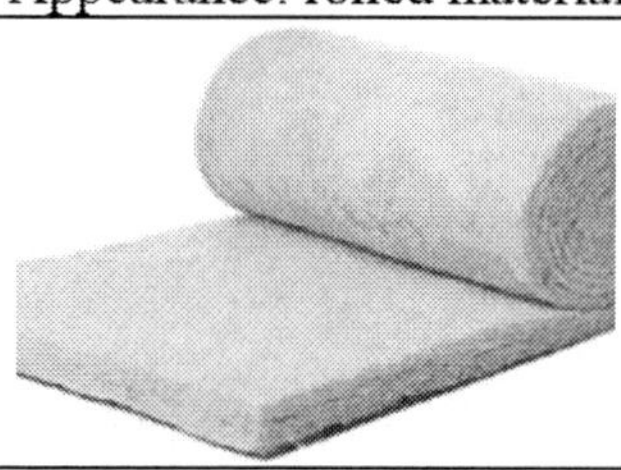	- dimensional stability to thermal variations; - low thermal conductivity; - increased resistance to aging and the action of aggressive environmental agents.	- low shock resistance; - reduced workability; - immersion / diffusion water absorption;

Recently, it has begun to be applied other thermal insulation systems based on pulverizing materials, such as closed cell polyurethane foam and organic foam (contains soybean oil, castor oil or reused synthetic material, recycled pet stain), table 3 [8] and thermal insulation paints.

Table 3. Organic ecological foam.

Appearance: sprayed layered	Advantages	Disadvantages
	- complete sealing of the coverage area; - recommended for easy application to curved surfaces; - very good adhesion to the support layer; - ecological material; - high durability.	- high degree of impermeability; - relatively low fire resistance; - high costs (compared to other thermal insulation systems).

3. Multicriteria comparative analysis

As presented above, there are advantages and disadvantages of using these thermal insulation systems. For a good evaluation and the choice of the most efficient thermo-system (to satisfy the requirements imposed for water tanks), a multicriteria comparative analysis (most commonly used in Romania) has been performed, based on a scoring system, taking into account the following criteria:

- cost of the investment;
- duration of execution;

- durability of the material;
- technological process;
- environmental impact;
- fire resistance;
- thermal performance.

The evaluation system used awarded points from 1 to 5, corresponding to the ratings from "inefficient" to "very effective".

For the parametric quantification, an overground water tank made of reinforced concrete (having a 5000 m^3 capacity) was analysed. From a structural point of view, on the outside, the perimeter wall was rehabilitated by post-tensioning, figure 3 (a) and the roof elements by the use of carbon fibre reinforced polymeric (CFRP) composite materials, figure 3 (b).

(a) (b)

Figure 3. Strengthening of the structural elements: (a) post-tensioning of the vertical wall perimeter, (b) R.C. beams strengthened with CFRP strips.

3.1. *Cost of the investment*

The quantification of the first two criteria (cost and duration) for the 3 methods of thermal insulation analysed has been achieved by creating the bill of quantities with the related resources (material, labor, equipment, transport) using the software program Intersoft Deviz Professional (using the market prices from 2019), figure 4.

From the point of view of the estimated cost of the investment (figure 4), by applying organic foam thermal rehabilitation has an investment value of approximately 70% higher compared to the use of extruded polystyrene, respectively 27.9% in the case of basaltic mineral wool.

This fact significantly influences the investor's decision regarding the costs and the duration of the new construction.

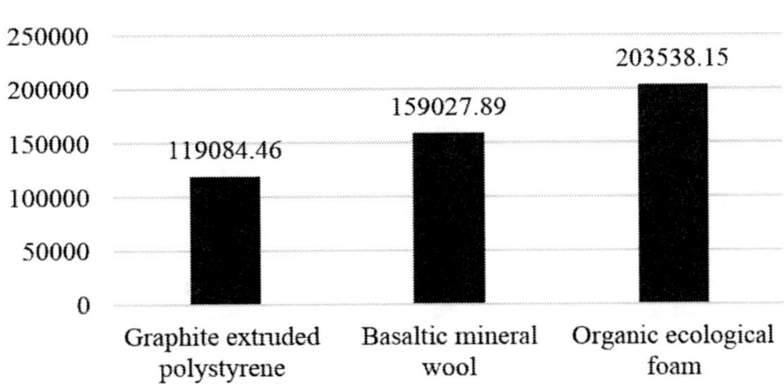

Figure 4. Comparative analysis according to the value of the investment

Depending on the costs involved for each of the thermal insulation systems, the following score was given, table 4:

Table 4. Cost of the investment.

Material	Degree of efficiency
1. Graphite extruded polystyrene	5*
2. Basaltic mineral wool	4*
3. Organic ecological foam	3*

3.2. Duration of execution
The estimated duration of execution, figure 5, in the case of each proposed thermal rehabilitation solution, from a percentage point of view varies significantly. Thus, in relation to the execution time of the use of an extruded polystyrene thermo-system, in the case of the application of the basaltic mineral wool it decreases by 23.2%, whereas, in the case of using the organic foam, the percentage is a considerable one, decreasing dramatically by 82.2%.

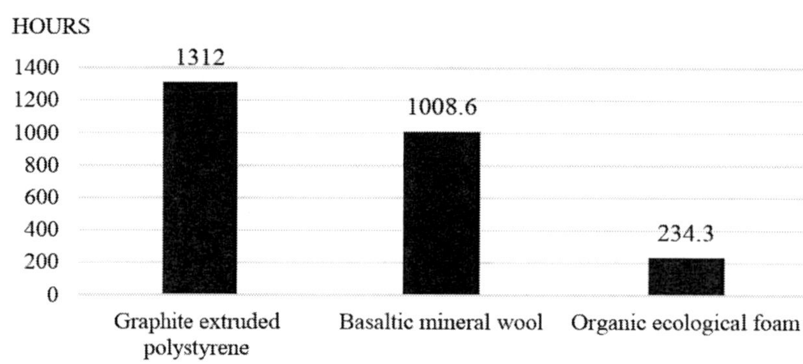

Figure 5. Comparative analysis according to the execution time

In this case, the following score was given, table 5:

Table 5. Duration of execution.

Material	Degree of efficiency
1. Graphite extruded polystyrene	1*
2. Basaltic mineral wool	2*
3. Organic ecological foam	5*

3.3. Durability of the material
Durability is the capacity of a material to maintain its properties for a long time without significant deterioration / degradation, having a low impact on the environment.

According to the analysed technical data sheets, the extruded polystyrene has an estimated life of about 25 years (considering that the technological steps have been properly respected).

Mineral wool has a much longer shelf life compared to extruded polystyrene, the correct thermo-system having a lifetime estimated between 25÷50 years, preserving its qualities while polystyrene degrades over time much faster, leading to the need for additional repair / maintenance work.

In the case of the thermal insulation system with organic foam, under normal operating conditions, the estimated service life is at least 40 years, no maintenance work required.

Based on the information gathered from the technical data sheets, the following score was obtained, table 6:

Table 6. Durability of the material.

Material	Degree of efficiency
1. Graphite extruded polystyrene	2*
2. Basaltic mineral wool	4*
3. Organic ecological foam	5*

3.4. Technological process

Technological procedure applied like thermo-insulation for a curved surfaced (eq. vertical walls of an r.c. water tank) impose qualified workmanship and a significant period to put in practice, because is a material from rigid plates. Unlike these, basaltic mineral wool is found on the market in the form of mattresses and is much more adaptable to the curved surfaces of insulation.

The insulation with organic foam represents an alternative to the one with polystyrene or mineral wool, in that it is applied as a mixture of two liquid materials and is applied by spraying in a layer of several centimeters on the outer face of the structural elements Depending on the thickness of the insulation to be made, a single layer or more layers of foam are applied [8].

In the case of constructions in operation, the thermal rehabilitation works are performed after the leak test of the tank has been performed. These are made mainly in the case of the overground / semi-buried tanks.

The quality of the thermal insulation works at the special structures of water storage, has a fundamental role because, by observing all the requirements, it ensures a long life of operation, without being necessary interventions for this purpose.

From the point of view of the complexity and efficiency of the technological process, the following score was given, table 7.

Table 7. Technological process.

Material	Degree of efficiency
1. Graphite extruded polystyrene	2*
2. Basaltic mineral wool	3*
3. Organic ecological foam	5*

3.5. Impact on the environment

In the manufacturing process of extruded polystyrene (98% air and 2% styrene), a quantity of gases is used that does not have relatively significant effects on the environment, compared to the basaltic mineral wool.

In Romania, organic ecological foam used in thermal insulation works contains only natural ingredients and is resistant to mold, moisture and fungi. This substance has no negative impact on the environment. From the point of view of the impact of the materials on the environment, table 8, the following score was awarded:

Table 8. Impact on the environment.

Material	Degree of efficiency
1. Graphite extruded polystyrene	1*
2. Basaltic mineral wool	2*
3. Organic ecological foam	5*

3.6. Fire resistance

In Romania, the fire safety measures of the constructions are regulated by the norm P118/1999. It determines according to the category of importance of the construction, the type of construction, destination and size, the strength and safety in operation of the construction in case of fire.

In the case of special constructions of water storage, for thermal insulation systems, the fire risk is defined by fire hazard categories, which express:

- categories A (BE3a), B (BE3b): very high risk of fire;
- category C (BE2): high risk of fire;
- category D (BE1a): average risk of fire;
- category E (BE1b): low risk of fire [9].

According to the data in the data sheets of the manufacturers of insulating materials and of the provisions of normative P118/1999, the following score was awarded according to the degree of efficiency, table 9.

Table 9. Fire resistance.

Material	Degree of efficiency
1. Graphite extruded polystyrene	1*
2. Basaltic mineral wool	5*
3. Organic ecological foam	3*

3.7. Thermal performance

The thermal insulation performance of a material depends on the thickness of the material and the thermal conductivity (mainly) and the water absorption capacity (in the secondary). A high thermal resistance is achieved by using thermal insulation materials with a thickness (d) as high as possible and a thermal conductivity (λ) as low as possible. For the analysed materials, according to the studied technical data sheets, their thermal conductivity falls within close limits, table 10, for the same thickness of the thermal insulation layer.

Table 10. Thermal conductivity coefficients.

Material	Thermal conductivity λ [W/mK]
1. Graphite extruded polystyrene	0.040 – 0.042
2. Basaltic mineral wool	0.035 – 0.038
3. Organic ecological foam	0.020 – 0.025

From the point of view of the thermal performance of the materials, the following score was given, table 11.

Table 11. Thermal performance.

Material	Degree of efficiency
1. Graphite extruded polystyrene	2*
2. Basaltic mineral wool	3*
3. Organic ecological foam	4*

Based on the criteria described above, the information was centralised, table 12, creating a ranking regarding the thermal insulation materials proposed in the comparative analysis.

Table 12. Multicriteria analysis – centralizing situation.

Descriptive criterion	Graphite extruded polystyrene	Basaltic mineral wool	Organic ecological foam
1. Cost of the investment;	5*	4*	3*
2. Duration of execution,	1*	2*	5*
3. Durability of the material;	2*	4*	5*
4. Technological process;	2*	3*	5*
5. Impact on the environment;	1*	2*	5*
6. Fire resistance;	1*	5*	3*
7. Thermal performance.	2*	3*	4*
TOTAL:	**14***	**23***	**30***

4. Conclusions regarding the comparative analysis

The aim of the multicriteria analysis was to quantify the proportions of advantage and/or disadvantage presented by each of the three thermal insulation materials commonly used in Romania, based on market information and the quotations drawn up, in order to select the most energy efficient and eco-friendly.

The most effective material proved to be organic ecological foam, which is recommended in terms of lead time, material durability and technological process.

The basaltic mineral wool is noted for its fire resistance (non-propagation of flame), the positive environmental impact being at a price / quality ratio in the average quantification class.

From the point of view of the costs involved and the impact on the environment (recycling process) extruded polystyrene is the ideal solution, but according to the awarded score, it is deficient in relation to the other criteria (except the technological process), thus occupying the last place.

In the application of the comparative analysis, the investor / beneficiary has the role to formulate and establish the net benefits of the proposed solutions. This leads to a high complexity of the comparison between gains / losses, because the value of today's investment (costs / benefits) is not directly compatible with the value over a certain period of time (for instance, 20 years).

References

[1] Law 10/1995 *regarding the quality in building processes*

[2] HG 766/1997 *Regulation on the building's behaviour in-time monitoring, intervention measures and post-use of constructions*

[3] Hobjilă V and Luca M 2000 *Complementary instructions to the design and expertise of some types of hydrotechnical constructions* (Iasi, Cermi Press) p 205 ISBN 973-8000-28-6

[4] SR EN ISO 14040:2002 *Environmental Management. Life-cycle Evaluation. Principles and working domains*

[5] Covatariu D and Filip A 2018 *A Comprehensive Approaching of the Water Tank's Durability Affecting Factor in Different Stages* (Bulletin of the Polytechnic Institute of Iasi) section: Construction – Architecture **64** (68) (4) p 142-156

[6] Zamfirescu D and Postelnicu T 2003 *Reinforced Concrete Durability* (Bucharest: Matrix Rom Press) p 4 ISBN 973-685-666-6

[7] SR EN 13164:2012 *Thermo-insulation products for buildings. Products made of extruded polystyrene foam*

[8] SR EN 13165:2003 *Thermo-insulation products for buildings. Products made of rigid polyurethane foam*

[9] P 118/1999 *Norms regarding Buildings Fire safety*

Planning risk management in case of a drinkable water tank facility operation

G Covatariu[1*], D Covatariu[2] and A Filip[3]
[1] Department of Structural Mechanics, Faculty of Civil Engineering and Building Services, "Gheorghe Asachi" Technical University of Iasi, Romania
[2] Department of Civil and Industrial Engineering, Faculty of Civil Engineering and Building Services, "Gheorghe Asachi" Technical University of Iasi, Romania
[3] Faculty of Civil Engineering and Building Services, "Gheorghe Asachi" Technical University of Iasi, Romania

[*] gabriela.covatariu@tuiasi.ro

Abstract. The continued assurance of the needs of drinking water and the proper quality is vital for ensuring good living conditions. As a result of this idea, the risk analysis of the technological process of providing drinking water in the case of water tanks and the creation of a risk management plan and its permanent pursuit can lead to a more efficient process. Water tanks are high importance facilities for communities and during their lifecycle, there are a lot of uncertainties that can occur and affect it. In order to have an efficient and correct facility management during the functioning period, performing a risk analysis of these events is mandatory for this kind of constructions. Creating a good risk management plan and keep following it during the objective lifetime keeps the qualitative indicators of the water into planned parameters and the low costs.

1. Introduction

In the case of special structures, such as water tanks, in order to obtain quality constructions, the achievement and maintenance of fundamental requirements (according to Law 10/1995) shall be compulsory fulfilled and maintained for the entire period of operation of the construction, as follows: Mechanical strength and stability; Fire safety; Hygiene, health and the environment; Safety and accessibility in service; Energy saving and thermal insulation; Sustainable use of natural resources [1]. Compliance with these requirements leads to the establishment of the quality assurance system. The quality system in construction is the ensemble of organizational structures, responsibilities, regulations, procedures and means, which competes in the realization of the quality of construction at all stages of conception, realization, exploitation and its post-using.

To this end, the quality system shall lead to the achievement and exploitation of appropriate quality constructions and shall be applied differently depending on the categories of building importance. An important component of quality management is risk management.

The risk is the "effect of uncertainty on objectives", according to ISO 31000 [2]. Its effect can have a positive or negative deviation from what is expected.

A centralised public system exploiting a water source is a major responsibility that has implications for public health and safety in densely populated urban areas. Nowadays, drinking water is an absolute necessity not only for consumption, sanitation, fire extinguishing, but also for industrial processes. The centralized water supply system consists in capturing and bringing water from the source, pumping stations, treatment stations, storage tanks (for creating an emergency reserve and the volume necessary to compensate for consumption variations) and the water distribution system to consumers.

Finding a sufficient source of water to be safely exploited and then distributed to consumers (with almost no interruption) requires reliable infrastructure.

In the period 1955 ÷ 1990, many water supply systems (capture and distribution networks, pumping stations, treatment stations, reservoirs, installations) have been executed in Romania, and which are currently at the end of the normed service period.

Water tanks are high importance facilities for communities and during their lifecycle, there are a lot of uncertainties that can occur and affect it. In order to have ~~an~~ efficient and correct facility management during the functioning period, performing a risk analysis of these events is mandatory for this kind of constructions. Creating a good risk management plan and keep following it during the objective lifetime keep the qualitative indicators of the water into planned parameters and the low costs.

For assuring a good quality of drinkable water, can be used the Stockholm framework, presented by the World Health Organisation conceived for assessment of risk for water-related microbiological hazards (figure 1).

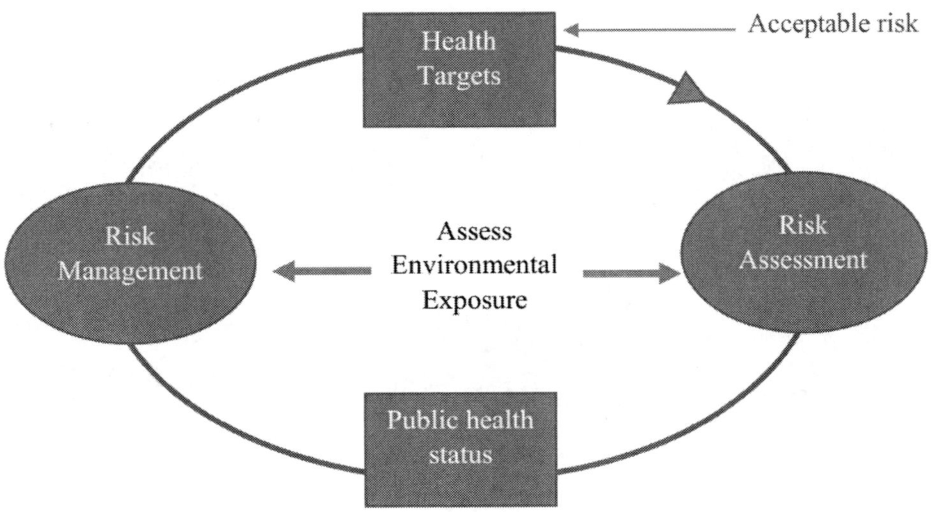

Figure 1. Simplified representation of the Stockholm Framework (WHO2001) [3].

2. Risk Managements Processes

According to PMI Practice Standard Project Risk Management [4] and PMBok 6th edition [7] the processes required to create ~~a~~ proper risk management for any project/structure.

a. *Plan risk management* is the process where are defined the scope and objectives of the risk management process. During this process are analysed historical data (in order to find previous risks and their data in similar structures), standards and documentation (to establish the parameters to measure and their acceptance level), tolerability criteria of the business to risks.

b. *Identify risks* is the process of identifying risks and the sources of risk and also documenting their characteristics; It is identified from the design, execution and, in particular, the exploitation phase, of the structure, as well as from the experience gained from the exploitation of similar structures.

c. *Qualitative analysis*: in the process where the risks are evaluated by the probability and impact and are ranked through risk index:

Risk index = expected loss/gain impact from an event x probability of the event occurring

It is well known that the first 20% of causes can determine 80% effects (according to Pareto principle)

so, it is important to analyse and find actions to avoid or mitigate it.

d. *Quantitative analysis* is the process of numerically analysing the combinations of effects of the identified risks and other sources of uncertainties.
e. *Plan risk responses:* Proposal for minimisation and/or avoidance/transfer of the identified risks and their possible effects, assign the risk owners (the persons that are accountable for any risk).
f. *Implement risk responses* is the process that follows the agreed-upon planned actions, only in case of risk occur.
g. *Monitor risks* is the process that tracks the identified risks, monitoring the agreed-upon risk responses, identifying and analysing new risks and if that risk process is efficient.

3. Case study

In Moldavian region of Romania, there is an important number of drinkable water tanks that due to lengthy exploitation and lack of a time-monitoring programme, in conjunction with the lack of maintenance/repair work (generated by minimum funds) has resulted in a significant decrease in sustainability characteristics.

For an appropriate risk analysis, a water tank was proposed as a case study. The analysed construction is an overland water reservoir (figure 2) made of pre-stressed concrete, having 5000 m^3 capacity. The construction was executed in the '80 and is still in operation, ensuring the water needs for the municipality of the Nicolina district of Iasi (RO).

Figure 2. Water tank - Miroslava, Iaşi county.

The water tank has a cylindrical shape, with an inner radius of 13.85 m and a useful height of 8.05 m. Inside there is a central reinforced concrete pillar with dimensions of 70×70 cm, supported by a continuous foundation (C18/22.5), with an upper chapter.

The wall of the tank is made up of prefabricated elements (C 25/30) of 17 cm thick, with a post-tensioned rub on the outside. At the top of the wall was made a reinforced concrete belt with 28 × 20 cm dimensions.

At the top, the tank was closed with prefabricated T-shaped roof elements (C 25/30), radially laid-down, supporting one end on the top of the wall and the other on the central pillar.

During the operation of the structure were carried out maintenance and repair work on the inside of the circular wall of the tank to diminish the corrosion process of the reinforcement. On the outside, periodic waterproofing works were carried out on the roof due to water infiltrations. The building also supported the effects of earthquakes during the exploitation period [9].

The list of the identified risks:

a. *Physical security* – operational risk in drinking water production: These risks could occur due to several unpredictable or accidental actions (like fire, blast, earthquake, containment loss, landslides), but could be avoided by a good structure's protection (from designing and execution stages) and by positioning as an isolated structure (as is possible far from populated areas).
b. *Supplying water quality* – is given by the water treatment stations, were the procedures to improve

the water quality (exploited from natural sources) are applied in order to achieve the quality exigencies for drinkable water (standards requirements)

c. *Resulted drinking water quality* – Water must be checked under the microbial, chemical, radiological and acceptability aspects [3]. If the water does not qualitatively correspond to the requirements, it must be treated. The water treatment process is carried out using a complex set of structures and installations designed to improve qualitative parameters. In Romania, the most used method of water treatment is made by chlorination, for technical and economic reasons. Water chlorination (figure 3) [8] is carried out through various processes (additives with normed chlorine doses and over-chlorination with increasing chlorine dose), and this method applies to both centralized and local water supply systems. [5, 6]

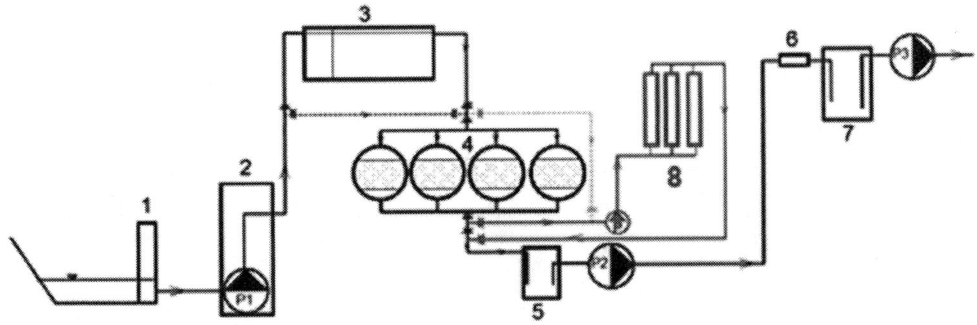

Figure 3. Chart of the water treatment procedure by chlorination
1. Pumping raw water; 2. Pre-chlorination; 3. Coagulation/flocculation;
4. Decantation; 5. Pumping process; 6. Filtering on multi-layer quartz layers;
7. Filtration on active charcoal layer; 8. Post-chlorination.

d. *Overflow/overfill* – is a minimal risk, due to the sensor network which monitors 24/24 the water volume in the water tank, and also due to the procedure that imposes that the upper liquid level never success a safety level (figure 4).

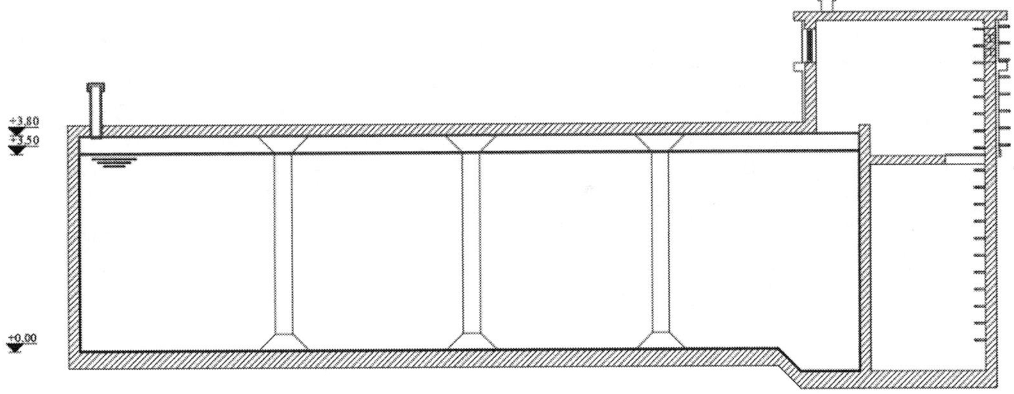

Figure 4. Water tank section.
Blue zone – water; Red zone – air (safety level)

e. *Piping / fitting failure* – is a risk which could occur, but is a minimum one due to the permanent monitors' activities (daily) and of the replacement services (stored in a nearby storehouse);

f. Internal / external corrosion

3.1 Internal corrosion action

It is caused by the attack of chlorine ions contained in water stored in the reservoir. The most common form of attack on the intimate structure of concrete is the action of chlorine in the treated water, the speed of the corrosion process being influenced by operating conditions, temperature, humidity, etc.

During the corrosion process produced by chlorine ions, the passive layer on the surface of the reinforcement is destroyed. Since the anodic area is reduced and the surface of the catheter is expanded over the entire surface of the reinforcement, there is a substantial reduction in the section of the reinforcement.

Due to the constant exploitation of the water tank, the operating environment promotes the triggering and evolution in various manners of the reinforcement corrosion in the structural elements (figure 5), thereby reducing the durability of the construction.

Figure 5. Internal corrosion effects (degraded reinforcements).

3.2 External corrosion action

At the wall's bottom, the accelerated corrosion process of the post-stressed reinforcement (figure 6.a.) was caused by the continuous infiltration of the water resulting from precipitation and the detachment of the torch due to excessive moisture at the base (figure 6.b.), as well as the succession of frost-thaw cycles.

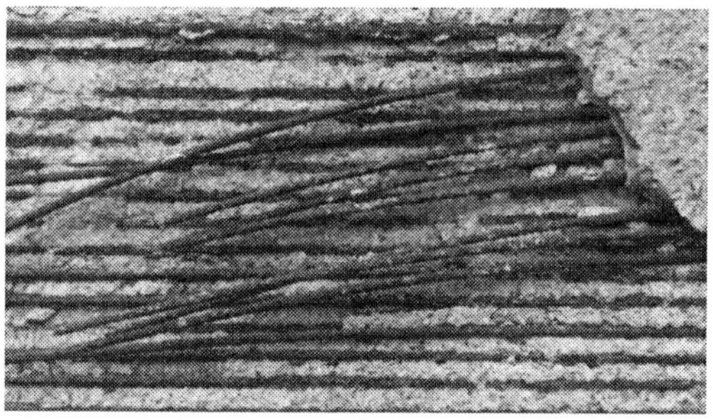
a)

b)

Figure 6. Degradations at the tank wall's bottom.

In the case of post-stressed reinforcement, there was no need to carry out further tests to determine mechanical characteristics, as it was found during the visual inspection that it was significantly affected by the process of necessary to replace it completely.

On the outside, at the top of the perimeter wall of the tank was observed the corrosion of the reinforcement and, in some cases, the rupture of the post-stressed reinforcement (figure 7 a.) and the detachment of the torch on large surfaces (figure 7 b.).

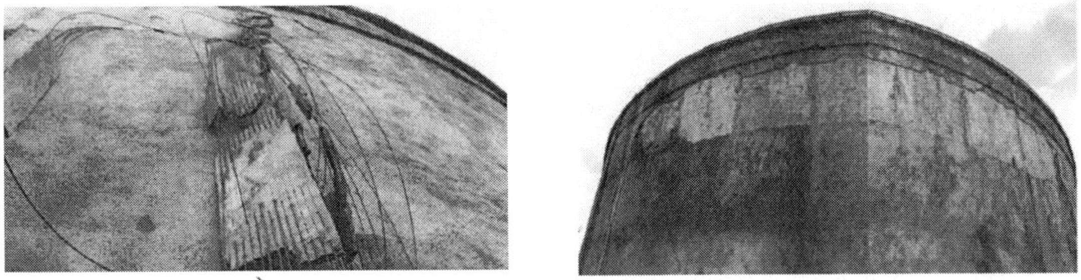

a) b)

Figure 7. Degradations at the tank wall's upper zones.

a. Structural failure

Due to the action of aggressive environmental agents (water infiltrations resulting from precipitation, freeze-thaw cycles, acid rains, strong winds, etc.), degradation processes have been amplified over the last 5 years.

On the outside of the roof elements, the interaction with aggressive environmental agents (solar radiation, water infiltration, vapor exchange), coupled with some execution errors (waterproofing works), led to the amplification of the process of degradation (figure 8).

a) b)

Figure 8. Structural damages: Basement level's excess moisture (a) and cracks (b).

b. Alarm failure

Whereas most water storage constructions have an advanced state of degradation as a result of particular operating conditions (chemical aggressiveness environments, high humidity, technological processes, etc.), it is often necessary to provide maintenance and repairs on construction objectives. The monitoring of functional parameters shall be intermittent and may be included in an automated system together with other constructions and installations.

The operation of these water storage structures cannot be achieved without strict and permanent supervision. To this end, it is necessary to investigate and verify the structural elements, as well as the precise determination of the functional parameters and processes of the hydrotechnical structure and installations, in respect to the compliance with the safety criterion in the exploitation of drinkable water tanks.

c. *Impact to the business and financial penalties* – in case of occurring of an uncertainty event in water tank functioning this can affect the community and can't assure the needed water capacity.
d. *Loss of equipment use / inventory* – failure of equipment used; represent a medium risk due to possibility to damage of the water tank's services, which may result in periods of stopping the water supply (in case of repair necessity).

e. *Natural resource damages* – It may be possible only in the event of a miss-operation during exploitation of a water source, in which case can be hypothetically infested at the source. An

IOP Conf. Series: Materials Science and Engineering **789** (2020) 012015 doi:10.1088/1757-899X/789/1/012015

impossible risk due to the automation and processes specific to the operation that does not allow the double-sense circuit (source⇨station and stations⇨source) but only one way (source⇨station).

f. *Criminal penalties* – in the event of a break-in, an attempt to destroy the facilities of a drinking water tank or, even, the attempted infestation of the stored liquid is severely penalised by the laws being considered by the jurisprudence of serious offenses. However, in practice, it is considered a minor risk because these objectives (drinking water tanks) are considered the objectives of strategic importance and are equipped with security and alarm devices, according to the laws of protection for the particular importance objectives.

g. *Loss of life* – possible risk, which may occur only in the case of non-conform operation of existing automation (very low risk) or possible workmanship accidents (falls from height, non-insurance, lack of protection for dangerous manipulation, access to prohibited places, etc.)

4. Risk assessment matrix

The qualitative risk assessment is made using the risk matrix, for the classification of risks.

One of the versions of the risk matrix is the one based on the probability of producing the risk and the impact of that risk in the event of its occurrence (in terms of the expenses necessary to solve the consequences).

The estimation of the probability and the impact of the risk in the case of the realization of the risk matrix is subjective and is usually performed based on an internal scale of the organization. The organization establishes what each level of probability or impact represents based on the information found in the database created from the previous works of the company.

In the proposed case study, the possible risks were evaluated in terms of probability and impact and it was realised the risk matrix, as follows:

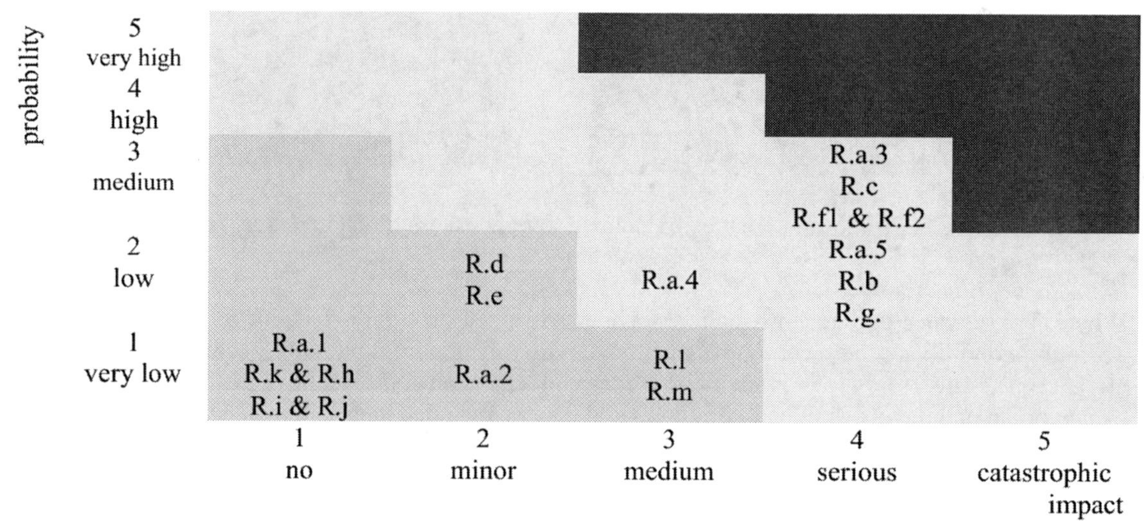

R.a.1 fire
R.a.2. blast
R.a.3. earthquake
R.a.4 containment loss
R.a.5 land slides
R.b. supplying water quality
R.c. resulted drinking water quality
R.d. overflow/overfill
R.e. piping / fitting failure
R.f.1 internal corrosion
R.f.2 external corrosion
R.g. structural failure
R.h - alarm failure
R.i. impact to the business and financial penalties
R.j. loss of equipment use/inventory
R.k. natural resource damages
R.l. criminal penalties
R.m. loss of life

In the studied case, the interpretation of the risk matrix it turns out that there are no risks in the red zone, which means that there is no high risk in our analysis, in the functioning of the water basin. In the area of medium risk, we have the risks of an earthquake, of a landslide, of corrosion, both internal and external, nonconforming quality of water, supplying or resulted from the tank, containment loss from the tank and structural failure of it. All risks need to be monitored throughout the operating period of the water tank, as the status of each risk can be changed at any time.

5. Final remarks

The Risk Assessment Matrix analysis can conclude that:

- here are no risks in the critical area (red zone);
- there are current risks (yellow zone) that need to be managed by planning the responses corresponding to the risks assessed and also, constantly monitored;
- there have been identified as lower index risks (green zone), which has to be monitored.

Creating a strategy to assure a good function of water pump & treatment stations, water transportation systems and water deposits (drinkable water tanks) it is of vital local and national importance for ensuring the health of the population.

The continued assurance of the needs of drinking water and the proper quality is vital for ensuring good living conditions. So that the risk analysis of the technological process of providing drinking water in the case of water tanks and the creation of a risk management plan and its permanent pursuit can lead to a more efficient process.

References

[1] Romanian Gov 1997 *HG 766/1997*
[2] ISO 2018 *ISO 31000:2018 – Principles and Guidelines on Implementation*
[3] WHO 2011 *Optimizing regulatory frameworks for safe and clean drinking-water*, www.who.int
[4] PMI 2009 *Practice Standard Project Risk Management*
[5] Romanian Gov 2002 *Law 458/2002 on the quality of drinking water* (with subsequent modifications and additions)
[6] WHO 2017 *Water quality — guidelines, standards and health: assessment of risk and risk management for water-related infectious disease* (Fewtrell, Bartram, 2001) 4th ed
[7] PMI 2017 *Project Management Book* PMBok 6th edition
[8] Giurconiu M, Mirel I, Carabeț A, Chivereanu D, Florescu C and Staniloiu C 2002 *Hydro utilities buildings and services* West Publishing House Timisoara
[9] Covatariu D and Filip A 2018 A Comprehensive Approaching of the Water Tank's Durability Affecting Factor in Different Stages *Bulletin of the Polytechnic Institute of Jassy* section Construction – Architecture **64** (68) 4 p 141-156

The thermal field analysis of the rotary kiln for the cement plants by means of the image processing techniques

A Danila[1*] and F L Tamas[2]

[1] Transilvania University of Brasov, Electrical Engineering and Computer Science, Brasov, Romania
[2] Transilvania University of Brasov, Faculty of Civil Engineering, Brasov, Romania

[*] adrian.danila@unitbv.ro

Abstract. The implementation of the image processing techniques for the analysis of the thermal field images captured with infrared scanners from the shield of the rotary kilns increase the quality of the information required for the optimization, monitoring and failure prevention of the kiln. In this work, authors presented a method that implement advanced algorithms such as the Radon transform, the segmentation and labeling of images for estimating the space-distributed, parametric model of the kiln's thermal field. The aim of the investigation is analyzing the similarities/differences between the thermal profile of the kiln and the kiln's maintenance events. The implementation of the proposed method proved that the correlations between the changes of the thermal profile and the renewal of the thermal coating of the kiln may be well emphasized. In this perspective, this method is suitable for building a thermal profile of the kiln during an entire maintenance time span.

1. Introduction

Since decades, the rotary kilns are used for cement and lime production. Recent estimations count that the cement production is set to increase between 12% and 23% by 2050. In addition to its economic importance, the cement manufacture rises environmental concerns due to the high energy consumption, the release of greenhouse gases, NO_x, SO_2, and particulates etc.

The enterprises in the cement industry spend a large amount of financial and technological resources for the protection of the environment aiming to reduce the greenhouse gas emissions and particulates according to the environmental regulations.

In this perspective, the improvement of the thermal efficiency of the kiln, the failure prevention of the thermal coat and the decrease of time required for the adjustments of the flame length at the startup of the process are specific measures that allow the reduction of specific costs and the improvement of the quality of the final product.

Considering the importance of the optimal rotary kilns operation, designers, manufacturers in the cement industry as well as researchers in the field continuously improve the methods and apparatus dedicated to the observation, prediction and control of the rotary kilns (figure 1).

Figure 1. The interior of the rotary kiln under investigation during maintenance.

The processes and phenomena that occur during the clinker manufacture are well documented; a referential book is [1]. In the technical literature, many articles are dedicated to the process control and monitoring of the rotary kiln in the cement industry.

The typical control strategies, meant to substitute the manual operator are [2]:

- the expert-based solutions;
- the model predictive control.

The expert-based methods, such as the fuzzy logic, proved useful for nonlinear processes (e.g. the burning process). In model predictive strategy, the plant dynamics are described by an explicit predictive model.

The model predictive control computes the process inputs and the control action to optimize the future plant behavior over the prediction horizon in accordance with a given objective function. The formulation is solved at each time-step using updated measurements from the process. Models such as black-box and gray-box may be used within the model predictive control formulation.

The method captures the local behavior of the plant as close to the reference as possible. In this work, the authors used a system identification approach and focused on the thermal processes within the rotary kiln. In a previous work [3] a time-domain flows' dependencies of the materials and particulate in a rotary kiln for solid waste reclamation were investigated. In this work a space-distributed model of the thermal field is taken into account, based on the method firstly implemented in [4].

The organization of this work is as follows. In Chapter 2, the implementation of the image processing method for the analysis of the kiln's thermal field from scanner images is presented. Two implementations of the proposed method, correlated with a maintenance event of a rotary kiln are presented in Chapter 3. In Chapter 4, the results and their interpretations are detailed.

2. Methods

2.1. The manufacture of the cement clinker

The rotary kilns are combustion chambers mainly used for cement manufacturing, solid and semisolid hazardous waste treatment [1].

The main part of the rotary kiln is an inclined cylinder which rotates slowly to accomplish mixing of the materials and to expose all surfaces and substances to heat and oxygen, figure 2.

The flame residence time is of the order of 2 to 5 seconds and the flame temperature is of over 2000°K. The material residence time within the kiln is typically greater than an hour. This feature allows maintaining the uniformity of the quality of the final product.

However, the quality of the product may be affected by various factors such as the variance of the inlet materials properties, the fuel, the status of the thermal insulation of the shield, as well as other physical and chemical factors.

In this respect, various researches were made in the area of the control and observation of the complex physical and chemical processes that occur within the rotary kilns during their operation [1], [5, 6].

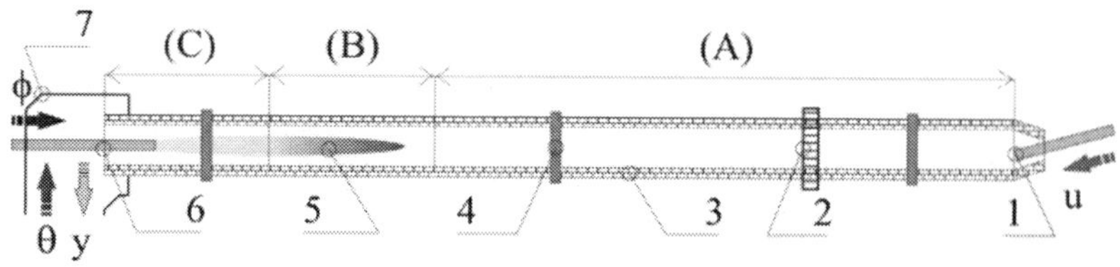

Figure 2. The schematic diagram of the cement rotary kiln thermal zones: (A) – Decomposition; (B) – Transition; (C) – Sintering. 1 – Feed In; 2 – Drive gear; 3 – Firebrick lining and steel shell; 4 – Tyre; 5 – Flame; 6 – Burner pipe; 7 – Kiln hood. u – Feed in; y – Clinker to cooler; Φ – Fuel; θ – Hot air from cooler.

2.2. Knowledge models vs. parametric models for the cement rotary kiln

The clinker manufacturing consists in the transformation of a mixture of raw materials (limestone – $CaCO_3$ and clay) into alite (Ca_3SiO_5). The chemical reactions required for the process evolves on stages onto three zones of the rotary kiln. In table 1 the space-distribution and the corresponding temperatures related to the chemical reactions occurring to the clinker manufacturing are presented.

Table 1. The clinker manufacturing in the rotary kiln. The correlations between the chemical reactions and their corresponding temperatures[a].

No.	Kiln zone	Chemical reactions	Temperature (°C)
1.	Decomposition		
		$CaCO_3 = CaO + CO_2$	600-900
		$CaO + Al_2O_3 = CaO \cdot Al_2O_3$	800
		$CaO + Fe_2O_3 = CaO \cdot Fe_2O_3$	800
		$CaO + CaO \cdot Fe_2O_3 = 2CaO \cdot Fe_2O_3$	800
		$3(CaO \cdot Al_2O_3) + 2CaO = 5CaO \cdot 3Al_2O_3$	900-950
2.	Transition		
		$2CaO + SiO_2 = 2CaO \cdot SiO_2$	1000
		$3(2CaO \cdot Fe_2O_3) + 5CaO \cdot 3Al_2O_3 + CaO = = 3(4CaO \cdot Al_2O_3 \cdot Fe_2O_3)$	1200-1300
		$5CaO \cdot 3Al_2O_3 + 4CaO = 3(3CaO \cdot Al_2O_3)$	1200-1300
3.	Sintering		
		$2CaO \cdot SiO_2 + CaO = 3CaO \cdot SiO_2$	1350-1450

[a] Ref. [1].

The science-based models of the clinker manufacturing process in the rotary kiln refer to the interaction between the ensemble of transport processes and chemical phenomena that occur into the kiln during operation. The transport processes refer to the bed phenomenon, the transverse and the axial motion of the bed. The physical phenomena refer to the free board aerodynamic phenomena, the granular flows in the kiln, the particle mixing and segregation, and the heat transfer processes. The chemical phenomena refer to the chemistry of the cement process and to the combustion chemistry of the fuel used. Documented knowledge-based models for these complex processes are presented in [1]. Although, the knowledge-based models are effective to the experimental data, the complex interactions that occur during the real-life manufacturing of the clinker into the kiln are difficult to predict by means of these models. The system identification approach and its related parametric time-domain models are more suitable for on-line observations, predictions and control, [3] due to the following reasons: (a) the time-domain models are related to the observed data and to predefined abstract classes of models that are not related to the nature of the processes depicted, (b) the models are adaptive in the sense that the values of the coefficients are estimated each time a new data set is acquired from the process and (c) by adopting a time- and space-distributed model for the process the modeling of the overall interactions between the processes and transport phenomena may be observed.

2.3. The implementation of the image processing techniques for estimating the parametric models of the rotary kiln

In the flowchart depicted in figure 3, the images of the kiln's thermal spectra acquired from the scanner – (3), are decomposed on levels of temperature – (4). The result of the decomposition is a set of binary images. Each binary image is treated in the loop (5) as follows: in the module (6) the regions at the given level of temperature are detected; each region within the image is labelled recorded – module (7).

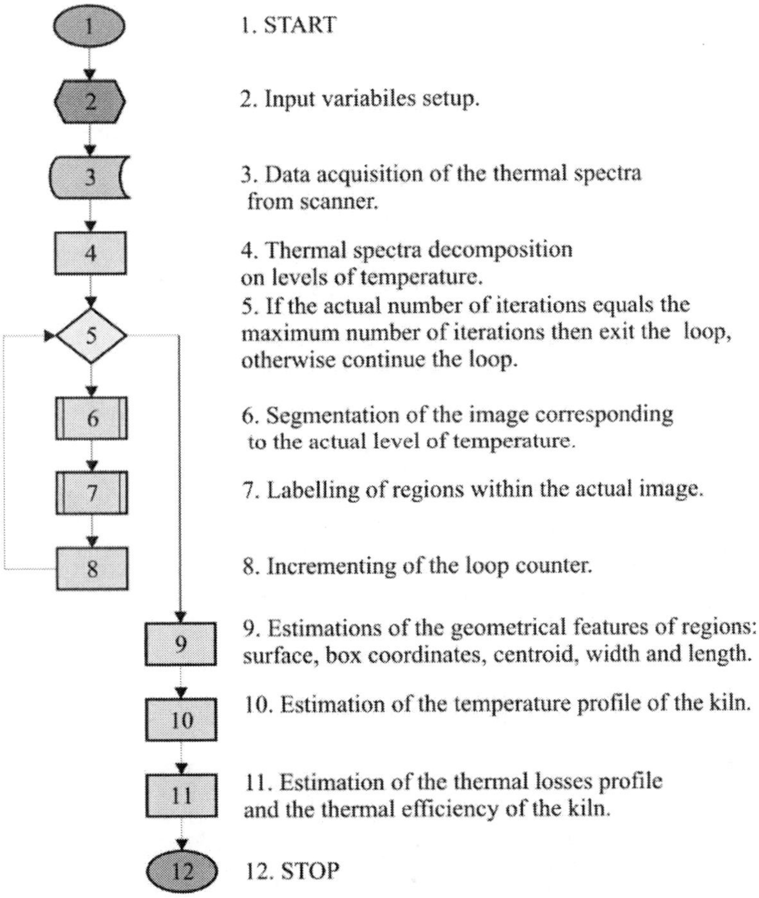

Figure 3. The flowchart of the proposed method and its software implementation.

After all levels of temperature are processed, the geometrical features of regions i.e. the surface coordinates, the width and the length of the outer box frames and the coordinates of the regions' centroids are estimated – module (9). These results are used in module (10) for the computations of the actual values of the 5 – order parametric model that depicts the thermal profile of the kiln – module (10). The resulting thermal model is used in module (11) to produce the estimate of the thermal losses and the thermal efficiency of the kiln. The thermal losses were estimated with the assumption that the heat transfer from the shield to the environment (air) is made by radiation.

3. The experiment

3.1. The experiment components, apparatus and software

The experiment was carried onto several sets of experimental scans acquired from the process at the "CRH Ciment Romania" – the Hoghiz Cement Factory. The technical features of the rotary kiln at the Hoghiz Cement Factory are given in table 2.

Table 2. The characteristics of the rotary kiln at "CRH Ciment Romania" – The Hoghiz Cement Factory.

Denomination	Units	Values
Kiln length	m	97.6
Kiln diameter	m	6.04
Kiln angular speed	rpm	2.1
Kiln fuel (a mix of coke, coal and powder waste)	t	1 to 12
Kiln production capability	t	3600

The shell of the kiln was monitored with the thermal scanner of type Infrared Kilnscan. The main features of the Infrared Kilscan scanner are: continuous full temperature map, with a true one-brick resolution, hot spots early warning before damage, monitoring and extension of refractory lifetime, reliable preventive maintenance scheduling, scanner's data direct linkage to the Plant Control System.

The inside kiln temperature was also monitored via a Pyroscan - Ruggedized HDR Pyrometric Camera. The data from the scanner was transmitted to the control room through an optic fiber link to avoid electromagnetic interference. The monitoring system was also provided with hot spot detection but produced poor predictions of the thermal process's evolution with time. The software applications were written in the Python environment [7].

3.2. The organization of the experiment

The organization of the experiment was as follows.

Observations of the thermal spectra at the surface of the kiln at two different moments of time were made. The first observation – Test 1 – was made before the deadline of the maintenance cycle of the kiln. The second observation – Test 2 – was made after the thermal insulation of the kiln had been renewed. In figure 4 an example of the thermal spectra image acquired from the thermal camera KilnScan captured within the tests is presented.

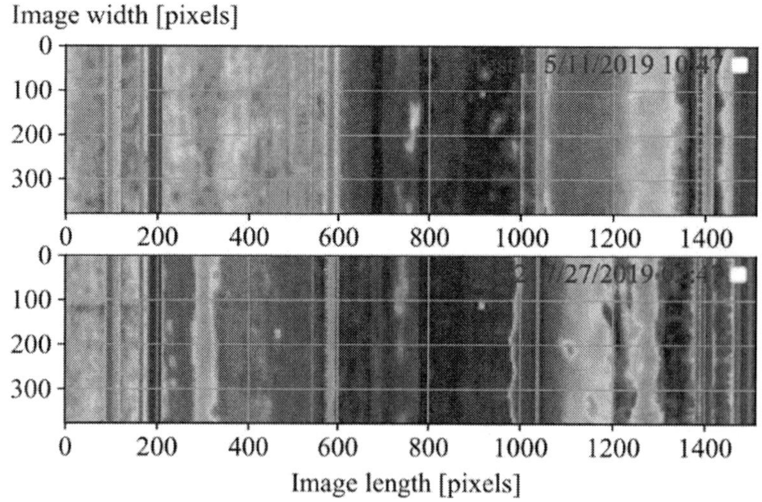

Figure 4. The image of the thermal field of the rotary kiln acquired with the KilnScan scanner.

The images acquired from the scanner with threshold on levels of 10 Celsius degrees as presented in figure 5.

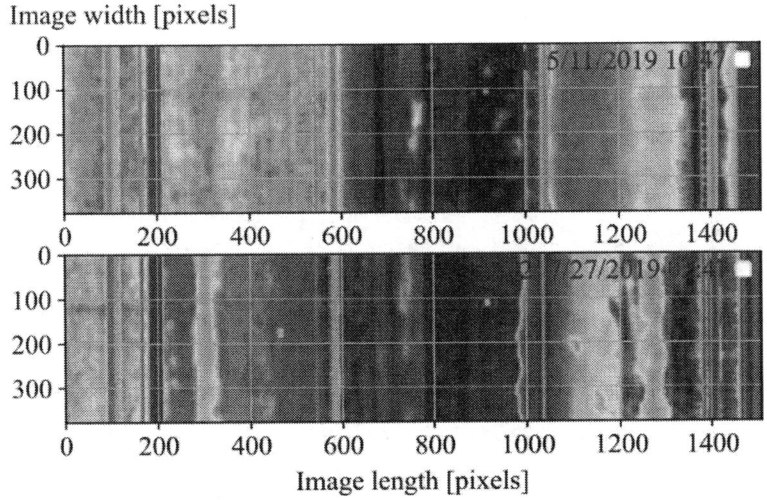

Figure 5. The image of the thermal field in figures 4 after a 10°C threshold was implemented.

For each level of temperature, the binary image was extracted and processed according with the procedure depicted in Chapter 2. The binary image at 180 Celsius degrees is presented in figure 6.

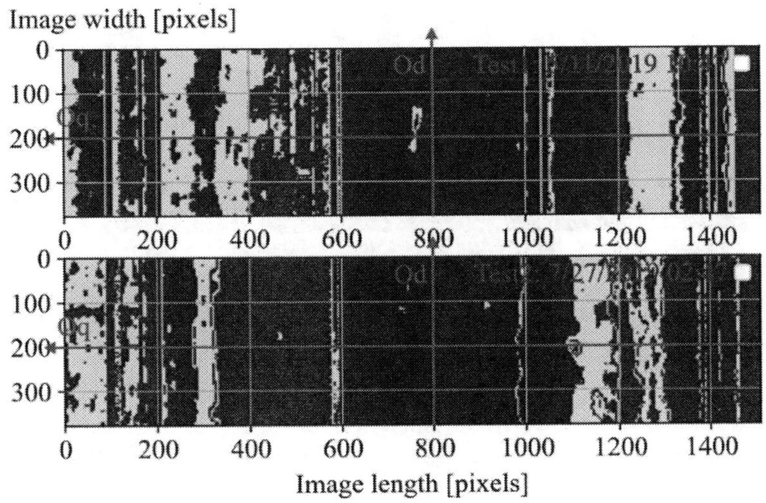

Figure 6. The 180°C threshold binary image of the thermal field of the rotary kiln.

4. Results and discussions

In this section, the interpretation of the image processing analysis is presented as follows:

- the analysis of the estimations of the Radon transform on the (Od) direction of images;
- the 5th – order polynomial approximation of the surface versus temperature dependency;
- the power-losses versus temperature estimate.

4.1. The estimations of the Radon transform on the (Od) direction of images

An example of the estimates of the Radon transforms from the acquired images is depicted in figure 7. The comparison of the estimates at test 1 and test 2, provides the followings:

- the levels of projections are of the form of random variables;
- the levels of the projections corresponding to the input side of the kiln (the left side of each image) are statistically higher than the levels corresponding to the output side of the kiln (the right side of each image);

- the levels of the projections corresponding to the middle area are in the middle range of values;
- the similarity between the projections levels at test 1 and test 2 may not be made at the samples level thus one must use a statistical approach.

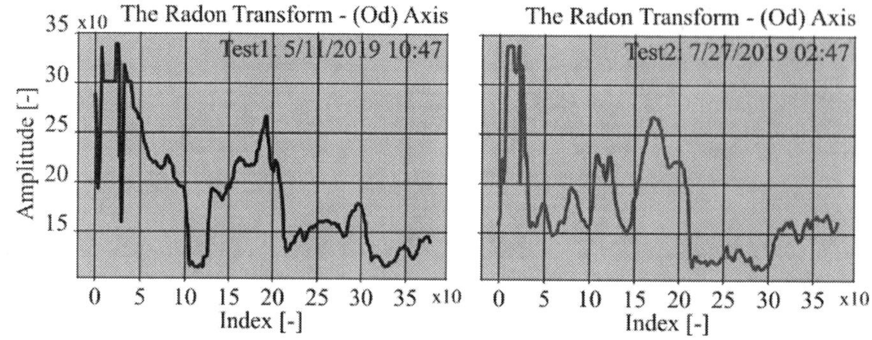

Figure 7. The Radon transforms onto the (Od) axes of images presented in figures 6. Test 1 – to the left and Test 2 – to the right.

To emphasize this idea, the histogram (or the empirical probability function) of the Radon transforms values for the Test 1 and the Test 2 are depicted in figure 8.

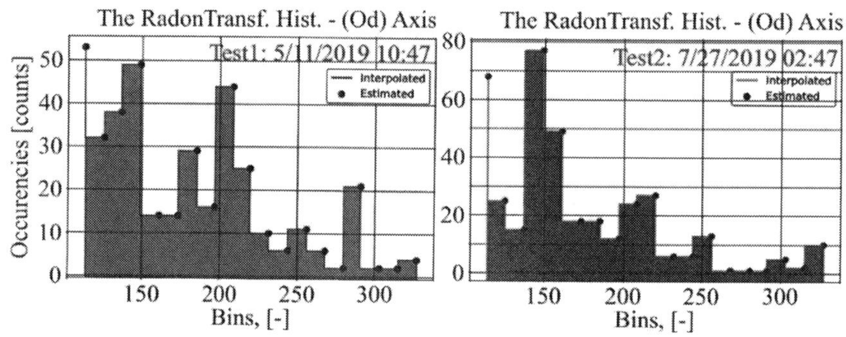

Figure 8. The Histograms of the Radon transforms from figure 7: Test 1 – to the left and Test 2 – to the right.

The histogram from Test 1 gives a sample mean of 19.89 and a variance of 16.62 and the histogram from Test 2 gives a sample mean of 19.89 too and a variance of 22.05. This result proves the intensities of projections are statistically higher in Test 2 than in Test 1.

4.2. The approximation of the surface versus the temperature dependencies

The surface versus temperature dependencies are depicted in figure 9.

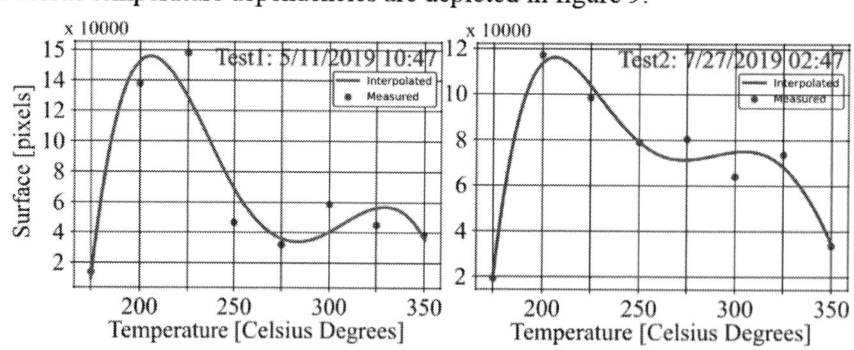

Figure 9. The dependencies of the surfaces versus the temperature: Test 1 – to the left and Test 2 – to the right.

IOP Conf. Series: Materials Science and Engineering **789** (2020) 012016 doi:10.1088/1757-899X/789/1/012016

The 5th-order polynomial approximation of the estimates features the same profile. The given polynomial and its coefficients may be interpreted as a parametric model of the thermal process in the kiln and thus may be used for predictions.

For both tests there are three local extremes: one local maximum around 200°C, one local minimum around 275°C and another local maximum around 350°C.

At Test 1, the amount of surface in the low range of temperature is higher and the amount of surface in the high range of temperature is lower than the corresponding ranges in the Test 2.

One may assess that the kiln was less heated during Test 1 than Test 2. However, the given profile proves that the improvement of the thermal insulation made before Test 2 has reduced the values of the temperature at the surface at the kiln.

A numerical comparison between the given tests regarding the local max/min values of temperature and surface is given in table 3.

Table 3. The dependencies of surfaces vs. temperature – the local max/min values[a].

No.	Kiln's zone	Decomposition		Transition		Sintering	
	Test No.	Temp. (°C)	Surface[b] (pixels)	Temp. (°C)	Surface[b] (pixels)	Temp. (°C)	Surface[b] (pixels)
1.	Test 1	211.75	152513	283.5	33939.7	329.0	56887.1
2.	Test 2	206.50	116312	273.0	71290.2	304.5	75099.4
3.	Test 1 vs Test 2	2.48%	23.74%	3.85%	-110.05%	7.45%	-32.01%

[a] The local max/min values are taken from figure 9;

[b] The pixels count is related to the dimensions of the image in figure 5.

One may observe there are similar max/min values of temperature at all regions of the kiln for both tests. However, the difference of surfaces of the transition region at Test1 and its corresponding at Test 2 is (-110.05)% of the surface at the transition region at Test 1.

The dependencies of the power losses at the surface of the kiln with respect to the temperature are depicted in figure 10.

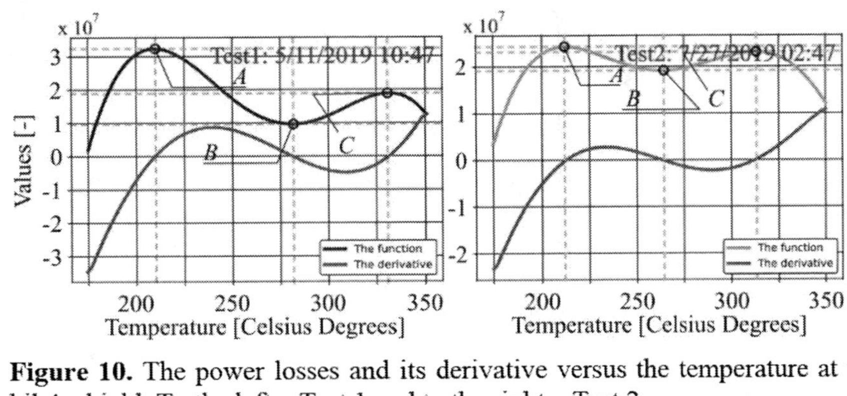

Figure 10. The power losses and its derivative versus the temperature at the kiln's shield. To the left – Test 1 and to the right – Test 2.

The estimations prove that the power losses at the sintering region decreased with (-23.45)% from Test 1 to Test 2 and the power losses at the transition region increased significantly with 99.59%. In the decomposition zone, the power losses also increased with 24.82%. The corresponding values of temperature didn't change significantly.

The integral of the power losses density with respect to the temperature estimates the total power losses of the kiln. The comparison proves that the efficiency of the kiln at Test 1 is lower than its counterpart at Test 2.

The local max/min values are presented in table 4.

Table 4. The dependency of the power-losses vs. temperature – the local max/min values[a].

No.	Kiln's zone	Decomposition		Transition		Sintering	
	Test No.	Temp. (°C)	Power (kW)	Temp. (°C)	Power (kW)	Temp. (°C)	Power (kW)
1.	Test 1	210.00	32373.60	281.75	9579.15	330.75	18790.10
2.	Test 2	211.75	24337.90	262.50	19118.90	313.25	23196.40
3.	Test 1 vs. Test 2	-0.83%	24.82%	6.83%	-99.59%	5.29%	-23.45%

[a] The local max/min values are from figure 10.

5. Conclusions

The cement manufacturing processes into the rotary kilns consists of a complex interaction between chemical, thermodynamic and mechanical processes. From the analysis point of view, the knowledge-based model of each component of this interaction is well documented in the literature. However, for online observations, monitoring and control, the above approach is less efficient. In this respect, the authors proposed implementing a parametric, polynomial model of the thermal power-losses surface distribution of the kiln. The values of the coefficients of the polynomial are determined with the least-squares algorithm in order to minimize the prediction errors.

Within this work, the proposed method was implemented onto observations of the thermal spectra of the rotary kiln at the "CRH Ciment Romania" – The Hoghiz Cement Factory. Two significant experiments – correlated with the maintenance cycle of the kiln – were taken into account. The comparison between the estimated power-losses of the kiln at the aforementioned tests, emphasized the possibility of detection of changes during the lifetime of the coating of the kiln. Further research intends to evaluate the power-losses profile of the kiln over an entire maintenance time-span that could allow evaluating the actual status of the thermal coating from measurements.

Acknowledgements
Authors wishing to acknowledge "CRH Ciment Romania" – The Hoghiz Cement Factory for its assistance, valuable expertise and the experimental data provided for this work.

References

[1] Boateng A A 2008 *Rotary Kilns Transport Phenomena and Transport Processes* (Amsterdam: Elsevier)
[2] Sahasrabudhe R, Sistu P, Sardar G and Gopinath R 2006 *IEEE Control Systems Magazine* **222** 56
[3] Danila A 2011 *Environmental Engineering and Management Journal Sustainable energy* **8** pp 1161-1168
[4] Danila A 2018 *Int. Conf. And Exposition on Electrical and Power Engineering* (Iasi: Technical University Gh. Asachi)
[5] Bányász Cs, Keviczky L and Vajk I 2003 *IEEE Control Systems Magazine* **222** 87
[6] Clarke D W 1988 *IEEE Control Systems Magazine* **222** 49
[7] Van der Walt S, Schönberger J L, Nunez-Iglesias J, Boulogne F, Warner J D, Yager N, Gouillart E, Yu T and the scikit-image contributors 2014 *scikit-image: Image processing in Python* PeerJ 2:e453

IOP Conf. Series: Materials Science and Engineering **789** (2020) 012017 doi:10.1088/1757-899X/789/1/012017

Numerical methods to find the slope at a point on map using level curves. Application in road designing

O Deaconu[1*] and A Deaconu[2]

[1] Department of Civil Engineering, Faculty of Civil Engineering, Transilvania University of Brasov

[2] Department of Computer Science, Transilvania University of Brasov

[*] ovideaconu@unitbv.ro

Abstract. The problem of determining the slope at a point on a map using level curves has practical applicability in designing the roads. This problem is reduced to the problem of finding the shortest length segment through a point that connects two coplanar lines. Two methods to solve this problem are presented: a fast approximate method and a slower but exact algorithmic method. The results are compared at the end.

1. Introduction

When designing a new road some points are fixed on the map (where the new road will pass). It is very important for designer to know as exact as possible the slope at each such point. Instead of sending workers on the field to measure the slope at different points these values can be calculated using the methods from this paper fast and with very good accuracy. The slope at a point can be determined using the closest two level curves to that point (figure 1). The slope is given by the shortest length segment that connects the two level curves. Since a level curve consists in consecutive coplanar segments of straight lines, our problem can be reduced to finding the shortest length segment that connects two coplanar line segments located on two consecutive level curves.

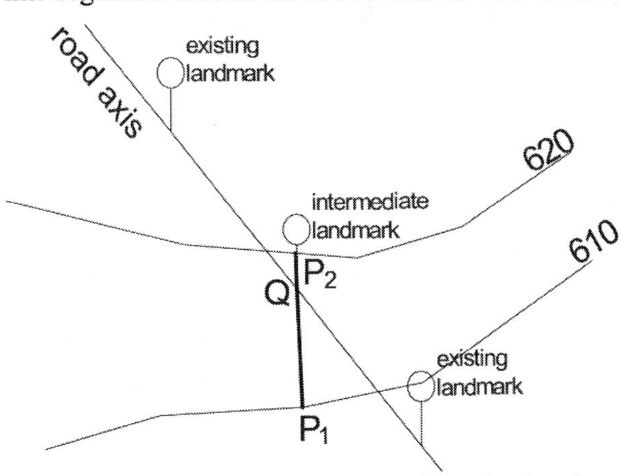

Figure 1. Finding the slope at a point using level curves.

This problem has also applications in railways, roads and bridges when circular curves are needed to draw as can be seen in figure 2 [1].

IOP Conf. Series: Materials Science and Engineering **789** (2020) 012017 doi:10.1088/1757-899X/789/1/012017

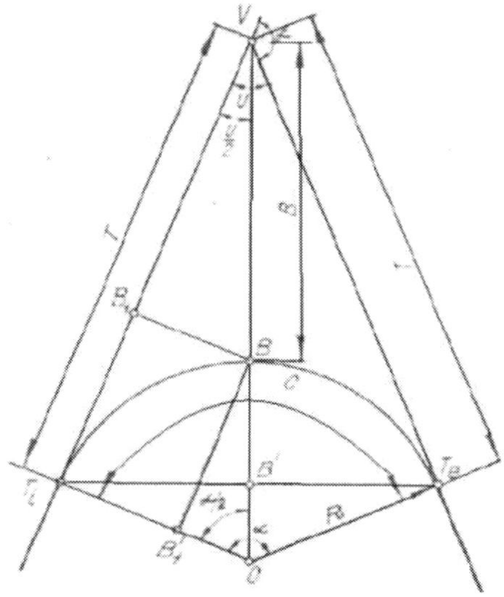

Figure 2. Drawing a circular curve.

We will show next how to calculate the shortest segment that passes through a given point Q and connects two coplanar lines d_1 and d_2 (figure 3). We shall denote this problem as SSTPC2L (Shortest Segment Through a Point that Connects 2 Lines problem). In figure 3 the points A_{11} and A_{12} determine the line d_1 and the points A_{21} and A_{22} determine the line d_2.

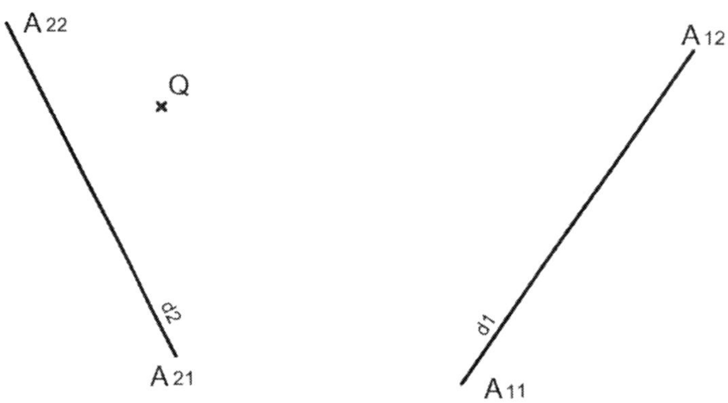

Figure 3. Connecting the point Q to the lines d_1 and d_2.

We present two methods of solving SSTPC2L: an approximate method presented in section 2 and an exact mathematical method presented in section 3. In section 4, the experimental one, six different positions of the Q points are presented for which the length of the connecting segments was determined by both methods, thus observing the variation of the error of the first method.

2. Approximate method

The method consists in drawing from the point Q the perpendiculars on the two given lines d_1 and, respectively, d_2. We construct a line passing through Q and it is parallel to the line determined by the feet of the two perpendiculars. The obtained line is denoted by d_3, and its intersection with the lines d_1 and d_2 is the searched segment $[P_1P_2]$ (approximate solution of SSTPC2L), like in figure 4.

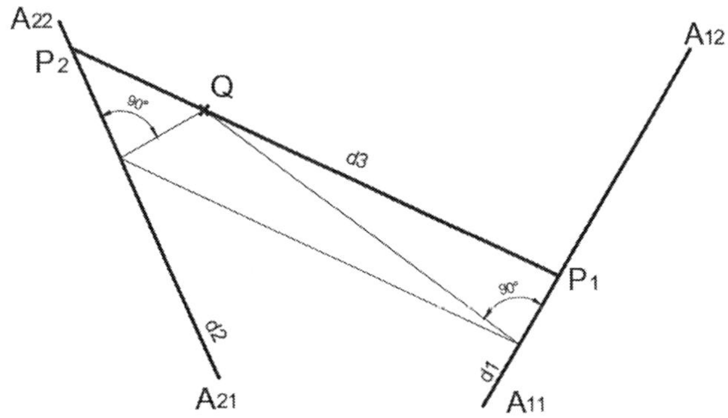

Figure 4. Approximate method of drawing the line d_3.

3. The mathematical method

In order to have simpler calculus we apply some initial transformations (rotations and translations) to the points A_{11}, A_{12}, A_{21}, A_{22} and Q so that the bisector of the angle between the lines d_1 and d_2 becomes parallel to Oy axis and the point Q is in the origin O, i.e., $Q = O(0,0)$.

The following 2 systems of equations have to be solved in order to find the points $P_i(x_i,y_i)$ ($i = 1,2$) where the lines d_i ($i = 1,2$) intersect with the line d:

$$\begin{cases} y = a_i x + b_i \\ y = ax \end{cases} \tag{1}$$

Since A_{i1}, A_{i2} define the line d_i ($i = 1,2$) in (1) we have:

$$a_i = \frac{(y_{i2} - y_{i1})}{(x_{i2} - x_{i1})} \tag{2}$$

$$b_i = y_{i1} - a_{i1} x_{i1}. \tag{3}$$

Solving the 2 systems of equations, the points $P_i(x_i,y_i)$ ($i = 1,2$) are found, where:

$$x_i = \frac{b_i}{a - a_i} \tag{4}$$

$$y_i = \frac{ab_i}{a - a_i} \tag{5}$$

The distance between P_1 and P_2 must be minimized, i.e.:

$$\min_a f(a) \tag{6}$$

where:

$$f(a) = (y_2 - y_1)^2 + (x_2 - x_1)^2 \tag{7}$$

By replacing (4) and (5) in (7) the following formula for $f(a)$ is obtained:

$$f(a) = \left(\frac{a_2 b_2}{a - a_2} - \frac{a_1 b_1}{a - a_1} + b_2 - b_1\right)^2 + \left(\frac{b_2}{a - a_2} - \frac{b_1}{a - a_1}\right)^2 \tag{8}$$

In order to find the value a_{min} where $f(a)$ is minimum, the equation:

$$f'(a) = 0 \tag{9}$$

must be solved, where:

$$f'(a) = 2\left(\frac{a_2b_2}{a-a_2} - \frac{a_1b_1}{a-a_1} + b_2 - b_1\right)\left(\frac{-a_2b_2}{(a-a_2)^2} + \frac{a_1b_1}{(a-a_1)^2}\right) + 2\left(\frac{b_2}{a-a_2} - \frac{b_1}{a-a_1}\right)\left(\frac{-b_2}{(a-a_2)^2} + \frac{b_1}{(a-a_1)^2}\right) \quad (10)$$

The value a is the slope of the line d_3 and, consequently, a_{min} must be searched in the interval given by the slopes of the perpendiculars QQ_1 and QQ_2 from Q on the lines d_1 and, respectively, d_2 (figure 5), i.e.:

$$a \in (a_L, a_R) \quad (11)$$

where:

$$a_L = \min\left\{\frac{-1}{a_1}, \frac{-1}{a_2}\right\} \quad (12)$$

$$a_R = \max\left\{\frac{-1}{a_1}, \frac{-1}{a_2}\right\} \quad (13)$$

Remark 1: Since the bisector of the angle between d_1 and d_2 is parallel to Oy axis, it can be seen that:

$$a_L = -a_R \quad (14)$$

In order to obtain the solution of a_{min} of the equation (9) a numerical method [2] can be used such as bisection method (interval halving method) or tangent method. We preferred bisection method [3] and we performed it on the interval defined in (11).
After a_{min} is calculated, the points P_i (i = 1,2) are obtained using (4) and (5) and the distance between these two points is:

$$\sqrt{f(a_{\min})} \quad (15)$$

In order to go back to the initial system of coordinates, the inverse initial transformations must be applied in reverse order to the points A_{11}, A_{12}, A_{21}, A_{22}, Q, P_1 and P_2. So, the following algorithm is obtained to solve the problem SSTPC2L:
Algorithm for SSTPC2L

- Transform the points A_{11}, A_{12}, A_{21}, A_{22} and Q so that the bisector of the angle between the lines d_1 and d_2 is parallel to Oy axis and the point Q is in the origin O;
- Find the solution a_{min} of the equation $f'(a) = 0$ (10) using bisection method on the interval (a_L, a_R) (11);
- Calculate the coordinates of the points P_i (i = 1,2) using (4) and (5), where $a = a_{min}$;
- Apply the inverse transformations in reverse order to the points A_{11}, A_{12}, A_{21}, A_{22}, Q, P_1 and P_2;
- Calculate the distance between P_1 and P_2 using (15).

We implemented the algorithm above in Visual C++ 2017 and in the bisection method we set the error err = 0.0001 for finding the value a_{min}. In figure 5 a graphically output of our program is presented.

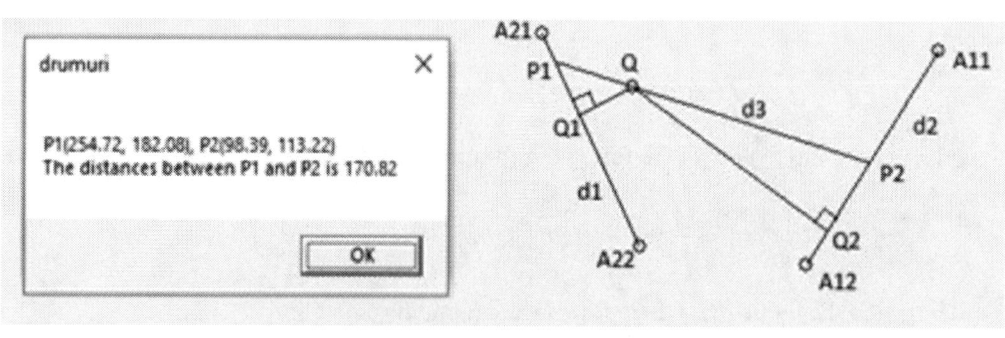

Figure 5. Output of the Visual C++ program.

4. Experiments

In order to compare the results of the two methods, two random lines d_1 and d_2 were chosen and several Q points were considered: near the bisector of the angle formed by d_1 and d_2, near the two straight lines d_1 and d_2 and other intermediate points.

In experiments the two lines d_1 and d_2 are defined by the points $A_{11}(200, 100)$ and $A_{12}(280, 220)$, and, respectively, $A_{21}(100, 110)$ and $A_{22}(40, 230)$. Six different Q points were chosen with the coordinates: $Q_1(70, 175)$, $Q_2(95, 200)$, $Q_3(156, 185)$, $Q_4(170, 210)$, $Q_5(210, 205)$ and $Q_6(250, 180)$ (figure 6).

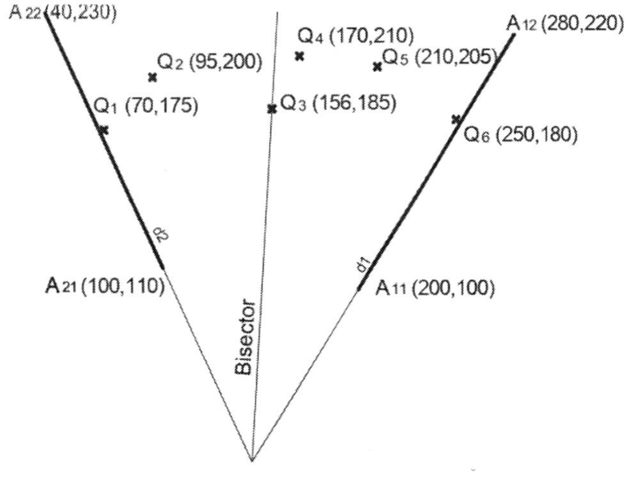

Figure 6. Different positions for Q point.

In figure 6 the 6 positions of the point Q with the obtained results are presented. P_1 and P_2 are the intersection points calculated with the first method. P'_1 and P'_2 are the intersection points calculated with the second method. With L was denoted the length of the segment obtained by the approximate method and with l was denoted the length of the segment obtained by the exact method.

As one can see in table 1 the first approximate method provides good enough solutions. Moreover, as expected, if the point Q is located near the bisector given by the lines d_1 and d_2 or near the segments d_1 and d_2 the solutions given by the two methods are very alike.

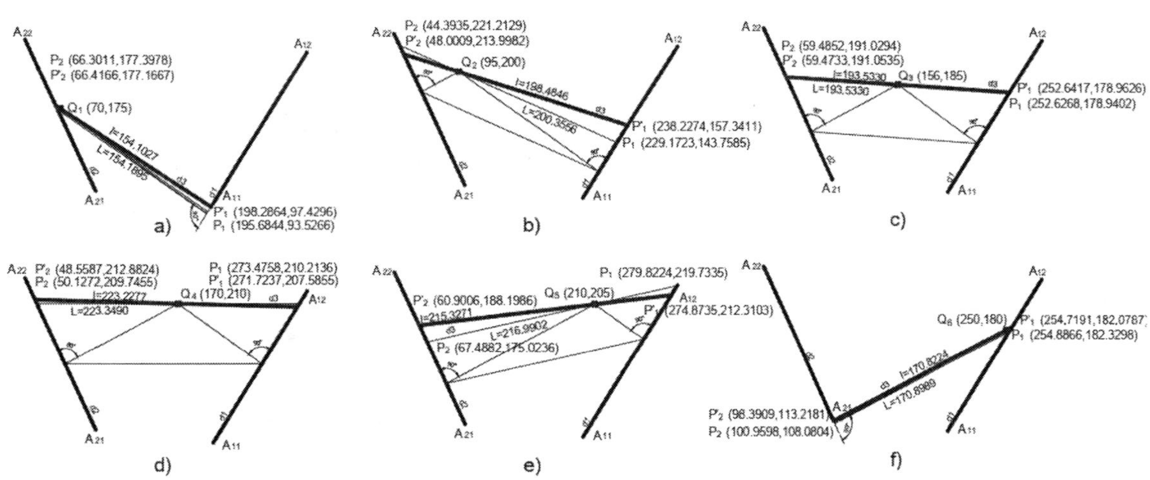

Figure 7. Various positions of point Q and obtained results.

Table 1: Comparison between the two methods.

Coordinate		P1	P2	P'1	P'2	L	l	L-l	Error (%)
Q1	x=	195.6844	66.3011	198.2864	66.4166	154.1895	154.1027	0.0868	0.0563
	y=	93.5266	177.3978	97.4296	177.1667				
Q2	x=	229.1723	44.3935	238.2274	48.0008	200.3556	198.4846	1.8710	0.9426
	y=	143.7585	221.2129	157.3411	213.9982				
Q3	x=	252.6268	59.4733	252.6417	59.4852	193.5330	193.5330	0.0000	0.0000
	y=	178.9402	191.0535	178.9626	191.0294				
Q4	x=	273.4758	50.1272	271.7237	48.5587	223.3490	223.2277	0.1213	0.0543
	y=	210.2136	209.7455	207.5855	212.8824				
Q5	x=	279.8224	67.4882	274.8735	60.9006	216.9902	215.3271	1.6631	0.7724
	y=	219.7335	175.0236	212.3102	188.1986				
Q6	x=	254.8866	100.9598	254.7191	98.3909	170.8989	170.8224	0.0765	0.0448
	y=	182.3298	108.0804	182.0787	113.2181				

5. Conclusions

In conclusion when designing a new road it is very important for designer to know as exact as possible the slope at some points of the future road. Instead of sending workers on the field to measure the slope at different points these values can be calculated with very good accuracy using the methods from this paper. We presented two methods to solve SSTPC2L, a fast and approximate one and an exact and slower one. The error of the first method is low (table 1). It has the advantage of being very fast and accurate enough. So, if we have many points for which SSTPC2L is applied the first method can be preferred. Although, if we look for an exact solution, the second method has to be used.

References

[1] Diaconu E, Dicu M and Racanel C 2006 *Căi de comunicații rutiere - principii de proiectare* (Roadways - design principles) Conspres Bucuresti
[2] Quarteroni A, Sacco R and Saleri F 2007 *Numerical Mathematics*, Springer
[3] https://www.mathwarehouse.com/calculus/continuity/continuity-bisection-method.php

Improvement of the concrete behaviour to sulphate corrosion using fly ash admixture collected by wet process

L I Diaconu[1*], M Rujanu[1], A C Diaconu[2], A A Şerbănoiu[1], D T Babor[1] and D Plian[1]

[1] Technical University "Gheorghe Asachi" from Iasi, Faculty of Civil Engineering and Building Services, Department of Concrete, Materials, Technology and Management, Romania
[2] Romanian Association of Earthquake Engineering

[*] liviagroll@yahoo.com

Abstract. For the concretes subjected to sulphate corrosion the use of the H II/A-S cement type is recommended. To improve the behaviour of these concretes to aggressive sulphate and magnesia corrosion, the use of the fly ash admixture collected by wet process for replacing the fine part of the aggregate is proposed. Because this type of corrosion is manifesting by expansion the analysis was conducted by measuring the dimensional modifications. In the Laboratory of the Faculty of Civil Engineering and Building Services of Iasi, an experimental program was conducted in order to assess the dimensional modification of the some samples prepared by concrete mixes based on belitic cements of type H II/A-S and embedding fly ash subjected to a combined sulphate and magnesia corrosion process using a mixed solution of Na_2SO_4 and $MgSO_4$ having certain concentrations. The prismatic concrete samples were preserved in distilled water and in the combined solution, and then after 28 and 90 days their dimensions were measured by the Graff device. In this paper is presented the behaviour of the concretes in the first life time and, as further research other samples will be tested for longer time.

1. Introduction

Concrete durability continues to be a subject of challenges for design professionals, and specialists of concrete. The durability of concrete structures is affected by many environmental factors, the sulphate corrosion being one of the most frequent and detrimental processes [1].

The sulphate corrosion is characterised by the expansion of the cement stone due to the formation of crystallized products that are containing a great number of crystallization water molecules, which normally will lead to volume increase.

This type of corrosion is also produced by the sodium sulphate, which is a soluble sulphate, often encountered in practice. This is reacting with the calcium hydroxide from the cement stone forming another calcium sulphate. The calcium sulphate is combining with the calcium hydro aluminate giving a complex salt crystallized with 31 water molecules, called ettringite. This tendency of crystals formation leads that in a first stage to be observed an increase of the cement stone's strength, an improve of the structural characteristics, so finally internal friction to appear, that in the end are leading to the decay of the concrete's structure [2].

So, the sulphate corrosion supposes the formation of more hard soluble compounds, having a great molecular volume which will deposit in the concrete's pores, capillaries and microcracks and can lead, in certain conditions to internal frictions appearance and concrete's decay [3].

The formation of the compact and impermeable structure of the concrete represents a main condition, and supposes that in the conditions of the use of some cement type having high characteristic strengths to sulphate aggressiveness, the increase of the compactness to be realize not only through the cement dosage and the water / cement ratio, but with fly ash admixture, as well. In the case of using of some active ashes from chemical point of view, these will react with the calcium hydroxide resulted from the reaction with water of the cement's mineralogical compounds, forming the calcium hydro silicates which are covering the ash granules favouring the formation of new gelic products. The previous researches highlighted favourable behaviour for the ash admixtures collected by wet process in proportions of 40% from the cement dosage [4].

The analysis of the concretes from the buildings main structures subjected to intense sulphate aggressive actions is conducted on concretes based on cement with high specific strengths to chemical aggressiveness, such as H II/A-S type cement, trying to highlight the favourable influence of the thermal power plant fly ash as compositional factor [5, 6].

The obtained concrete has to present high specific strengths to aggressive actions, because at least in the first stage, after the placing, when the concrete's porosity is greater, the reaction between the aggressive solution and the cement stone's compounds can lead to an intense corrosion process in the initial stage [7]. Also it has to be considered that the sulphate corrosion effect can occur by expansion of the affected structure and in this study are suggested two testing terms, at 28 and 90 days, in order to highlight the action of the aggressive agent.

Taking into account that the suggested cement for the research is of the type H II/A-S 32.5 composite cement, the evolution of the sulphate corrosion was studied on four concrete mixes containing this type of cement in different dosages and different thermal power plant dosages, using a high concentration of the aggressive agent (2% Na_2SO_4 and 1% $MgSO_4$).

2. Experimental program

In the Laboratory of the Faculty of Civil Engineering and Building Services of Iasi, an experimental program was conducted in order to assess the dimensional modification of the some samples prepared by concrete mixes based on belitic cements of type H II/A-S and embedding fly ash subjected to a combined sulphate and magnesia corrosion process using a mixed solution of Na_2SO_4 and $MgSO_4$ having certain concentrations.

The experimental program was conducted according to the requirements of the actual Standards (NE 012/1:2007 [8] and NE 012/2:2010) [9] referring at the realization of concretes in aggressive environments and that are requiring limits for the compositional factors and, in the same time, are recommending function of the exposure class of the concrete, even the cement type that is required for use. Taking into account the limitations imposed by the Standards four concrete mixes were realized.

2.1. Materials

To realize the concrete mixes, for different values of the compounds dosages, the following materials were used:

- river aggregate – the maxim dimension;
- distillate water;
- cement – type H II/A-S 32.5;
- thermal power plant fly ash collected by wet process according to SR EN 450-1:2012 [10];
- solutions of Na_2SO_4 and $MgSO_4$ with concentration of 1000 mg/dm^3, SO_4^{2-} ions.

2.2. Concrete mixes

For different values of the component dosages the following compositions were realized and are presented in table 1.

Table 1. Concrete mixes.

Concrete mixes	Component dosages					Slump (cm)
	Aggregates (kg/m^3)	Cement (kg/m^3)	Ash (kg/m^3)	Water (l/m^3)	W/C+Ash	
R1	1800	250	125	210	0.56	S3 (7.0)
R2	1800	100	200	223	0.74	S3 (8.5)
R3	1800	200	200	215	0.55	S3 (8.5)
R4	1800	300	150	216	0.49	S3 (8.5)

2.3. Experimental set up

Prismatic samples having 7x7x28 cm sizes were prepared. Three of each concrete mix were immersed and preserved in water up to the testing moment as witness samples.

The other three samples of each concrete mix were immersed in a 2% concentration Na_2SO_4 solution mixed with a 1% concentration $MgSO_4$ solution for 28 and 90 days as shown in figure 1.

Figure 1. Samples preservation.

After the preserving time the dimensional modifications were measured.

The determination expansion due to the corrosion phenomenon was realized according to the SR EN 2833-2009. This standard establishes the test methods for the determination of hardened concrete volume changes of and of the strains during the determination constant humidity and temperature conditions using the micro dial gauge and the Graff equipment [11]. The Graff equipment is illustrated in figure 2.

Figure 2. Graff equipment.

3. Experimental results and discussions

According to the compositional characteristics of the four concrete mixes, after the samples were subjected to corrosion the following results were registered.

The results are presented in the table 2.

Table 2. Registered results.

Concrete mixes	Aggressive environment, concentration (%)	W/(C+Ash) ratio	Expansion mm/m	
			28 days	90 days
R1	2% Na_2SO_4 + 1% $MgSO_4$	0.56	0.000	0.000
R2		0.74	0.070	0.070
R3		0.55	0.035	0.035
R4		0.49	0.035	0.035

The following graphic representation, figure 3, shows a clearer appreciation of the concrete's behavior to sulphate corrosion.

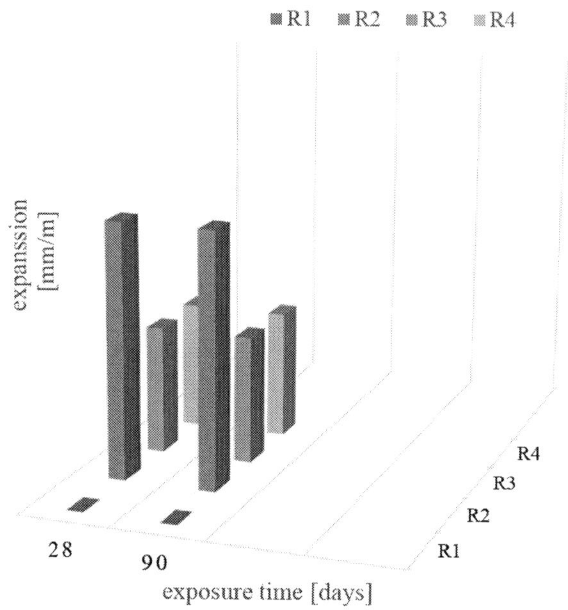

Figure 3. Expansion variation of the four receipts in conditions of combined sulphate and magnesia corrosion.

Analyzing the experimental results completed by the graphic illustration can be concluded the following:

- the mix 1 presents the best behaviour for an ash proportion of 50% from the cement dosage C=250 kg/m^3;
- mixes 3 and 4 present for the both testing terms a favourable behaviour because the expansion values are very low due to the fact that the cement presents strength besides the action of SO_4^{2-} ions, phenomenon which is due to the reduced contain of aluminates from the cement and the compactness is greater determined by the ash admixture in proportions that are shown in the research;
- mix 2 presents a weaker behaviour but under the accepted limit for the expansion and is due the reduced cement contain, which represents the main element regarding the sulphate chemical attack;
- it can be considered that the mix 1 presents the best behaviour in the conditions of the experimental program because the values of the expansion in intense concentration conditions are presenting values below the accepted limits for the mentioned exposure times.

IOP Conf. Series: Materials Science and Engineering **789** (2020) 012018 doi:10.1088/1757-899X/789/1/012018

4. Conclusions

The use of the fly ash collected by wet process is improving the concretes behaviour to sulphate corrosion by processes which are increasing the compactness and decreasing the permeability.

The use of the fly ash collected by wet process which is improving the concretes behaviour to sulphate corrosion has many economic effects by ashes recycling and replacing the fine part of the aggregate.

The first experimental results are showing the favourable behaviour of the concretes based on fly ash to sulphate aggressive actions in the first part of life.

A further testing will highlight the favourable behaviour of the concretes based on fly ash subjected to sulphate corrosion for a longer life time.

References

[1] Georgescu D, Apostu A and Gavrilescu R 2009 Descriptive and performance approach to ensure the concrete durability *Romanian Journal of Materials* **39** (4) 307

[2] Neville A M 2003 *The Properties of Concrete* (Technical Publishing House, Bucharest)

[3] Georgescu D and Gavrilescu R 2013 Modern approaches to ensure concrete performances regarding durability *Romanian Journal of Materials* **43** (2) 119

[4] Georgescu D P 2001 *Guide for manufacturing durable concrete according to the National application Annex of SR EN 2061. Durability classes* (ed Everest Bucharest)

[5] ACI Committee 232.2R-03 2003 *Use of fly ash in concrete* (American Concrete Institute Farmington Hills Michigan)

[6] Ştefan (Oancea) I, Bărbuţă M, Budescu M, Mihai P, Banu O M and Ţăranu N 2018 Particularities regarding the mechanical behaviour of some types of sustainable concrete mixes with waste materials *Romanian Journal of Materials* 236

[7] Muntean R and Cazacu C 2013 Considerations regarding floors execution and rehabilitation using ferrocement Journal of Applied Engineering Sciences **3** (2) *The annals of the University from Oradea* **3** (2) 79

[8] NE 012/1:2007 *Normativ pentru producerea betonului şi executarea lucrărilor din beton, beton armat şi beton precomprimat. Partea 1: Producerea betonului (Norm for production and execution works of concrete, reinforced concrete and restressed concrete – Part 1: Concrete production)*

[9] NE 012/2:2010 *Normativ pentru producerea şi executarea lucrărilor din beton, beton armat şi beton precomprimat. Partea 2: Executarea lucrărilor din beton (Norm for production and execution works of concrete, reinforced concrete and restressed concrete – Part 2: Execution of concrete works)*

[10] SR EN 450 1:2012 *Fly ash for concrete - Part 1: Definition, specifications and confomity criteria*

[11] SR 2833:2009 *Tests on concretes. Determination of the axial shrinkage of hardened concrete*

Considerations about fire behaviour of an unprotected wood elements according to Romanian Code SR EN 1995-1-2-2004

R Erbaşu[1*] and D Ţăpuşi[1]

[1] Department of Civil Engineering, Technical University of Civil Engineering Bucharest, Romania

[*] ruxi@erbasu.ro

Abstract. Increased interest in wood as construction material has led to the development of research on improving physical and mechanical characteristics of different wood species or various types of processed wood. Since wood construction materials are combustible, it is necessary to improve their fire behavior. This is achieved by treating the surface or the mass of materials using flame retardant substances, which can delay the ignition and can reduce the speed of propagation of fire. Although its benefits: specific weight, thermal conductivity, chemical stability, good vibration and shock absorption, good dimensional stability against temperature, ease of processing, etc., wood has the great disadvantage that it will ignite easily, being a cellulosic material.

1. Introduction

According to Romanian Code SR EN 13943 [1], the fire behavior means changing or maintaining physical and chemical properties of a product exposed to fire. The concept includes both the reaction to fire and the fire resistance.

The reaction to fire is defined as the behavior of a material which by its own decomposition, supplies a fire to which is exposed under specified conditions.

The concept of fire resistance is the property of a building element to keep, for a given time interval, the required stability, the tightness and / or the thermal insulation.

The fire tests allow the evaluation of the physical and chemical parameters that characterize the construction and to establish the behavior of the structures and the structural elements at the complex action of fire by estimating measures corresponding to the different limit states of fire resistance.

The theoretical research has been developed in parallel with the experimental research, pursuing the development of calculation models and methods that expand the field of knowledge in this area. The theoretical approach allows the calculation of the behavior in case of fire of large elements and structures, for which experimental tests are difficult to perform. The process of simulation through calculation of the behavior of structures at the action of fire can be a more economical way of investigation.

The essential requirement, „*fire safety*", will be fulfilled by taking measures and establishing rules for locating, designing, constructing and operating the buildings, installations and facilities, as well as establishing performance criteria and levels in terms of fire for the respective constructions.

The purpose of this paper is to determine the difference between two approaches in case of fire design: the results obtained through the numerical calculation according Romanian Code SR EN 1995-

1-2-2004 [2] and the series of experimental results obtained after performing two sets of fire resistance tests on one floor and one roof built with glued laminated timber beams [3].

For the analytical modelling the SAP2000 finite element method computer program was used and strength checks were performed at various time intervals, up to failure. The computations respect the provision of Romanian Code SR EN 1995 - 1- 2 - 2004 [2].

2. General presentation of the experiment

Initial set-up data for the experiment [3]:

- wood strength class: GL24; beams section: 140x167x7000 mm;
- permanent load on the roof beam 45 kg/m; permanent load on the floor beams 35 kg/m.

Exposure to fire was made on three sides and a span of 2.50m. 24 thermocouples were installed on the element surface and in the furnace, for measuring the temperature during the test, as shown in the figure below. The wood is not fire proofed.

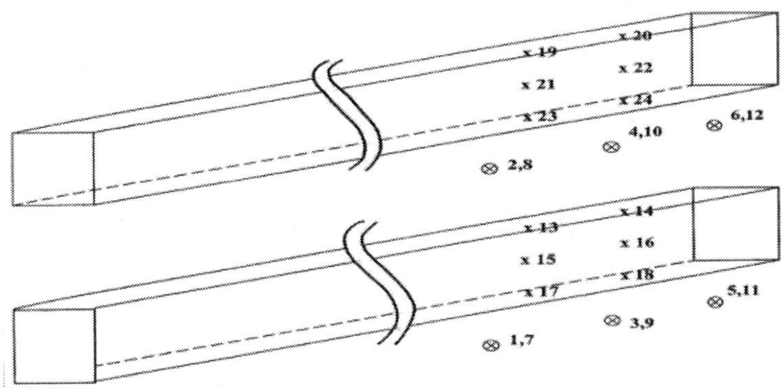

Figure 1. Thermocouples positioning

The temperature/time furnace heating diagrams for both tests are the following:

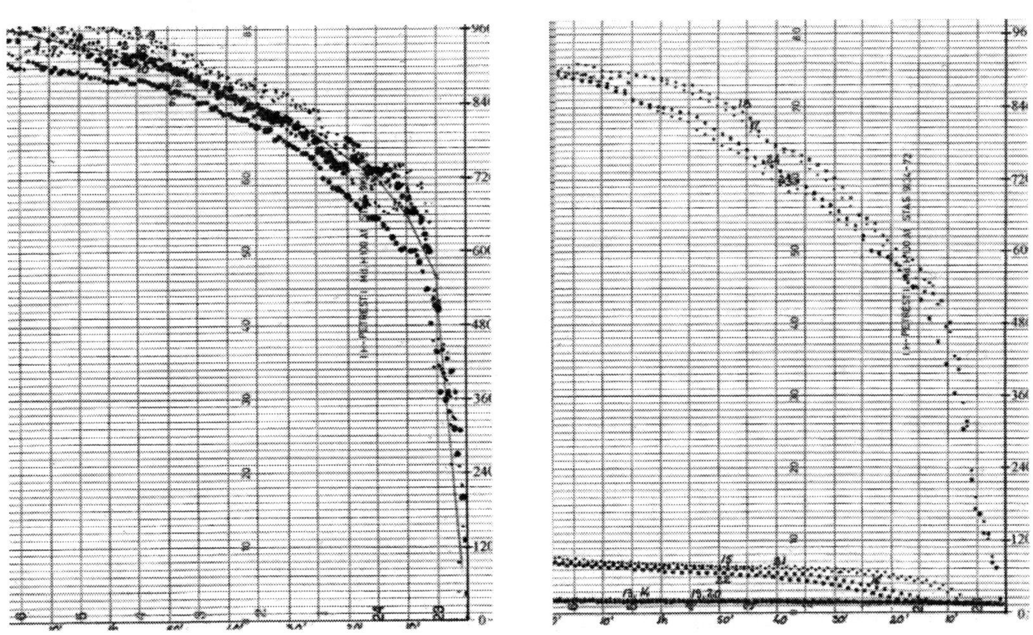

Figure 2. Temperature/time diagram for the two experimental tests

The observations during the experiment described above and that are relevant to this study are the following:

- at time t = 18 min, the roof beam has a deflection of 1.00 cm and the floor beam of 0.5 cm;
- at time t = 25 min, floor beam has a deflection of 1.00 cm;
- at time t = 66 min, both beams show a deflection of 2.00 cm;
- at time t = 85 min, the test is stopped.

The experimental results show that the two beams were subjected to fire for 85 minutes. According to the code SR EN 13501-2 [4], the wooden beams fulfilled the condition of resistance up to 60 minutes, under the conditions of loading to which they were subjected, being framed in the resistance class R60.

3. Calculation of the beams according to SR EN 1995-1 - 2/2004 [2]

According to the obtained information in the roof beam experiment, this was modeled in SAP2000 program, loads being applied on three sides of the beam and strength checks were performed at different time interval, up to failure.

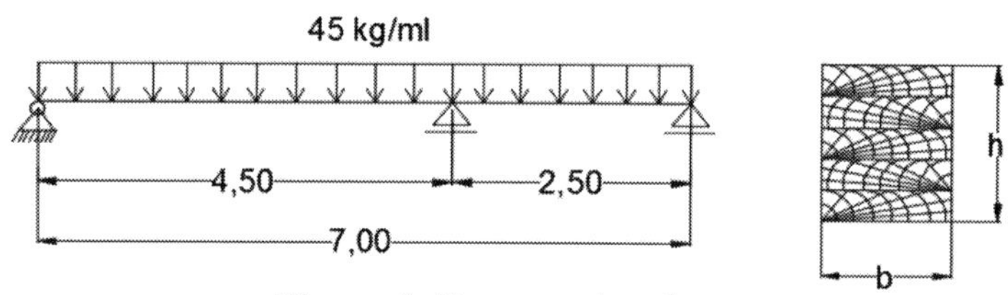

Figure 3. Beam static scheme

Beam section: b = 140 mm, h = 167 mm, span = 7.00 m

Taking into account the ISO 834 [4] time-temperature curve the model temperatures were determined:

$$T = 345\log_{10}(8t + 1) + T_0(°C) \tag{1}$$

$$T = 20°C \tag{2}$$

where

T - hot gas temperature in the furnace fire test, in the vicinity of the element exposed to fire

t - during fire exposure (min)

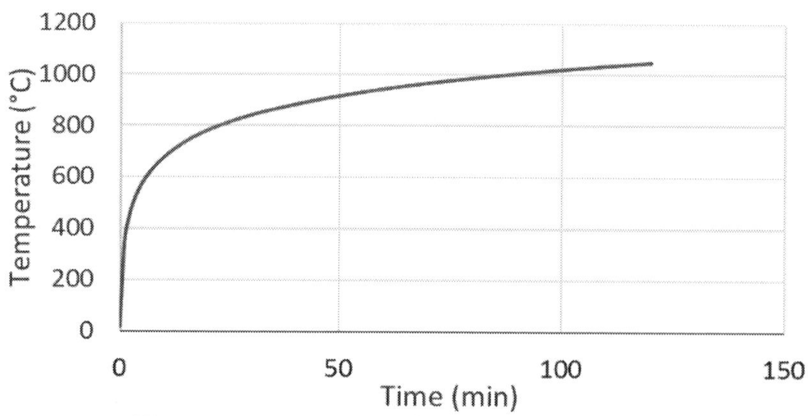

Figure 4. Temperature Time Curve - ISO 834 [4]

Fire design was performed using the reduced cross section method, according to SR EN 1995 – 1-2 – 2004 [2].

For this computation method the duration of fire resistance depends on the resilience of non charring section.

Calculation of the design bearing capacity with reduced cross-section beam $M_{rd,r}$.

$$M_{Rd,r} = W_r f_{d,fi} \tag{3}$$

$$f_{d,fi} = k_{mod,fi} \frac{f_{20}}{\gamma_{M,fi}} \tag{4}$$

$$f_{20} = k_{fi} f_k \tag{5}$$

$$f_{d,fi} = k_{mod,fi} \frac{k_{fi} f_k}{\gamma_{M,fi}} \tag{6}$$

where

$f_{d,fi}$ - value of the resistance to fire;
$k_{mod,fi}$ - fire-factor modification;
f_{20} - 20% fractile resistance at normal temperature;
f_k - resistance characteristic value;
$\gamma_{M,fi}$ - partial safety coefficient of wood to fire action.

$$W_r = \frac{b_r\, h_r^2}{6} \tag{7}$$

W_r - strength modulus of the non-charring section
b_r - width of cross section
h_r - depth of cross section

The fire resistance checks were made for different time intervals from t = 18 min to t = 85 min, according to the experimental data.

The SAP 2000 analysis led to the design sectional moments $M_{Ed,fi}$ for the considered time intervals.

Table 1. Check of the roof beam from $t = 0$ to $t = 85$ min.

Time (min)	Beam Section b_r (mm)	Beam Section h_r (mm)	Sup Temp (°C)	Inf Temp (°C)	f_{max} (mm)	$M_{Rd,r}$ (kNm)	$M_{Ed,fi}$ (kNm)	Check
18	102.2	148.1	20	766	8.29	10.31	5.65	ok
20	98	146	20	781	8.916	9.61	5.58	ok
25	91	142.5	20	815	10	8.50	5.47	ok
30	84	139	20	842	11.345	7.47	5.27	ok
35	77	135.5	20	865	12.6	6.50	5.01	ok
40	70	132	20	885	13.96	5.61	4.69	ok
45	63	128.5	20	902	15.33	4.79	4.32	ok
50	56	125	20	918	16.7	4.03	3.92	ok
55	49	121.5	20	932	18.1	3.33	3.48	Beam fails
60	42	118	20	945	19.5	2.69	3.02	Beam fails
66	33.6	113.8	20	960	21	2.00	2.46	Beam fails
70	28	111	20	968	21.9	1.59	2.08	Beam fails
75	21	107.5	20	979	21.57	1.12	1.62	Beam fails
80	14	104	20	988	22	0.70	1.17	Beam fails
85	7	100.5	20	997	22.78	0.33	0.74	Beam fails

Thus, it appears from the above calculation that the beam fulfils the resistance conditions up to time t = 50 min, 10 minutes earlier than in the experimental case, this showing the fact that the analytical computational offers a safety interval to cover the unpredicted risks that may appear in case of real service conditions.

According to the Code SR EN 13501-2 [4], there will be a different framing of the wooden beams in fire resistance classes: R30 – for the analytical computational and R60 for the experimental case.

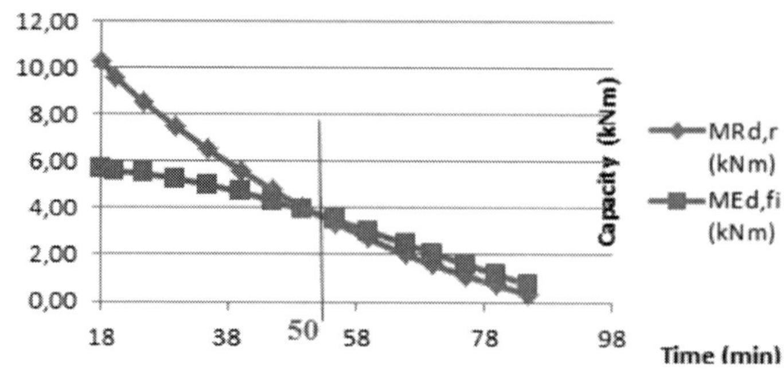

Figure 5. Capacity- requirement graph of the roof beam.

Following the previous example, the floor beam was modelled in SAP2000, according to information from the experiment and checked until failure.

Table 2. Check of the floor beam from t = 0 to t = 85min.

Time (min)	b_r (mm)	h_r (mm)	Sup Temp (° C)	Inf Temp (° C)	f_{max} (mm)	$M_{Rd,r}$ (kNm)	$M_{Ed,fi}$ (kNm)	Check
18	102.2	148.1	20	766	8.22	10.31	5.48	ok
20	98	146	20	781	8.84	9.61	5.42	ok
25	91	142.5	20	815	10	8.50	5.31	ok
30	84	139	20	842	11.27	7.47	5.13	ok
35	77	135.5	20	865	11.29	6.50	4.87	ok
40	70	132	20	885	13.91	5.61	4.56	ok
45	63	128.5	20	902	15.3	4.79	4.2	ok
50	56	125	20	918	16.7	4.03	3.8	ok
55	49	121.5	20	932	18.1	3.33	3.36	Beam fails
60	42	118	20	945	19.62	2.69	2.92	Beam fails
66	33.6	113.8	20	960	21.27	2.00	2.36	Beam fails
70	28	111	20	968	22.2	1.59	1.99	Beam fails
75	21	107.5	20	979	23.15	1.12	1.53	Beam fails
80	14	104	20	988	23.17	0.70	1.08	Beam fails
85	7	100.5	20	997	23.2	0.33	0.66	Beam fails

As in the case of roof beams, the results of the experimental test and the analytical calculation obtained for the floor beam are maintained: the beam is framed in the class of resistance R60 – according to experimental tests and in the resistance class R30 by the analytical calculations.

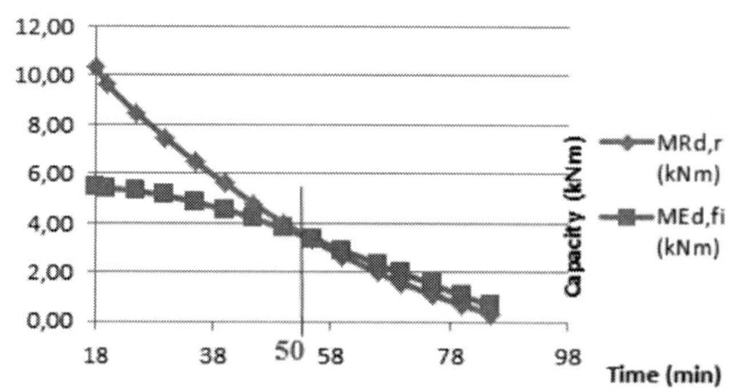

Figure 6. Capacity- requirement graph of the floor beam.

4. Fireproofing of wood

The process of improving the fire behavior of combustible materials is called *fireproofing*. The fire retardant does not exclude the ignition and combustion of the material, but it gives an improved fire behavior over a certain period of time.

Improvement of the fire behavior of wood elements can be done by protecting the surfaces of the elements with: (i) fireproof cladding, made by using one or more layers of wood-based panels, a layer of gypsum or clay shale fibers, (ii) flame retardant substances applied to the surface (heat-foaming paints) or in the mass of the element.

According to SR EN 1995-1-2 [2] computations, when using protection made of wood panels a minimum of 13.3 mm of timber panel or plywood is necessary to delay the charring starting time with 15 min meaning that one 15 mm thickness protective panel of wood is sufficient to offer satisfying fireproofing

Considering the flame retardant variant with the gypsum coating, in order to delay the charring initiation time by 15 minutes, a minimum of 11 mm gypsum board panel is needed.

5. Conclusions

Although its benefits: specific gravity, thermal conductivity, chemical stability, good vibration and shock absorption, good dimensional stability against temperature, ease of processing, etc., wood has the great disadvantage that it will easily ignite, being a cellulosic material.

Wood, even if is a combustible material, performs well in terms of structural resistance to fire. As explained in this study, when exposed to fire wood retains its strength for a longer period of time.

In this study, one aims to achieve the comparison between the results of the fire tests performed on two beams of glued laminated timber and the results of the analytical calculation, performed according to the Codes in force in Romania in compliance with the experimental conditions.

By comparing the results obtained from the experiments with those obtained by the analytical calculation, on the unprotected structural elements, it resulted that, regarding the fire resistance, the analytical calculation leads to more restrictive results, compared to the experimental approach.

Fireproofing is one of the requirements for fire safety that lead to improving the fire resistance of a building.

References

[1] SR EN ISO 13943 2011 *Fire security. Vocabulary*
[2] SR EN 1995-1-2 2004 *Design of timber structures. General - Structural fire*
[3] Test report no. 931 2005 S.C. GLULAM S.A.
[4] SR EN 13501-2 2010 *Fire classification of products and building element*

Development of experimental designs of the integrated heater for the disposal of low-potential waste heat of ventilation emissions

V S Ezhov[1], **N S Semicheva**[1], **A P Burtsev**[1*], **D A Ermakov**[2] **and N S Perepelitsa**[1]

[1] Department of HVAC, Southwest State University, Kursk, Russia
[2] Department of electro supply, Southwest State University, Kursk, Russia

[*] ap_burtsev@mail.ru

Abstract. The aim of the study is to develop an experimental design of a complex air heater, conduct and analysis of the experiments, as well as to determine the main characteristics of the thermo-electric generator during the utilization of low-potential heat-La exhaust gases. The experimental setup consists of two units – thermoelectric generator operating on the principle of cross-heat exchanger heat exchange for heat recovery with parallel warmed of the supply air, which is supplied in the form of a mixture in the burner boiler unit and adsorber unit, filled blast furnace slag for the purification of waste gases from nitrogen oxides, c and carbon.

1. Introduction

The actual problem of development of fuel and energy and housing and communal complexes of the Russian Federation is the problem of energy saving and environmental safety. The solution of these problems will ensure a steady increase in the efficiency of heat generation and distribution from the source to the final consumer. Closely related to this is the economic growth of the country, the improvement of the environmental situation and the safety of the population [1].

One of the main directions of increasing the efficiency of heat-generating plants is the use of devices and sets of equipment to reduce the heat content and temperature of smoke gases, reduce emissions of flue gases and, as a consequence, increase environmental safety adjacent to the boiler area. To achieve these goals, the paper proposes to use the effect of thermoelectricity, which provides Autonomous power supply of anticorrosive equipment and intensification of the process of adsorption of harmful components by granular blast-furnace slag in a complex air heater [2].

Application in boiler equipment [3] of deep heat recovery or installations of use of latent heat of vaporization of the leaving flue gases (for example, contact heat exchangers, air heaters) increases the General efficiency of a boiler room by 1-3%.

Air heaters [4] make it possible to utilize heat emissions from boilers by contact of flue gases and blast air supplied from the upper zone of the boiler (air heaters of various types increase fuel economy by 1-4%) [5].

For the implementation of increasing the degree of heat recovery the resection of the gases used, the effect of thermoelectricity is a phenomenon of direct conversion of heat into electricity in conductors by direct heating and cooling of junctions of two conductors passing current. In the first approximation, the appearance of thermo-pods depends only on the material of conductors and the temperatures of hot and cold junctions [6].

For this purpose, the Department of Heat, gas and water Supply of the SWSU developed several innovation structures of air heaters [7-10], the main scientific novelty of which is the utilization of low-potential heat of exhaust gases with temperature up to 150°C and direct generation of electricity due to the effect of thermoelectricity.

2. Research methods

Air taken by the heater from the laboratory room was used as a working medium. Tests on heat exchange at counterflow in the thermal-electric section were carried out for a number of fixed speeds and air consumption at the experimental unit [10].

A fragment of a thermoelectric generator of an autonomous air heater with an indication of the main structural dimensions on figure 1 [10]. The red arrows indicate the flow of hot air from the air heater located on the left side. Blue circles indicate the flow of cold air from the axial fan located at the bottom. In this case, the flow direction is from bottom to top. (to the observer).

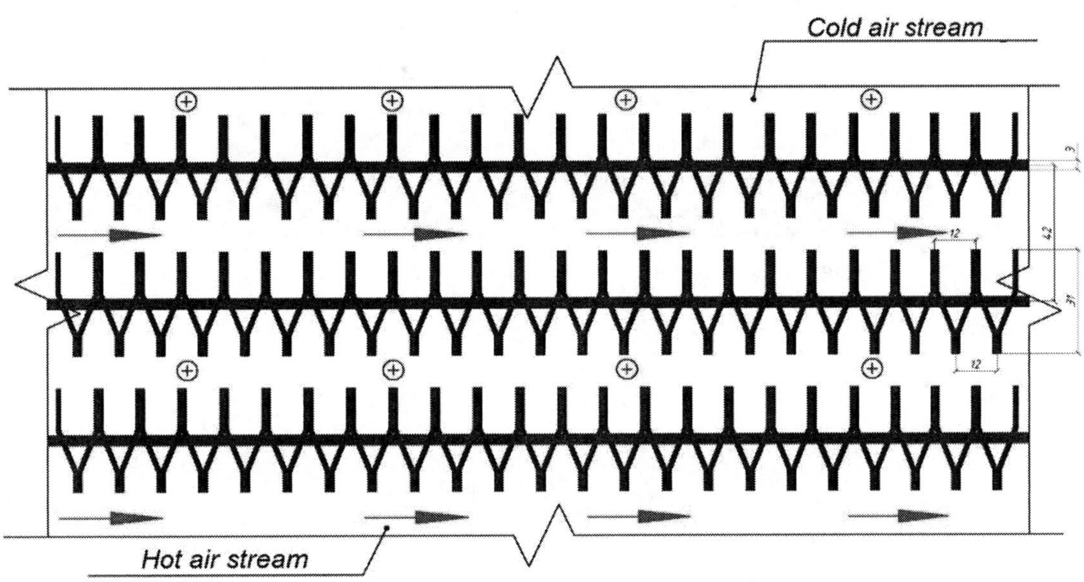

Figure 1. – Fragment of a thermoelectric generator.

The main characteristics of thermionic elements made of chrome (M1) and copper (M2) are: coefficient of thermo-EMF $\alpha = 12.97 \cdot 10^{-3}$ V/K; q-factor $=2.8 \cdot 10^{-3}$ K^{-1} and conductivity $\sigma = 8 \cdot 104$ Ohm-$1 \cdot m^{-1}$ [8].

The flow temperature was determined as the arithmetic mean value of mercury thermometer readings at the points of entry and exit from the channels of thermoelectric sections [9].

The average air velocity in the channels of thermoelectric sections was determined by the expression:

$$\overline{\omega} = c_н \cdot \omega_{анем} \quad (1)$$

$c_н$ is the coefficient of flow unevenness

The total heat perception of the air flow was determined from the heat balance equation, [12]:

$$Q' = \overline{\gamma}_в \cdot \overline{G}_{кан} \cdot (t'_{вых} - t'_{вх}) \cdot 10^3 \quad (2)$$

where $\overline{\gamma}_{in} = c_{in} \cdot \rho_{in}$; c_{in}, ρ_{in} heat capacity and density of the air flow at its average temperature, kJ/(kg° C), kg/m^3; $\overline{G}_{chn}$ air flow rate in the channel , $\overline{G}_{chn} = \overline{\omega} \cdot F_{chn}$, m^3/h; t'_{in}, t'_{ext} air temperature at the inlet and outlet of the channel of the heat exchange element, for the passage of the heated air, °C; F_{chn} – the cross-sectional area of the channel, determined from the geometric dimensions of the channel, m.

A general view of the experimental setup on figure 2.

Figure 2. Scheme of the experimental setup. 1 – housing; 2 – pipe for the inlet of the heating medium; 3 – construction Hairdryer; 4 – axial fan; 5 – stand for the fan; 6 – LEDs.

To determine the heat transfer coefficient, the method of stationary heat flow was used, in which Newton-Richman law is used [12]:

$$dQ' = \alpha_{F_{cm}} \left(t'_{в.cm} - t'\right) dF_{cm}, °C; \tag{3}$$

where $t'_{in.sd}$, t' - the temperature of the heat-emitting surface and the air, respectively, °C; F_{sd} – the area of the heat-emitting surface, m^2.

In that case, if all the values included in equation (3) are related to small elements of the body surface, the average coefficient of heat transfer from the heat-emitting surface was determined for each individual experiment, respectively, from the expression [11]:

$$\overline{\alpha}' = \frac{\frac{1}{L}\int_0^L q(x)dx}{\frac{1}{L}\int_0^L \left(t'_{в.cm} - t'\right)dx} = \frac{Q'}{F_{cm \cdot}\left(\overline{t}'_{в.cm} - \overline{t}'\right)}, \text{Вт/(м}^2 \cdot \text{°C)}; \tag{4}$$

F_{sd} - the calculated heat transfer surface of the heat exchange surface, m^2; $\overline{t}'$ - the average temperature of the heated air flow in the channel, determined from the ratio, °C:

$$\overline{t}' = \frac{t'_{вх} + t'_{вых}}{2}, °C \tag{5}$$

where $\bar{t}_{in.sd}$ - the average temperature of the heat exchange surface from the heated air flow, determined by the formula (6), °C:

$$\bar{t}'_{в.ст} = \bar{t}_{ст} - \frac{Q'}{\frac{\lambda_{ст}}{\delta_{ст}}}, °C \tag{6}$$

where δ_{sd} - the thickness of the heat transfer surface 2 mm; λ_{sd} - the coefficient of thermal conductivity of the material of the heat transfer surface of heating, W/(m·° C); $\bar{t}_{sd}$ - the average wall temperature determined by the formula (7), °C:

$$\bar{t}_{ст} = \frac{\bar{t}' + \bar{t}''}{2}, °C \tag{7}$$

where $\bar{t}''$ - the average temperature of the heating air in the channel, determined by (8) similarly (5), °C.

$$\bar{t}'' = \frac{t''_{вх} + t''_{вых}}{2}, °C; \tag{8}$$

where t''_{in}, t''_{ext} - the air temperature at the inlet and outlet of the channel of the heat exchange element, for the passage of the heating air, °C.

The average value of the coefficient [11] of heat transfer for each experiment was determined from the following ratio:

$$\bar{k} = \frac{Q'}{F_{ст} \cdot \Delta\bar{t}}, Вт/(м^2 \cdot °C); \tag{9}$$

where $\Delta\bar{t}$ - the average temperature head (°C), determined at $\frac{\Delta t_б}{\Delta t_м} \le 1,7$, with sufficient accuracy, as the arithmetic mean temperature difference according to the formula:

$$\Delta\bar{t} = \frac{\Delta t_б + \Delta t_м}{2}, °C; \tag{10}$$

where $\Delta t_б, \Delta t_м$ - the temperature difference between the media at the other end of the heating surface, respectively, °C

$$\Delta t_б = t''_{вх} - t'_{вых}, °C; \tag{11}$$

$$\Delta t_б = t''_{вых} - t'_{вх}, °C. \tag{12}$$

The amount of heat received from the heating air was determined for each experiment as,

$$Q'' = \bar{\gamma}_в \cdot \bar{V}_{кан} \cdot \left(t''_{вх} - t''_{вых}\right) \cdot 10^3, Вт. \tag{13}$$

The average heat transfer coefficient [11] from the heating air flow to the heat transfer surface was also determined for each individual experiment, respectively, from the expression:

$$\bar{\alpha}'' = \frac{Q''}{F_{ст} \cdot \left(\bar{t}'' - \bar{t}''_{в.ст}\right)}, Вт/(м^2 \cdot °C); \tag{14}$$

F_{sd} - the calculated heat transfer surface of the heat exchange surface (wall); $F_{sd} = 0.049$ m^2; $\overline{t}''$ - the average temperature of the heated air flow in the channel, °C; $\overline{t}''_{in.sd}$ - the average temperature of the heat exchange surface from the heated air flow, determined by the formula (15), °C:

$$\overline{t}''_{в.ст} = \frac{Q''}{\frac{\lambda_{ст}}{\delta_{ст}}} + \overline{t}_{ст}, \text{°C}; \tag{15}$$

where $\overline{t}_{sd}$ - the average wall temperature, °C.

The average value of the heat transfer coefficient [12] for each experiment was determined from the following ratio

$$\overline{k} = \frac{1}{\frac{1}{\overline{\alpha}'} + \frac{\delta_{ст}}{\lambda_{ст}} + \frac{1}{\overline{\alpha}''}}, \text{Вт/(м}^2 \cdot \text{°C)}. \tag{16}$$

Table 1 shows the results of experimental studies followed by the determination of the main characteristics of the thermoelectric generator.

Table 1. Results of experimental studies.

Name of the measured value	Designation	Dimension	The number of the series of experiments		
			1	2	3
The temperature of the heated air at the inlet to the channel	$t'_{вх}$	°C	25.4	25.4	25.4
Temperature of heated air at the outlet of the channel	$t'_{вых}$	°C	27.7	28.1	29.2
Heating air temperature at the inlet of the channel	$t''_{вх}$	°C	81.7	150.4	178.7
Heating air temperature at the outlet of the channel	$t''_{вых}$	°C	42.8	55.1	63.8
Air velocity in the channel for the passage of heated air	$\overline{\omega}_{наг}$	m/s	5.1	5.1	5.1
Air velocity in the channel for the passage of heating air	$\overline{\omega}_{гр}$	m/s	1.1	1.1	1.1
Average temperature of heated air	$\overline{t}'$	°C	26.6	26.8	27.3
Thermal perception of heated air flow in one channel	Q'	W	552	660	990
The heat transfer coefficient from the wall of the channel to heated air	$\overline{\alpha}'$	W/(m²·K)	104.3	58.5	72.8
Voltage in an electrical circuit	V	V	1.14	2.02	2.26
Amperage	I	A	0.356	0.631	0.728
Power	N	W	0.406	1.275	1.645

Figure 3 graphically shows the dependence of the generated electric power on the heat flux perceived from hot air by thermoelectric sections in an autonomous air heater.

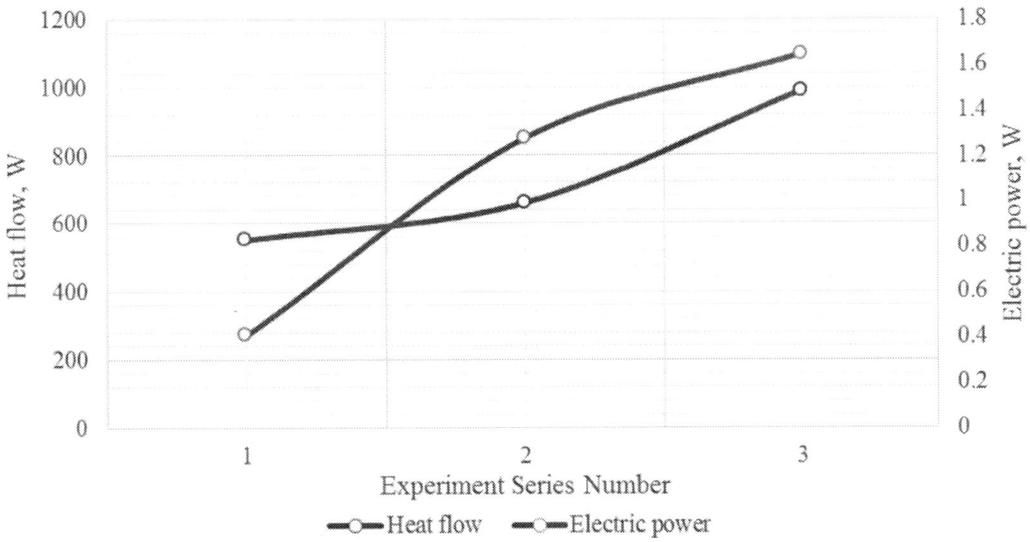

Figure 3. Schedule of dependence of generated electric power on heat flow in thermoelectric generator.

3. Conclusions

The main conclusions of this article are:

- an experimental design of the air heater is developed, which has a thermoelectric EMF source operating as a result of the passing conversion of heat into electricity, which allows to utilize the low-grade heat of the exhaust gases from 140°C to 60°C;
- experimental studies to determine the basic parameters of the thermoelectric generator at cross-heat exchange between the exhaust gases and cold air at the laboratory installation for heat recovery of the exhaust gas;
- as a result of cross-heat exchange, the discharge gas temperature at the outlet of the complex air heater is reduced by 40-60%, while an increase in the efficiency of the boiler unit by 1-1.5% is recorded as a result of an increase in the supply air temperature in the boiler burner device from 20°C to 55°C.

References

[1] Ezhov V S, Semicheva N E, Burcev A P and Popova M E 2018 Razrabotka effektivnogo kompleksnogo korrozionnoustojchivogo vozduhopodogrevatelya (Development of an effective integrated corrosion-resistant air heater), *5-j Mezhdunarodnoj molodezhnoj nauchnoj konferencii* (5th International Youth Scientific Conference) Vol **2**, p 292–295

[2] Ezhov V S and Shik D A 2018 Kompleksnyj vozduhopodogrevatel' dlya nagreva vozduha i ochistki dymovyh gazov (Integrated air heater for heating air and flue gas cleaning), 2-j *Vserossijskoj nauchnoj konferencii perspektivnyh razrabotok molodyh uchenyh* (2nd All-Russian Scientific Conference of Promising Developments of Young Scientists) Vol **3**, p 79–83

[3] Ezhov V S, Semicheva N E and Telegin A A 2018 Razrabotka meropriyatij po povysheniyu KPD teplogeneriruyushchih ustanovok na TEC SZR (Development of measures to increase the efficiency of heat generating plants at the CHPP of SZR) Kursk, *3-j Mezhdunarodnoj nauchnoj konferencii studentov i molodyh uchenyh*. (3rd International Scientific Conference of Students and Young Scientists.) Vol **4** p 187–191

[4] Ezhov V S, ZHmakin V A and Burcev A P 2018 Povyshenie effektivnosti sistemy ochistki dymovyh gazov kotel'noj na prirodnom gaze (Improving the efficiency of a flue gas cleaning system for a natural gas boiler) Kursk, 3-j *Mezhdunarodnoj nauchnoj konferencii studentov i*

molodyh uchenyh (3rd International Scientific Conference of Students and Young Scientists) Vol **4** p 196–199

[5] Ezhov V S, Semicheva N E and Zabelin I S 2018 Kotel'naya s korrozionnostojkim teploobmennym oborudovaniem (Boiler room with corrosion-resistant heat exchange equipment), *6-j Mezhdunarodnoj molodezhnoj nauchnoj konferencii* (6th International Youth Scientific Conference) Vol. **4**, p 343–348 DOI 10.1088/1755-1315/72/1/012022

[6] Ezhov V S, Pavlov S V, Burcev A P and Brezhnev A V 2018 Modelirovanie pryamogo preobrazovaniya tepla v elektrichestvo dlya energosnabzheniya ITP, BST (Simulation of direct conversion of heat into electricity for power supply of ITP, BST) *Byulleten' stroitel'noj tekhniki* (Bulletin of construction equipment) no. **3**, p 48–50

[7] Ezhov V S and Semicheva N E 2019 Stekloblochnyj vozduhopodogrevatel'-elektrogenerator (Cinder block Air heater'-electric generator) RF patent. №2592938

[8] Ezhov V S 2019 Kompleksnyj vozduhopodogrevatel (Integrated air heater) RF patent. №2595289

[9] Ezhov V S and Burcev A.P. 2019 Kompleksnyj korrozionnoustojchivyj vozduhopodogrevatel (Integrated Corrosion Resistant Air Heater) RF patent. №2691896

[10] Ezhov V S, Semicheva N E and Burcev A P, Emel'yanov S G 2019 Avtonomnyj vozduhopodogrevatel (Freestanding air heater), RF patent. №2705193

[11] Vladimir Y, Natalia S and Brezhnev A 2018 Characterization of Thermoelectric Generators for Cathodic Protection of Pipelines of the City Heating, *Adv. Intell. Syst. Comput.* Vol. **2** p 670–678

IOP Conf. Series: Materials Science and Engineering **789** (2020) 012021 doi:10.1088/1757-899X/789/1/012021

The mechanical properties and the freeze-thaw resistance of an ecological micro concrete made with sunflower stalks

C M Gradinaru[1], A A Serbanoiu[1]* and R Muntean[2]

[1] "Gheorghe Asachi" Technical University, Faculty of Civil Engineering and Building Services, Iasi, Romania
[2] Transilvania University of Brasov, Faculty of Civil Engineering, Brasov, Romania

* serbanoiu.adrian@tuiasi.ro

Abstract. The aim of this study was to analyze the density, compressive strength, tensile strength and the freeze-thaw resistance of a cement micro concrete made with sunflower stalks that replaced 20%, 35%, 50%, 60%, 80% and 100% of the sand from its composition. The results showed, in general, a decrease of the analyzed properties, but the use of 20% and 35% sunflower stalk aggregates led to the improvement of the splitting tensile strength. A significant advantage of this type of micro concrete was its density, being obtained a lightweight building material, with up to 822 kg/m^3.

1. Introduction

Currently, an important goal of researchers is to find alternative solutions to build cheaper, more environmentally friendly and at the same time using sustainable materials. Sustainable materials are generally those made from natural or recycled raw materials, whose production has a reduced impact on the environment, requiring reduced amounts of energy and non-renewable resources. While the solution of recycled raw materials has been extensively studied, research into alternative variants of renewable raw materials is at the beginning. In general, renewable raw materials are by-products or waste of plant or animal origin.

Ecological concrete is most often the result of using a waste category material as a partial or total replacement of cement, sand and/or aggregates or that concrete whose production process does not significantly affect the environment. The ecological concrete is a revolutionary material that has a very low environmental pollution effect compared to the conventional concrete.

Excessive use of aggregates causes depletion of these natural resources, and reckless mining activities to extract these materials could lead to environmental problems, such as landscape damage and disruption of the ecosystem, water, soil and air pollution [1, 2].

Construction materials made by using annual plants can have the appropriate properties to meet modern design requirements such as sustainability in construction, efficiency in operation and carbon footprint reduction. The high porosity of the plant aggregates and their low apparent density recommend them for use in the composition of the concrete, giving it improved qualities of thermal and acoustic insulation compared to the classical concrete.

The vegetable aggregates have a high flexibility which translates into a non-fragile plastic and elastic behavior, increased ductility beyond the point of maximum mechanical strength and a high deformability under pressure.

The use of vegetal aggregates such as sunflower stems in concrete has advantages in terms of a combination between the use of agricultural waste, low gas emissions and their local availability. Their high oxygen content makes them a good thermal insulator. By chopping, sunflower stems can be used as aggregates in lightweight concrete production, being a sustainable alternative solution for currently marketed variants, such as expanded clay or expanded polystyrene, which are highly energy-consuming.

Aggregates from sunflower stems are ultra-light, with a bulk density of 105±2 kg/m^3 and a water content of 9.4% measured in a room at (20±2)°C and (35±5)% relative humidity. Similar results were obtained for hemp (103±2 kg/m^3) and sunflower stem (105±2 kg/m^3) [3]. By comparison, the relative density of wood particles is 305.8±66 kg/m^3 [3]. The internal porosity of the sunflower stem is very high and tends to enhance the absorption of water in the sunflower aggregates.

The aim of this study was to analyze the density, compressive strength, tensile strength and the freeze-thaw resistance of a micro concrete made with sunflower stalks that replaced 20%, 35%, 50%, 60%, 80% and 100% of the sand from its composition.

2. Materials and methods

The present research aimed to investigate the effects of sun flower aggregates on the micro concrete mechanical properties. For this purpose, starting from a reference micro concrete (RC) strength class 16/20, micro concrete recipes with 20%, 35%, 50%, 65%, 80% and 100% vol. mineral aggregates replacement sunflower stalks granules were developed.

2.1. Materials

The RC contained the following components:

- Portland cement CEM II/A-LL 42.5R, with granulated blast furnace slag and limestone;
- natural sand (0-4 mm diameter) and gravel (class 4-8 mm) extracted from riverbeds;
- water as to ensure a water/cement ratio of 0.52;
- additives: a policarboxilateter based superplasticizer and a rhodanid based accelerator; the superplasticizer was used to obtain a smaller water/cement ratio and to improve the mortar workability; the accelerator was used to enhance the cement hydration process, necessary especially in the case of concrete with vegetal aggregates.

The study involved the development of 7 micro concrete compositions, as follows:

- 1 recipe of reference micro concrete , RC, with 320.19 kg/m^3 cement and 706.77 l/m^3 mineral aggregates with maximum diameter of 8 mm, which served as a starting and comparison basis for micro concrete compositions made with sunflower stalk aggregates as a partial and total replacement material for mineral aggregates.
- 6 micro concrete recipes with sunflower aggregates (granules smaller than 5mm and fibers smaller than 25 mm obtained by sunflower stalks shredding and then treated with sodium silicate solution) as a replacement material for conventional aggregates by 20%, 35%, 50%, 65%, 80% and 100%, noted with SFC20, SFC35, SFC50, SFC65, SFC80 and SFC100, respectively.

2.2. Methods

2.2.1. Determination of the apparent density of the fresh and hardened micro concrete. The apparent density measuring of the fresh and hardened micro concrete was determined according to [4] and [5] and consisted in determining the mass of a micro concrete sample and relating it to the volume of the respective sample in compacted state.

For the determination, a watertight metal container of cylindrical or parallelepiped shape (including the molds for the molding of the tests on mechanical strengths) having the smallest dimension equal to at least four times the maximum size of the concrete aggregate granules was used. As a result, it was taken the arithmetic mean of 3 determinations, rounded with max. ±5 kg/m^3.

2.2.2. Determination of compressive strength. Micro concrete specimens were tested under mono-axial compression, with a hydraulic concrete press that carries a uniformly distributed load on the surface of the specimen, figure 1, according to [6]. The specimens were loaded perpendicularly, in the casting direction casting, figure 2.

For the evaluation of the compressive strength, the average of three tests was calculated. The value of the compressive strength was determined using the formula (1) [6]:

$$f_c = \frac{F_{max}}{A_C} \left[\frac{N}{mm^2}\right] \tag{1}$$

where F – maximum force applied to the specimen, expressed in [N]; A_c – the compressed area perpendicular to the direction of application of the load, expressed in [mm^2].

Figure 1. Micro concrete testing for compressive strength.

2.2.3. Determination of flexural tensile strength. The determination of the flexural tensile strength was done on prisms simply recessed, subjected to bending by applying a concentrated force in the middle of the opening, according to the standard. The rupture occurred in the maximum bending moment section, through a crack that appeared in the stretched area, under the concentrated force, splitting into two tubes, figure 2.

According to [7], the flexural tensile strength, f_{cf}, is given by the relation (2):

$$f_{cf} = \frac{3}{2} \cdot \frac{F_b \cdot l}{d_1 \cdot d_2^2} \left[\frac{N}{mm^2}\right] \tag{2}$$

where F_b – breaking force, expressed in [N]; l – distance between the two supports, expressed in [mm]; d_1, d_2 – the average width, respectively the average height of the cross-section, expressed in [mm];

Figure 2. Specimens tested for flexural tensile strength.

2.2.4. Determination of the splitting tensile strength. The determination of tensile strength by splitting was performed on cylinders with a diameter of 100 mm and a length of 200 mm. This test consisted of compressing a specimen after two diametrically opposed generators, as it is seen in figure 3.

Figure 3. Micro concrete samples testing to splitting tensile strength.

Splitting tensile strength (f_{ct}) was calculated using the formula (3) [8]:

$$f_{ct} = \frac{2F_b}{\pi bh} \left[\frac{N}{mm^2}\right] \tag{3}$$

where Fb – breaking force, expressed in [N]; π – value of pi number; b – the average width of the breaking cross section, expressed in [mm]; h – the average height of the breaking cross section, expressed in [mm].

The tests were performed according to the [8]. For testing, a hydraulic press was used, recording the value of the breaking force. The breaking force was applied perpendicular to the direction of casting of the concrete, continuously and evenly, until breaking. The average of the results of three tests was calculated and then interpreted.

2.2.5. Determination of the micro concrete's resistance to freeze-thaw cycles. The determination of the freeze-thaw resistance of the micro concrete samples (figure 4 and figure 5) was carried out in accordance with the provisions of [6] and [9]. This test was performed by measuring the variation of the compressive strength of the micro concrete samples (figure 6). To perform the tests, 10 cm cube-shaped specimens were used.

The working method required that all micro concrete specimens be placed in the water bath at (20±5)°C, four days before the actual testing began.

The water level was initially up to 1/4 of the height of the specimen, after 24 hours it increased to 1/2 of the height of the specimens, then up to 3/4, followed by 3 days after the initial introduction into the water the specimens were at least 20 mm in totality below the water level. To achieve full saturation, they were kept completely immersed in water for 24 hours.

Figure 4. Concrete samples held in refrigerated enclosure at (-17)°C.

Figure 5. Micro concrete samples kept in a water bath at (20±5)°C.

The air temperature in the refrigerator was adjusted so that it gradually dropped in about 1 h to (-17)°C, and then constant for 3 h. After the freeze cycle, the thaw begun by completely immersing the samples in water, at a temperature of (20±5)°C, for 4 hours.

Samples were subjected to 50 freeze-thaw cycles.

The loss of compressive strength (η) was determined by the formula (4) [9]:

$$\eta = \frac{R_m - R_t}{R_m} \cdot 100 \quad (4)$$

where R_m – the average compressive strength of the reference samples [N/mm²]; R_t – the average compressive strength of the samples subjected to freeze-thaw cycles [N/mm²].

Figure 6. Compressive strength testing of the micro concrete samples subjected to 50 freeze-thaw cycles.

3. Experimental results

3.1. The micro concrete density

Figure 7 graphically represents the values of the density of micro concrete compositions, recorded at the time of casting and after the maturation period of 28 days. As for the density of micro concrete compositions with sunflower aggregates, it registered significant decrease, both at casting and after 28 days, between 11% and 60%.

For a percentage of mineral aggregates replacement with vegetable aggregates of over 35%, lightweight micro concrete compositions were obtained.

The increase of the percentage of plant aggregates caused the density decrease from 1845.3 kg/m^3 to 822 kg/m^3 for micro concrete made only with plant aggregates.

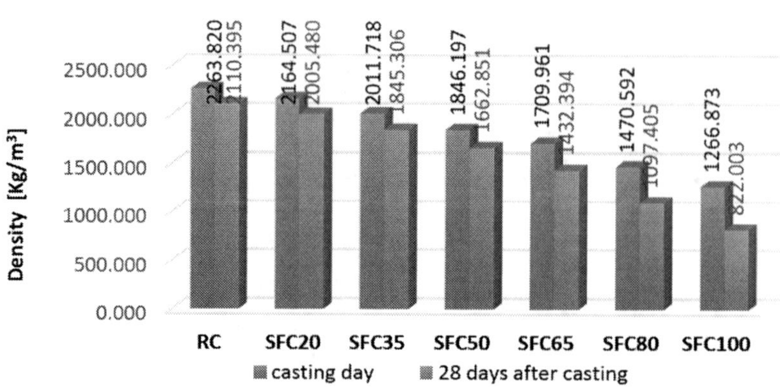

Figure 7. Micro concrete density in fresh and dry state [kg/m^3].

3.2. Micro concrete compressive strength

Figure 8 exhibits how the values of compressive strengths of the mortar samples tested at the age of 28 days are plotted. Micro concrete with vegetable aggregates showed significant decreases in compressive strength, being closely related to their density. For a percentage of replacement of mineral aggregates up to 50%, compressive strengths between 14.8% and 5 N/mm^2 were recorded. A replacement of 20% of the mineral aggregates led to a decrease of this parameter by around 24% of the obtained concrete, while a 100% replacement value to around 95% decrease. The micro concrete with 50% sunflower aggregates obtained a compressive strength value of 5.48 N/mm^2, very similar with Autoclaved Cellular Concrete.

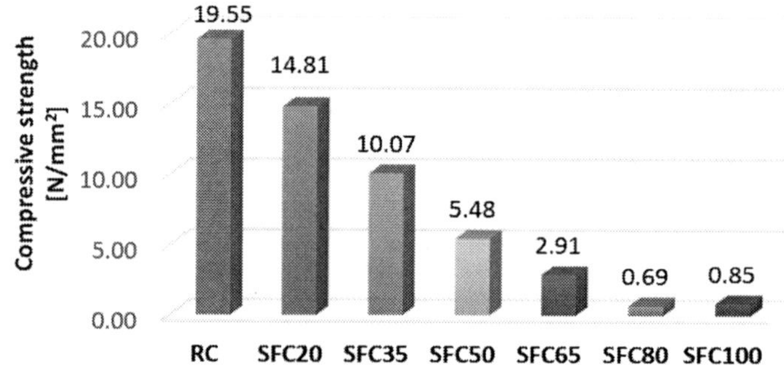

Figure 8. Compressive strength of the micro concrete at 28 days [N/mm^2].

3.3. Micro concrete flexural tensile strength

The values of flexural tensile strength are presented in figure 9. Flexural tensile strength registered a 17.48% decrease in case of SFC20, 49.50% in case of SFC50 and 90.21% in case of SFC100.

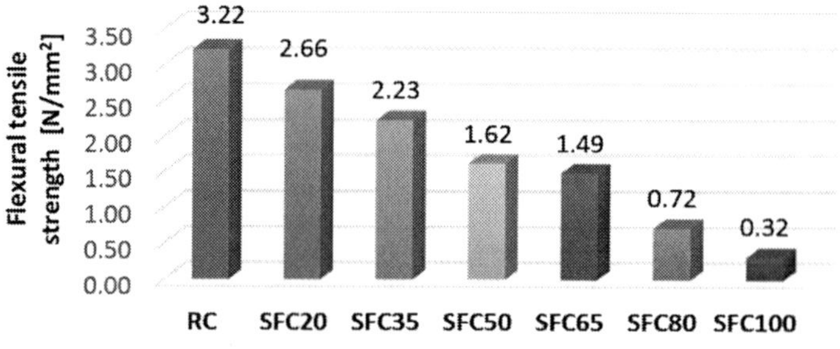

Figure 9. Flexural tensile strength of the micro concrete [N/mm^2].

3.4. Micro concrete splitting tensile strength
The values of splitting tensile strength are presented in figure 10. The 20% and 35% of the mineral aggregates' replacement had positive influence over the splitting tensile strength of these micro concrete compositions. A value of 50% of sunflower aggregates led to a small decrease of this parameter by around 14%, and 65% to a decrease of 50%. The micro concrete compositions with 80% and 100% of plant aggregates led to a very strong decrease of the splitting tensile strength of around 90%.

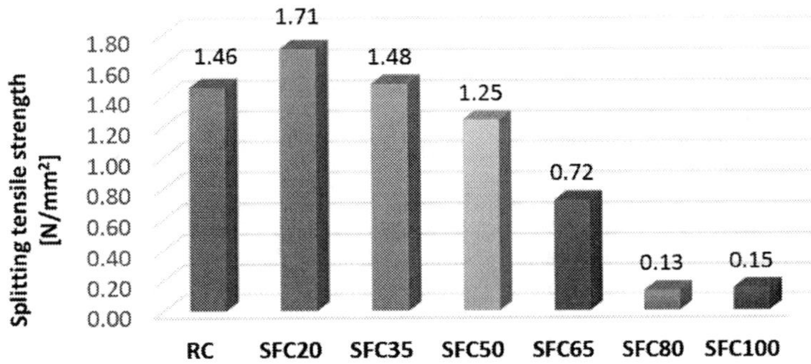

Figure 10. Splitting tensile strength of the micro concrete [N/mm^2].

3.5. Freeze-thaw resistance of the micro concrete
Figure 11 indicates the values registered at the compressive test by the micro concrete samples subjected to 50 freeze-thaw cycles, comparatively to each of their reference samples that were kept in relative constant temperature and humidity.

The sunflower stalk aggregates improved the freeze-thaw resistance in the case of 20% using. While RC compressive strength decreased by 22.80% as compared to their reference samples, SFC20 registered a decrease of this parameter by 8.34%. SFC35 and SFC50 had a compressive strength decrease of 31.68% and 39.17%, respectively, with around 9% and 16% smaller than RC. The micro concrete compositions with over 65% of sunflower aggregates did not resist to the 50 cycles of freeze-thaw.

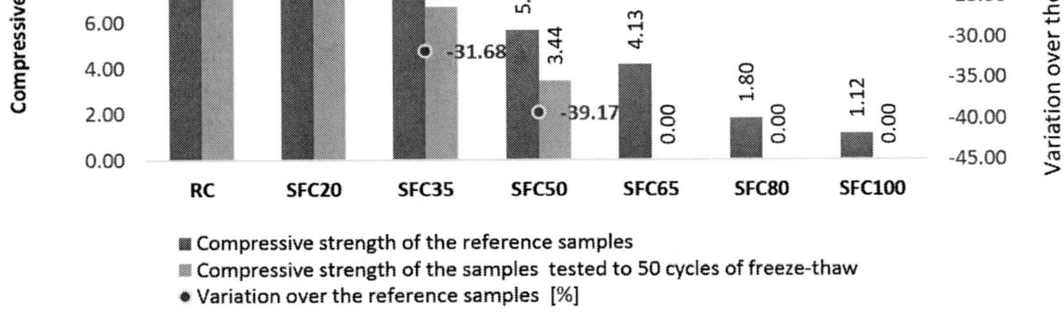

Figure 11. Freeze-thaw resistance of the micro concrete [N/mm^2].

4. Conclusions

The experimental tests revealed that the sand replacement from the reference micro concrete composition with sunflower stalks granules led to the following conclusions:

- a lightweight micro concrete was obtained for replacement volumes bigger than 35%, with densities between 1845.30 kg/m^3 up to 822 kg/m^3 in the case of total replacement.
- compressive strength of the samples decreased along with plant aggregates increasing quantity. A 50% replacement of the sand led to an almost 75% smaller value of this parameter.
- in case of the flexural tensile strength, a decrease resulted as well , SFC50 registering around 50% smaller value than RC.
- the sunflower granules led to an improvement of the splitting tensile strength in the case of 20% and 35% replacement rate due to the presence of the sunflower stalk fibers; till this replacement level, these fibers had an positive influence on this parameter
- in the case of SFC 20, the tested samples obtained a higher resistance to the 50 freeze-thaw cycles, the compressive strength decrease comparative to their own reference sample being by 14.46% smaller than RC.

Acknowledgement

Special thanks to the Sika Romania representatives for their technical support and for providing the necessary additives in order to accomplish this research.

This work was supported by a grant of the Romanian Ministry of Research and Innovation, CCCDI – UEFISCDI, project number PN-III-P2-2.1-CI-2018-1194, 199CI / 2018, within PNCDI III.

References

[1] Grădinaru C M, Şerbănoiu A A, Babor D T, Sârbu G C, Petrescu-Mag I V and Grădinaru A C 2019 When Agricultural Waste Transforms into an Environmentally Friendly Material: The Case of Green Concrete as Alternative to Natural Resources Depletion *Journal of Agricultural and Environmental Ethics* 77-93

[2] Mo K H, Alengaram U J, Jumaat M Z, Yap S P and Lee S C 2016 Green concrete partially comprised of farming waste residues: a review *Journal of Cleaner Production* **117** 122-138

[3] Nozahic V and Amziane S 2012 Influence of sunflower aggregates surface treatments on physical properties and adhesion with a mineral binder *Composites* **43** (A) 1837–1849

[4] SR EN 12350-6:2019 *Testing fresh concrete. Density*

[5] SR EN 12390-3:2019 *Testing hardened concrete. Compressive strength of test specimens*

[6] SR EN 12390-5:2019 *Testing hardened concrete. Flexural strength of test specimens*

[7] SR EN 12390-6:2010 *Testing hardened concrete. Tensile splitting of test specimens*

[8] SR EN 12390-7:2019 *Testing hardened concrete. Density of hardened concrete*

[9] SR EN 3518:2009 *The determination of freeze-thaw resistance by measuring the variation of the compressive strength and / or the modulus of relative dynamic elasticity*

Considerations on the use of ultra-high performance concrete for water tanks

A Grigorean[1*], L I Diaconu[1] and F L Tamas[2]

[1] Technical University "Gheorghe Asachi" from Iasi, Faculty of Civil Engineering and Building Services, Romania
[2] Transilvania University of Brasov, Faculty of Civil Engineering, Brasov, Romania

[*] grigoreanadrian@yahoo.com

Abstract. The development of ultra-high strength and high performance concrete is a relatively new field in the profile industry. The last 150 years have shown that, despite getting more and more resistance, the practical applications have been too many times due to innovations in materials science. This phenomenon seems to be caused by the increased costs per unit of volume, as resistance increases, as well as a cautious retention of using new materials in practical applications. The present study aims to be an encouragement of practical structural applications by presenting an Ultra High Performance Concrete (UHPC), developed with locally available materials. As it is known, concrete is the main building material worldwide. In the last decades, his leadership position has been strengthened by the constant concern of the researchers and by obtaining the various types of concrete with increasing performances from the point of view of the fresh and hardened state, obtaining special structural applications.

1. Introduction

In 2007, within the program PN II -IDI- ID_1053 / 2007-2010, "ULTRAPERFORMANT CONCRETE - ENVIRONMENTAL ECOLOGY", approached the theoretical and experimental study of the ultra-high performance concrete, regarding the realization sustainable concrete, capable of maximum protection of the environment. Slim concrete elements were obtained which, for the same capable moment, are comparable with those of steel in terms of mass and geometric dimensions tested at bending and shearing [1].

At the Technical University of Cluj-Napoca, high-strength reinforced concrete has been researched since 1988. Experimental studies were carried out following the achievement of predetermined ductility factors [2] for frame nodes made with dispersed reinforced concrete (a fiber volume of at least 1% is sufficient to carry out reductions of the transverse reinforcement), for constructions located in areas with high seismicity [3].

In France, Lafarge, in collaboration with the manufacturer of chemical compounds Rhodia (www. Rhodia.com) and with the construction division of the Bouygues corporation (www.bouygues.fr), have developed a UHPC marketed under the name Ductal. It incorporates metallic or organic fibers that lead to the improved ductility of the composite material. Tests have shown that the compressive strength increases about 6-8 times (about 230 MPa without traditional reinforcement) compared to a regular concrete (RC), while the bending resistance has values of about 30-60 MPa.

A challenge for any practitioner is to use new materials and emerging technologies, in the absence of an explicit consecration at the level of the incident norm. To meet this challenge, several action groups have initiated, in recent years, transitional guidance, until the introduction of this new material in the design norm. The following list, organized by country, presents the following guidance resources for engineers:

- France - in March 1999, at the request of the Scientific and Technical Commission of the Association of Civil Engineers of France, the development of an interim code for UHPC structures began. It was published in French and English in January 2002.
- Australia - with the support of the construction company VSL (Australia), the University of New South Wales, Australia, has published "Design Code for BPR Pre-Compressed Beams" in accordance with the methodology and spirit of the national code Australian Code AS3600-1994. This paper presents examples of design and guidance to obtain specific resistance to compression, bending, cutting force and torsion, as well as checking the crack opening, limiting arrow, fire resistance, fatigue resistance, loss of tension from pre-compression and related anchors [4].
- Japan - in sept. 2004, the Japanese Society of Civil Engineers published "Recommendations for the Design and Construction of Structures with UHPC – pre-norm". The document includes design recommendations, rules for selecting material properties for the design, testing and obtaining the desired durability, execution technologies and examples of bridges built in the country. It is estimated that an English version will be available soon [4].
- United States – in 2002, the Federal Highway Administration (FHWA) initiated a research project in collaboration with the Massachusetts Institute of Technology (MIT) to optimize highway bridges by using UHPC. This study was based on the experience accumulated at MIT since 1999, respectively the similar one accumulated by FHWA since 2000, when a study of the potential UHPC for use on bridges began. Following this collaboration was published the report CEE Report R03-01, "Models for optimizing bridge beams made with UHPC on highways". The report presents the design strategy of a brittle-plastic material, with an elasto-plastic composite fibre reinforcement. Also, there is a comparison between this model, based on the maximum crack opening, and the design rules [4].

The first research programs that introduced the terminology of "Ultra-High Performance Concrete" were initiated in 1985. The adopted definition uses the criterion of "compressive strength" and imposes a lower limit of at least 150 MPa, Schmidt and Fehling (2005). For those UHPCs, which use constituents with very small granular dimensions (powders) and therefore have higher strengths, the term "Concrete with Reactive Powders" (RPC) is preferred [3].

2. Technical requirements

The first part of the present research aimed at achieving high-performance concrete compositions with reactive powders (RPC), at which the resistance is between 150-200 N/mm^2.

The researches focused on two types of ultra-high performance concrete (UHPC) with reactive powders (RPC):

- concrete with and without fibers, having the medium consistency class;
- concrete with and without fibers with a workability, respectively the higher consistency class.

An important element is the selection of materials in order to obtain the concrete forms. A challenge is to succeed in achieving the performance with locally available materials.

The special properties of RPC are obtained with a high dosage of Portland cement, a small W/B ratio (water / binder), possible through the use of the latest generation of superplasticizers in large dosages, the use of high-binding pyrolastic additives – ultrafine silica and the incorporation of steel fibers as dispersed reinforcement [4].

By using ultrafine silica, a by-product of the metallurgical industry that is often considered to be a partial or added replacement for Portland cement, it achieves energy savings, reduces gas (CO_2) emissions in cement production and, finally, an environmentally friendly concrete with positive effects in terms of compressive strength and sustainability. Thus, the waste of one industry can be the raw material for another industry. The materials used as concrete additives, having binding properties, bear, in the specialized literature, the names of additional constituents with cementitious characteristics [5].

The aggregates used are quartz sands with a maximum diameter of 1.2 mm. Although the composite mixture has an appearance close to that of a mortar matrix, it will be referred to as concrete.

The Portland cement used comes from the company Lafarge Romania, the last generation superplasticizer of polycarboxylate type is offered by BASF Romania, the metallic steel fibers with dimension of Ø0.4/25 mm (diameter / length), with bent ends and Ø0.175/6 mm - straight, are purchased from BAUM CAS FIBRES Romania, the ultrafine silica type "Elkem Microsilica Grade 940 U undensified" is offered by the company BASF Romania, and the quartz sands come from the Career company Bega Minerale Industriale Aghires [5].

At the beginning of the researches, it was started from the realization of the high compactness of the mixtures taking into account the Apollonian distribution of specific to the reactive powder concretes, the principle being that each smaller particle fills the space between the larger particles.

The mixtures were made with a mixer with forced mixture and with predetermined times (DIEM mixer with a capacity of up to 139 liters).

The principles underlying the composition of RPC are:

- increasing the homogeneity by eliminating the coarse aggregate;
- increasing the density in compacted state by optimizing the granular distribution of the mixture;
- improving the microstructure by applying a heat treatment after hardening;
- improving the ductility by incorporating steel fibers;
- establishing the technology as easy to apply in practice [5]

3. Concrete mixtures with and without fibers

In the first stage of the research, fiber-free mixtures were made, with different amounts of cement but especially of different types and different dosages of superplasticizing additives. The results were less favorable from the point of view of the resistance and the obtained workings.

The second phase of the research followed, in which mixtures were made with and without fibers, with different percentages of polycarboxylate ACE 440 BASF type superplasticizing additive and different dosages of long steel fibers. The third phase followed, in which the long fiber mixture was resumed - the most successful in the second stage (OE 23), with a short fiber mixture but also a hybrid one (1/2 fibers), long steel + 1/2 short steel fibers) [5].

These mixtures are considered representative, they fulfill the performance from the point of view of the concrete in fresh and hardened state. The quantities of the components of the recipes are related to the cement unit.

3.1. Fresh characteristics of fiber-free and long-fiber concrete mixtures - Stage II

The ME 22 mixture, also called the Witness, with the best performances, is the most fluid and falls in the second spreading class of spraying (SF2) for a self-compacting concrete, but when determining the viscosity, the values of time T500 (s) and the leakage through the V-hull exceeded the requirements. A self-compacting concrete has to meet all three requirements, therefore ME 22 is not self-compacting but concrete with very good workability. The strengthened characteristics of concrete mixtures with and without the addition of steel fibers. Are presented the values of the compressive strengths of the samples kept in water or subjected to the heat treatment immediately after removing the formwork, the samples being sealed at the casting face, by brushing with an insulating film based on Baumit type BA2 resins.

The tests were performed on 71 mm side samples and the results represent the average of 6 tests.

It is observed that the heat treated concrete up to the age of 6 days (1 day in molding and 5 days heat treatment), has higher values of the compressive strength compared to the one kept in the water until the age of the test [5].

The resistance to compression at the age of 28 days, in the case of heat treated concrete, increases very little or in some cases there is even a decrease in values compared to those from 6 days.

In concrete kept in water up to the age of 28 days, it is observed that the resistance increases over time until the last test age (28 days).

3.2. The strengthened characteristics of concrete mixtures with the addition of long steel fibers
The favorable influence of the steel fibers on the compressive strengths is observed for both modes of subsequent treatment. The true contribution of the fibers and an important feature of the dispersed reinforced concrete is reflected in the results of the bending resistance test.

3.3. Parallel between the characteristics of concrete without fibers, of those with long, short fibers and with hybrid steel fibers
The ME 22 mixture is considered a control for long, short fiber reinforced concrete and for hybrid fiber reinforced concrete, based on the OE 23 mixture, which has the best characteristics in fresh and hardened state. The hybrid fiber blend is made with 50% long fibers (Ø4/25 mm) and 50% short fibers (Ø0.175/6 mm). The fiber-free concrete scraping and scraping have values similar to those of hybrid concrete, and the long-fiber concrete has lower values, yet it is very workable. It is observed, from the values of squeezing and spreading, that long fibers influence the workability, even in smaller quantities [6].

The temperature is influenced by the large amount of cement in the composition, the amount of additive and, in some cases, the duration of mixing the concrete [5].

The occluded air content is normal for such a dense concrete, from which the air bubbles come out very hard. Analyzing the values, it is observed that the mixtures with the addition of steel fiber have a lower content of occluded air than the control concrete - without fibers. The elimination of the occluded air is obtained by vibration. The occluded air content may have much lower values for such concrete (BUIP). It has a high resistance to repeated cycles of frost - thawing, because, following the heat treatment, the microstructure of the concrete is improved by applying an optimum temperature of subsequent treatment (min. 90°C), which activates the reactivity of the microsilica and leads to a reduction in porosity [7].

Analyzing the values, it is found that, in a fresh state, the characteristics of the hybrid fiber concrete (OE 23 hybrid) are approximately the same as those of the fiber-free composition (ME 22), but better than the long fiber composition (OE 23) with fibers L, and the compressive strengths are almost identical to the compositions made only with fibers [8].

In conducting this experimental program on hybrid fiber compositions, we aimed to improve the working characteristics of ultra-high performance concrete and their behavior at complex service demands.

4. Consideration on the use of Ultra High Performance Concrete in tanks
In the particular of two 5000 m^3 water tanks (figure 1), for the intangible reserve required in case of fire, practically built in the same time, one in Buzau (noted REZ 1) and one in Bacau (noted REZ 2), we used Ultra-High Performant Concrete class C50/60 with short metallic fibers at REZ 2 and normal concrete from project class C25/30 at REZ 1. Their characteristics are according to table 1, where ND means not determined.

Figure 1. Water tank in Buzau (left) and in Bacau (right) – second phase of casting.

Table 1. Characteristics of concrete with and without fibers in fresh and hardened state.

	UHPC with heat treated fibers					Normal concrete (for REZ 1)
	ME22 with fibers	OE 23 with long fibers	OE 23 with long fibers	OE 23 with long fibers	OE 23 with long fibers	C25/30 XC4+XF1 D_{max} 16 mm, S3
Characteristics UHPC	Stage II	Stage II	Stage III with extension of mixing			
Water / Cement	0.22	0.23	0.23	0.23	0.23	0.55
Water / Binder	0.18	0.18	0.18	0.18	0.18	ND
Settling (mm)	265	247	270	270	260	ND
Spread (mm) after - 10 minutes	560	428	560	565	545	ND
- 60 minutes	-	-	508	515	500	ND
Temperatures (°C)	26.7	29.5	32.5	32.28	35.3	34.8
Occluded air (%)	4.60	4.56	4.50	4.35	4.40	5.5
Density of fresh concrete (kg/m^3)	2260	2400	2395	2437	2413	2430
Recipe density calculation (kg/m^3)	2289	2410	2410	2410	2410	2440
f_{cm} 1 day (MPa)	65.1	77.9	87.8	80.2	84.2	12.1
f_{cm} 7 days (MPa)	131	172	171.3	169.3	172.8	24.4
f_{cm} 28 days (MPa)	132	173	-	157.2	172.4	44.3

Analyzing the test values on cubes with 71 mm of the table 1, we notice a significant increase in the workability of the concrete. This is because the additive ACE 440 makes its effect more and more visible over time. With the increase of the mixing time of the concrete, of the workability, its temperature also increases, theory denied in the case of the other mixtures made with predetermined times. The increase in temperature is due to the high dosage of additive and the heat of hydration of CEM I 52.5R [8].

The value of the resistance obtained on the smaller cubes as size will always be higher and the starting mathematical relation for the equivalence of the average resistances on cubes of different dimensions is: $f_{c,\,cub200} = 0.95 \cdot f_{c\,,\,cub150} = 0.92 \cdot f_{c,\,cub100}$ [9].

The fresh concrete takes the form of the pattern, faithfully copying its imprints, reinforcing the statement related to the very good workability of these concretes, for a better filling of the pattern.

The classic technology at REZ 1, provided a four stage casting of the walls in order to eliminate the effects of the contractions and avoid the cracking of the concrete. At each casting, 3x6 samples (14.1 cm cubes) were taken, and the tests were performed at 1, 7 and 28 days, on 6 cubes.

The obtained results for REZ 1 (normal casting) reveal some disadvantages:

- the working joints (of the 4 casting stages) required special treatment like expandable cords, joint band, and careful vibration. After casting, due to cracking, corrections were made with special insulation materials to prevent leakage.
- The processing time of casted concrete (for maintaining humidity) it was four days, compared to the case of UHPC, where the time was two days.
- Subsequent works (filling, technological tests) started ten days after the last casting (stage 4) respectively 25 days after first casting (stage 1) compared to six days from the casting in the case of UHPC.

The obtained results for REZ 2 (experimental), when the casting has been done in one stage, reveal:

- the costs of renting the formwork and the costs of manpower were 60% lower;
- the costs with the concrete itself were 18% higher;
- the execution time was 50% lower than the classical solution;
- no cracks appeared;
- since the required strength has been reached 10 days earlier, the technological samples and the commissioning were also outdated by 10 days.

IOP Conf. Series: Materials Science and Engineering **789** (2020) 012022 doi:10.1088/1757-899X/789/1/012022

5. Conclusion

- An UHPC concrete with high initial compressive strength (>170 MPa) was obtained.
- Existence of fiber influence on physical-mechanical characteristics.
- Designing the compositions of a high performance concrete (HPC) and ultra-high performance concrete (UHPC) is essential, as it will determine its structural performance.
- Obtaining these concrete is not possible without the use of materials with higher characteristics and preliminary tests.
- There are no standardized compositions – each application requires a characteristic recipe.

Some of the advantages in using UHPC for making concrete tanks are:

- complete replacement of transverse reinforcement (stirrups) or passive reinforcement (for pre-compressed concrete);
- making slimmer sections and larger openings for the same load-bearing capacity;
- increased productivity through faster reuse of formwork (high resistance in the first days);
- reduction of workmanship, transportation and materials for making formwork;
- higher rigidity for structural elements;
- low maintenance costs over the life of the structure;
- applications for prefabricated and monolithic elements;
- maximum durability for aggressive agents of any kind due to very low porosity;
- obtaining an apparent surface that does not require further finishing.
- the realization of ultra-high performance concrete by using mainly local materials is a remarkable performance in the field of structural concrete.
- brings important economic and qualitative advantages.
- however, much more attention needs to be paid to all the phases of design, transport and commissioning, because the errors are having a significant weight in the final product quality (UHPC).

References

[1] Magureanu C 2007-2010 Program PN II – IDEI – ID_1053 / 2007-2010 *Betoane Ultraperformante (Ecologia Mediului) Ultra-Performance Concrete (Environmental Ecology)* p 44

[2] Olariu I and Ioani A 1991 *Precast reinforced concrete frames with steel fibre concrete joints* (ACI Structural Engineering Review) p 175

[3] Olariu I, Ioani A and Poienar N 1988 Steel fiber reinforced ductile joints, (Proceedings of Ninth World Conference on Earthquake Engineering, Tokyo-Kyoto) vol **IV** p 28 657

[4] Schmidt M and Fehling E 2005 Entwicklung, Dauerhaftigkeit und Berechnung Ultrahochfester Betone (UHPC) Forschungsbericht DFG FE 497/1-1 *Structural Materials and Engineering Series* No. Kassel University Press ISBN 3-89958-108-3) p 19-29

[5] Lee N P and Chisholm D H 2005 *Reactive Powder Concrete* (study Report SR 146 Ltd. Judgeford, New Zealand) p 14 24 34 39

[6] Markovic I 2006 *High performance hybrid fibre concrete* (Development and Utilisation)

[7] Cheyrezy et al 1995 *Microstructural Analysis of RPC*

[8] Magureanu C, Heghes B, Corbu O, Szilágyi H and SoSa I 2008 Behaviour of high and ultra high performance fibre reinforced concrete (8HSC/HPC Symposium Tokyo) p 44

[9] Rossi P, Renwez St and Guerrier F 1996 Les bétons a ultra-hautes performances fibrés: formulation, comportement mécanique, modélisation (Colloque International Francophone sur les Bétons Renforcés de Fibres Métalliques Toulouse) p 217

IOP Conf. Series: Materials Science and Engineering **789** (2020) 012023 doi:10.1088/1757-899X/789/1/012023

Characteristics of buildings and seismic vulnerability assessment for the specific area of the city of Osijek

M Hadzima-Nyarko[1*], G Pavić[1], B Bulajić[2] and D Kukaras[3]

[1] Faculty of Civil Engineering and Architecture, University of Josip Juraj Strossmayer, Osijek, Croatia
[2] Faculty of Civil Engineering, University of Novi Sad, Novi Sad, Serbia
[3] Faculty of Civil Engineering Subotica, University of Novi Sad, Novi Sad, Serbia

[*] mhadzima@gfos.hr

Abstract. Seismic risk is a measure of the damage that is expected in a specific area and time interval. It is based on the level of seismicity of locations (hazard), the resistance of buildings (vulnerability), and the type, quality and quantity of exposed assets (exposure). It can be measured by the expected economic loss, loss of life or damage to property size. Reducing the seismic risk comprises three phases: assessment, planning and implementation. The risk assessment requires a multidisciplinary approach that takes into account the expected physical damage (damage to buildings and economic losses), as well as social, organizational and institutional factors. At the city level, a risk assessment should start assessing the physical damage as a result of basic tools and the link between risk and vulnerability values of buildings. In this paper the creation of a database of existing buildings, classification of buildings and a procedure for determining the building typology that prevail in the city of Osijek is described. Based on the formed database, it is possible to implement certain seismic vulnerability methods.

1. Buildings database for the city of Osijek

An initial step in assessing the vulnerability of all types of buildings potentially exposed to earthquakes is creating a database of buildings.

Although building databases are extremely important, data collection for them is rarely part of the established collection processes for data of interest to society [1].

The assessment of the exposure of and damage to buildings in the Republic of Croatia is aggravated by the fact that there is no city that has a database with information on the number, types and characteristics of existing buildings.

That is why, at the Faculty of Civil Engineering and Architecture in Osijek, a project of creation of the database of buildings for the city of Osijek has been started. The goal of this project is to produce reliable and useful information on attributes that describe the structural and other characteristics of city buildings, and which will satisfy the needs for accurate and high-quality earthquake and other risk assessments for the city of Osijek.

Building database also provides data needed for quantifying damage in monetary units [2].

A high-quality building database for a particular area of research implies, first and foremost, the availability of complete and detailed information on the geographical location of buildings, their total surface area and their structural characteristics. However, due to common problems such as inaccessibility or lack of official data and statistics to assign relevant characteristics and parameters, approximate procedures are often used.

The process of creating a database requires the collaboration of numerous experts in various fields of science as well as civil engineering practitioners. Gathering all the necessary building data requires finding and analyzing building documentation, field visual data collection and measurements, and the final analysis and documentation of the collected data.

Numerous methods have been developed to estimate earthquake exposure and damage to buildings. The reliability and precision of these methods depends on the classification of the buildings, which must cover all the individual features of the buildings in a uniform manner.

Although the use of the already developed classifications (PAGER-STR, HAZUS, GEM, etc.) is common, their use in geographic environments for which they have not been developed may cause epistemic uncertainties in a later vulnerability model [3].

That is why the building database for the city of Osijek, at the building level only certain, valuable attributes are being collected, as the most likely related to certain existing building typologies in the city and which allow for risk calculation for each investigated building according to its compatibility in relation to one or more construction.

The building database of the city of Osijek contains more than 1000 buildings so far and continues to be completed and upgraded.

2. Structural systems Classification

By identifying the recurring characteristics, and analyzing similarities and changes in the structural features of existing buildings in the city of Osijek, the structural systems of the existing buildings in Osijek are classified as follows:

- M5 – unreinforced masonry structures with flexible floors (old bricks),
- M6 – unreinforced masonry structures with rigid floors,
- M7 – masonry structures with horizontal and vertical ties (confined masonry),
- RC1 – reinforced concrete frames,
- RC4 – shear walls (without E.R.D.).

In adopting the classification, guidelines and designations from Giovinazzi [4] were respected and used.

The years when buildings were built and the basic structural and performance requirements in building regulations at the time when they were built, were the basic criteria we used to classify buildings into adopted classes. For the buildings in the current database, figure 1 shows percentages of certain structural system classes.

The data shown in figure 1 indicate that 98% of the buildings that are so far included in the Osijek's building database, are masonry structures.

The significant percentage of masonry in the city and the complex and extremely nonlinear behavior of masonry during the earthquake justify the creation and existence of this database and reduce the fact that relatively few low intensity earthquakes have been recorded in the city and the surrounding area. The database represents the basis for reliable assessment of the earthquake vulnerability for all types of buildings in the city, and hence for the reliable assessment of seismic risk [5].

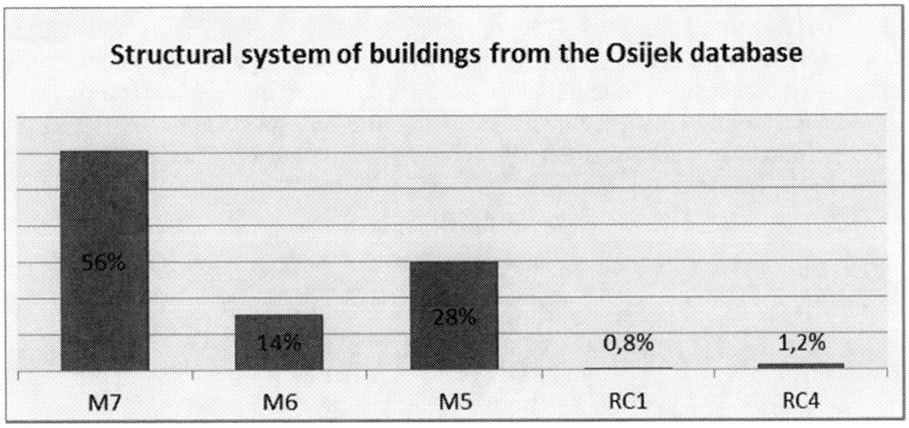

Figure 1. The proportion of different structural systems in the current building database for the city of Osijek.

A series of seismic microzonation studies made for the north-western Balkans [6-10] and a recent study of strong earthquake ground motion in the same region [11], all show that deep geological sediments strongly affect at the severity of both longer period waves. The Osijek is located in the valley, near the river Drava, and the deep geological deposits beneath it are comprised of the rocky, sandy and muddy soils with infrequently high groundwater levels. These deposits can be significantly disturbed by distant earthquakes and cause damage to the housing stock, especially to masonry buildings.

3. Features that describe the buildings in the database

The earthquake resistance of a building is influenced by some characteristics more than others, e.g. the lateral resistance of the structure and the predominant material by which the structure is built (brick, reinforced concrete, steel, etc.). Therefore, selecting the relevant parameters that will best describe the construction and other features of the building is the initial step in the process of creating a building database.

The Global Earthquake Model (GEM Basic Building Taxonomy v2.0) recommends 13 building characteristics that, alone or in synergy, can significantly affect the behavior of buildings during an earthquake [12]. However, in reality it is almost impossible to identify all of those characteristics.

When selecting the characteristics that will describe the existing buildings of the city of Osijek, the recommendations from the GEM were taken into account, but their number was reduced due to the lack of availability of some of them, and their selection was adjusted to the specific characteristics of the construction technology and materials from which the existing buildings in the city were built.

A form was created for the field data collection, and a mobile application was later developed to collect the following building data:

- building location information: address, cadastral number,
- the position of the building in relation to the particular city block,
- general information: the purpose of the building, the year of construction and/or reconstruction, (if any), and the number of persons living in the household,
- information on the geometric characteristics of the building: floor plan dimensions, net and gross floor planes, floor plan blueprints, floor level, floor space, and total height,
- regularity in floor plan and height,
- information on the main structural system of the building,
- information on building materials used,
- information on the roof structure and cover.

According to Salgado-Galvez *et al.* [13], the characteristic that can significantly influence potential earthquake damage to buildings is the age of the structure. In addition, this characteristic can be used to determine the used design codes and to define the type of load-bearing structures.

Given the characteristic construction technologies, the time periods during which buildings from the city of Osijek database were constructed are divided into 4 groups – see figure 2.

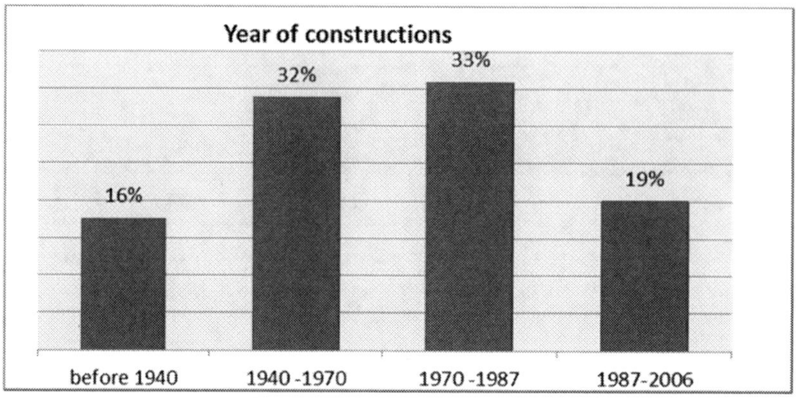

Figure 2. The proportion of buildings in the current database for the city of Osijek with respect to year of construction.

The data collected so far on the year of construction of the buildings in the database show that the largest number of buildings was built between 1970 and 1987, and the smallest number of buildings in the period prior to 1940.

Among the most critical features in terms of the expected earthquake behavior of buildings are structural irregularities (floor plans and/or verticals).

González Herrera and Gómez Soberón [14] studied the devastating earthquakes of 1980-2003 and their impact on residential buildings. Their research has shown a great cause and effect relationship between the degree of damage to a building (due to an earthquake) and structural irregularities in the affected buildings.

Eurocode 8 [15] also stipulates that the structural system of buildings must be approximately symmetrical with respect to the two orthogonal axes to ensure sufficient lateral rigidity and proper mass distribution. In addition, the building must have a compact configuration plan and the corners or edges of the entrance, if any, should not exceed 5% of the floor area.

The buildings covered by the database are mostly rectangular. With respect to the regularity of the ground plan, 75% of buildings have the regular floor plan and 25% have the irregular floor plan, as shown in figure 3.

An important characteristic for damage modeling is the position of the buildings in the city block. Specifically, in densely populated urban centers, buildings are at risk of impacting each other, and more damage can be expected on buildings located at the end of a series of buildings and at corners.

The fundamental period of vibration can have a significant impact on the behavior of a building during an earthquake. The height of the building is one of the characteristics that determines this dynamic property.

Higher buildings are characterized by longer vibration periods and are usually considered flexible; however, this does not mean that they will suffer less damage than low or lighter buildings.

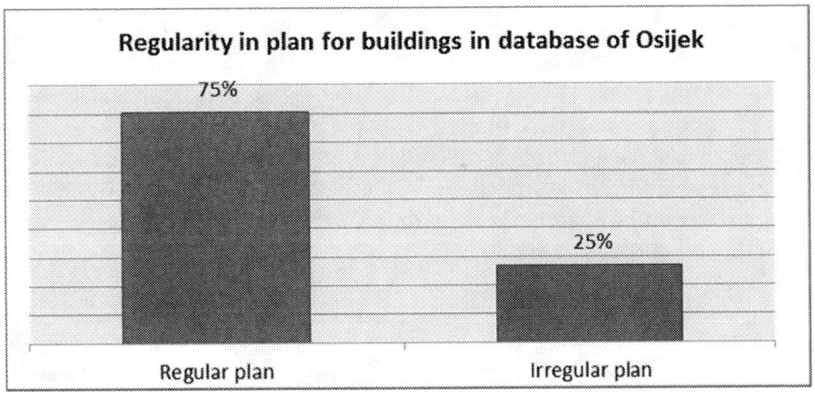

Figure 3. The proportion of buildings in the current database for the city of Osijek with respect to regularity in plan.

In addition to determining the earthquake hazard, the earthquake behavior of high and low buildings is also influenced by the characteristics of the underlying soil, and not only by the first 30 m of the soil profile, but also by the deep geological layers [11].

Numerous modern reinforced concrete high rise buildings were destroyed in the 1985 Mexico City earthquake due to amplified vibration in soft soil deposits and because of the resonance caused by a predominant shaking frequency of about 2 sec.

Most of the buildings listed so far in the Osijek building database are lower buildings, generally up to 4 stories high.

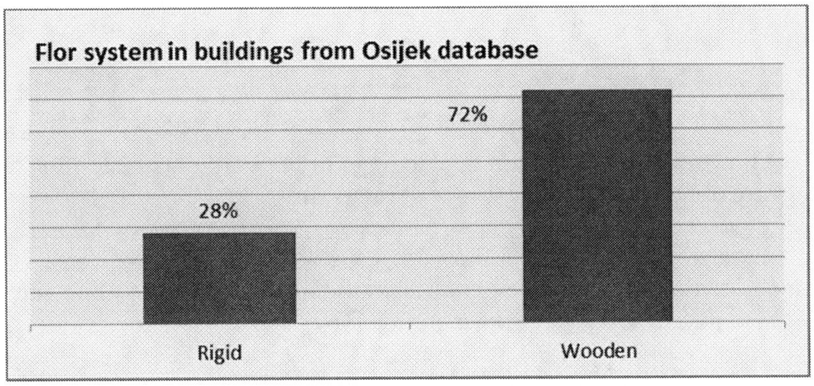

Figure 4. The proportion of buildings in the current database for the city of Osijek with respect to floor systems.

Good earthquake transmission during earthquake action is ensured by the floor system that is sufficiently rigid in its plane. The floor system must prevent the walls from being separated and ensure that during an earthquake, the building acts as a single spatial structure with well-connected elements. Softer floors (i.e. wooden floors) often cannot be considered absolutely solid in their plane, so the transmission of earthquake forces onto all vertical elements is not entirely ensured.

In the city of Osijek database, floor systems are classified as rigid or wooden floor systems, and their percentage distribution in the current database is shown in figure 4.

The parameter that has a strong influence both on the behavior of the structural elements and on the entire structure during earthquakes is the construction material or the type of building material from which the building was made. Buildings that are completely identical in terms of floor planes, number of floors, structural type, etc., but made of different types of materials, will have significantly different behavior even for the same intensity of earthquake ground shaking. The worst earthquake behavior can be expected in the case of unreinforced masonry and reinforced concrete frames with masonry fill, while reinforced concrete buildings with shear walls, steel frame buildings and wood buildings can be expected to show very good performance during the same earthquake.

For the city of Osijek database purposes, the construction material data are collected on site, from project documentation, based on knowledge of the building users or by evaluation based on knowledge of building age and researcher's experience.

The walls of the base building are made of brick and/or concrete. Using a brick block is the most common: 922 buildings, or 86%, were built by brick blocks. The ordinary (most commonly used) brick was used in 128 buildings, which is 12% of the total number of buildings. Concrete walls were used in 15 buildings or 1%, and concrete blocks in 5 buildings or 0.5% of the total number of buildings. In the database there were also 5 masonry buildings which were constructed of raw brick (in Tvrđa and Upper Town), and that makes 0.5% of all buildings (figure 5).

The walls of the building from the Osijek base are mostly made of brick elements – see figure 5. The most commonly used are brick blocks in 86% of buildings and ordinary brick in 12% of buildings. In the oldest parts of the city (in Tvrđa and Upper Town), the walls of the building are made of adobe.

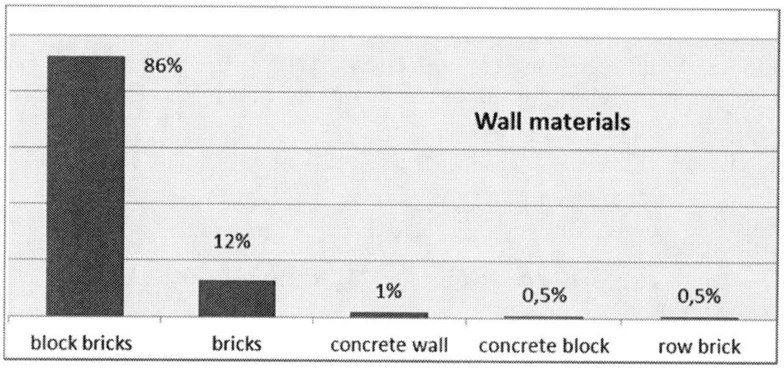

Figure 5. The proportion of buildings in the current database for the city of Osijek with respect to wall materials.

The data on buildings collected so far shows that the average gross floor area is 769 m^2 and the average floor area is 226 m^2. The total height of the buildings is between 4.48 m and 37.3 m, and the average height is 8.46 m. These data are shown in figure 6.

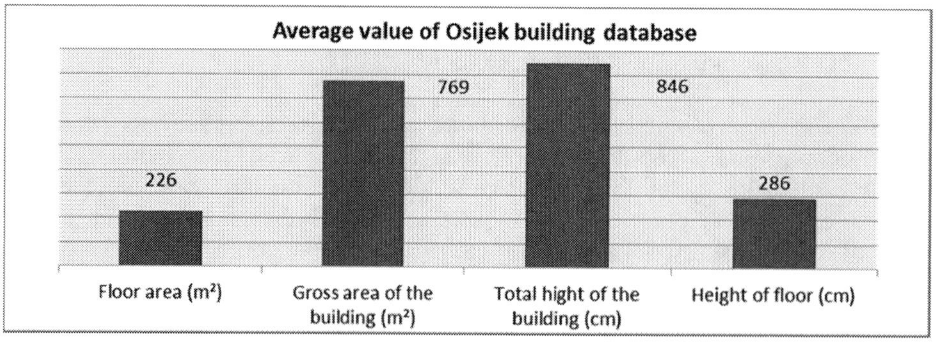

Figure 6. The proportion of buildings in the current database for the city of Osijek with respect to average dimensions.

So far, the data have been collected for buildings in the following city areas: Tvrđa, larger parts of the Upper Town, New Town (Sjenjak), some parts of the Industrial District, Retfala and South II, and the current database currently covers at least the whole Lower Town area – see figure 7.

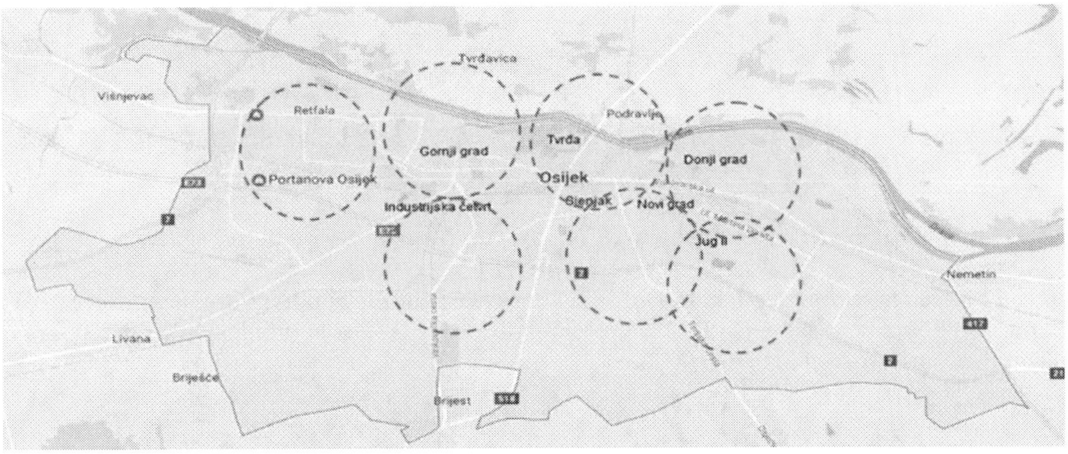

Figure 7. Parts of the city of Osijek that are currently included in the database.

4. Seismic vulnerability assessment for the selected city block

4.1. Characteristics of the buildings from the selected city block

In this article, a seismic vulnerability assessment is applied for the selected city block. The observed area includes 69 buildings in the wider city center of Osijek (figure 8), of which: 47 are residential buildings, 14 residential/commercial buildings and 7 commercial buildings. The necessary data on the buildings were collected by field work, searching on the Internet and visiting the State Archives of the City of Osijek.

Figure 8. Selected area for which the seismic method is applied in this study.

Most of the buildings are single-storey buildings with wooden floors, and their structural system is a masonry structure. The buildings are old, built with a traditional approach, while bricks are used as the main building material. Figure 9 shows an example of a typical unreinforced masonry building with a flexible floor structure.

Figure 9. Typical building from the city block analysed in this study.

It can be seen from figure 10 that in this city block, 96% of the buildings are unreinforced masonry buildings with wooden floors (M5), 3% are unreinforced masonry buildings with rigid floors (M6), while only 1% is confined masonry (M7).

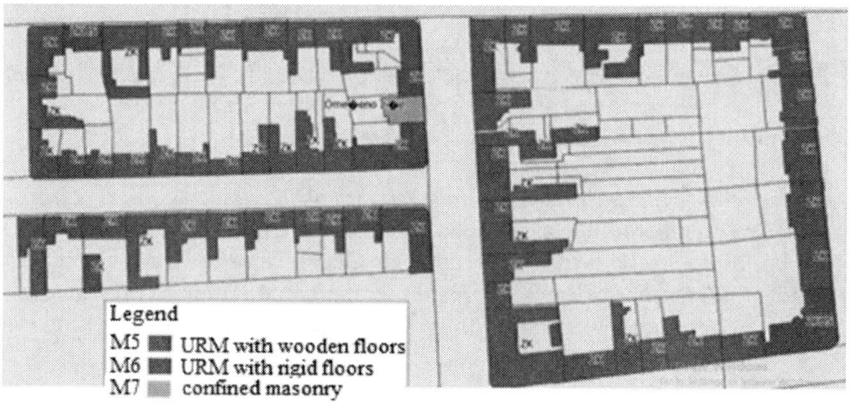

Figure 10. Structural system of the buildings in the city block that is analyzed in this study.

4.2. Macro seismic method

Macro seismic method, developed by Giovinazzi and Lagomarsino [4], [16] from the definition provided by the European Macro seismic scale EMS-98 [17], is also called "indirect" since the relationship between seismic action and the response is obtained through the vulnerability index. Vulnerability is measured in terms of a vulnerability index V and a ductility index Q, both evaluated considering the building typology and its constructive features [16].

The correlation between the seismic input and the expected damage, as a function of the assessed vulnerability, is expressed in terms of vulnerability curves described by a closed analytical function [16]:

$$\mu_D = 2.5\left[1 + tanh\left(\frac{I + 6.25 \times V_I - 13.1}{Q}\right)\right], \quad (1)$$

where:

I – the macroseismic intensity;

V_I – the vulnerability index;

Q – the ductility index; the proposed value is 2.3 for residential buildings.

The vulnerability index V_I of every building considers the regional vulnerability modifier and the behavior modifier. The behaviour modifiers take into account the following characteristics of each building:

- building irregularities,
- number of floors,
- vertical irregularity,
- state of preservation,
- aggregate building position etc.

The total behavior modifier for a single building is the sum of the individual values.

The calculated vulnerability index for every building can be connected with most probable damage state (table 1).

Table 1. Damage states and mean damage index values [18]

Mean damage index intervals	Most probable damage state	EMS-98 Damage Grade
0-0.5	None	D0
0.5-1.5	Slight	D1 (Grade 1)
1.5-2.5	Moderate	D2 (Grade 2)
2.5-3.5	Substantial to heavy	D3 (Grade 3)
3.5-4.5	Very heavy	D4 (Grade 4)
4.5-5.0	Destruction	D5 (Grade 5)

The calculated vulnerability indices for the buildings of the selected city block, connected with most probable damage state, are presented in figure 11 for two levels of intensity, VIII and IX.

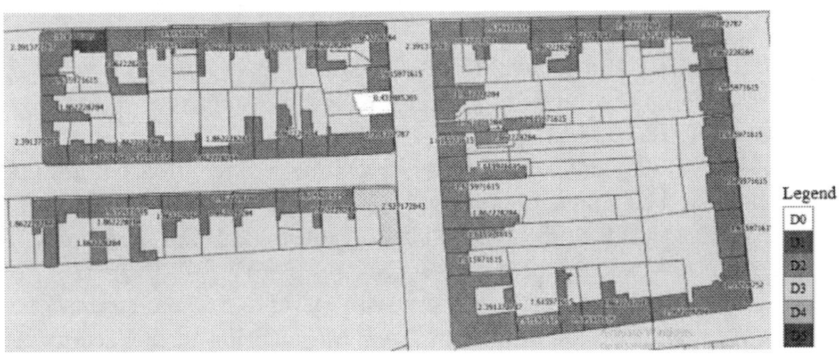

(a)

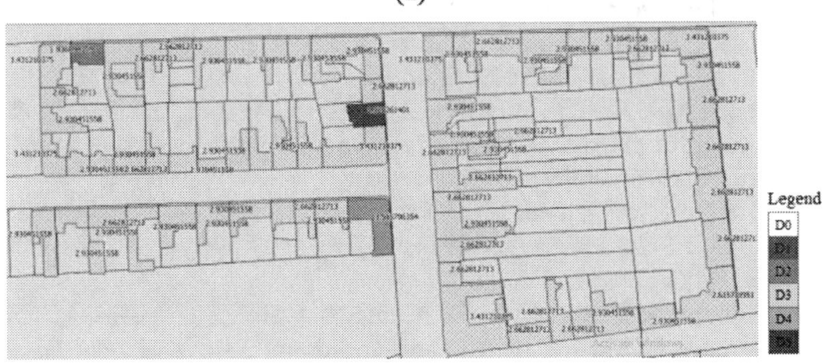

(b)

Figure 11. Most probable damage states obtained for intensities: a) VIII, b) IX, for the city block that is analyzed in this study – the same colors that are used in table 1 are used here to indicate different damage states.

5. Conclusions

Seismic vulnerability assessment, as a main component of earthquake risk, can only be performed if there is exposure, i.e. a database of buildings. In this paper, the characteristics of the buildings database of the city Osijek is presented, which, so far, contains more than 1,000 buildings. The dominant structural systems, that is, the typology of buildings, the characteristics of the buildings in the data base are also shown: number of floors, planar and vertical regularity, position in the block, year of construction, etc. are collected. The collected data on buildings, visual inspection, and the available documentation in the National Archives are classified in the database and analyzed. The purpose of the database is to enable an assessment of the seismic vulnerability for the existing city buildings.

In the example of the selected Osijek city block, the goal was to assess seismic vulnerability. Therefore, the vulnerability was conducted and estimated for the several blocks of the city Osijek by an empirical earthquake vulnerability estimation method, the macroseismic method. The macroseismic method calculates a vulnerability index, which is a function of the most likely value of an index of damage of a type of building, a regional and a behaviour modifier. The influence of the behavior modifier is expressed in the increase in the vulnerability index. Based on the calculated vulnerability indices, the mean damage grade of each building was calculated for two levels of seismic intensity. The mean damage grade can be correlated with the most likely damage state according to EMS-98. According to the obtained results, the calculation of the macroseismic method show that the unreinforced masonry structures with wooden floors for earthquake of intensity VIII will suffer moderate damages (most likely damage state), and for earthquake of intensity IX those buildings will suffer substantial to heavy damages. Due to similarities in existing building types, these conclusions can be extended in other regions of Croatia, especially in northern and eastern parts of the country.

References

[1] Dell'Acqua F, Gamba P and Jaiswal K 2013 Spatial aspects of building and population exposure data and their implications for global earthquake exposure modeling *Nat. Hazards* **68/3** 1291-1309 DOI: 10.1007/s11069-012-0241-2

[2] Barbat A H, Moya FY and Canas J 1996 Damage Scenarios Simulation for Seismic Risk Assessment in Urban Zones *Earth. Spect.* **12/3** 371–394

[3] Jaiswal K, Wald D and Porter K 2010 A global building inventory for earthquake loss estimation and risk management *Earthq. Spect.* **26** 731–748 DOI: 10.1193/1.3450316

[4] Giovinazzi S 2005 The vulnerability assessment and the damage scenario in seismic risk analysis. *Ph.D. Thesis*. Department of Civil Engineering of the Technical University Carolo-Wilhelmina at Braunschweig and Faculty of Engineering Department of Civil Engineering of University of the Florence

[5] Hadzima-Nyarko M and Kalman Šipoš T 2017 Insights from existing earthquake loss assessment research in Croatia *Earthquakes and structures* **13/4** 365–375

[6] Lee V W, Manić M I, Bulajić B Đ, Herak D, Herak M and Trifunac M D 2015 Microzonation of Banja Luka for performance-based earthquake-resistant design *Soil. Dyn. Earthq. Eng.* **78** 71–88 DOI 10.1016/j.soildyn.2014.06.035

[7] Lee V W, Trifunac M D, Bulajić B Đ, Manić M I, Herak D and Herak M 2017 Seismic microzoning of Belgrade *Soil. Dyn. Earthq. Eng.* **97** 395–412 DOI: 10.1016/j.soildyn.2017.02.002

[8] Lee V W, Trifunac M D, Bulajić B Đ, Manić M I, Herak D, Herak M, Dimov G and Gičev V 2017 Seismic microzoning of Štip in Macedonia *Soil. Dyn. Earthq. Eng.* **98**, 54–66 DOI 10.1016/j.soildyn.2017.04.003

[9] Lee V W, Trifunac M D, Bulajić B Đ, Manić M I, Herak D, Herak M and Dimov G 2017 Seismic microzoning in Skopje, Macedonia *Soil. Dyn. Earthq. Eng.* **98** 166–182 DOI 10.1016/j.soildyn.2017.04.007

[10] Bulajić B Đ, Bajić S and Stojnić N 2018 The effects of geological surroundings on earthquake-

induced snow avalanche prone areas in the Kopaonik region *Cold Regions Science and Technology* **149** 29–45 DOI 10.1016/ j.coldregions. 2018.02.005

[11] Bulajić B Đ, Manić M I and Lađinović Đ 2013 Effects of shallow and deep geology on seismic hazard estimates – a case study of pseudo-acceleration response spectra for the north-western Balkans *Nat. Hazards* **69** 573–588 DOI 10.1007/s11069-013-0726-7

[12] Brzev S, Scawthorn C, Charleson AW, Allen L, Greene M, Jaiswal K and Silva V 2013 GEM Building Taxonomy Version 2.0, Exposure modelling *Technical report* 2013-02 v1.0.0.

[13] Salgado-Gálvez M A, Cardona OD, Carreño M L and Barbat A H 2015 Probabilistic seismic hazard and risk assessment in Spain, National and local level case studies, *Monografía CIMNE IS* **69**, ISBN: 978-84-993307-7-3 Depósito legal: B-2889-2015, Monografías de Ingeniería Sísmica Editor Barbat AH

[14] González Herrera R and Gómez Soberón C 2008 Influence of Plan Irregularity of Buildings, *Proc. of the 14th World Conference on Earthquake Engineering* Beijing October 12-17

[15] EN 1998-1 2004 Eurorcode 8: Design of Structures for Earthquake Resistance. Part 1: General Rules, Seismic Actions and Rules for Buildings. Bruxeles: European Committee for Standardization

[16] Lagomarsino S and Giovinazzi S 2006 Macroseismic and mechanical models for the vulnerability and damage assessment of current buildings *Bull. Earthq. Eng.* **4** 415–44

[17] Gottfried, M., Roger M.W., Schwarz, J., Stucchi, and M. Cahiers 1998 European Macroseismic Scale (EMS-98) Ed. Grünthal, Centre Européen de Géod ynamique et de Séismologie, vol. **15** Luxemburg

[18] Lantada N, Irizarry J, Barbat AH, Goula X, Roca A, Susagna T and Pujades LG 2010 Seismic hazard and risk scenarios for Barcelona, Spain, using the Risk-UE vulnerability index method *Bul. Earthq. Eng.* **8** 201–229

Investigations regarding the degradation state of a Romanian country road

D Hoha[1*], N Țăranu[1], D Ungureanu[1], V Boboc[1], D N Isopescu[1], E-L Pleşcan[2], A Boboc[1], M C Scutaru[1] and L I Domnița[1]

[1] "Gheorghe Asachi" Technical University, Iasi, Romania,
[2] Transilvania University of Brasov, Faculty of Civil Engineering, Brasov, Romania

[*] didi.hoha@gmail.com

Abstract. This study is aimed to identify and to describe some of the most common types of pavement degradations that may develop on some Romanian county roads. According to the road safety management system, The County Direction of Road and Bridge Administration has the responsibility to ensure the traffic conditions, by preventing or repairing any degradation that may have adverse effects on the ride's safety and comfort. The most common types of degradations that can be identified on the Romanian county roads are: block cracks, longitudinal / transverse cracks, joint stepping, rocking, joint sealant defects and local declines. The process of interventions for restoring the road safety condition is continuous and span through the entire lifecycle of the structure (planning, design, build, operate, maintain and decommission). In order to illustrate the degradation state and to identify the most appropriate type of intervention required to restore the road safety condition, a case study is presented. The latter consists in a county road located in the Northern part of Romania (Suceava County), designated with the route number DJ 177A. The viability of the road was evaluated based on visual inspections and laboratory testing of its component materials. Based on the experimentally results several intervention works were suggested.

1. Introduction

The performance of road pavements degradation with the impact of the traffic volume and the effect of the environmental climate, in case of heavy traffic and severe environmental conditions, the road structure is usually designed as a concrete pavement, also referred as rigid pavement. This type of road structure is considered a long-lasting solution, since it possesses several advantages when compared to regular bituminous pavement [1, 2]. For example, the concrete pavements are suitable for road traffic with large point loads and they also withstand diesel spillage and other aggressive materials [3]. Moreover, the rigid pavements are suitable for cases where sub-grade strength is low and the road is exposed to high temperature. However, due to the evaporation and the hydration processes, the humidity of the concrete cannot be maintained in a close range. This phenomenon causes a negative pressure in the capillary pores and thus, produces drying shrinkage and cracks inside the concrete element [4].

According to the Romanian National Norm CD 155-2001 [5] the assessment of the concrete pavement performances is performed based on the value of the degradation evaluation index. The latter accounts for various degradations, including block cracks, longitudinal and transverse cracks, joint stepping, rocking, surface spelling and raveling. These degradations are, usually, caused by the gradually accumulated damage of the pavement constituent materials, under the combined action of traffic load, humidity, light and temperature, in a significant period of time [6-10].

In order to illustrate the degradation state and to identify the most appropriate type of intervention required to restore the road safety condition, a case study is presented. The latter consists in a county road located in the Northern part of Romania (Suceava County), designated with the route number DJ 177A. The road starts from the Frasin town, near the bridge over the Suha River and connects two major roads of national interest DN 17 and DN 17B. It should be also noted that the DJ 177A connects the towns of Frasin and Broşteni (where the headquarters of several wood processing companies are located) and crosses the settlements of Frasin, Doroteia, Stulpicani, Ostra, Tarniţa, Holda, cumulating a total length of 43.615 km. Thus, this road serves as the main route for the transportation of significant wood and timber cargos, between local producers and retailers. The geometry of the road consists in a 6 m carriageway (two driving lanes of 3 m width) bordered by two ballast shoulders, each having a width of 1 m. The degradation evaluation index was assessed by both visual inspections and laboratory testing of the pavement materials, including uniaxial compression test, indirect tensile test and static modulus of elasticity evaluation.

2. Pavement degradation index

In most of the cases, the evaluation index of concrete pavement serviceability for maintenance focus on the features related to driving safety, such as: the bearing capacity, the deflection state, the surface condition, the riding quality and comfort, the joints condition, and the skidding resistance [11, 12]. These six aspects of concrete pavement performance can be easily evaluated with automatic test equipment, such as deflectometers and bearing plates. However, the results provided by these devices represent the strength of the entire pavement structure and cannot be limited to the structural layers of interest.

Therefore, the exact bearing capability of each individual course can be determined only through laboratory tests, performed on the pavement materials. In this manner, the evaluation index of the concrete pavement can be properly determined, by quantifying the contribution of each component of the structure.

2.1. Identification of the homogeneous sections for site investigation

According to the Romanian Norm AND 547-2013, the homogeneous sections for site investigations are determined by taking into account the following parameters: the traffic characteristics, the type of the road structure, the hydrologic type, the geometry of the transverse profile and the level of degradation (defined empirically by the grading good, medium or bad). The site evaluation sector located between km 0 + 000 and km 2 + 520 was selected to determine the evaluation index of the surface course of the county road DJ 177A. According to the design documentation, the road structure is homogeneous throughout the entire site investigation sector.

2.2. Degradation state of the site investigation sector

The evaluation of the degradation state of the concrete pavement was performed according to the provisions given in the Romanian Norm AND 547/2013. The concrete pavement of the county road DJ 177A consists in a thin-plate structure exposed to the external environment for 54 years. Due to various traffic loads and severe environmental conditions, several types of degradation were developed. The most relevant degradations that were identified during the visual inspection stage are described in the following.

2.2.1. Joint sealant degradation. This degradation consists in cracks that are developed in the seal material, creating a break path for external materials and substances to penetrate into the joints. The joint degradations do not disturb the road traffic, but favor the occurrence and development of several failure mechanisms. This is due to the atmospheric water, which can seep through the joints into the lower road courses and into the foundation ground. The joint sealant degradation of the county road DJ 177A is illustrated in figure 1.

2.2.2. Longitudinal cracks. This type of concrete pavement degradation consists in a large crack (over 3 cm), located at the longitudinal joints level. Similar to the joint sealant degradation, the longitudinal cracks can lead to water infiltrations in the lower structural courses. The longitudinal cracks developed on the county road DJ 177A are illustrated in figure 2.

Figure 1. Joint sealant degradation on the county road DJ 177A.

Figure 2. Longitudinal cracks developed on the county road DJ 177A.

2.2.3. Slab settlement/joint stepping degradation. The slab settlement degradation is characterized by the development of a level variation between the boundaries of two adjacent slabs. A concrete pavement is considered to have an unacceptable settlement when the differences between the longitudinal or the transverse levels among two neighbour slabs is greater than 5 mm. The slab settlement degradation occurred on the county road DJ 177A is illustrated in figure 3.

2.2.4. Surface spalling/Ruptures. Usually, these degradations consist in a break down or a disintegration of the slab surface at edges, joints and corners. The surface spalling degradation can also be developed as a continuous path of cracks with variable depth that can even reach the thickness of the surface course. The ruptures occur either on small surfaces, or extended ones, along the entire length of the transverse joints. The surface ruptures developed on the county road DJ 177A are illustrated in figure 4.

Figure 3. Slab settlement/joint stepping developed on the county road DJ 177A.

Figure 4. Surface spalling developed on the county road DJ 177A.

2.2.5. Transverse cracks. This degradation consists in a crack that develops transversely across the pavement/slab. Depending on the opening level, the cracks can be classified as active or passive. Active cracks are considered to have openings which increase or decrease with more than 0.5 mm at a daily variation of the concrete temperature above 10 °C.

The passive cracks are considered those whose openings remain almost constant as the concrete temperature varies. The transverse cracks that were identified on the county road DJ 177A are illustrated in figure 5.

2.2.6. Longitudinal cracks. This degradation consists in an unconnected crack that develops longitudinally along the pavement/slab. Typically, it occurs due to the settlement of the foundation layers, or due to the transverse contractions of the concrete slabs (triggered by uneven load bearing capacity throughout the carriageway width). The longitudinal cracks that were identified on the county road DJ 177A are illustrated in figure 6.

Figure 5. Transverse cracks identified on the county road DJ 177A.

Figure 6. Longitudinal cracks identified on the county road DJ 177A.

2.2.7. Corner cracks. This degradation consists in several diagonally arranged cracks, in a right-angled triangle pattern, whose hypotenuse connects a joint or a transverse crack with a longitudinal joint or a slab edge. These degradations can occur frequently on both sides of two adjacent slabs. The corner cracks that were identified on the county road DJ 177A are illustrated in figure 7.

2.2.8. Potholes. The potholes that develop in concrete pavements are characterized by a roundly shaped cavity, having depths greater than 3 cm. Generally, the potholes develop in heterogeneous segregated concrete pavements, by gradually material removal due to repeated traffic. The potholes that were identified on the county road DJ 177A are illustrated in figure 8.

Figure 7. Corner cracks developed on the county road DJ 177A.

Figure 8. Potholes identified on the county road DJ 177A.

2.3. Evaluation of the pavement degradation index

The pavement degradations were grouped by representative types, and then evaluated according to the provisions given in the Romanian Norm CD 155/2001. The surface degradation state is characterized by the pavement degradation index (ID=14.7), calculated with the following equation.

$$\text{ID} = number\ of\ degraded\ slabs\ (NDS) \cdot total\ number\ of\ slabs\ on\ a\ lane\ (TNS)^{-1} \quad (1)$$

$$\text{NDS} = \text{D1} + 0.5 \cdot \text{D2} + (0.5 \cdot D3 \cdot N) \cdot S^{-1} + 0.3 \cdot \text{D4} \quad (2)$$

where N = the number of slabs on a lane = 4 pcs.; S = the sector surface measured on a lane = 90 m^2; D1 = the number of slabs damaged by settlements = 1 pcs.; D2 = the number of slabs damaged by filled potholes and crocodile cracks = 2 pcs.; D3 = the area of the surface damaged by longitudinal, transverse and corner cracks = 52.2 m^2; D4 = the area of the surface damaged by ravelling = 0.

$$\text{NDS} = 1 + 0.5 \cdot 2 + (0.5 \cdot 52.2 \cdot 4) \cdot 90^{-1} + 0.3 \cdot 0 \quad (3)$$

$$\text{NDS} = 3.16\ \text{pcs.} \sim 3\ \text{pcs} \quad (4)$$

2.4. The surface degradation rating criteria

The surface degradation rating criteria is established based on the pavement degradation index (ID), as presented in table 1 (CD 155-2001) [5]. As it can be observed, for the value of the pavement degradation index (ID) of 14.7, the surface degradation is considered to be "bad". According to the Romanian Norm CD 155-2001, Appendix 7, mandatory maintenance works are required, generally consisting in laying either minimum two bituminous layers, or an equivalent concrete layer.

Table 1. The surface degradation rating criteria (CD 155-2001) [5].

Surface Degradation Rating	ID Value
BAD	> 13
MEDIUM	7.5...13.0
GOOD	5.0...7.5

VERY GOOD	< 5

3. Uniaxial compression test

The uniaxial compression test is a well-established testing practice, which enables the evaluation of the compressive strength of various pavement materials [13].

Also, through repeated uniaxial compressive tests, the static modulus of elasticity can be determined. For this work, 8 cylindrical core specimens (4 for compressive tests and 4 for elasticity modulus evaluation) were extracted from the existing concrete course of the county road DJ 177A [14, 15]. The compressive tests were performed using a universal testing machine, Zwick Roell 1000 kN. The force-displacement variations are graphically depicted in figures 9-12. The ultimate compressive strength and the corresponding moduli of elasticity are listed in table 2.

Table 2. Compressive strengths and elasticity moduli.

No. crt.	h (*cm*)	Φ (*cm*)	Test type	F_{max} (*kN*)	E_c (*MPa*)	f_c (*MPa*)
1	19.20	9.10	Compressive strength	183.60	-	28.23
2	19.50	9.40	Compressive strength	250.00	-	36.02
3	18.50	9.30	Compressive strength	282.60	-	41.60
4	18.80	9.30	Compressive strength	211.60	-	31.15
5	19.90	9.40	Modulus of elasticity	-	33543.5	-
6	20.20	9.30	Modulus of elasticity	-	28394.8	-
7	19.80	9.30	Modulus of elasticity	-	43209.4	-
8	19.20	9.30	Modulus of elasticity	-	35493.4	-

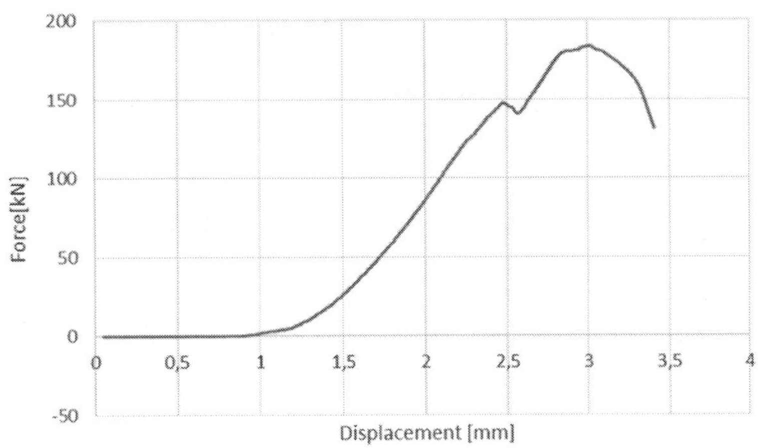

Figure 9. Core sample 1. Force-displacement variation.

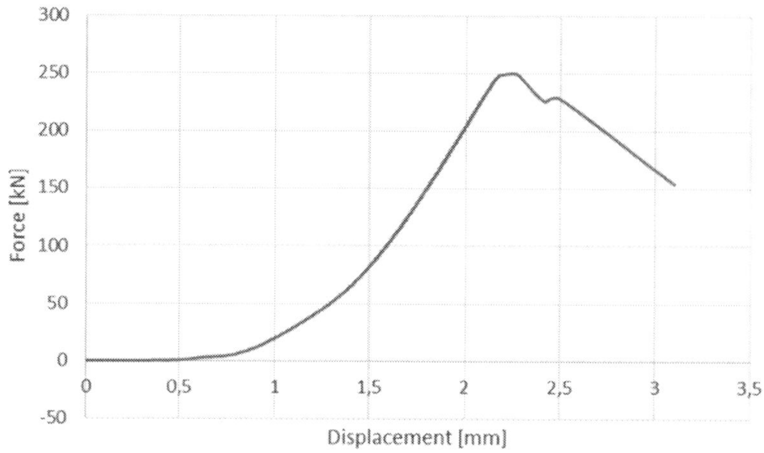

Figure 10. Core sample 2. Force-displacement variation.

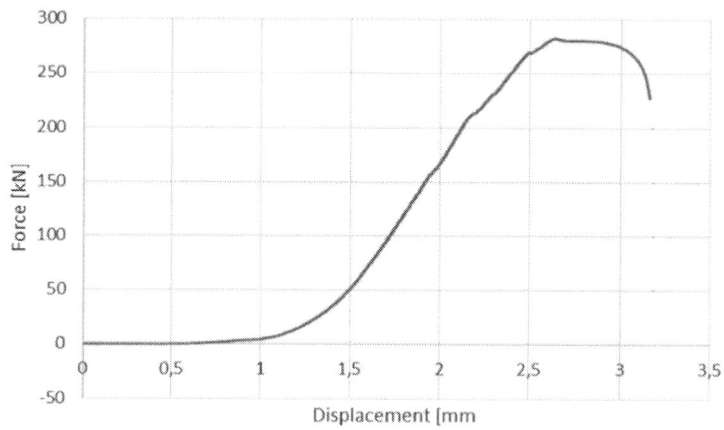

Figure 11. Core sample 3. Force-displacement variation.

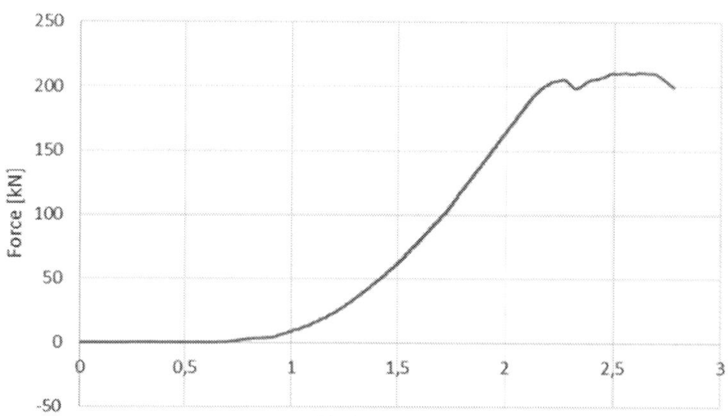

Figure 12. Core sample 4. Force-displacement variation.

4. Indirect tensile test

The indirect tensile test is a commonly used test, performed to support the pavement design and to evaluate the courses tensile characteristics. Moreover, the indirect tensile strength is often used to evaluate the fatigue stress at the bottom of the continuous courses. Most importantly, the cores extracted from existing concrete pavements, even those consisting in relatively thin layers, can be tested directly in the laboratory, by applying this test method. For this work, 4 cylindrical core specimens were extracted from the existing concrete course of the county road DJ 177A [16]. The indirect tensile tests were performed using a universal testing machine, Zwick Roell 1000 kN. The ultimate tensile forces and the ultimate tensile strengths are listed in table 3.

Table 3. Ultimate tensile forces and strengths.

No. crt.	h (*cm*)	Φ (*cm*)	Test type	F_{max} (*kN*)	E_c (*MPa*)	f_{ct} (*MPa*)
1	17.50	9.30	Tensile strength	72.00	-	2.82
2	17.80	9.40	Tensile strength	82.40	-	3.14
3	17.30	9.30	Tensile strength	95.00	-	3.76
4	15.00	9.30	Tensile strength	91.20	-	4.16

5. Conclusions

This work presents various investigations, performed both on site and on laboratory tests aiming to evaluate the state of a county road. Based on the visual inspection findings, it can be concluded that the road was inadequately exploited and the maintenance works that were conducted during the service life period were unsatisfactory. The degradations that were identified on the road surface cover a large spectrum of defects, which can negatively influence the ride comfort and the road safety conditions. Since the road administration did not perform the proper type of intervention works (only some asphalt laying on isolated surfaces), the required works to remedy the degradations are complex and expensive. The possible solutions consist in laying either minimum two bituminous layers, or an equivalent concrete layer. The pronounced variations of the elasticity moduli, of the compressive and tensile strengths indicates that the traffic was not consistent, most likely fluctuating between periods with heavy and light volumes. Also, the scattering of results indicates that the maintenance works were executed on isolated portions of the concrete pavement. If the maintenance works would have been executed on time and on the entire road surface, the mechanical characteristics of the course constituent materials would have ranged in a considerable narrow interval.

References

[1] Florida Department of Transportation, Office of Design, Pavement Management Section 2019 *Rigid pavement design manual*
[2] Arizona Department of Transportation, Roadway Engineering Group, Pavement Design Section 2017 *Pavement design manual*
[3] National Concrete Pavement Technology Center, Iowa State University, Institute for Transportation 2018 *Guide for Concrete Pavement Distress Assessments and Solutions: Identification, Causes, Prevention and Repair*
[4] Yang J, Guo Y, Shen A, Chen Z, Qin X and Zhao M 2019 Research on drying shrinkage deformation and cracking risk of pavement concrete internally cured by SAPs *Construction and Building Materials* **227** 116
[5] CD 155-2001 *Technical Instructions on determining the technical condition of modern roads*
[6] Jiang Z, Ren P, Zhang H, Xie T and Qiu Y 2009 Explore and analyze for failure process of asphalt mixtures *International Conference on Transportation Engineering* pp 3032
[7] Cao M, Li Y and Fang M 2012 *To discuss the asphalt pavement's early damages* Civil Eng. Urban Plann. pp 795
[8] Chen D, Bilyeu J, Scullion T, Nazarian S and Chiu C 2006 Failure investigation of a foamed-asphalt highway project *J. Infrastruct. Syst.* **12** 33

[9] Lu S 2012 Fatigue equation of asphalt mixture considering the influence of loading rate *Eng. Mech* **29** 276
[10] Gongyun L, Xiaoming H and Dunzhu B S 2008 Anti-ultraviolet aging tests of asphalts adapting to environment in Tibetan Plateau of China *J. Southeast Univ.* **24** 503
[11] Andrei R, Puslau E and Plescan C 2010 Evaluation of Performance of Flexible Pavement by Using Specific Pavement Condition Indicators *National Congress of Roads and Bridges*
[12] Andrei R, Durmusoglu Z, Boboc V and Puslau E 2010 Assessment of road pavement condition, based on the combined concept of chance, change and entropy *WSEAS Transactions on Information Science and Applications*
[13] Gao Y, Geng D, Huang X and Li G 2017 Degradation evaluation index of asphalt pavement based on mechanical performance of asphalt mixture *Constr. and Building Mater.* **140** 75
[14] * * * SR EN 12390-3/2009 *Testing hardened concrete. Part 3: Compressive strength of specimens*
[15] * * * SR EN 12390-13/2014 *Testing hardened concrete. Part 13: Determination of the secant modulus of elasticity in compression*
[16] * * * SR EN 12390-6/2010 *Testing hardened concrete. Part 6: Stress resistance by splitting test pieces*

Study on the adsorption and purification performance of runner with different staged regeneration modes

H Hao[1*], X Y Song[1] and P Xin[1]

[1] School of Municipal and Environmental Engineering, Shenyang Jianzhu University, Shenyang, China

[*] 552046581@qq.com

Abstract. In this paper, the hierarchical regeneration mode of three-zone, four-zone and two-stage regeneration of the wheel is firstly determined. Silica gel is selected as the adsorbent, and toluene is selected as the VOC pollutant. Visual Basic software is used to establish the coupling adsorption model of heat, humidity and VOC pollutants in the staged-regeneration wheel. The dehumidification and VOC adsorption performance of the graded regenerated wheel with different influencing factors were compared, and the variation rules of the dehumidification and toluene adsorption concentration of the staged-regeneration wheel with different influencing factors were analyzed. The results show that, when the regeneration energy consumption of the runner is the same, the dehumidification capacity of the two-stage runner is the highest, the concentration of toluene adsorbed by the three-zone runner is the highest, and the dehumidification capacity and the concentration of pollutants adsorbed by the two-zone runner are the lowest. When the capacity of dehumidification is the same, the regeneration energy consumption of two-stage runner is the lowest, and the regeneration energy consumption of two-zone runner is the highest. With the increase of adsorption capacity, the lower the regeneration temperature is, the more significant the energy saving effect will be. When the capacity of dehumidification is the same, the concentration of pollutants adsorbed by the runner in the three regions is the highest. In addition, with the increase of air entrance temperature and humidity, the dehumidification capacity of the two-stage runner changes the most, and the adsorption concentration of toluene of the two-stage runner changes the most. When the concentration of toluene in the treatment zone was different, it had no effect on the dehumidification effect of the runner, and the concentration of toluene adsorbed by the runner in the third zone changed the most.

1. Introduction

With the continuous advancement of science and technology and the rapid development of society in China, the level of national economy has been improved in recent years, and people's living and working environment have also undergone significant changes. In addition to daily life, people's working environment has gradually tended to be mainly indoor. Studies have shown that people spend more time than 87% [1] of the whole day in indoor space, accounting for most of the proportion. Therefore, in order not to affect people's health and work efficiency, the indoor environment requirements of modern buildings are constantly improving, and gradually pursue to make people more comfortable living and working environment.

As an air conditioning system with new operating mode, the runner dehumidification air conditioning system can not only meet the requirements of indoor temperature and humidity, but also absorb indoor VOC, which has been paid more and more attention [2, 3]. Runner dehumidifier has strong dehumidification capacity, large dehumidification capacity, and continuous operation [4]. Continuous, humidity controllable, compact structure, energy saving and material saving etc. However, the regeneration process of adsorbent requires higher temperature, which is also the main energy consumption in the process of dehumidification and purification of runner. The graded regeneration method has been proven to reduce the regeneration temperature requirements of the runner and improve the thermal performance of the system. Collier and Cohen [5] put forward the hierarchical graded regeneration runner dehumidification air conditioning system. Its principle is to add a preheating zone before the regeneration zone and use silica gel as desiccant. Through simulation and experimental research, the system can significantly improve the COP of the air conditioning system. Scholars of Shanghai Jiaotong University [6, 7] proposed two kinds of two-stage runner dehumidification system. Because of the different regeneration and treatment zones of the runner, it is divided into TSDC system and OTSDC system. The results show that the regeneration temperature of the two systems obviously decreases when the treatment air entrance conditions are the same, and the thermal COP of the two systems are both above 1.0. Schultz [8] used direct solar radiation to regenerate the desiccant, and the results showed that the system had a lower COP than the system using hot air to regenerate the desiccant. Han Wang [9] puts forward a new technology which combines the graded regeneration runner with evaporative cooling technology. Compared with the general composite air conditioning system, the regeneration energy consumption of the new composite air conditioning system is reduced by about 42% and the COP is increased by about 78%. Jia Chunxia [10] made a desiccant rotor using a silica-based composite desiccant material, and experimentally analyzed the dynamic and steady-state dehumidification characteristics of the desiccant rotor dehumidifier. The experimental results show that the dehumidification advantage of the composite runner is more obvious in low humidity climate conditions. For purification, Hines [11] et al. tested the adsorption effect of different adsorbents, and selected polar molecules such as water, toluene, formaldehyde, 1,1,1-trichloroethylene and non-polar molecules such as carbon dioxide and radon, which proved that silica gel has the ability to absorb water and other molecules simultaneously. Aristov [12] respectively prepared composite adsorbents modified with $CaCl_2$ silica gel and modified silica gel with Li Br. The prepared adsorbents were used in air intake devices, and their adsorption properties and amounts were experimentally tested. The regeneration temperature of the adsorbent is lower than 90°C, which can greatly reduce the consumption of regeneration heat energy. Wei Zhang [13] utilizes the ability of silica gel to adsorb water and other molecules at the same time. Using silica gel as desiccant, the adsorption capacity of silica gel runner to indoor VOC pollutants was verified by experiments. In this paper, Visual Basic software is used to establish the coupled adsorption model of heat, humidity and VOC pollutants in the graded regeneration runner. For different staged regeneration modes, the dehumidification capacity and VOC adsorption performance of the runner under different influencing factors were compared.

2. Establishment of model

2.1. Establishment of physical model

The traditional dehumidification runner is divided into two zones: dehumidification zone and regeneration zone. In order to reduce the renewable energy consumption of runners and improve the energy utilization rate, the dehumidification runners in this paper adopt the hierarchical regeneration mode, which are three-zone runners, four-zone runners and two-stage runners composed of two ordinary runners.

The specific regional division is shown in figure 1. The three-zone runner is mainly divided into three parts: dehumidification zone, preheating regeneration zone and heating regeneration zone, which can realize energy cascade utilization and reduce high-grade energy consumption. For the four-zone and two-stage runner, it is mainly composed of dehumidification zone and regeneration zone. For the four-zone and two-stage runners, it mainly consists of dehumidification zone and regeneration zone.

This dehumidification method with intermediate cooling can make the runner close to isothermal dehumidification and increase the dehumidification capacity of the runner. Assuming that all air passages in the runner have the same specifications, the flow direction of treated air and regenerated air is reverse flow.

When the entrance air state is stable, the heat and mass transfer state of the runner in any air passage in the same radial direction is the same, and at different times when the runner rotates, each airflow passage must pass through the same state, the Lagrangian coordinate system is established by taking any airflow passage of the runner as the research object, and the working state of the entire runner at a certain moment is described by using a state of the airflow passage at different times in one rotation cycle.

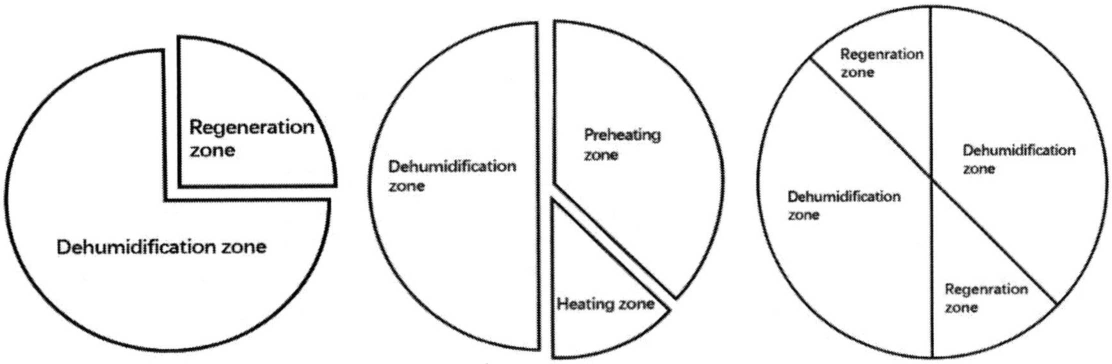

Figure 1. The zone division of desiccant staged-regeneration wheel.

2.2. Establishment of mathematical model

- Energy conservation equation in air

$$\frac{\partial t}{\partial \tau}+\omega\frac{\partial t}{\partial \phi}+\frac{m_i}{\rho_i}\frac{\partial t}{\partial z}=\frac{\alpha F_v(t_d-t)}{\rho_i f_s(C_{pa}+DC_{pv}+C\cdot C_{pm})} \quad (1)$$

- Water mass conservation equation in air

$$\frac{\partial D}{\partial \tau}+\omega\frac{\partial D}{\partial \phi}+\frac{m_i}{\rho_i}\frac{\partial D}{\partial z}=\frac{k_y F_v}{\rho_i f_s}(D_d-D) \quad (2)$$

- VOC mass conservation equation in air

$$\frac{\partial C}{\partial \tau}+\omega\frac{\partial C}{\partial \phi}+\frac{m_i}{\rho_i}\frac{\partial C}{\partial z}=\frac{k_y F_v}{\rho_i f_s}(G-C) \quad (3)$$

- Energy conservation equation of adsorbent and matrix materials

$$\frac{\partial t_d}{\partial \tau}+\omega\frac{\partial t_d}{\partial \phi}-\frac{\lambda_d(1-f_s)}{[M_z C_{pz}+M_d(C_{pd}+WC_{ps}+SC_{pn})]}\frac{\partial^2 t_d}{\partial z^2}=\frac{\alpha F_v(t-t_d)+k_y F_v Q(D-D_d)}{[M_z C_{pz}+M_d(C_{pd}+WC_{ps}+SC_{pn}]} \quad (4)$$

- Equation of water mass conservation on adsorbent surface

$$\frac{\partial W}{\partial \tau}+\omega\frac{\partial W}{\partial \phi}-D_e(1-f_s)\frac{\partial^2 W}{\partial z^2}=\frac{k_y F_v}{M_d}(D-D_d) \quad (5)$$

- VOC mass conservation equation on adsorbent surface

$$\frac{\partial S}{\partial \tau}+\omega\frac{\partial S}{\partial \phi}-D_{ev}(1-f_s)\frac{\partial^2 S}{\partial z^2}=\frac{k_y F_v}{M_d}(C-G) \tag{6}$$

- Adsorption equilibrium equation of moisture and VOC in runner

$$D_d=\frac{0.622\cdot\varphi\cdot P_s}{B-\varphi\cdot P_s} \tag{7}$$

$$\varphi=\frac{W\cdot R}{W_{max}\left[1-\frac{W(1-R)}{W_{max}}\right]} \tag{8}$$

$$S=H_i\cdot G\cdot\rho_i \tag{9}$$

2.3. Preliminary validation of the model

At present, the experimental research on the graded regenerative runner only stays in the dehumidification stage, and the VOC adsorption characteristics of the graded regenerative runner are not available. In order to preliminarily verify the reliability of the model, only the dehumidification capacity of the graded regenerative runner is verified. For four-zone and two-stage runners, the values of air state parameters and physical parameters input in the program are given in reference [14]. The values of specific parameters are shown in table 1.

Table 1. The parameters of the body and the adsorbent in the four-zone and wheel.

Various parameters of the runner			
Treat air entrance moisture content (g/kg)	14.3	Regeneration air entrance moisture content (g/kg)	14.5
Runner thickness (m)	0.1	Runner diameter (m)	0.4
Runner speed (r/h)	8	Processing temperature (°C)	35
Processing air flow rate (m/s)	2.5	Regenerated air velocity (m/s)	2.5
Specific heat of silica gel at constant pressure (J/kg·K)	921	Density of silica gel (kg/m^3)	1201
Specific heat of ceramic matrix at constant pressure (J/kg·K)	880	Density of ceramic matrix (kg/m^3)	625
Specific heat of lithium chloride at constant pressure (J/kg·K)	3000	Volume ratio of lithium chloride in compound desiccant	15
Pore size of composite adsorbent (mm)	4	Proportion of airflow passage to cross-sectional zone	0.816

Because the adsorbent in the paper is silica gel and the adsorbent in the literature is composite desiccant, the physical properties of the adsorbent, such as its comprehensive density, constant pressure specific heat capacity and thermal conductivity, are simulated. The simulated results are compared with the experimental results, and the results are shown below.

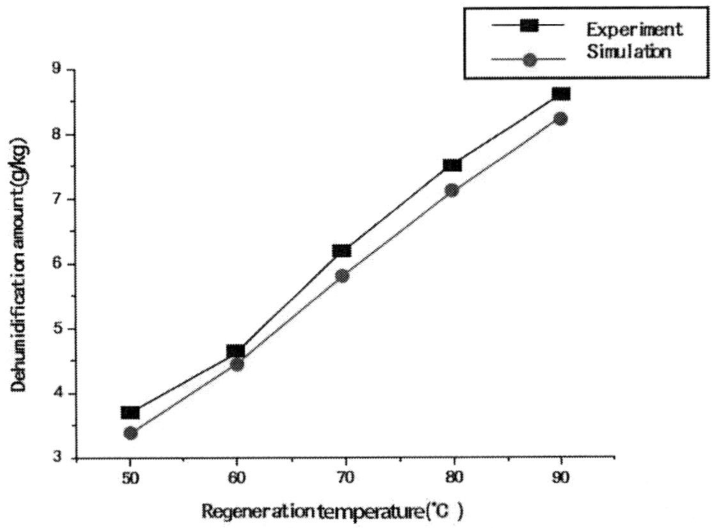

Figure 2. The experimental value and analog value of dehumidification capacity at the regeneration temperature variation of four-zone wheel.

The above figure shows the experimental and simulation results of four-zone runner dehumidification at different regeneration temperatures. From the figure, we can see that the difference between simulation and experiment is very small. When the regeneration temperature is 80°C, the corresponding error is the largest. The difference is about 0.413 g/kg. After calculation, it is 5.51%, and the error is controlled within 10%. This shows that although the characteristics of VOC adsorption are taken into account in the model, the effect of dehumidification is not significant, it is in the acceptable range, the mathematical model is basically accurate.

For the two-stage runner, the parameters in reference [14] are taken as follows: the moisture content of treated air entrance is 15 g/kg, the moisture content of regenerated air entrance is 11 g/kg, and the regeneration temperature is 70°C. The physical parameters of the other parts of the runner are shown in table 1. The simulated results are compared with the experimental results. The results are as follows:

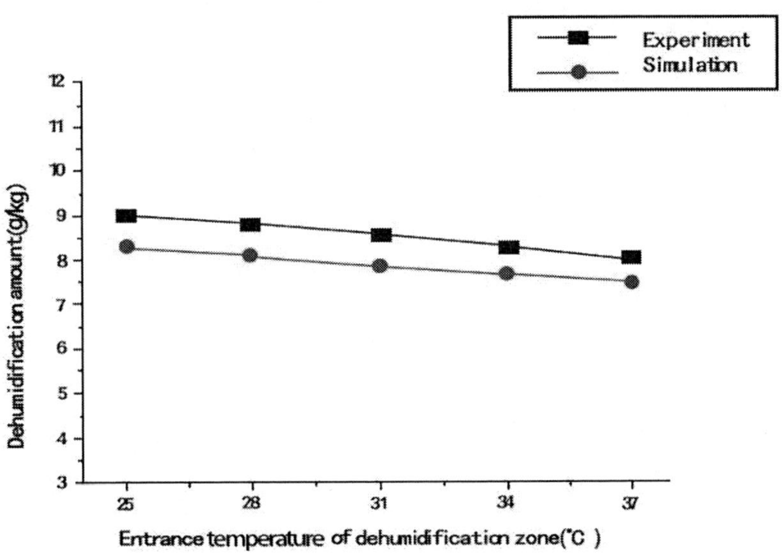

Figure 3. The experimental value and analog value of dehumidification capacity at the entrance temperature variation of two-staged wheel.

The figure above shows the experimental and simulated results of dehumidification capacity of two-stage runners at different entrance temperatures in the dehumidification zone. As can be seen from the figure above, when the entrance temperature is 31°C, the corresponding error is the largest. The difference between the results is about 0.76 g/kg. After calculation, it is 8.83%. The error is controlled within 10%. The model meets the accuracy requirements. Therefore, the mathematical model established is basically accurate.

3. Comparative analysis of runner performance under different staged regeneration modes

3.1. The value of initial parameters

For different staged graded regeneration runners, the dehumidification and adsorption performance of the runners are also different. For the convenience of comparison, this paper refers to the air heat and humidity parameters in reference [15], and simulates them as standard operating conditions. VOC pollutants choose toluene. In order to make the research meaningful, the concentration of toluene entering the room should be higher than that in the room. Referring to China's indoor air quality standards, the concentration of toluene should not be higher than 0.2 mg/m^3 in all indoor air pollutants. Therefore, the toluene concentration of imported air in this paper is 500 μg/m^3. The specific parameters are as shown in table 2.

Table 2. Table of wheel value under standard condition.

	Two-zone	Three-zone	Four-zone	Two-stage
Runner thickness (m)	0.2	0.2	0.2	0.2
Runner speed (r/h)	15	15	15	15
Specific heat of adsorbing materials (J/kg·k)	921	921	921	921
Import humidity of treated air (g/kg)	13	13	13	13
entrance temperature of treated air (°C)	30	30	30	30
VOC concentration at import for air treatment (μg/m^3)	500	500	500	500
entrance humidity of regenerated air (g/kg)	13	13	13	13
entrance temperature of regenerated air (°C)	70	70	70	70
entrance temperature of preheated air (°C)	—	45	—	—
entrance humidity of preheated air (g/kg)	—	13	—	—
Secondary entrance temperature of treated air (°C)	—	—	40	—
Secondary entrance humidity of treated air (g/kg)	—	—	9	—
VOC concentration at secondary import of treated Air (μg/m^3)	—	—	200	—
Air handling capacity (m^3/h)	150	150	150	300
Air volume of regenerated air (m^3/h)	50	100	50	100
Preheated air volume (m^3/h)	—	50	—	—
Handling Air Speed (m/s)	1.78	2.67	3.56	3.55
Wind speed of regenerated air (m/s)	1.78	2.67	3.56	3.55
Preheated air velocity (m/s)	—	2.67	—	—
Processing zone Angle (°)	270	180	135	270
Angle of regeneration zone (°)	90	120	45	90

3.2. Performance comparison of runners under different staged regeneration modes
Through simulation, the outlet air temperature, humidity and VOC toluene concentration of regenerated runners under standard conditions are obtained. In order to analyze the adsorption characteristics of runners more intuitively, a comparative analysis between runners is needed. In order to facilitate comparison, the parameters of air state in the simulation process are as follows: the entrance temperature of the treatment zone is 30°C, the entrance humidity is 13 g/kg, the entrance toluene concentration is 500 μg/m^3; the entrance temperature of the regeneration zone is 70°C, the entrance humidity is 13 g/kg, and the entrance toluene concentration is 10 μg/m^3. The dehumidification capacity and VOC concentration of the runner are compared.

3.2.1. Comparison of dehumidification capacity and toluene adsorption concentration at the same regeneration temperature. The figure below shows the comparison of dehumidification capacity between the runner and the two-zone runner under different staged regeneration modes with the same regeneration energy consumption. According to the graph, the dehumidification capacity of two-stage runner is 3.48 g/kg, that of three-stage runner is 5.49 g/kg, that of four-stage runner is 5.78 g/kg, and that of two-stage runner is 5.84 g/kg.

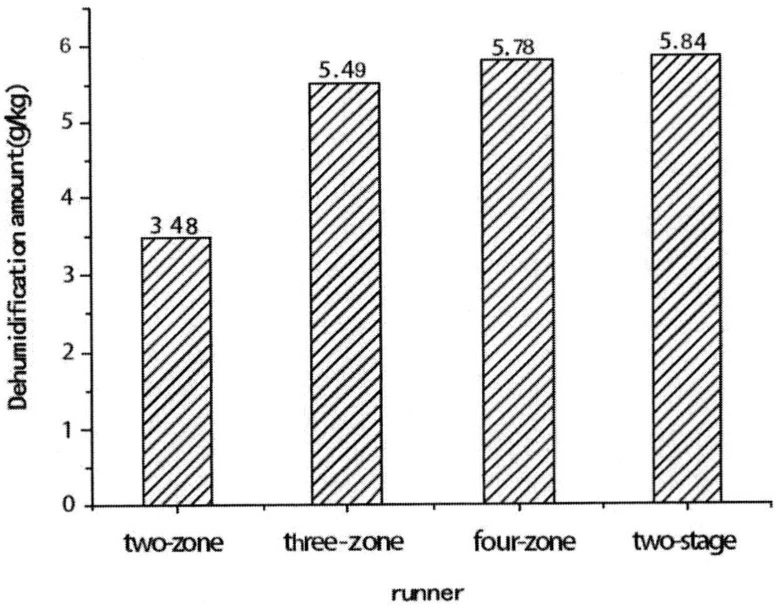

Figure 4. The dehumidification capacity of the wheel under the same regeneration temperature.

For the four-zone and two-stage runners, when the adsorbent adsorbs water, it releases heat of adsorption, and the temperature of treated air increases. When the air temperature rises to a certain extent, the mass transfer coefficient decreases and the dehumidification process slows down.

For two-zone and three-zone runners, because of the existence of a preheating zone in three-zone runners, the consumption of high-grade energy is reduced, the temperature of adsorbent is lowered, and the equilibrium vapor partial pressure of air on adsorbent surface is lowered.

When the entrance humidity of the treated air is the same, the partial pressure difference between the treated air and the surface air of the adsorbent increases, and the driving force of dehumidification increases.

Therefore, the dehumidification capacity of the three-zone runner is larger than that of the two-zone runner, and the dehumidification capacity of the two-zone runner is the smallest.

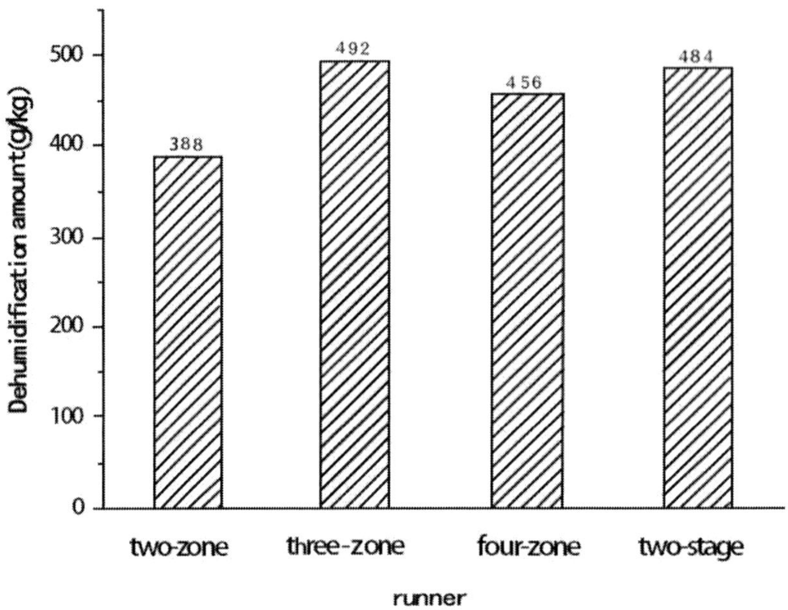

Figure 5. The adsorb toluene capacity of the wheel under the same regeneration temperature.

It can be seen that although the toluene concentration of the second zone runner is the lowest, it can also meet the toluene concentration limit specified in the standard. Because the content of pollutants in the air is less than one thousandth of the moisture content, VOC concentration in the air is very low.

The mass transfer process of pollutants is similar to that of water. The adsorption effect of runner on VOC pollutants in air depends on the desorption time of adsorbent in regeneration zone. On the treatment side, adsorbents absorb moisture and pollutants from the air and then send them to the regeneration zone.

On the regeneration side, high temperature air flowing through the runner will take away most of the water and pollutants on the adsorbent surface, but there are also a small capacity of water and pollutants remaining on the adsorbent surface due to limited time and low regeneration temperature. At the same time, because regenerated air preferentially desorbs the water on the adsorbent surface, the regeneration time needed to satisfy the simultaneous adsorption of water and pollutants on the runner is relatively longer, in order to ensure that the pollutants on the adsorbent surface achieve better desorption effect and keep the adsorption effect unchanged. Increasing the residence time of the runner in the regeneration zone can be achieved by reducing the rotating speed of the runner or increasing the angle of the regeneration zone.

On the one hand, because of the preheating zone of the three-zone runner, the desorption time is longer. On the other hand, because the three-zone runner has the smallest dehumidification capacity, the adsorbent has more time to desorb and regenerate, so the concentration of pollutants adsorbed by the three-zone runner is the highest.

3.2.2. Comparison of regeneration temperature and adsorption toluene concentration at the same dehumidification volume. The below figure shows the regeneration temperature and the concentration of toluene adsorbed by the runner at different dehumidification rates.

When the dehumidification capacity is 4 g/kg, the regeneration temperature of the two-zone runner is 78°C, the adsorption toluene concentration is 403 μg/m^3, the regeneration temperature of the three-zone runner is 55.5°C, and the adsorption toluene concentration is 487 μg/m^3; the regeneration temperature of the four-zone runner is 55°C, the adsorption toluene concentration is 431 μg/m^3; the regeneration temperature of the two-stage runner is 54.5°C, and the adsorption toluene concentration is 457 μg/m³; the regeneration energy consumption of three-zone, four-zone and two-stage runners is 28.8%, 29.5% and 30.1% less than that of two-zone runners respectively.

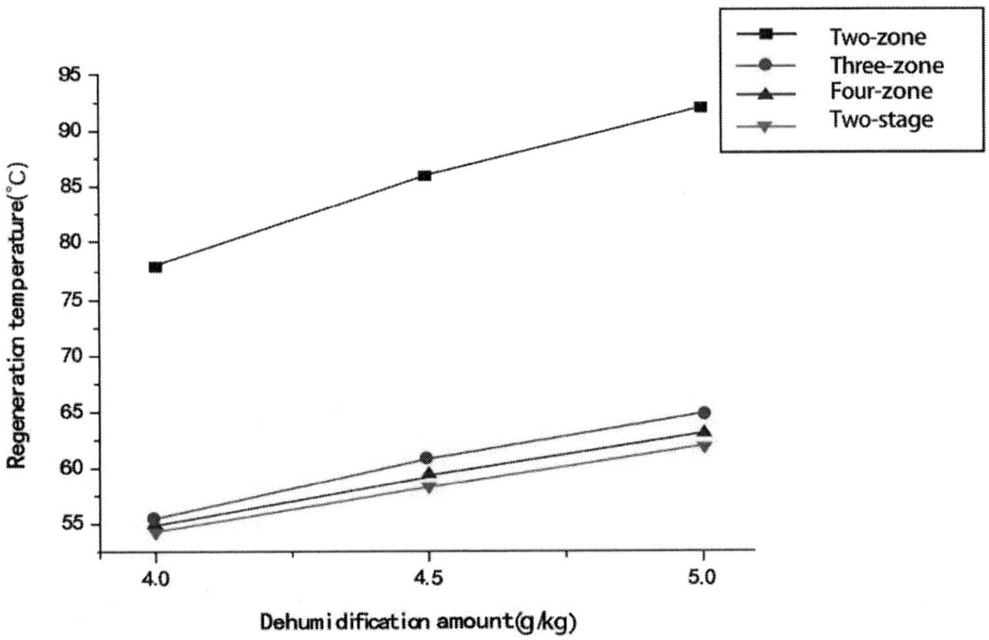

Figure 6. The regeneration temperature of wheel under the same dehumidification.

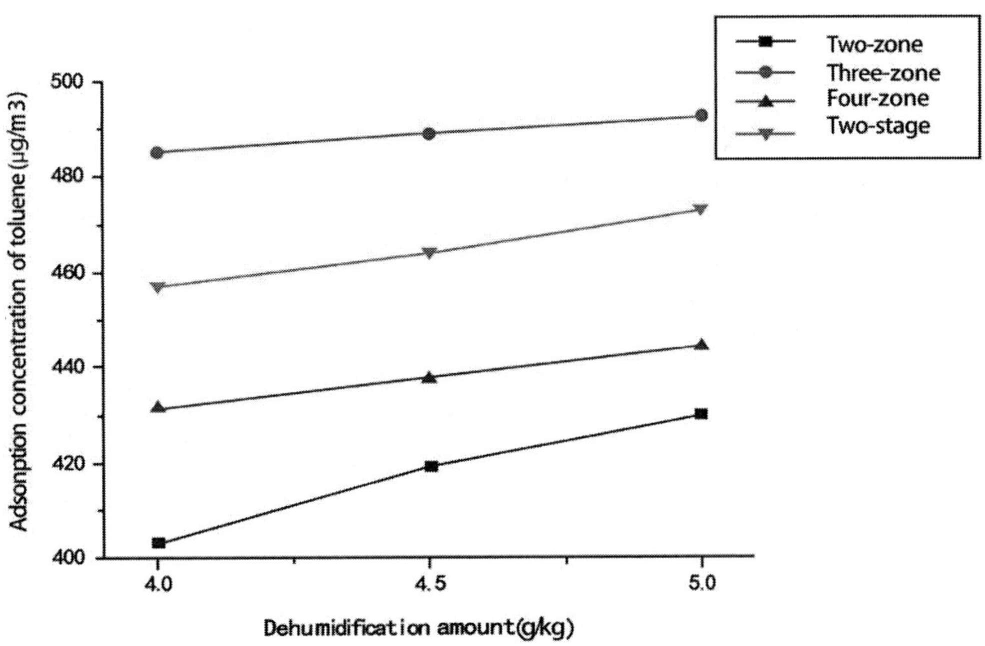

Figure 7. The adsorb toluene capacity of the wheel under the same dehumidification.

Under the condition of 4.5 g/kg dehumidification, the regeneration temperature of the two-zone runner is 86°C, the concentration of toluene adsorbed is 419 μg/m³, the regeneration temperature of the three-zone runner is 61°C, the concentration of toluene adsorbed is 489 μg/m³, the regeneration temperature of the four-zone runner is 59.5°C, the concentration of toluene adsorbed is 438 μg/m³, and the regeneration temperature of the two-stage runner is 58.5°C, the concentration of toluene adsorbed is 464 μg/m^3. The regeneration energy consumption of three-zone, four-zone and two-stage runners is 29.1%, 30.8% and 32.0% less than that of two-zone runners, respectively. Under the condition of 5 g/kg dehumidification, the regeneration temperature of the two-zone runner is 92°C, the concentration of toluene adsorbed is 430 μg/m^3, the regeneration temperature of the three-zone runner is 65°C, the concentration of toluene adsorbed is 491 μg/m^3, the regeneration temperature of the four-zone runner is 63.2°C, the concentration of toluene adsorbed is 446 μg/m^3, and the regeneration temperature of the two-stage runner is 62°C. The concentration of toluene is 472 μg/m^3. The regeneration energy consumption of three-zone, four-zone and two-stage runners is 29.3%, 31.3% and 32.6% less than that of two-zone runners, respectively.

For a certain graded runner, with the increase of dehumidification, the energy-saving effect of the graded runner is more significant than that of the two-zone runner. With the increase of desiccant capacity, the greater the relative humidity of air on the adsorbent surface, the greater the equilibrium vapor partial pressure of air on the adsorbent surface, the greater the power of mass transfer between the adsorbent and the regenerated air, the better the desorption effect and the lower the growth rate of regeneration temperature, so the greater the desiccant capacity, the better the energy-saving of the runner.

With the increase of dehumidification, the lower the regeneration temperature is, the slower the growth rate of regeneration temperature is. In other words, the greater the capacity of dehumidification, the more remarkable the energy-saving effect of two-stage runner is than that of four-zone runner, and the more remarkable the energy-saving effect of four-zone runner is than that of three-zone runner. Because when the temperature and humidity of air entrance are the same, the vapor partial pressure of wet air remains unchanged. With the increase of dehumidification capacity, the relative humidity of air on adsorbent surface increases, the equilibrium vapor partial pressure on adsorbent surface increases, which reduces the vapor partial pressure difference between treated air and adsorbent surface, and makes the dehumidification rate lower and lower. The higher the regeneration temperature, the higher the heat exchange between the air and the adsorbent. The temperature of the adsorbent is $T_{two\text{-}zone} > T_{three\text{-}zone} > T_{four\text{-}zone} > T_{four\text{-}zone}$, and the saturated water vapor partial pressure $P_{two\text{-}zone} > P_{three\text{-}zone} > P_{four\text{-}zone} > P_{four\text{-}zone}$, the equilibrium water vapor partial pressure difference between the treated air and the adsorbent surface air is $P_{two\text{-}zone} < P_{three\text{-}zone} < P_{four\text{-}zone} < P_{two\text{-}stage}$. Thereby, the dehumidification effect of the two-stage runner is promoted, so in order to achieve the same dehumidification effect, the growth rate of the regeneration temperature required for the two-zone runner is the largest. Therefore, for the graded regenerative runners in different modes, the regeneration rate of the three-zone runners is the largest, the energy-saving effect is the slowest, the growth rate of the two-stage runners is the smallest, and the energy-saving effect is the most significant.

In terms of purification, the concentration of toluene adsorbed by three-zone runner is the highest, while that by two-zone runner is the lowest. For the adsorption process of pollutants, the adsorption characteristics are similar to those of the above analysis. When the capacity of dehumidification is the same, the adsorption capacity of adsorbent is most affected by the regeneration time, because the regeneration angle of the three-zone runner is the largest and the desorption time is the longest, so the higher the concentration of adsorbed pollutants. In addition, when the VOC concentration and other state parameters of the treated air in the runner remain unchanged, the regeneration temperature of the runner increases with the increase of dehumidification, which enhances the desorption effect between the regenerated air and the adsorbent surface, so the ability of the runner to adsorb pollutants increases.

As can be seen from the figure above, the increase speed of toluene concentration in the two-zone runner is the largest, while that in the three-zone runner is smaller than that in the four-zone runner and the two-stage runner. From the above analysis, with the increase of dehumidification, the ability of runner to adsorb pollutants increases. The higher the concentration of adsorbent adsorbing pollutants, the higher the partial pressure of adsorbent surface compounds. As a result, the absorptive potential energy of treated air and adsorbent surface decreases, and the driving force of adsorption decreases. So, the concentration of toluene adsorbed by three-zone runner increases slowly.

4. Analysis of influencing factors on dehumidification and purification performance of runner

4.1. Effect of treatment air entrance temperature

As can be seen from the figure below, the dehumidification capacity of the runner decreases with the increase of the entrance air temperature in the treatment zone.

When the entrance temperature of treated air rises, the heat transfer between air and adsorbent surface decreases, which leads to the increase of adsorbent surface temperature and the increase of saturated vapor partial pressure on adsorbent surface.

As a result, the partial pressure difference between the treated air and the surface air of the adsorbent decreases, and the adsorption effect becomes worse, so the dehumidification capacity of the runner becomes lower and lower.

With the increase of entrance air temperature, the change of dehumidification capacity of two-stage runner is greater, and the change of dehumidification capacity of two-zone runner is the smallest.

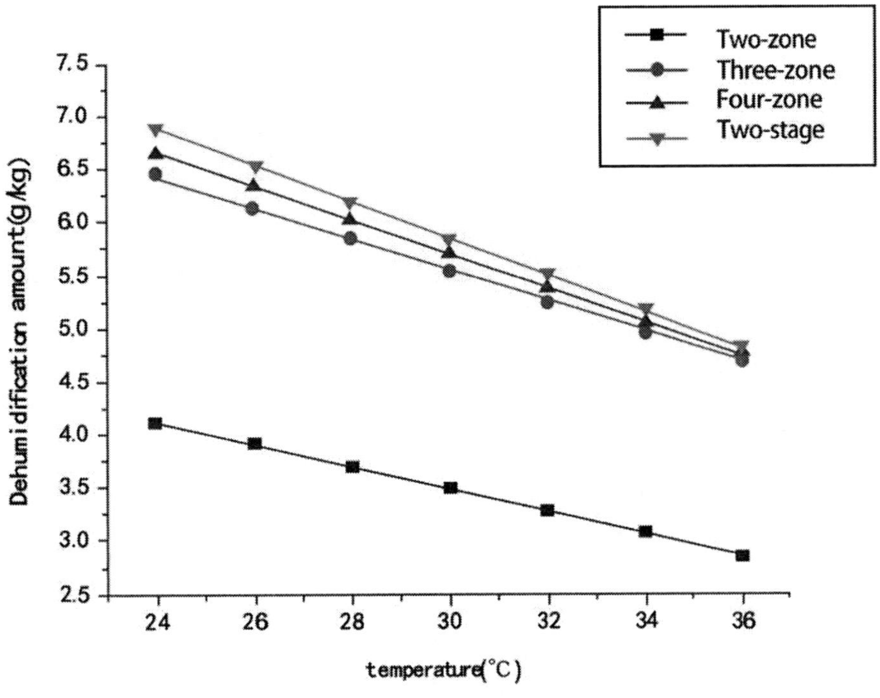

Figure 8. The dehumidification capacity of wheel under the different entrance temperature of processing zone.

When the entrance temperature of treated air rises, the saturated vapor partial pressure on the adsorbent surface increases; the smaller the dehumidification capacity, the greater the relative humidity of treated air, the closer the partial pressure of water vapor in air is to saturation, and the greater the difference between the partial pressure of water vapor on the air and the adsorbent surface, the greater the driving force of water transfer, and the greater the removal of water vapor partial pressure, the decrease of humidity is inversely proportional, so the lower the dehumidification, the slower the dehumidification decreases.

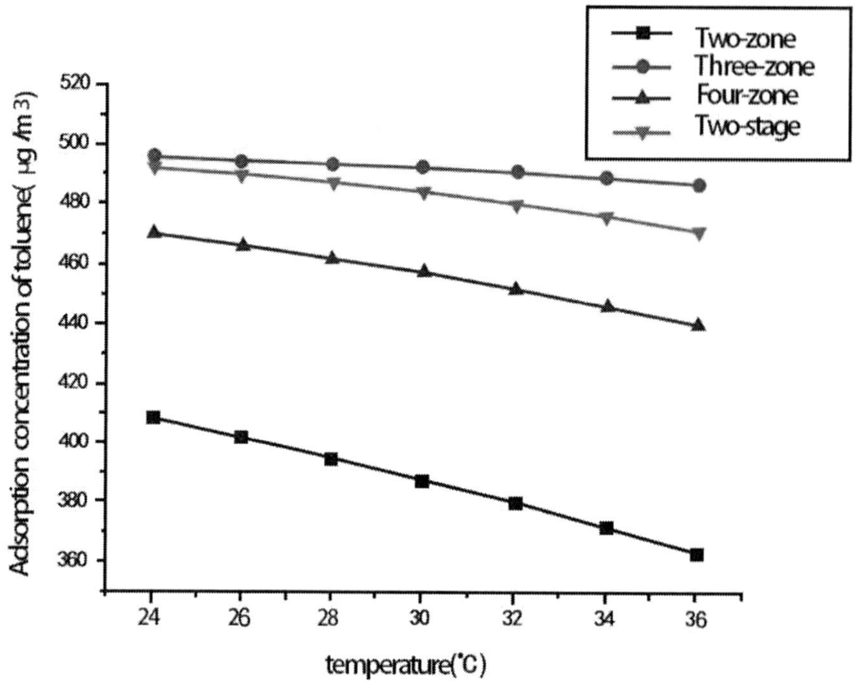

Figure 9. The adsorb toluene capacity of wheel under the different entrance temperature of processing zone.

As can be seen from the above figure, the concentration of toluene adsorbed by the runner decreases with the increase of the entrance air temperature in the treatment zone. When the entrance temperature of the treated air rises, the surface temperature of the adsorbent increases, which increases the equilibrium concentration of the adsorbent surface, reduces the potential energy of the treated air and the adsorbent surface and weakens the mass transfer process, so the lower the concentration of the adsorbent adsorbing toluene.

With the increase of entrance temperature of treated air, the concentration of toluene adsorbed by three-zone runner changed the least and that by two-zone runner changed the most. When the entrance temperature of treated air rises, the concentration of toluene adsorbed by runner decreases, and the concentration of toluene in treated air increases.

When the regenerated air state remains unchanged, the higher the concentration of toluene adsorbed, the more sufficient the adsorbent desorption is, the lower the equilibrium concentration of toluene on the adsorbent surface is, thus, the greater the concentration difference between the treated air and the adsorbent surface, the stronger the adsorbent's ability to adsorb pollutants. Therefore, the higher the concentration of toluene adsorbed by the runner, the slower the reduction rate is.

4.2. Treatment of the effect of air entrance humidity.

As can be seen from the figure below, the dehumidification capacity of the runner increases with the increase of air humidity at the entrance of the treatment zone.

When the entrance temperature of treated air is unchanged, the increase of entrance air humidity will increase the partial pressure of water vapor in the air, so that the differential pressure of water vapor on the surface of air and adsorbent will also increase, and the adsorption effect will be better, so the dehumidification capacity will be higher and higher.

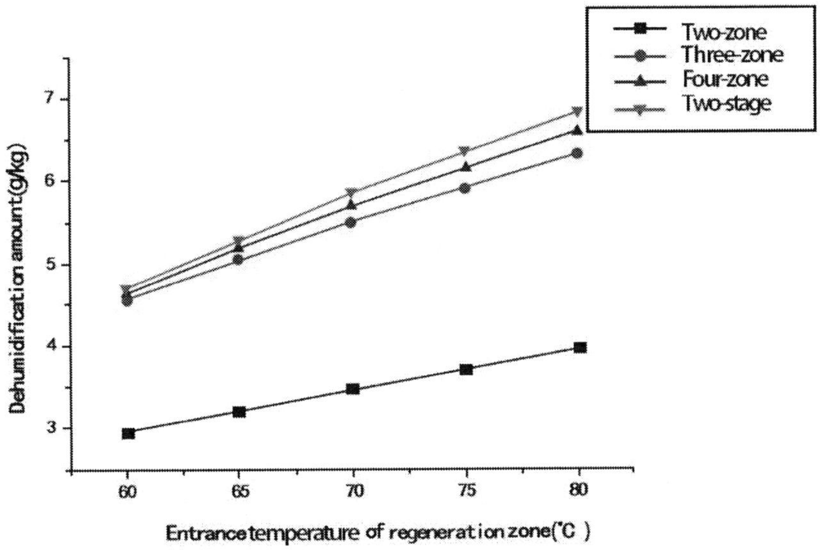

Figure 10. The dehumidification capacity of wheel under the different entrance humidity of processing zone.

With the increase of entrance air humidity, the dehumidification capacity of two-stage runner changes the most and that of two-zone runner changes the least. When the entrance humidity of treated air rises, the dehumidification capacity of runner increases, which decreases the partial pressure of water vapor in treated air. The smaller the dehumidification capacity, the lower the heat released by adsorbent in runner, the lower the average temperature of adsorbent surface, and the lower the saturated partial pressure of water vapor on adsorbent surface, therefore, the smaller the difference of partial pressure of water vapor between treated air and adsorbent surface air is, the smaller the driving force of dehumidification is. Therefore, with the increase of humidity at the entrance of treated air, the lower the dehumidification capacity is, the slower the change rate is.

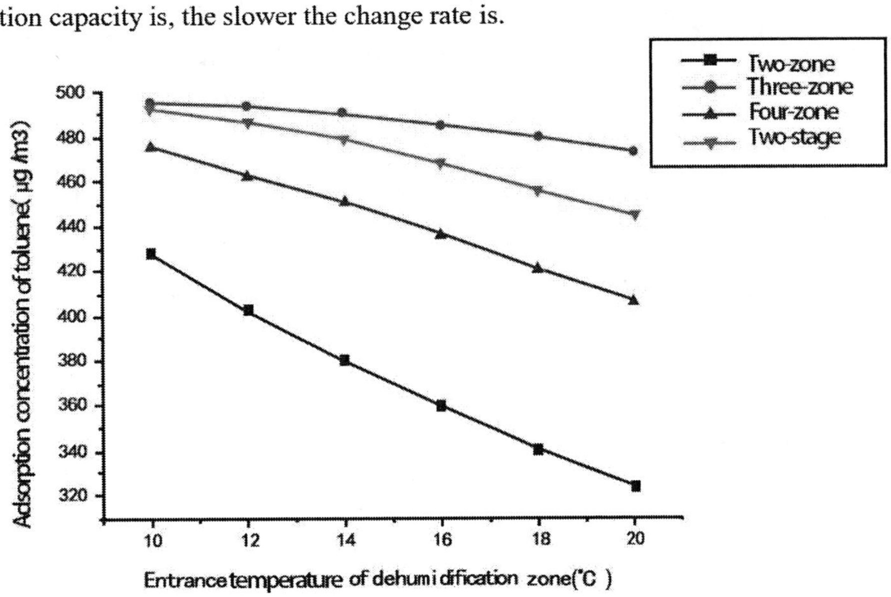

Figure 11. The adsorb toluene capacity of wheel under the different entrance humidity of processing zone.

As can be seen from the figure above, the concentration of toluene adsorbed by runner decreases with the increase of air humidity at the entrance of the treatment zone.

When the humidity of the treated air entrance increases, the capacity of dehumidification of the runner increases, the heat released by the adsorbent increases, the average temperature of the adsorbent increases, and the equilibrium concentration of pollutants on the adsorbent surface increases. As a result, the adsorption potential energy of treated air and adsorbent surface decreases and the mass transfer process is weakened, so the concentration of toluene adsorbed becomes lower and lower.

With the increase of air entrance humidity, the concentration of toluene adsorbed by the three-zone runner changed the least and the two-zone runner changed the most. When the humidity of air entrance increases, the concentration of toluene adsorbed by runner decreases, and the concentration of toluene in treated air increases.

When the regenerated air state remains unchanged, the higher the concentration of toluene adsorbed, indicating that the more adequately adsorbent desorption, the lower the equilibrium concentration of toluene on adsorbent surface. Thus, the greater the concentration difference between the treated air and the adsorbent surface, the stronger the adsorbent's ability to adsorb pollutants. Therefore, the higher the concentration of toluene adsorbed by the runner, the slower the reduction rate is.

4.3. Effect of entrance temperature of regenerated air

As can be seen from the figure above, the dehumidification capacity of the runner increases with the increase of the entrance temperature of regenerated air.

When the moisture content in the air remains unchanged, when the entrance temperature of regenerated air rises, the partial pressure of water vapor in the air does not change, but at the same time the partial pressure of saturated water vapor increases, which makes the relative humidity of regenerated air smaller, enhances the purification effect of regenerated air, and thus increases the ability of adsorbent to absorb water. Therefore, the dehumidification capacity of the runner is getting higher and higher.

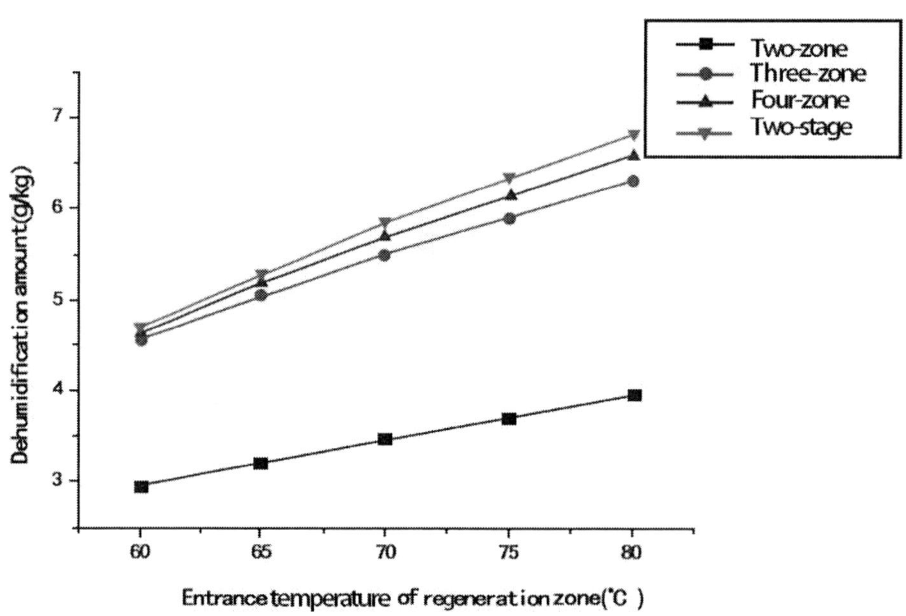

Figure 12. The dehumidification capacity of wheel under the different entrance temperature of regeneration zone.

With the increase of entrance temperature of regenerated air, the change of dehumidification capacity of two-stage runner is the biggest and that of two-zone runner is the smallest.

When the entrance temperature of regenerated air rises, the partial pressure of water vapor in the air remains unchanged; the greater the dehumidification capacity, the higher the saturated partial pressure of water vapor in the adsorbent, which increases the partial pressure difference between regenerated air and the surface air of the adsorbent.

The desorption effect is obvious, thus enhancing the driving force of water transfer in the treatment zone. So, the faster the capacity of dehumidification increases.

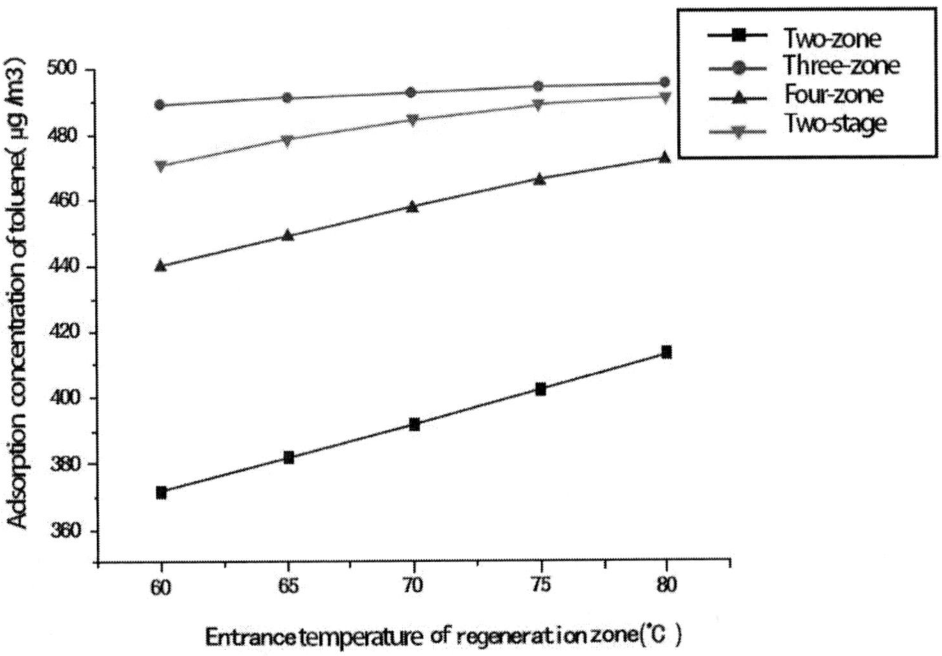

Figure 13. The adsorb toluene capacity of wheel under the different entrance temperature of regeneration zone.

As can be seen from the above figure, the concentration of toluene adsorbed by runner increases with the increase of air temperature at the entrance of regeneration zone.

When the entrance temperature of regenerated air rises, the equilibrium concentration of pollutants on the adsorbent surface rises, and the adsorbent is fully regenerated, that is, the concentration of pollutants on the adsorbent surface decreases, thus increasing the adsorption potential energy on the treated air and adsorbent surface and accelerating the mass transfer process, so the concentration of adsorbent adsorbing toluene is increasing.

With the increase of the entrance temperature of regenerated air, the concentration of toluene adsorbed by the three-zone runner changed the least and that by the two-zone runner changed the most. When the entrance temperature of the regeneration air rises, the concentration of toluene adsorbed by the runner becomes higher and higher, the concentration of toluene in the air decreases, and the partial pressure of the compound decreases.

The greater the concentration of adsorbent adsorbing pollutants, the larger the partial pressure of the adsorbent surface compound, the smaller the adsorption potential energy between the treatment air and the surface air of the adsorbent, and the smaller the driving force of adsorption, the slower the increase.

4.4. Effect of regeneration air entrance humidity

As can be seen from the figure below, the dehumidification capacity of the runner decreases with the increase of the humidity of the regenerated air entrance.

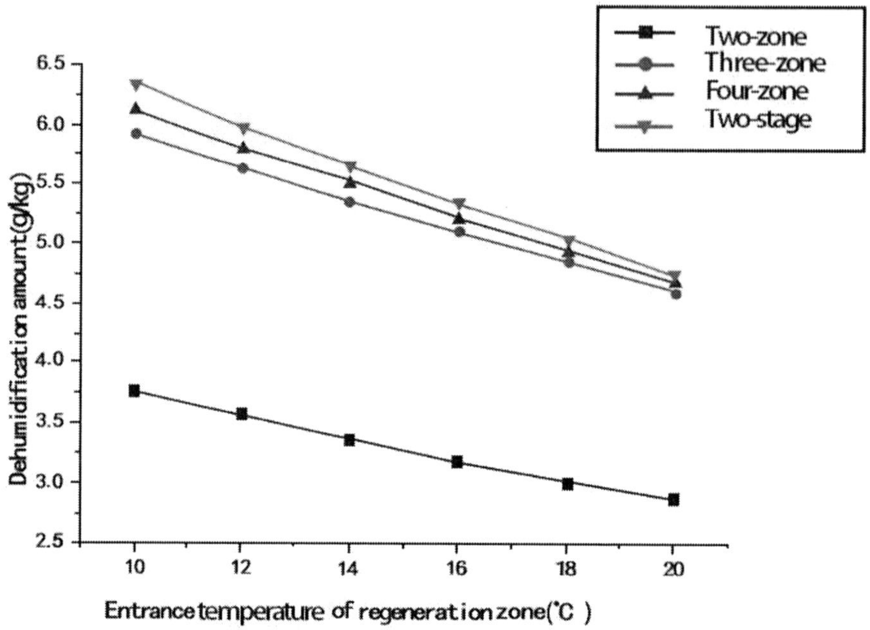

Figure 14. The dehumidification capacity of wheel under the different entrance humidity of regeneration zone.

When the entrance temperature of regenerated air is unchanged, the higher the humidity of regenerated air is, the higher the relative humidity is, and the greater the partial pressure of water vapor in air. When the partial pressure of saturated water vapor on the adsorbent surface is unchanged, the differential pressure of water vapor between regenerated air and the air on the adsorbent surface becomes smaller, which leads to the deterioration of desorption ability. As the humidity of regenerated air increases, the dehumidification capacity of runner becomes lower and lower. With the increase of the humidity of the regenerated air entrance, the dehumidification capacity of the two-stage runner varies the most, and that of the two-zone runner varies the least. When the entrance humidity of regenerated air increases, the relative humidity of regenerated air increases, and the partial pressure of water vapor in air also increases. The smaller the dehumidification capacity, the lower the heat released by adsorbent in the runner, the lower the average surface temperature of adsorbent, and the lower the saturated vapor partial pressure on the surface of adsorbent. Therefore, the smaller the partial pressure difference between the regenerated air and the adsorbent surface is, the smaller the driving force of the desorption process is, and the slower the desorption speed is, so the desiccant capacity of the runner is reduced.

As can be seen from the figure below, the concentration of toluene adsorbed by the runner increases with the increase of air humidity at the entrance of the regeneration zone. When the humidity of regenerated air increases, the desiccant capacity of runner decreases, the average temperature of adsorbent surface decreases, and the equilibrium concentration of pollutants on adsorbent surface decreases, thus increasing the adsorption potential energy between treated air and adsorbent surface, so the concentration of toluene adsorbed by runner becomes higher and higher. With the increase of entrance humidity of regenerated air, the concentration of toluene adsorbed by three-zone runner changed the least, and that by two-zone runner changed the most. When the entrance humidity of regenerated air increases, the concentration of toluene adsorbed by runner increases, the concentration of toluene in air decreases, and the partial pressure of compounds decreases. The higher the concentration of pollutants adsorbed by adsorbents, the higher the partial pressure of compounds on the adsorbent surface, the less the adsorption potential energy between the treated air and the adsorbent surface air, and the less the driving force of adsorption, so the slower the speed of adsorbing pollutants on the runner.

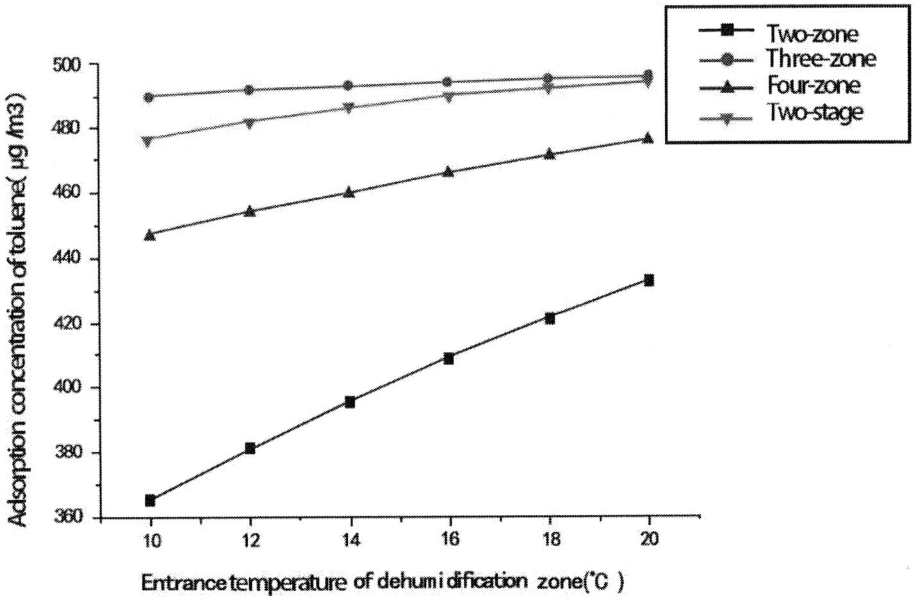

Figure 15. The adsorb toluene capacity of wheel under the different entrance humidity of regeneration zone.

4.5. Effect of toluene concentration at air entrance

As can be seen from the below figure, with the increase of toluene concentration at the entrance of the dehumidification zone, the dehumidification capacity of the runner remains basically unchanged. Because the adsorbent occupies the main position in the process of adsorbing water and VOC, that is, the adsorbent adsorbs water preferentially, so the change of VOC concentration in air has little effect on dehumidification.

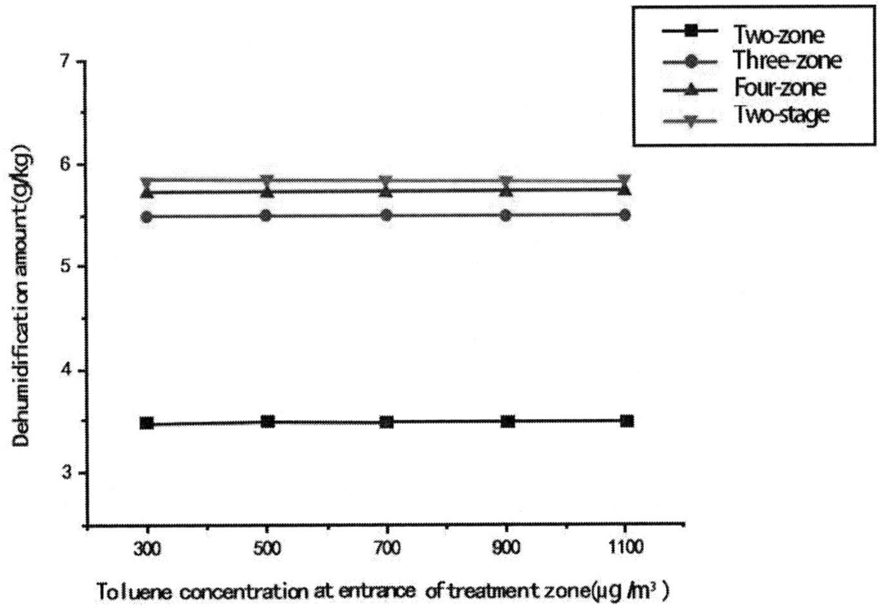

Figure 16. The dehumidification of wheel under the different entrance toluene concentration of processing air.

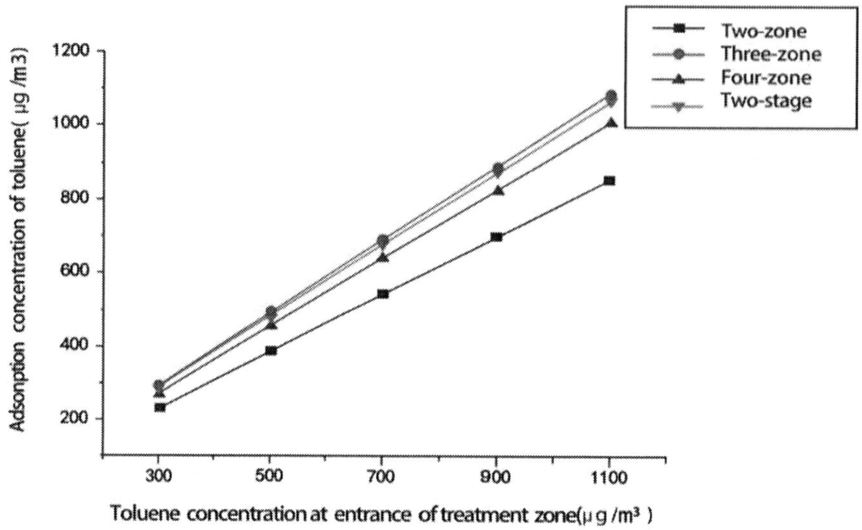

Figure 17. The adsorb capacity of wheel under the different entrance toluene concentration of processing air.

As can be seen from the figure above, the concentration of toluene adsorbed by the runner increases with the increase of toluene concentration at the entrance of the treatment zone.

When the temperature and humidity of the treated air entrance remain unchanged, the increasing concentration of imported toluene will increase the adsorption potential energy of the treated air and the adsorbent surface, which will increase the concentration of the adsorbed toluene. In addition, the concentration of toluene adsorbed by silica gel was proportional to the entrance concentration of toluene, and the concentration of toluene adsorbed by three-zone runner changed most.

According to the results of reference [12], when the concentration of indoor pollutants meets the range specified in the Air Quality Standard, the adsorption capacity of the runner does not reach saturation, then the adsorption characteristics of the adsorbent materials on the runner remain unchanged. From the above analysis, under the same conditions, the concentration of toluene adsorbed by the three-zone runner is the largest, so the change speed of the concentration of toluene adsorbed is the largest.

5. Conclusion

In this paper, toluene is chosen as pollutant to study. According to the established mathematical model, the changes of graded regeneration runners in different stages are compared with those in the second zone when the dehumidification capacity (regeneration temperature) is the same.

At the same time, for different staged graded regeneration runners, the dehumidification capacity and VOC adsorption performance of the runners under different influencing factors are compared, and the variation law of the runners under different influencing factors is analyzed.

The main conclusions are as follows:

- when the renewable energy consumption of the runner is the same, the two-stage runner has the largest dehumidification capacity, the three-zone runner has the highest concentration of toluene, and the two-zone runner has the lowest dehumidification capacity and the concentration of pollutants adsorbed. When the dehumidification capacity is the same, the regeneration energy consumption of the two-stage runner is the lowest and that of the two-zone runner is the highest, and the lower the regeneration temperature is, the more remarkable the energy saving effect is. When the capacity of dehumidification is the same, the concentration of pollutants adsorbed by the three-zone runner is the highest, and with the increase of the capacity of dehumidification, the ability of the runner to adsorb pollutants increases. The higher the concentration of pollutants adsorbed by the runner, the slower the increase rate is.

- With the increase of the entrance air temperature of the runner, the desiccant capacity of the adsorbent becomes lower and lower. The desiccant capacity of the two-stage runner changes the most and that of the two-zone runner changes the least. At the same time, the concentration of toluene adsorbed by the runner decreases more slowly, and that of the three-zone runner decreases the fastest, and that of the two-zone runner. With the increase of the humidity of the process air entrance, the dehumidification capacity of the runner is getting larger and larger, among which the change of the two-stage runner is the largest, the change of the runner of the second zone is the smallest; the concentration of the toluene adsorbed by the runner is getting lower and lower, and the toluene of the third zone is adsorbing toluene. The change in concentration was minimal, and the concentration of toluene adsorbed in the second zone is the largest.
- With the increase of regenerated air temperature, the desiccant capacity of runner is getting higher and higher. The desiccant capacity of two-stage runner changes the most, and that of two-zone runner changes the least. The concentration of toluene adsorbed by runner is getting higher and higher, and the concentration of toluene adsorbed by three-zone runner changes the least and that of two-zone runner changes the most. As the humidity of regenerated air increases, the dehumidification capacity of runner decreases. The two-stage runner has the greatest change and the two-zone runner has the smallest change. The concentration of toluene adsorbed by runner is getting higher and higher, among which the concentration of toluene adsorbed by three-zone runner has the smallest change and that of toluene adsorbed by two-zone runner has the greatest change. Therefore, according to the specific requirements of indoor environment, appropriate regeneration temperature should be selected to reduce the regeneration energy consumption of runners.
- When the toluene concentration in the runner treatment zone is different, the desiccant capacity of the graded regeneration runner remains basically unchanged, and the concentration of toluene adsorbed by the runner becomes higher and higher. The concentration of toluene adsorbed by the runner in the three regions is the highest, and the concentration of toluene adsorbed by the runner is also the largest.

Acknowledgments
This work is supported by the Ministry of Housing and Urban-Rural Development project2017-R1-006. In addition, I really appreciate my tutor and my friends who really gave me so much help. What should I do is just working hard on my subject and then achieving great success.

References
[1] Klepeis N E, Nelson W C and Ott W R2001 *J. Exp Anal Environ Epid* **11** pp 231-252
[2] Yanjun D and Ruzhu W 2003 *J. Engineering Thermophysics* **24** pp 205-507
[3] Yasu Z and Peilin C 1999 *J. HVAC* **29** pp64-66
[4] Lizhi Z 2005 *MA. Eng. Dehumidification technology*
[5] Collier R K and Cohen B M 1991 *J. ASME Sol Energy ENGR* **113** pp 157-163
[6] Tianshu G, Yong L and Ruzhu W 2008 *J. Energy* **33** pp 1807-1815
[7] Tianshu G, Yong L and Ruzhu W 2009 *Int J. Ref* **32** pp 498-508
[8] Waugaman D G, Kini A and Kettleborough C F 1993 *J. Energy Res. Techology* **115** pp 1-8
[9] Han W2013 *PhD. Chongqing Univ*
[10] Chunxia J 2006 *PhD. Shanghai Jiaotong Univ*
[11] Hines A L and Ghosh T K 1992 *J. GRI Conter* **5089** pp 246-1821
[12] Aristov Y I and Tokarev M M 1996. *J. Reaction Kinetics and Catalysis Letters* **59** pp 325-333
[13] Wei Z 2007 *PhD. Tianjin Univ*
[14] Tianshu G 2008 *PhD. Shanghai Jiaotong Univ*
[15] Hong H 2006. *PhD. Tianjin Univ*

Study on regional applicability of dehumidification air conditioning system with different stage regeneration modes

H Hao[1*], Y D Zhang[1] and P Xin[1]

[1] School of Municipal and Environmental Engineering, Shenyang Jianzhu University, Shenyang, China

[*] 944946338@qq. com

Abstract. This article also put forward staged-regeneration wheel of the air conditioning system with condensing heat recovery technology, according to the climate zones in China, using the DesT software simulation in different areas of thermal and moisture load in a villa, and regeneration of the runner desiccant air conditioning system performance calculation, the results show that the compared with the two area runner desiccant air conditioning system, hierarchical regeneration runner desiccant air conditioning system performance are improved, the system of the refrigerant mass flow is reduced, the cooling load of the evaporator is reduced, the total load also significantly reduce energy consumption, air conditioning systems exergy efficiency improved; For regions with low temperature, low humidity or moderate humidity, the desiccant air conditioning system of three-areas graded regeneration wheel is more energy saving. In temperature and humidity relatively moderate areas, classification of renewable energy saving effect of the wheel were similar, but the four-areas wheel exergy efficiency obviously is higher than other wheel; For regions where both temperature and humidity are relatively high, the energy saving of the air conditioning system of the rotary wheel in the three-areas is not obvious, and the energy saving effect of the two-stage rotary wheel is the most significant.

1. Introduction

In recent years, the rotary dehumidification air conditioning system as a new air conditioning system in the operating mode, using the adsorption principle of the adsorbent, can separately handle the heat load and the wet load of the air conditioning system, not only fundamentally reducing the energy consumption of the air conditioning system, but also it solves the problem of mold contamination of fresh air in traditional air conditioning systems and has gradually received attention. However, higher temperatures are required during the regeneration of the adsorbent, which is also the most important energy consumption in the runner dehumidification purification process.

The graded regeneration method has been proven to reduce the regeneration temperature requirements of the runner and improve the thermal performance of the system [1]. At present, domestic and foreign scholars have studied the graded regenerative dehumidification air conditioning system. Compared with the two-zone rotary dehumidification air conditioning system, the graded regenerative dehumidification air conditioning system has obvious energy-saving effect. Collier and Cohen [2] proposed the concept of a graded regenerative wheel. The principle is to add a preheating zone in front of the regeneration zone.

The system uses silica gel as a desiccant. After simulation and experimental research, the system can significantly improve the air conditioning system COP. Meckler [3] designed a system in which a two-stage runner is coupled to a conventional air conditioner.

The system first uses the full heat exchange runner to cool the treated air, and bears part of the latent heat load, and then sends the treated air to the dehumidification runner for further dehumidification. The research results show that the dehumidification capacity of the full heat exchange runner can reach 30% to 50%, and no extra heat is needed during processing; Henning [4] also studied a similar two-stage rotor dehumidification system. In China, Ge Tianshu [5] proposed the concept of a two-rotor two-stage dehumidification air conditioning system. Compared with the ordinary runner, the two-stage runner has better adsorption performance, and under standard operating conditions, both the regeneration temperature and the helium efficiency of the two-stage system are improved; Wang Han [6] proposed a new hybrid air-conditioning system that combines a graded regenerative rotor and evaporative cooling technology.

Compared with the general composite air-conditioning system, the regenerative energy consumption of the graded regenerative composite system is reduced by about 42%, COP increased by about 78%. This paper intends to construct three dehumidification air conditioning systems that couple the grading regeneration runner with the condensing heat recovery technology. According to China's building climate zoning, seven typical cities are selected, and DesT software is used to calculate the heat and humidity load of a villa in different regions.

The performance parameters of three kinds of graded regenerative dehumidification air conditioning systems are calculated separately. Compared with the two-zone rotary dehumidification air conditioning system, the regional applicability of different graded regenerative dehumidification air conditioning systems is analyzed.

2. Workflow of graded regenerative dehumidification air conditioning system

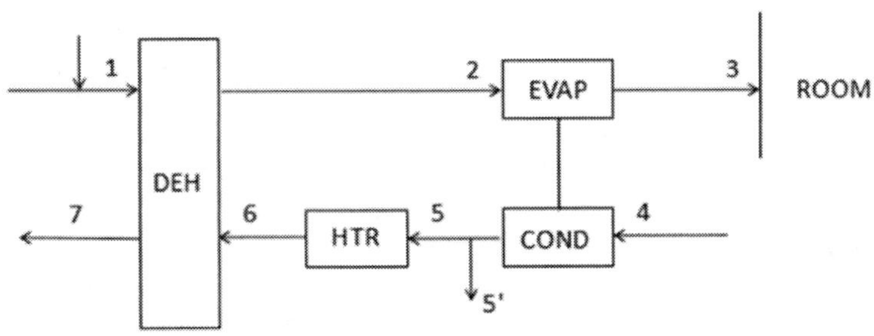

Figure 1. The dehumidifying air-conditioning systems of two-areas rotating wheel.

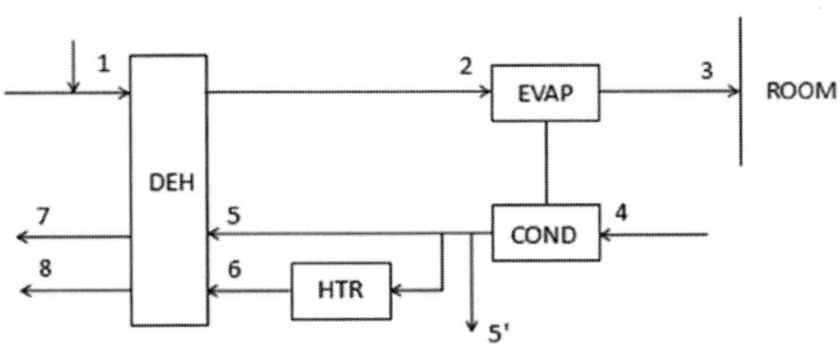

Figure 2. The dehumidifying air-conditioning systems of three-areas rotating wheel.

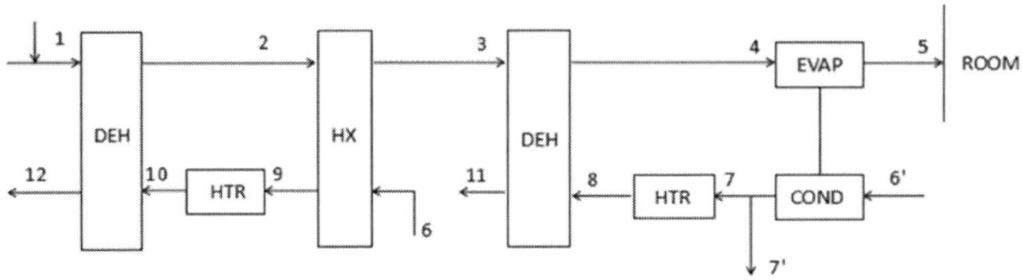

Figure 3. The dehumidifying air-conditioning systems of two-stage rotating wheel.

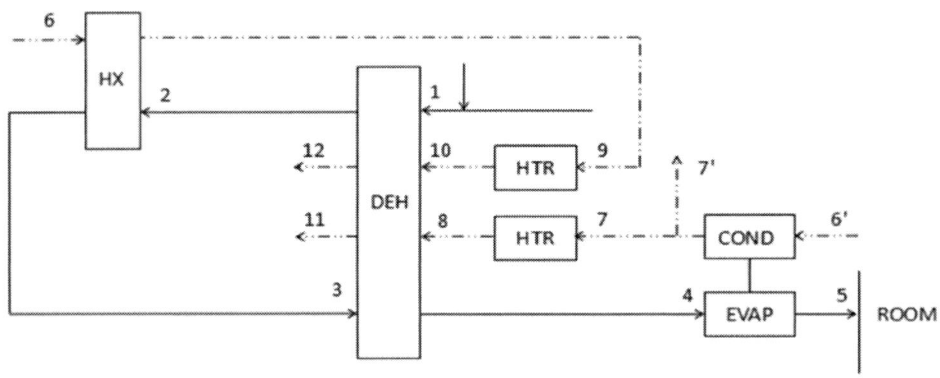

Figure 4. The dehumidifying air-conditioning systems of four-areas rotating wheel.

The figure above shows the working principle of the staged regenerative dehumidification air conditioning system. Among them DEH is a Wheel dehumidifier, EVAP is an evaporator, COND is a condenser, and HTR is a regenerative heater. The treatment air first enters the runner on the dehumidification side to perform dehumidification and temperature rise, and then is sent to the sensible heat exchanger to cool down, and after that the cooled air is sent to the Wheel dehumidifier again, and is processed to the Supply air condition by the evaporative cooler.

On the regeneration side, for the three-zone runner, the air passing through the condenser is divided into two strands, one is sent to the preheating zone of the runner, and one is sent to the regenerative heater to be heated to the regeneration temperature required by the system; for the four-zone and two-stage runners, the regeneration air is divided into two stands.

One is heated by the condenser and sent to the regenerative heater for heating, and the other is heated by the sensible heat exchanger and sent to the regenerative heater for heating. In this paper, the dehumidification air conditioning system chooses R22 as the refrigeration cycle working medium. It has been shown that the outlet air temperature of the condenser of the system is not higher than 55°C, which cannot meet the requirements of adsorbent regeneration. Therefore, an auxiliary heat source is needed to heat the regeneration air [7].

3. Mathematical description of system performance parameters

Mixed state point of fresh air and return air:

$$h_1 = \frac{G_w h_w + G_N h_N}{G_w + G_N} \tag{1}$$

$$d_1 = \frac{G_w d_w + G_N d_N}{G_w + G_N} \tag{2}$$

In the formula: G_W - mass flow of fresh air outside, kg/h; G_N - mass flow of indoor return air, kg/h; h_W - the enthalpy of fresh air outside, kJ/kg; h_N - the enthalpy of indoor return air, kJ/kg; d_W - moisture content of fresh air outside, g/kg; d_N - moisture content of indoor return air, g/kg.

Sensible heat exchanger:

Dehumidification area:

$$t_{A2} = t_{A1} - \varepsilon_{hx}(t_{A1} - t_B) \tag{3}$$

$$d_{A2} = d_{A1} \tag{4}$$

Regeneration Zone:

$$t_{B2} = t_B + \varepsilon_{hx}(t_{A1} - t_B) \tag{5}$$

$$d_{B2} = d_B \tag{6}$$

In the formula, ε_{hx} - heat-transfer coefficient, value 0.9; t_{A2}, d_{A2} - air temperature and humidity at the outlet of the sensible heat exchanger on the dehumidification side; t_{A1}, d_{A1} - air temperature and humidity at the inlet of the sensible heat exchanger on the dehumidification side; t_{B2}, d_{B2} - air temperature and humidity at the outlet of the sensible heat exchanger on the regeneration side; t_B, d_B - air temperature and humidity at the inlet of the direct evaporative cooler on the regeneration side.

Evaporation temperature and condensation temperature [8]:

$$Q_{evap} = M_k(h_{A2} - h_{A1}) \tag{7}$$

$$t_{cond} = t_w + 15 \tag{8}$$

$$t_{evap} = t_{A_2s} - \frac{t_{A_2s} - t_{os}}{0.94E_s} \tag{9}$$

In the formula, M_k - mass flow through the evaporator air, kg/s; h_{A1}, h_{A2} - the air enthalpy of the inlet and outlet of the evaporator, kJ/kg; t_W - outdoor air temperature, °C; t_{cond} - condensing temperature of the condenser, °C; t_{evap} - evaporation temperature of the evaporator, °C; t_{A2s}, t_{os} - wet bulb temperature of air inlet and outlet of evaporator, °C; E_s - the wet bulb temperature efficiency of the.

Refrigeration system:

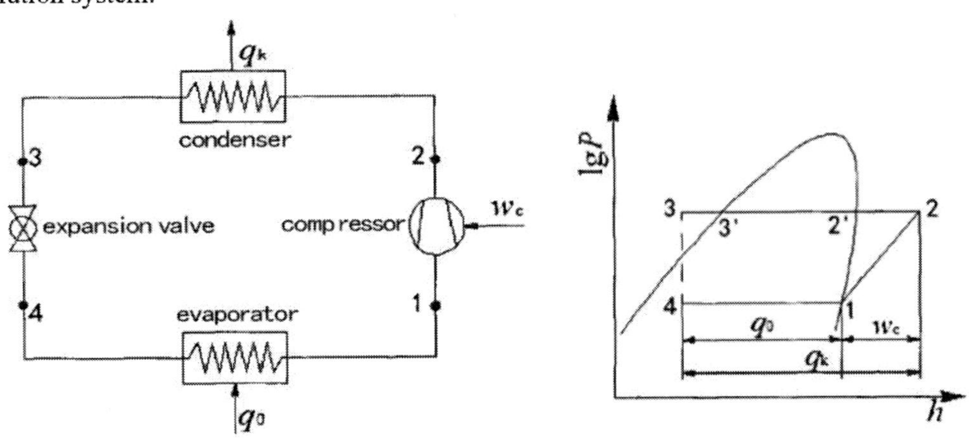

Figure 5. Theoretical cycle of a refrigeration system.

The figure shows:

Cooling capacity per unit mass refrigerant q_0:

$$q_0 = h_1 - h_4 \tag{10}$$

Condenser heat per unit mass refrigerant q_k:

$$q_k = h_2 - h_3 \tag{11}$$

Unit power consumed when compressor compresses unit mass of refrigerant w_c:

$$w_c = h_2 - h_1 \tag{12}$$

Compressor theoretical power consumption P_{th}:

$$P_{th} = M_r (h_2 - h_1) \tag{13}$$

Condensing heat ϕ_k:

$$\phi_k = M_r (h_2 - h_3) \tag{14}$$

In the formula, M_r - mass flow of refrigerant, kg/s; h_1 - enthalpy of refrigerant at the compressor inlet, kJ/kg; h_2 - enthalpy of refrigerant at the compressor outlet, kJ/kg; h_3 - the enthalpy of refrigerant at the condenser outlet, kJ/kg; h_4 - enthalpy of refrigerant at evaporator inlet, kJ/kg.

Regeneration energy consumption of air conditioning systems:

$$Q_r = m_r c (t_2 - t_1) \tag{15}$$

In the formula, Q_r - regenerative energy consumption of the system, kW; m_r - mass flow of regenerative air, kg/s; c - specific heat of regeneration air, kJ/(kg·°C); t_2 - the temperature of the regeneration air at the runner outlet, °C; t_1 - the temperature of the regenerated air at the runner inlet, °C.

3.1. Exergic loss and exergic efficiency of grading rotary dehumidification air conditioning system

In this paper, the parameters in the atmospheric environment are selected as reference points for the calculation of exergic loss and the analysis of exergic efficiency.

The atmospheric pressure is P_0, the thermodynamic temperature of air is T_0, and the saturated moisture content W_0 at air temperature is used as the reference parameter, Then the calculation formula of exergy under different wet air states (P, T, W) is as follows:

$$e = (c_{pa} + wc_{pv})T_0\left(\frac{T}{T_0} - 1 - \ln\frac{T}{T_0}\right) + (1 + 1.608w)R_a T_0 \ln\frac{p}{p_0} + R_a T_0\left[(1 + 1.608w)\ln\frac{1 + 1.608w_0}{1 + 1.608w} + 1.608w \cdot \ln\frac{w}{w_0}\right] \tag{16}$$

among them, R_a - dry air gas constant, value in the text is 0.287 kJ/kg/K; C_{pa} - constant pressure specific heat of dry air, value in the text is 1.003 kJ/kg/K; C_{pv} - constant pressure specific heat of water vapor, value in the text is 1.872 kJ/kg/K.

As can be seen from the above equation, exergy of wet air can be divided into three parts, the first term is heat exergy of wet air, expressed with e_{th}; the second item is mechanics exergy, expressed with e_{me}, When the system under study is open and constant pressure, this term is ignored; the third item is exergy of wet air, also known as diffusion exergy, expressed with e_{ch}.

IOP Conf. Series: Materials Science and Engineering **789** (2020) 012026 doi:10.1088/1757-899X/789/1/012026

$$e = e_{th} + e_{me} + e_{ch} \tag{17}$$

According to the above we can know, the graded runner composite air conditioning system studied in this paper uses the state of the air in the atmosphere as a reference point to establish the exergic equilibrium equation.

Because the system inlet specific exergic is larger than the system outlet specific exergic, therefore can be expressed as:

$$e_{in} - e_{out} \geq 0 \tag{18}$$

The system's specific exergic loss i is defined as:

$$i = e_{in} - e_{out} \tag{19}$$

The system exergic efficiency η is defined as:

$$\eta = \frac{e_{out}}{e_{in}} = 1 - \frac{i}{e_{in}} \tag{20}$$

4. Examples and analysis

4.1. Regional selection

China's climate zoning is determined in accordance with the current national standard <Thermal design code for civil building> GB50176, taking the coldest month average temperature as the main indicator, it is mainly divided into five climatic zones: severe cold region, cold region, hot summer and cold winter region, hot summer and warm winter region, and mild region.

At present, in the severe cold and cold regions of China, according to the differences in the number of heating days (HDD18) and the number of air conditioning days (CDD26) in a year, the existing <Design Standard for Energy Efficiency of Residential Buildings in Severe Cold and Cold Zones> [9] divides the two large building climate regions in detail, the severe cold regions are mainly divided into three parts, including severe cold zone A, zone B and zone C; divide the cold area into 2 parts, including cold areas A and B.

The specific division of each climate sub-area is shown in table 1. Among them, the hot summer and cold winter regions and the hot summer and warm winter regions respectively choose representative Chongqing and Guangzhou as the basis for calculation.

In summary, according to the relevant division basis, seven typical cities were selected for research, namely Mohe, Harbin, Shenyang, Dalian, Tianjin, Chongqing and Guangzhou. The outdoor design state parameters of different regions are shown in table 2.

Table 1. Climate classification in severe cold and cold regions.

Climatic region	Climate area	Partition basis	Typical city
severe cold region	Severe Cold Area A	6000≤HDD18	Mohe
	Severe Cold Area B	5000≤HDD18<6000	Harbin
	Severe Cold Area C	3800≤HDD18<5000	Shenyang
cold region	Cold Area A	2000≤HDD18<3800 CDD26≤90	Dalian
	Cold Area B	2000≤HDD18<3800 90<CDD26	Tianjin

Table 2. Outdoor design status parameters in different areas.

location	Outdoor design air dry bulb temperature in summer (°C)	Outdoor design wet bulb temperature in summer (°C)	Relative humidity	Moisture content of outdoor design air in summer (g/kg)	Outdoor design air velocity in summer (m/s)
Mohe	29.1	20.8	57%	13.8	1.9
Harbin	30.7	23.9	62%	16.5	3.2
Shenyang	31.5	25.3	65%	18.2	2.6
Dalian	29.0	24.9	71%	17.3	4.1
Tianjin	33.9	26.8	63%	20.4	2.2
Chongqing	35.5	26.5	59%	20.8	1.5
Guangzhou	34.2	27.8	68%	22.4	1.7

4.2. Simulation calculation

The villa has two floors above ground with a land area of 122 m², a total construction area of 198.5 m² and a building height of 6.25 m, The building simulation analysis of the project includes thermal parameters of the enclosure structure, room ventilation, room heat disturbance, etc, uses DesT software to fully simulate the hourly cooling load of buildings in different areas; the villa simulation overview is shown in figure 6:

Figure 6. The simulation diagram of building.

Table 3 shows the heat and humidity load of air conditioners in different regions. The wet load mainly considers the moisture production of indoor personnel and equipment.

Table 3. The table of load calculation in different areas.

location	Mohe	Harbin	Shenyang	Dalian	Tianjin	Chong qing	Guang zhou
cooling load (kW)	12.7	13.2	14.3	14.8	14.9	15.1	16.5
humidity load (kg/h)	11.9	11.9	11.9	11.9	11.9	11.9	11.9

The air treated by the runner in the air conditioning system is a mixture of outdoor fresh air and indoor return air. The ratio of return air volume to fresh air volume is 4:1, and the value of fresh air state parameters is shown in table 2, it is assumed that the air volume to be processed is constant, the state parameter of the air supply is determined by the heat and humidity load in the building, and the return air state parameter is the indoor air design state parameter. In addition, the air distribution ratio of the heating zone and the preheating zone in the three-zone rotary dehumidification air conditioning system is 1:1. In order to meet the requirements of comfort air conditioning, according to the provisions of GBJ 50019-2003, the designed indoor air temperature is 26°C, and the designed indoor air humidity is 10.2 g/kg.

Table 4. The performance comparison of air conditioning systems in different regions.

	Mohe	Harbin	Shenyang	Dalian	Tianjin	Chongqing	Guangzhou
Zone 2 rotary air conditioning system							
Refrigerant mass flow rate kg/s	0.141	0.160	0.177	0.165	0.184	0.196	0.206
Evaporator cooling load kW	21.3	23.5	26.2	24.8	29.2	29.5	31.1
Compressor energy consumption	2.62	2.98	3.45	3.18	3.62	3.70	3.94
Regenerative load	11.23	13.98	15.23	14.59	19.82	23.56	26.12
exergic loss	1.02	1.33	1.54	1.47	1.66	1.84	1.87
exergic efficiency	62%	61%	57%	60%	55%	53%	51%
Zone 3 rotary air conditioning system							
Refrigerant mass flow rate kg/s	0.134	0.153	0.172	0.163	0.188	0.192	0.203
Evaporator cooling load kW	20.3	23.3	26.1	24.9	28.6	29.2	30.9
Compressor energy consumption	2.55	2.91	3.44	3.10	3.57	3.65	3.86
Regenerative load	5.74	7.58	8.54	7.88	11.02	13.26	15.18
exergic loss	0.74	1.04	1.29	1.12	1.46	1.48	1.70
exergic efficiency	71%	68%	62%	67%	59%	55%	52%
Zone 4 rotary air conditioning system							
Refrigerant mass flow rate kg/s	0.094	0.107	0.118	0.116	0.133	0.142	0.153
Evaporator cooling load kW	14.6	16.2	17.8	17.0	20.8	22.0	23.4
Compressor energy consumption	1.78	2.09	2.34	2.20	2.56	2.75	2.98
Regenerative load	8.03	7.97	8.78	8.70	9.35	9.62	10.14
exergic loss	0.34	0.35	0.43	0.37	0.61	0.78	0.82
exergic efficiency	87%	86%	83%	86%	76%	73%	72%
Two-stage rotary air conditioning system							
Refrigerant mass flow rate kg/s	0.094	0.106	0.118	0.113	0.135	0.144	0.155
Evaporator cooling load kW	14.3	16.1	17.9	17.1	20.5	21.9	23.5
Compressor energy consumption	1.79	2.01	2.36	2.14	2.57	2.74	2.95
Regenerative load	7.47	7.88	8.22	8.20	8.90	9.57	9.93
exergic loss	0.34	0.49	0.64	0.53	0.68	0.72	0.74
exergic efficiency	83%	81%	79%	80%	79%	77%	76%

The above table shows the performance calculation results of the grading regenerative rotary composite air conditioning system under several typical climatic conditions in China.

Under the seven air conditioning conditions, compared with the two-zone rotary dehumidification air conditioning system, the mass flow, regenerative load and compressor energy consumption of the grading regenerative air conditioning system are significantly reduced, and the exergic efficiency is improved. In the graded dehumidification air conditioning system, all the wet loads in the air are borne by the runner, but a part of the heat is released while dehumidifying, and the heat load in the system is shared by the sensible heat exchanger and the evaporator. Therefore, the energy consumption of the compressor is significantly reduced; at the same time, due to the existence of the sensible heat exchanger, the temperature of the runner outlet is reduced, so the evaporator refrigeration load is significantly reduced.

When the indoor environmental parameters have the same value, the outlet temperature of the runner is related to the parameters of the outdoor environment. The higher the outdoor air temperature, the larger the cooling load and the higher the outlet temperature of the runner. Therefore, in order to meet the indoor air supply requirements, evaporative condensation system needs to consume more energy and can't achieve energy saving effect; when the humidity in the air is high, the capacity of the adsorbent to adsorb moisture in the runner is relatively high, and the greater the dehumidification amount of the runner, the greater the latent heat load, so the higher the required regeneration temperature, the greater the energy consumption for regeneration, at the same time, the air temperature at the exit of the runner has increased, the sensible heat load of the vapor compression system has increased, and the total load energy consumption has also increased.

In summary, the total load energy consumption of the composite air-conditioning system is related to the temperature and humidity of the outdoor environment. Although the graded regeneration runner has good energy-saving effects, the specific energy saving needs to be analyzed according to the outdoor environment and the dehumidification performance of the runner.

In Mohe, where the temperature and humidity are relatively low, the mass flow rate of the refrigerant in the two-zone rotary air conditioning system is 0.141 kg/s, the energy consumption of the compressor is 2.62 kW, the renewable energy consumption is 11.23 kW, exergy efficiency is 62%, Compared with the two-zone rotary dehumidification air-conditioning system, the mass flow rate of the three-zone rotary refrigerant is not much different from that of the compressor. The refrigerant mass flow rate of the four-zone and two-stage composite rotary dehumidification system is reduced by 33.3%. Compressor energy consumption decreased by 32.1% and 31.7% respectively. The exergic efficiency of the three-zone, four-zone and two-stage runners increased by 14.5%, 40.3% and 33.8% respectively, and the regeneration load decreased by 48.9%, 27.6% and 33.5% respectively.

Compared with the outdoor temperature and humidity in Mohe, Dalian, where the outdoor temperature is similar and the humidity is relatively high, the mass flow of refrigerant of the rotary air-conditioning system in zone 2 is 0.165 kg/s, the energy consumption of compressor is 3.18 kW, the regenerative energy consumption is 14.59 kW, and the exergic efficiency is 60%. Compared with the two-zone rotary dehumidification air-conditioning system, the refrigerant mass flow rate of the four-zone and two-stage composite rotor dehumidification system is reduced by about 26.7% and 31.5%, and the compressor energy consumption is reduced by 30.8% and 32.7% respectively. The exergic efficiency of the four-zone and two-stage runners increased by 11.7%, 43.3% and 33.3% respectively, and the renewable energy consumption decreased by 46.0%, 40.0% and 43.8% respectively.

From the perspective of energy consumption, the energy saving effect of the three-zone runner is more obvious when the temperature is lower, and the energy saving effect of the three-zone runner system is most significant in the region with low humidity, and with the increase of humidity, the energy saving effect of the four-zone and two-stage runners is becoming more and more obvious, according to the analysis, when the dehumidification amount is low, the grading runner can meet the dehumidification requirements at a lower regeneration temperature. For the three-zone runner, due to the presence of the preheating zone, the consumption of high-grade thermal energy is reduced, so the regeneration load is also relatively low. Therefore, for areas with lower temperatures, lower humidity or milder temperatures, the three-zone grading regenerative air conditioning system is relatively more energy efficient.

In Harbin and Shenyang areas where temperature and humidity are relatively mild, the temperature and humidity of Harbin are lower than that of Shenyang, for the Harbin area, the mass flow rate of the refrigerant in the second-zone rotary air-conditioning system is 0.160 kg/s, the energy consumption of the compressor is 2.98 kW, the energy consumption for regeneration is 13.98 kW, and the efficiency of exergic is 61%, compared with the two-zone rotary dehumidification air-conditioning system, the mass flow rate of the three-zone rotary refrigerant is similar to that of the compressor, and the refrigerant mass flow of the four-zone and two-stage composite rotary dehumidification system is reduced by about 33.1%, compressor energy consumption decreased by 29.9% and 32.5% respectively, the exergic efficiency of the three-zone, four-zone and two-stage runners increased by 11.4%, 41.0% and 32.8% respectively, and the regeneration load decreased by 45.8%, 43.0% and 43.6% respectively.

For the Shenyang area, the mass flow rate of the refrigerant in the second-zone rotary air-conditioning system is 0.177 kg/s, the energy consumption of the compressor is 3.45 kW, the energy consumption for regeneration is 15.23 kW, and the efficiency of the exergic is 57%, compared with the two-zone rotary dehumidification air-conditioning system, the mass flow rate of the three-zone rotary refrigerant is not much different from that of the compressor, the refrigerant mass flow rate of the four-zone and two-stage composite rotary dehumidification system is reduced by 33.3%, compressor energy consumption decreased by 32.2% and 31.6% respectively. The exergic efficiency of the three-zone, four-zone and two-stage runners increased by 8.7%, 45.6% and 38.5% respectively, and the regeneration load decreased by 43.9%, 42.3% and 46.0% respectively. From the perspective of energy consumption, in the areas where the temperature and humidity are relatively mild, the energy saving effects of the three-zone, four-zone and two-level systems are not much different. The analysis suggests that in areas where the temperature and humidity are not very high, the dehumidification and energy consumption of the grading regeneration runners are not much different, but the exergic efficiency of the four-zone runners is significantly higher than that of the other runners.

For the Tianjin, Chongqing and Guangzhou regions where the temperature and humidity are relatively high, the specific analysis is as follows: for the Tianjin area, the mass flow rate of the refrigerant in the second-zone rotary air-conditioning system is 0.184 kg/s, the energy consumption of the compressor is 3.62 kW, the energy consumption for regeneration is 19.82 kW, and the efficiency of exergic is 55%, compared with the two-zone rotary dehumidification air-conditioning system, the refrigerant mass flow rate of the four-zone and two-stage composite rotor dehumidification system is reduced by 27.7% and 26.6%, and the compressor energy consumption is reduced by about 29.3% and 29.0%. The exergic efficiency of the four-zone and two-stage runners increased by 7.3%, 38.2% and 43.6% respectively, and the regeneration load decreased by 44.4%, 52.8% and 55.0% respectively. For the Chongqing area, the refrigerant flow rate of the second-zone rotary air-conditioning system is 0.196 kg/s, the compressor energy consumption is 3.70 kW, the regenerative energy consumption is 23.56 kW, and the exergic efficiency is 53%, compared with the two-zone rotary dehumidification air-conditioning system, the refrigerant mass flow rate of the four-zone and two-stage composite rotor dehumidification system is reduced by about 26.5%, and the compressor energy consumption is reduced by about 25.7% respectively, the exergic efficiency of the three-zone, four-zone and two-stage runners increased by 5.7%, 37.7% and 45.3% respectively, and the regeneration load decreased by 43.7%, 59.2% and 59.3% respectively.

For the Guangzhou area, the mass flow rate of the refrigerant in the second-zone rotary air-conditioning system is 0.206 kg/s, the energy consumption of the compressor is 3.94 kW, the energy consumption for regeneration is 26.12 kW, and the efficiency of exergic is 51%, compared with the two-zone rotary dehumidification air-conditioning system, the refrigerant mass flow rate of the four-zone and two-stage composite rotor dehumidification system is reduced by about 25.0%, and the compressor energy consumption is reduced by 24.3% and 25.1% respectively, the exergic efficiency of the two-stage runners increased by 5.9%, 37.2% and 49.0% respectively, and the regeneration load decreased by 41.9%, 61.2% and 61.9% respectively. For areas with relatively high temperature and humidity, the dehumidification performance of the three-zone runner decreases. With the increase of dehumidification, the energy saving of the three-zone runner system is less obvious. The energy-saving effect of the two-stage runner dehumidification air-conditioning system is the most significant.

5. Conclusion

This paper proposes the concept of combining the regenerative runner and the condensing heat recovery technology into the composite rotor dehumidification air conditioning system. The main conclusions are as follows:

- compared with the two-zone rotary dehumidification air conditioning system, the performance of the staged regenerative dehumidification air conditioning system is improved; the refrigerant mass flow rate in the system is reduced, the refrigeration load of the evaporator is reduced, the total load energy consumption is also significantly reduced, and the exergic efficiency of the air conditioning system is also improved;
- the energy consumption of the staged regenerative dehumidification air conditioning system is greatly affected by the outdoor environment. For areas with low temperature, relatively low humidity or milder temperature, the three-zone grading regenerative air conditioning system is relatively more energy efficient than the two-zone rotary dehumidification air conditioning system.
- In areas where temperature and humidity are relatively mild, the energy-saving effect of the grading regeneration runner is not much different, but the exergic efficiency of the four-zone runner is significantly higher than that of the other runners;

For areas with relatively high temperature and humidity, the energy saving of the three-zone rotary air conditioning system is not obvious, and the energy saving effect of the two-stage runner is the most significant.

Acknowledgments

This work is supported by the Ministry of Housing and Urban-Rural Development project (2017-R1-006). In addition, I really appreciate my tutor and my friends who really gave me so much help. I should work hard on my subject and achieve great success.

References

[1] Ying Y, Han W, Wei Z and Zhao D 2013 *J. Cryogenics Supercond* **1** pp 41-44
[2] Collier R K and Cohen B M 1991 *ASME J. Sol Energy ENGR* **113** pp 157-163
[3] Meckler G 1989 *J. ASHRAE Trans* **95** pp 1116-1123
[4] Henning H M 2007 *J. Appl Therm Eng* **27** pp 1734-1749
[5] Tianshu G 2008 *PhD. Shanghai Jiaotong Univ*
[6] Han W 2013 *PhD. Chongqing Univ*
[7] Hong H, Yufeng Z and Guobing Z 2008 *J. Shenyang Jianzhu Univ* **24** pp 675-678
[8] Yufeng Z, Hong H and Lei F 2008 *J. Acta Energy Sol Sinica* **24** pp 675-678
[9] JGJ 26—2010 *S. Design standard for energy efficiency of residential buildings in severe cold and cold zones*

IOP Conf. Series: Materials Science and Engineering **789** (2020) 012027 doi:10.1088/1757-899X/789/1/012027

Design of flow measurement system based on LabVIEW

H Hao[1*] and H L Zhang[1]

[1] School of Municipal and Environmental Engineering, Shenyang Jianzhu University, Shenyang, China

[*] 291074584@qq.com

Abstract. The traditional flow meter is installed in the pipeline, not only to increase initial investment and cause stress loss to the system, but also to use a large amount of manpower and material resources in the data recording. In some areas that are not easily measured, such as buried in the ground, it can be difficult to measure. In order to solve these problems, based on the LabVIEW2013 as platforms and the hardware of sensor and data acquisition card, combined with the serial communication program to design the flow measurement system, getting traffic through the information collected data and the data acquisition card to convey in the program, get traffic after processing, according to flow to calculate the hydraulic balance, judging by the imbalance rate and the degree of hydraulic balance pipeline system is in a state of hydraulic balance. At last, through the actual test analysis, it is concluded that the flow measurement system for the conclusion of practical application, it not only can be used for pipe flow measurement and equilibrium state judgment, and also to have certain security on accuracy.

1. Introduction

The concept of virtual instrument was first proposed by the America company National Instruments (NI) in 1986 and has been developed and applied at high speed [1], investigating its reason, mainly is the computer ability to the powerful data processing, and then combined with the data acquisition card or bus and other related hardware and programming software and developed a kind of convenient and visual instruments.

Virtual instrument is the most essential is the meaning of "software is instrument", it has the following characteristics: strong running environment, software and hardware resource sharing, short development cycle, less cost and visualization programming languages and display interface etc [2]. LabVIEW is a through the use of graphic symbol, function and connection for programming software platform, it uses a programming language called graphical programming language, namely the G language, using the G language compared with traditional programming languages, short time-consuming, programming efficiency has improved greatly.

Computer, hardware, and LabVIEW combined into a virtual instrument can fully configured, however the cost is only a small part of the traditional instruments, even when changes to the requirement of instrument, it can be modified in an instant.

Therefore, the flow measurement system based on LabVIEW will have great development prospect and application scale in the field of flow measurement.

2. System composition

The entire virtual instrument system consists of two parts: hardware and software design.

The hardware foundation includes differential pressure sensor, data acquisition card and serial port communication program, etc. By comparing the advantages and disadvantages and using range of a variety of flow measurement method, combining the application of the virtual instrument conditions, this paper uses the differential pressure type flow measurement methods, and selection of the differential pressure type sensor in the flow sensor, throttle type models for BC69, its technical indicators as shown in table 1.

Choose the data acquisition card (DAQ) [3] for PCI8025, physical channel number is 4 channel, the input method of analog quantity is Double-end analog revenue, 12 of the 1 D/A channels and switch input and output channels all 16 road, resolution is 14, programmable gain 1 times, 16 k word FIFO memory, the sampling rate of up to 1.6 MHz, trigger type has edge trigger and pulses (level) trigger, the nonlinear error is plus or minus 1 LSB, system measuring accuracy is 0.05%, operating temperature range for - 40□ ~ + 85□. Serial communication programs mainly include the initialization of serial port and serial port data transmission, and the initialization of serial port is mainly used to set baud rate.

In the communication, the serial port communication baud rate is 9600, which uses the query method to send the data in the program design, the flow chart of serial communication program as shown in figure 1.

Table 1. Performance indicators of throttle flow sensor.

range	Working voltage	The output voltage	precision	Allow overload
0~5000 kPa	24VDC	0~5 V	±0.5% FS	300% FS

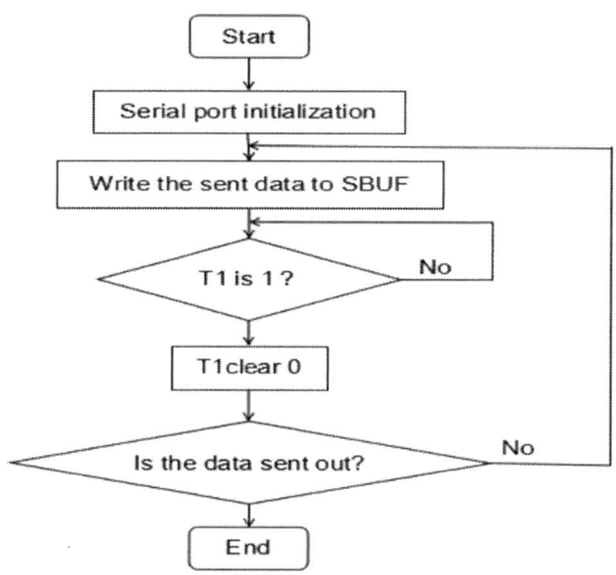

Figure 1. The flow chart of serial communication program.

System software design using LabVIEW platform, a whole program [4] written in it composed by one or more of the children of the virtual instrument, virtual instrument in the following paper will be referred to as VI, each VI is constitute by three parts: the Front Panel , visual interface, program Block Diagram - the background program show and ICONS/connection port - connection hub [5].

System design in this paper not only includes the flow measurement, and even after the measured flow on the basis of these data get pipeline imbalance rate and hydraulic balance to reflect the piping system is whether or not to achieve hydraulic balance state.

Main program includes three parts: login module, real-time data and historical data, mainly through the serial port to receive the 8-bit digital quantity conversion and acquisition by lower machine to send

4 road A/D, through data processing to calculate the flow rate value, and the degree of hydraulic balance, unbalance rate can be shown up on the front panel at the same time. In addition, achieving the function of alarming, data storage and data acquisition parameter setting, etc.

3. The software design

3.1. Custom control and subVI design

Before editing the VI program may customize some controls that have specific request, on the custom control panel, it is a front panel that similar to the VI, according to user needs to be set, but the difference is not the program block diagram. In order to make the whole interface clearer and more intuitive, in this article, the mainly customized controls of the alarm lamp, starting collection, flow rate and hydraulic balance are made in real time.

As shown in figure 2, the self-definition setting of the control is shown in real-time display of alarm lamp, starting collection, flow rate and hydraulic balance.

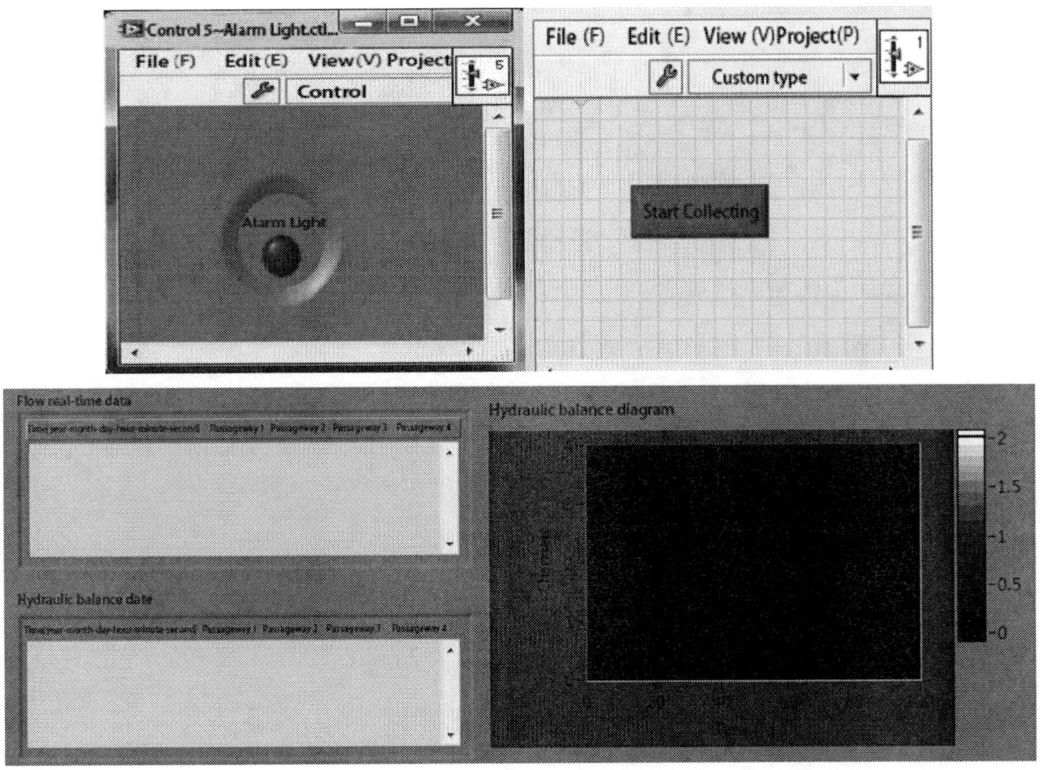

Figure 2. The customized controls.

In addition to customize control, it is written to the subVI of the call before the main program is written. The subVI, which is set up in the paper, has the flow meter operator VI, the hydraulic equilibrium calculation subVI, the alarm subVI, and TAB sub VI, etc, which can be called in the main VI. The following is a presentation of the part VI, figure 3 is the flow meter operator VI's front panel and program block diagram, flow meter operator program block diagram of the writing is according to the principle of flow measurement of differential pressure type, and on the various parameters input and run after receiving the differential signal can get flow measurements.

Figure 4 is the pipeline resistance loss subVI and imbalance rate calculation subVI program, the concept of programming is based on unbalance rate, namely the difference value of pipeline working pressure and the pressure loss of the calculation of line and the ratio of the pipeline working pressure.

Figure 5 is hydraulic balance calculation VI's front panel and program block diagram, programming is also derived from the concept of hydraulic balance, in the case of known design flow to calculate the hydraulic balance.

SubVI programming is completed, in order to make the main VI successful call, in the upper right corner to select the mode of connection port and establish connection with input, output controls one by one, from the white line to non-ferrous wire frame, can be connected to the main VI to call.

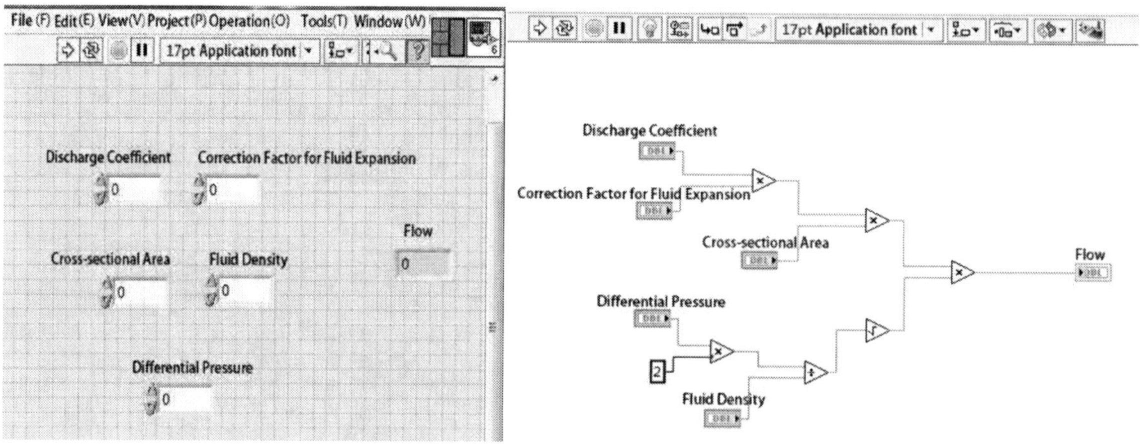

Figure 3. The front panel and block diagram of the sub-VI-flow calculation.

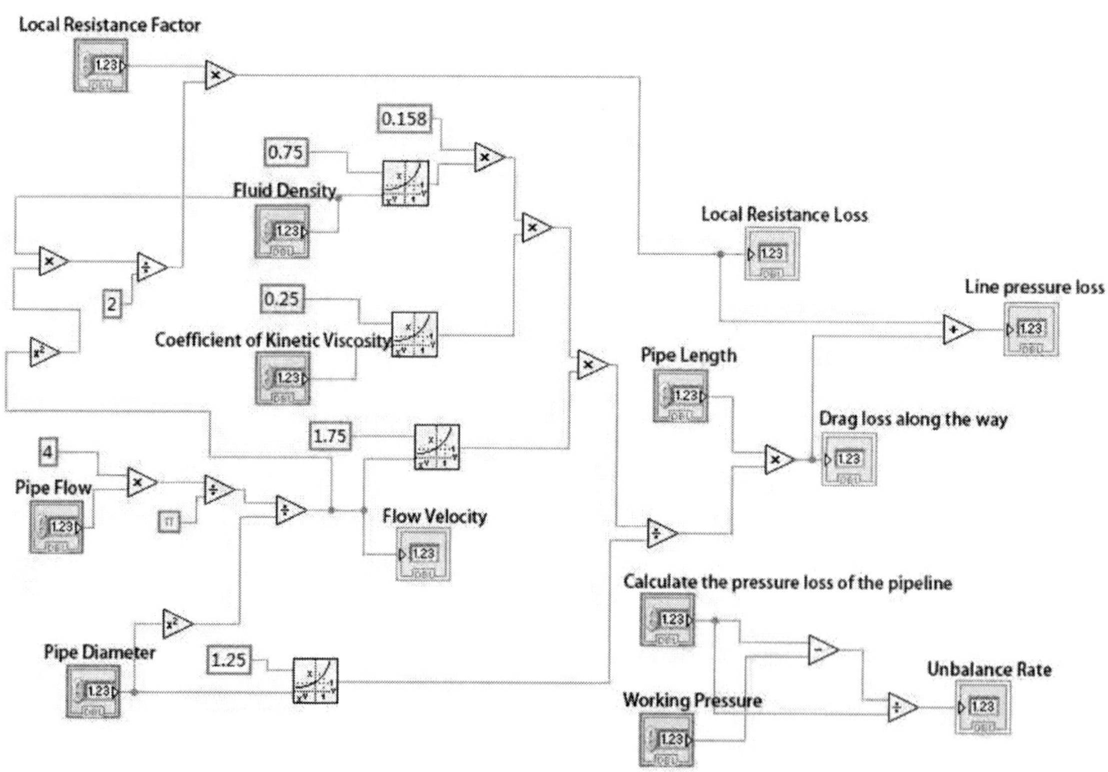

Figure 4. The block diagram of the sub-VI-unbalance rate calculation.

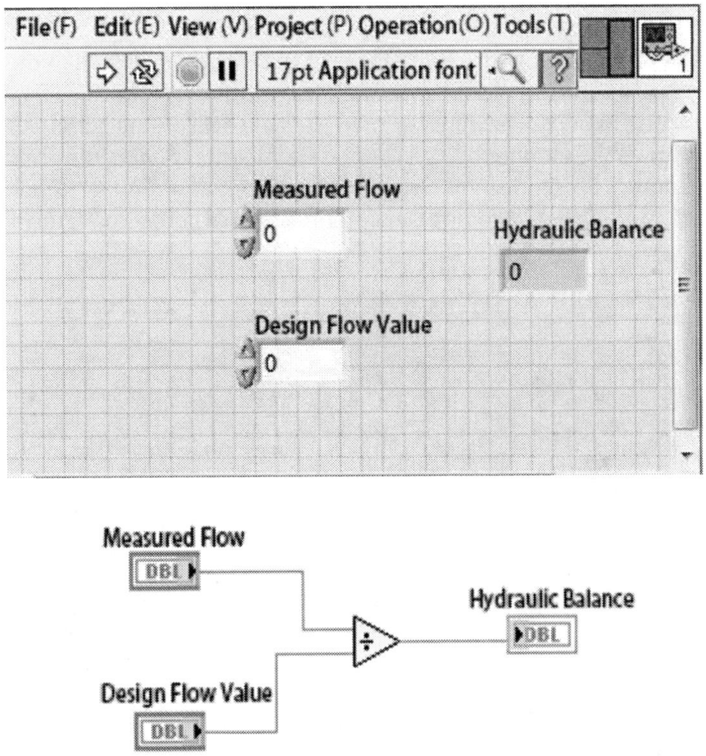

Figure 5. Front panel and block diagram of the sub-VI-Hydraulic balance calculation.

3.2. The login module

Login interface can stop the non-operation personnel operating system, only the prescribed operating personnel can perform relevant operation, not only can ensure the safety of the system, at the same time greatly reduce adverse consequences due to operator's wrong operation, so setting up the login module is necessary.

When operators enter the user name equal to the set value, we have set the program execution down into the structure of the second condition, the input password value equals the value again, clicking on the login confirmation, system login successful. Otherwise, the wrong password login, the system could not log in.

Tab controls make login interface and real-time data and history query interface form a complete system well-organized, in other words, does not perform the login interface program, will not be able to perform at the back of the program, which greatly increases the security and confidentiality of the system [6].

3.3. Data collection and processing module

Data collection and processing module is the display interface of real-time data and the query storage interface of historical data.

The real-time data window Is mainly used to implement 4 road traffic data real-time display, real-time flow curve display, unbalanced rate and degree of hydraulic balance figure display, and can set the configuration of the serial port in this interface, such as baud rate, data bits, stop bits and parity bit; in addition, it can set the rate of flow collection, the upper and lower limit of alarm flow and the storage path of selecting data.

It can start and stop data collection at the same time. In addition, when the flow value exceeds the design limit, the alarm indicator will blink and make an alarm sound.

IOP Conf. Series: Materials Science and Engineering **789** (2020) 012027 doi:10.1088/1757-899X/789/1/012027

Figure 6 shows the front panel of the real-time data interface.

Figure 7 shows the corresponding program block diagram, including VISA configuration serial node, which can be conveniently configured for the parameters of the serial port.

The serial buffer reads the node, which can read the data sent to the serial data buffer by the node; property node, which reads the number of bytes in the serial port; connect the above nodes correctly and combine the while structure and condition structure to carry out continuous reading of the data sent by a machine; by using the data frame decoding data of the obtained serial port data, the corresponding data of each channel is obtained.

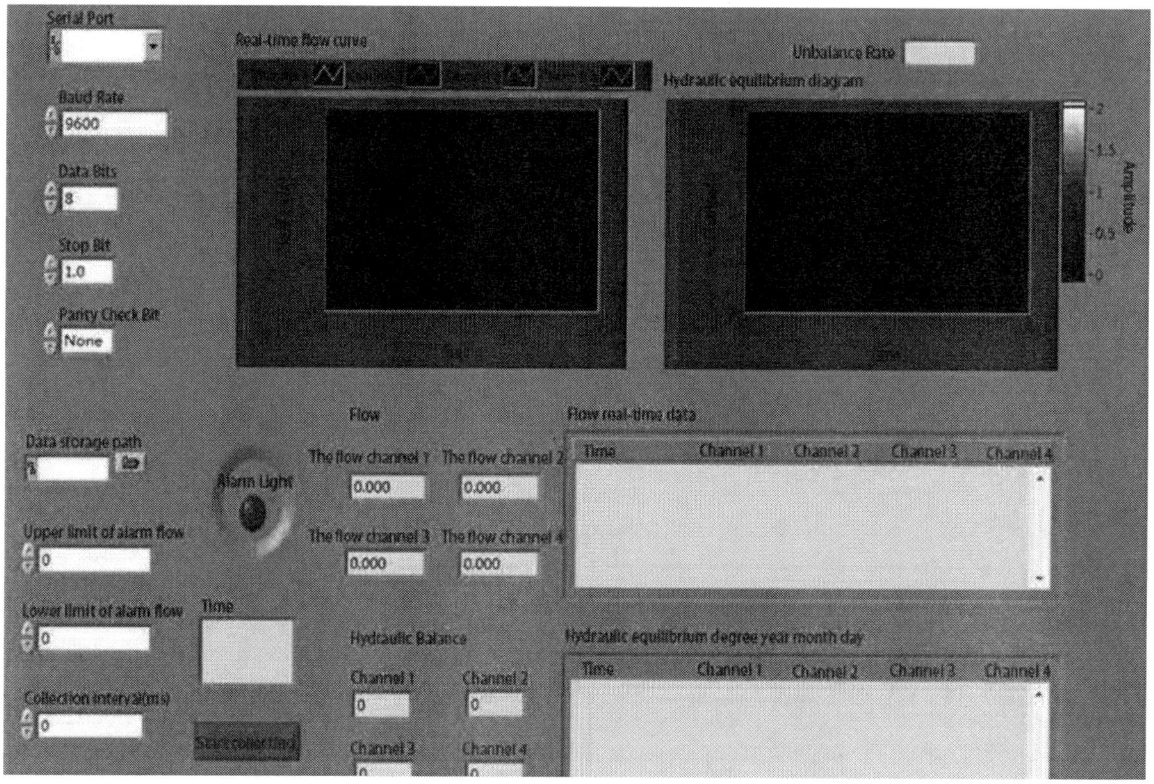

Figure 6. The front panel of real-time data interface.

Each channel data after processing, data can be entered to the computer interface for display, with the waveform chart to display the real-time flow curve, imbalance rate displayed in numerical form, with the strength of the chart color depth to represent the extent of the actual flow deviates from the design flow, namely hydraulic balance, at the same time of each channel is shown on the interface flow numerical and hydraulic balance; data storage is done primarily by writing to a text file.

The design by pressing the start button to read the data stored in the file, and on the historical data interface shows the historical data, including historical flow graph, historical hydraulic balance figure and historical time corresponding flow value.

For the convenience of data analysis, program flowchart is a specialized in the data is used to read history, its program design as shown in figure 8, the corresponding block diagram of the front panel and the process of the historical data interface are similar to the front panel showing the real-time data interface in figure 6.

IOP Conf. Series: Materials Science and Engineering **789** (2020) 012027 doi:10.1088/1757-899X/789/1/012027

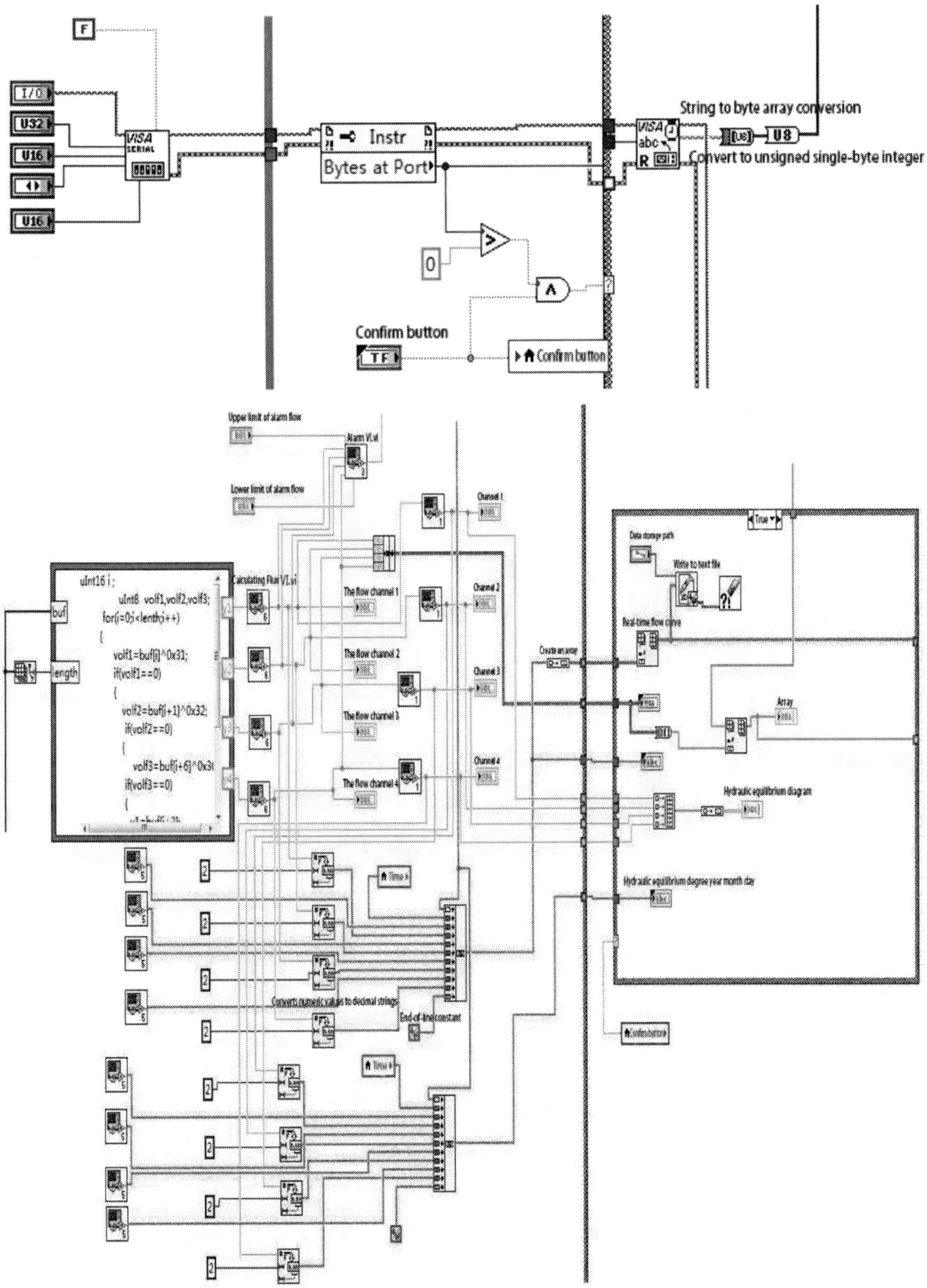

Figure 7. The program diagram of real-time data interface.

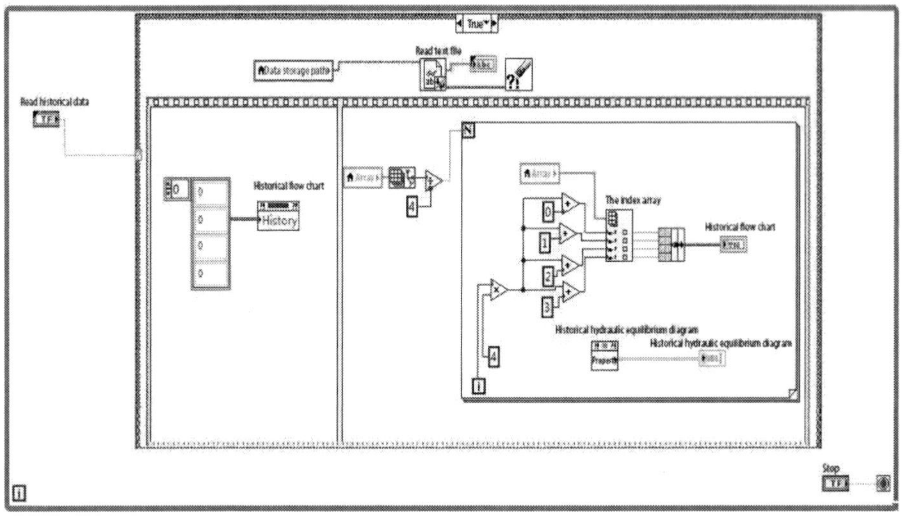

Figure 8. The program diagram of historical data to read.

4. System test analysis

The test validation of the flow measurement system based on LabVIEW mainly consists of two aspects: one is whether can get a precise flow value, the other aspect is whether the hydraulic balance and pipeline imbalance can be obtained after the measured flow data.

In the process of testing, we mainly is shown in figure 9 pipe for the differential pressure signal acquisition, A, B, C, D respectively represent 4 users, in figures 1-4 installation of differential pressure sensor, sampling the differential information of channel 1 - %4, through the data acquisition card plug wire connection, sending information to the PC, running can get data processing of the collected information, so get the numerical flow, in order to compare with the measured results of the flow measurement system, we reference literature [7] were measured flow numerical using ultrasonic flow meter, and the error between the two and the actual design flow is calculated respectively.

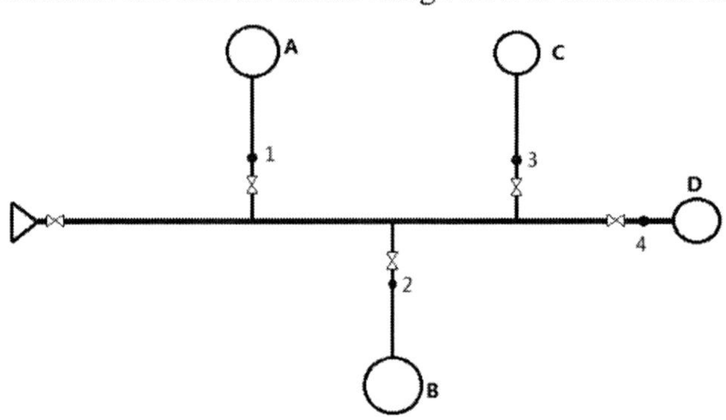

Figure 9. The measurement pipeline in experiment.

Through practical test and analysis can be concluded that the measured values of flow measurement system and the measured value of the ultrasonic flowmeter is very closed, and because of the flow measurement system in the process of flow values obtained only with the differential signal has direct relation, avoiding a lot of interference from outside world , such as the environment and human factors, so its measuring accuracy is higher, it can completely satisfy the measurement requirements, hydraulic balance and imbalance rate are also intuitive display correctly, and reflect the pipeline operation state of equilibrium. It can also record, query and analyze historical data.

5. Conclusion

Through research, this paper designed the flow measurement system based on LabVIEW, the hardware foundation based on differential pressure sensor, data acquisition card, setting up a serial port communication, the design of the software of the flow measurement system, adopts the modular concept.

On the basis of custom controls and establish a subVI completed the design of the main program, the system can continuous collect signal sent by sensors, and after further to deal with these signals, achieving data computing, storage and read, and a series of processing and the flow measurement; by measuring the flow value, the hydraulic balance ratio and the pipe imbalance rate are calculated to determine whether the pipeline is in a state of equilibrium.

Test validation for the flow measurement system, not only can be flow numerical display accurate, but also can meet the measuring accuracy, so the flow measurement system can be applied practically, especially in some occasions is not easy to measure such as the buried pipe flow, and it can be judged by imbalance rate and the degree of hydraulic balance state of the hydraulic balance of the pipeline system.

Acknowledgments

This work is supported by the Ministry of Housing and Urban-Rural Development project. In addition, I really appreciate my tutor and my friends who really gave me so much help.

References

[1] Lian H Z and Zhao Y J 2010 *J. Inst. Meas* **2** pp 21-23
[2] Zhou Y 2008 *J. China SciTechnol Inno Guide* **25** pp 11-12
[3] Liu J Y, Zhang S Y and Zhang Y J 2011 *J. Coal mine safety* **42** pp 84-86
[4] Fan L N 2001 *Inst user* **8** pp 5-9
[5] Chen X H 2007 *J. First Edi* Beijing Tsinghua Univpress
[6] Zhi J M 2017 *MA. Eng.* Shenyang Jianzhu Univ
[7] Shi W J 2014 *MA. Eng.* Xi 'an Petroleum Univ

Computer modeling and simulation - tool for design of buildings and its systems

M Horneț[1*], R D Todor[2], I L Cîrstolovean[1], D C Năstac[1] and P Mizgan[1]

[1] Transilvania University of Brasov, Romania, Department of Buildings Services, Brasov
[2] Transilvania University of Brasov, Romania, Marketing, Turism and International Affairs Department, Brasov

[*] mircea.hornet@unitbv.ro

Abstract. This paper presents a real example of optimizing the design of buildings and the systems they serve, through computer modeling and simulation. It highlights the capabilities of the DesignBuilder simulation software used as a design tool for buildings and systems for heating, ventilation and air conditioning. The paper also includes discussions and conclusions about how simulation results can provide important information for building design. The building for which the research presented here were done is the Transilvana University of Brasov Development Research Institute (Romania).

1. Introduction

Reducing energy consumption for the operation of buildings, of any kind, is a wish of today's modern economy, both nationally and globally. That is why solving, as accurately as possible, the problems of heat transfer and fluid flow is of major importance in the various applications in the field of civil engineering [1].

It is well known that currently one of the best techniques available for prediction, in the design phase of the actual thermal behavior and energy consumption of a building, is computer modeling and simulation. [2]. This technique, initially used mainly in research, has become increasingly used in current design practice.

With this technique a building is treated and optimized in its entirety, not as a sum of components and systems optimized separately. Also, the simulation has the advantage that in addition to the structure itself of the building, it can also include the interior and exterior environment (of the building), and can also deal with problems related to the classical energy systems of the building, or those of renewable energy [2]. At the same time, with this technique, several solutions regarding the geometry of the building, the construction of the tire or of the heating/cooling (air conditioning) systems can be analyzed in detail and compared with each other.

However results are highly dependent on the knowledge of the person setting up the model and interpreting the results. This is a problem in extending the use of CFD in building design, because CFD software becomes easier to use by people with little understanding of the mathematical model underlying it [3].

In the present research, we used the DesignBuilder software tool. This is a unique software tool for creating and evaluating building projects. It provides a wealth of data on environmental performance, such as: maximum summer temperatures, comfort conditions, energy consumption, carbon emissions, as well as dimensions of HVAC components. DesignBuilder uses the latest EnergyPlus simulation engine to calculate building performance and is suitable for use by architects, energy consultants and construction service engineers. Among the typical uses of this software can be mentioned [4]:

- Calculation of temperature and pressure distribution in and around buildings, using CFD;
- Evaluating a wide range of options for the facade, for the effect on overheating, energy consumption and visual appearance;
- HVAC design, including sizing of heating and cooling equipment;
- Thermal simulation of naturally ventilated buildings;
- Viewing the location of the solar shading *in situ*;
- Verification of optimal use of natural light.

To carry out the simulations, either real time data for the respective site or, in the absence of them, weather data of the reference year climate, for the locality in which the building is located are required. The simulations of the present research were performed on the basis of hourly values for five meteorological variables, provided by a weather station in Brasov.

Also, in choosing the design solution we also considered the studies performed by our colleagues from the marketing department, regarding the life cycle cost analysis in a building design [5].

2. The building for which the research was carried out

The building for which the research presented in this paper was carried out is actually an ensemble of 12 identical buildings (designed as research laboratories), two access and connection spaces between the buildings (called spines) and a centrally located atrium that forms the Transilvana University of Brasov Development Research Institute (ICDT). The way the 12 buildings are located is shown in figure 1 below. The dimensions of this ensemble are: 180 m long, 78 m wide and between 4.20 m and 7.40 m high, the atrium having the height of 10.20 m.

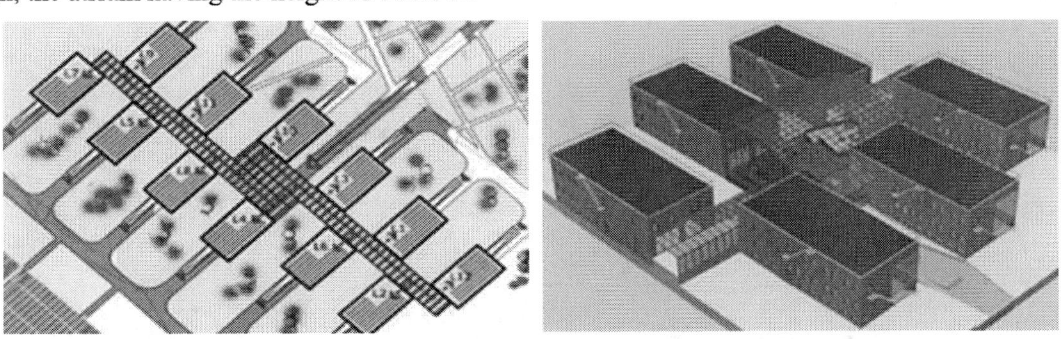

Figure 1. The access and connection spaces between the 12 buildings and DesignBuilder model of part of the building.

Each of the 12 buildings is composed of basement, ground floor and floor, and has a built area of 1440 m^2 each. The strength structure of the building is made of reinforced, longitudinal and transverse reinforced concrete columns and frames. The outer walls are made of BCA GBC 50 with a thickness of 20 cm and they are clad with aluminum panels, which have a polystyrene insulation with a thickness of 10 cm. The horizontal elements of the resistance structure are made of monolithic reinforced concrete slabs of 15 cm thick. The roof is of terrace type and has a mineral wool insulation with a thickness of 40 cm. [6].

The buildings are oriented with the main facades to the south and were designed with two glazed curtain walls: one on the south facade, with an area of about 90 m^2 and the other on the west (respectively east), with an area of about 36 m^2. The buildings are not sheltered.

The two connecting spaces between the buildings (Spinele) have an area of 480 m^2, each, the Atrium has an area of 630 m^2, resulting in a total useful area of the whole built, of 18870 m^2. The side walls and the roof of the Spines and the Atrium - with the exception of those common with the laboratories - are glazed surfaces.

3. Research results

Here is presented how, through the carried out researches, the optimal solutions were chosen for two of the important problems whose solution depends, to a large extent, the energy consumption of the building assemblies, namely:

- choosing the best solution for glazed surfaces;
- choosing the optimal solution for heating systems.

In addition to solving these specific problems, from the research carried out with the help of simulations, other important conclusions and indications have been drawn regarding: the interior environment of the buildings (in the cold and hot seasons), the maximum heating and cooling tasks, the operation and control of HVAC systems etc.

3.1. Choosing the best solution for glazed surfaces

Whereas:

- large portions of the external walls are designed as glazed;
- in Romania, the solar thermal inputs during the summer, through a triple glazed surface are about 6.5 MJ/m² - for the south orientation and 5.5 MJ/m² - for the west orientation (which significantly influences the overall energy of the building and its performances, the first part of the study aimed at choosing the best solution for glazed surfaces [6].

As the architectural design mode and the degree of glazing differ from one type of building to another, the study was carried out for each type of building separately:

- laboratories;
- spines;
- atrium.

Thus, for each type of building the annual energy consumption for heating (and cooling) was compared for four variants (double glassing surface, triple glassing surface and two different types of glass). The results of this part of the study are shown in figure 2, below.

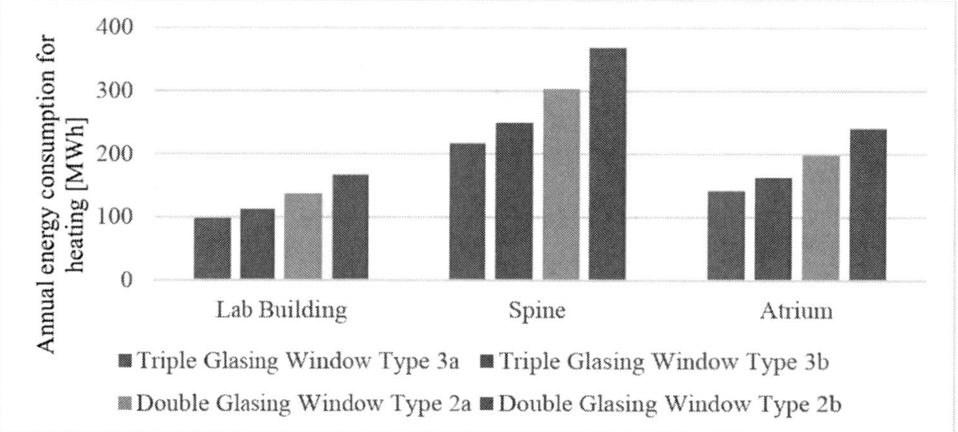

Figure 2. Annual energy consumption for heating depending on the type of glazing.

Following the study, for all the glazed facades and all the windows, a very high performance triple glazing system was chosen, whose specific characteristics are: low solar energy transmittance (32.4%), low overall heat gain factor (32.8%) and high visible transmittance (58.2%).

3.2. Choosing the optimal solution for heating systems

Whereas:

- the way of architectural conception differs from one type of building to another;
- the destination, occupancy and interior comfort parameters are different, also, the study was performed for each type of building separately: laboratories, spines and atrium.

Regarding the laboratories (in which the indoor temperature in the cold season must be 20° C) we considered the following:

- the structure of the strength structure with concrete pilars and frames, with horizontal elements consisting of monolithic reinforced concrete slabs, which constitutes a structure with an important thermal mass;
- the fact that the laboratories do not have a high degree of occupancy, which means that they do not need major fresh air exchanges;
- the fact that although they are important, the solar thermal inputs during the hot season are not as great as in the case of the spines and the atrium, which means that we can achieve the indoor climate and without having an air conditioning system, which would be a great energy consumer.

Therefore, we chose to analyze two solutions for the heating system:

- Thermoactive building systems (TABS);
- heating system with static radiators.

Regarding the spines and the atrium (in which the indoor temperature in the cold season must be 18° C) we considered the following:

- the fact that the solar thermal inputs during the hot season are very high, which means that a cooling system is needed;
- the fact that these spaces do not have a high occupancy, which means that they do not need major fresh air exchanges;
- the fact that they do not need to be cooled during the cold season.

Therefore, we chose to analyze two solutions for the heating system:

- TABS;
- heating system with hot air, with which to carry out the cooling in the hot season. The results of this part of the study are shown in figure 3, below.

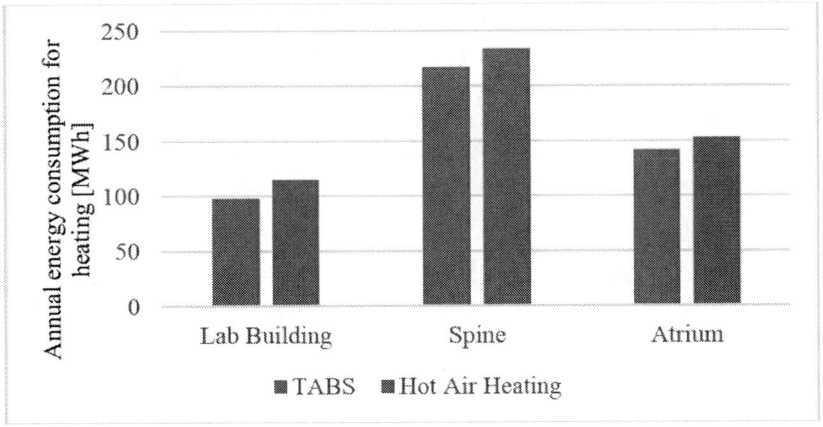

Figure 3. Annual energy consumption for heating depending on the heating system.

The primary purpose of this project was to ensure that the buildings as a whole have the least energy consumption for heating (and cooling). According to this objective, and based on the results of the study carried out (presented in the figure above), we have chosen the best solutions for the heating (and cooling) systems of the whole building. As for the laboratories, we opted for heating with a TABS, whose energy consumption is lower due to the use of the important thermal mass of the building. Regarding the cooling, given the relatively low temperatures (the location is located in a mountainous area) of the outside air in the hot season, we opted for natural ventilation and night cooling, with the help of operable windows. Regarding the spines and the atrium, an heating system with air would respond better to the indoor climate, both in the hot and cold season, but would involve a high energy consumption. Considering:

- that they are not continuously occupied spaces, but only connection spaces between the buildings, and that in the cold season we must ensure a lower indoor air temperature (18°C);
- that in the atrium it is necessary to realize the comfort conditions, in the cold season, only for an area occupied with the height of (2.50-3.00) m;
- the relatively low temperatures of the outside air during the hot season, which favors the cooling by natural ventilation and a very good night cooling, we opted, for the spines and the atrium, also for heating with a TABS.

Obviously, energy saving was an important factor in choosing this solution. In addition, the owner's intention was also taken into account in the future to develop a system of ground/water heat pumps (with wells drilled to a depth of 100 meters), which would provide the full amount of thermal aget needed to heat the institute's buildings. We considered that in this situation, during the hot season, with the help of the TABS, a passive cooling with cooled water is carried out in the drilling holes of the heat pump system. Opting for this solution I accepted that, both in the cold and hot season, short periods of time can occur when the thermal comfort could suffer.

4. Conclusions

Today, an important trend in the design of buildings and the systems that serve them, is that it starts from the requirements of integration of the latest technologies in the field of construction, and less from the primary need to design, through modeling and simulation, energy-efficient buildings.

Even so, there are many design teams in the field of heating, ventilation and air conditioning (HVAC), who are aware of the benefits, in terms of energy performance and environmental protection, of this design technique, and therefore use it. We believe that, in the near future, by introducing design standards based on energy performance, construction design will increasingly use the technique of building design through computer modeling and simulation. In the present paper, a real application related to the optimization of the design of the buildings and the systems that serve them was presented, through computer modeling and simulation. The DesignBuilder software, which we have used here, has the ability to provide all the data related to the thermal behavior of the building being designed, the conditions of its interior environment, to help the designers in adopting the most efficient solutions. Of course, it is not easy to do design using this technique. This requires a thorough knowledge of the software, a certain experience in its use and in the interpretation and validation of the data it provides. It also takes more time, which makes this technique somewhat costly.

References

[1] Clarke J A 2011 *Energy Simulation in Building Design*. 2nd edition London, New York Routledge
[2] Tian H et al 2018 Simulation-Based Decision Support Tools in the Early Design Stages of a Green Building-A Review *Sustainability-Open Access Journal* **10** (10) 3696 ISSN 2071-1050
[3] Harvie G 2019 *Computational fluid dynamics for buildings* Designingbuildings.co.uk
[4] www.designbuilder.co.uk
[5] Todor R D et al 2017 Implementing the life cycle cost analysis in a building design *Advanced Engineering Forum* Vol.21
[6] Horneț M et al 2018 Using the earth's natural potential for heating a building *Journal of applied engineering sciences* vol. **8** (21) issue 1 ISSN 2247-3769

IOP Conf. Series: Materials Science and Engineering **789** (2020) 012029 doi:10.1088/1757-899X/789/1/012029

Establishing workability asphalt mixtures

S C Iriciuc[1], N I Gălușcă[1] and C Pleșcan[2*]

[1] "Gheorghe Asachi" Technical University of Iasi, Faculty of Civil Engineering and Building Services, Romania
[2] Transilvania University of Brasov, Faculty of Civil Engineering, Romania

[*] plescan.costel@unitbv.ro

Abstract. The article shows the need to improve repair technologies for bituminous pavements especially during the cold season. As a result of laboratory tests, it was found the need for other attempts. These are attempts on storable asphalt mixtures workability at various temperatures for the installation period, but also for the storage period. Through the laboratory tests was shown the complex interaction between mechanical behaviour of asphalt mixtures and variability factors acting on the road structure, this complex framework is determined by fundamental mechanical properties that influence long-term performance.

1. Introduction

To ensure the safe conduct of road traffic in any season is necessary to ensure continuity of maintenance and repair of pavement surface throughout the year.

Bituminous pavements degradation repair in the season, the time period between 15 October and 15 March with temperatures generally below 10 °C, when there are no precipitations filling with storable asphalt mixtures is recommended [1].

Holes filling of the bituminous street pavements are necessary for ensuring traffic safety and the prevention of holes development under the combined action of traffic and water [2].

2. Workability of storable asphalt mixtures workability

Workability is the property of asphalt mixtures which allows the production, handling, placing and compaction of asphalt mixtures with minimal energy consumption [3, 4].

This test method is used to generate information on potential characteristics handling, laying, compaction and performance of asphalt mixtures.

This test method is also used on storable asphalt mixtures for control improvement over storage of these materials, timing and method of packing, climate conditions of storage and application.

3. Measuring devices

Workability measuring device was designed in the road laboratory of the Faculty of Civil Science Engineering.

This device is composed of a mixer used to make building materials, mixer produced by Humboldt HL200, an electronic device manufactured by UNI-TREND Technology for measuring the electric intensity and a computer on which is installed the software for data interpretation recorded by electronic measuring device of the electric intensity, figure 1.

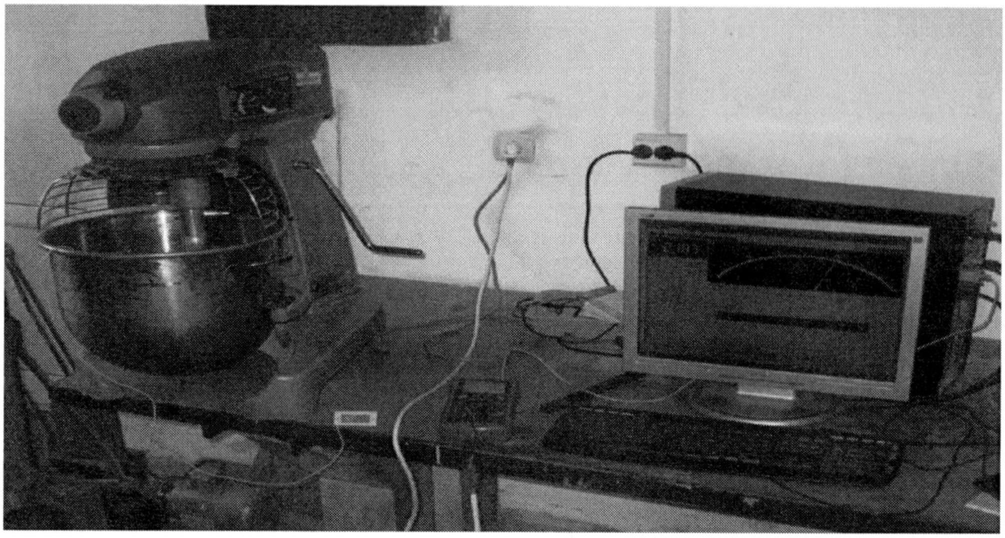

Figure 1. Workability measuring device.

HL200 mixer used is equipped with a timer that can be set during the mixing of mixture, a stirrer with transmission speeds that can provide 3 speeds of 107 198 and 361 rpm plus an addition lower speed necessary for liquid material composition of 53 rpm. Mixing vessel has a capacity of 20 litters and is made of aluminium.

Determination of the storable asphalt mixture workability is made as follows:

- storable asphalt mixture is placed in the mixing bowl;
- the mixing bowl is place in the mixer device.
- the machine ensured to carry out tests under normal safety conditions;
- the thermometer is placed to display mixture temperature during the determination;
- the speed and duration of mixing is established;
- the engine is turned on. When pallets are rotating the mixture from the vessel is opposing.
- the engine needs more power which leads to an increase of current intensity.
- the current intensity is measured with an amper measuring device and represents current intensity required for asphalt mixture mixing;
- measurement results are automatically transmitted to a computer and interpreted.

4. Workability of laboratory studied storable asphalt mixtures

In laboratory were studied four storable asphalt mixtures:

- control asphalt mixture (M0) without flow agent;
- storable asphalt mixtures with bitumen mixed with a flow agent (M1) – the flow agent used is an oil called STAROIL GL30;
- storable asphalt mixtures with bitumen mixed with a flow agent (M2) – the flow agent used is a solvent(white-spirt);
- storable asphalt mixtures RRD standard (M3).

To study the workability of M1, M2 and M3 storable asphalt mixtures comparisons have been made between these mixtures and control mixture (M0). The tests on M1, M2 and M3 mixtures aimed the same workability as the control mixture (without flow agent) at the laying time, at 120 °C (laying temperature of hot asphalt mixtures).

Were made three batches of each control mixture, 8 kg each, and were put in the mixing vessel. When the mixture temperature reached 120 ° C, the first speed started (107 RPM) for 1 min registering values of intensity within 5 seconds. In the end was given three results for each batch of control mixture. The test was repeated at 110 °C, 100 °C and 90 °C registering three intensity values for the three temperatures.

In table 1 values have been calculated by averaging between each batch of the four temperatures.

Were made three batches, 8 kg each, for M1 mixture (Storable asphalt mixtures with bitumen mixed with a flow agent – the flow agent used is oil called STAROIL GL30) and for M2 mixture (Storable asphalt mixtures with bitumen mixed with a flow agent – the flow agent used is white-spirit). The two storable asphalt mixtures were tested at 20 °C, 15 °C, 10 °C, 5 °C, 0 °C, -5 °C, -10 °C.

Table 1. Values of the electric intensity for M0 mixture.

Batch	Electric intensity at 120 °C, A	Electric intensity at 110 °C, A	Electric intensity at 100 °C, A	Electric intensity at 90 °C, A
I	2.149	2.183	2.278	2.454
II	2.141	2.181	2.276	2.457
III	2.145	2.185	2.280	2.456
Average	2.145	2.183	2.278	2.456

In tables 2 and 3 values were calculated by averaging between the values obtained for each batch to determine the four temperatures, for M1 and M2 mixtures.

Table 2. Values of the electric intensity for M1 mixture.

Batch	I -10 °C A	I -5 °C A	I 0 °C A	I 5 °C A	I 10 °C A	I 15 °C A	I 20 °C A
I	2.919	2.652	2.399	2.271	2.244	2.162	2.138
II	2.917	2.725	2.405	2.280	2.241	2.169	2.132
III	2.914	2.730	2.413	2.278	2.250	2.159	2.131
Average	2.917	2.702	2.406	2.276	2.245	2.163	2.134

Table 3. Values of the electric intensity for M2 mixture.

Batch	I -10 °C A	I -5 °C A	I 0 °C A	I 5 °C A	I 10 °C A	I 15 °C A	I 20 °C A
I	3.121	2.986	2.621	2.371	2.291	2.219	2.182
II	3.130	3.002	2.625	2.365	2.287	2.228	2.186
III	3.112	2.995	2.617	2.367	2.274	2.221	2.177
Average	3.121	2.994	2.621	2.368	2.284	2.223	2.182

The storable asphalt mixtures RRD standard (M3), a storable asphalt mixture ready prepared and packaged in the bucket was warm up for reaching the desired temperature.

The storable asphalt mixtures RRD standard (M3) was tested at 20 °C, 15 °C, 10 °C, 5 °C, 0 °C, -5 °C, -10 °C.

In tables 4 values were calculated by averaging between the values obtained for each batch to determine the four temperatures

Table 4. Values of the electric intensity for M3 mixture.

Batch	I -10 °C A	I -5 °C A	I 0 °C A	I 5 °C A	I 10 °C A	I 15 °C A	I 20 °C A
I	2.727	2.427	2.325	2.276	2.250	2.176	2.155
II	2.728	2.433	2.316	2.274	2.246	2.175	2.149
Average	2.728	2.430	2.321	2.275	2.248	2.176	2.152

In figure 2 is the workability variation of asphalt mixtures according to the temperature and the temperature range, considered optimal, that can be achieved under optimal laying and compacting conditions for each mixture type.

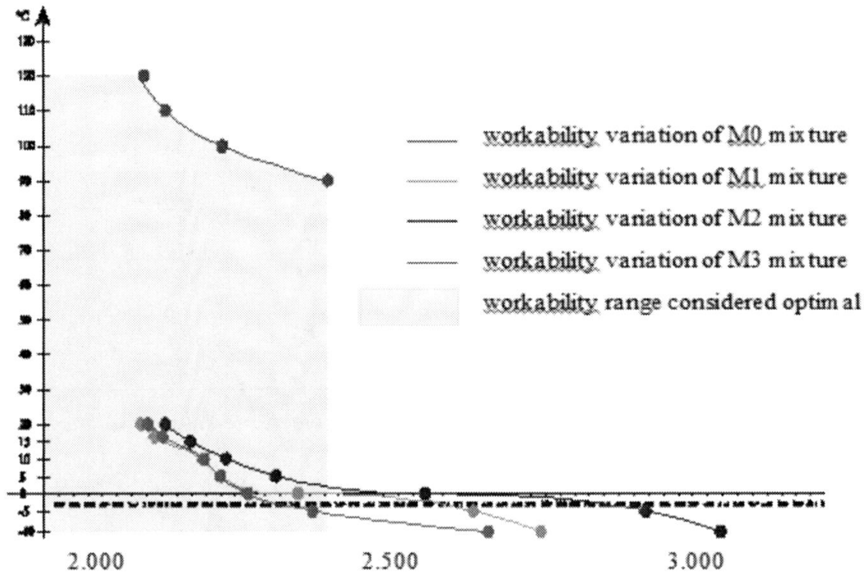

Figure 2. Storable asphalt mixtures workability variation according to the temperature.

5. Conclusions

Having established the storable asphalt mixtures workability in the temperature 10^0 C - 20^0 C range, were observed the following:

- M1 mixture meets the required workability for laying and compaction in optimal quality conditions, up to a temperature of 0 °C;
- M2 mixture meets the required workability for laying and compaction in optimal quality conditions, up to a temperature of 5 °C;
- M3 mixture meets the required workability for laying and compaction in optimal quality conditions, up to a temperature of -5 °C.

From a comparison of the three mixtures M1, M2 and M3 was found that although we have close granulometric curve, the difference is made by the flow agent. The flow agent for M3 mixture provides workability at low temperatures.

References

[1] Bambang I S and Kariantoni G 2005 Workability and Resilient Modulus of Asphalt Concrete Mixtures Containing Flaky Aggregates Shape *(Journal of the Eastern Asia Society for Transportation Studies)* **6** 1302
[2] Gimiga G R *Improvement of the technologies of repair of the isolated degradations in the bituminous garments,* PhD thesis Technical University "Gheorghe Asachi" of Iasi
[3] Salah M K, Mohd Y A R and Ahmad K A *Development of Workability Measuring device for Asphalt Mixture using transducer by means of Energy*
[4] Iriciuc S C, Găluşcă N I and Pleşcan C, Steelworks slag aggregate in asphalt mixtures *Bulletin of the Transilvania University of Brasov Vol. 11 (60) 102*

IOP Conf. Series: Materials Science and Engineering **789** (2020) 012030 doi:10.1088/1757-899X/789/1/012030

Research on hysteretic behavior of external T-stiffener box-column I-beam space connections

L Jin[1], L P Chen[1] and Y Zhang[1*]

[1] School of Civil Engineering, Shenyang Jianzhu University, Shenyang, China

[*] ZY_JZDX@163.com

Abstract. The bearing capacity and deformation performance of external T-stiffener box-column to I-beam space connections were analyzed in this paper. A great deal of full-path analysis of space connections has been carried out by using finite element program ABAQUS with different parameters, which are box-column section width, I-beam flange width, section height of I-beam, flange width and web length of T- stiffener. FEM is proposed for numerical simulation under cycling loading based on the mechanical properties of materials. The result show that the hysteresis curve is the spindle shape and relatively full of T-stiffener box-column I-beam space connections. The ductility coefficient μ is larger than 4.0 and equivalent viscous damping coefficient h_e is larger than 0.2 of the space connections, which have been met the limitation of the specification requirements. The bearing capacity and deformation performance of external T-stiffener box-column I-beam space connections can be improved with the increasing of T- stiffener flange width and web length effectively. Due to the fine performance of ductility, energy dissipation and seismic performance, the external T-stiffener box-column I-beam space connections can be proposed used in multi-layer steel structure engineering widely.

1. The first section in your paper

Box-column for steel structures are generally welded by four steel plates [1], as shown in figure 1. Generally, in order to improve the lateral stiffness and torsion resistance of box-column, an internal partition is provided in the connection between box-column and I-beam at a corresponding position [2].

However, the actual welding process between the inner partition and the four steel plates is the greatest difficulty [3]. In view of the problem, the actual welding process is time-consuming, laborious and the quality of the weld is not guaranteed.

The external T-stiffener box-column I-beam connections can be used to avoid the welding difficulties of box-columns during welding [4-6]. The relevant research shows that the external reinforcement is simpler, faster and more economical than the traditional method of adding diaphragm in the column, and the T-external reinforcement connections have enough strength, rigidity and rotation capacity [7-9]. The T-stiffener box-column I-beam connections is shown in figure 2 [10-12].

At present, most of the researches are focused on plane connections, for empty connections the research of space connections is relatively few [13-14]. Due to the uncertainty of the direction of seismic force, the stress of the middle column joint is characterized by spatial stress. Therefore, in this paper, T-stiffener box-column I-beam space connections is selected for research, as shown in figure 3. The hysteretic behavior of spatial connections is studied by changing the parameters of box columns, I-beams and T-stiffeners.

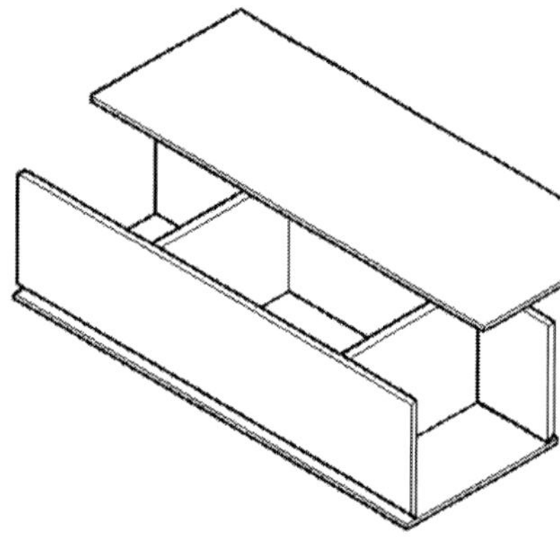

Figure 1. Box-column.

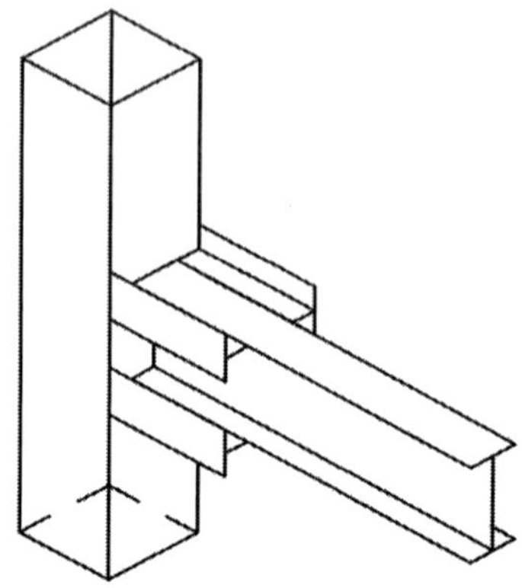

Figure 2. T-stiffener box-column I-beam connections.

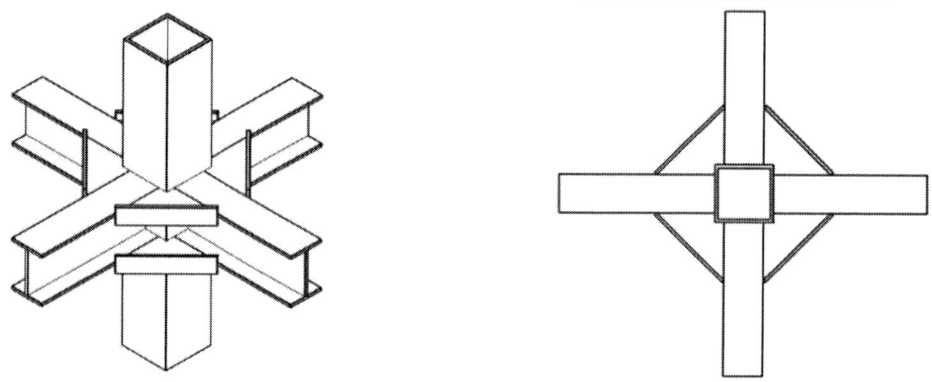

(a) Node axonometric drawing. (b) Node top view.

Figure 3. T-stiffener box-column I-beam space connections geometry model.

2. Finite element model

2.1. Connection design

The finite element program ABAQUS is used to build the modelling [15], the connection geometry model and loading method are shown in figure 4. Steel Q345B is selected and other properties of the material are shown in the table 1.

The constitutive relation is selected as a linear elastic elastoplastic mechanical model of double-fold line. The steel stress-strain relationship is shown in figure 5.

The external T-stiffener box-column I-beam space connections is referred to as ST in the following content. Eight kinds of working conditions for simulation analysis were designed. The working condition number is shown in figure 6.

Table 1. Mechanical properties [16].

material	Yield stress fy/MPa	Tensile strength (MPa)	Elasticity modulus (10^5MPa)	Poisson's ratio
Q345B	375.7	544.5	2.06	0.3
artifacts	Length of I-beam (mm)		Height of box column (mm)	
ST	1250		750 mm above and below each box column	

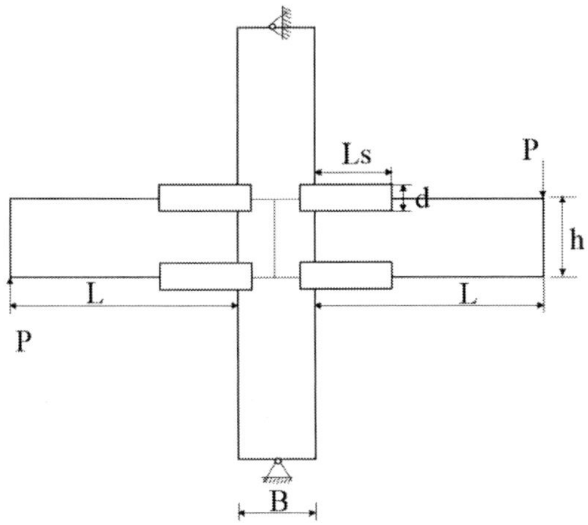

Figure 4. Geometry model and loading method.

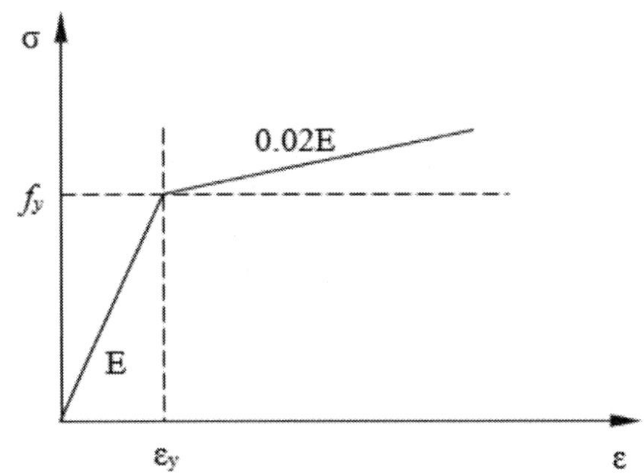

Figure 5. Steel stress-strain relationship.

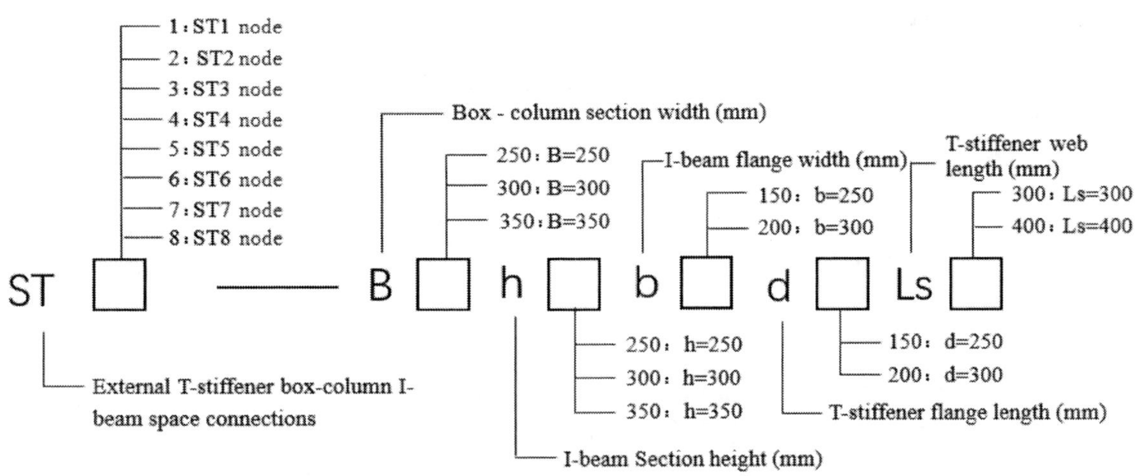

Figure 6. Working condition number.

2.2. Establishment of finite element model

According to table 1 dimensions of beams and columns, the numerical model of spatial connection is established, the T-stiffener box-column I-beam space connections finite element model is shown in figure 7.

The selection of mesh shape and global size of the finite element model will directly affect the calculation cost and accuracy of the calculation results. By comparison, when the overall dimensions of box column, I-beam and T-stiffener are 60 mm, 40 mm and 30 mm respectively, the calculation accuracy and calculation time are appropriate.

Tie restraint relationship were used between box column, I-beam and T-stiffener, reciprocating load is applied by displacement control at beam end [17-19].

Three stages of loading Δ/ 3, 2 Δ/ 3 and Δ before yield, one cycle per stage, after yielding, cycle three times for each stage according to 2 Δ, 3 Δ, 4 Δ, 5 Δ. The loading curve is shown in figure 8.

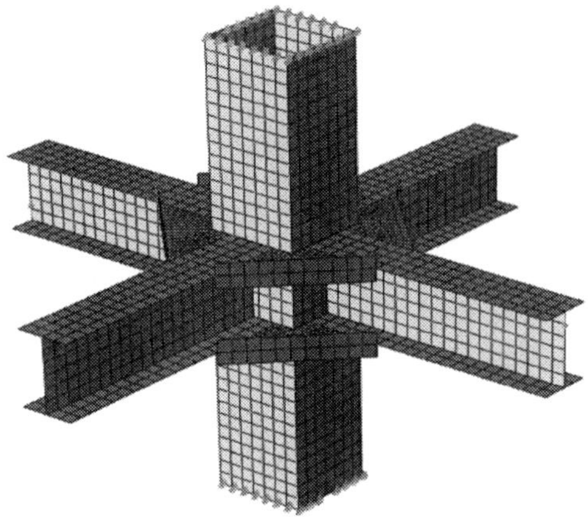

Figure 7. Finite element model.

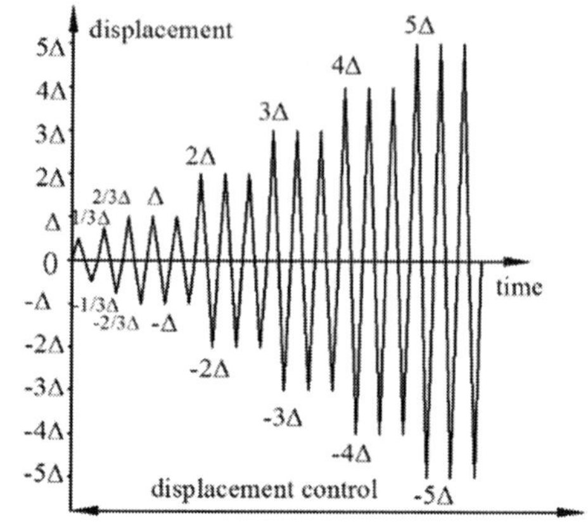

Figure 8. Typical cyclic loading sequence.

3. Hysteresis performance

3.1. Hysteresis curve

The numerical model was established refer to the above steel constitutive. The working condition design numbers of the 8 space connections are shown in table 2. The hysteresis curves of the end of beam is shown in figure 9.

Table 2. Design condition.

condition	number	condition	number
1	ST1-B250 h250 b150 d100 Ls300	2	ST2-B250 h350 b150 d100 Ls300
3	ST3-B300 h350 b200 d100 Ls300	4	ST4-B350 h300 b200 d100 Ls300
5	ST5-B350 h350 b200 d100 Ls300	6	ST6-B300 h350 b150 d100 Ls300
7	ST7-B300 h350 b200 d160 Ls300	8	ST8-B300 h350 b200 d100 Ls400

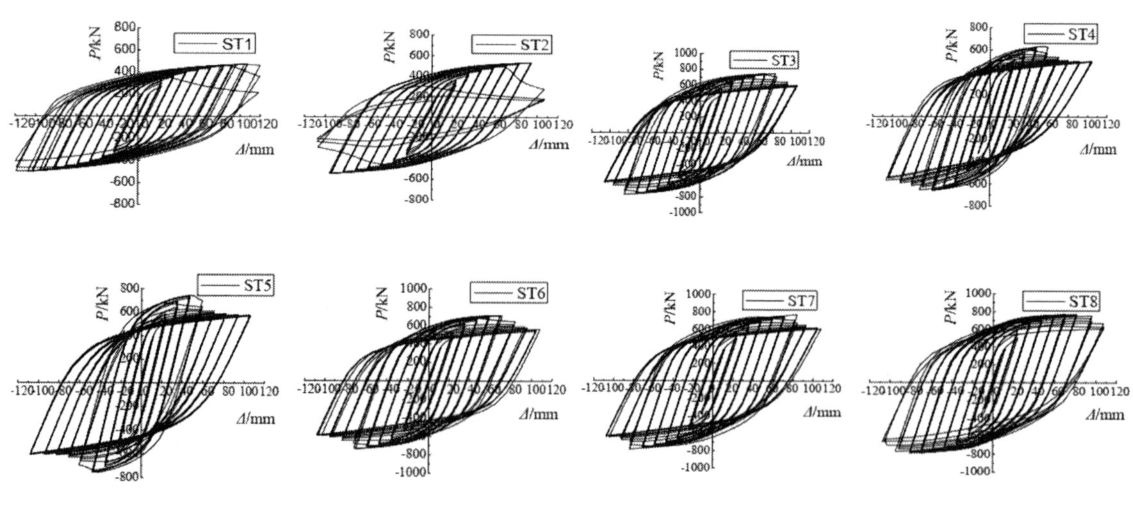

Figure 9. Hysteresis curves.

The result show that the beam and column dimensions of connection ST1 and ST2 are much smaller than others. As a result, its bearing capacity decreases, but its ductility increases. Connection ST1- ST8 has a decrease in stiffness under cyclic loading.

However, the hysteresis curve becomes fuller with the increasing of the cross-section width of the box column, the height of the I-beam and the length of the stiffener web. Hysteresis performance has been improved significantly.

3.2. Skeleton curve

The figure 10 represents the skeleton curves of the external T-stiffener box-column I-beam space connections. As shown in figure 10(a), with the section width of the box column increasing from 250 mm to 350 mm, the bearing capacity of connection ST6 is increased by 61.41% compared with connection ST2. The bearing capacity of connection ST5 is increased by 7.67% compared with connection ST3. As shown in figure 10(b), with the increasing of the height of the I-beam, the bearing capacity of connection ST2 is increased by 8.28% compared with connection ST1. The bearing capacity of connection ST5 is increased by 22.93% compared with connection ST4. As shown in figure 10(c) and figure 10(d), compared with connection ST3, the increasing of T-stiffeners' flange width and web length leads to the load capacity of connection ST7 and ST8 increase by 0.83% and 1.14%, respectively.

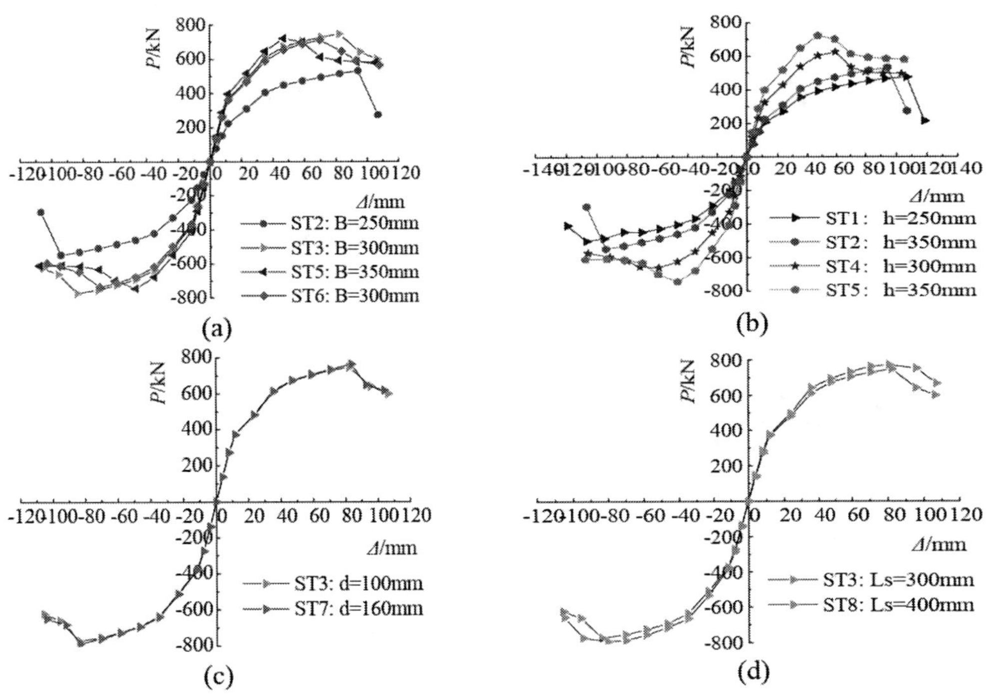

Figure 10. Skeleton curves.

3.3. Stiffness degradation

The figure 11 are the stiffness degradation curves of the external T-stiffener box-column I-beam space connections. As shown in the stiffness degradation curves and the table 3.

Since connection ST1 and connection ST2 are smaller size than others, it can be seen that the initial stiffness of the connection ST1 and ST2 is about 1.5 times lower than the initial stiffness of others as the same displacement. The stiffness of the connection ST5 is increased by 21.55% compared with connection ST4, the stiffness of the connection ST8 is 5.65% higher than that of connection ST3.

Increasing the section width of box column and the web length of T-stiffener can improve the stiffness of space connections effectively.

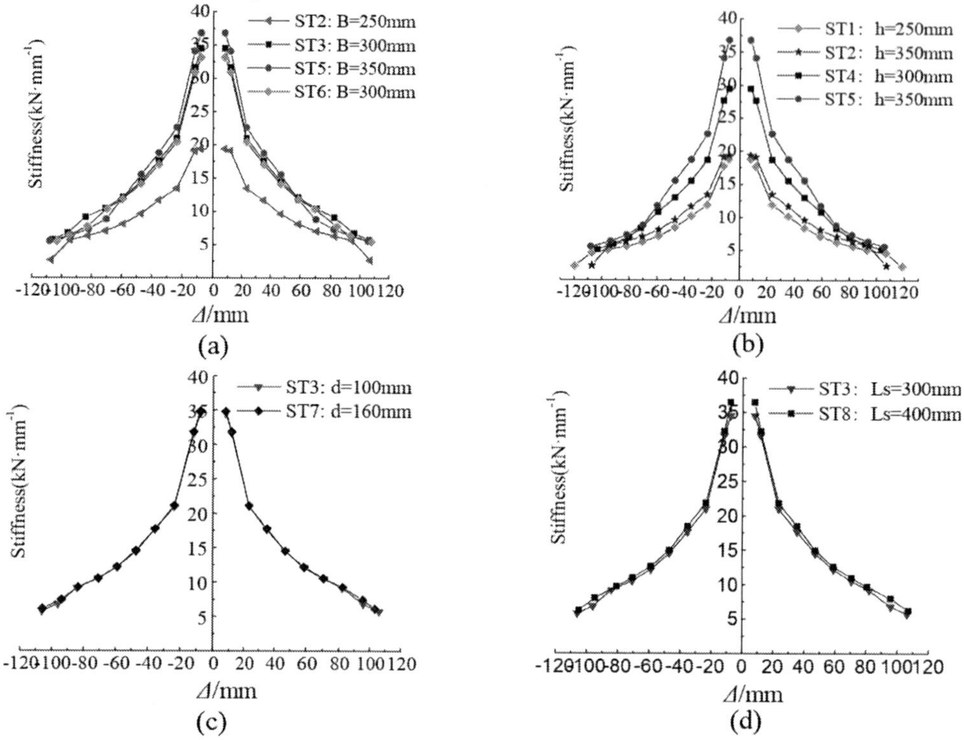

Figure 11. Stiffness degradation curves.

Table 3. Cyclic stiffness of each connection.

condition	Cyclic stiffness (N/mm)									
	K_1	K_2	K_3	K_4	K_5	K_6	K_7	K_8	K_9	K_{10}
ST1	18.774	17.686	11.903	10.212	8.428	7.149	6.238	5.576	5.041	2.611
ST2	19.385	19.083	13.467	11.679	9.637	8.120	7.076	6.319	5.717	2.655
ST3	34.503	31.639	20.984	17.615	14.466	12.117	10.476	9.143	6.793	5.756
ST4	29.452	27.684	18.663	15.525	12.983	10.812	8.351	6.990	5.952	5.091
ST5	35.798	34.075	22.592	18.743	15.542	11.761	8.832	7.340	6.362	5.575
ST6	33.136	30.897	20.437	17.025	14.118	11.863	10.368	7.753	6.391	5.488
ST7	34.947	31.908	21.176	17.790	14.528	12.190	10.568	9.272	7.186	6.090
ST8	36.431	32.261	21.882	18.476	14.933	12.596	10.957	9.668	7.988	6.232

3.4. Ductility

The calculation of the carrying capacity and ductility coefficient μ of the space connections ST1-ST8 and other performance indicators are shown in table 4.

With the increasing of flange width of T-stiffener from 100 mm to 160 mm and length of T-tiffener web from 300 mm to 400 mm, the space connections ductility coefficient increased by 5.69% and 3.09%. The ductility coefficient μ is larger than 4.0 of each condition and can meet the requirement of seismic specifications.

Table 4. Node carrying capacity and deformability.

Condition	Yield capacity Fy/kN	Yield displacement δy/mm	Ultimate capacity Fu/kN	Limit displacement δu/mm	Ductility coefficient μ
ST1	207.066	11.786	475.049	106.751	9.057
ST2	224.211	11.786	531.440	94.293	8.000
ST3	370.333	11.789	746.921	80.1413	6.797
ST4	324.351	11.785	625.164	59.306	5.032
ST5	398.732	11.782	721.154	47.345	4.018
ST6	361.889	11.786	710.112	69.818	5.924
ST7	373.416	11.785	764.652	84.659	7.184
ST8	375.536	11.786	769.999	80.771	6.853

3.5. Energy consumption

The equivalent viscous damping coefficient h_e and the energy dissipation coefficient E of the peak hysteresis loop of each condition are shown in table 5.

According to the energy dissipation coefficient E in the table 5, through the comparison of the two groups of connections ST2, ST6 and connections ST3, ST5. It can be seen that the energy consumption capacity of the enlarged box-column section width is increased by about 1.18% and 3.91% respectively. With the cross-section height of the I-beam increasing, for example, from connection ST1 to ST2, from connection ST4 to ST5, it can be seen that the energy consumption capacity is increased by 1.33% and 6.36% respectively.

Comparing the energy dissipation coefficient E of the connection ST3 and ST6, it is known that the energy consumption capacity of the flange of the I-beam is increased by 7.89%.

As shown in table 5, figure 12 and figure 13, connection ST3, ST7 and ST8, with increasing width of the flange of the T-stiffener and the length of the web of the T-stiffener, resulting in an increase in energy consumption of 2.59% and 6.79% respectively.

The equivalent viscous damping coefficient h_e of the eight space connections is larger than 0.2, indicating that this type of space connection can meet the requirements of seismic design energy performance.

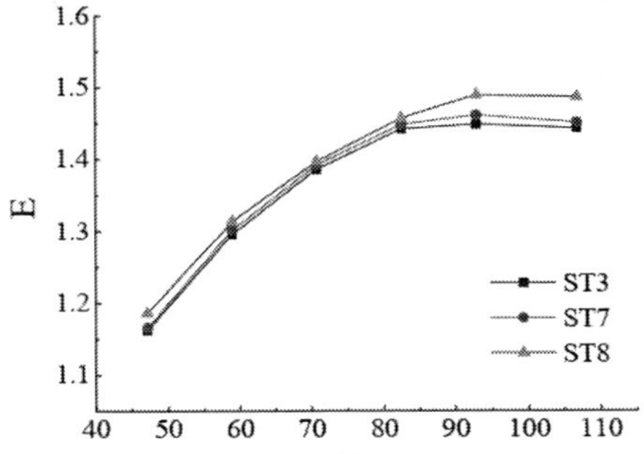

Figure 12. E-displacement curves.

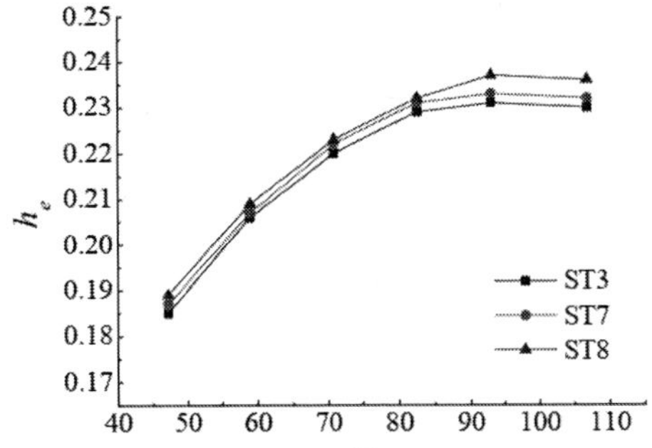

Figure 13. h_e-displacement curves.

Table 5. Energy dissipation coefficients.

Dissipation coefficient	ST1	ST2	ST3	ST4	ST5	ST6	ST7	ST8
h_e	0.240	0.263	0.230	0.223	0.237	0.246	0.231	0.237

E	1.508	1.528	1.443	1.400	1.489	1.546	1.461	1.486

Conclusions

The performance of the external T-stiffener box-column I-beam space connections have been studied. The research focuses on the influence of changing the section width of the box column, the width of the I-beam flange, the height of the section, the width of the T-stiffener flange and the length of the web on the bearing capacity and deformation performance. The following conclusions are got:

- the hysteresis curves of the T-stiffener box-column I-beam space connections are the full spindle shape. The performance of deformation is strong in the elastoplastic stage;
- the ductility coefficient μ of the space joints with T-stiffener is larger than 4.0 and meets the requirement of seismic specifications;
- the stiffness decreases significantly with the increasing of the displacement. The rate of descent has a rapid decrease followed by a slow decrease.
- The equivalent viscous damping coefficient h_e is larger than 0.2 and energy dissipation coefficient E is larger than 1.4.

Acknowledgement

The work presented in this paper was supported by the natural foundation guidance program of Liaoning province, China (2019-ZD-0678) and the program for "Xing Liao Talent" of Liaoning province, China (XLYC1807188).

References

[1] Mao H Jia B L and Bu Y 2011 *J. Electric Welding Machine* **41** pp 53-57
[2] Deylami A and Gholipou R M 2011 *J. Procedia Engineering* **2011** pp 3252-3259
[3] Zhao X B, Zheng Z K and Pan J F 2016 *J. Steel Construction* **31** pp 92-95
[4] Wang Y, Liu Y, Wang P and Gao P 2011 *J. Progress in Steel Building Structures* **13** pp 1-7
[5] Shanmugam N E 1997 *J. Journal of Structures* **17** pp 36-60
[6] Shanmugam N E and Ting L C 1995 *J. Journal of Structural Engineering* **121** pp 824-830
[7] Yang X D 2011 *D. Xian University of Architecture and Technology*
[8] Nia Z S, Mazroi A, Ghassemieh M and Pezeshki H 2014 *J. Earthq Eng & Eng Vib* **13** pp 717-29
[9] Kato B 2014 *J. Journal of the Structure Division* **108** pp 343-360
[10] Wang M, Shi Y and Wang Y 2013 *J. Constructional Steel Research* **90** pp 140-152
[11] Mirghaderi S R, Torabian S and Keshavarzi F 2011 *J. Engineering structures* **2011** pp 2034-2048
[12] Jin L 2006 *D. Harbin Institute of Technology*
[13] Shao Y S, Jin L, Xie L L and Zhang Y C 2009 *J. Journal of Harbin Institute of Technology* **41** pp 20-25
[14] Li F X and Wang X W 2018 *J. Steel Construction* **33** pp 98-102
[15] Xu P Z, Yang L, Zhu Y G and Jing H K 2018 *J. Steel Construction* **33** pp 93-97
[16] Zhang Z X 2015 *D. China University of Mining and Technology*
[17] Liu E H 2017 *D. Chang an University*
[18] Wei K F, Li J F, Liu S X, Li X and Chen H 2018 *J. Steel Construction* **33** pp 17-20
[19] Xu Z G, Zhang Z B and Deng C G 2017 *J. Sichuan Building Science* **43** pp 75-79

Dynamic characteristics of CLT panels: computer modelling and simulations

L Kozarić[1], M Vojnić Purčar[1*], Z Tekic[1], M Bešević[1] and S Zivković[1]

[1] Faculty of Civil Engineering Subotica, University of NoviSad, Serbia

[*] vojnicmartina@gmail.com

Abstract. The fundamental vibration frequency and deflections due to the unit static force of the cross-laminated timber panels of different heights and spans were calculated. The calculation of the effective flexural rigidity was performed by the Gamma method. Obtained values are compared with the values obtained by modeling the construction in the Ansys software package. Three types of floor panels were analyzed with the same combinations of layer thicknesses in the cross section, but with different number of layers. The panels are 4, 5 and 6 meters long and 1 meter wide. Based on the results, recommendations have been made in accordance with current criteria that define the acceptable behavior of lightweight floor structures in relation to vibrations caused by human action.

1. Introduction

Cross laminated timber (CLT) is a modern product that greatly improved the physical properties of traditional wood as building material [1]. CLT is made of controlled dried wooden elements – layers of uniform width, free of defects (knots, resin etc.). By removing all of the defects and cross-gluing the layers it is possible to produce the material that has more uniform mechanical properties than the traditional wood.

A CLT cross section contains a minimum of three cross-glued layers, but most often five or seven. Alternate layers are mutually perpendicular; however, it is possible to double certain layers so that greater strength can be achieved in a desired direction.

Today the use of CLT in multi-family houses is increasing and there is a need to develop better performing CLT floors in order to increase the comfort of the residents. Floor vibration is significant problem for these structures. Because they are light and flexible, vibration is a source of discomfort in the use of this floor systems and its major cause are dynamic movements produced by human activities, such as walking.

This paper deals with numerical analysis of dynamic behaviour of CLT floor structures. Advantages and possible shortcomings will be pointed out.

2. Serviceability limit state of vibration

Vibration of timber floors is a problem of serviceability limit state which refers to discomfort from the aspect of use, at loads that regularly exist in everyday use of the structure.

The feeling of discomfort when vibrating is individual and varies from person to person. It is well known that vibrations that are disturbing to one person may not be to another. This makes it very difficult to determine the threshold of human acceptability. The standards and literature find various proposed criteria that a timber floors must satisfy in designing to ensure its acceptable behavior due to dynamic loading, the most commonly used combination is the criterion of the minimum value of the fundamental frequency and the criterion of the maximum allowable deflection due to unit static forces, CLT Handbook [2], which is:

$$\frac{f}{d^{0.7}} \geq 13 \quad \text{or} \quad d \leq \frac{f^{1.43}}{39} \tag{1}$$

where:

f - fundamental frequency (Hz),

d - maximum deflection due to static force $F = 1\ kN$ (mm).

3. Materials and methods

The effective stiffness, fundamental frequencies and maximum deflection due to static force for three simply supported floor CLT panels were analyzed. Spans were varied from 4.0 m to 6.0 m. The height of the panel is varied from 9.8 cm to 22.6cm depending on number of the layers in the cross section of the panel (figure 1).

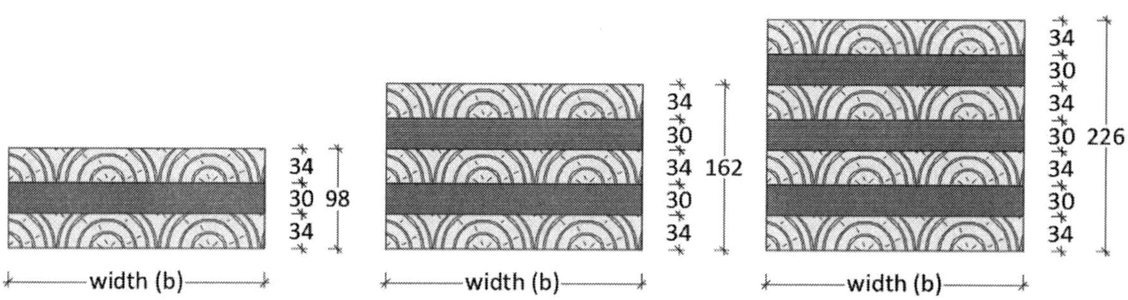

Figure 1. Geometric characteristics of the CLT panels FP1, FP2 and FP3.

The mechanical characteristics of the longitudinal and lateral layers of the timber floors are shown in table 1. The shear of the longitudinal layers is neglected because the ratio of the span to the height of the CLT panels is greater than 30 [2]. The effective stiffness is calculated by the Gamma method [3] and the mean density for C24 wood class is 420 kg/m^3 and for C18 is 380 kg / m^3.

Table 1. Mechanical properties of the longitudinal and lateral layers.

Layers	Strength classes	f_m (MPa)	E_0 (MPa)	E_{90} (MPa)	G_0 (MPa)	G_R (MPa)
Longitudinal	C24	24	11000	370	690	69
Lateral	C18	18	9000	300	560	56

The fundamental frequency of a simply supported timber floor can be calculated analytically using the equation:

$$f = \frac{\pi}{2l^2}\sqrt{\frac{(EI)_{ef}}{\gamma A}} \tag{2}$$

where: l - span of timber floor (m); $(EI)_{ef}$ - effective panel stiffness 1.0 m wide in the direction of span (Nm^2); γ - volume of wood (kg/m^3); A - area of the cross section (m^2).

The maximum deflection due to the unit static force expressed in mm can be determined by the following equation:

$$d = \frac{1000 P l^3}{48(EI)_{ef}} \tag{3}$$

where: $P = 1000$ N.

The finite element method was applied using Ansys 14.0 computer software. The timber floor is modeled as a 2D laminate model consisting of layers, which represent the third dimension of the model. Construction modeling was performed using elements from the Ansys library. A 4-node two-dimensional element SHEL181 was selected. It was used as a 2D element which thickness is defined with the stacking sequence of the layers. An element is defined by four nodes and six degrees of freedom in each node, three translations and three rotations. The basic input data for this element are: modulus of elasticity, Poisson's ratio (coefficient) and volume of wood. The finite element is 0.125 x 0.25 m in size (figure 2). The adopted mesh was determined by an iterative procedure, until the results of two successive steps differed by less than 1%.

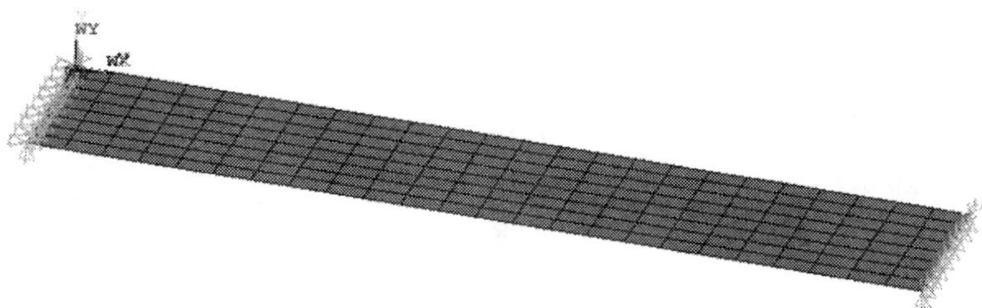

Figure 2. Arrangement of nodes and finite elements of the CLT panel FP1 with 4,0 m span.

4. Results and discussion

The floor fundamental frequency and the maximum deflection due to the unit static force are calculated according the CLT Handbook [2]. Figure 3 shows the first vibrational shape of the CLT panel with 4.0 m span. The results are shown in table 2, 3 and 4. Also, the corresponding limiting values are given.

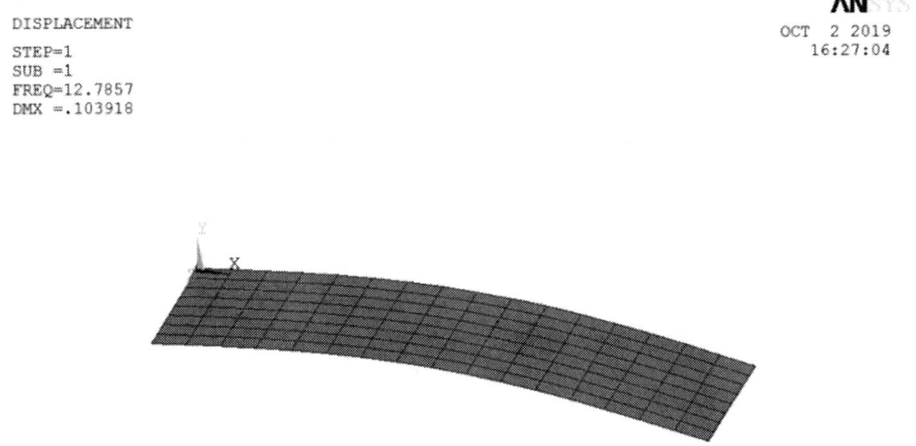

Figure 3. First characteristic mode shape of CLT panel FP1 with 4.0 m span.

IOP Conf. Series: Materials Science and Engineering **789** (2020) 012031 doi:10.1088/1757-899X/789/1/012031

Table 2. The fundamental frequency of the CLT panels.

CLT		f_{min} = 10 Hz [4]		
		The fundamental frequency f (Hz)		
		l= 4.0 m	l= 5.0 m	l= 6.0 m
FP1	Analytical	12.70	**8.22**	**5.74**
	Ansys	12.79	**8.26**	**5.76**
FP2	Analytical	18.74	12.23	**8.58**
	Ansys	18.79	12.27	**8.61**
FP3	Analytical	24.91	16.28	11.42
	Ansys	23.98	15.87	11.23

Table 3. The unit point load deflection of the CLT panels.

CLT		d_{max} ≤1.15 mm [4]	d_{max} ≤ 1.01 mm [4]	d_{max} ≤ 0.82 mm [4]
		The unit point load deflection		d (mm)
		l= 4.0 m	l= 5.0 m	l= 5.0 m
FP1	Analytical	**1.69**	**3.23**	**5.52**
	Ansys	**1.70**	**3.27**	**5.54**
FP2	Analytical	0.47	0.88	**1.49**
	Ansys	0.48	0.89	**1.51**
FP3	Analytical	0.19	0.36	0.60
	Ansys	0.21	0.38	0.62

Table 4. Combined design criterion.

CLT		$\frac{f}{d^{0.7}} \geq 13$		
		Combined design criterion		
		l= 4.0 m	l= 5.0 m	l= 6.0 m
FP1	Analytical	**8.80**	**3.62**	**1.74**
	Ansys	**8.82**	**3.60**	**1.74**
FP2	Analytical	31.79	13.37	**6.49**
	Ansys	31.41	13.31	**6.45**
FP3	Analytical	79.66	33.28	16.33
	Ansys	71.50	31.24	15.69

The results in tables 2 and 3 show a decrease in the value of the fundamental frequency of the CLT panel with increasing its span, while its value changes significantly with the change in the number of layers in the cross section. The modal analysis in the Ansys software package confirmed the analytically obtained values, which indicates very high precision of the analytical expressions.

If the value of 10 Hz was adopted for the required minimum fundamental frequency of the light floor structure, which corresponds to the more stringent recommendations in the available literature, CLT panels of type FP1 with spans 5.0 m and 6.0 m would not be acceptable from the standpoint of serviceability limit state, as well as panels of type FP2 with span 6.0 m. But if the value of 8 Hz [5, 6] was accepted as the minimum required fundamental frequency of CLT panels, spans of the FP1 panel could range up to 5.0 m, and for FP2 panels even 6.0 m.

However, the criterion of maximum allowed deflection due to the unit static force and combined design criterion whose results are shown in table 3 and 4 are much stricter when evaluating the dynamic acceptability of CLT floor panels. According to these criteria, only FP3 panels would not cause discomfort to the user in their daily activities, regardless of the span tested. FP2 type panels meet the prescribed requirement for the span of 4.0 m and 5.0 m, while FP1 type panels do not meet the prescribed requirement in any tested case.

It is evident that the increase in the number of layers placed in the longitudinal direction significantly affects the dynamic behavior of the panels as the values of the criteria change with their heights, and it is difficult to assess whether or not the structure satisfies based only on one parameter.

5. Conclusions

The presented results of the analyzed floor CLT panels show that they can be very successfully applied in the floor structures of residential and commercial buildings, provided that an adequate dynamic calculation is carried out. The combined design criterion and the unit point load deflection limit are significantly stricter compared to the fundamental frequency limit therefore it is advisable to use multiple recommendations and guidelines when evaluating the acceptability of a structure from a dynamic point of view.

Acknowledgement

The present work has been supported by The Ministry of Education and Science of the Republic of Serbia (Project No. ON174027).

References

[1] Kozarić Lj, Prokić A and Bešević M 2015 Unakrsno lamelirani drveni elementi u savremenim drvenim konstrukcijama zgrada (Cross-laminated wooden elements in contemporary wooden structures of buildings) *Građevinski materijali i konstrukcije* (Building Materials and Structures) **4**: 51-69

[2] Karacabeyli E and Douglas B 2013 *CLT Handbook: Cross-laminated Timber* FPInnovations

[3] Kozarić Lj, Prokić A, Bešević M and Vojnić Purčar M 2016 Proračun ugiba unakrsno lameliranog drvenog međuspratnog panela (Calculation of the deflection of a cross laminated wooden floor panel) *Zbornik radova Građevinskog fakulteta* (Proceedings of the Faculty of Civil Engineering) **30** 63-72

[4] Allen,D E and Pernica G, 1998 Control of flor vibration. *Construction Technology*, **22** p 124

[5] Smith I and Chui Y H 1988 Design of lightweight wooden floors to avoid human discomfort *Canadian Journal of Civil Engineering* **15** p 254-262

[6] EN 1995-1-1 2010 Design of timber structures. General rules for buildings 2012

Contributions to the consolidation of wooden floors to structures with load-bearing masonry walls

A V Lapuste[1*] **and H A Mociran**[1]

[1] Technical University of Cluj-Napoca, Faculty of Civil Engineering, Cluj Napoca, Romania

[*] adina.lapuste@mecon.utcluj.ro

Abstract. The consolidation of wooden floors is a common intervention in the practice of building rehabilitation. The necessity of this type of works comes from the sensitivity of wood to biological degradation, the existence of a significant number of heritage buildings in which the construction of the slabs was done without ensuring compliance with the requirements of the normal exploitation (limiting the deformations), the occurrence in operation of the need to increase the capable loads. It is required, in case of consolidation at earthquake of the buildings with wooden floors, the creation of rigid slab (rigid in horizontal plane). The consolidation solutions that can be adopted are different, trying to respond to the diversity of situations/particularities which are encountered in practice. The paper briefly reviews the solutions currently used for the rehabilitation of wooden floors, then presenting a solution for the consolidation of a wooden floor belonging to an intermediate floor acted by gravitational actions, respectively a solution for achieving the rigid slab of an attic, of a multi-storey building.

1. Introduction

In the paper, problems of rehabilitation of wooden floors in multi-story masonry buildings are addressed. The analysis done for some situations encountered in the practice of rehabilitation is useful by the way of approaching the problem and the way of treating the deficiencies that make the rehabilitation necessary.

The rehabilitation / consolidation of the wooden floors is done when the necessity of increasing the bearing capacity / stiffness to the gravitational actions or to obtain the effect of rigid slab to horizontal actions.

The wooden floors can be damaged over time by biological attack on the construction (insect-fungal attacks), dislocation of the supports due to mechanical vibrations, weakening the connection of the floor with the other structural elements, etc.

The need to rehabilitate the slabs results from the decrease of their performance at gravitational actions, as a result of the degradation, or due to the increase of the performance requirements given by a possible change of function.

2. Current practice in the field

2.1. Increasing of bearing capacity for gravitational actions

In current practice, the increase in the bearing capacity and the rigidity of the slabs at gravitational actions are realized in several variants:

Working from bottom up, repairing each wooden beam in order to increase its strength and reduce its deformations. Following this procedure, the following methods are used:

- steel beams are placed between the wooden beams, at half distance between them;
- the arrangement of a pair of steel profiles (double T, U, angles or cold rolled profiles) as shown in figure 1 (a, b, c);
- solutions can be adopted to consolidate the beam over its entire length through especially designed steel structures, as in figure 2 [3].

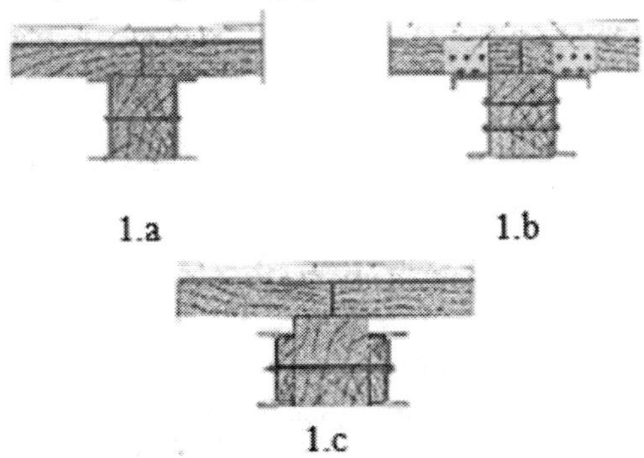

Figure 1. Rehabilitation of a wooden floor from the bottom up. 1.a) U steel profile, 1.b) steel composed profile, 1.c) double T steel profile [3].

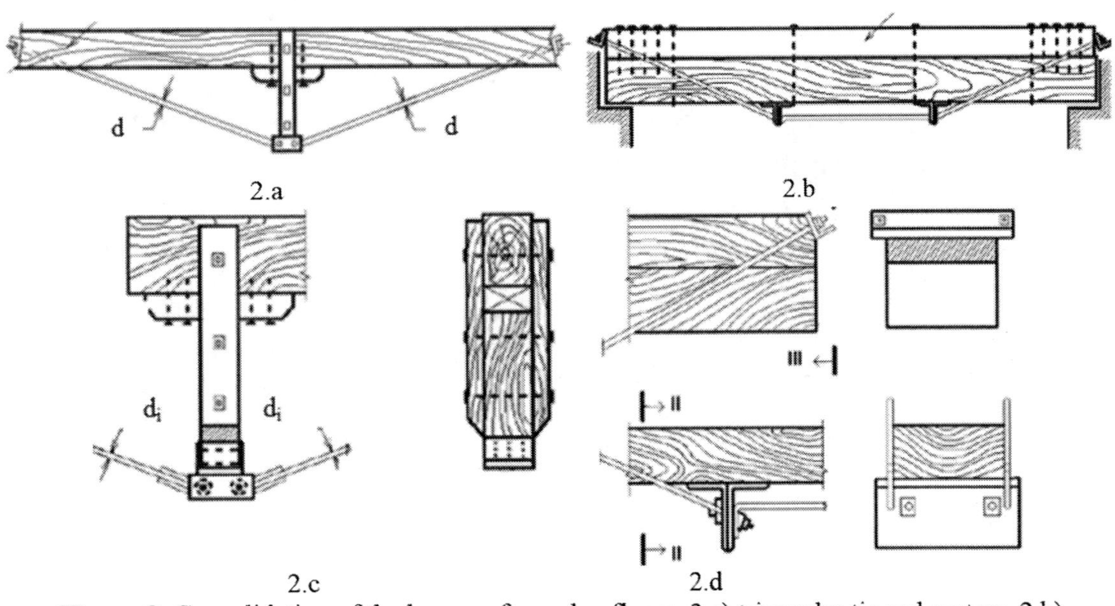

Figure 2. Consolidation of the beams of wooden floors, 2.a) triangular tie-rod system, 2.b) trapezoidal tie-rod system , 2.c) anchor triangular tie-rod system 2.d) design/anchor trapezoidal tie-rod system [2].

2.2. Increase of bearing capacity for horizontal actions

At the same time, there is the necessity to ensure the behaviour of the floor to horizontal actions, as a rigid slab. This is usually not done on traditional wooden floors and interventions are needed to achieve this behaviour.

The more frequently used variants of transforming a floor into a rigid slab are:

2.2.1. The execution of mixed wood-reinforced concrete floors. The solution involves the execution of a reinforced concrete slab over the existing wooden floor, ensuring connections between the wooden structural elements of the existing floor and the concrete slab, through metal connectors as in figure 3. The solution can be used only if the existing flooring of the wooden floor can be replaced and the additional weight brought by the concrete slab can be taken over by the structure of the building.

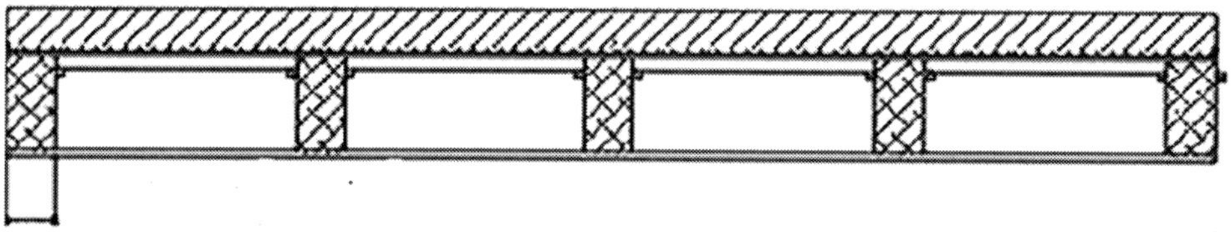

Figure 3. Reinforced concrete slab over the wooden floor [4].

2.2.2. We can achieve a rigid slab effect using reinforced concrete tie beams and a double planking of boards at 45 degrees. The problem of connecting the reinforced concrete tie beams and the planking of boards arises. The reinforced concrete can be connected to the wood by means of metal parts like in figure 4.

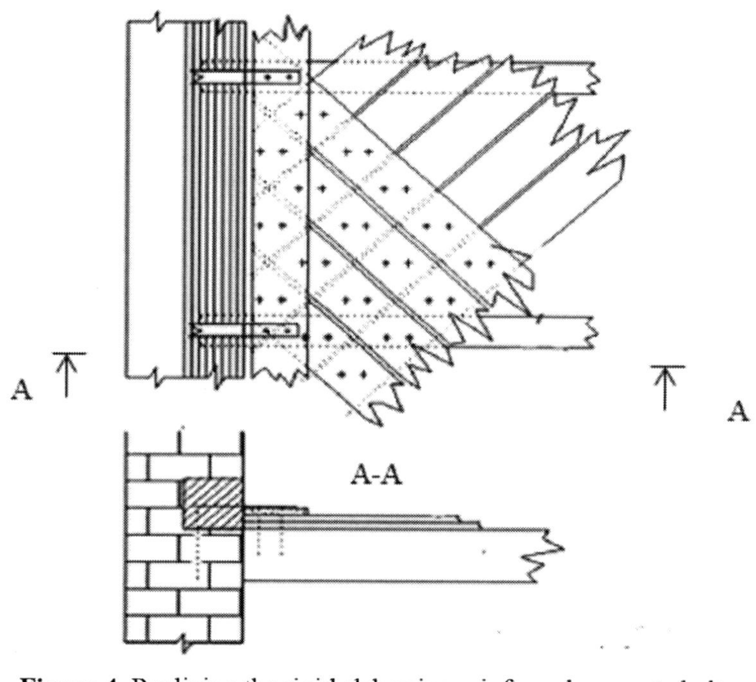

Figure 4. Realizing the rigid slab using reinforced concrete belts and a double plank of boards [4].

3. Contributions regarding the consolidation of wooden floors

3.1. Increase of the bearing capacity for gravitational actions

It is proposed to consolidate a wooden floor that, under gravitational actions, has developed deflections far beyond the limit ones. The consolidation is applied, due to the imposed conditions, at the bottom of the slab. The large deformations of the floor beams do not allow the use, for consolidation, of steel elements to be attached to the floor beams.

It is proposed to intercalate between the wooden beams of the slab some steel beams from HEB profiles, located tangentially, in the middle of the opening, at the structural floor. At the bottom of these elements and arranged perpendicularly on them, two metal profiles are suspended which will become supports for the wooden beams of the slab. The support will be ensured by the introduction of hardwood feathe

The proposed solution preserves the state of deformation of the floor is presented in figure 5, but prevents the increase of deformations. The dimensioning of the system is made by conditions of deformability and load bearing capacity of the slab. The direction of unloading of the floor on the wall is kept.

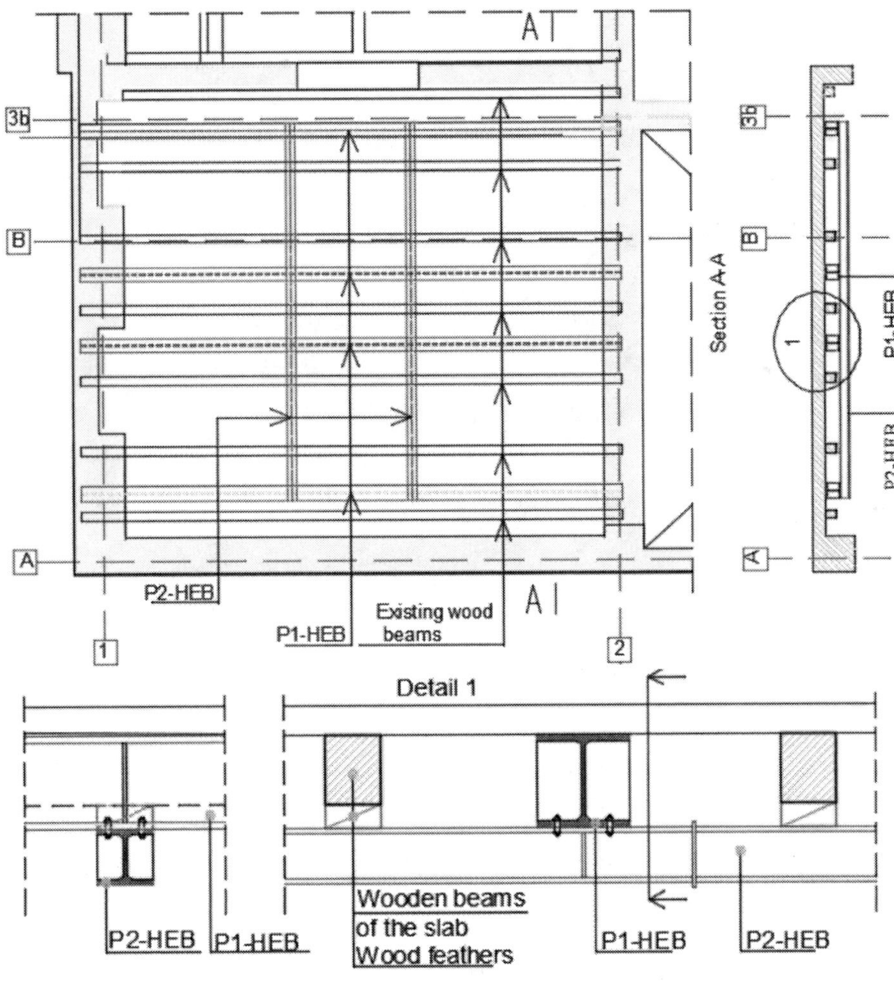

Figure 5. Increase of the bearing capacity of a slab subjected to gravitational

actions by consolidate with additional steel elements.

3.2. Increase of bearing capacity for horizontal actions

The aim is to make a rigid slab at the floor level of the attic, in a building with a structure with load-bearing walls and wooden slabs.

For this purpose, the chords of the main truss (included in the slab structure, consisting of adjacent beams) are reinforced with UPE steel profiles arranged on the two faces of the section and supported (by means of concrete bearings) on the longitudinal walls. The reinforced concrete bearings are anchored to the walls by means of rods inserted into holes made in masonry (diameter min. 30 mm, depth min. 800 mm) and fixed with injected mortar. At the bottom of the wooden floor, along the walls on which it is supported, UPE metal profiles fixed by the lower sole of the strings for strengthening the chords are arranged. These ensure the increase of the length of the wooden beams of the slab and together with the profiles for strengthening the chords and the round steel diagonals arranged in a X shape create a rigid horizontal system. The steel elements UPE placed along the walls, at the bottom of the slab, also act as tie rods of the respective walls, ensuring their anchoring at the ends.

The following are the constructive solutions that can be adopted in this case are shown in figures 6 and 7.

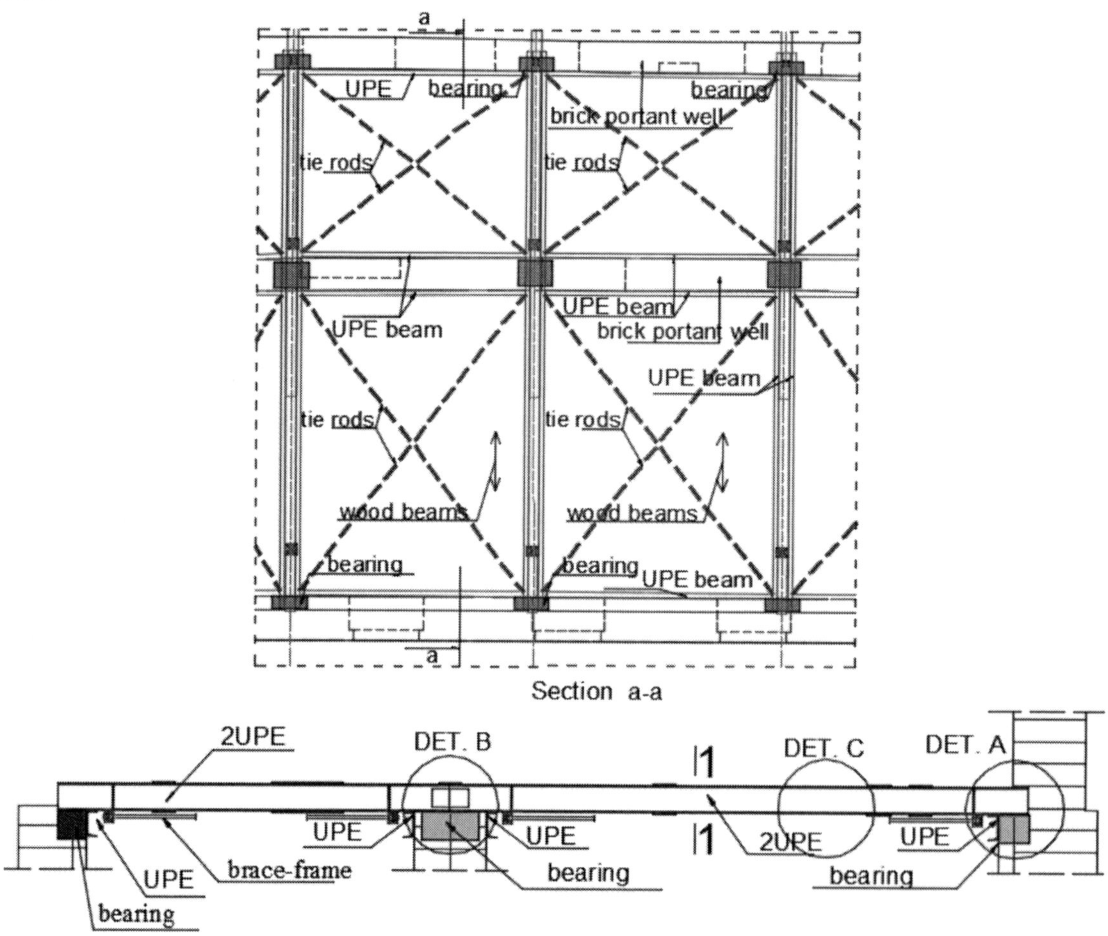

Figure 6. Slab plane. Disposal of the rigid elements in plane.

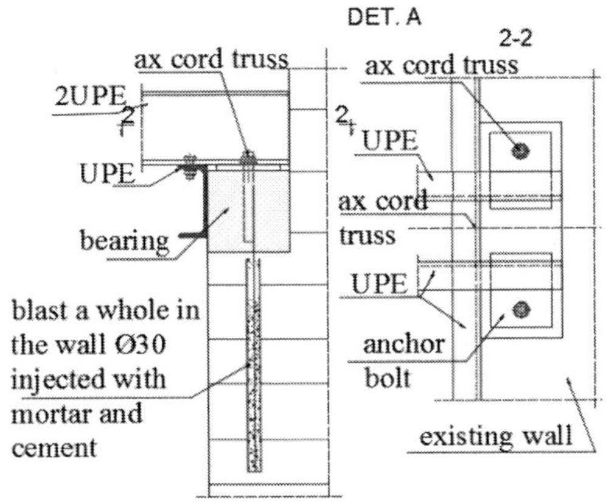

Figure 7.a Constructive detail. Support detail – steel confection on marginal wall.

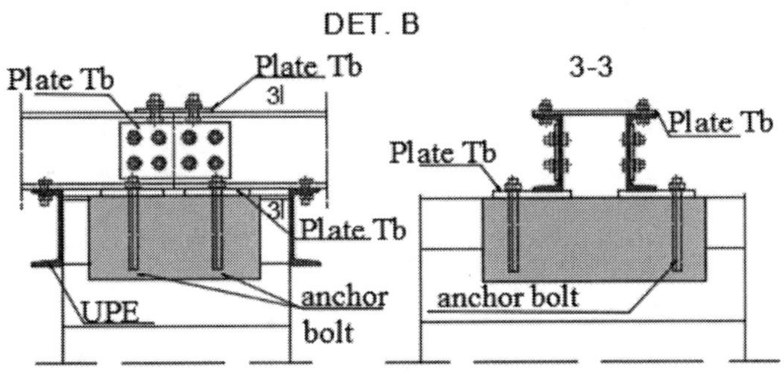

Figure 7.b Constructive details. Support detail/continued steel confection on the internal wall

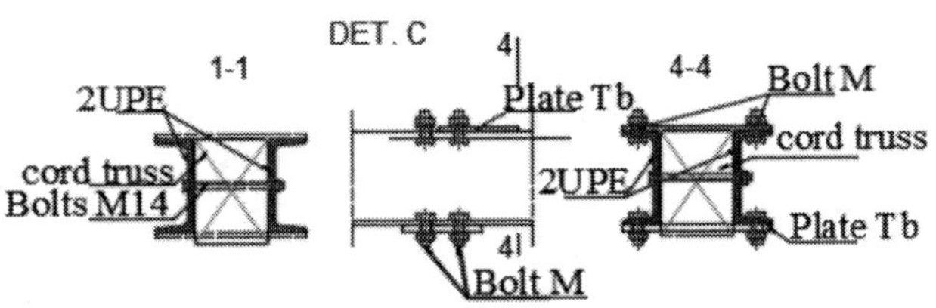

Figure 7.c Constructive details. fixed detail steel profiles with the wooden beam. Lateral stiffening assemble steel profiles.

4. Conclusions

The presented solutions try to contribute to the development / diversification of the intervention modalities for the rehabilitation / consolidation of the wooden floors.

Proposed structural technique for the retrofitting of wooden slabs subjected to gravity loads is formulated for interventions at the inferior part of the slab only. Dimensioning of steel consolidation elements is undertaken in a way that the strains and stresses in the slab's wooden elements are

IOP Conf. Series: Materials Science and Engineering **789** (2020) 012032 doi:10.1088/1757-899X/789/1/012032

controlled under post-retrofitting loads. The slab's behavior on the vertical structural elements does not undergoes any adjustment. Another important aspect is the fact that the solution does not involve important building technology.

According to the suggested solution, an accomplishment of rigid slab effect in the case of wooden floor slabs assures a compound effect of vertical structural components (masonry walls) horizontally loaded on condition that the system fully connects with the masonry walls. Recommended system also provides the support of main trusses of the framework by transforming its chords in elements capable of undertaking gravity loads. They become in this case the horizontal elements of an attic. The abovementioned solution increases the attic's rehabilitation process.

References

[1] P100-1-2013 Cod de proiectare seismică – partea I – Prevederi de proiectare pentru clădiri (P100-1-2013 Seismic design code – part I – Design provisions for buildings)

[2] P100-3-2018 Cod de proiectare seismica – partea a III-a – Prevederi pentru evaluarea seismica cladirilor existente (P 100-3-2018 Seismic design code – part III - Prescription for seismic evaluation of the existing buildings)

[3] Dogariu A 2004 Soluții constructive cu materiale metalice pentru consolidarea și / sau reabilitarea structurilor din zidărie sau beton armat, Teză de doctorat (Dogariu A 2004 Constructive solutions with steel materials for the consolidation and / or rehabilitation of the structures of masonry or reinforced concrete, PhD thesis)

[4] Spoială L T 2007 Posibilități de transformare a planşeelor de lemn în diafragme plane orizontale (şaibe) rigide, Revista AICPS, 1/2007 (Spoială L T 2007 Possibilities of transforming wooden floors into horizontal flat diaphragms – rigid slabs, AICPS Magazine, 1/2007

IOP Conf. Series: Materials Science and Engineering **789** (2020) 012033 doi:10.1088/1757-899X/789/1/012033

Analysis of influence of subway station deep foundation pit excavation on adjacent bridge piles

B Li[1,2*], W Sun[2] and G K Li[2]

[1] School of Civil Engineering, Shenyang Jianzhu University, Shenyang, China
[2] Key Laboratory of Disaster Prevention and Control of Liaoning Province, Shenyang Jianzhu University, Shenyang, China

[*] bingleesy@163.com

Abstract. Based on the engineering background of the deep foundation pit construction of Shenyang metro Line 10, the numerical simulation software Midas the GTS NX, and the three-dimensional integral calculation model considering the interaction of foundation pit supporting structure, surrounding soil and viaduct is established, and the numerical simulation value and the measured data are compared and analyzed, the applicability of the model is verified, the influence of foundation pit excavation on the deformation of pile foundation is studied, and the optimization scheme is given. The construction plan of Simulation + forecast is established to guide the practical construction of the project. The results show that when the depth of the pile foundation is 1.5 times the depth of the foundation pit, and the horizontal displacement of the pile foundation is within 1/2 of the control value, this is the optimal depth of the pile foundation. This is the minimum standard safe distance when the pier is 1/2 from the depth of the foundation pit.

1. Introduction

Excavation of deep foundation pit will disturb the surrounding soil, cause horizontal displacement of the enclosure structure, vertical settlement of soil in the range of 1 times excavation depth around the foundation pit.

The deep foundation pit project is a very complicated problem of geotechnical engineering which involving soil properties, supporting structures, groundwater, etc.

When the deep foundation pit is under construction, it will affect the stability of the adjacent viaducts and cause safety hazards.

Now the site monitoring of deep foundation pit construction is used as the background. Considering the mutual coupling relationship between the deformation of the support structure of the foundation pit and the vertical deformation of the bridge, a three-dimensional model was built using MIDAS GTS NX. The model includes foundation pits, supporting structures and viaducts, simulate the entire excavation process of foundation pit, analyze the simulation results, understand the impact of deep foundation pit excavation on the viaduct, and then feed back to the site construction to solve the actual engineering problems and guide similar engineering construction [1].

2. Engineering background

Shenyang Metro North Daying Street Station is the transfer station for Metro Line 10 and Line 4. The main body of the station adopts the construction method of open pit excavation. Bored piles + steel supports are used as supporting structures, the method of precipitation is precipitation outside the pit. The excavation depth of the main structure of the foundation pit is 20.5 m. The length of the retaining pile is 26.6 m. The viaduct is located 9.0 m to the east of Line 10 of Beidaying Street Station.

The pile foundation of the viaduct is a bored pile with a buried depth of 28 m. To ensure that vehicles on the viaduct can pass through during the construction of the subway, monitor the horizontal displacement of the pile body of enclosure pile in the deep underground layer and understand the horizontal deformation of the enclosure pile.

Compare the numerical simulation results with the actual monitoring results to verify the rationality of the mode.

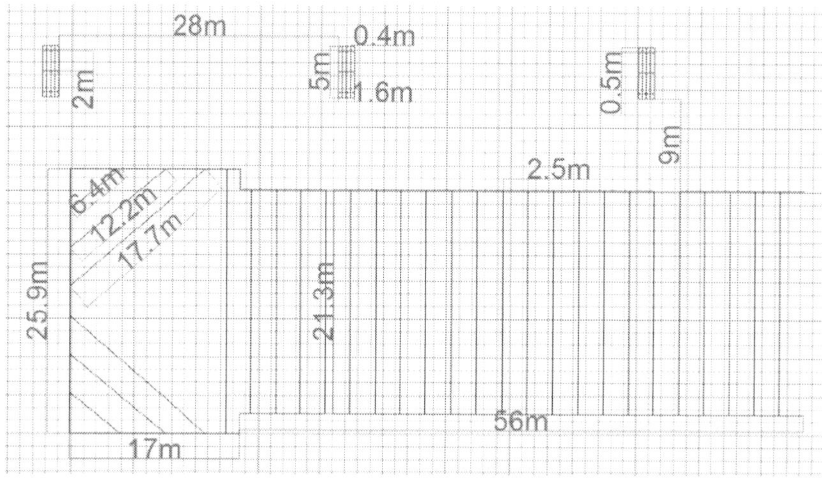

Figure 1. The North Street station layout.

3. Numerical Simulation

3.1. Constitutive model

Using Mohr-Coulonb criterion for numerical simulation in foundation pit construction stage [2].

Mohr-Coulonb criterion:

According to the Mohr criterion, the damage is generally expressed by the following equation:

$$|\tau| = f(\sigma) \tag{1}$$

The simplest Moir's damage envelope is a straight line, and the straight line envelope equation is as follows:

$$|\tau| = c + \sigma \tan \varphi \tag{2}$$

The meaning of each item in the above formula is as follows:

φ - effective internal friction angle,

σ - effective normal stress on the failure surface,

τ - shear stress on the failure surface (kPa),

c - effective cohesion (kPa).

In 1900, Mohr according to the stress circle of the ultimate stress state when the strength is destroyed, an envelope curve tangent to the stress circle is obtained.

The expression of the curve equation is:

$$\frac{\sigma_1 - \sigma_3}{2} - \frac{\sigma_1 + \sigma_3}{2}\sin\varphi - c\cos\varphi = 0 \tag{3}$$

Moore-Coulomb is expressed by principal stress ($\sigma_1 \geq \sigma_2 \geq \sigma_3$) , and the equation of the failure surface is as follows:

$$\sigma_1 \frac{1 - \sin\varphi}{2c\cos\varphi} - \sigma_3 \frac{1 + \sin\varphi}{2c\cos\varphi} = 1 \tag{4}$$

From (4), we can know that the ultimate shear stress τ in any plane is only related to the normal stress σ in the same plane.

According to this calculation model, the following assumptions are made [3]:

- steel supports and other components of the same material are isotropic materials;
- in the analysis of the construction phase, the influence of the water level is not considered;
- underground continuous wall is a linear elastic model, and there is no relative slippage between the wall and the soil;
- the ground surface around the foundation pit is set as a uniform load and the load value is 50 kPa;
- in the analysis process, the influence of the excavation time and space on the calculation results is not considered [4].

3.2. The establishment of a numerical model

The actual soil structure for excavation of foundation pits is shown in table 1:

Table 1. Key soil parameters.

Serial number	Geological stratification	γ (*kN/m³*)	C (*kPa*)	Φ (°)	E (*MPa*)
1	Miscellaneous fill	18.4	33.8	11.4	3.5
2	Silver sand	17.4	0	23.5	11.0
3	Round	21.0	0	35.0	26.2
4	Round gravel	21.8	0	36.4	32.2
5	Gravel sand	20.8	0	36.0	29.0

Due to the excavation of the foundation pit in layers. Before each layer of soil is excavated, the water will drop below the soil layer. The simulation did not consider water.

Drilled cast-in-place piles are the main components of the enclosure structure. Mainly bear the lateral pressure of soil and lateral pressure of steel support. This value simulation, in order to facilitate the establishment of the enclosure structure model and improve the accuracy of the calculation, an underground continuous wall with equal stiffness are replaced by cast-in-place piles in the simulation.

In this simulation, in this simulation, the wall unit simulation using the plate unit simulation is more in line with the actual situation of the foundation pit excavation.

In the GTS simulation, the underground continuous wall is adopted rectangular plate element to simulate the enclosure structure. Assume that the bored pile diameter is D, pile spacing is t.

Then a single pile should be equivalent to wall-type underground continuous wall with a length of D + t.

If the equivalent thickness of the ground wall after the equivalent is h, from the principle of equal stiffness can be obtained as follows:

$$\frac{1}{12}(D+t)h^2 = \frac{1}{64}\pi D^4 \tag{5}$$

$$h = 0.589D^{1.5}\sqrt{\frac{1}{1+\frac{t}{D}}} \tag{6}$$

In the excavation of the foundation pit, after consulting the drawings, the diameter of the enclosure pile D = 1.0 m and the pile spacing t = 0.4 m. after calculation, h = 0.50 m can be obtained.

Therefore, the equivalent underground continuous wall with thickness h = 0.50 m is used in the simulation of the enclosure pile. underground continuous wall adopts linear elastic model.

The concrete strength grade is C30, the elastic modulus of concrete is E = 30 GPa, Poisson's ratio is 0.2. The main steps for building a model using Midas GTS software are as follows:

(1) Drawing a 2D plan using Autodesk CAD - first define the size and location of the foundation and pier and the soil and structure within the affected area according to the project overview, then draw the flat shape in Autodesk CAD and save the file as ".dxf" format. Midas GTS "Extend" command - after importing the plane figure into the GTS software, Extend the "line" into a solid through the "Extend" command, it consists of three parts: the foundation pit, the pier and the surrounding soil within the affected area.

(2) Midas GTS "Extension" command: after the flat graphics are imported into the GTS software, the "line" is expanded into entities by the "expand" command, including the foundation pit, the pier and the surrounding soil within the influence range.

(3) Boolean Operation - Embedded Command: the overlapped part of the entity generated by the extended command, by embedding commands, the foundation pit and the bridge pile body are embedded in the soil within the scope of influence. So that no overlapping parts appear.

(4) The division of the grid: use 3D automatic meshing, use the Hybrid Grid Builder to mesh all the entities step by step, while defining the grid group properties separately. At the same time define the grid group attributes. Divide grids from inside to outside and from small to large. The quality of the mesh has a great influence on the results of the finite element analysis. The calculation accuracy has a direct relationship with the mesh quality. The finer the mesh, the higher the accuracy of the calculation. In this simulation of the number of values. The soil is a flexible material and the enclosure structure is a rigid material. Therefore, the grid in the excavation area of the pit is properly encrypted, and the farther away from the pit is, the sparser it is. Check the mesh topology after meshing to determine if the meshes are coupled [5].

(5) "Extended grid" command: the "Extended Grid" command - expands the foundation pit, the pier and the surrounding soil to the negative direction of the z-axis, respectively. According to the excavation sequence of the foundation pit, the information on the expansion of the foundation pit, the pier and the surrounding soil is consistent, so that the foundation pit analysis can be performed accurately. After the expansion grid is complete click on "Check command" to check the mesh topology. The free faces are all orange, indicating that the expanded grid is correct and will prepare for the subsequent definition of the construction phase.

(6) After "Extract" command: after the mesh is completed, use the extraction command to divide the underground continuous wall and bridge pile grid, and define the grid group attributes. At the same time, 1D grids such as steel supports and ring beams are divided, and steel supports and ring beams are extracted from the nodes to 1D.

(7) The foundation model is established, define the construction phase to simulate the process of foundation pit excavation. In the construction phase, the excavation, lining and support structure installation and erection of the soil are mainly accomplished through the "activation" and "passivation" of the grid group elements, boundary conditions, and static loads. Define the construction phase as shown in table 2 below.

Table 2. Numerical simulation of construction condition tables.

Construction state	Simulation content
Project Status 1	Initial Stress Analysis
Project Status 2	Building diaphragm wall
Project Status 3	Excavate the first layer of soil to -2.50 m
Project Status 4	Set up the first steel support at -2.00 m, excavate the second layer of soil to -8.90 m
Project Status 5	Set up a second steel support at -8.40 m and excavate the third floor to -12.40 m
Project Status 6	Set up the third steel support at -11.90 m, excavate the fourth floor to -16.70 meters
Project Status 7	The fourth steel support is erected at -16.2 meters and the fifth layer of soil is excavated to 20.50 meters

Figure 2 is a three-dimensional numerical model of the foundation pit and bridge. Figure 3 is a schematic diagram of enclosure pile, steel support and bridge pile foundations. The model is cuboid soil which overall size of 110.0 m by 106.0 m by 42.0 m. The soil is divided into 5 layers according to geological conditions.

The simulated foundation pit is in the central position of the overall model. The green part of the model is the surface monitoring point. The standard section of the foundation pit is 21.3 m wide, and the viaduct is 9.0 m outside the foundation pit.

The bridge piers and bridge piles are automatically coupled with the soil, and the boundary conditions of the foundation are set, and then the displacement of the pile foundation is cleared.

The "self- weight" and "auto-constraint" functions in the software were used to simulate the model's deadweight and boundary conditions.

Then calculate and analyze the model. Study the influence of excavation of deep excavation on the bridge pile foundation. In order to make the excavation simulation process almost reasonable. The role of groundwater and soil consolidation and creep are ignored in calculations [6].

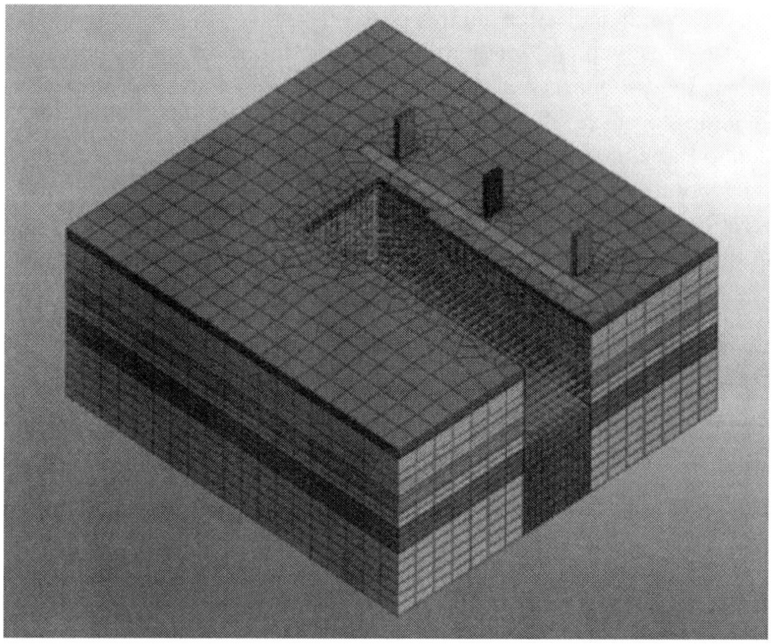

Figure 2. Schematic diagram of the whole model.

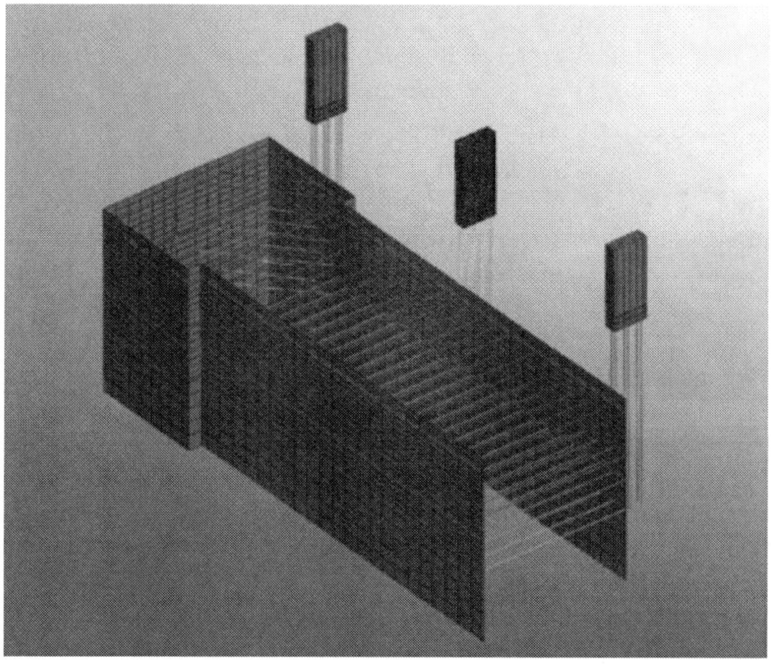

Figure 3. Support structure and bridge model.

3.3. Simulate the horizontal displacement of the enclosure structure

Deep excavation will certainly disturb the surrounding soil during excavation. enclosure structure will produce horizontal deformation, figure 4 is a horizontal displacement deformation cloud chart of the retaining pile.

IOP Conf. Series: Materials Science and Engineering **789** (2020) 012033 doi:10.1088/1757-899X/789/1/012033

The enclosure structure near one side of the bridge is inclined to the inside of the foundation pit, and horizontal deformation occurs from the top to the middle and lower parts [7].

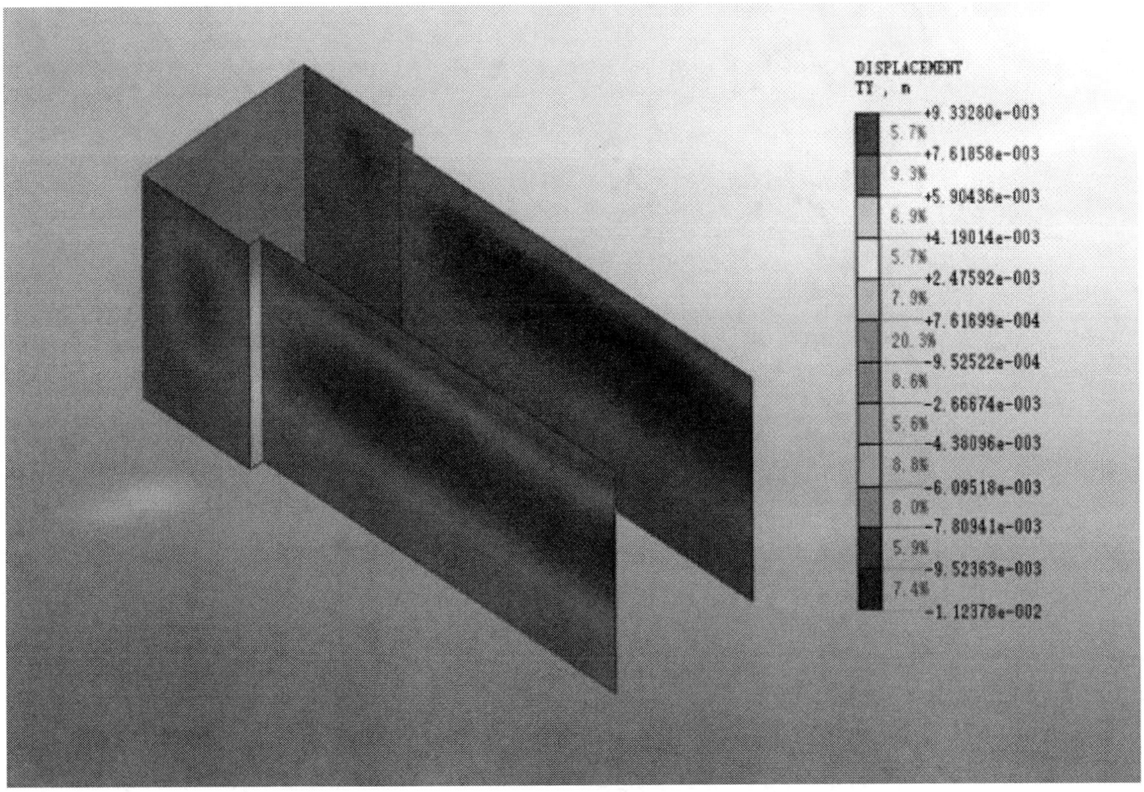

Figure 4. Enclosure wall horizontal displacement nephogram Y direction.

Figure 4 shows that the horizontal displacement of the wall of the enclosure wall near the side of the bridge is greatly deformed at the middle and upper part of 10.0 m.

The surrounding soil will inevitably loosen. At the same time, the bridge pile produces horizontal deformation towards the inside of the foundation pit.

Figure 5 shows the measured values of field monitoring. The maximum is 9.52 mm.

Figure 6 shows the numerical simulation.

The combination diagram shows that the deformation of the enclosure pile is toward the inside of the foundation pit during the initial process of the entire foundation pit. And the amount of deformation is related to the excavation depth of the foundation pit.

The excavation of the foundation pit is completed, and the enclosure piles are bow-shaped deformation [8]. The maximum displacement of the enclosure pile is at the upper middle position of the enclosure pile.

The numerical simulation results are similar to the on-site monitoring data trends, prove the rationality of the three-dimensional model.

Therefore, the numerical simulation can effectively predict the deformation of the enclosed structure and the bridge pile foundation during the construction of the deep foundation pit, and guide the construction on site.

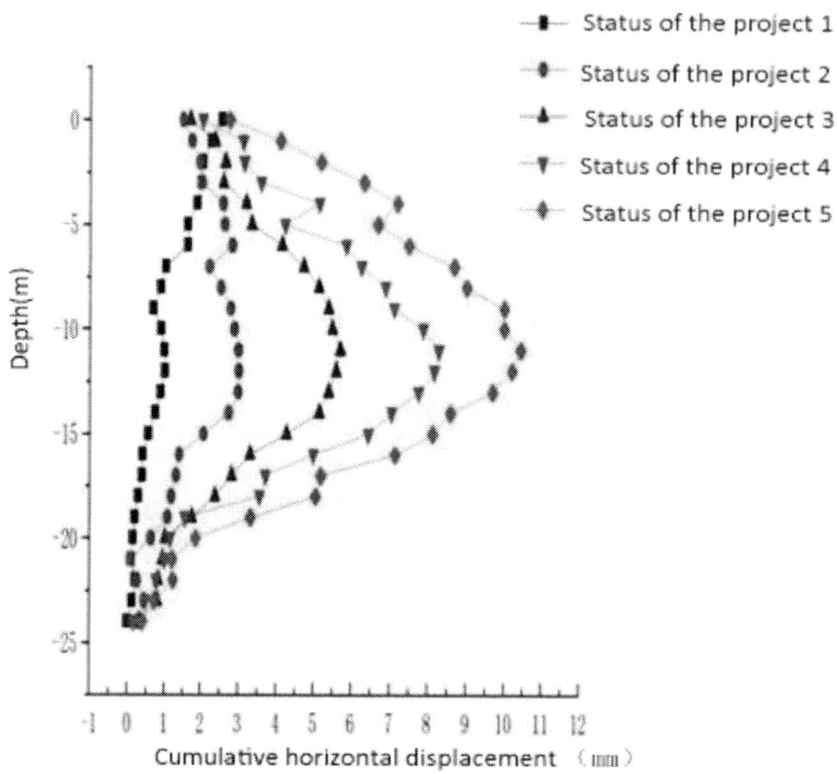

Figure 5. Site monitoring of CX-9 enclosure pile.

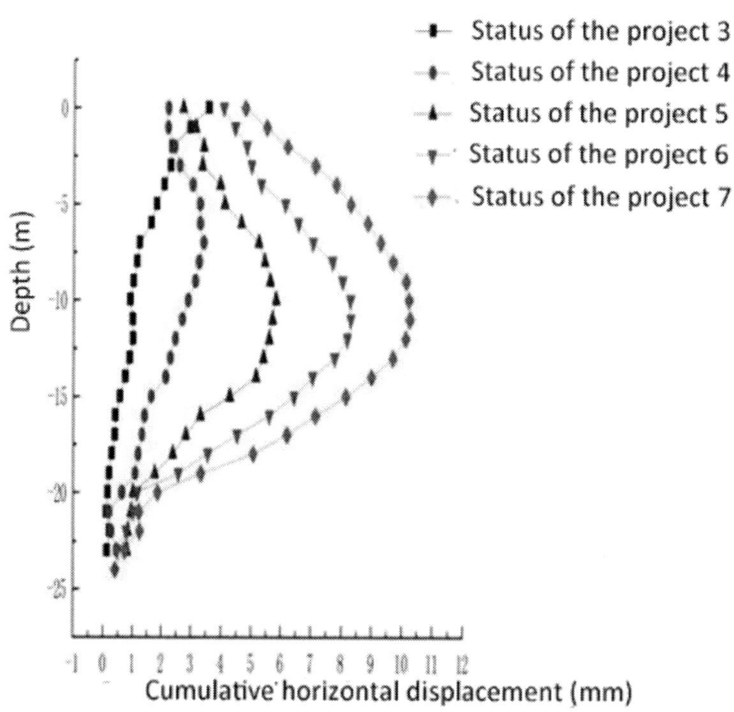

Figure 6. Numerical simulation of CX-9 enclosure pile.

3.4. Analysis of horizontal displacement of bridge piles

When the soil deforms, the bridge pile will also be affected. The safety of the bridge pile directly affects the safety of the viaduct.

Figure 7 shows the horizontal deformation displacement diagram of the pile foundation. The overall deformation of the pile foundation is also "bow-shaped" with the maximum displacement in the middle of the pile foundation. The maximum displacement value is 6.42 mm and the pile foundation controlling value is 10 mm. Under the conditions of the foundation pit construction, the deformation of the pile foundation is in a stable state. Through the horizontal deformation diagram of retaining piles and bridge piles, it can be seen that the changes show similar trends.

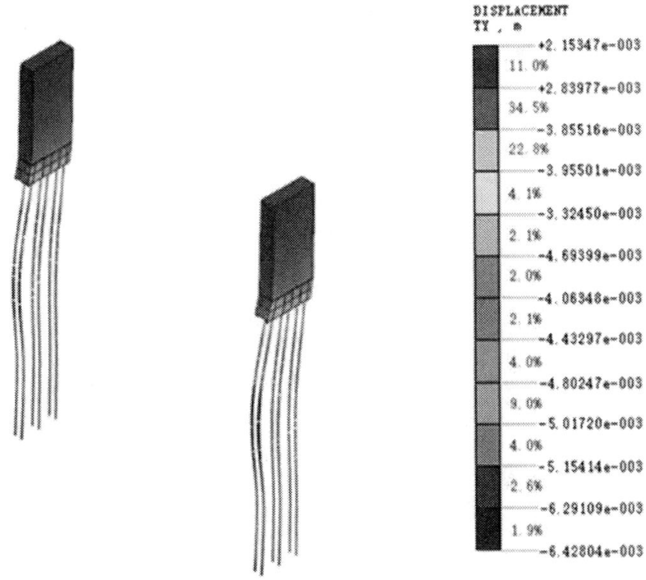

Figure 7. Horizontal deformation of pile body of bridge pile.

4. Influence of deep foundation pit construction on pile body

4.1. At the same time of excavation of deep foundation pits, the influence of different depths of buried soil of pile foundation on horizontal displacement of pile foundation is analyzed

The depth of grave of the bridge pile foundation has a certain influence on the deformation of the pile foundation.

Analyze the relationship between the depth of pile foundation and the depth of foundation pit, and then analyze their influence on the horizontal deformation of pile foundation, and the optimum depth of the pile foundation is obtained.

Combining the actual engineering to establish a corresponding three-dimensional numerical model, Maintain the excavation conditions of foundation pits, the depth of the buried enclosure pile, the structural parameters of the soil and the parameters of the enclosure structure remain unchanged. Divide the depth of pile foundation into six categories: 21.0 m, 24.0 m, 27.0 m, 30.0 m, 33.0 m, and 36.0 m. Through the numerical simulation calculation, the deformation curve of the pile foundation is obtained, and then analyze the influence of pile foundation depth on horizontal deformation of pile foundation.

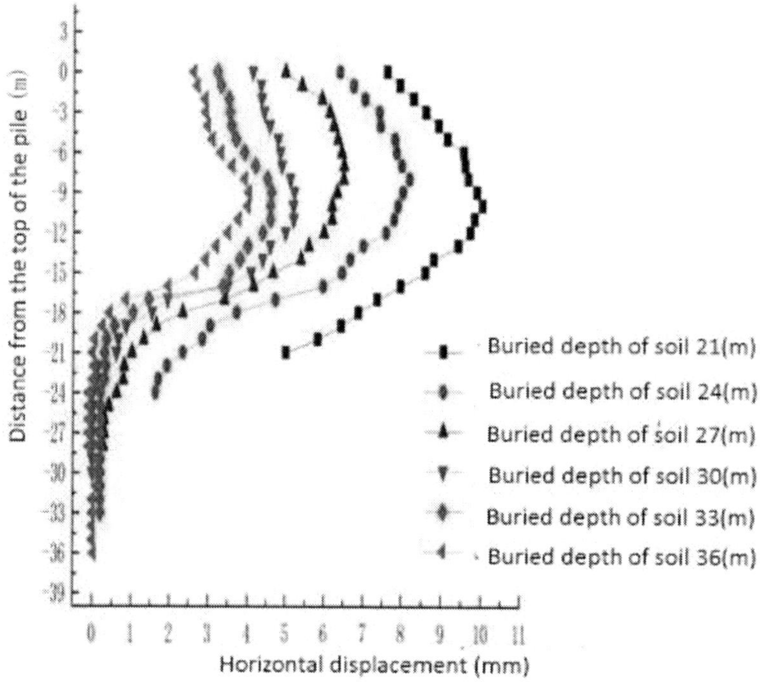

Figure 8. The horizontal displacement curve of pile foundation of different depth of pile foundation.

As shown in figure 8, during the actual excavation of the foundation pit project, the different depths of the pile foundations have some influence on the horizontal displacement of the pile foundation. When the depth of the pile foundation is 21.0 m, the maximum displacement of the pile foundation is 10.2 mm.

When the depth of pile foundation is more than 30.0 m, the maximum horizontal displacement of the pile does not decrease significantly with the depth of pile foundation. The change of the horizontal displacement of pile foundation is reduced, and the variables are all within 1mm. It shows that when the soil depth of the pile foundation of the bridge is 30.0 m, it is approximately 1.5 times the depth of the foundation pit, and the horizontal displacement of the pile foundation is within 1/2 of the controlling value.

The controlling value is 10 mm, which is the optimum depth of the pile foundation. Under this engineering background and project profile, when the soil depth of the pile foundation is 2 times the depth of the foundation pit, the influence of excavation of the deep foundation pit on the horizontal deformation of the pile foundation is negligible.

4.2. Analysis of the influence of different distances to the foundation pit adjacent to the pile foundation

Set up a three-dimensional construction numerical model for deep foundation pits and bridges, study the effect of different distances from the bridge pier on the horizontal displacement of adjacent piles, and maintain the parameters of the soil and the parameters of the enclosure.

The distance from the bridge pier to the edge of the foundation pit is 5.0 m, 8.0 m, 10.0 m, 13.0 m, 16.0 m and 20.0 m respectively. The excavation depth is constant and the construction conditions are not changed. Analyze the effect of different distances on the horizontal deformation of the pile foundation and obtain the safety distance

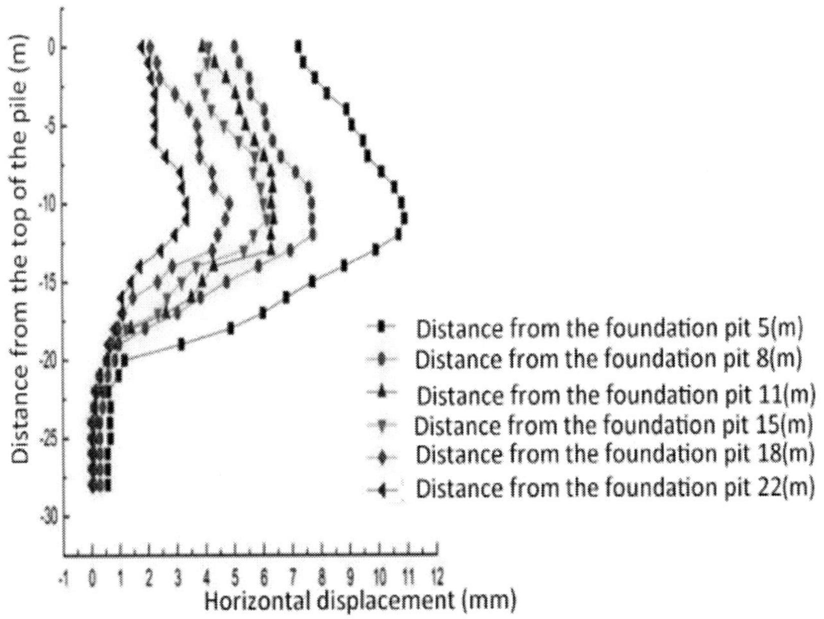

Figure 9. Horizontal deformation of pile base at different distance from foundation pit.

As can be seen in figure 9, as the distance from the pile foundation to the edge of the foundation pit increases, the overall horizontal displacement of the pile foundation gradually decreases. When the bridge pier is 5.0 m away from the edge of the foundation pit, the maximum horizontal displacement of the pile foundation is 12.24 mm.

When the pier is 22.0 m away from the edge of the foundation pit, the maximum horizontal displacement of the pile is 3.87 mm. When the distance is more than 22.0 m, it is almost considered that this distance has no effect on the bridge pile foundation. When the distance is 8m, the settlement of pile foundation has reached the control value.

Therefore, when the distance between the pier and the edge of the foundation pit is less than 8 m, corresponding protection measures must be taken. This shows that when the distance from the bridge to the foundation pit is less than 1/2 of the depth of the foundation pit, it is necessary to strengthen construction monitoring or adopt some protective measures.

4.3. Construction optimization and protection measures

Based on the above research contents, the design scheme is optimized based on the original design scheme. Based on the premise of ensuring the safety of adjacent foundation pits, the distance between foundation pits and bridge piers and the soil depth of pile foundations were optimized. See table 3 for a list of optimization parameters for the construction of foundation pits.

Modify the previous model and re-simulate the calculation. Modify the previous model and re-simulate the calculation.

Table 3. Parameters optimization of shield construction.

Optimization project	original plan	optimization
The distance between pile foundation and foundation pit	9.0 m	14.0 m
Pile foundation depth	28.0 m	34.0 m

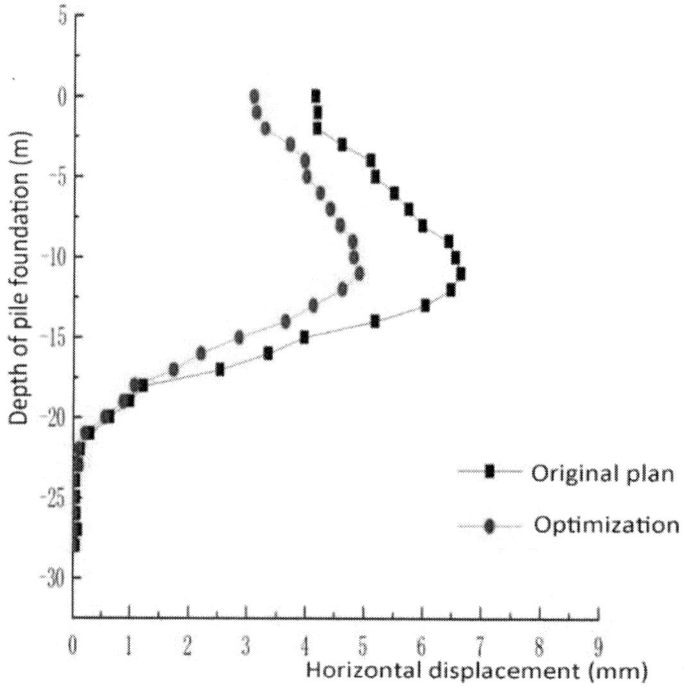

Figure 10. Comparison of horizontal displacement curves of the original plan and the optimization plan.

The numerical simulation method was used to analyze the optimized design scheme. As can be seen from figure 10, the overall horizontal displacement of the optimization scheme becomes smaller, the curve is flatter than the original scheme, and the fluctuation range is small. The maximum value of the horizontal displacement of the optimized pile foundation is 4.90 mm, which is less than 1/2 control value and can be considered to achieve the optimization effect. In the subsequent actual projects, the on-site design plan is compared with the optimized plan. Then through numerical analysis for calculation and analysis, to predict the possible hazards of construction, effective control the safety of construction workers and the construction progress.

The commonly used method of soil reinforcement in the project [9] is to protect the pile foundation. Soil reinforcement has two methods.

The first type is to reinforce the surrounding soil of the foundation pit, increase the strength of the surrounding soil, and thus reduce the deformation of the soil to achieve the role of protecting the pile foundation.

The second type is the foundation of the reinforcement pile foundation to improve the foundation bearing strength of the pile foundation. Inhibit the deformation of the pile foundation and protect the safety of the viaduct. Soil reinforcement methods include pre-grouting method, freezing method, and high-pressure spin-spray mixing method [10]. This project adopts the pre-grouting method and uses cement paste to reinforce the surrounding soil of the foundation pit, to increase the strength of the surrounding soil. Reduce the disturbance of the surrounding soil to achieve a stable pile foundation.

5. Conclusion

This paper takes the deep foundation pit project of the station of No. 10 line of Beidaying Street Station in Shenyang as the research background, combined with on-site monitoring. Using the finite element analysis software MIDAS GTS NX to establish three-dimensional models of deep foundation pits, supporting structures and viaducts, the influence of deep foundation pit excavation on the bridge pile foundation under different conditions was studied and the conclusions are as follows:

- when the depth of the piled foundation of the bridge is 30.0 m, it is about 1.5 times the depth of the foundation pit; the horizontal displacement of the pile foundation is within 1/2 of the control value, which is the optimum depth of the pile foundation;
- when the distance between the pier and the pit is 1/2 times the depth of the pit, it is the minimum safety distance; less than this distance, we must strengthen construction monitoring or take some protective measures;
- the data monitored at the site often lags behind actual changes; therefore, based on the actual project, this paper uses numerical simulation method to simulate the deep excavation and establish a "simulation-prediction" information construction program.

References:

[1] Ruan Y F, Xu C M and Zhang L H 2011 *J. Safety Science and Technology of China* **7** pp 73-77
[2] Dnan L L, YU L and Sun Y 2011 *J. Safety Science and Technology of China* **7** pp 38-43
[3] Cai F and Ugai K 2001 *J. Hong Kong* 305-310
[4] Lin Y, Hu C M, Li Y H and Zhang J 2010 *A. Construction Technology* **39** pp 18-21
[5] Yang B, Xi Z R and Pan S H 2013 *A. Hydroelectric Energy Scienc.* **31** pp 143-146
[6] Zhang Y C, Yang G H and Yao J 2010 *J. Chinese Journal of Geotechnical Engineering* **32** pp 109-115
[7] Yum J H, Kima S S and Kima D Y 2005 *J. Journal of Photochemistry and Photobiology A: Chemistry* **171** pp 269-273
[8] Liu J, Yao H L and Ren J X 2010 *J. Rock and Soil Mechanics* **31** pp 456-461
[9] Tan Y C, Tang Y R and Peng J Q 2009 *A. Urban Rail Transit Study* **12** pp 21-24
[10] Xie Z Y and Bai H C 2011 *B. Sichuan Building Materials* **4** pp 238-242

IOP Conf. Series: Materials Science and Engineering **789** (2020) 012034 doi:10.1088/1757-899X/789/1/012034

Tool wear analysis of multi-layer diamond coated tools for cutting white marble

F Lu [1*], H X Li [1], C Gabor [2], Y Y Wang [1] and J B Zhao [1]

[1] School of Mechanical Engineering, Shenyang Jianzhu University, Shenyang, China
[2] Transilvania University of Brasov, Materials Science Department, Brasov, Romania

[*] lufeng72@126.com

Abstract. Study on the tool wear of cutting white marble by multi-layer diamond coated tool, and discussion of the wear characteristics and failure mechanism of the tool are presented in this paper. Multi-layer diamond coating was prepared by hot filament chemical vapor deposition (HFCVD). It was deposited on the surface of cemented carbide ball-end milling tool. The surface appearance of the tool was analyzed by an optical microscope. The grain state was analyzed by scanning electron microscopy. Comparing the surface morphology of the cutting for 1h, the area where the coating peeled off became larger when the cutting tool was cut for 2 hours, and the falling range was expanded along the cutting-edge direction while the flank surface was expanded, and the tool joint wear was severe. The composition of diamond grains was investigated by Raman spectroscopy. Comparing the surface morphology of the cutting for 1h, the area where the coating peeled off became larger when the cutting tool was cut for 2 hours, and the falling range was expanded along the cutting-edge direction while the flank surface was expanded, and the tool joint wear was severe.

1. Introduction

Natural stone is a typical brittle material. It is mainly composed of $CaCO_3$, $MgCO_3$ and SiO_2. It is basically insoluble in water and is a hard-white fine-grained marble. As one of the widely used stone materials in construction materials, natural stone is very easy to engrave because of its solid and delicate texture. In the course of stone processing, commonly used tools such as diamond sintered tools, electroplated diamond tools, often have shortcomings such as low processing precision, high manufacturing cost, high wear rate, easy peeling etc, which cannot meet daily needs [1-4].

The diamond coated tool is a new type of cutting tool that can effectively improve the wear resistance and surface hardness of the tool. To a certain extent it can extend the service life of the tool.

At present, the research on the cutting of diamond coated tools in the metal field is widely studied. However, there is less research on the application of diamond tools in the field of stone processing.

In this paper, hot filament chemical vapor deposition (HFCVD) was used to prepare a diamond coating on the surface of cemented carbide tools. The wear mechanism of multi-layer diamond coated tools during cutting process was studied, which provided a theoretical basis for the preparation of diamond-coated tools for brittle rigid materials [5-7].

2. Test

A WC-Co (YG6) cemented carbide milling cutter was used as a substrate material, and a multilayer diamond coating was prepared by the method of hot filament chemical vapor deposition.

Figure 1 shows the system of hot filament chemical vapor deposition. The vacuum unit of the system provides a good environment for diamond growth. The cooling device uses running water to remove excess heat from the substrate and the chamber wall to keep the surface temperature of the device constant. The gas supply system supplies the hydrogen and hydrocarbon gases required for diamond growth. It provides a gas-free, non-contaminating reaction environment. And the equipment can set the parameters required for the test such as gas flow ratio, chamber pressure and substrate temperature.

Figure 1. Hot filament chemical vapor deposition system.

The test uses silk as a hot filament for heating. The distance between the hot filaments is 10mm, and the pressure in the reaction chamber is evacuated to 0.5 Pa. Subsequently, a mixed gas is introduced for carbonization. This process is to avoid wasting too much carbon source on the hot wire, which would result in producing graphite with insufficient carbon source. The substrate temperature is 800□, the hot filament temperature is 2400□, the gas pressure is maintained at 5 kPa, the H_2 flow rate is 800 sccm, and the CH_4 flow rate is 8.40 sccm, respectively. The ratio of methane to hydrogen is adjusted at two concentration values of 1% and 5% to prepare a composite film system in which two films are alternately stacked [8].

Multi-layer diamond coating was prepared by hot filament chemical vapor deposition method, and the film system with three layers was formed by alternately superposing the coating. This test kept the total thickness of the film uniform, about 7 μm. The adjusted carbon source concentration during deposition is 1% to form a bottom diamond film. The carbon source concentration is increased to 5% by changing the proportion of methane to hydrogen, and the diamond film is continuously deposited to cover the underlying diamond film. The top diamond film and the underlying diamond film grow at the same carbon source concentration, covering the second layer. Finally, a multilayer diamond coating having three layers is produced.

A diamond coating having a three-layer coating was prepared, and the carbon source concentration from the substrate was 1%, 5%, and 1%, successively. And the total thickness is 7μm. The surface state of diamond-coated tools of this structure without cutting is compared with the surface state after 1 hour and 2 hours after cutting white marble.

3. Test results and analysis

The surface morphology of the film system and the wear surface of the tool surface were observed by Zeiss Super 55 field emission scanning electron microscope. The surface composition of the diamond coating was analyzed by using a Horiba, Lab Ram HR laser Raman spectrometer.

3.1. Raman spectroscopy of diamond coating

Figure 2 represents a Raman spectrum analysis of diamond film. The diamond film has obvious characteristic peaks at 1337 cm^{-1}, which means that there are much sp3 hybrid components in the nano-diamond film. The characteristic peak of diamond is shifted from 1332 cm^{-1} to 1337 cm^{-1} due to the existence of residual stress, and its residual stress is expressed as compressive stress [9]. The scattering peak of the amorphous carbon component is generally between 1350^{-1} cm and 1600^{-1} cm [10]. In the figure 2, the diamond film exhibits a broad scattering peak at 1491 cm^{-1}, indicating that there are some crystal defects such as dense grain boundaries and grain defects in the film. It may be derived from the C-H bond in trans-polyacetylene, which has a greater impact on the quality of the diamond film [11].

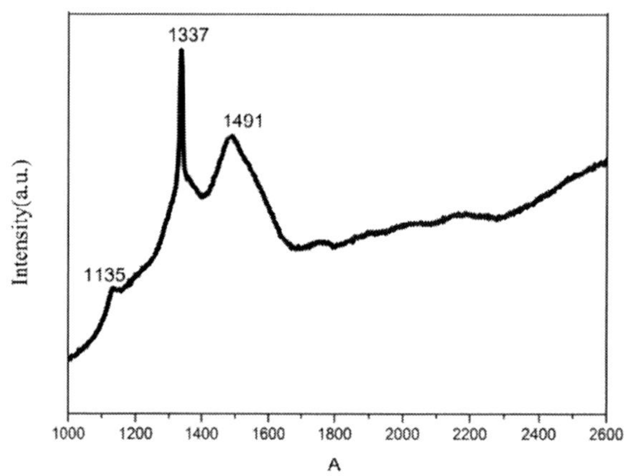

Figure 2. Raman spectroscopy of diamond coating.

This paper explores the changes of the cutter head and the cutting edge of the diamond coated tool during the cutting test. The cutting material is white marble stone with a cutting speed of 3000r/min, a feed rate of 300mm/min and a cutting depth of 1mm. During the cutting process, the stone and the tool flank surface impact and friction, and the failure mode is mainly abrasive wear. It is generally believed that the abrasive wear morphology of diamond is divided into five forms: complete, smooth, microscopic, macroscopic, and shedding [12].

3.2. Analysis of surface morphology of diamond coated tools

Figure 3 shows the macroscopic morphology of the blade tip of the diamond coated tool under the optical microscope, and figure 4 shows the macroscopic morphology of the blade of the diamond coated tool under the optical microscope. Figure 3 (a) and figure 4 (a) show the surface topography of the diamond-coated tool when it is not cut, from which it can be seen that the coating at the tip is completely and evenly coated on the surface of the tool, and the surface of the tool is intact.

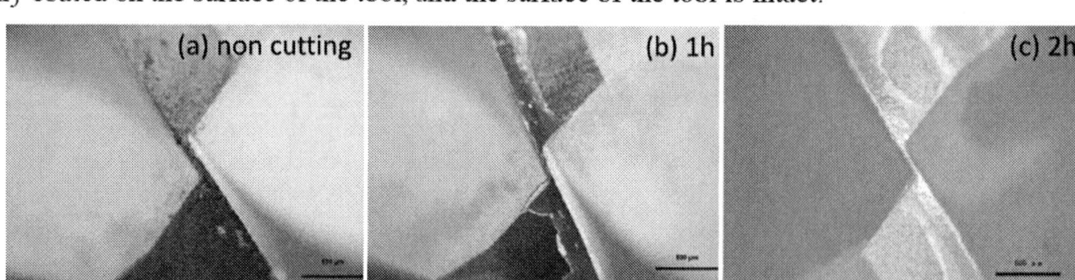

Figure 3. Optical microscopy of the surface wear profile of the tip of diamond coated tool.

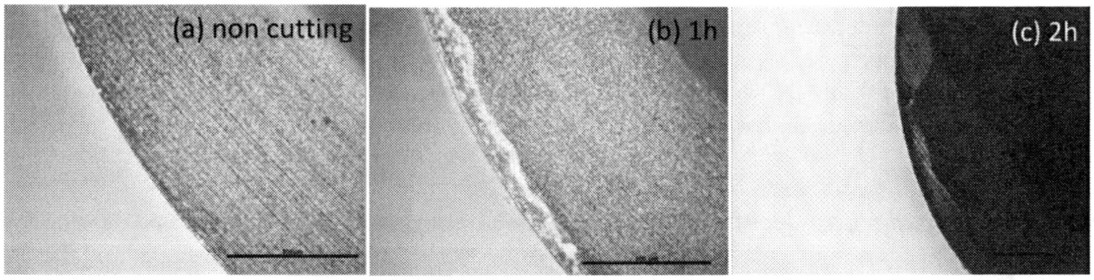

Figure 4. Optical microscope of surface wear profile at the edge of diamond coated tool.

From the figure of the diamond-coated tool for 1h, it is obvious that there is trace of coating peeling off. And the position of falling off is mainly at the flank along the cutting edge. Comparing the surface topography of 1h cutting, it can be seen that after the diamond coating tool is cut for 2h, the area of the coating falling off becomes larger, and the falling range extends in the flank surface along the cutting-edge direction. The reason for this phenomenon is that the high-speed rotation of the milling cutter and the continuous collision and wear of the stone eventually lead to the coating falling off. As the cutting time increases, the amount of wear increases and the coating falls off [13-15].

3.3. Microscopic topography of diamond coated tool surface

Figure 5 is a scanning electron micrograph of the wear state at the blade edge. Figure 5(a) shows the diamond coated tool without cutting, the grain size is evenly covered on the tool base, and the surface is flat. Figure 5(b) is a topographical view after 1h cutting.

Figure 5. Scanning electron micrograph of the edge of the diamond coated tool.

It can be seen that the diamond coating covered by the surface of the substrate has a large area of grain detachment and the side flank is bonded. There are two main reasons for the large area of the coating to fall off. First, the effect of the Co removal treatment is not obvious during the preparation of the coating tool [16-17].

The time to remove cobalt is too short, and the graphitization is serious during the growth of the diamond particles. If the time is too long, the strength of the base material is lowered. The mechanical occlusion effect of the surface of the substrate and the diamond particles is not good, and the bonding force is not strong. When subjected to large impact and friction, large-area peeling occurs. Second, the diamond coating is constantly subjected to large impact forces during the cutting process. And the expansion coefficients of diamond and tool differ greatly.

High thermal stress and cutting force reduce the fatigue strength of the coating, resulting in large area shedding. Figure 5(c) shows the morphology after 2h cutting, and the bond wear is severe. The main cause of bond wear is high-speed extrusion and friction between the tool and the substrate, resulting in very high heat of cutting. High cutting temperature, the more severe the bond wear. High cutting temperatures cause the fatigue strength of the diamond film to decrease, and micro cracks also appear.

When the micro cracks continue to expand to a limit, the coating will fall off. Micro pits appear during the cutting process [18]. As the cutting progresses, fine chips and abrasive particles enter the pit. This makes the larger diameter micro-pit structure cause secondary cutting, resulting in greater cutting force. The impurities accumulated in the pits are constantly increasing.

3.4. Diamond coating tool blade surface topography

Figure 6 shows a scanning electron micrograph of the cutting edge of a diamond coated tool for 2h. Obvious detachment and notch at the blade edge were observed under the microscope, and the damage area was large. In addition to cutting force and cutting temperature, there is a reason that Acid-base etching removes surface Co, which has a certain influence on the hardness and toughness of the tool itself during the preparation of the coating [19-20].

Adhesive chips are distributed at the edge of the blade because the cutting tape tends to form a high temperature region, and the abrasive grains of the diamond coating die are easily affected by the high temperature during the falling process. This situation will affect the waste generated by the cutting stone cannot be discharged, and will reduce the surface quality of the processed stone, processing accuracy and efficiency have also been greatly affected.

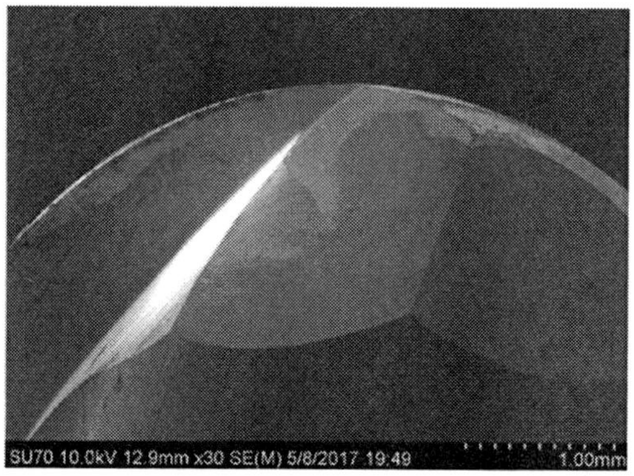

Figure 6. Scanning electron micrograph of blade at 2h cutting of diamond coated tool.

4. Conclusions

When the diamond coated tool is cutting for 1h, the position where the coating falls off is mainly at the flank surface along the cutting edge direction. Comparing the surface topography of 1h cutting, it can be seen that when it is cutting for 2h, the area of the coating falling off becomes larger, and the falling range extends in the flank surface along the cutting edge direction.

A large area of the diamond coating covered by the surface of the tool substrate cut for 1h is peeled off, and the side surface is bonded. When it is cutting for 2h, the increase of tool bond wear is mainly due to the high cutting pressure and friction between the tool and the base to generate high cutting heat. The higher the cutting temperature, the more severe the bond wear. The high temperature region formed by cutting tends to adhere the abrasive particles and also reduces the fatigue strength of the coating.

Cutting force is reduced, cutting environment is bad, resulting in a decrease cutting efficiency. During the pre-treatment process, the acid-base etching method will affect the hardness and toughness of the tool itself. The etching time is too long, the tool quality is degraded, and large gaps are prone to occur. If the etching time is too short, Co cannot be removed, which promotes the formation of graphite impurities and reduces the quality of the tool.

Sintered diamond tools and electroplated diamond tool are common cutting tools. Electroplated diamond tool is more severely worn on the bottom than the frank when cutting marble. So, the frank is the main cutting surface. It is a limitation for cutting process. Sintered diamond tool needs to choose pre-alloy powder with good wear resistance.

At the same time, it needs to match type, concentration, materials of diamond. This can lead limitation in the choice of materials. In order to reduce the limitation of work, the CVD diamond-coated tool is studied in this article.

Acknowledgments
This study was supported by the Liaoning Natural Science Foundation (20170540757), Program for Science and Technology Innovation Talents of Young and middle-aged of Liaoning (SYSCXRC2017002).

References
[1] Wu Y H, Yan G Y and Zhao D H 2018 *J. Machine Tools and Automated Machining Technology* **5** p144
[2] Skordaris G, Bouzakis K D, Charalampous P, Kotsanis T, Bouzakis E and Lemmer O 2016 *J. CIRP Annals* **65** p101
[3] Kun Z and Gyurika I 2013 *J. International Review of Applied Sciences and engineering* **4** p63
[4] Huang Q, Yu D, Xu B, Hu W, Ma Y, Wang Y and Tian Y 2014 *J. Nature* **510** p250
[5] Wu Y H, Yan G Y and Zhao D H 2017 *J. Lubrication & Sealing* **42** p54
[6] Dumpala R, Kumar N, Kumaran C R, Dash S, Ramamoorthy B and Ramachandra Rao M S 2014 *J. Diamond and Related Materials* **44** p71
[7] Sein H, Ahmed W, Jackson M, Woodwards R and Polini R 2004 *J. Thin Solid Films* **447** p455
[8] Suzuki H, Okada M, Asai W, Sumiya H, Harano K, Yamagata Y and Miura K 2017 *J. CIRP Annals* **66** p93
[9] Salgueiredo E, Amaral M and Almeida F A 2013 *J. Surface and Coatings Technology* **236** p380
[10] Zhang S Q, Wang Tand Tang Y B 2017 *J. Integrated Technology* **6** p10
[11] Tang Y, Li Y S, Yang Q and Hirose A 2010 *J. Diamond and Related Materials* **19** p496
[12] Bolshakov A P, Ralchenko V G, Yurov V Y, Popovich A F, Antonova I A, Khomich A A, Ashkinazi E E, Ryzhkov S G, Vlasov A V and Khomich A V 2016 *J. Diamond and Related Materials* **62** p49
[13] Wang C C, Wang X C and Sun F H 2018 *J. Surface and Coatings Technology* **353** p49
[14] Cheng C, Xue P and Xue F 2018 *J. Carbon* **130** p215
[15] Çalışkan H, Kurbanoğlu C, Panjan P, Čekada M and Kramar D 2013 *J. Tribology International* **62** p215
[16] Lü F X, Tang W Z and Song J H 2004 *J. Metal Processing: Thermal Processing* **6** p62
[17] Park K H, Beal A, Kwon P and Lantrip J 2011 *J. Wear* **271** p2826
[18] Nakamura M, Sumomogi T and Endo T 2003 *J. Surface and Coatings Technology* **169** p743
[19] Chen N, Shen B, Yang G and Sun F 2013 *J. Applied Surface Science* **265** p850
[20] Aslantas K, Onur Ö, İsmail U and Büyüksağş I S 2009 *J. Advanced Manufacturing Processes* **24** p1423

Researches on energy conversion of municipal waste by plasma decomposition for energy-efficiency in civil engineering

M L Lupu[1*] D N Isopescu[1] I Cucoş[1] I Antonescu[1] S G Maxineasa[1] and I-R Baciu[1]

[1] "Gheorghe Asachi" Technical University of Iasi, Faculty of Civil Engineering and Building Services, Iasi, Romania

[*] mariuslucianlupu@gmail.com

Abstract. This paper presents the research, technologies and testing equipment developed for the energy plasma conversion of municipal waste in order to enhance energy-efficiency in the civil engineering domain. The purpose of this work is the environmentally friendly improving the efficiency of waste conversion processes by using controlled plasma decomposition reactors developed by the authors, which generate syngas, electrical and thermal energy which can be used in civil engineering.

1. Introduction and study objectives

In 1960 the world population was 3 billion, in 2019 is over 7.7 billion and it is expected to be 8.1 billion in 2025 [1]. The growth of the global population mixed with economic evolution led to rapid urbanization and industrialization, which automatically changed the consumption pattern of the population that finally head up to the rapid increase of Municipal Solid Waste (MSW) at alarming rates [5, 16, 24]. A lot of countries adopted Waste To Energy (WTE) technologies for efficient management of large quantities of waste to produce energy [5, 6]. A foresight made by the International Renewable Energy Agency revealed that the world has the potential of generating approximately 13 GigaWatt of energy from the WTE sector alone [2, 3, 7]. WTE technologies have been considerably modernized in the last years, in 2012, the USA alone generated 14.5 million MWh of electricity from 84 WTE facilities [4].

The general principles of waste management are: the waste prevention to reduce emissions, the reduction of hazardous substances in material streams and the increase resource efficiency, the preparation for reuse that involves checking, cleaning, or recovery by which products or components of products, which have became waste could be prepared for reuse, and, also, the energy recovery from waste and other recovery activities. The energy production from waste (WTE) involves burning waste and using energy content of the waste to produce electricity or to obtain heat and power, heat is used for various services. The disposal of waste in landfills in ecological storage is the least desirable option in waste management hierarchy haveing the most negative effects on environment and human health [8, 9, 10, 21, 25, 26].

Classic technologies for the energetic processing and recovery of recyclable materials from municipal waste are [11, 23]:

IOP Conf. Series: Materials Science and Engineering **789** (2020) 012035 doi:10.1088/1757-899X/789/1/012035

a) Selective collection/sorting and recycling followed by controlled storage of waste in green landfills, which means a sum of technologies for the reuse of materials obtained from waste, to become the raw material for various economic processes. This is characterized by [27]:

- solid wastes are obtained without the energetic contribution and with negative influence on the environment;
- a mixture of combustible gases (CH_4, CO, H_2, C) is obtained with low purity, low calorific value and a high percentage of dangerous accompanying gases [5, 12];
- the technological installations are voluminous, and they require large areas for installation, they are high energy consumption, the duration of the technological process is very high and the volume of fuel gas obtained is small.

b) Thermochemical processes which are characterized by:

- direct combustion - it produces thermal energy, a large volume of ash and inert materials, a high percentage of dangerous gases with a negative impact on the environment;
- gasification - obtaining a gas mixture of H_2, CO, CH_2 with reduced calorific value, a high volume of inert materials;
- pyrolysis/incineration - the thermal process that takes place in the absence of oxygen, produces combustible gases and thermal energy (CH_4, CO, H_2, C obtained from the decomposition of biological material at 700 °C according to the chemical reaction $C_6H_{10}O_2 \rightarrow CH_4 + 2CO + 3H_2O + 3C$ [5].

2. Material and methods

Modern technologies for energy recovery of municipal waste as a renewable energy source are represented by Plasma gasification, which is a V - generation technology through [13]:

- a mixture of combustible gases H_2, CO, CH_2, CH_3, CH_4 (syngas) is produced with high purity and calorific value [14];
- a vitrified mineral waste with a volume between 5-10% with minimal impact on the environment, hardness and properties similar to ceramic materials (use in construction) [15][28];
- the resulted syngas can be used to obtain raw materials in the field of the chemical industry, fuels or directly converted into electricity and heat [16];
- the technological installations are compact, and the technological processes are carried out in short periods of time, while the volume of fuel gas obtained is high [16].

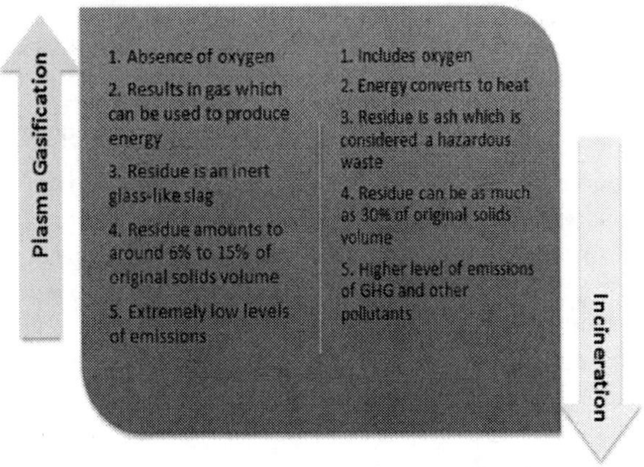

Figure 1. Comparative analysis between plasma gasification and pyrolysis/incineration.

These modern technologies are defined as:

a) Westinghouse plasma technology - plants use plasma torches on gas, equipment is high working temperatures which are about 4,000 °C, electricity consumption is high, processed waste requires humidity reduction, syngas obtained has a high percentage (5-10%) of gases dangerous for the environment with low calorific power, the volume of inert material obtained is high [17, 18].

b) Plasma hydrogen technology obtained by dissociation of water in plasma jet - the installations are compact of small size, with working temperatures between 10,000-20,000 °C, do not require drying of waste, the syngas obtained has a high percentage of combustible gases $H_2 \approx$ 50-55%, $CO \approx$ 20-25%, $CO_2 \approx$ 15%, and accompanying gases are arround 1-5% with minimal environmental impact [17, 18, 19].

The following layout (figure 2) describes the overall configuration and the main components of a new technological developed installation, and the prototype is a result of a scientific research project. The necessary utilities are:

- power grid with an installed power of 210 kW at 380 V, 50 Hz,
- running water with a flow of at least 2 mc/h.

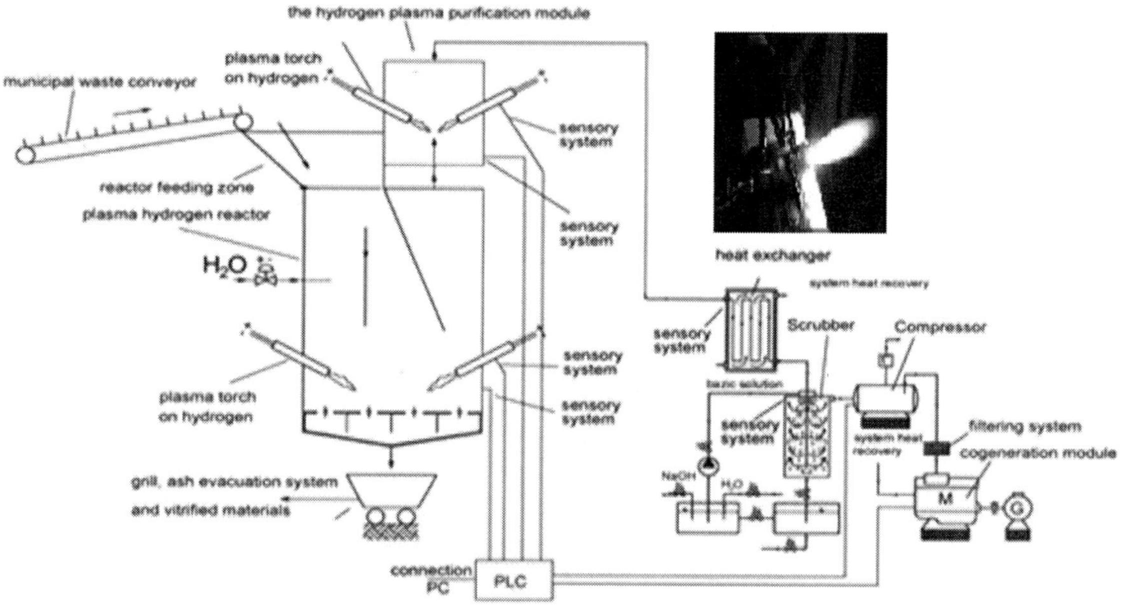

Figure 2. The schematic diagram of equipment for hydrogen plasma conversion of municipal waste.

The main objective of the research was the design of an eco-innovative plasma-based technology for the recovery of waste as a renewable energy source, improvement of the technical and economic performance of the plasma installation for the waste processing by upgrading some components of the system (plasmas, adaptive thermal conduction processes from the plasma reactor enclosure, plasma gas purification system), and optimization of the syngas production technology by controlling the plasma conversion of waste depending on the results obtained from the modeling and simulation of thermo-chemical processes with the help of Chemcad specialized software [20].

The new technological system converts into synthetic gas any solid municipal waste using high-temperature plasma technology and hydrogen technology. Synthesis gas can be used to generate energy, liquid fuels or other sustainable energy sources. It operates in a continuous mode using as a source of energy only to initiate the process of the electricity network, minimizing the energy consumption by means of its own cogeneration system. The heat produced in the process is recovered at various points in the plant and is used to obtain the heating agent needed to supply an external district heating system [20].

Electricity produced through another cogeneration module may be supplied to a local power grid or national system [22].

The designed installation has the following technical features, presented in table 1:

Table 1. Raw - energy balance

Entry	Exit
raw material (organic substance) = 410 kg/h humidity = 35% - caloric power = 13.9 Mj/kg - water as vapor = 65 kg/h	slag (vitrified) = 8.2 kg/h inert material = 141.8 kg/h residual heat of 0.007 MWt
electricity: 0,200 MWe/h of which: - plasma torque = 0.142 MWe/h - water / solid separator = 0.050 MWE/h - organic substance carrier in the reactor = 0.005 MWe/h - measurement and control systems = 0.003 MWe/h	singas = 325 kg/h, which represents 1,533 MWt chemical energy with 10.87 Mj/m^3 calorific value

In table 2 is presented the balance conversion of thermal energy into electricity.

Table 2. Energy balance

Step 1 - energy conversion	Step 2 - energy conversion
The cogeneration system with a *single-turbine and a 34% efficiency generator produces 0.521 MWe/h of which: 0,200 MWe/h is for the operation of plasma torches*, separator, sludge conveyor and control-control system.	The steam turbine receives 0.74 MWt at 90% efficiency, operates an electric generator with a 30% efficiency that produces 0.222 MWe/h. *Thermal energy recovered:* *- in the single wash and neutralization module = 0.231 MWt* *- boiler 1 (steam turbine) = 0.273 MWt of which 0.143 MWt for torches and plasma reactor* *- boiler 2 (heat / steam exchanger recovered from the steam turbine) = 0.443 MWt*
Electricity available to internal consumers or the public grid: *0.521 MWe + 0.222MWe = 0.743 MWe;* of which 0.200 MWe consumed for the operation of the plasma installation and 0.543 MWe for sending in the network. *Total thermal energy recovered:* *0.231 MWt + 0.273 MWt + 0.443 MWt = 0.947 MWt* of which are available for external heating to consumers = 0.804 MWt	

These technical parameters were compared with the parameters resulting from the modeling and simulation of thermo-chemical processes of plasma waste processing with Chemcad resulting in a small variation between them. Figure 3 presents the modeling and simulation of the main reactor with plasma on hydrogen.

The sludge from in the designed installation is largely eliminated, the resulting vitrified material is inert and in minimal quantity, easy to recycle with the possibility of recovering heavy metals.

The installation allows the reduction of greenhouse gas emissions:

- reduction up to 40% CO_2 and 100% CH_4, efficient filtration without toxic gas emissions;
- reduction of the quantity of SO_2, SO_3, and NO_x (acid rain factors);
- reduction of dioxins and furans.

The average operating time of the plasmas is approx. 3,000 h (24 h/day) maintenance requires the replacement of anodes and cathodes from the plasma construction (replacement is a simple assumption to replace compact equipment without removing the reactor). The installation works completely automated, online, from a distance without the intervention of an operator.

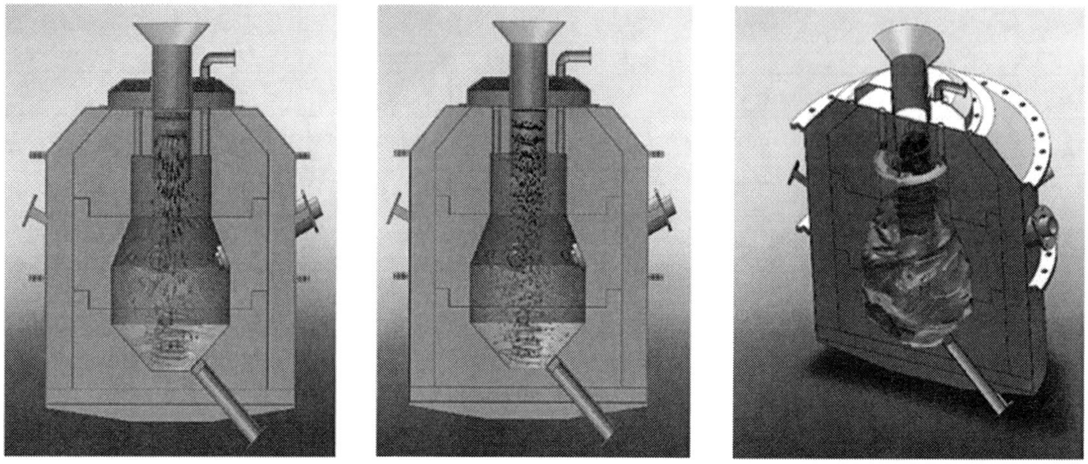

a) Simulation of the hot gas flow in the plasma reactor.

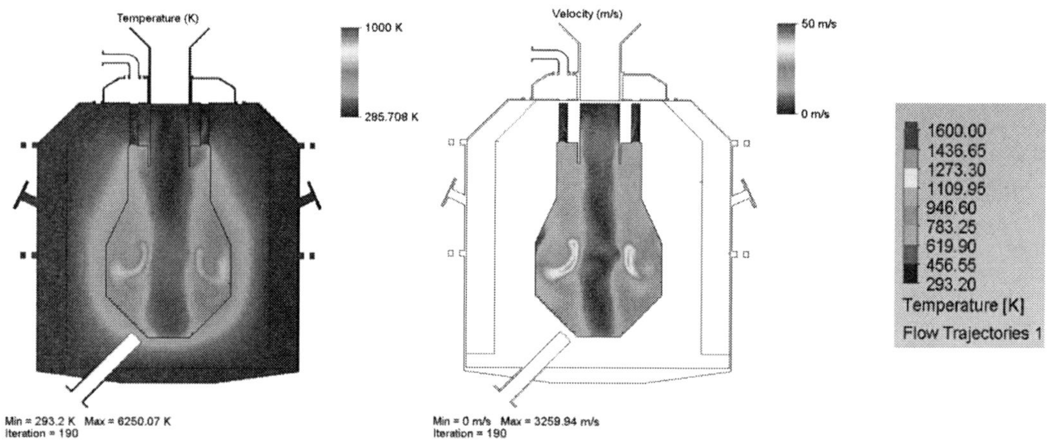

b) Simulation of the temperature in the plasma reactor and the speed of gas displacement.

Figure 3. Modeling and simulation of the main reactor with plasma on hydrogen

3. Conclusions

The goal of the paper was to describe the efficiency of waste conversion processes by using controlled plasma decomposition reactors developed in the research, which generate syngas, electrical and thermal energy that can be used in civil engineering as treatment of a large variety of wastes (municipal solid wastes, construction wastes, heavy oil, used car tires, medical wastes etc).

By using the plasma technologies different environmental problems can be resolved:

- reducing substantially the waste going to landfill sites;
- reducing CO_2 emissions;
- reducing the usage of fossil fuels;
- safe destruction of medical and numerous hazardous waste.

Beside syngas and heat, another important product of the plasma decomposition process is vitrified slag, which potentially can be used in the construction sector.

Acknowledgments
This paper was realized with the support of project EFECON – Eco-innovative products and technologies for energy efficiency in construction, POC/71/1/4 - Knowledge Transfer Partnership, Cod MySMIS: 105524, ID: P_40_295, Project co-financed by the European Regional Development Fund.

References
[1] United Nations 2019 World Population Prospects 2019 *Department of Economic and Social Affairs*
[2] Whiteman A, Esparrago J, Rinke T and Arkhipova I 2016 Renewable Energy Statistics 2016 *International Renewable Energy Agency*
[3] Kumar A and Samadder S R 2007 A review on technological options of waste to energy for effective management of municipal solid waste, *WASTE MANAGE.*, **69**, 407–22
[4] Michaels T 2014 Directory of Waste-To-Energy Facilities *Energy Recovery Council*
[5] Gomez A, Rani E, Amutha D, Cheeseman C R, Deegan D, Wise M, and Boccaccini A R 2009 Thermal plasma technology for the treatment of wastes: A critical review *J. Hazard. Mat.* **161** (2–3) 614-26
[6] Commission Of The European Communities 2006 Action Plan for Energy Efficiency, *EUR-Lex*
[7] Commission Of The European Communities 2006 Renewable Energy Road Map, *EUR-Lex*
[8] Commission Of The European Communities 2006 Sustainable power generation from fossil fuels, *EUR-Lex*
[9] Commission Of The European Communities 2006 Towards a European Strategic Energy Technology Plan, *EUR-Lex*
[10] Commission Of The European Communities 2007 An energy policy for Europe, *EUR-Lex*.
[11] Demirbas A 2005 Potential applications of renewable energy sources, biomass combustion problems in boiler power systems and combustion related environmental issues *PROG. ENERG. COMBUST.* **31** 171–92
[12] Van Loo S and Koppejan J 2008 Handbook of Biomass Combustion and Co-firing *Earthscan*.
[13] Sørensen B 2017 Renewable Energy. Its physics, engineering, use, environmental impacts, economy and planning aspects *Elsevier Academic Press*
[14] Ni M, Leung D Y C, Leung M K H, and Sumathy K 2006 An overview of hydrogen production from biomass *FUEL PROCESS TECHNOL,* **87** 461 – 72
[15] Umberto A 2012 Process and technological aspects of municipal solid waste gasification. A review *WASTE MANAGE.* **32 (4)** 625-39
[16] Dinu M, Socol C, and Marinas M 2005 Mechanisms of convergence and cohesion *Economic Publishing House*
[17] Akishev Y, Aponin G, Balakirev A, Grushin M, Karalnik V, Petryakov A, and Trushkin N 2013 Spatial-temporal development of a plasma sheet in a surface dielectric barrier discharge powered by a step voltage of moderate duration *PLASMA SOURCES SCI. T.*, 22
[18] Fridman A, and Kennedy L 2004 Plasma Physics and Engineering *Taylor & Francis*
[19] Matveev I 2015 Plasma Assisted Combustion, Gasification, And Pollution Control *Outskirts Press*, 2
[20] Matveev I, and Rosocha L 2010 Guest Editorial: Classification of plasma systems for plasma-assisted combustion *IEEE T. PLASMA SCI.* **38 (12)** 3257-264
[21] Matveev I, Matveeva S, and Kirchuk E 2009 Plasma assisted combustion technologies *European Combustion Meeting* 4
[22] Klimov A, Moralev I, Tolkunov I, and Matveev I 2008 Plasma assisted ignition and combustion of hydrocarbon fuel in high-speed airflow by HF streamer discharge, *International Workshop and Exhibition on Plasma Assisted Combustion* **4** 32-3

[23] Kazemi S 2013 Use of recycled plastics in wood plastic composites – A review, *WASTE MANAGE.* **33 (9)** 1898-905
[24] Patel M L, and C J 2012 Plasma Gasification: A Sustainable Solution for the Municipal Solid Waste Management in the State of Madhya Pradesh, India *J ENVIRON SCI,* 3
[25] Tendler M, Rutberg P, Van Oost G 2005 Plasma Based Waste Treatment and Energy Production *PLASMA PHYS. CONTR. F.*, 47
[26] Springer M D, Burns W C, and Barkley T 1996 Apparatus and Method Fortreating Hazardous Waste *United States Patent No 5.534.659*
[27] Leal-Quirós E 2004 Plasma Processing of Municipal Solid Waste *BRAZ. J. PHYS.* **34** 1587-593
[28] Mountouris A, Voutsas E, Tassios D 2008, Plasma Gasification of Sewage Sludge: Process Development and Energy Optimization *ENERG. CONVERS. MANAGE.*, 49

Optimized design of air inlet devices based on environmental analysis of a broiler house model

Y X Ma[1*] and H F Zou[1]

[1] School of Municipal and Environmental Engineering, Shenyang Jianzhu University, Shenyang, China

[*] 1040993118@qq.com

Abstract. A typical broiler house in the cold region of northern China was selected as the research object of this study. The environment of the broiler house was studied in depth by using the Airpak software. Then, an optimized design of air inlet devices for this typical broiler house in the cold region, under the condition of transverse ventilation, was made on basis of two influence factors: the length of the flow guiding device, and the direction of airflow. The optimized air inlet devices helped to improve the airflow in the broiler house, hence changing environmental factors such as internal temperature distribution, wind speed distribution, and carbon dioxide distribution. The length of the ideal flow guiding device should be approximately 1 m and not more than 2 m. Influence by the direction of the air inlet devices was affected by length of the guided airflow. Increase in the direction of the air inlet device led to rise of the body plane temperature of the chickens and the overall wind speed, while carbon dioxide distribution remained basically unchanged. The best direction of airflow (3) should be 30°～45°.

1. Introduction

In the cold regions of northern China, temperature is high in summer, the winter is long and severely cold, while the spring and autumn are short, causing the broiler houses closed design [1-2]. In this study, a typical broiler house in the cold regions of northern China was selected as the research object. The air inlet devices commonly used in broiler houses in northern China were studied and optimized using simulation software.

J A Osorio et al. conducted numerical simulation on a naturally ventilated chicken house in Brazil and obtained the distribution of wind speed inside and outside the chicken house and the distribution of harmful gas concentration.

Brian D. Luck et al. developed a scalable environmental assessment system for wind speed measurements and applied it to a commercial chicken house. Wisate et al. simulated a chicken coop with a size of 120 (length) 14 (width) 2.5 (height) m and equipped with multiple intake air system by CFD technology, analyzed the distribution of ventilation rate, temperature and wind speed of the internal air flow, and compared the results with experimental tests.

The experimental results showed that:

1) equipped with multiple air inlet system, the ventilation rate in the region was increased by 57%;

2) the average wind speed in the house increased by 25%, but the fluctuation value decreased;

3) CFD simulation results are very close to the experimental results, and CFD can completely replace the experimental results within the allowable error range.

In-bok LEE et al. simulated the wind speed field of a naturally ventilated chicken house by using CFD software's RNG model and k-two-path model respectively, and then tested the actual wind speed field of the chicken house. They concluded that the RNG model was the closest to the experimental results, with an error of -6.2%.

Menegali et al. selected broiler chickens as 3-week old and conducted a timed test and analysis on the air quality (carbon dioxide, ammonia and oxygen distribution) of broiler chicken houses under the conditions of minimum ventilation volume in winter, including positive pressure ventilation, natural ventilation and negative pressure ventilation. Research and analysis show that both positive pressure ventilation and negative pressure ventilation provide sufficient oxygen, and ensure that harmful gas, ammonia, carbon monoxide, etc. are within the standard range. However, for these three ventilation forms, the distribution level of ammonia and carbon monoxide in the negative pressure ventilation form is the highest under the minimum ventilation system.

The airflow structure of the broiler house was improved by optimizing the direction and length of airflow in the flow guiding device, thereby improving environmental factors such as temperature distribution, wind speed distribution and carbon dioxide distribution.

Despite numerous differences between the air inlet devices of broiler houses at home and abroad with respect to style, size, installation form and control mode, their functions and principles are basically the same in the following three main aspects:

- providing necessary volume of fresh air for the broiler house which can be controlled and adjusted during different seasons [3-4];
- implementing varying forms of ventilation in the broiler house during different seasons, such as transverse ventilation in winter [4-5]; and
- following the basic principle of submerged jet (referred to as jet) in airflow structure of air inlet device. The difference, however, is that most closed-style broiler houses abroad adopt wall-attached jet, while domestic broiler houses exclusively rely on jet from air inlet devices [6-7].

2. Jet and airflow structure of transverse ventilation

2.1. Jet

Jet refers to the flow of air formed by the injection through a nozzle, orifice, or slit [8]. Figure 1 is a turbulent jet structure diagram.

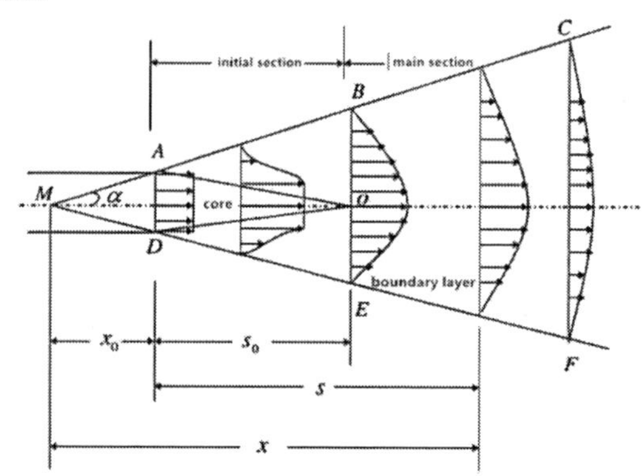

Figure 1. Jet structure diagram.

The following formula [9] can be obtained from the jet theory:

$$\frac{R}{r_0} = \frac{x_0 + s}{x_0} = 3.4a\bar{x} \tag{1}$$

Where: R - jet radius; r_0 –section radius; a - jet diffusion angle; x_0–jet range from the nozzle section; s- turbulence coefficient, taking 0.076 ~ 0.08; $\bar{x}$ - dimensionless distance from the pole, whose expression is as follows:

$$\bar{x} = \frac{x_0 + s}{r_0} \quad (2)$$

Through the jet theory, the range of the air inlet device in the broiler house can be theoretically calculated. Then, a theoretical quantitative basis is provided for the optimization scheme of the air inlet device through theoretical analysis.

2.2. Airflow structure of transverse ventilation

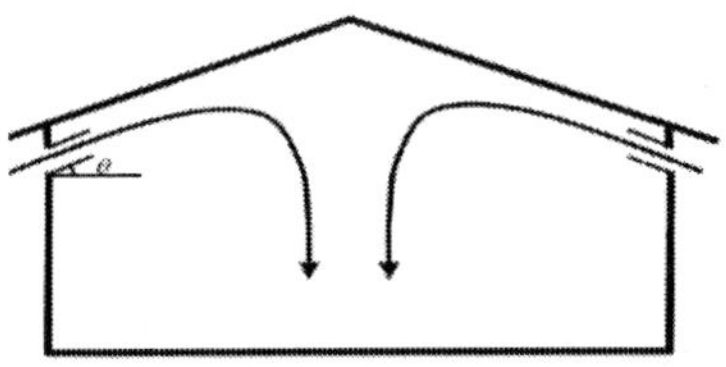

Figure 2. Schematic diagram of airflow in X cross section of transverse ventilation.

The low-temperature fresh air outside enters the broiler house through the air inlet device, and continues to rise while flowing transversely under influence of the jet. Then, after reaching the highest point, the air sinks and reaches the body plane of the chickens.

An abroad broiler house adopts a rectangular vent, which is located close to the ceiling. Therefore, the fresh air enters through the air inlet and forms a wall-attached jet, before sinking at the middle of the ceiling. Meanwhile, due to consideration of costs of the air inlet device, a domestic broiler house in cold region adopts the method of small vent with high wind speed, so that, in the absence of wall-attached jet, the airflow can reach the middle of the house before sinking [10, 11].

The house air inlet device is shown below:

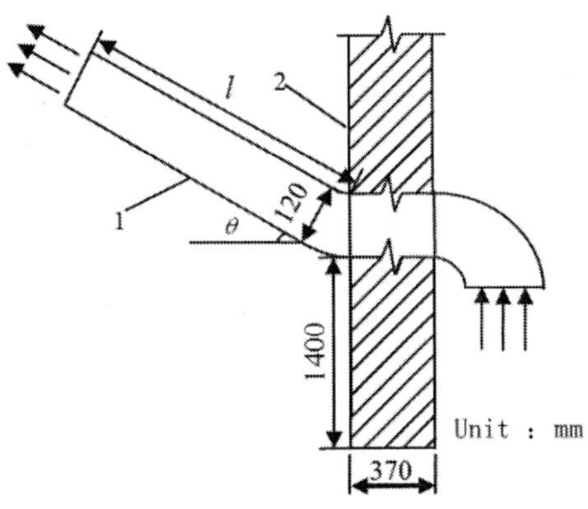

Figure 3. Air inlet device diagram of the broiler house.

Where l is the length of the flow guiding device, and θ is the angle between the flow guiding device and the transverse plane, i.e. direction of the airflow.

Regardless of interaction between multiple air inlet devices (including issues regarding installation spacing, single- or double-sided installation, etc.), as to the air inlet device itself, the length of the flow guiding device and the direction of the airflow are two major factors that influence the environment inside the broiler house. These factors may cause the incoming air to travel a longer distance before reaching the body plane of the chickens, or accelerate the airflow inside the broiler house, thereby increasing the heat exchange between the low-temperature fresh air and the internal air and saving energy while avoiding direct blow of cold air at the chickens. Altering the conditions of the air inlet device can influence the temperature, whilst changing the wind speed simultaneously. Therefore, the pros and cons should be considered when adopting transverse ventilation in winter [12, 13]. This study, through studying of the air inlet device, intends to find a reasonable length of the device and the direction of the airflow, which provides a theoretical basis for the standardization of the air inlet device.

3. Research on air inlet guiding device

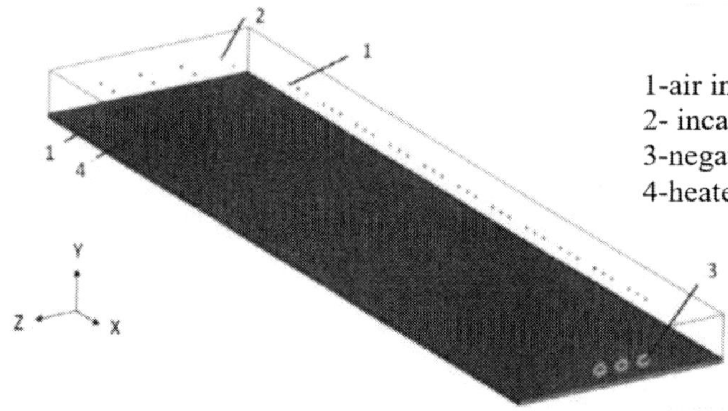

Figure 4. Model of the broiler house.

The theoretical model of the broiler house in this study substantially adopted relevant data and conditions of the tested broiler house. In order to simplify the model, certain parameters were approximately replaced in this study, on condition that such replacement had basically no influence on results of the simulation. A total of 11,307 broilers were farmed inside the broiler house, which was a closed space whose size was 112 m (X) x 13 m (Z) x 2.95 m (Y). The broiler house model is as shown in figure 4.

3.1. Study on the guide length of the inlet device

During winter, under the condition of transverse ventilation, flow guiding devices are commonly installed on basis of the air inlets in order to promote airflow. Thereby, the temperature distribution, wind speed and carbon dioxide distribution in the broiler house tend to become more homogeneous as a whole [14].

For the broiler house with a span length of 13 m, the height of the air inlet device is 1.4 m, and the guide direction is 30°C. Under these conditions, the guide lengths were set at 0 m, 1 m, and 2 m respectively. The effect of the length of the guiding device on the temperature field, wind speed and carbon dioxide distribution of the chicken body plane was studied.

3.1.1. Simulated analysis of temperature distribution in the broiler house with different guide lengths. As shown in table 1, the temperature distribution trends under different guide lengths are basically the same. At the initial section (X = 0 m to X = 20 m), the temperature was relatively low due to effect by the intrusion and infiltration of cold air. At the middle section (X = 20 m to X = 95 m), the temperature in the area close to the heater was relatively high, while the temperature in the area away from the heater was relatively low, due to effect by the air inlet device and the location of the heater. At the end section (X = 95 m to X = 112 m), the temperature was relatively homogeneous, which was close to the average temperature of the chicken body plane, due to the mixed airflow. When in the absence of flow guiding device, i.e. l = 0 m, the average temperature of the chicken body plane was 23.7°C. When l = 1 m, the average temperature of the chicken body plane was 25.1°C. When l = 2 m, the average temperature of the chicken body plane was 25°C. When in the absence of flow guiding device, the cold airflow entering the broiler house directly mixed with the internal air and sank to the chicken body plane. Because the flow guiding device increased the distance from the cold airflow to the plane of the chicken body while reducing the pressure on the airflow by the heater, the heat exchange process was strengthened. If the airflow guiding device was excessively long, though the convective heat transfer process allowing the airflow to enter the house would be strengthened, the mixing process of cold and hot airflows, which was a more direct heat transfer process, would be shortened.

Table 1. Temperature distribution statistics (°C) of the broiler house with different guide length (*l*) (Section Y = 0.25 m).

X	*l*	Z = 0 m	3 m	5 m	7 m	9 m	11 m	13 m
Initial segment 0~20 m	0 m	22	21	23	23	24	25	26
	1 m	23	23	24	27	26	26	26
	2 m	23	21	24	27	26	25	26
Middle section 20~95 m	0 m	22	22	23	25	26	28	31
	1 m	27	23	25	28	29	28	25
	2 m	27	23	25	28	29	28	25
End section 95~112 m	0 m	23	22	24	24	24	26	27
	1 m	27	23	26	27	27	27	26
	2 m	27	23	27	28	28	28	27

As can be seen by comparing the data in the table, in terms of temperature, the effect of a flow guiding device was superior to that without a flow guiding device. An appropriate guide length ensured the temperature inside the boiler house while maintaining the homogeneity of temperature of the chicken body plane. If the guide length was too short, temperature of the chicken body plane would be too low. If the guide length was too long, though the average temperature of the chicken body plane would be basically unchanged, temperature in certain low-temperature areas would become even lower, while temperature in certain high-temperature areas would become even higher, thus increasing the construction costs.

According to the requirements of temperature control in the broiler house during winter, and due to the above analysis, in terms of temperature, the optimal length of flow guiding device for a typical broiler house in cold regions should be approximately 1 m.

3.1.2. Simulated analysis of wind speed distribution in the broiler house with different guide lengths.

Table 2. Statistical table of wind speed distribution (*m/s*) of the broiler house with different guide lengths (*l*) (section Y = 0.25 m).

X	*l*	Z = 0 m	Z = 3 m	5 m	7 m	9 m	11 m	13 m
Initial segment 0~20 m	0 m	0.05	0.25	0.1	0.15	0.15	0.25	0.45
	1 m	0.1	0.3	0.15	0.2	0.25	0.35	0.5
	2 m	0.1	0.3	0.15	0.2	0.3	0.4	0.5
Middle section 20~95 m	0 m	0.05	0.25	0.15	0.2	0.25	0.3	0.45
	1 m	0.1	0.35	0.25	0.3	0.3	0.4	0.5
	2 m	0.1	0.3	0.2	0.3	0.4	0.5	0.6
End section 95~11 2m	0 m	0.2	0.25	0.25	0.25	0.25	0.3	0.45
	1 m	0.15	0.3	0.3	0.35	0.35	0.4	0.5
	2m	0.15	0.35	0.35	0.35	0.4	0.45	0.5

As shown in table 2, with increase in the length of the flow guiding device, the wind speed in the broiler house was enhanced. Ventilation conditions in winter required that the wind speed to be less than 0.3 m/s, which was different from the requirement in summer.

When the lengths of the flow guiding device were between 0 m and 1 m, respectively, the wind speed of the chicken body plane basically met this requirement. However, when the length was 2 m, as could be known from the simulation process, the wind speed in the area of the broiler house from when X = 65 m to X = 112 m exceeded the requirement standard, and the areas of original wind speed exceeding the standard increased when closer to the heater. When l = 0~1 m, the wind speed requirements of the broiler house was met. Therefore, when equipping a typical broiler house with a flow guiding device, its length should be less than 2 m.

3.1.3. Simulated analysis of carbon dioxide distribution in the broiler house with different guide lengths.

Table 3. Level of carbon dioxide of chicken body in the broiler house with different guide lengths (l) (section Y = 0.257 m).

Guide Length	l = 0 m	l = 1 m	l = 2 m
Carbon dioxide average (mg/m3)	5008	4910	4910
Carbon dioxide maximum (mg/m3)	7145	6750	6866
Carbon dioxide minimum (mg/m3)	3301	3539	3673

As shown in table 3, when in the absence of flow guiding device, the mean carbon dioxide of the chicken body plane was 5008 mg/ m^3, with a maximum value of 7145 mg/m^3. When a flow guiding device had been installed, the mean carbon dioxide value of the chicken body plane was 4910 mg/ m^3, and the requisite fresh air in the broiler house came from the air inlet devices.

In terms of ventilation, the addition of flow guiding device, in essence, altered the location of the air inlet. Hence, influence by the span length of the broiler house was reduced, and the location of the air inlet was heightened, thereby improving the airflow and the heat exchange in the airflow. As can be seen from table 5, the addition of the flow guiding device helped improve the homogeneity of carbon dioxide distribution on the chicken body plane.

However, it did not reduce the overall carbon dioxide value of the chicken body plane. In this study, the reasonable length of the flow guiding device is about 1 *m*.

3.2. Study on the direction of airflow

The direction of airflow is another important parameter of the flow guiding device. Essentially, on the one hand, it influences the height of the air inlet jointly with the flow guiding device; on the other hand, it alters the path of the incoming cold airflow after it passes through the air inlet [15].

For small vents, the effect of the direction of airflow is affected by the guide length of the airflow guiding device. The longer the guide length, the more prominent the effect of the direction of airflow. For instance, when the guide length equals 0, the direction of airflow virtually no effect on the temperature field, wind speed and carbon dioxide distribution of the chicken body, so long as the value of direction is $0° \leq \theta \leq 90°$.

While considering the height limit of a typical broiler house, in order to better study the influence of the direction of airflow on the temperature field, wind speed and carbon dioxide distribution of the chicken body, this study performed simulation processes based on three direction values, i.e. $\theta = 30°$, 45°, 60° respectively, on condition that l = 1 *m*. The simulation process was consistent with the above studies on the height of the inlet and the guide length.

The simulation data and results were as follows:

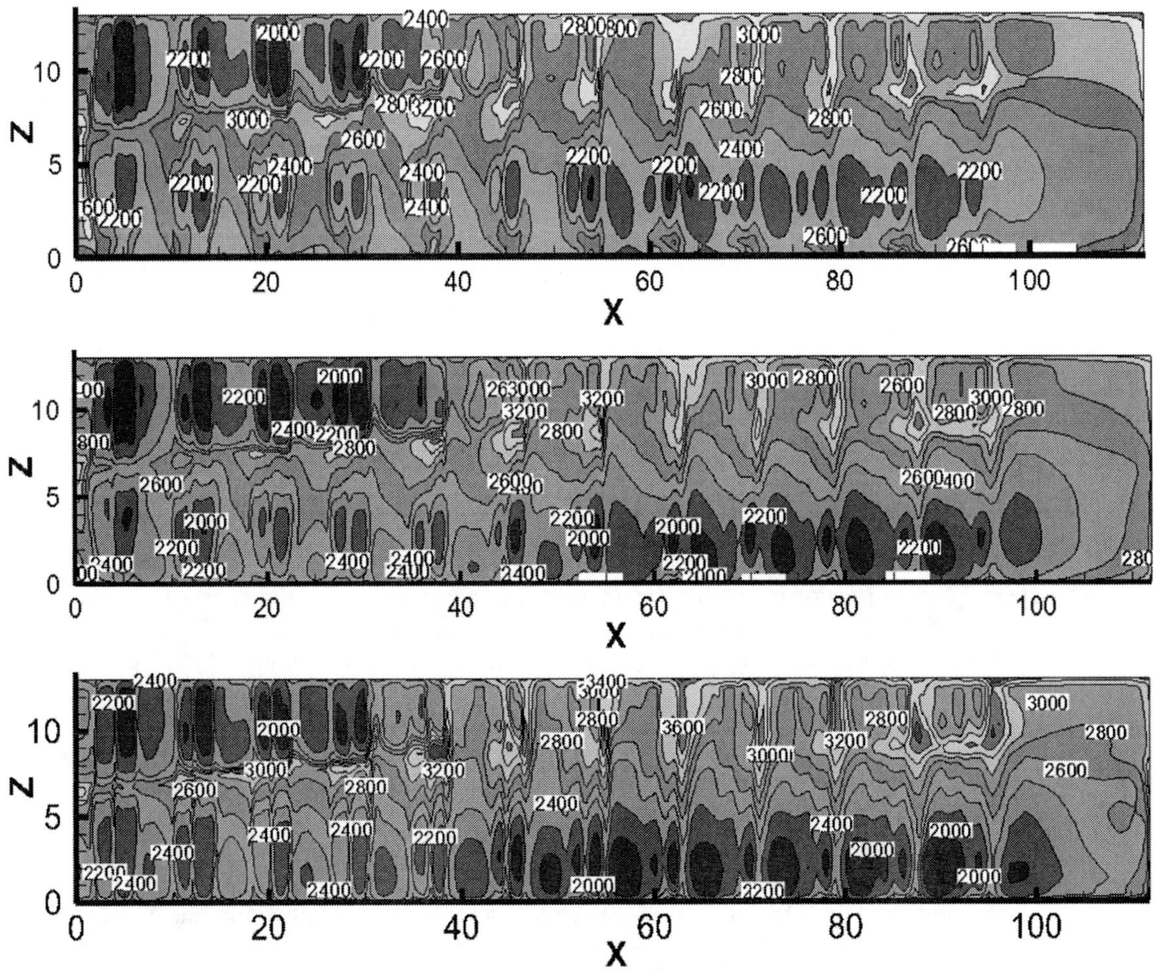

Figure 5. Carbon dioxide distribution map of Y = 0.25 section.

Table 4. Temperature distribution statistics table (°C) of the broiler house with different direction of airflow (section Y = 0.25 m).

θ (°)		Z = 0 m	3 m	5 m	7 m	9 m	11 m	13 m
Initial section 0~20 m	30	23	24	25	26	27	26	24
	45	23	24	25	27	27	26	24
	60	23	24	26	28.5	28.5	27	24
Middle section 20~95 m	30	26	24	26	28	28	27	25
	45	24	25	26.5	28	28	27	25
	60	25	26	27.5	28	28	28	27
End 95~112 m	30	26	25	26	27	27	26.5	26
	45	25	25	26	27	27	26.5	26
	60	28	26.5	27	28	28	28	27

Table 5. Wind speed distribution statistics table (m/s) for the broiler house with different direction of airflow (Y = 0.25 m).

θ (°)		Z = 0 m	3 m	5 m	7 m	9 m	11 m	13 m
Initial segment 0~20 m	30	0.15	0.25	0.15	0.2	0.25	0.35	0.4
	45	0.15	0.2	0.15	0.2	0.25	0.4	0.5
	60	0.2	0.35	0.2	0.2	0.25	0.4	0.5
Middle section 20~95 m	30	0.2	0.2	0.2	0.3	0.3	0.4	0.5
	45	0.25	0.25	0.25	0.3	0.3	0.4	0.5
	60	0.2	0.35	0.3	0.3	0.35	0.4	0.6
End 95~112 m	30	0.15	0.25	0.3	0.35	0.35	0.35	0.4
	45	0.15	0.25	0.3	0.35	0.35	0.35	0.4
	60	0.2	0.35	0.35	0.35	0.4	0.4	0.4

The simulation results showed that:

- increase of the direction of airflow within a certain range (30° to 45°) did not result in significant change in the temperature of the chicken body plane, and the temperature basically reached the standard. However, increase in the direction of airflow exceeding 45° would cause the temperature of the chicken body plane to rise with the increase of the direction of airflow, and the temperature of most areas of the broiler house exceeded the standard. By respectively studying the temperature of the chicken body plane in three direction values of 30°, 45° and 60°, the optimal direction of airflow in this study should be between 30° and 45° in terms of temperature.
- When the direction of airflow (θ) equaled 30° and 45° respectively, the wind speed of the chicken body plane basically met the requirements; high wind speed mainly concentrated on the side of the heater and the fan towards the end; while when θ equaled 60°, the wind speed in certain areas of the chicken body plane in the broiler house exceeded the standard and failed to meet the ventilation requirements, which was undesirable.
- As the direction of airflow increased, the carbon dioxide distribution remained basically unchanged. The distribution in certain low carbon dioxide regions was lower, and that of certain high carbon dioxide regions was higher. According to the relevant standards of ventilation, when the angle was 30°~45°, the carbon dioxide values basically met the requirements; when the direction of airflow was 60°, the carbon dioxide value of certain regions exceeded the standard. Therefore, according to carbon dioxide distribution, the optimal angle should be 30° ~ 45°.

4. Conclusion

In this study, Airpak software was used to study the influence of the air inlet device on the environment of a typical broiler house in cold region under the condition of transverse ventilation, including the guide length and the direction of airflow. Conclusions are as follows:

- increase in the guide length of the air inlet device helped to enhance the mean temperature of the chicken body plane and the airflow in the broiler house, thereby making the carbon dioxide of the chicken body plane more homogeneous. When the length of the flow guiding device was 0~1 m, the effect of the flow guiding device was particularly obvious; when the length of the guiding flow was 1~2 m, the rise in temperature was not obvious, and the wind speed in certain regions might exceed the standard. Therefore, the optimal length of the flow guiding device should be about 1 m and not more than 2 m.
- The effect of the direction of airflow of the air inlet device was affected by the length of the airflow. Increase in the direction of airflow caused the chicken body plane temperature and the

overall wind speed to rise, while the carbon dioxide distribution remained basically unchanged. When the direction of airflow was 30° ~ 45°, the rise in temperature was not obvious, while the rise in wind speed was more obvious. When the direction of airflow was 45° ~ 60°, the rises in temperature and wind speed were both obvious. However, as to this study, when the direction of airflow was 60° ($\theta = 60°$), the overall temperature exceeded the standard and the wind speed in certain regions also exceeded the standard. Therefore, the optimal flow direction in this study (θ) should be 30° ~ 45°.

- This study performed more detailed research on the air inlet device based on relevant existing studies. However, there are still some shortcomings and defects: firstly, the influence between the air inlet devices was not considered, such as the spacing of the air inlet devices and the staggered arrangement of the air inlet devices on both sides. Secondly, the study on the direction of airflow and the guide length of the air guiding device can be further refined. For instance, the control group in the study on direction of airflow selected θ as 30°, 45° and 60° respectively. On basis of this study, a further comparative study can be performed when selecting θ as 30°, 35°, 40° and 45° respectively.

The air inlet device in a typical domestic broiler house is different from the common rectangular air inlet. This study on the air inlet device through CFD software has certain reference values for the design and construction of typical broiler houses in cold regions, and for the standardization of broiler houses.

References

[1] Huichun Z 2017 *J. Chinese Journal of Animal Husbandry and Veterinary Medicine* p 94
[2] Jinguo N, Lizhen S, Binlin R and Binbin W 2016 *J. China National Poultry* pp 43-46
[3] Xuanjiang Y 2015 *J. Study on Environmental Parameters and Correlation of Summer and Winter in An Eight-Story Caged Egg House D* p 136
[4] Kejia Z 2014 *D. Comprehensive Evaluation of Environmental Parameter Control in Modern Super-Large-Scale Laying Hens in Winter and Spring* p 66
[5] X.-J.n L 2011 *J. Transactions of the ASABE* May-June pp 1059-1068
[6] Casey K D, Gates R S, Wheeler E F, Hongwei X, Zajaczkowski J L, Topper P A and Yi L 2004 *C. Proceedings of the Air and Waste Management Association's Annual Meeting and Exhibition, Proceedings of the A and WMA's 97th Annual Conference and Exhibition* pp 2095-2102
[7] Bustamante E, Calvet S, Fernando E, Torres, Antonio G and Hospitaler A 2017 J. *Biosystems Engineering* pp 55-68
[8] Henian L, Jing L and Fluid M 2016 *M. China Construction Industry Press*
[9] Xueping G and Higher F M 2009 *M. Tianjin University Press*
[10] Van C, Doan Z, Milan K and Pavel 2017 *C. Engineering for Rural Development, 16th International Scientific Conference Engineering for Rural Development, Proceedings* pp 461-466
[11] Purswell J L, Luck B D, Davis and Jeremiah D 2014 *J. Applied Engineering in Agriculture* pp 471-475
[12] Yu Z, Hui C, Yongzhi B, Chong W and Renlu H 2017 *J. China Poultry* pp 41-45
[13] Haidong S 2011 *J. Poultry Science* pp 23-25
[14] Xiaoming J, Yanlin Z, Xingguang Z, Li H, Fanghuan L, Jing W, Hui L and Min L 2017 *J. Journal of Huazhong Agricultural University* pp 113-120
[15] Blanes-Vidal G V, Balasch E and Torres A 2008 *J. Biosystems Engineering* pp 105-116

Catastrophic landslide stabilization on a restricted area through a combined solution in Bran, Brasov County

M Măntulescu[1] and I Tuns[1*]

[1] Transilvania University of Brasov, Faculty of Civil Engineering, Brasov, Romania

[*] ioan.tuns@unitbv.ro

Abstract. The paper presents a combined method for the rehabilitation of a road on a restricted area by reinforced soil above a double sheet of piles. The principle is illustrated by a case study of a landslide in Bran, Brasov county. Due to the restriction imposed by the limit of properties we have to find out the way to stabilize a small part of the landslide and in the mean time to let the efforts induced by the rest of the sliding mass to act. To determine the status of efforts the problem on a restrained area without stabilizing the rest of the landslide which is much bigger we have considered that there are lateral surcharges due to the movement of the sliding mass.

1. Introduction

This article was brought about by a landslide that has affected properties, utilities and communications in the village of Bran, Brasov county. The main problem is that because of the huge dimensions of the phenomenon that passes through many more private properties it was not possible to stop the causes, rather to limit some effects that concern the whole community.

Landslides has affected the street Bradul Inalt from Sohodol district on the length of 50 meters, with moving a compartment (the northern one) by about 6 meters horizontally and 3-4 meters vertically (figure 1).

1.1. Technical premises

From a technical point of view the problem becomes more complicated than other similar situations where the solution is probably routine, because of the restrictions imposed by the legal limitations. As a result, the landslide cannot be stabilized by eliminating the causes.

They are for the most part outside the public domain where the street is and for which the government funds were allocated.

Thereby only the effects were covered, but keeping in mind that the phenomenon is dynamic and ongoing, efforts required by the next ground movements were taken in consideration.

The area affected by movements is very large, so the field investigations did not cover the entire surface. The landslide is regressive.

The major triggering cause is the erosive action of the downstream creek, located about 500 m away. The destabilization of the foot of the slope has brought ground movements, the surface has become irregular, and cracks appeared, favoring water stagnation and water infiltration in terrain.

The highly varied lithology in which the cohesive material alternates with the non-cohesive one (with high permeability) has permitted groundwater circulation, creating areas of minimal resistance.

(a)

(b) (c)

Figure 1. (a) Local effect of the landslide; (b) A building destroyed by the landslide; (c) The same building before the landslide.

1.2. Existing data

On the left side of the street (north) there was a sewerage system beneath the concrete ribbon and on the left side of the street there was at that time a gas pipe of 120 mm diameter.

The slip surface with a possible circular shape in the plane has a maximum arrow of about 5 meters from the right edge of the road, endangering the safe operation of the upstream tennis court (about 2 meters distance) and the construction in the adjacent area.

The slip has affected the communal road on a length of about 50 meters owned by the Bran City Hall, as well as two buildings, privately owned, downstream. In fact, the sliding is much broader, having a length of over 500 meters, over a width of 50-100 meters in the north-north-eastern direction of the road to the affected area (figure 2).

The instability phenomena were reported at the level of the road and neighboring constructions from 2013-2014 and accentuated in the summer of 2016 when the street became impracticable.

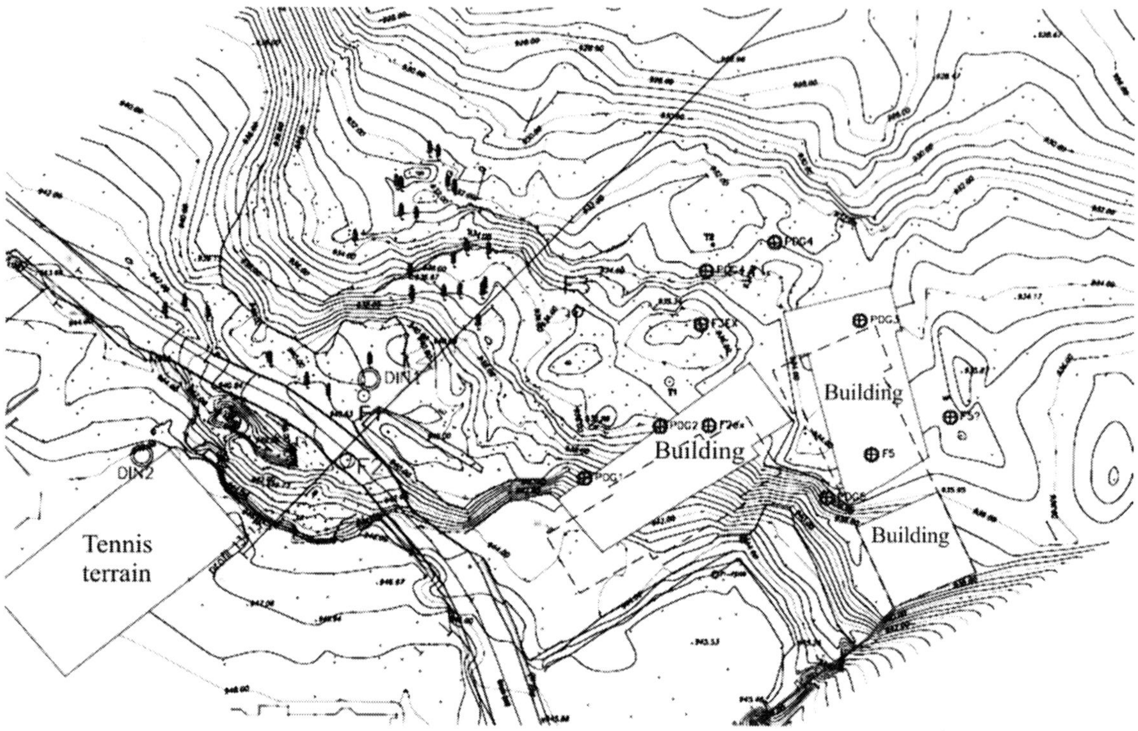

Figure 2. Detail of the landslide plan (contours are at every 0.5 m).

1.3. Geological-geomorphological contexts

From a geographic point of view, Bran is located in the depression headway Bran-Rucar, which is highest in the Giuvala Pass from Fundata commune (1235 m), a depression corridor guarded on the south-eastern side of the Bucegi-Leaota Mountains (2504 m) and the Piatra Craiului Mountains to the west.

The studied area belongs geologically to the Oriental Carpathians, and geographically in the Rucar-Bran corridor, belonging to the Meridional Carpathians. This depression corridor corresponds to a vast tectonic depression with the north-south direction situated between the Bucegi mountains in the east and the Fagaras massif in the west.

The Bran Branch, to which the studied area belongs, is an intermountain depression in which "Sohodol Piedmont" is one of the high piedmonts. The site itself is located above the Turcu valley, where deposits from the alteration of the base rock, low-cemented sandstones, is developing. The "Bradul Inalt" hill area is part of an intermountain basin of the Oriental Carpathians, constituted as a patch of Miocene sedimentary formations built from sandy clays over which a dacitic tuff follows, and then a gray quartz sandstone.

Surface deposits are generally made of cohesive materials and various sands with gravel and boulders.

2. Geotechnical data

Soil investigations was made by drilling with continuous coring and collecting undisturbed samples and also by dynamic penetrations.

According to the groundwater level also, the slip surfaces are situated at 6 m, and 9 m depth from the ground level at that time (figure 3 a).

The results of the investigations revealed a complex lithology, in which cohesive soil alternates with non-cohesive material (figure 3 b).

Here are the results of the most significant boreholes:

Borehole - F1:

- 0.00 – 0.60 m fill (gravel, silt, topsoil);
- 0.60 m – 2.50 m sandy clay with gravel, yellow-grey with low plasticity and consistency, wet;
- 2.50 m – 5.50 m sandy clay, yellowish, plastic consistent;
- 5.50 m – 8.20 m clayey sand yellowish with lens of reddish sand and rare gravel, wet;
- 8.20 m – 8.60 m micaceous sand, yellowish-brown, saturated;
- 8.60 m – 9.00 m clayey silt with gravel yellowish-brown, plastic stiff with low moisture;
- 9.00 m – 10.30 m clayey sand, grey with gravel, compact;
- 10.30 m – 10.60 m silty sand, brown, with gravel;
- 10.60 m – 11.10 m gravel with sand light brown.

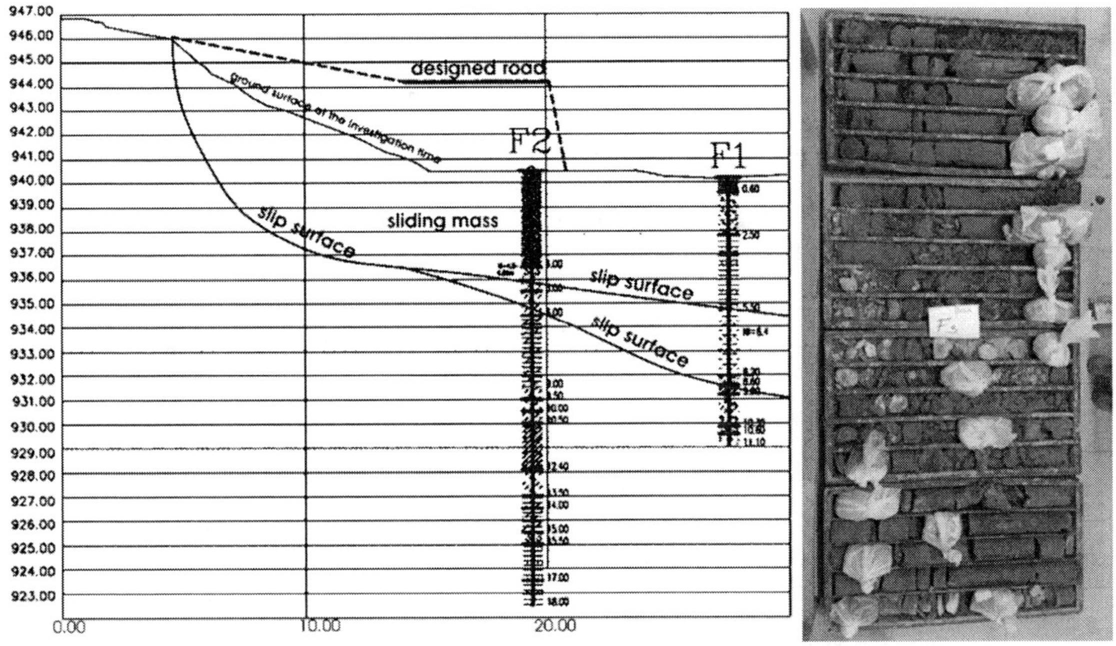

(a) (b)

Figure 3. (a) Simplified landslide profile; (b) Samples from borehole F3.

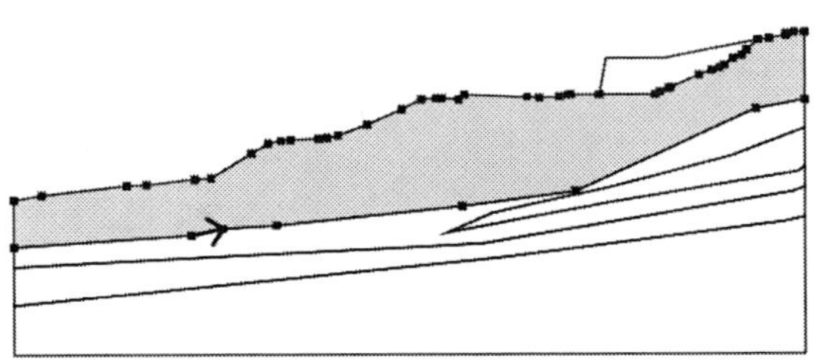

Figure 4. Meshing profile with highlighted sliding mass (with GEO5).

Slope stability (figure 4) was determined with Geo 5 software by analytical methods. Presuming only that the road will be restored, the safety factor was inacceptable (less than one).

On the next step it was calculated adding a double sheet of piles; the piles dimensions were adjusted until it has been obtained a safety factor over 1.5.

The length of the piles was calculated on the ground pressure plus traffic surcharge, without counting the passive resistance of the soil below the road.

This lack of resistance was counted because the sliding phenomenon is much larger and it is possible to go on in the future. The system with a double sheet of piles was conceived according to [1], and over it a blanket made by reinforced soil.

3. Conclusions

General stability analysis (figure 5) assuming that the slip surface has a multi-planar shape. The analysis (made with Geo 5 and Geo-Slope software) on the existing situation calculated a safety factor equal to $Fs^{static} = 1.02$.

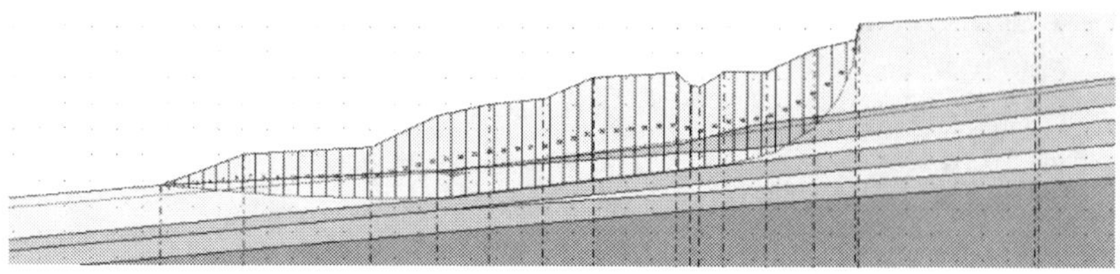

Figure 5. Analysis on the existing situation.

So the condition of verification, even in analytical way, under static conditions is not satisfactory and it is required to ensure the overall stability of the slope. After re-profiling the slope to correct the existing road, the safety factor decreases. It is necessary to carry out retaining works down the road at the limit of the existing platform in order not to require additional expropriations. Thus, a retaining work was provided from pilots drilled down the road on two rows of pilots to reach the safety factor Fs = min. 1.3 (figure 6). The re-profiling of the downhill slope will be made with a slope of 1:6 and upstream of the road will be re-profiled with a slope of 1:3. In this case the safety factor will increase up to $Fs^{static} = 1.39$.

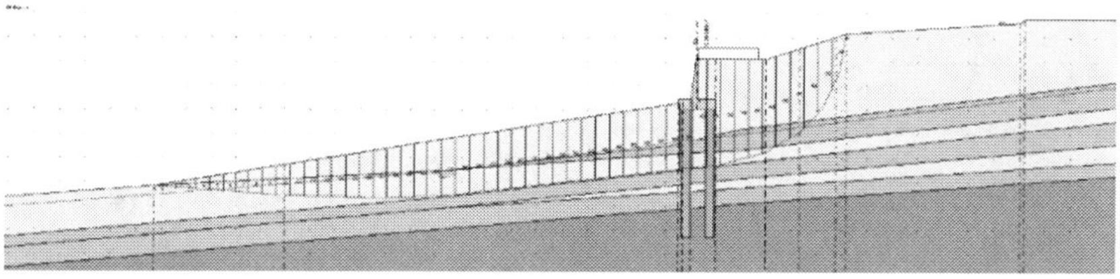

Figure 6. Analysis on the existing situation with a retaining structure.

Analyzing the general stability of the versant in seismic conditions according to SR EN 1998-125/2004 for approach 1, group 2, were applied the following partial coefficients: $\gamma_\varphi = 1.25$ (partial coefficient for the internal friction angle of the material) and $\gamma_c = 1.25$ (partial coefficient of the cohesion of the material); as a result the safety factor was $fs_{seism} = 0.72$, much less than the acceptable $F_s^{adm} > 1.0$. General stability analysis by cylindrical - circular slip surfaces (figure 7 and figure 8) has led us to lower values of the safety factor $F_s^{seism} = 0.66 < F_s^{adm} = 1.0$.

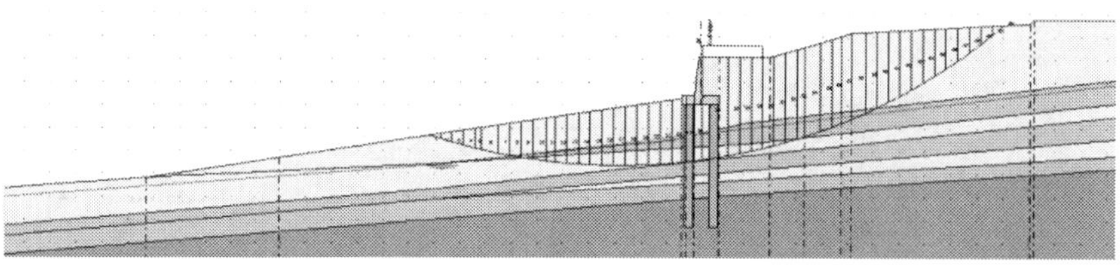

Figure 7. General stability analysis by cylindrical - circular slip surfaces.

Figure 8. Analysis of downstream stability of support after re-profiling assuming a cylindrical - circular slip surface.

The slope is not stable downstream of the retaining structure. To ensure the stability of the slope upstream of the supporting work it must be able to take the thrust of E = 1150 KN. The main solution was to increase the pile's dimensions.

In these reasons, to accommodate additional loads that were induced to execution by downstream compaction, there were provided two rows of piles embedded in the base rock layer with an average length of 18.00 meters in the chess board with 1.50 m diameter, at a distance of at least 3.50 cm apart and 2.0 m between strings (figure 9).

(a) (b) (c)

Figure 9. Aspects during and after the executions: (a) checking of pile's continuity with down-hole method; (b) introducing steel reinforcement cage of the piles (c) beam's reinforcement with anchors casing; (d) panoramic view of finished structure; (e) rehabilitated road segment.

The so described position of the piles is also in agreement with the [2] report's conclusions. The reinforced soil wad calculated according to [3]. The geogrill is 4 meters large, enough to act like a support structure (figure 10).

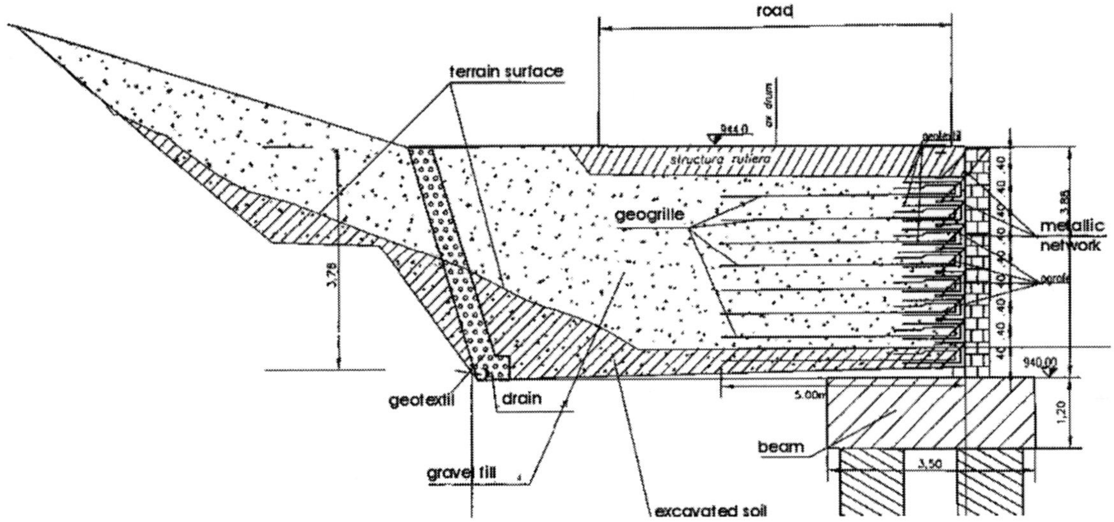

Figure 10. Typical profile through stabilization's structure.

The continuity of the piles was checked by with down-hole method, according to [4].

The forces that act on the slope have been limited to a narrow area for now. In any case, the sheet of piles has a positive effect in many ways, in addition to the stability of the road. Thus, according to [5] the vibrations transmitted by the traffic will be attenuated and it will act as a stiff wave barrier. The efficacy is determined by the depth and the stiffness contrast with soil.

Assuming the results described in this study, traffic-induced vibrations will be significantly attenuated over a distance of at least 50 meters; adding to it the effort's shadow behind the piles as is mentioned in [6] and the restoring of the slope geometry over that distance by uniforming the surface, results in an increase in stability in the significant area. However, these parameters have not been included in the calculation of the safety factor, they are implicitly found in the supra-unitary fraction.

Because of the unpredictability of the events, a monitoring program was set up to raise alerts of potential displacements that go over the reasonable limits. To mitigate unfavorable developments, additional solutions have been provided that can be added to the existing ones (anchors, horizontal drains). A monitoring system was also provided according to a schedule for the next two years. Topographic benchmarks were mounted on the beam of the piles and measurements were carried out which up to this date did not indicate any movement; however, we consider it too early to be able to draw conclusions on the behavior of the structure.

For this reason, casing anchors included in the pile's beam were included. Depending on the results during monitoring time, anchors will be installed.

However, the sliding phenomenon remains active downstream, which requires that steps be taken in the future to stabilize the entire slope by completing a second stabilization step. This will be achieved by restoring the slope geometry, squeezing the sliding mass and possibly introducing other support systems at a later stage of the present. Without these measures, in time, the measures envisaged in this study could also be affected.

References

[1] NP 123-2010 *Normativ privind proiectarea geotehnică a fundațiilor pe piloți (Normative on geotechnical design of foundations on piles)*

[2] Poulos H G 2011 Design of reinforcing piles to increase slope stability *Canadian Geotechnical Journal*

[3] GP 093-06 *Ghid proiectarea structurilor de pământ armat cu materiale geosintetice și metalice (Guide to the design of reinforced soil structures with geosynthetic and metallic materials)*

[4] NP 137-2014 *Normativ pentru evaluarea in situ a rezistentei betonului din construcțiile existente (Normative for the assessment of in situ resistance of concrete in existing constructions)*

[5] Dijckmans A, Ekblad A, Smekal A, Degrande G and Lombaert G 2016 Efficacy of a sheet pile wall as a wave barrier for railway induced ground vibration *Soil Dynamics and Earthquake Engineering*

[6] Hai L, Lusu N, Muhannad T S and Raich A 2015 Interaction between Laterally Loaded Pile and Surrounding Soil *Journal of Geotechnical and Geoenvironmental Engineering*

IOP Conf. Series: Materials Science and Engineering **789** (2020) 012038 doi:10.1088/1757-899X/789/1/012038

Friendly environment solution for a suffusion case

M Măntulescu[1*]

[1] Transilvania University of Brasov, Faculty of Civil Engineering, Brasov, Romania

[*] mmantulescu@unitbv.ro

Abstract. The paper describes about solving the phenomenon of suffusion in the foundation ground of a building. The soil was investigated with direct and indirect methods (drilling and dynamic penetration, electric tomography, seismic refraction). The filling of the cavity was made with a low permeability or practically impermeable material, homogeneous in granularity, with approximately the same stiffness as the surrounding soil.

1. Introduction

Suffusion phenomena represent a major risk for engineering construction. The frequency with which they affect the constructions is relatively low compared to other geotechnical risks (earthquakes, landslides etc.). If the phenomenon of suffusion in artificial dikes is not taken into account, where they represent the main danger, the appearance of suffusion crater occurs mainly in the urban center causing emotions. According to the publication, they are more common, which indicates that the causes of the triggering are mostly anthropogenic [1]. Probably for this reason these phenomena are not explicitly included in the hazard area studies. Consequently, addressing the phenomenon, both in terms of causes and effects, and in terms of solutions, is not systematized in literature, and the resolutions of different situations have an artisanal aspect. This article presents an extreme situation caused by such a phenomenon and describes the stages of investigating causes, uncertainties and feasible solutions.

2. About the suffuses phenomenon

Originally suffusion means when something slowly spreads throughout something else. With some different nuances in the soil case, there are several ways in which this phenomenon can manifest and as a result it can be described under different names according to the mode of manifestation, figure 1.

Internal erosion:

- Concentrate flow
- Regressive erosion
- Contact erosion
- Suffusion

However, they all share some common elements:

- Migration occurs under the action of a hydrodynamic current
- The migrating material leaves voids behind; those voids are moving in the opposite direction to the migrant material.

In the case of the last two forms of void migration (contact erosion and suffusion) there are at least two different entities, by granularity in the case of soils: the first one is migrating inside of the second.

Voids genesis is conditioned by three factors:

- The existence of a nucleus of voids, which usually has a different cause than the occurrence of the phenomenon of suffusion;
- Spatial ratio between the two components;
- The ratio of the current force to the granularity of the migrant component.

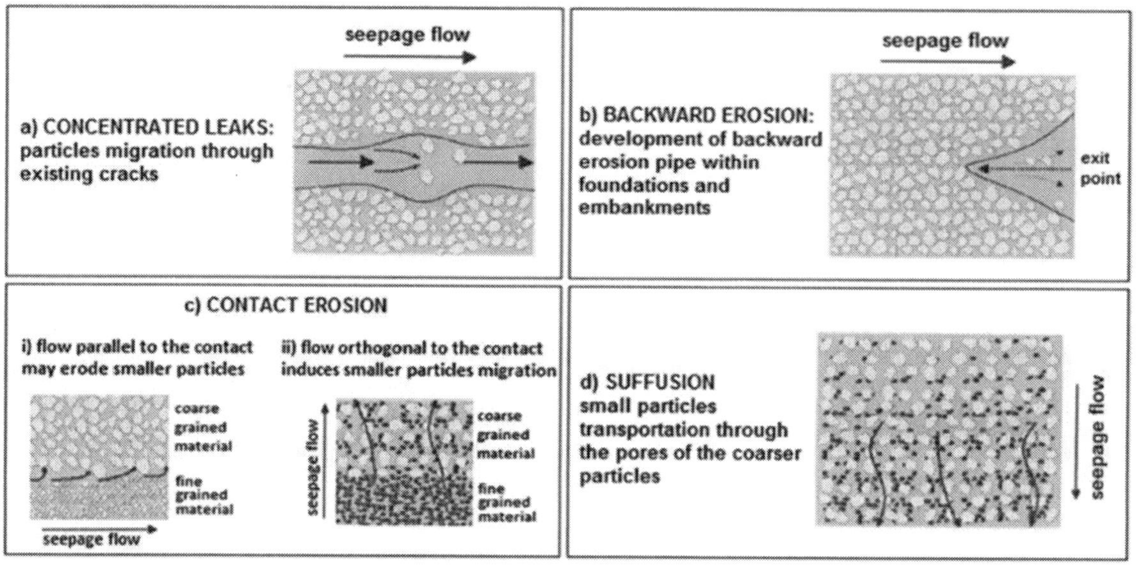

Figure 1. Different types of internal erosion or suffusion [2].

For the case of internal erosion and concentrated flow, the current force of water must be determined by overcoming the critical hydraulic gradient.

In case of the suffusion, the hydraulic gradient must not necessarily reach the value of a critical hydraulic gradient calculated for the whole soil's layers.

Its value must be able to train only the smallest particles provided that there is a gap in the continuity of the particle size curve.

This particle gauge, namely the gap correspondent, will represent the free space through which the fine particles will circulate.

In other words, the cumulative curve must have a horizontal plane, which would indicate the existence of the corresponding gaps, whereby it is possible to migrate the fine particles among the coarse particles.

Several formulas are known in the literature to indicate a limit. Granulometry criterion for the phenomenon so called "Grain Size Criteria"; the limit state equation can be written in the following general form [3]:

$$Z = f(Cu)\, d_{85,F} - d_{15,S} \quad (1)$$

where $d_{85,F}$ is a diameter such that 85% of the grain's diameter of the filling are smaller than this size, $d_{15,S}$ is a diameter such that 15% of the grain's diameter of the skeleton are smaller than this size, f(Cu) is function of uniformity coefficient.

The positive value of Z gives the stage of the safety of this simple system. Another example is the Terzaghi criterion (Terzaghi and Peck1948) with f(Cu) = const. = 4~5, including some factor of safety and C(cu) = 9, which is only valid for uniform material and filter combinations:

$$Z = 9\, d_{85,F}\, d_{15,S} \quad (2)$$

Or another criterion for recognizing land that presents a danger of suffocation is given by the relationship (A.N. Patraşev, Gh Pradevnîi 1963 – [4]):

$$\alpha = \frac{d_3}{d_7} \geq (0.32 + 0.16 \cdot \frac{d_{60}}{d_{10}}) \cdot \sqrt[6]{\frac{d_{60}}{d_{10}}} \cdot \frac{n}{1-n} \quad (3)$$

where d_3, d_7 have the same meaning as described above.

In addition to the granulometry criterion, an additional element may emerge in the formation of the funnels suffuses to trigger the activation of the phenomenon, namely the collapse of the roof of the initial void under its own weight.

The conditions in which these situations arise are studied in detail in particular in the case of underground excavations, mine galleries, tunnels etc.

3. Geological situations where naturally occurring phenomena of suffuses may occur

In the whole literature, the existence of these naturally occurring creeks is related to a limestone base in which some karstic voids occur at some point, and the above layer, usually limestone, fall into this void under its own weight (figure 2).

Rather, the phenomenon is mentioned in connection with loessoid, macroporous formations, but this hypothesis should not be considered in the studied area.

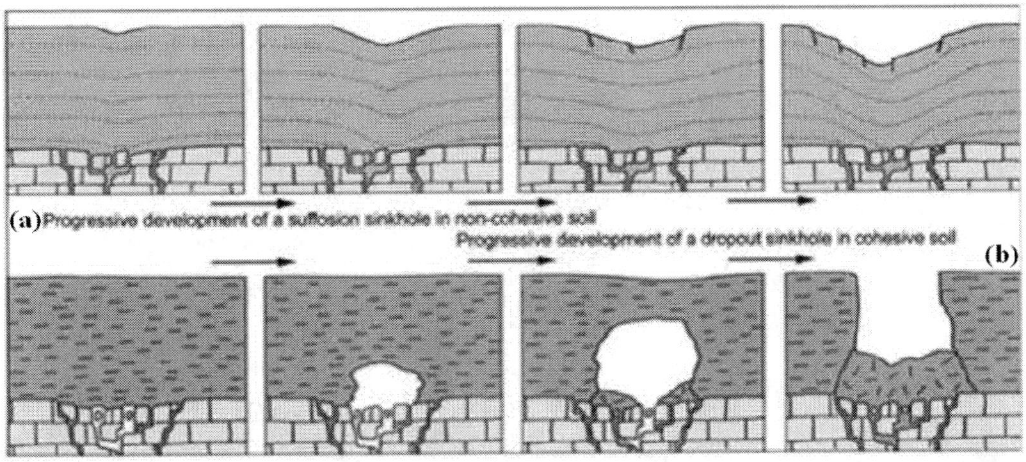

Figure 2. Evolution of suffuses cone in the case of a fine (a) or coarse (b) layer over limestone [5].

Once the deep gap is formed, it is possible for the fine particles in the roof to be trapped by a descending stream of water, and the gap to migrate in reverse direction to this current in a regressive way.

Another hypothesis, less probable but possible in the case studied, is that of an internal regression erosion, namely: precipitation infiltration waters accumulate at the base of the alteration bark and form a current between the base rock and the prologue deposits capable of trains and fine particles.

In urban areas, the phenomenon was manifested mainly by defective municipal networks (sewers, water supply pipes).

In this respect, there are spectacular examples of enormous sinkholes in cities, with depths of hundreds of meters sometimes and with diameters of meters or tens of meters.

4. History of the emergence and specific development of the cavity in Brasov

The studied area comprises an area in a residential district at the base of the south-eastern slope of Tampa Mountain, in the city of Brasov (figure 3).

On the land there are several blocks of flats which have a height regime D + P + 2, P + 3E or D + P + 4E. The constructions were put into use in the years 1982-1985 and so far they have generally behaved well (figure 4).

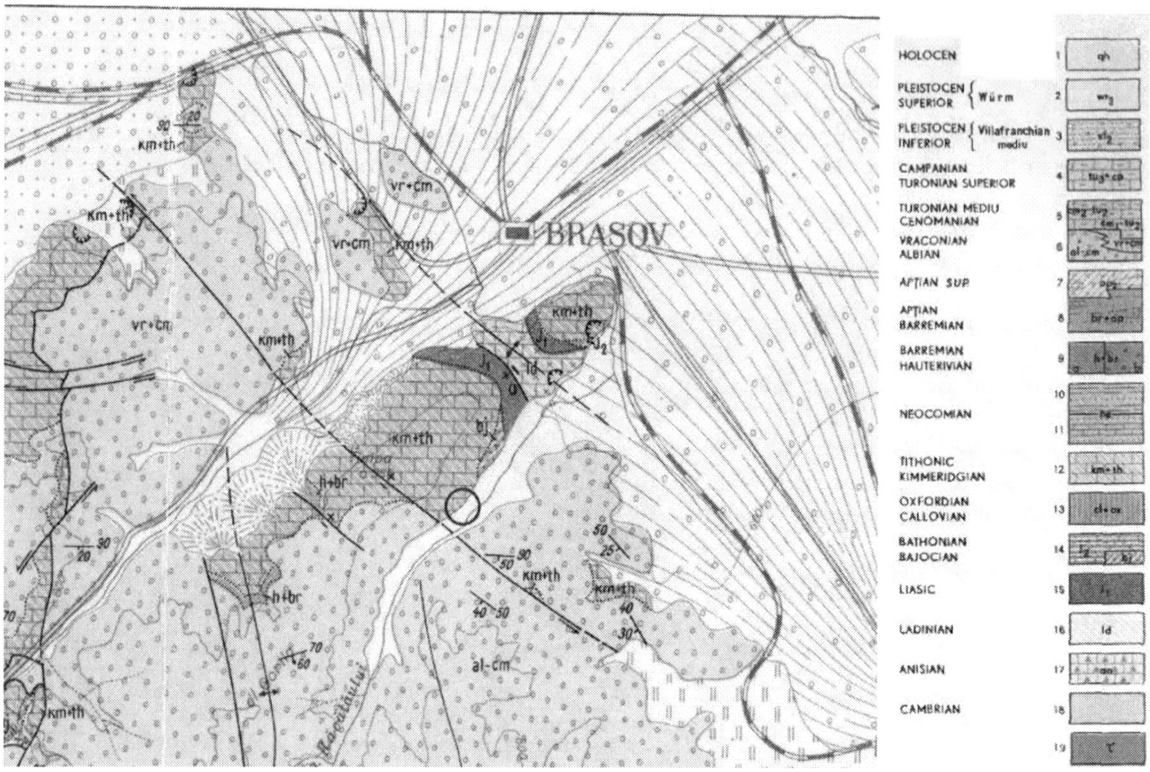

Figure 3. Geological map of the area (scale 1:50000, according to 1972 Geological Institute of Romania).

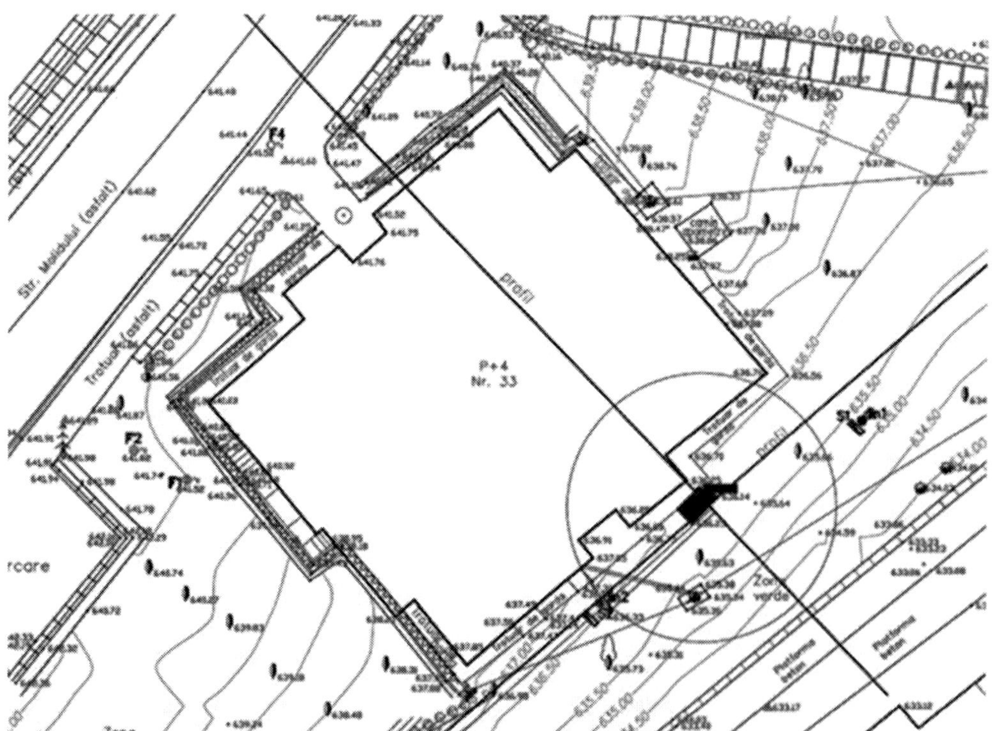

Figure 4. Situation plan.

IOP Conf. Series: Materials Science and Engineering **789** (2020) 012038 doi:10.1088/1757-899X/789/1/012038

The exception made the foundation ground of the block; so in the summer of 2017 the tenants observed the existence of a cavern of 4x4x4.5 m under the foundation on the downstream facade of the block. On the southern side of the block appeared in a cavern with a depth of approx. 4.20 m to the surface (figure 5).

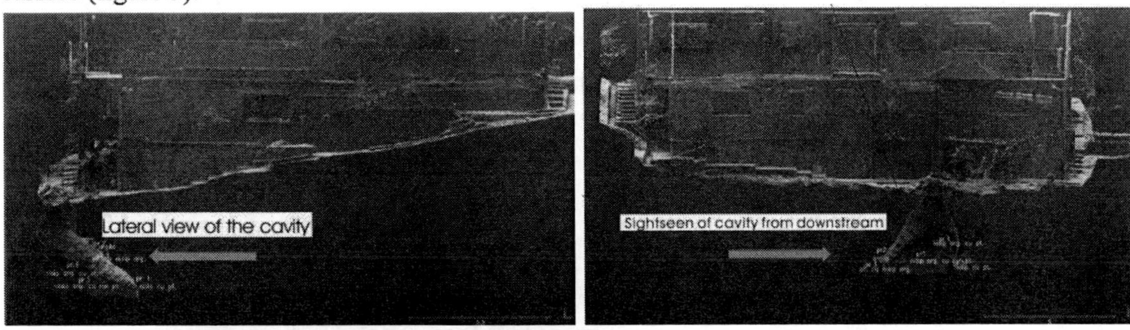

Figure 5. Profiles with the affected block resulting from the scan; the foundations and the position of the cavity are highlighted.

It was shaped like a flat-bottomed pear with a diameter of approx. 4 m and a depth of 4.20 m below the block foundation. The lateral walls of the cavern were circular in shape, and they were smooth, wallowed with concentric layers, probably due to the precipitation of some salts (figure 6).

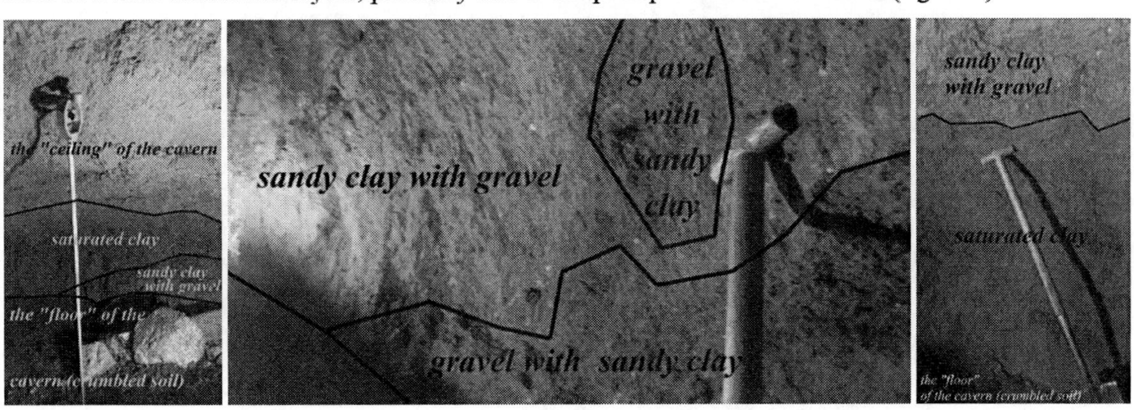

(a) (b) (c)

Figure 6. Cave Pictures in December 2017: (a) the northern wall of the cavern, under the block; (b) the "floor" of the cavern is about 50 cm higher than in September; (c) the western wall of the cavern.

Later, the cavern reduced its size by about 50% lower due to the slowness of the material on the southern, downstream wall. The sidewalks in the North-East corner are left with a maximum of (7-10) cm, relative to the thermal insulation that was executed 3-4 years ago.

4.1. Considerations on geological-geomorphological context from the point of view of the suffusion
The studied site belongs geomorphological to the piedmont on the southern slope of Tampa Mountain.

Piedmont was formed by the junction of some manure cones, resulting from the deposition of the disaggregated material and transported by the torrents.

As a result these proluvial deposits have a cross-stratification in which alternate layers with very varied granularity (boulders, clays, sandy sandy gravel sands). Over them lay a deluvial clay layer, which flows downstream.

The bedrock, meaning a Tithonian-Barremian age limestone of Jurassic lies at much greater depths of tens, perhaps over a hundred meters (figure 7).

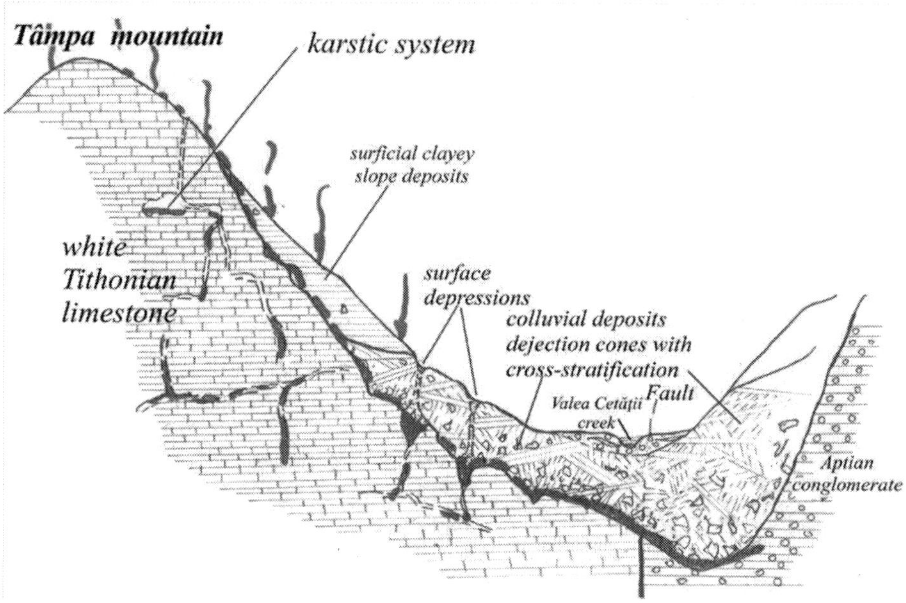

Figure 7. Geological profile through Valea Cetatii (extended karstic system is hypothetically; until now in real way it has been investigated only few meters on the top of the mountain).

The Jurassic reef limestone forming the Tampa Mountain allows the development of such karst voids. On Tampa there are at least two entrances in a karst system, near the Tampa Saddle, at a much higher altitude than the investigated area, with approx. 200 m. The system was not explored, but from the accounts of those who entered the first portion and stopped at the mouth of some vertical wells, they seem to have a size of at least the order of tens of meters. An additional argument is the existence of the cave on the Rasnoavei Valley, developed in the same type of limestone, which is very large in the order of hundreds of meters long, with impressive halls, true cathedrals inside. Investigations to determine the causes and magnitude of the phenomenon. Some images from the cavity investigation are presented in figure 8 and the geoelectric profile in figure 9.

Figure 8. Images during cavity investigation.

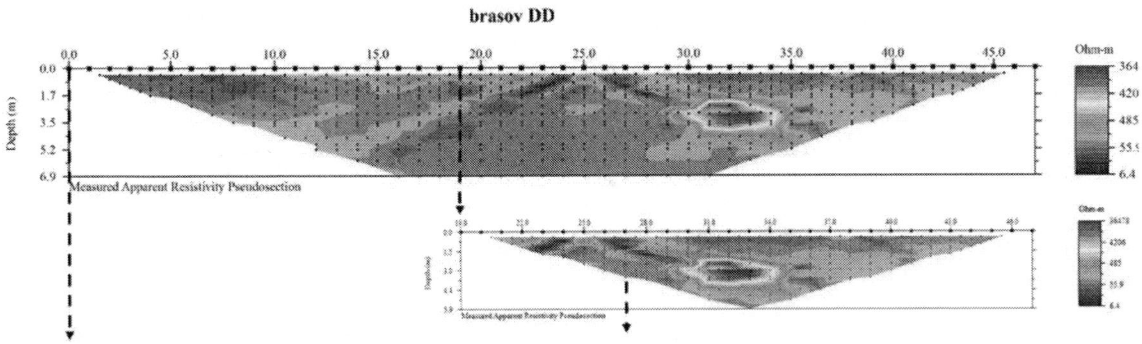

Figure 9. Geoelectric profile parallel to the building facade. Red areas indicate high electrical resistivity; the position of the cavity is observed at one third of the length to the right.

For deciphering the conditions in which the phenomenon occurred, they were made:

- 4 continuous drilling,
- 10 reconnaissance drills,
- two open ditches;
- three electrical tomography profiles;
- 3 Lefranc tests to determine permeability in situ;
- 5 samples from the cavern were harvested,
- the cavern was scanned,
- sewer networks in the area were video investigated.

5. Comments on the results of the investigations

In principle, in the upstream area, a clayey cohesive complex has been found to surface up to approx. 10 m after which all the drillings (except F 1 that stopped above) penetrated into a coarse material made of gravel or limestone blocks and submerged rounded conglomerate in a sandy matrix. It cannot be said that the bedrock has been reached. The soil permeability coefficient determined by the Lefranc method was: in the F1 drilling depth of 10 m (in slurry clay) $K = 5.05 \times 10^{-5}$ cm/sec. in F2 drilling with 3.55 m deep (in slurry clay) $K = 6.2 \times 10^{-6}$ cm/sec. In the F3 drilling depth 10.0 m (in clayey matrix groove) $K = 7.82 \times 10^{-6}$ cm/sec. Geophysical investigations, meaning electric tomography, have highlighted areas with low resistivity that could indicate the existence of other drought gaps, or the "channel" through which the water with the washed material protrudes into a collector. Reconnaissance drills made after the electrical tomography (about 10 in number) did not confirm the existence of any void. Low electrical resistance areas have been found to be coarser soil.

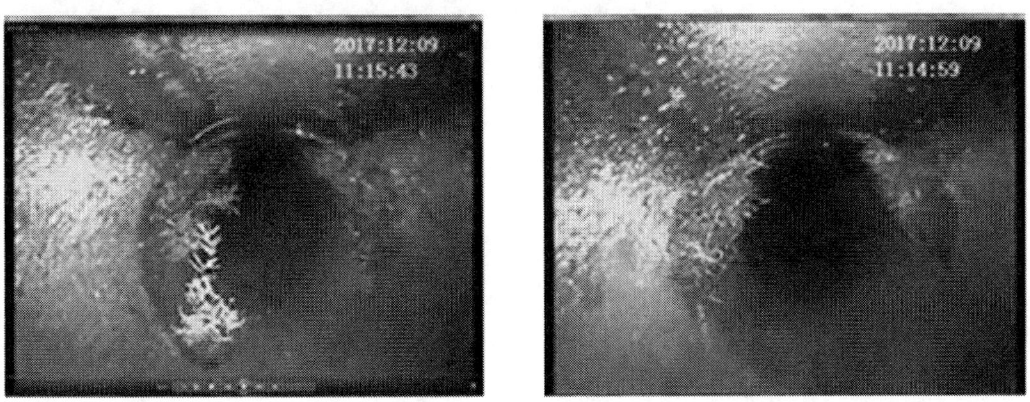

Figure 10. Imperfect sewer connections on the CM4-CM3 sewer; the roots penetration and aggregates deposited on the tube (the images are rotated differently).

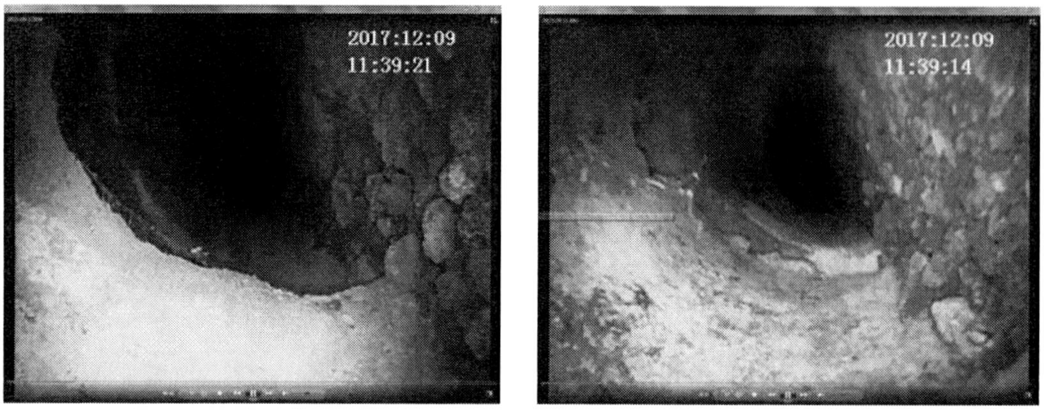

Figure 11. Images from the video inspection of the drain under the block; aggregate deposits are deposited on about half of the drainage section and precipitation of salts on the walls (the images are rotated 90° in a trigonometric sense).

The video investigation of the sewerage networks led to the following observations: The drain that crosses the building is half filled with washed soil (figure 10 and 11).

When it was filled with water (about 3 cubic meters), most of the volume followed its normal course, reaching the opposite dormitory, the eastern one, then the downstream sanitation (about 30 seconds). However, some of the water crossed the foundations of the block, and rolled over the basement floor to the room above the cavity. Moreover, this room has the floor left 30-40 cm halfway out of the block.

As can be seen in figure 12 there are satisfied from the granulometry point of view of the conditions mentioned in the literature. Thus, the granularity of the silty-clay matrix of gravel, where over 70% of the coarse material (sand and gravel) and 19,27% (colloidal) clay are observed, compared with only 9.4% of silt, which represents the "gap" mentioned above.

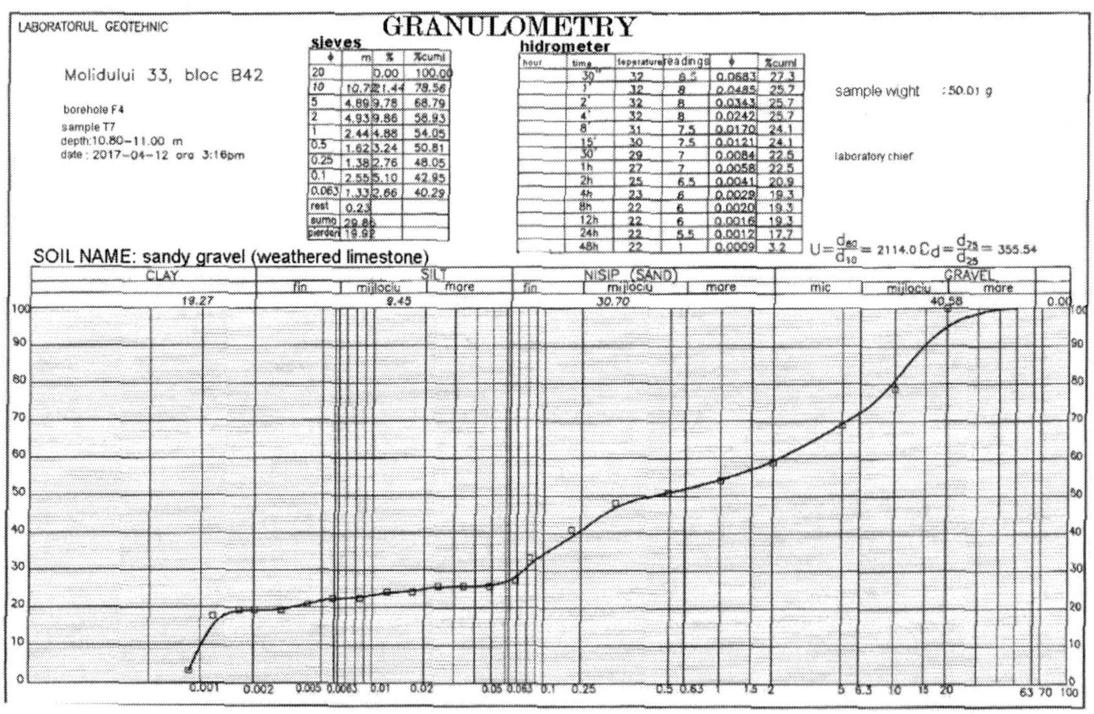

Figure 12. Example of a granulometry test performed on the claystone matrix.

Granulometry analyzes carried out on the harvested samples show that this criterion is fulfilled both for the red clayey complex of deluvial genesis and for the proluvial deposits. Another argument is the existence of the aggregates (of the earth) in the sewer pipes, which can represent the material in the cavern.

The amount of aggregate in the tube is not very significant as long as it can be reduced by washing. The joints between the sewer pipes are faulty. They see almost every joint as the roots of the plant penetrated through the joints.

The networks are made of concrete tubes in a one-meter section, which, besides the problems of execution, seem to have been involved in the motion of the earth, enough to create significant joints for the process of suffuses.

6. Solutions adopted and verifying their sustainability

As a result of the emergency work, the block was secured by improving the adjacent ground by means of injection of material (bentonite solution, cement, sodium silicate) and cavity filling.

The safety of the block followed the following steps:

- Cavity filling;
- Obtaining the drain and injecting waterproof material into the drain;
- Creating a sealing veil downstream of the block;
- Execution of drainage works upstream of the block and rehabilitation of the sewage system;
- Exterior landscaping around the block.

The filling of the cavity was made with a low permeability or practically impermeable material, homogeneous in granulometry, with approximately the same rigidity as the surrounding soil.

A cork on the surface was created to fill the cavity and then injected with downstream material at a reduced pressure (maximum two atm). After approximately filling the filling cavity, injecting material from the basement, the drain was blocked at both ends with concrete plugs after which a suspension of bentonite was injected with 2-3 atm. pressure. Pressure limitation is required to avoid penetration of the subsoil with injections.

A sealing veil was formed from a series of grouting injections (bentonite and clay) at a distance of one meter to a minimum depth of 9 m. The arrangement of the injection line is parallel to the block at a distance of 2 m from the facade.

Parallel to the upstream facade, at a distance of one meter, a drain with a minimum depth of 2.5 m will be executed.

The drain behind the supporting wall near the pavement of the street, including the barbacans, will be restored.

The sewer network has been redesigned so that the rainwater will be removed from the block and the joints will be secure.

During the execution, we found the following:

- Large suspension losses in some injectors I4 (4454 l), I5 (4651l), I6 (4192 l) and I7;
- at injector I7 between 7-12 m depth – total suspension loss (zero pressure bar pressure); the injection was resumed for 5 consecutive days, and an injection pressure of 4 bar was achieved after a total consumption of 4847 l of suspension (1/3 C/A);
- Two dynamic heavy penetrations were made up to 16 m deep, resulting that the soil is loose, poorly consolidated, with gaps made by washing / moving fine particles.

The work has been visually controlled and verified by laboratory tests and "in situ" tests (dynamic penetration) and it has been ascertained that the required requirements have been met.

The material was inserted into the cavern in elementary thicknesses of 25 cm thick, from the surface downwards.

At least two oblique tubes were mounted from the surface to the bottom of the cavern. After hardening of the material cast in the cavern, a new amount of 3 atm. to 5 atm. was injected at 2-7 days until refusal.

7. Conclusions

The investigations carried out have not intercepted other caverns or churning funnels. Besides, it would have been very unlikely that such a supposed goal would hit a survey. However, the results indicate that there are obvious prerequisites for their existence, especially in terms of dynamic penetration.

Figure 13 shows the number of blows (penetration resistance) v depth. It can be noticed that after the surface filling layer, resulting from landscaping and general excavation between depths of approx. 3-6 m was encountered a layer of coarse deluvial clay with a relatively high resistance to dynamic penetration. It is then observed a decrease in penetration resistance, because at a depth of 15-16 m there is a minimum (2-4 strokes/10 cm). Normally in a uniform stratification, unaffected by other factors, the penetration resistance increases with the depth on the one hand due to the consolidation of the lower layers under the load of the top of the earth column and on the other hand due to the rotation of the rods, which is particularly significant in the case of the clays. A statistical processing of these penetrations indicates at a depth of 13-16 m a pore index of 0.8-1.25, which is very much for a normally consolidated ground.

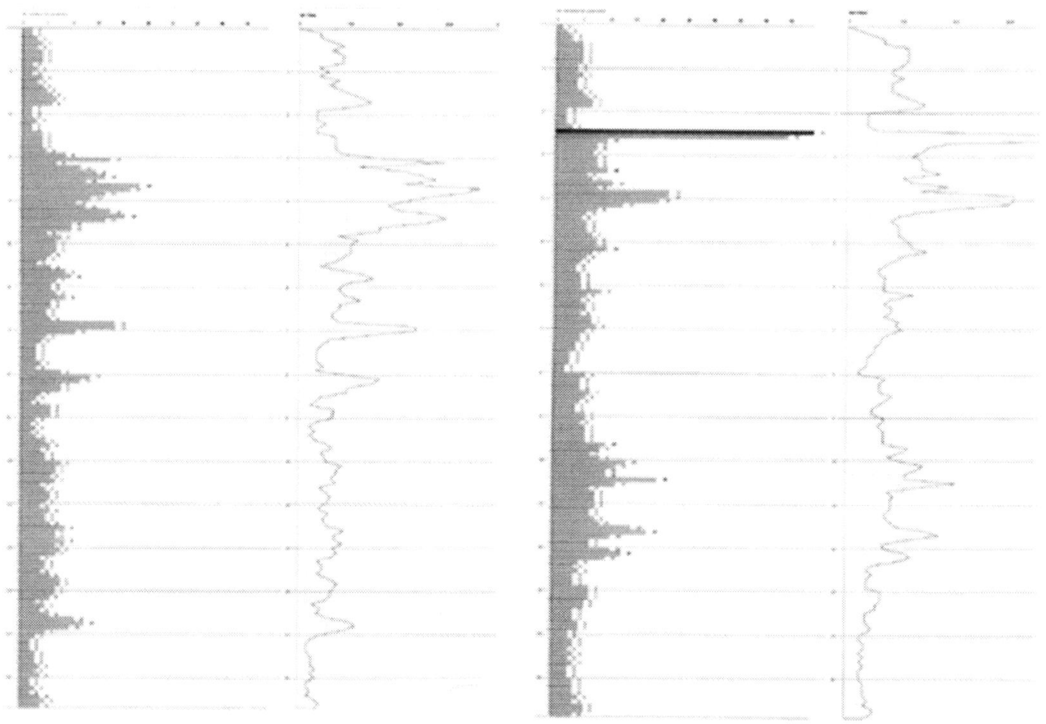

Figure 13. Dynamic penetration resistance charts performed by Geosond S.A. (no. of strokes vs. depth).

The study of the old topographical plans, which formed the basis of the urban plans on which the district was built, drew attention to the existence of a number of 44 funnels, assumed by existing suffuses, at the base of the southern slope of Tampa, between the slopes and the valley axis. We believe that their appearance is a natural and not anthropic phenomenon.

Acknowledgments
Author wish to acknowledge colleagues who has contribute to this work as follows:
Transylvania University of Brasov: Cristian Năstac, Iosif Vorovencii, Mihai Niță, Cătălin Lixăndroiu;
Glob Consult International S.R.L.: Onica Medeșan, Alexandru Partene, Suceveanu George;
Geotesting S.R. L.: Adrian Diaconu, Ioan Scolaru;
Geo-Tech S.R.L.: Istvan Szekely;
Geosond S.R.L: Petre Uță;

References

[1] Lin K and Takahashi A 2014 Experimental investigations on suffusion characteristics and its mechanical consequences on saturated cohesionless soil *Soils and Foundations* 2014 **54** (4):713–730

[2] Federico F 2017 Particle Migration Phenomena Related to Hydromechanical Effects at Contact between Different Materials in Embankment Dams in *Granular Materials*

[3] Salehi Sadaghiani M. R. and Witt K. J. 2011 *Variability of the grain size distribution of a soil related to suffusion* ed Vogt, Schuppener, Straub & Bräu Bundesanstalt für Wasserbau

[4] Popa A and Ilieş N M 2009 *Consolidarea fundaţiilor (Foundation consolidation)* Ed U T Press-Cluj Napoca

[5] https://www.usbr.gov/ssle/damsafety/risk/BestPractices/Chapters/I-2-20150401.pdf

Thermal analysis of a structure made by using cold formed steel sections

S G Maxineasa[1*], D N Isopescu[1], I R Baciu[1], M L Lupu[1] and T C Dragan[2]

[1] "Gheorghe Asachi" Technical University of Iasi, Faculty of Civil Engineering and Building Services, Iasi, Romania
[2] SC Next Studio SRL, Miroslava, Romania

[*] sebastian.maxineasa@tuiasi.ro

Abstract. In the last decades, man-made environmental burdens have led to a significant increase of negative effects felt by the human race. Thus, some industrial sectors have developed and implemented a series of solutions with the goal of minimising their negative ecological impact. Taking into account that the construction sector is one of the most important economic players in every national economy, one that has a tremendous negative impact over the environment, in the last years, various recommendations and limitations have been drawn in order to reduce the consumption of non-renewable energy. In order to constantly reduce the impact of the built environment over the natural one, civil engineers should develop and promote sustainable solutions that have a direct effect on improving the environmental performances of the built environment. One such solution is the use of cold formed steel sections (CFS) for building new structures. This paper analyses the linear thermal transmittance of the thermal bridge formed by the vertical steel studs inside the external walls. The results show that these steel elements can be used in order to build energy efficient buildings.

1. Introduction

In the last decades, the issues that are related to the negative effects of the global warming phenomena have significantly increased the level of awareness of how the human race is negatively influencing the state of the Earth's ecosystem. Another current significant environmental problem, one that is going to affect the development of future generations, is represented by the alarming levels of natural resources consumption. Nowadays, the amount of natural resources that are consumed every year exceeds the Earth's capacity of renewing the stock of raw materials by 75% [1]. Taking into account that all the ecological problems we are facing are a direct result of our daily activities, different solutions have to be developed and applied in all economic activities at the global scale in order to reduce the load over the natural environment.

Due to the important volume of greenhouse gases emissions into the atmosphere as well as the tremendous amounts of consumed raw materials, the construction sector is considered to have one of the most significant roles in achieving the primary goals of sustainable development at the global scale. Out of the entire quantity of raw materials consumed worldwide, almost 60% is used in the construction sector. At the same, the activities specific to this industry are responsible for emitting more than 40% of the total amount of greenhouse gasses, and for producing around 25% of worldwide waste. Another significant environmental issue concerning this industry is reflected by the fact that more than 60% of the extracted natural resources are used in the built environment [2-7]. Therefore, the built environment is exerting a significant negative load over the natural one, being one of the most pollutant economic activities. In this respect, all construction sector specialists should increasingly promote solutions with the goal of significantly improving the environmental performances of the construction sector.

The environmental footprint of the building industry is highly influenced by the type and amount of construction materials that are used in order to satisfy the need of a secure and protective indoor space. Thus, optimizing the structural systems and the materials used in building new constructions can reduce the present alarming rates of environmental burdens resulted from all the activities that are characteristic to this industry. Another important influence over the overall ecological impact of the construction sector is represented by the way we use the buildings. There are different studies showing that from the entire life cycle of a construction, one that is not energy efficient, the operation or usage stage is responsible for between 80 and 90% of the total environmental impact of that building [8-12]. By taking this into account, along the years, various solutions have been used in order to reduce this negative environmental impact. Nevertheless, efforts should be intensified in order to improve the energy efficiency of the buildings and thus, improving the overall environmental performances of the construction sector.

The environmental impact over the operation phase from the life cycle of a building is mainly influenced by the level of energy consumption and greenhouse gases emissions used in view of satisfying the optimal indoor condition for the occupants. Related to this issue, at the present moment, the passive house standard is considered to recommend the most restrictive values regarding the thermal performances of a building.

A passive house should have a very thick thermal insulation, highly insulated windows frames, an airtight building envelope, and a high performance ventilation system with efficient heat and/or energy recovering. One of the most important characteristics of this type of buildings is that in order to achieve the passive house limitations, the thermal bridge free design and construction principle must be used. This is a must when it comes to achieving a building with high energy performance, which is formed by using envelope elements (i.e. external walls, floor slabs and roof areas) with a value range from 0.10 W/(m^{2}K) to 0.15 W/(m^{2}K) for the thermal transmittance coefficient (these values are valid for the Central European climate) [12].

Therefore, by taking into account all the above, it is justified to support the idea that reducing the ecological footprint of the built environment represents a very important problem that can be resolved by considering innovative approaches. The use of cold formed sections (CFS) could represent a solution for reducing the environmental impact of the construction sector by building new structures that satisfy the passive house regulations.

The aim of the present paper is to analyse if in the case of a building made by using CFS profiles, the thermal bridges formed inside the exterior walls by the vertical metal studs have an influence over the thermal bridge free design principle of a passive house.

2. Case study

In order to achieve the goal of the study, the authors have used the principles presented in the C 107-2005 Romanian normative [13] and in the international standard ISO 10211:2017 [14] to assess if the value of the linear thermal transfer coefficient (ψ), characteristic to the thermal bridges formed by the vertical metal studs inside the exterior walls, is lower or equal with 0.01 W/(mK). If this restriction is satisfied, then the thermal bridge free design can be used. In order to correctly dimension the thickness of the thermal insulating materials, with respect to the passive house principles, the authors have considered the value range from 0.10 W/(m^{2}K) to 0.15 W/(m^{2}K) for the thermal transmittance coefficient (U). Taking into account that for evaluating the linear thermal transfer coefficient, the value of the specific unidirectional thermal resistance (R) of the considered building envelope element is needed [13], the above mentioned U values have been translated in R values (table 1). The computation of the specific unidirectional thermal resistance has been completed according to Romanian normative C107-2005 by using the equation (1):

$$U = \frac{1}{R}\left[W/\left(m^2K\right)\right] \Rightarrow R = \frac{1}{U}\left[\left(m^2K\right)/W\right] \tag{1}$$

where R is the specific unidirectional thermal resistance (m^2K/W).

Table 1. U-R equivalence.

U (W/m^2K)	R (m^2K/W)
0.10	10
0.15	6.67

Figure 1 presents the construction detail of the exterior wall that has been put under analysis. According to C107-2005 [13], before determining the value of the linear thermal transmittance (also known as linear thermal transfer, according to C107-2005), ψ, it is necessary to determine the specific unidirectional thermal resistance of the element (R).

The R value represents the sum of the thermal unidirectional resistances of the construction element's layers and the superficial resistance to the radiative and convective heat transfer of the inner and outer surfaces [13], [15-17]. Table 2 shows the computation of the R value of the considered envelope's element. Equation 2 has been used in order to determine the specific unidirectional thermal resistance [13]:

$$R = \frac{1}{\alpha_i} + \sum\frac{d_k}{\lambda_k} + \frac{1}{\alpha_e}\left[\left(m^2K\right)/W\right] \tag{2}$$

where α_i and α_e are the convective and radiative heat transfer coefficients of the internal and external surfaces, in $W/(m^2K)$; d_k represents the thickness of the layer "k" (m); λ_k represents the thermal conductivity of the layer "k" (W/mK).

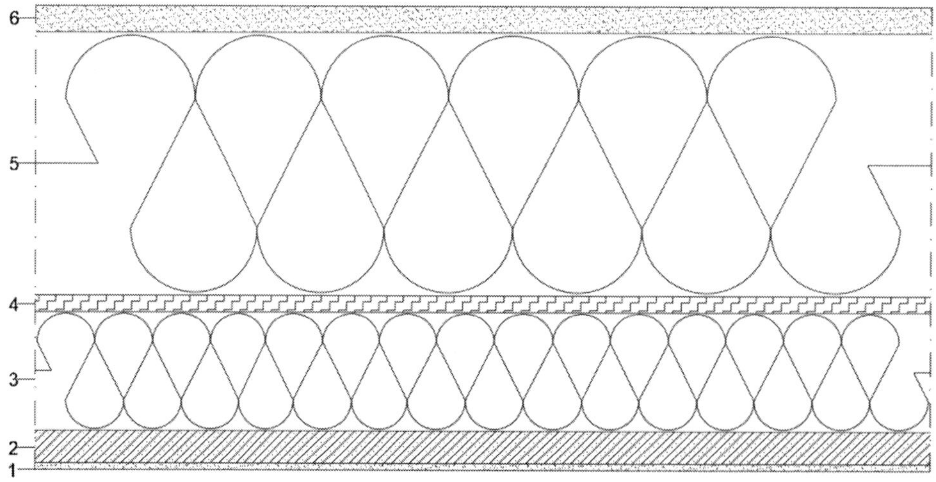

Figure 1. Material layers of the analysed construction element.

As mentioned above, the thickness of the thermal insulating materials has been dimensioned by considering the R values that are recommended to be satisfied by a passive house situated in a European climate. By analysing table 2, it can be observed that the achieved value is between the imposed values.

IOP Conf. Series: Materials Science and Engineering **789** (2020) 012039 doi:10.1088/1757-899X/789/1/012039

Table 2. U-R equivalence.

Nr.	Layer	Thickness (d) (m)	Thermal conductivity (λ) (W/mK)	Layers' thermal resistance (R) (m^2K/W)	Thermal resistance (R) (m^2K/W)
1.	Internal plaster	0.005	0.87	0.006	8.503
2.	Plasterboard	0.025	0.21	0.12	
3.	Rock wool (1)	0.09	0.04	2.25	
4.	OSB panel	0.0125	0.204	0.06	
5.	Rock wool (2)	0.2	0.034	5.88	
6.	External plaster	0.02	0.87	0.02	
External surface resistance $1/\alpha_e$=1/24=0.042 m^2K/W					
Internal surface resistance $1/\alpha_i$=1/8=0.125 m^2K/W					

3. Thermal bridge analysis

According to ISO 10211:2017 [14] a thermal bridge represents a "part of the building envelope where the otherwise uniform thermal resistance is significantly changed by full or partial penetration of the building envelope by materials with a different thermal conductivity, and/or a change in thickness of the fabric, and/or a difference between internal and external areas, such as occur at wall/floor/ceiling junctions".

The same standard defines the linear thermal transmittance (ψ) as the "heat flow in steady-state compared to a reference heat flow rate calculated disregarding the thermal bridge, divided by length and by the temperature difference between the environments on the either side of a linear thermal bridge" [14]. In order to compute the value of ψ, the authors have used equation (3) [13]:

$$\psi = \frac{\phi}{\Delta T} - \frac{B}{R}\left[W/(mK)\right] \tag{3}$$

where Φ represents the heat flow rate (W/m); ΔT represents the difference between the interior and exterior temperatures (K); B represents the width of the modelled linear thermal bridge (m); R represents the specific unidirectional thermal resistance of the element (m^2K/W).

Figure 2 presents the considered thermal bridge that is formed by the vertical steel studs in the interior of the analysed construction detail (i.e. external wall).

The value of the heat flow rate, necessary to compute the influence of the linear thermal transmittance, has been determined by using the RDM 6 software.

Table 3 details the characteristics of the materials used for the finite element assessment. Also, in order to complete the model, the authors have considered T_e=-18°C for the exterior temperature boundary conditions (characteristic for climatic zone III) and T_i=20°C for the interior temperature.

Table 3. Properties of the materials considered in the analysis.

Material (Layer)	Thermal conductivity (λ) (W/mK)	Bulk density (ρ) (kg/m^3)	Specific heat capacity (c) (J/kgK)
Interior plaster	0.87	1700	840
Plasterboard	0.21	1100	840
Rock wool (1)	0.04	70	750
OSB panel	0.204	600	2510
Rock wool (2)	0.034	60	750
Exterior plaster	0.87	1700	840
Steel profiles	58	7850	480

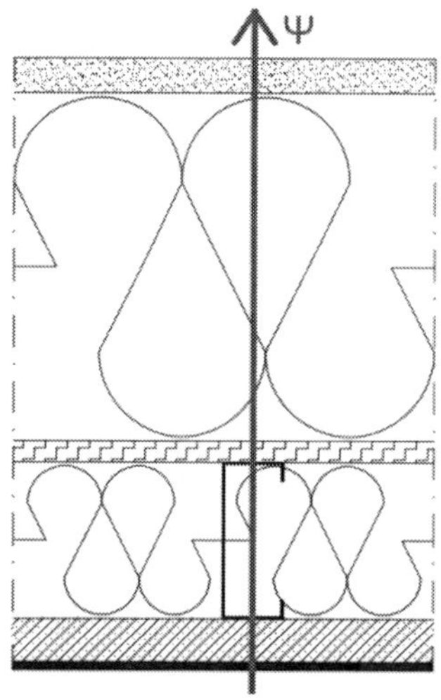

Figure 2. Analysed thermal bridge.

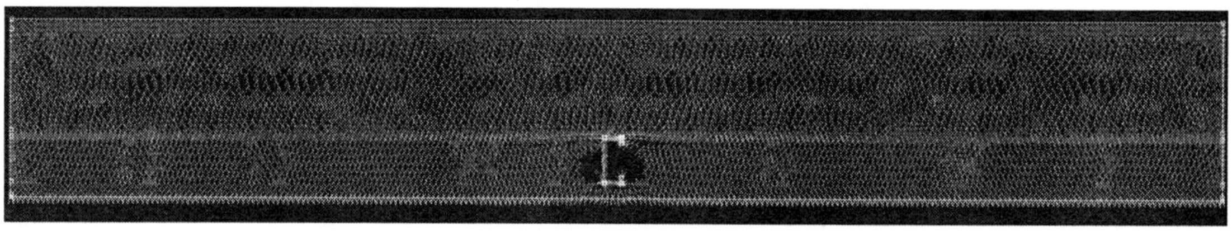

Figure 3. Heat flow of the analysed thermal bridge.

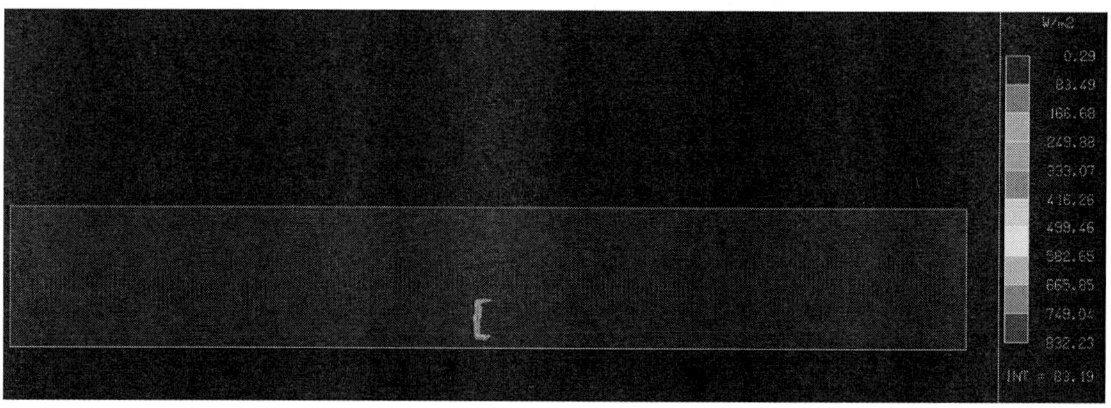

Figure 4. Isovalues of the heat flow of the analysed thermal bridge.

Figures 3 and 4 show the heat flow resulted for the considered thermal bridge. It can clearly be observed that the vertical steel stud forms a thermal bridge that has an important influence over the assessed building envelope's element.

Analysing the figures above, we can also see that the construction detail has been properly developed due to the fact that important values of the heat flow are only registered in the vertical thermal bridge.

Table 4 presents the computation of the linear thermal transmittance (ψ). The result clearly shows that the influence of the analysed thermal bridge can be neglected in determining the thermal resistance of the building envelope's element. The achieved value is lower than the limit of 0.01 W/(mK) imposed by the passive house standard in order to apply the principle of thermal bridge free design and construction.

Table 4. Computation of the linear thermal transmittance.

Heat flow rate (Φ) (W/m)	Temperature difference (ΔT) (K)	Width of the thermal bridge (B) (m)	Thermal resistance (R) (m^2K/W)	Linear thermal transmittance (W/mK)
10.96	38	2.4	8.503	0.006

4. Conclusions

The present state of the natural environment imposes the development and promotion of a series of innovative solutions that can help reduce the ecological footprint of our daily activities. From the consumption rates of natural resources to the volume of greenhouse gases emissions into the atmosphere, year after year, alarming values regarding the impact over the Earth's ecosystem are registered. While all economic activities have an important negative effect over the environment, the construction sector is considered to be the most pollutant. In the last decade, different solutions have been implemented in order to reduce the impact of the built environment over the natural one, but these efforts must be supplemented in order to significantly improve the ecological performances of this sector. In the near future, it is necessary to reduce the load over the environment of the operation stage from the life cycle of a construction by using different solutions that are related to high energy efficient buildings.

One such solution is represented by the use of the passive house standard which is considered to be one of the most restrictive norms related to the energy consumption of a building. This standard imposes the use of the free thermal bridge design and construction principle, which means that if the value of the linear thermal transmittance of a thermal bridge is lower than a certain value, the effect of that thermal bridge is neglected in the computation of the thermal transmittance of a building envelope's element. The goal of the present paper was to assess if in the case of using cold formed steel sections, the value of the linear thermal transmittance of the thermal bridge formed by the vertical steel studs inside of an external wall is lower than the one imposed by the passive house standard. The resulted value shows that the ψ value is lower than the limit imposed for considering a free thermal bridge design. Thus, the CFS profiles can be used for high energy efficient buildings.

Acknowledgments

This paper was elaborated with the support of the "Ecoinnovative Products and Technologies for Energy Efficiency in Constructions – EFECON" re-search grant, project ID P_40_295/105524, Program co-financed by the European Regional Development Fund through Operational Program Competitiveness 2014-2020.

References

[1] Global Footprint Network 2019 https://www.overshootday.org/

[2] Maxineasa S G, Entuc I S, Taranu N, Florenta I and Secu A 2018 *J. Clean. Prod.* **175** 164-175

[3] Maxineasa S G and Taranu N 2018 Life cycle analysis of strengthening concrete beams with FRP *Eco-efficient Repair and Rehabilitation of Concrete Infrastructure* ed F Pacheco-Torgal et al (Duxford: Woodhead Publishing) pp 673-721

[4] Messari-Becker L, Bollinger K and Grohmann M 2013 Life-cycle assessment as a planning tool for sustainable buildings *Life-Cycle and Sustainability of Civil Infrastructure Systems. Proceedings of the Third International Symposium on Life-Cycle Engineering (IALCCE 2012)* A Strauss et al (CRC Press) pp 1558-1562

[5] Bribian I Z, A Capilla V and Uson A A 2011 *Build. Environ.* 46 1133-1140

[6] Ding G K C 2014 Life cycle assessment (LCA) of sustainable building materials: an overview *Eco-Efficient construction and building materials. Life Cycle Assessment (LCA) Eco-Labelling and Case Studies* ed F Pacheco-Torgal et al (Cambridge: Woodhead Publishing Limited) pp 38-62

IOP Conf. Series: Materials Science and Engineering **789** (2020) 012039 doi:10.1088/1757-899X/789/1/012039

[7] Pacheco-Torgal F and Labrincha J A 2013 *Constr. Build. Mater.* **40** 729-737
[8] Maxineasa S G 2015 Composite and hybrid solutions for sustainable development in constructions PhD Thesis (in Romanian)
[9] Adalberth K, Almgren A and Petersen E H 2001 *International Journal of Low Energy and Sustainable Buildings* **2** 21
[10] Huberman N and Pearlmutter D 2008 *Energ. Buildings* **40** (5) 837-848
[11] Ortiz O, Pasqualino J C, Diez G and Castells F 2010 *Resour. Conserv. Recy.* **54** 832-840
[12] *Active for more comfort: Passive House. Information for property developers, contractors and clients* 2018 Third Edition. International Passive House Association (iPHA) Darmstadt
[13] C107-2005 *Normative Regarding the Calculation of Thermic Properties of Building Construction Elements* (in Romanian) Including the Ministerial order no. 386/2016 regarding the modification of C107-2005
[14] ISO 10211:2017 *Thermal bridges I building construction – Heat flows and surface temperatures – Detailed calculations* International Organization for Standardization
[15] Pruteanu M, Maxineasa S G, Vasilache M and Taranu N 2014 *Bulletin of the Polytechnic Institute of Iasi* Fasc **1** 31-42
[16] Baciu I-R et al 2019 *IOP Conf. Ser.: Mater. Sci. Eng.* **586** 012007
[17] Isopescu D N et al 2017 *IOP Conf. Ser.: Mater. Sci. Eng.* **209** 012076

IOP Conf. Series: Materials Science and Engineering **789** (2020) 012040 doi:10.1088/1757-899X/789/1/012040

Effectiveness of tuned mass dampers for improved seismic performance of dual eccentrically braced steel frames

H A Mociran[1*] and A V Lăpuşte[1]

[1] Technical University of Cluj-Napoca, Faculty of Civil Engineering, Cluj-Napoca, Romania

* horatiu.mociran@mecon.utcluj.ro

Abstract. Tuned mass dampers (TMDs) have been widely used for the control of seismic response of various new or existing buildings, successfully. However, the use of TMDs in the case of dual eccentrically braced steel frame structures has not been found in the technical literature. The paper appraises numerically the efficiency of TMDs in enhancing the seismic response of dual eccentrically braced steel frame structures and investigates the influence of mass of the TMD on the level of the response reduction achieved. To achieve these goals, a typical mid-rise five story steel frame structure with perimeter dual eccentrically braced frames, located in Bucharest, has been considered. Then, bidirectional TMDs with three mass ratios 0.04, 0.06 and 0.08 have been added on the top floor of the building, without redesign the frame members. Comparative nonlinear time history analyses of the unequipped and equipped structure have been performed with accelerograms of Vrancea 1977 type. Numerical results (peak base shears, peak inter-story drifts ratio and peak absolute story accelerations) show the efficiency of TMDs in mitigating the seismic response of dual eccentrically braced steel frames.

1. Introduction

Dual eccentrically braced steel frames are constituted by moment resisting frames (MRFs) and eccentrically braced frames with short links (EBFs). These frames combine the main advantages of MRFs (high energy dissipation capacity) and EBFs (high lateral stiffness), while minimizing their disadvantages (significant inter-story drifts at the upper stories of MRFs and great story shears at the lower stories of EBFs) [1].

A TMD is one of the most reliable passive devices for reducing the response of buildings subjected to wind and earthquakes, due to the advantage of an easy installation, non-interference with vertical and horizontal load paths and low maintenance costs [2, 3].

A TMD consists of a mass, spring and a damper which is attached to a story of the structure with the aim of reducing the structural response. The frequency of the TMD is tuned to the first natural frequency of the structure. When that frequency is excited, the TMD will resonate out of phase with the structural movement. The TMD dissipates energy, when its mass oscillates relative to the structure [4, 5].

The current TMD originates from a vibration absorber device, invented by Frahm in 1909 [6]. Over the last few decades, TMDs have been installed in many buildings such as Centerpoint Tower (Sidney, Australia), Citicorp Center (New York, USA), John Hancock (Boston, USA), Burj Al Arab (Dubai, UAE) etc. – to mitigate wind-induced vibrations and Taipei 101 (Taipei, Taiwan) – to mitigate the seismic response of building [3, 7].

Several papers have derived the optimum parameters of tuned mass dampers that reduce the seismic response of single and multiple degree of freedom structures [7, 8, 9]. The parameters of TMDs have been developed considering different optimization criteria of the structure: minimum displacement of the structure, minimum velocity of the structure, minimum acceleration of the structure etc. [6].

Studies regarding the optimal placement of the TMDs in the structure recommend locating the damper at the top floor, where the displacement amplitude of the fundamental mode is highest [8].

Several researchers have investigated numerically or by experimental studies the effectiveness of TMDs in mitigating the seismic response of various types of structures (symmetrical or asymmetrical - steel or reinforced concrete moment resisting frames) [2, 3, 8, 9 10, 11, 12, 13].

To the knowledge of the authors, this paper represents the first study regarding the seismic protection of dual eccentrically braced steel frame structures by using TMDs.

2. Description of the structure

The seismic performance of dual eccentrically braced frames equipped with TMDs has been investigated by numerical simulations on five story office building, located in Bucharest.

The building has a square plan (18x18 m), as shown in figure 1. The structure is formed by two groups of four planar frames oriented along two orthogonal directions. The perimeter moment resisting frames (MRFs) are eccentrically braced in the middle bay, being dual frames (figure 2), while the internal frames are MRFs. The floors are treated as rigid diaphragms, their thickness being 15 cm.

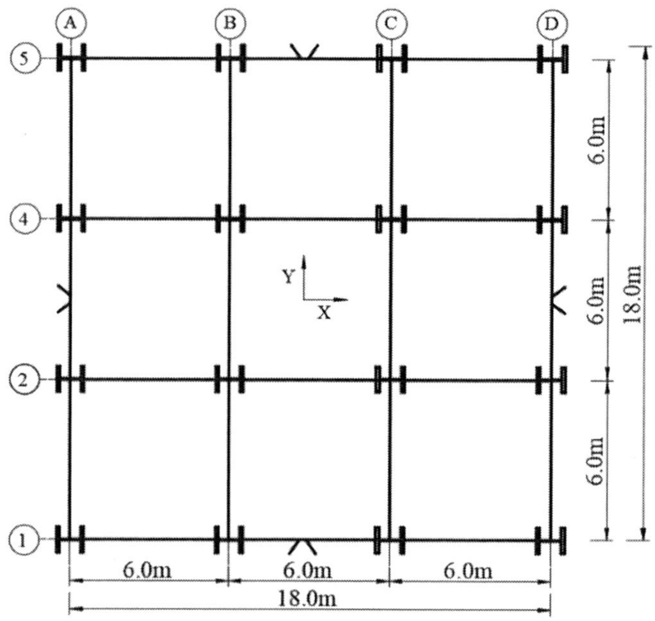

Figure 1. Plan view of the building.

The structure has been designed according to P100-1 (2013) and EN 1993-1-1 (2005) requirements, considering a dissipative behaviour with ductility class high (DCH) [14, 15].

Two limit states have been checked:

- Ultimate Limit State (ULS) - for an earthquake with a reference return period of 225 years;
- Serviceability Limit State (SLS) – for an earthquake with a reference return period of 40 years, considering the inter-story drift limit as 0.5% of the story height.

Modal response spectrum analysis has been performed independently for the ground motion in two horizontal directions. The main characteristics of the design spectrum are:

- peak ground acceleration: a_g=0.30g;
- corner period: T_c=1.6 s;
- behaviour factor: q=6 (dual eccentrically braced steel frames with DCH).

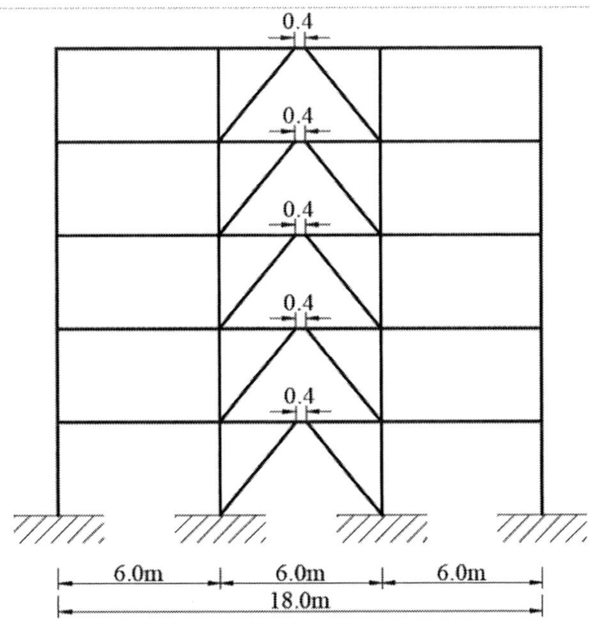

Figure 2. Elevation of a perimeter dual frame.

The overstrength of the structural system, Ω_T=2.5, has been used for the design of non-dissipative members.

Steel grade S235 has been used for dissipative members (links and beams), while steel grade S355 for the other members.

The characteristic value of dead load is 5.5 kN/m^2, except the roof level where it is 5.5 kN/m^2, while the characteristic value of live loads is 2 kN/m^2 for all floors.

The structural member sections are tabulated in Tables 1-4.

Table 1. Member cross sections for frames from axes 1 and 4.

Story	Exterior columns	Interior columns	Current beams	Middle bay beams	Braces
4-5	HEM 160	HEM 160	IPE 240	IPE 400	HEB 200
1-3	HEM 160	HEM 260	IPE 240	IPE 600	HEB 240

Table 2. Member cross sections for frames from axes 2 and 3.

Story	Exterior columns	Interior columns	Beams	Middle bay beams
4-5	HEM 160	HEM 220	IPE 300	IPE 300
1-3	HEM 260	HEM 280	IPE 330	IPE 330

Table 3. Member cross sections for frames from axes A and D.

Story	Current beams	Middle bay beams	Braces
4-5	IPE 240	IPE 400	HEB 200
1-3	IPE 240	IPE 600	HEB 240

Table 4. Member cross sections for frames from axes B and C.

Story	Interior columns	Current and middle bay beams
4-5	HEM 220	IPE 300
1-3	HEM 280	IPE 330

Then, the structure has been equipped with bidirectional TMDs, located on the top floor, without redesigning its members.

Mass ratio μ is defined as [6]:

$$\mu = \frac{m_d}{M_{tot}} \tag{1}$$

where m_d is the damper mass and M_{tot} is the total mass of unequipped structure.

The influence of mass of the TMDs on the seismic response has been studied for three mass ratios: 0.04, 0.06 and 0.08 [16, 17].

The design of the TMD involves specifying the mass, stiffness and damping coefficient. Den Hartog method has been used to compute these parameters [18, 19, 20]:

$$\alpha_{opt} = \frac{1}{1+\mu}\sqrt{\frac{2-\mu}{2}} \tag{2}$$

where α_{opt} is the optimum frequency ratio;

$$\zeta_{opt} = \sqrt{\frac{3\mu}{8(1+\mu)}}\sqrt{\frac{2}{2-\mu}} \tag{3}$$

where ζ_{opt} is the optimum damping ratio of TMD;

$$k_d = 4\pi^2\mu\alpha_{opt}^2\frac{m_d}{T^2} \tag{4}$$

where k_d is the stiffness of TMD;

$$c_d = 4\pi\mu\alpha_{opt}\zeta_{opt}\frac{m_d}{T} \tag{5}$$

where c_d is the damping coefficient of TMD.

3. Numerical analyses and discussions

To assess the influence of the TMD mass on the seismic response of the dual eccentrically braced frames, nonlinear time history analyses have been performed by the program SAP 2000. This is a structural analysis program with capabilities for nonlinear dynamic analysis of structures equipped with energy dissipation devices. The researches have shown that results obtained by SAP 2000 for structures equipped with energy dissipation devices are in good agreement with the results obtained by ANSYS and with experimental results [21]. This is the reason why SAP 2000 has been used for the numerical studies.

The seismic loading is represented by seven sets of semi-artificial accelerograms of Vrancea 1977 type (NS and EW components). The seismic performance of the structures has been analyzed for Ultimate Limit States (ULS) and Serviceability Limit States (DLS).

Peak base shears, which represent the maximum lateral forces that appear at the base of the structure during an earthquake; peak inter-story drift ratios, that are related with non-structural drift sensitive components damages; and peak absolute story accelerations that are related with non-structural acceleration sensitive components damages, have been selected as response parameters [22]. Due to space considerations, the response parameters are graphically represented only for the dual eccentrically braced frames along the X direction. Comments refer to dual frames from both directions.

3.1. Base shears

As an example, base shear response history in the case of equipped structure with mass ratio of 0.08, at ULS, along X direction, for one set of semi-artificial accelerograms of Vrancea 1977 type, compatible with the response spectrum for Bucharest, is depicted in Figure 3. For this set of semi-artificial accelerograms, the peak value of base shear, at ULS, along X direction, is extracted from the graph. Similarly, in the case of equipped structure with mass ratio of 0.08, at ULS, along X direction, the peak base shears are extracted from the corresponding graphs of the other 6 sets of semi-artificial accelerograms. Finally, the mean of the 7 values is computed. The same procedure has been applied for all considered cases (unequipped and equipped), for all response parameters.

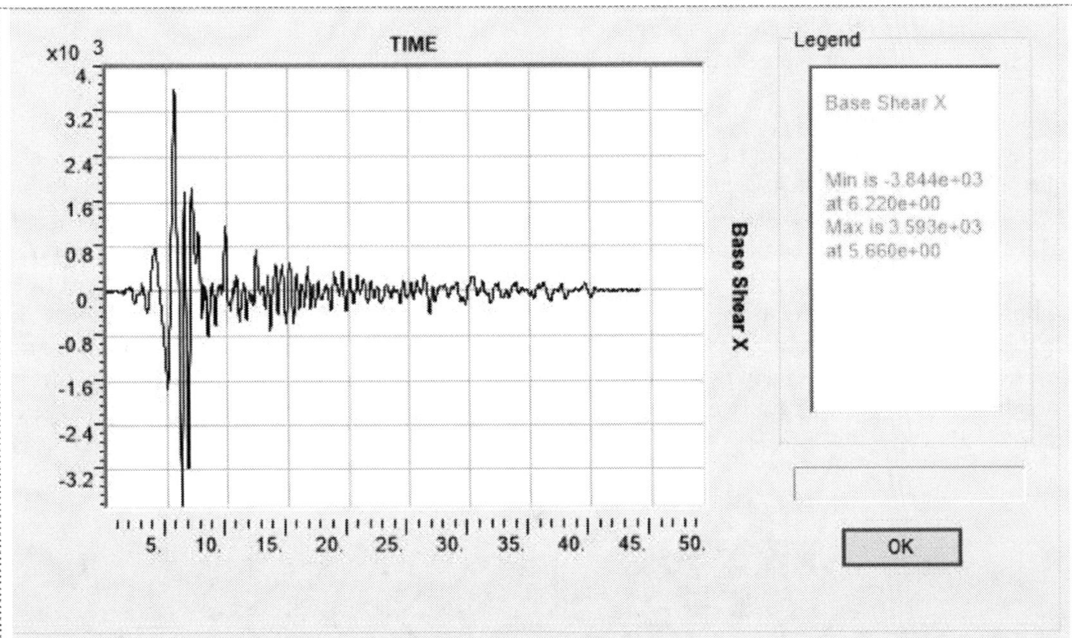

Figure 3. Base shear response history in the case of equipped structure with mass ratio of 0.08, at ULS, in X direction, obtained for one set of accelerograms of Vrancea 1977 type.

Figure 4 shows the average value (of the seven-time history analyses) of the maximum values of the base shears at ULS, of the equipped and unequipped structure, along X direction.

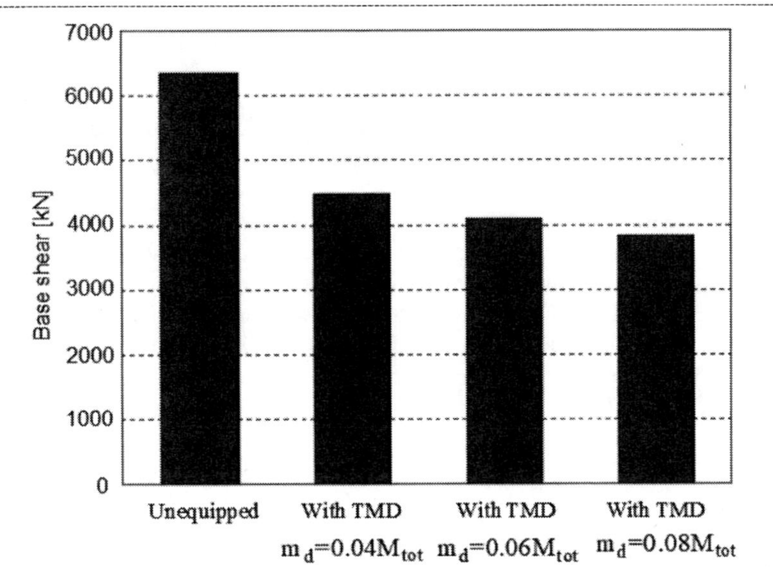

Figure 4. Base shears at ULS, in X direction.

It can be observed that by adding TMDs to the structures, the values of the maximum base shears have been significantly decreased in all cases. The reductions are slightly higher in the Y direction.

In the X direction, the peak base shear has been reduced by 29.4% in the case of the equipped structure with mass ratio of 0.04, by 35.72% in the case of the structure with mass ratio of 0.06 and by

39.65% in the case of the structure with mass ratio of 0.08, compared to the unequipped structure. In the direction Y, the reductions have ranged from 31.55% to 41.16%.

3.2. Inter-story drift ratios

Figure 5 shows the distribution of the peak inter-story drift ratios over the height of structure, at SLS, for the perimeter moment dual eccentrically braced frame, along X direction.

By equipping the structure with TMD, the peak inter-story drifts along X direction have been reduced by 27.32%-29.11% for the structure with mass ratio of 0.04, by 34.05%-35.27% for the structure with mass ratio of 0.06, and by 37.90%-39.55% for the structure with mass ratio of 0.08, compared to the unequipped structure. The distribution of inter-story drift ratios over the height of structure is similar along the Y direction, the reductions being slightly higher. Along Y direction, the reductions have ranged from 27.65%-41.11%. It is worth mentioning that on both directions, reductions of inter-story drifts over the height of structure are approximately constant, slightly higher values being registered at the lower floors.

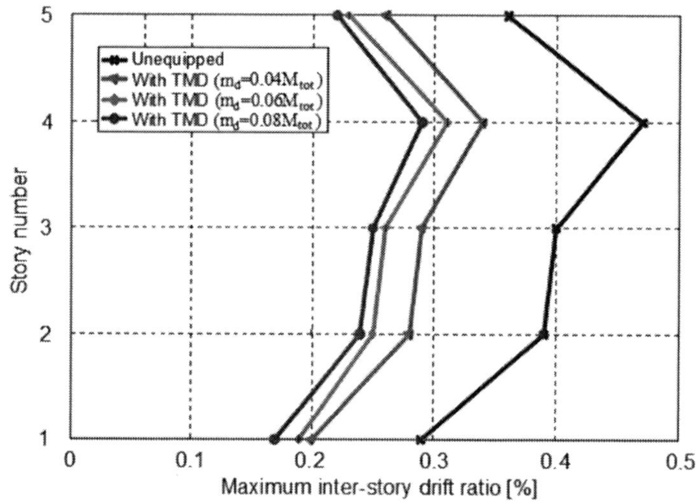

Figure 5. Inter-story drift ratios at SLS, in X direction.

3.3. Absolute story accelerations

Figure 6 illustrates the distribution of the peak absolute story accelerations over the height of structure, at ULS, for the perimeter moment dual eccentrically braced frame, along X direction.

In the case of unequipped structures, it can be observed an increase of lateral accelerations from the lower floors to the upper floors. By equipping the structure with TMDs, the maximum lateral accelerations along X direction have been reduced by 17.45%-20.12% for the structure with mass ratio of 0.04, by 23.58%-25.47% for the structure with mass ratio of 0.06, and by 27.70%-29.58% for the structure with mass ratio of 0.08, compared to the unequipped structure. Slightly higher reductions have been obtained along Y direction. These reductions have ranged from 18.7% to 31.3%. The distribution of the reductions over the frame height is approximately constant at all floors, for all equipped structures.

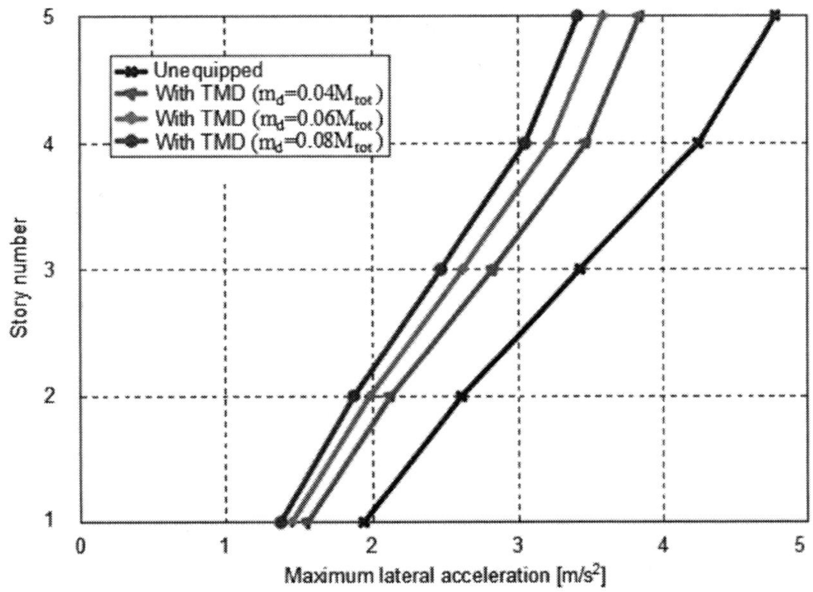

Figure 6. Absolute story accelerations at ULS in X direction.

4. Conclusions

A 3D five story structure with perimeter dual eccentrically braced frames, located in Bucharest, has been chosen for this study. Then, the building has been equipped at the top floor with bidirectional TMDs with three mass ratios 0.04, 0.06 and 0.08, without redesigning the structural members. The influence of mass ratios on the seismic response as well the efficiency of the TMDs in enhancing the seismic response of bare structure has been investigated through nonlinear time history analyses.

The main findings from this research are summarized as follows:

1. The analysed seismic response parameters (peak base shears, peak inter-story drift ratios and peak absolute lateral story accelerations) have been reduced for all equipped structures, in orthogonal directions X and Y, in comparison with the bare structure. The reductions are higher, as the mass ratios are higher.
2. Base shears have been reduced by 29.4%-31.55% for mass ratio 0.04, by 35.72%-37.31% for mass ratio 0.06 and by 39.65%-41.16% for mass ratio 0.08.
3. Inter-story drift ratios have been reduced by 27.32%-29.82.55% for mass ratio 0.04, by 34.05%-36.30% for mass ratio 0.06 and by 37.90%-41.11% for mass ratio 0.08.
4. Lateral absolute story accelerations have been reduced by 17.45%-21.4% for mass ratio 0.04, by 23.58%-26.80% for mass ratio 0.06 and by 27.70%-31.30% for mass ratio 0.08.

References

[1] Mociran H A 2016 Seismic performance of dual eccentrically braced steel frames equipped with fluid viscous dampers *Proc. of the 6th Int. Conf. on Struct. Eng., Mech. and Comput (Cape Town)* p 271
[2] Chey MH 2014 Study of TMD–controlled Reinforced Concrete Frame Structure under Code-based Artificial Earthquake *Rev. Tec. Ing. Univ. Zulia* **37** 1
[3] Hossain MA, Miah MS, Miah MJ and Hossain MM 2019 Performance evaluation of a newly developed translational tuned mass damper *Int. J. of Str. and Civ. Eng. Res.* **8** 2
[4] Elias S and Matsagar V 2017 Research developments in vibration control of structures using passive tuned mass dampers *Annual Rev. in Contr.* **44** 129
[5] Elias S and Matsagar V 2017 Research developments in vibration control of structures using passive tuned mass dampers *Annual Rev. in Contr.* **44** 129

[6] Soong TT and Dargush GF 1997 *Passive Energy Dissipation Systems in Structural Engineering* (Chichester: John Wiley&Sons) p 356

[7] Zahrai SM and Ghannadi A 2006 Effectiveness of Tuned Mass Dampers in Mitigating Seismic Vibration of Multi-Story Buildings *Proc. of the 7th Int. Cong. on Civ. Eng., (Teheran)*

[8] Sadek F, Mohraz B, Taylor AW and Chung R 1997 A methodology for estimating the parameters of tuned mass dampers for seismic applications *Earth. Eng. and Str. Dyn.* **26**

[9] Farghaly AA, 2012 Optimum design of TMD system for tall buildings *Int. J. Opt. Civil Eng.* **2** 4

[10] Praven JV and Anusha HP, 2018 A comparison study of RC framed structures using passive control devices as soft storey *Int. J. Eng. Res.&Tech.* **7** 05

[11] Nateghi F and Tarbali K, 2015 The effect of tuned mass damper on seismic response of building frames with uncertain structural characteristics *Int. J. of Eng* **28** 5

[12] Priya VRS and Gopisiddappa D, 2018 Effectiveness of tuned mass dampers in vibration control of multi-storied buildings *Int. Res. J. of Eng. And Tech.* **05** 6

[13] Patwa R and Maru S, 2018 Comparative study of seismic analysis of dampers in asymmetrical RC frame building *Int. Res. J. for Res. In App. Sc.&Eng. Tech* **6** 8

[14] P100-1/2013 Seismic design code-Part 1: Design rules for buildings (in Romanian), Bucharest.

[15] EN1993-1-1:2005 Eurocode 3: Design of steel structures-Part 1-1: General rules and rules for buildings, CEN-European Committee for Standardization, Brussels

[16] Mociran HA 2014 Earthquake response of steel frame structures – structures equipped with tuned mass dampers *Proc. of the 7th Europ. Conf. on Steel and Composite struct. (Naples)* p653

[17] Rana R and Soong TT 1998 Parametric study and simplified design of tuned mass dampers *Eng. Struct.* **20** 193

[18] Den Hartog JP 1956 *Mechanical Vibrations* (New York: McGraw-Hill)

[19] Elias S and Matsagar V 2017 Effectiveness of tuned mass dampers in seismic response control of isolated bridges including soil-structure interaction *Lat. Am. J. solids struct.* **14**

[20] Lin CC, Hu CM, Wang and Hu RY 1994 Vibration control effectiveness of passive tuned mass dampers, *J. of the Ch. Inst. of Eng. J.* **17**

[21] Scheller J and Constantinou MC 1999 *Response History Analysis of Structures and Energy Dissipation Systems: Verification Examples for Program SAP2000 -Technical Report MCEER-99-0002* (Buffalo: MCEER) p 112

[22] Taghavi S and Miranda E 2004 Estimation of seismic acceleration demands in building components *Proc. of 13th World Conf. on Earthquake Engineering (Vancouver)* p16

IOP Conf. Series: Materials Science and Engineering **789** (2020) 012041 doi:10.1088/1757-899X/789/1/012041

Multicriterial optimization of topo-geodetic techniques for behaviour monitoring in service stage of civil engineering structures

D-I Morariu[1*], D Lepădatu[1] and I Şerbănoiu[1]

[1] Faculty of Civil Engineering and Building Services, "Gheorghe Asachi" Technical University, Iasi, Romania

* morariudianaioana@yahoo.com

Abstract. Optimization of civil engineering systems is a complex process that includes both theoretical and practical knowledge of experts from different disciplines. Optimization is defined as a rigorous process by which the methods corresponding to an engineering problem are analyzed and selected. The objective of this scientific paper is to define an optimal system / technique in monitoring the behaviour of civil engineering structures by topographic methods using the multi-criterial decision theory, by analyzing, combining and comparing some topographic methods and techniques used in monitoring the behavior of objectives by identifying decisional criteria (quantitative and qualitative) and applying one-or multi-criteria decisions. Finally, by choosing quantitative and qualitative criteria, we can establish the optimal techniques that correspond to the monitoring activity of the behaviour in service stage of civil engineering structures. The necessity for this optimal system that uses an algorithm is to reduce the risks and uncertainties that may arise from the monitoring process of buildings. Also, the results obtained must fulfil a certain accuracy, time and cost conditions, as well as their extent for other monitoring activities in topographical measurements.

1. Introduction

Specialized civil engineering structures are structures that require at least a scheduled, if not permanent, monitoring to be able to identify in a timely manner structural degradations that may lead to unwanted collapse and implicit loss of human life. Thus, depending on the size and importance of the monitored behavior of the objective in service stage involves significant costs that depend on several factors [1]. Multicriteria analysis is a complex investigation method by which the multidimensional definition space is searched for the combination that satisfies user requirements for efficient optimization of a process or phenomenon [2].

The multi-criteria decision provides a systematic methodology that combines these objectives sometimes conflicting [3] and can generate a compromise solution closer to the reality of complex decision-making issues that are almost impossible to solve [4]. Multicriteria decision-making methods, among which we can mention the Promethee method [5], are a necessary tool, but valuable in the decision-making process [6]. Using this method of multicriteria analysis [7] and optimization in the present paper, we were able to quantify in an intelligent manner and decide according to the scenarios, the method that meets the imposed criteria and constraints, sometimes quite difficult to accomplish.

2. Visual Promethee – multicriterial analysis software

Visual Promethee Gaia is a virtual program to help with multi-criteria decisions that has implemented the Promethee and Gaia concept. The software was developed under the supervision of Professor Mareshal Bertrand of the Solvay School of Economics and Management in Brussels, Belgium. In 1984, Bertrand Mareschal's first Promethee program was implemented differently from today's version and was difficult to adapt to computers.

Later, the software was ported to IBM computers, providing the basis for PromCalc [8]. PromCalc software was developed around 1990 by Bertrand Mareshal and Jean-Pierre Brans, being the only interactive and graphical program in the field of multicriteria decision making. With the development of Windows 95 and 98 at the end of the 90's, a new program emerged in the decision technique: Decision Lab 2000.

In 2010, D-Sight was created by Yves de Smet at the University of Brussels and is available for use. The development of Visual Promethee (figure 1) began in 2012 to provide a replacement for the Decision Lab program. A strong emphasis is placed on the quality and consistency of the user interface, the visualization of the aspects and the ease of use of the software, implementing the latest and most advanced developments in multicriteria decisions. The program is available in 4 versions: Demo Edition, Academic Edition, Business and Online Edition [8].

PROMETHEE
METHODS

Figure 1. Visual Promethee Gaia software interface.

The Visual Promethee Gaia program classifies criterias in two ways: quantitative (economic, technical, time) and qualitative (numerical value scale).

The advantages of using Visual Promethee Gaia software are: the easy use of the program and its applicability in vast areas, the setting of scenarios according to the importance of the group of criterias, the graphical representation of the criteria in the multicriteria decision that can provide clarity on choosing the most optimal solution.

3. Results and discussions

The case study in this scientific paper uses multicriteria analysis based on concrete data. The information on which the three scenarios were based on come from monitoring through topographic methods different types of civil engineering structures. Moreover, the tools presented in the scenarios below contain details from the observations the authors had made during topographic activity. Knowing the particularity of each topographic instrument is essential when the operator wants to obtain an optimized system. The multicriteria analysis in this part of the paper was organized on two sets of criteria: technical and qualitative.

Technical criteria include: precision, measurement distance, memory space and time. On the other hand, qualitative criteria are represented by: price, resources and autonomy. Thus, we have drawn up three selection scenarios on the optimal method / technique that meets the requirements of the structure's monitoring process. Each group of criteria introduced into the Visual Promethee Gaia processing program has been assigned ratios of importance.

3.1. Scenario I

Scenario 1 was performed taking into account the absolute value of precision against the other criteria listed above. Accuracy influences directly the results of the monitoring, the better the observations of spatial displacements will have improved accuracy.

Table 1. Scenario 1-Accuracy.

Scenario 1	Level	Total Station	GPS	Laser scanner	UAV
Accuracy (mm)	7	15	30	20	30
Time (Hours)	4	2	1	2	1
Meas. Distanc (m)	100	1000	2000	200	50
Int. memory (Pts)	1000	17000	25000	50000	70000
Price (RON)	1000	1500	2000	3000	5000
Res. (operators)	3	2	1	1	1
Auton. (Hours)	4	5	6	6	1

The modern intelligent techniques presented in the figure 2 with which we can monitor engineering structures over time are: the level, the robotic total station, Global Positioning System (GPS), the laser scanner and the UAV. The data was introducing in the Visual Promethee Gaia program, where we obtained the following results (figure 2):

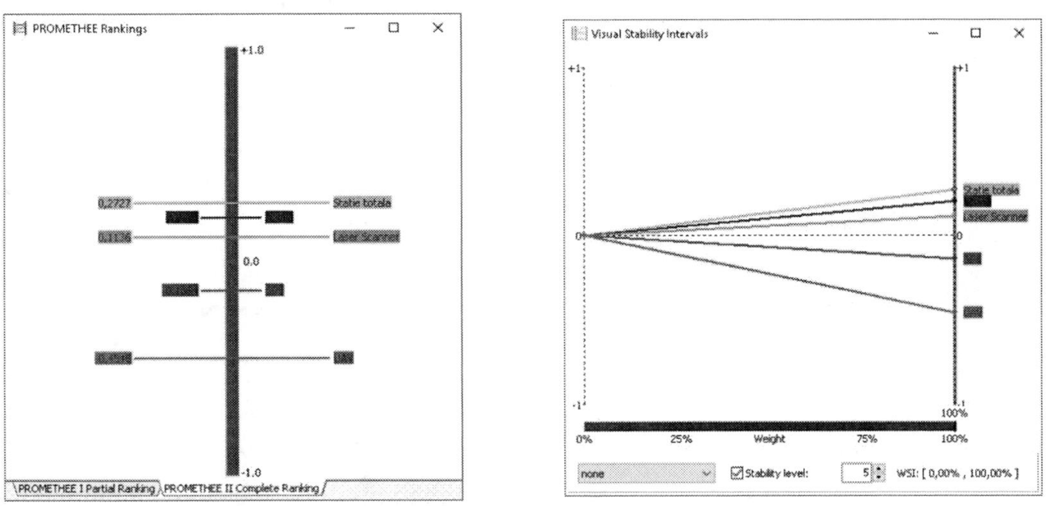

Ranching Stability

Figure 2. Scenario I – Maximum accuracy.

3.2. Scenario II

Scenario 2 is rendered by the influence of time value. Time depends on several factors: measurement distance, storage space, resources and autonomy of topographic tools. The robotic total station, laser scanner and UAV have time values assigned to the process of the monitored structure for at least one hour. This also depends on the size of the target, visibility to all landmarks and weather conditions (wind, rain, excessive heat). In the case of unmanned aerial vehicle flight, these factors are important because the process can only take place in circumstances beneficial to this type of technique.

Table 2. Scenario 2-Minimum time.

Scenario 2	Level	Total Station	GPS	Laser scanner	UAV
Accuracy (mm)	3	4	10	5	5
Time (Hours)	4	1	1	1	1
Meas. Distanc (m)	100	1000	2000	100	50
Int. memory (Pts)	1000	20000	30000	30000	50000
Price (RON)	1000	1500	1500	2000	2500
Res. (operators)	2	2	1	1	1
Auton. (Hours)	5	6	6	6	1

Also, the number of resources for laser scanning and UAV is minimum an operator and the autonomy is maximum of 6 hours, during which the spatial of the engineering structure can be recorded with the total station and laser scanning. The data processed using the Promethee method were materialized in graphical form (figure 3).

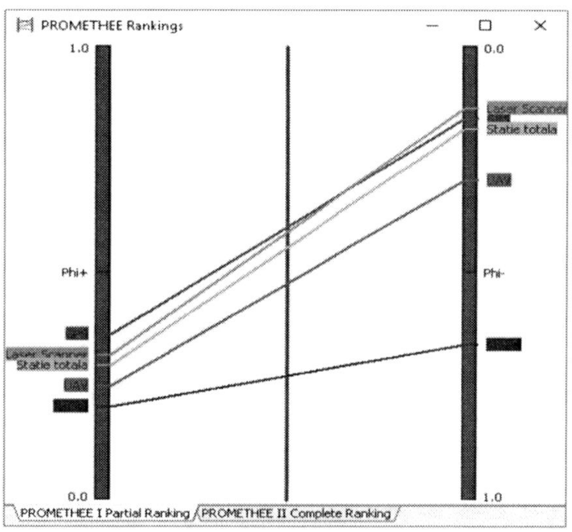

Ranching

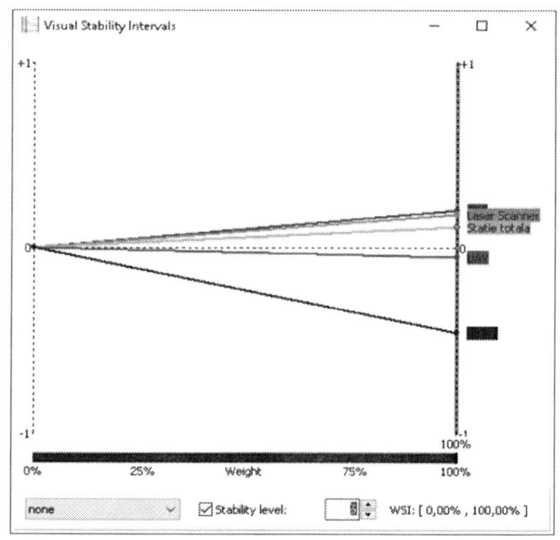

Stability

Figure 3. Results processed with Promethee method for Scenario II.

As against Scenario 1, the results of the vertical spatial displacements, where the accuracy was taken into account, Scenario 2 was based on the value of time. Thus, the laser scanner is the optimal technique to monitor the behavior of a civil engineering structures in service stage according to the time criterion. The values obtained in both charts: Promethee I Partial Ranking and Visual Stability Intervals classify the method on the first place, followed by the small difference of the robotic total station.

3.3. Scenario III

Scenario 3 was based on the introduction of the seven quantitative and qualitative criteria, but the results were obtained on the basis of the Global Cost criterion. This includes: the number of operators, the precision of the equipment and the time allocated to the determination of the spatial displacements. Following the latest software analysis we have obtained the following results:

Table 3. Scenario 3-Global Cost.

Scenario 3	Level	Total Station	GPS	Laser scanner	UAV
Accuracy (mm)	4	5	10	5	15
Time (Hours)	3	2	1	1	1
Meas. Distanc (m)	100	1500	2000	150	50
Int. memory (Pts)	1000	20000	30000	50000	60000
Price (RON)	1000	2000	3000	5000	6000
Res. (operators)	2	2	1	1	1
Auton. (Hours)	4	6	6	6	1

Following study 3 of the global cost of time monitoring of spatial displacements, we can see that the total robotic station corresponds to the minimum process cost criterion. The other two techniques are at a very large difference from the total station.

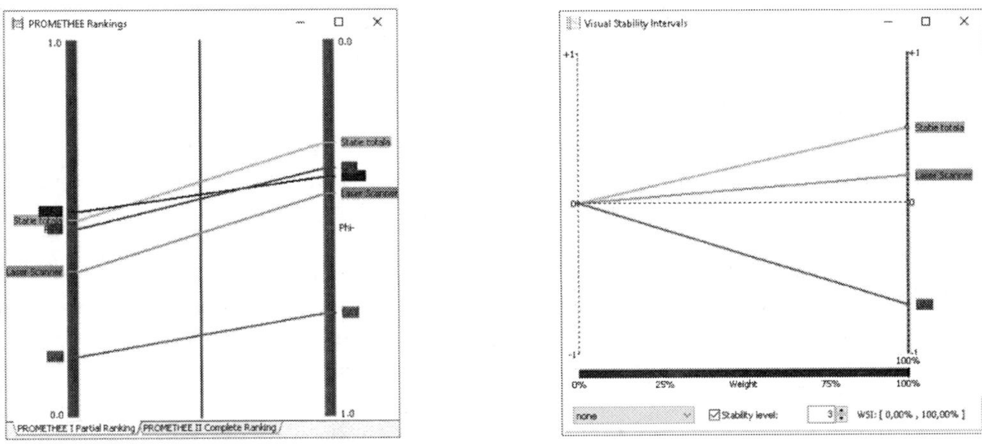

Ranching Stability

Figure 4. Results processed with Promethee method for Scenario III.

Finally, we compared the three scenarios to be able to apply the decision criteria for the optimal technique for monitoring engineering structures. Each scenario had the same criteria introduced, but the values differed in the precision, time and total costs. Also, the results were influenced by the measure distance, number of operators, autonomy and memory space.

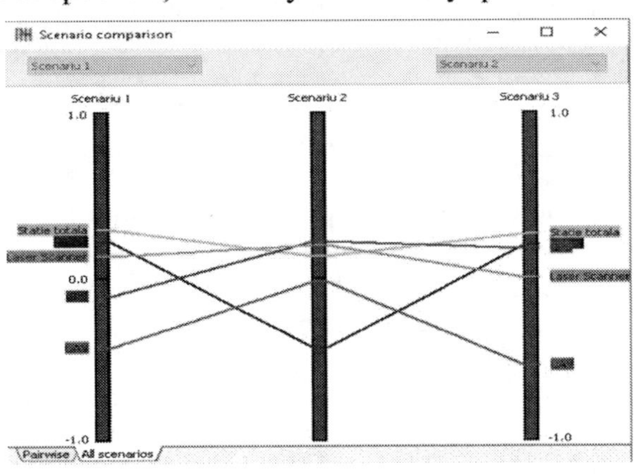

Figure 5. Comparative analyses for the three scenarios.

Thus, after analyzing the three situations, it is noticed that the total robotic station corresponds to the criteria: precision, time and cost. On the other hand, level can be used to monitor a civil engineering structure over time, depending on the purpose of the building and its size. Also, operator experience and professional software is required to achieve high accuracy results. The tools used in the process of monitoring are what we call smart technology optimization as being the latest generation instruments and definitely contribute to increasing the efficiency and accuracy of the monitoring process in time of civil engineering structures in service stage. We can add that this process is a mandatory one for the structures of the type presented (vital importance class) and also the precision and accuracy of the results can contribute in a significant manner to the disaster prevention caused by the loss of structural stability due to the undesirable appearance of some structural displacements that are not admissible.

4. Conclusions

Multicriteria optimization is an advanced analysis and optimization method that allows the user to obtain the desired response while taking into account several parameters that ultimately lead to a compromise between conditions or criteria that may be antagonistic, making it difficult to establish an alternative technique efficient to solve such problems. Thus, with this method several objective functions have to be optimized at the same time, and it is possible to determine the combination of control parameters of a process or phenomenon that simultaneously maximizes one and minimizes others. The process is based on two types of information: the partial importance of the criteria and the characteristics of the decision when the user compares the yield offered by each criterion.

In this paper we performed a multicriteria optimization using the Promethee method and Visual Promethee Gaia program through which we managed to rank in a professional manner a complex problem that allowed us to take into account the control parameters of the topo-geodetic monitoring process in the process of studying the behavior of civil structures in service stage and constraints that we could not quantify with other optimization methods. The final comparison showed that the robotic total station and the level are the most efficient tool in user-defined conditions and level of precision.

Within the monitoring in time of a civil engineering structure in service stage, the system will be optimized according to the accuracy, that is an important condition for the precision of the results, the time allotted to the operations necessary to obtain the results, but also their costs. The method comprises several objective functions that must be optimized at the same time and the combination of the control parameters of a process or phenomenon can be determined which simultaneously maximizes one and minimizes the other.

References

[1] Morariu D I and Lepadatu D 2018 Advanced method for station point control accuracy to monitor the behaviour in service stage of civil engineering structures using geodetic satellite technology *Proceedings of Eighth – International – Conference – On – Advances – in – Civil – Structural and Mechanical-Engineering Paris* pp 22-26

[2] Matthias E Figueira J R and Salvatore G 2010 Trends in Multiple Criteria Decision Analysis *Springer New York Dordrecht Heidelberg* London

[3] Teodoriu G and Serbanoiu I 2013 Decision analysis to evaluate thermal comfort in residential buildings *Romanian Journal of Civil Engineering* Vol. **4** pp 282-287.

[4] Seddiki M Anouche K Bennadji A and Boateng P 2016 A multi-criteria group decision-making method for the thermal renovation of masonry buildings: The case of Algeria *Journal of Energy and Buildings* Vol.**129** pp 471–483

[5] Taillandier P and Stinckwich S Using the PROMETHEE multi-criteria decision making methodto define new exploration strategies for rescue robots *IEEE International Symposium on Safety Security and Rescue Robotics Kyoto* Japan pp 321 - 326

[6] Radulescu C Z 2015 *Multicriteria Decision Methods and Applications in Sustainable Development*

[7] Zopounidis C and Pardalos P M 2010 *Handbook of Multicriteria Analysis* Springer edition ISBN-10: 3540928278
[8] Visual Promethee Gaia Manual – VPS 2013 (Available from: VP Solutions)

Assessment of code-based design approach of buckling restrained braced frames in P100-1/2013 using pushover analysis

E Mosnoi[1*], A Stratan[1] and T F Voica[1]

[1] Politehnica University of Timisoara, Faculty of Civil Engineering, Department of Steel Structures and Structural Mechanics, Romania

[*] eujen.mosnoi@student.upt.ro

Abstract. Buckling restrained braced frames (BRBFs) are a relatively new structural system, characterized by high seismic energy dissipation. The Romanian seismic design code P100-1/2013 has recently introduced design provisions for buckling restrained braced frames. According to the code, the dissipative components are designed for the seismic forces reduced by behavior factor q, while the non-dissipative components are capacity-designed for seismic forces amplified with an over-strength factor Ω^T. The objective is to assess the code-based capacity design procedure using the nonlinear static analysis. Three structural configurations are considered: two BRBF configurations and a dual BRBF configuration. The structures are first designed following the code procedure, and then analyzed using a pushover analysis, using a numerical model of the buckling restrained braces calibrated on experimental data.

1. Introduction

It is known that conventional brace members utilized to reduce the lateral displacements on structural buildings, when subjected to compressive forces exhibit buckling deformation and their hysteretic behavior is unsymmetrical in tension and compression. To solve this, engineers developed a new type of braces, the buckling-restrained brace (BRB), which due to its construction exhibits the same behavior in both tension and compression.

The buckling-restrained braced frame (BRBF) system differs from a conventionally brace frame in that the BRB braces are capable to yield in both tension and compression and are more ductile.

Sabelli, Mahin and Chang [1] showed that the seismic performance of BRBFs have comparable or better behavior than conventionally braces.

Uang and Kiggins [2] studied whether the use of a backup moment frame system, which provides a restoring force mechanism, in a dual system can minimize residual deformations.

Studies about design of BRBFs implementation in Romania were made by Stratan et all [3].

The objective of this paper is to perform an assessment for three structure configurations: a dual antiseismic structural system consisting from moment resisting frames (MRF) combined with buckling restrained braced frames (BRBF) and two buckling restrained braced frames (BRBF) with base of central columns fixed or pinned using pushover analysis. To achieve this we developed a finite model of the structure and calibrated the numeric model of BRB using an experimental model of BRB from IMSER project [6] data.

2. Prototype structure

2.1. Location

The location for this structure is city of Timisoara, Romania. The ground motion for this place has an acceleration of 0.20 g. The seismic action to be taken into account for the "Life Safety requirement" corresponds for a return period of 225 years and has a probability of exceedance of 20 % in 50 years.

2.2. Structural solution

As it can be seen in Figure the structures has the following particularities: Plan dimensions - 37.50 x 22.50 m, Spans on transversal direction - 3 x 7.50 m, Spans on longitudinal direction - 5 x 7.50 m, Story height - 3 x 3.50 m, Category - Office areas, Type of steel elements utilized in the structure - European hot-rolled steel sections, Steel grade for elements - S355. For exterior walls is utilized lightweight glass curtain wall system. The partition walls in interior are fabricated from gypsum boards fastened with screws on a self-supporting lightweight metal structure with infilling thermal insulation.

As Figure presents, we analyzed three structures. The lateral load resisting system consists from one inverted V buckling restrained braced frame (BRBF) for each of them. The BRBF is used in "X" direction in axes "1" and "6", and in "Y" direction in axes "A" and "D". First analyzed structure is a dual moment resisting frame (MRF) combined with BRBF - (D-BRBF) system.

For second and third structure, the MRFs are replaced with pinned beams, the lateral resisting system is only the BRBF. The second structure consists from BRBF with central columns fixed - (BRBF-F). The third structure consists from BRBF with central columns pinned - (BRBF-P).

The other frames are designed only for gravity loading. The secondary beams are considered pinned to the main beams. For gravity loading was considered a 0.5 m console on the perimeter contour of the structure, plan dimensions for gravity loading being - 38.50 x 23.50 m.

Due to the fact the structure is regular, the structural elastic analysis has been performed on a 2D model for half of the structure and only the frame from axis 1 was analyzed.

The gravity load resisting system is formed from a reinforced concrete slab cast on a profiled steel sheeting. The floor is supported on main and secondary beams.

The last floor is designed as an open space terrace. The structure has a compact shape in the plane, the floor made from reinforced concrete has sufficient stiffness in its plane so can provide the diaphragm effect at each level. As a consequence, the structure is regulated in plane in accordance with the requirements of paragraph 4.4.3.2 of Romanian seismic code P100-1 [5].

The analyzed structure is monotonous on vertically, with no discontinuities, with insignificant variation of rigidity and vertical resistance.

The mases applied on the construction are distributed relatively uniform, falling within the 50% limit of variation between adjacent levels. Consequently, the structure is vertically regulated as required by paragraph 4.4.3.3 of P100-1 [5].

The sections of structural elements are presented in table 1.

Table 1. Hot-rolled sections of frames.

Structure type	Level	Braced frame beams	Unbraced frame beams	BRB sections [mm]	Central columns	Marginal columns
	3	HE240A	IPE360	15x50	HE260B	HE220B
D-BRBF	2	HE240A	IPE360	20x65	HE260B	HE220B
	1	HE240A	IPE360	20x70	HE260B	HE220B
	3	HE240A	IPE400	15x60	HE220B	HE200B
BRBF-F	2	HE240A	IPE400	20x65	HE220B	HE200B
	1	HE240A	IPE400	20x75	HE220B	HE200B
	3	HE240A	IPE400	15x60	HE240B	HE200B
BRBF-P	2	HE240A	IPE400	20x65	HE240B	HE200B
	1	HE240A	IPE400	20x75	HE240B	HE200B

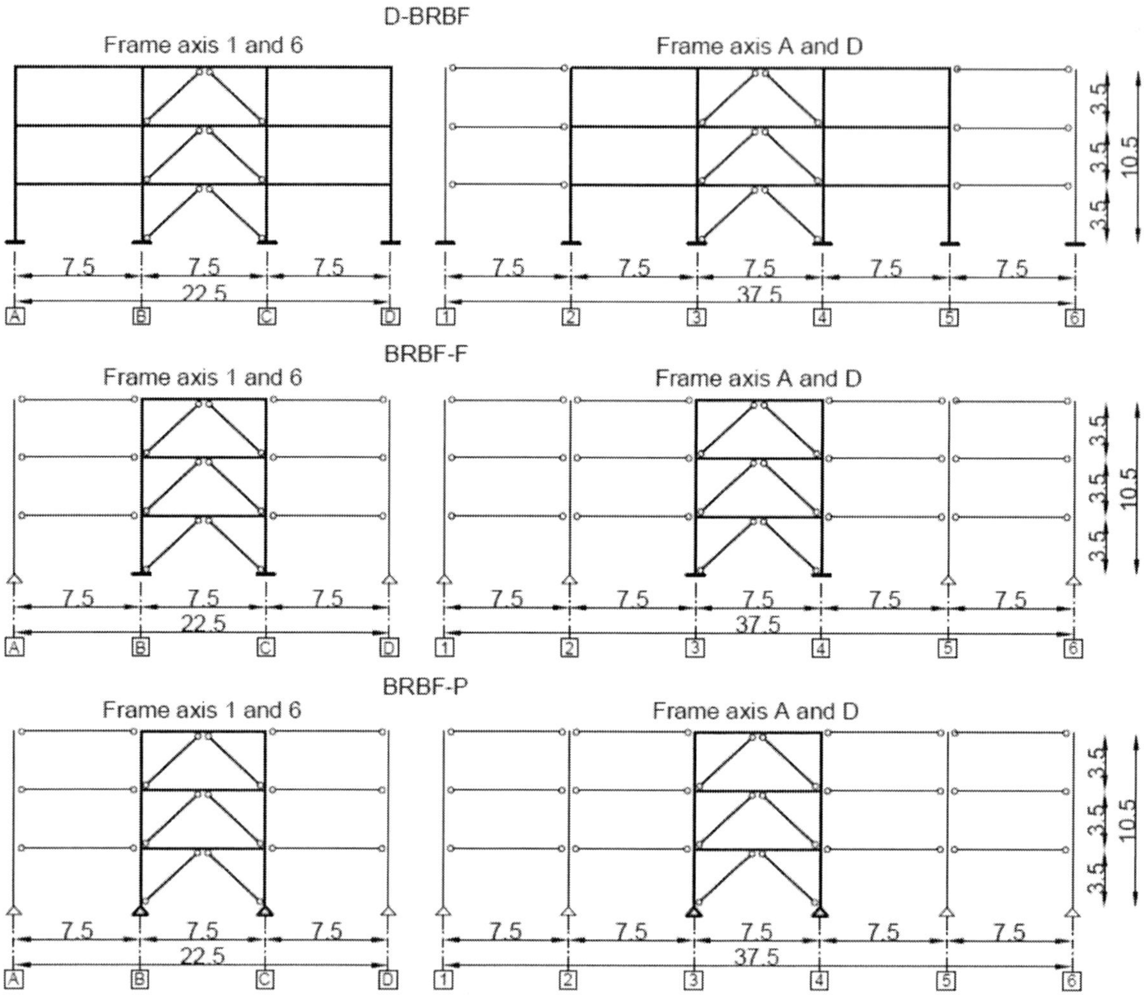

Figure 1. Side views of the structure.

2.3. Design of the structure

The elastic design is performed for ultimate and serviceability limit states. The norm according to which the design combinations are considered is CR0 [4].

The check of structure elements begins with persistent design situation. After this the structure elements are checked in seismic design situation.

In seismic design situation, first are checked dissipative members (buckling restrained braces (BRB), and unbraced frame beams for dual MRF+BRBF - (D-BRBF) system), (buckling restrained braces (BRB) for BRBF-F and BRBF-P systems). In second order are checked non-dissipative members (braced frame beams and columns). The non-dissipative members are designed having an over-strength with respect to the buckling restrained braces, and unbraced frame beams for D-BRBF system, and with respect to the buckling restrained braces for BRBF-F and BRBF-P systems.

For dual MRF+BRBF - (D-BRBF) system, according to paragraph 6.10.2 (2) from P100 [5], the unbraced frames, located in the bracing direction of the building has to be able to undertake at least 25% of the seismic action in the hypothesis in which bracing frames came out of work.

Accordingly to this, moment resisting frames (MRF) should be able to resist at least 25 % of the total seismic force using general formula:

$$F_y^{MRF} \geq 0.25 \cdot \left(F_y^{MRF} + F_y^{BRBF}\right) \quad (1)$$

Using the formulas below can be determined the duality of this particular structure:

$$2F_y^{MRF} + F_y^{BRBF} = F_y^{DUAL} \quad (2)$$

$$2F_y^{MRF} \geq 0.25 \cdot F_y^{DUAL} \quad (3)$$

where F_y^{MRF} - is the resistance of MRF; F_y^{BRBF} - is the resistance of BRBF; F_y^{DUAL} - is the total resistance of MRF+BRBF structure.

The duality checks for the MRF+BRBF (D-BRBF) structure is represented in table 2.

Table 2. Duality check.

Level	Beam	Brace	F_y^{BRBF} (kN)	F_y^{DUAL} (kN)	$2F_y^{MRF}$ (kN)	$0.25F_y^{DUAL}$ (kN)	Duality (%)
3	IPE360	15x50	389.28	802.70	413.42	200.67	51.50
2	IPE360	20x65	674.76	1088.18	413.42	272.04	37.99
1	IPE360	20x70	726.66	1140.09	413.42	285.02	36.26

3. The numerical modelling

3.1. Design of the structure of the BRB

The nonlinear behavior of the BRB is very important, and in order to have a high level of confidence in nonlinear computation it is necessary to calibrate the BRB according to an experimental model.

For numerical BRB behavior was utilized the experimental data for a BRB model from IMSER project [6]. The geometric data of experimental model are according to figure 2.

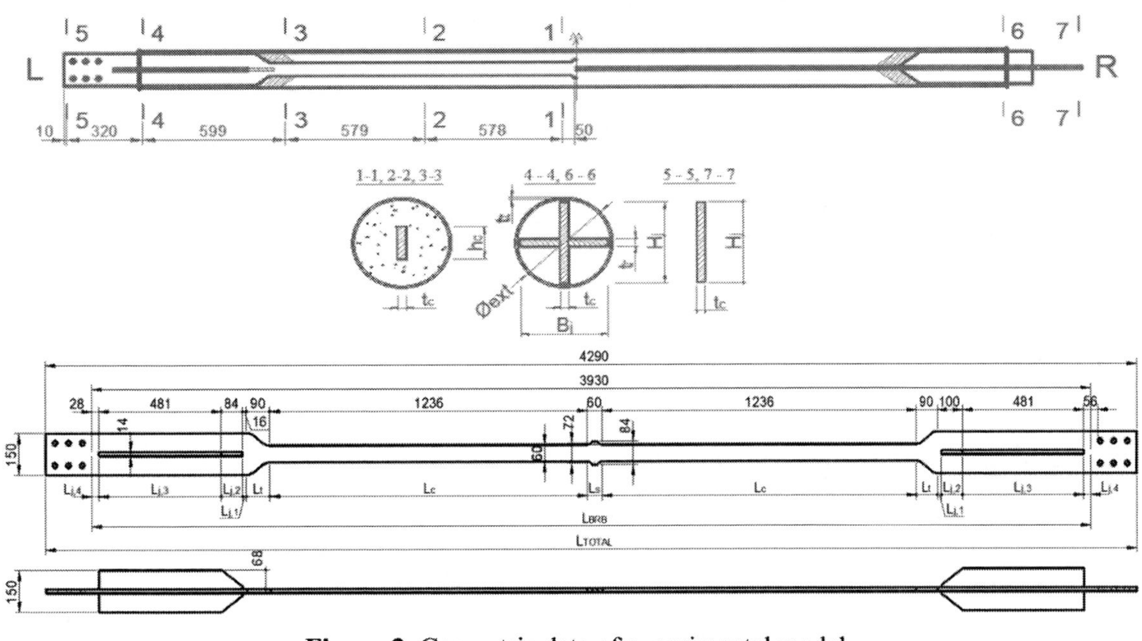

Figure 2. Geometric data of experimental model.

Due to the fact that the core of the BRB is composed of several segments with different sections, the effective stiffness of BRB - K_{ech} is determined according to formula:

$$K_{ech} = (\sum(Ki)^{-1} + \sum(Ke)^{-1} + \sum(Kt)^{-1} + \sum(Kp)^{-1})^{-1} \quad (4)$$

where: Ki- stiffness of joint zone; Ke- stiffness of elastic zone; Kt- stiffness of transition zone; Kp- stiffness of plastic zone.

For non-elastic calculus of buckling restrained braces, to take into account that the cross-sectional area varies along the length of the BRB, the equivalent stiffness of the member with constant section must be corrected by multiplying with the k factor:

$$k = K_{ech} \cdot L_n/(EA_p) \tag{5}$$

where A_p – plastic area of BRB steel core; $L_n = 5129.57\ mm$ - joint to joint length of buckling restrained brace.

The modelling of equivalent stiffness of BRB by an equivalent bar of constant section can be achieved by modifying the Young's modulus ($E=210000\ N/mm^2$) of the steel assigned to the section: $E_{ech} = k \cdot E = 315000\ N/mm^2$.

According to experimental data obtained from IMSER project [6], the value of effective stiffness of buckling restrained braces factor is: $k = 1.5$.

The force-deformation diagram of buckling restrained brace (BRB) is shown in figure 3.

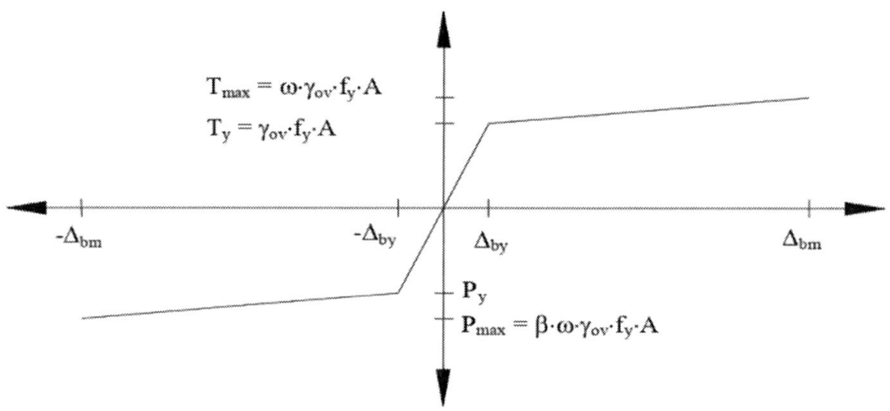

Figure 3. Force-deformation diagram of BRB.

The yield capacity at tension and compression:

$$T_y = P_y = \gamma_{OV} \cdot f_y \cdot A_p \tag{6}$$

Corrected capacity at compression according to P100-1-2013, paragraph 6.11.2(2):

$$P_{max} = \beta \cdot \omega \cdot \gamma_{OV} \cdot f_y \cdot A_p \tag{7}$$

Corrected capacity at tension according to P100-1-2013, paragraph 6.11.2(3):

$$T_{max} = \omega \cdot \gamma_{OV} \cdot f_y \cdot A_p \tag{8}$$

The utilized values of β, ω factor according to IMSER project [6] are: $\omega = 1.45, \beta = 1.17$.

The yielding capacity according to IMSER project [6] tests: $\gamma_{OV} \cdot f_y = 397{,}8\ N/mm^2$.

The deformation capacity in compression (Δ_{bm}^-) and deformation in tension (Δ_{bm}^+) are calculated:

$$\Delta_{bm} = \varepsilon_{c.max(\%)} \cdot L_n \cdot (100\%)^{-1} \tag{9}$$

for (Δ_{bm}^-) and (Δ_{bm}^+) was adopted $\varepsilon_{c.max} = 2.6\%$ of total BRB length - L_n.

The yield capacity deformation (Δ_{by}) is calculated:

$$\Delta_{by} = T_y \cdot L_n \cdot (E_{ech} \cdot A_p)^{-1} \tag{10}$$

The BRB behavior input data for model calibration was according to table 3.

Table 3. BRB calibration input values.

A_p (mm2)	P_{max} (kN)	T_{max} (N)	T_y (kN)	Δ_{by} (m)	Δ_{bm}^{-} (m)	Δ_{bm}^{+} (m)
840	566.889	484.520	334.152	0.00648	0.1334	0.1334

The numeric calibration of BRB behavior was made in a finite element model build in SAP2000 software which include a truss with length $L = L_n = 5129.57\ mm$. Using the parameters above was assigned a plastic axial P hinge at half-length of the brace.

The characteristics of BRB hinge introduced in SAP2000 software for numeric calibration, according to Table.

The comparison between experimental and numerical curves of BRB is presented in figure 4.

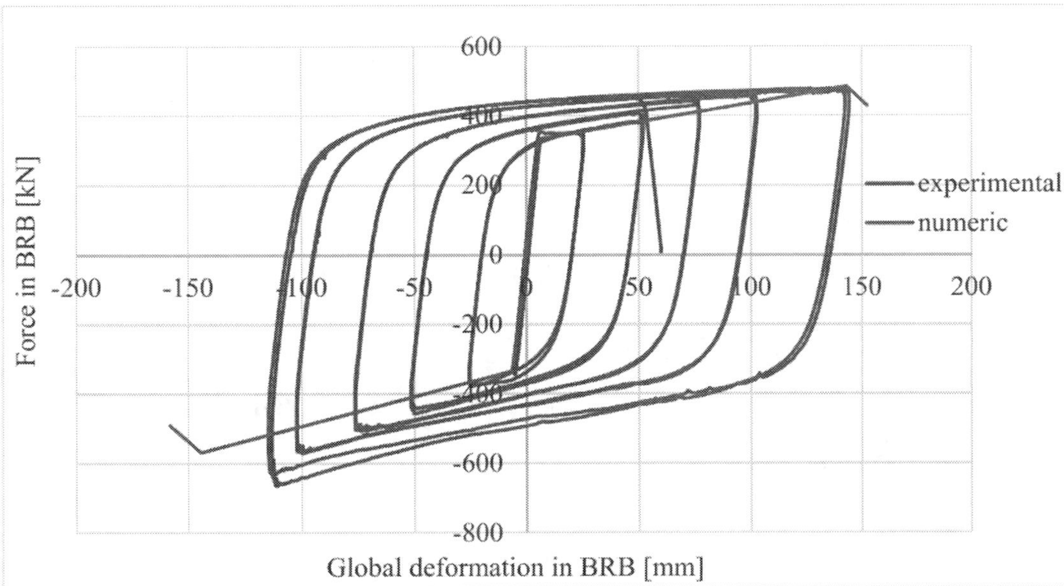

Figure 4. Force-deformation curves of experimental and numerical BRB model.

The cyclic experimental behavior model is quite similar with monotonic numerical model. The numerical monotonic model is used for non-linear static analysis.

3.2. Numerical model of the structure

The nonlinear model of the structure is a 2D model. Second order effects acting on interior gravity frames were modelled using a leaning column. The leaning column is considered pinned at each level and is linked to the structure by a rigid diaphragm constraint.

Concentrated plasticity was used to model the nonlinear response of the structure. The assigned plastic hinges for beams are type M3, for columns are type P-M3, and are assigned at each end of frames. The assigned plastic hinges for BRB braces are type P, at half-length of the brace.

4. Non-linear static analysis

4.1. Non-linear static analysis procedure

The non-linear static analysis is a two-step analysis. In the first step, the gravitational loads are applied on the structure. The lateral loads are applied in the second step.

4.1.1. Attribution of modal distribution of forces.

The static non-linear analysis is performing by use of a modal distribution of forces.

In the modal distribution of forces, forces are computed in function of the first mode of vibration of the structure. The forces are calculated with formula:

$$F_i = m_i \cdot \phi_i \tag{11}$$

where: F_i is the force to the assigned on each storey; m_i is the mass of each storey; ϕ_i is the deformation of each story.

The forces from modal and uniform distribution are shown in table 4.

Table 4. Forces for modal distribution.

Structure type	Modal				
	Level	$m_i[t]$	$\phi_i[-]$	$F_i[t]$	$F_i[-]$
	3	310.33	1.000	310.331	1.000
D-BRBF	2	298.34	0.682	203.593	0.656
	1	298.34	0.336	100.321	0.323
	3	310.33	1.000	310.331	1.000
BRBF-F	2	298.34	0.686	204.532	0.659
	1	298.34	0.329	98.201	0.316
	3	310.33	1.000	310.331	1.000
BRBF-P	2	298.34	0.692	206.595	0.666
	1	298.34	0.345	102.967	0.332

4.1.2. Attribution of plastic hinges.

The behavior input data for BRB braces plastic hinges due to experimental model calibration was according to table 5.

Table 5. BRB plastic hinges input values in SAP2000 model.

Structure type	Level	Hinge name	A_p (mm2)	P_{max}(kN)	T_{max}(N)	T_y(kN)	Δb_y(m)	Δ_{bm}^-(m)	Δ_{bm}^+(m)
	3	FH3	750	506.151	432.608	298.35	0.006479	0.1334	0.1334
D-BRBF	2	FH2	1300	877.328	749.853	517.14	0.006479	0.1334	0.1334
	1	FH1	1400	944.815	807.534	556.92	0.006479	0.1334	0.1334
	3	FH3	900	607.381	519.129	358.02	0.006479	0.1334	0.1334
BRBF-F	2	FH2	1300	877.328	749.853	517.14	0.006479	0.1334	0.1334
	1	FH1	1500	1012.3	865.215	596.7	0.006479	0.1334	0.1334
	3	FH3	900	607.381	519.129	358.02	0.006479	0.1334	0.1334
BRBF-P	2	FH2	1300	877.328	749.853	517.14	0.006479	0.1334	0.1334
	1	FH1	1500	1012.3	865.215	596.7	0.006479	0.1334	0.1334

4.2. Structural performance

4.2.1. Target displacement. The static non-linear second order analysis using $P - \Delta$ effects were performed. The target displacements (Td) were computed for ultimate limit state (ULS) according to N2 method [7].

Were analyzed the demand deformations in the elements of the structure at the calculated target displacement and were compared with allowed deformations in terms of radians for plastic rotations at the end of the beams and columns, and in terms of linear deformation at half-length of buckling restrained braces.

The plastic hinges for D-BRBF, BRBF-F and BRBF-P structures for modal distribution of forces at target displacement (Td) deformation in an upper point at level three and comparison with allowed deformations are presented in table 6 and respectively table 7 and table 8.

Table 6. Result of pushover analysis for D-BRBF structure.

Target displacement -158.20 [mm]			
Acceptance criteria - ULS			
Images of model at the target displacement			

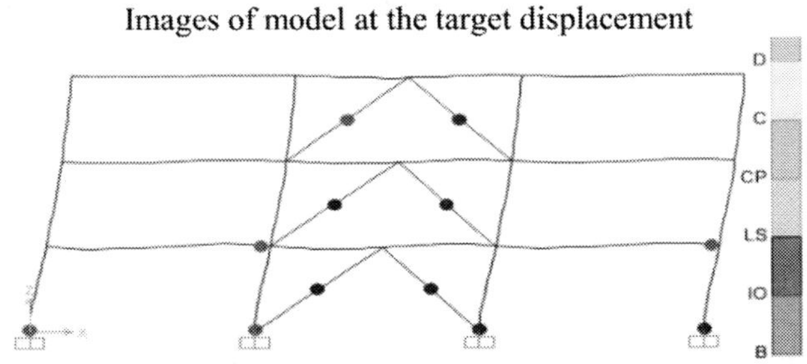

Demand vs. Capacity in terms of plastic rotations per element type [mrad]			
Beam		Column	
7.09 <	84.6	5.69 <	24.93
Demand vs. Capacity in terms of deformations per element type [mm]			
Brace			
41.3 <		133.40	

Table 7. Result of pushover analysis for BRBF-F structure.

Target displacement -159.46 [mm]			
Acceptance criteria - ULS			
Images of model at the target displacement			

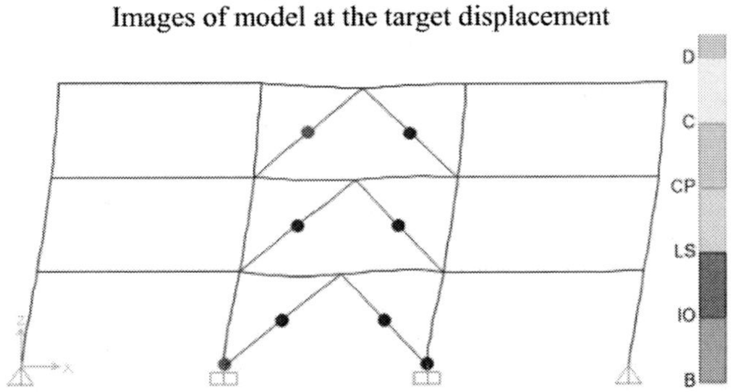

Demand vs. Capacity in terms of plastic rotations per element type [mrad]			
Beam		Column	
0 <	10.13	9.65 <	12.24
Demand vs. Capacity in terms of deformations per element type [mm]			
Brace			
47.3 <		133.40	

Table 8. Result of pushover analysis for BRBF-P structure.

Target displacement -159.01 [mm]
Acceptance criteria - ULS
Images of model at the target displacement

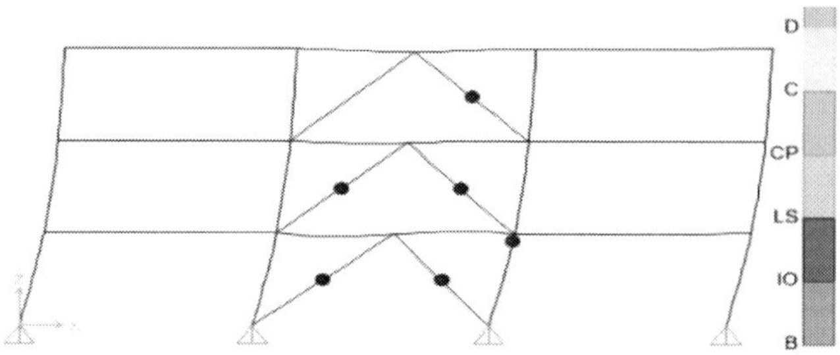

Demand vs. Capacity in terms of plastic rotations per element type [mrad]			
Beam		Column	
0 <	10.13	2.95 <	11.25
Demand vs. Capacity in terms of deformations per element type [mm]			
Brace			
62.4 <	133.40		

The results of pushover analysis and determination of target displacements for ultimate limit states (ULS) using N2 method for each of structures are presented in figure 5. The failure of all structures takes place along with failure of BRBs.

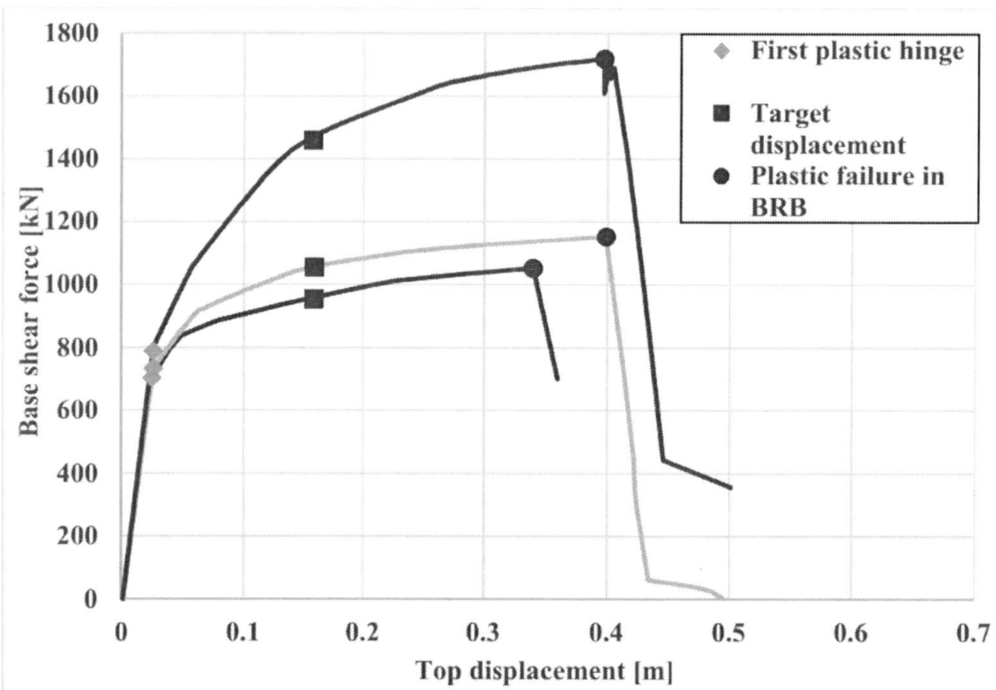

Figure 5. Pushover curves and target displacement at ULS.

From pushover analysis using N2 method we can see that al structures have a good behavior and member's deformations accomplishes the ULS deformations criteria. A better behavior has D-BRBF structure which is more resistant. BRBF-F and BRBF-P structures have nearly similar behavior, but BRBF-F structure has a better ductility –same ductility as D-BRBF. In mater of ductility D-BRBF and BRBF-F shows better behavior, the D-BRBF structure failed at top displacement d=2.5Td, BRBF-F structure failed at top displacement d=2.5Td, but BRBF-P structure failed at top displacement d=2.1Td, where Td is target displacement. In mater of base shear force resistance D-BRBF structure is more resistant than BRBF-F structure with 32.9%, and then BRBF-P structure with 38.8%. BRBF-F structure is more resistant than BRBF-P structure with 8.76%.

4.2.2. Analysis of over strength factor Ω^T. In this chapter we analyzed the over strength factor Ω^T resulted from analyses as a ratio between maximum pushover base shear force and base shear force calculated according to lateral force method in P100/1 [5].

The design Ω^T factor is an over-strength factor used to calculate the non-dissipative components of the structure according to P100/1 [5].

The over-strength factor Ω^T for buckling restrained braces (BRB) is calculated according to paragraph 6.11.5 (1) from P100/1 [5] with relation:

$$\Omega^T = \beta \cdot \omega \cdot \gamma_{OV} \cdot \Omega^N \qquad (12)$$

where $\Omega^N = \min_{1 \le i \le n} \Omega_i^N$, and $\Omega_i^N = N_{pl,Rd,i} \cdot (N_{Ed,i})^{-1}$; $\gamma_{OV} = 1.25$, -the over-resistance factor for S355 steel grade according to paragraph 6.2 (5) from P100/1 [5]; β, and ω were described in section 3.

The over-strength factor Ω^T for unbraced frame beams (MRF) is calculated according to paragraph 6.6.3 (1) from P100/1 [5] with relation:

$$\Omega^T = 1.1 \cdot \gamma_{OV} \cdot \Omega^M \qquad (13)$$

where $\Omega^M = \min_{1 \le i \le n} \Omega_i^M$, and $\Omega_i^M = M_{pl,Rd,i} \cdot (M_{Ed,i})^{-1}$.

- *D-BRBF structure*

From elastic calculus for D-BRBF structure was obtained $\Omega_{BRBF}^T = 2.37$ from primary dissipative members - BRBs and $\Omega_{MRF}^T = 2.76$ from secondary dissipative members - beams of MRFs. It was adopted the most detrimental over-strength factor - $\Omega_{D-BRBF}^T = \Omega_{MRF}^T = 2.76$ for checking of non-dissipative members of the D-BRBF structure.

- *BRBF-F structure*

From elastic calculus for BRBF-F structure was obtained $\Omega_{BRBF-F}^T = \Omega_{BRBF}^T = 2.33$ from primary dissipative members – BRBs. The secondary dissipative members does not exist in this structure due to the fact that the beams of unbraced frames are pinned.

- *BRBF-P structure*

From elastic calculus for BRBF-P structure was obtained $\Omega_{BRBF-P}^T = \Omega_{BRBF}^T = 2.25$ from primary dissipative members – BRBs. The secondary dissipative members do not exist for this structure.

The results of pushover analysis for ULS is represented in figure 6 which represents the base shear force vs. top displacement curve of structure for modal distribution of forces at ultimate limit states (ULS), considering the failure of BRB.

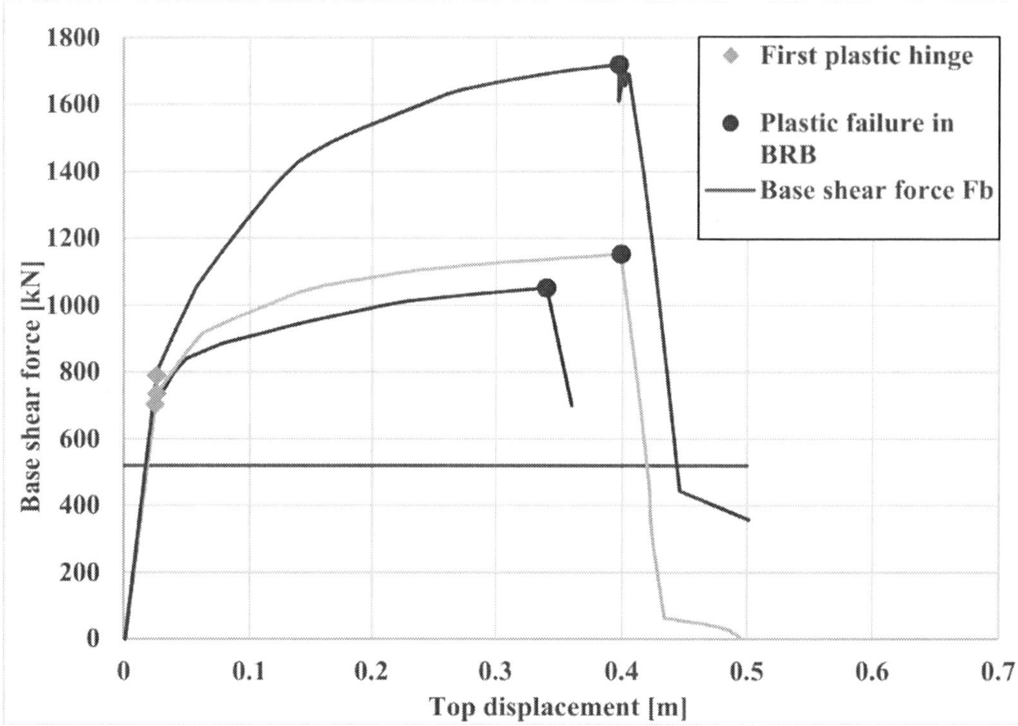

Figure 6. Pushover curves and base shear force.

The Base shear force F_b was computed according to lateral force method from P100/1 [5], paragraph 4.5.3.2.2.

Calculated base shear force is $F_b = 519\ kN$ for D-BRBF, BRBF-F and BRBF-P structures.

The Ω factor is calculated with formula:

$$\Omega^T = F_{max.push} \cdot (F_b)^{-1} \tag{14}$$

where $F_{max.push}$-The maximum force from pushover curve; F_b- Base shear force calculated according to paragraph 4.5.3.2.2 from P100/1 [5].

The results obtained for calculus of over strength factor of Ω^T calculated after pushover analysis, and using P100/1 code approach are presented in table 9.

Table 9. Calculus of over strength factor of Ω^T from pushover analysis and using P100/1 code approach.

	D-BRBF	BRBF-F	BRBF-P
Pushover	3.31	2.22	2.03
Code	2.76	2.33	2.25

The difference in percent's between elastic and pushover analysis is: for D-BRBF – (+16.6%), for BRBF-F – (-4.72%) and for BRBF-P – (-9.77%).

The analysis showed that there are differences between code based calculation of over strength factor of Ω^T and over strength factor of Ω^T calculated after pushover analysis.

For BRBF-F and BRBF-P are relatively small differences - (-5…-10%), but for D-BRBF structure is a 17% difference. This shows that code based calculus of over strength factor Ω^T for this type of dual MRF-BRBF structure is underestimated, and this type of dual MRF+BRBF structures needs more research.

IOP Conf. Series: Materials Science and Engineering **789** (2020) 012042 doi:10.1088/1757-899X/789/1/012042

4.2.3. Analysis of forces. In this chapter we analyzed the maximum efforts (P, M, V) in elastic analysis from most loaded non-dissipative members (columns and beams of buckling restrained braced frames) from non-dissipative seismic combination and compared with efforts in pushover analysis curves at maximum force response.

- *D-BRBF structure*

The efforts from elastic and pushover analysis for D-BRBF structure is presented in table 10.

Table 10. The efforts from elastic and pushover analysis for D-BRBF.

		P (kN)	M (kNm)	V (kN)
Floor 1 column	Elastic analysis	-1403.2	-165.0	-83.6
	Pushover analysis	-1235.3	-460.7	189.1
Floor 1 BRBF beam	Elastic analysis	-642.0	-85.9	57.2
	Pushover analysis	-635.2	256.5	133.8

The ratio of efforts is computed with formula:

$$f = F_{max.push} \cdot (F_{elastic})^{-1} \tag{15}$$

where $F_{max.push}$-The maximum effort from pushover analysis; $F_{elastic}$- The maximum effort from elastic analysis from non-dissipative seismic combination.

The results for floor 1 column are presented in figure 7 (a) and for floor 1 BRBF beam in figure 7 (b).

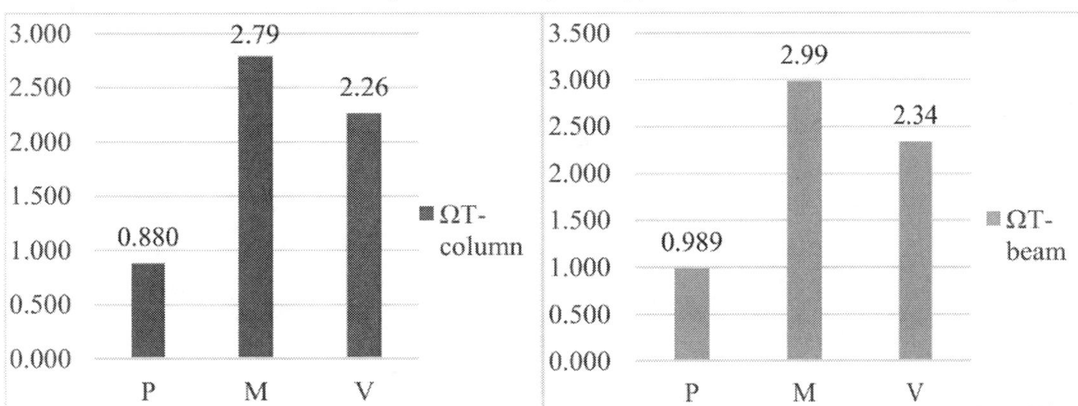

(a) (b)

Figure 7. The ratio of efforts f for column (a) and for BRBF beam (b) in D-BRBF structure.

- *BRBF-F structure*

The efforts from elastic and pushover analysis for BRBF-F structure is presented in table 11.

Table 11. The efforts from elastic and pushover analysis for BRBF-F.

		P (kN)	M (kNm)	V (kN)
Floor 1 column	Elastic analysis	-1377.9	72.9	32.4
	Pushover analysis	-1554.3	-194.5	23.0
Floor 1 BRBF beam	Elastic analysis	-589.3	-63.9	66.3
	Pushover analysis	-670.7	247.6	132.4

The ratio of efforts is computed with formula (15).

The results for floor 1 column are presented in figure 8 (a) and for floor 1 BRBF beam in figure 8 (b).

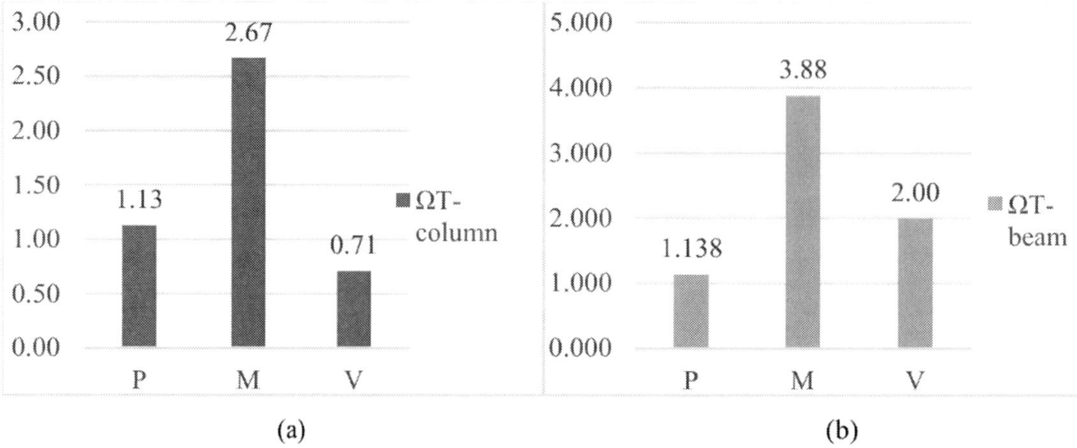

Figure 8. The ratio of efforts f for column (a) and for BRBF beam (b) in BRBF-F structure.

- *BRBF-P structure*

The efforts from elastic and pushover analysis for BRBF-F structure is presented in table 12.

Table 12. The efforts from elastic and pushover analysis for BRBF-P.

		P (kN)	M (kNm)	V (kN)
Floor 1 column	Elastic analysis	-1347.3	-21.0	6.0
	Pushover analysis	-1475.8	-167.0	-34.2
Floor 1 BRBF beam	Elastic analysis	-597.4	-58.2	63.5
	Pushover analysis	-679.3	250.4	133.5

The ratio of efforts is computed with formula (15).
The results for floor 1 column are presented in figure 9 (a) and for floor 1 BRBF beam in figure 9 (b).

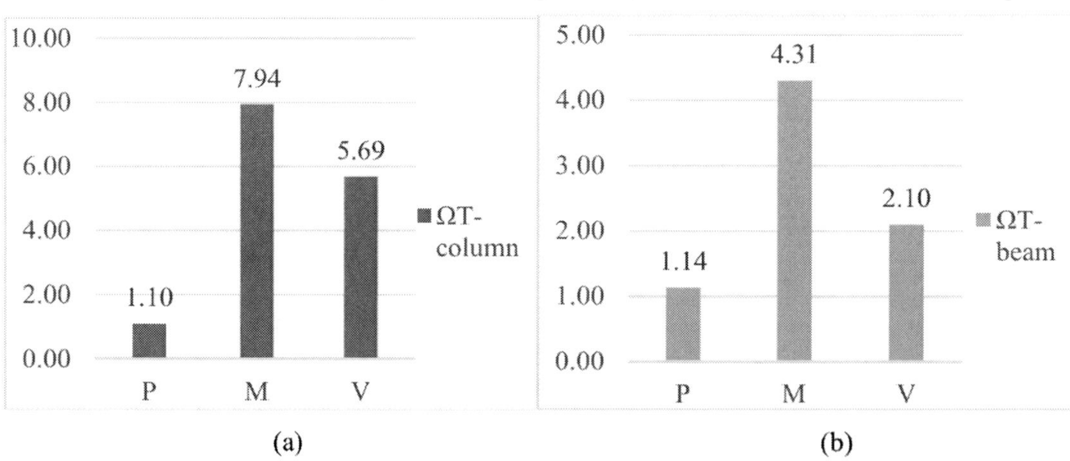

Figure 9. The ratio of efforts f for column (a) and for BRBF beam (b) in BRBF-P structure.

From this analysis we can conclude that differences between efforts of non-dissipative members from elastic analysis and efforts from pushover analysis at maximum failure force for the same non-dissipative members are quite different. For D-BRBF structure the differences of axial forces (P) are nearly the similar, most differences are for bending moment (M) - $\Omega_M^T = 2.79$ for column and $\Omega_M^T = 2.99$ for beam, and for shear force (V) - $\Omega_V^T = 2.26$ for column and $\Omega_V^T = 2.34$ for beam.

For BRBF-F structure the differences of axial forces (P) are nearly the similar, most differences are for bending moment (M) - $\Omega_M^T = 2.67$ for column and $\Omega_M^T = 3.88$ for beam, and for shear force (V) - $\Omega_V^T = 0.71$ for column and $\Omega_V^T = 2.0$ for beam.

For BRBF-P structure the differences of axial forces (P) are nearly the similar, most differences are for bending moment (M) - $\Omega_M^T = 7.94$ for column and $\Omega_M^T = 4.31$ for beam, and for shear force (V) - $\Omega_V^T = 5.69$ for column and $\Omega_V^T = 2.10$ for beam.

The differences for axial force for *D-BRBF* structure are:

Axial force: (-12.0 %) for column, and (-1.1 %) for beam. Bending moment and shear force have big differences.

For *BRBF-F* structure: Axial force- 11.3 % for column, and 12.1 % for beam. Bending moment and shear force have big differences except shear force for column which is (-29.1 %).

For *BRBF-P* structure: Axial force- 8.7 % for column, and 12.0 % for beam. Bending moment and shear force have big differences.

We can observe that efforts that varies the most are bending moment (M) and shear force (V) for all structures, but axial forces (P) remain quite similar. The BRBF-P structure has the most important differences of M and V efforts.

5. Conclusions

The results show that in mater of resistance the D-BRBF structure is more resistant. In mater of ductility, - the D-BRBF and BRBF-F structures has better ductility than BRBF-P structure. The total weight of frames is 30.67 t for D-BRBF, 29.75 t for BRBF-F and 30.73 t for BRBF-P structure, this shows that dual MRF+BRBF structure has nearly the same weight as BRBF structure, but instead is more resistant.

The pushover analysis according to N2 method showed that structures fulfils all deformations requirements at target displacement for ULS. The analysis of over strength factor Ω^T as a ratio between maximum pushover base shear force and base shear force calculated according to lateral force method in P100/1 showed a relatively small difference ~(-5...-10 %) for BRBF-F and BRBF-P structures, but a 17 % difference for D-BRBF structure. This shows that code based calculus of over strength factor Ω^T for this type of dual MRF-BRBF structure is underestimated, and needs more research.

The analysis of forces in non-dissipative members shows that the code approach provides a reasonable approximation for axial forces, but generally underestimates by large amount the bending moment and shear force demands. Consequently, as observed from pushover analysis, the resistance of non-dissipative members is insufficient for preventing their plastic hinging. Improved design approach is necessary, and will be addressed in future studies.

References

[1] Sabelli R, Mahin S and Chang C 2003 Seismic Demands on Steel Braced Frame Buildings with Buckling-Restrained Braces *Engineering Structures* Vol. **25** 655-666

[2] Uang C M and Kiggins S 2003 Reducing residual drift of buckling-restrained braced frames as a dual system *Proceedings of the International Workshop on Steel and Concrete Composite Construction (IWSCCC-2003)*, Oct. 8-9, Taipei, Taiwan: 189-198

[3] Stratan A, Zub C, Dogariu A, Florea D and Dubina D 2017 Design recomandations for buckling restrained braced frames – in Romanian, (*Editura Orizonturi Universitare*, Timisoara)

[4] * * * CR0-2012 *Romanian Design Code. The bases of designing structures*

[5] * * * P100-1 2013 *Romanian Seismic Design Code. Part 1: Design Provisions for Buildings*

[6] IMSER project nr. 99/2014 *Implementation of buckling restrained braces in the practice of antiseismic design in Romania*. Politehnica University of Timisoara, Romania

[7] Fajfar P A 2000 Nonlinear analysis method for performance-based seismic design *Earthq Spectra* **16(3)** 573–92

Approach, understanding, needs and integration of the BIM concept in the Romanian modern society

R Muntean[1], G C Chițonu[1*], D Radu[1], C E Cazacu[1] and T Gălățanu[1]

[1] Faculty of Civil Engineering, Transilvania University of Brasov, Brasov, Romania

* gabriela.chitonu@unitbv.ro

Abstract. The aim of this study is to look at the level of knowledge about BIM (Building Information Modelling) in 3 important constructions sectors: higher education institutions (HEI) - companies – public bodies. In reference to the studies and surveys made in Romania by a team from Transilvania University of Brasov, in the frame of an Erasmus+ research project - UrbanBIM, it is reasonable to hypothesize that most of the people surveyed did not understand what BIM really is, and some were unable to identify the differences between BIM and CAD (Computer-aided design). Making population surveys and analysing the data allow for determining the methodologies for future implementation of BIM technologies and improving the educational training. The concept understating shows the preponderance of using BIM technology in the construction sector, developed on extended levels, starting from urban planning, geographic information system (GIS), architecture, engineering and construction (AEC), building management and leading to life cycle analysis (LCA) and energy efficiency (EE).

1. Introduction

Beyond the basic definition of BIM which is explained in only three words: Building Information Modelling, the concept in itself is an extensive approach to computational work to serve a large scale of purposes. BIM is a technological approach which allows for further efficiency enhancements for economical and sustainable development across major sectors in the modern societies.

Since 1963, when the first CAD program proposed a graphical interface – Sketchpad, continued with GIS technology, further followed in 1982 by 2D CAD, and in 1987 highlighted by ArchiCAD, the software systems have created the foundation for actual BIM. From the beginning the CAD software has been used in projecting mainly for the Building Industry and has continued to develop as a first generation from 2D to 3D modelling [1, 2].

The concept of BIM has collected a history of innovation from the United States, Central and Northern Europe and all the way to Japan [1].

CAD technology used by architects and engineers, inter-connecting the same projects and providing computational-design deliverables for the building sector, is an essential tool which allows for planers, contractors and others stakeholders to get key information delivered in a more easily interpretable form.

The building industry and the software companies, being in a continuous development and economical competition, have continuously worked to build for themselves the framework for disseminating their products. Using and improving the assisted design, software, CAD, construction planning systems for technological products, they succeed to develop strategies in regard of Industrial-marketing and Building-process management [3].

Each sector of modern society can have its own definition for BIM. The level of understanding is connected with the level of society needs. The building industry sector has a most acute determination for developing methods which are bringing better management – control and faster results with less error in the building process or more efficiency in exploitation during the life cycle assessment.

The companies in the field of architecture, engineering and construction (AEC) which are the first technical participants involved in the construction process have great incentives as users of BIM tools and should be the first to develop and disseminate the progress in the field.

Building Information Modelling is the digital representation of the physical and functional characteristics of a facility - shared knowledge resource for information about it forming a reliable basis for decisions during its life cycle, from earliest conception to demolition. Technology is transforming the way that buildings and infrastructure are designed, constructed and operated [4].

The next important aspect in the BIM implementation is the educational system which can benefit by refining the fields and introducing basic training.

The public sector, government and smaller public administration entities are other users who can greatly benefit from having complete or almost complete information relating to projects or cases stored in one system and housed in one easy to access portal. This provides ease of use, allows for reduced time and resources in finding what one needs to be working on, which in turn allows for significant economic benefits, and easier record keeping.

2. Approach of BIM in Romania

In the last 30 years, Romania has been developed in multiple directions. One of the main problems, which the modern society is actually facing, is the cooperation and collaboration. The essence of BIM technology is to gather together different fields and different actors. The administrative - public sector is in continuous transformation. The EU Regulations are integrated in the public system.

The needs for efficiency required in different parts of the modern world like the United States, Western-Central and Northern Europe developed the concept AECOM (Architectural, Engineering, Construction/Contracting, Operating and Maintenance) which leads to collaborative integrated and centered model with the inclusion of the Life-cycle Analysis for buildings using BIM (Building Information Modelling/Management) and IT technologies [4].

In this regard, the technological systems, like 3D modelling, the collaboration and coordination, whether on-line or off-line, information management, mobility and 4D, 5D ,6D, 7D libraries, have been improved and issued and today BIM + IPDD (Integrated Project Design & Delivery) is available. The only solution to the challenges is collaboration [2]. Following the global expansion of the world construction industry, Romania is also seeking to bring forward global construction projects and the need for effective tools such as (BIM) for information management. At the Romanian Standards Association (ASRO - Romanian national organism for standardization) level, a technical committee CT334-Building Construction, Performance, Building Durability has been set up as a European-level CEN / TC 442 Building Information Modelling (BIM) Technical Committee.

At the beginning of 2019, ASRO published a new set of international standards to permit BIM to develop within projects and across borders, bringing benefits to the entire industry. Providing a framework for information management to collaborate with BIM, ASRO published the first two parts of ISO 19650 - Organization and digitization of information about buildings and civil engineering works, including building information modelling (BIM) - Information management using building information modelling (which was developed on the basis of British Standard BS 1192) [4].

3. Surveys and analysis about BIM in Romania

In order to improve BIM implementation to different professional levels and to identify the needs of the modern society, a team from Transilvania University of Brasov, in the frame of an Erasmus+ research project, "UrbanBIM – Innovative Educational Integration of Urban Planning based on BIM-GIS technologies and focused on Circular Economy Challenges", has undertaken surveys in different sectors of the Romanian community: educational, public-administrative and professional companies.

The UrbanBIM project has the main goal to develop a collaborative platform based on urban planning BIM and evaluating the challenges related to LCA (Life Cycle Analysis) of urban construction materials, in order to be used as training material for both HEI students and professional sectors.

A set of specific questionnaires has been supplied by all the partners in the UrbanBIM research project: from Romania, Transilvania University of Brasov (UTBv) and Romania Green Building Council Association (RoGBC), from Spain, Universidad de Sevilla (USE) and Asociación Empresarial de Investigación Centro Tecnológico del Mármol, Piedra y Materiales (CTM) and from Poland, Politechnika Warszawska (WUT) and Datacomp sp. z o.o. (Datacomp).

The preponderance of the people questioned is linked with building industry, namely youth and adults under 55 years of age. The survey seeks the input of population regarding future methods for implementation and creating a basis for an advanced BIM curriculum for the construction sector which should be used for higher education institutions in the teaching process.

The gender balance for the respondents in the Building sector is about 65% male and 35% female and the distribution in the fields shows: 44% engineering/building engineering, 10% architecture, 15% public administration, 22% construction work and 9% others.

The BIM concept is more used by the professionals in architecture/urbanism 21%, building engineering 31%, engineering 31%, construction 26%, the administrative sector is closer linked with urbanism.

The survey, in the first part, answers the following questions:

- Have you received BIM training? What kind of training?
- What is the proportion of people in your organization/studies center who use BIM?

All the answers from the three sectors, HEI, private companies and public bodies, indicate that more than 55% didn't receive any type of BIM training, 16% yes, self-taught, 23% yes, induction course, 2% yes, master course and less than 1% yes, expert course. The public bodies received the less BIM information compared with HEI and companies (figure 1).

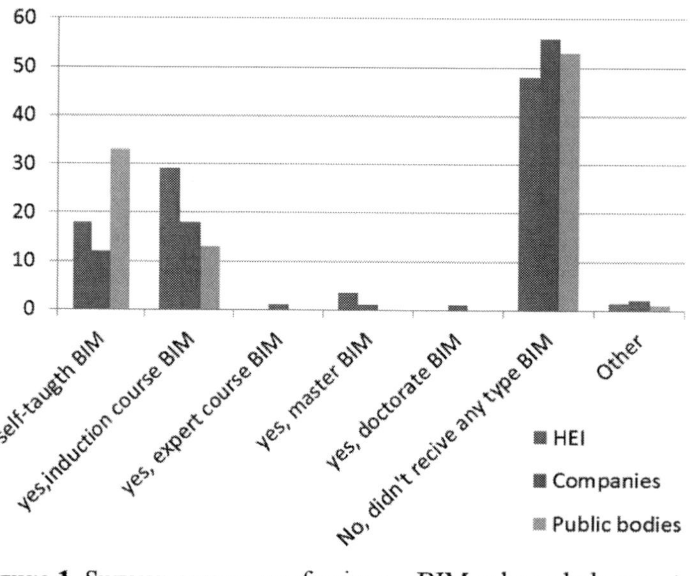

Figure 1. Survey: groups, professions – BIM acknowledgement.

The responses show that the basic level of general BIM knowledge is actually low, but the professions linked directly with the building sector have access to high technologies due to the work contracts, being mainly employees in big companies. The BIM concept is more popular among young people and seems easier to implement in the higher education institutions. Some employers currently use BIM and seek students capable of managing BIM, but do not ask for software expertise or previous training. The "social benefits" of BIM are necessary to be gradually implemented, social benefits meaning the sociological, behavioural, collaborative, psychological, and motivational benefits [4].

Educational institutions do not have BIM strategies in their curriculum. The training is more self-taught and once becoming employed in a building company as software user, some will receive training. Choosing BIM in projecting fields leads to efficiency, data analysis, lower management risk, quality control, national coverage.

BIM is more known among younger employees. The majority of people using BIM do not have extra acknowledgement about the advantages and the efficiency of the system in addition to the knowledge that they acquired by using BIM for no more than 5 - 6 years.

The benefits of using BIM are widely shared. The dominant observation was to create projects of better quality and the possibility of minimizing errors - both design and implementation, as it is shown in figure 2.

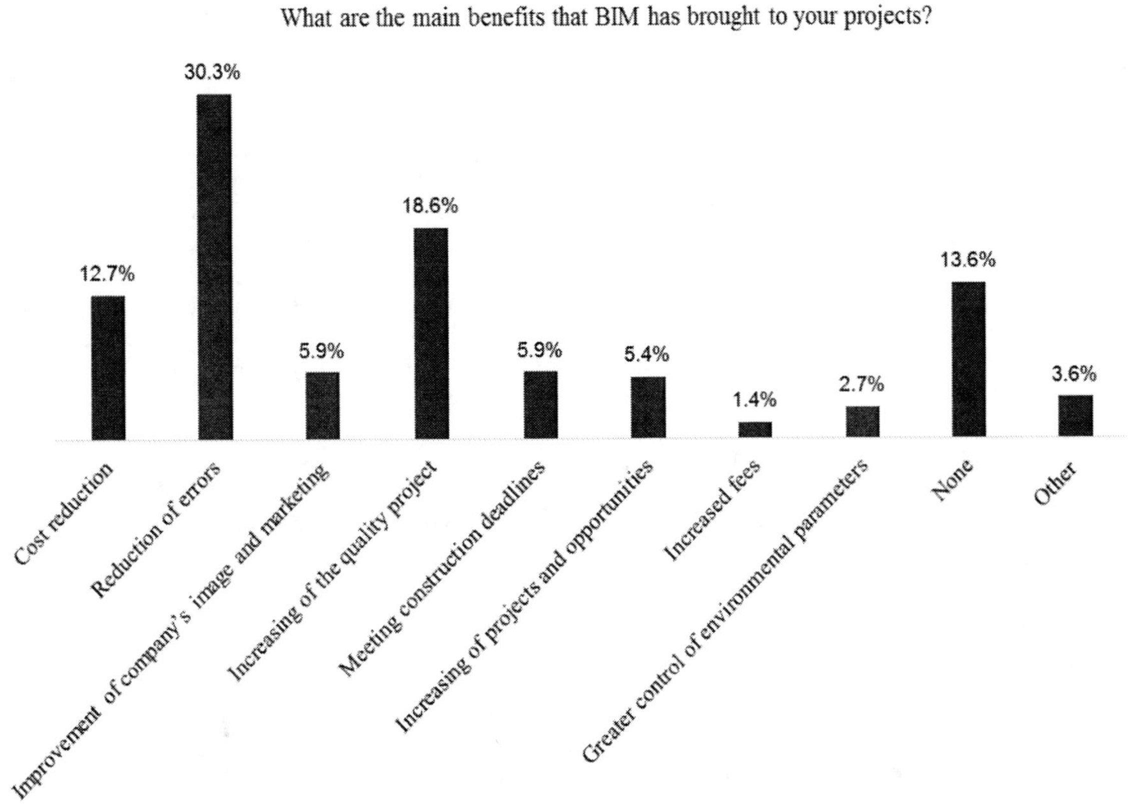

Figure 2. Survey: BIM implementation.

Construction programs seek to integrate BIM depending on the size of the company. However, small companies which have been using CAD 2D and 3D modelling, mostly in civil engineering and architecture, declare that they had contact with BIM. Over 60% of the respondents think that BIM will replace CAD in a long or medium term, but at the same time, about 27% assume that BIM will not replace CAD (figure 3).

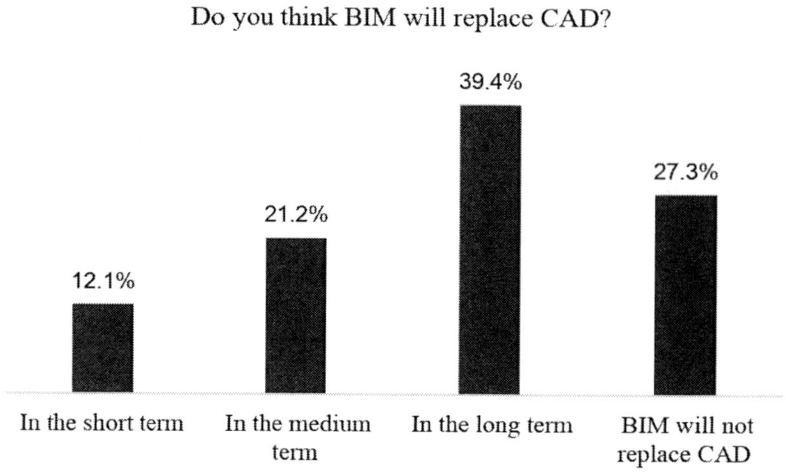

Figure 3. Survey about BIM / CAD.

The specialists in the architectural and construction industry are involved in the preparation of projects using modelling information 3D, 4D, 5D, 6D connected with the BIM, but BIM is more often used by larger companies [5].

Just about every source, including industry leaders, has its own definition of BIM, which can leave users with a few misconceptions: BIM is just for architects, BIM is just a design tool or BIM comes in a box [1].

The lack of standardized BIM will be covered in the coming years. The actual survey shows that in addition to the lack of information, training and the high costs of licences, there is a limited demand for BIM technologies in the Romanian construction market. This is not encouraged by the existing requirements of the administrative system for urban planning (figure 4).

The fact that Romania is an EU country and it is part of the Smart City – map Projects (there are over 300 projects) leads to intelligent development for the future urban planning [6].

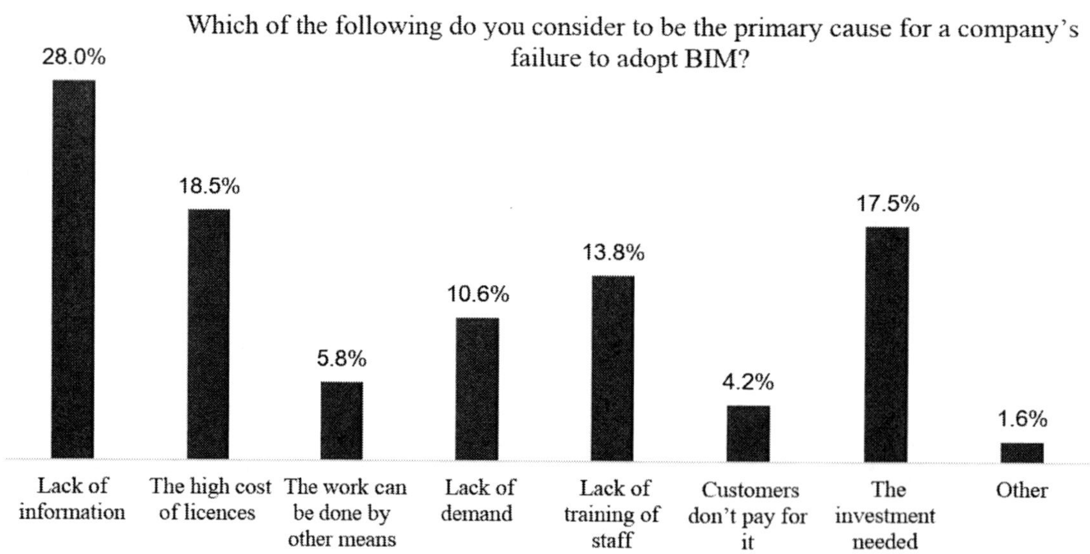

Figure 4. Survey: BIM - failures in implementing.

The Romanian Association for Smart City and Mobility has launched with the strategic support of DELL EMC, Smart City projects in several counties in Romania, which involve new technologies in intelligent development with relevant directions in monitoring the environment, the air quality, analysing the urban traffic, public lighting control, also creating communities connections [7].

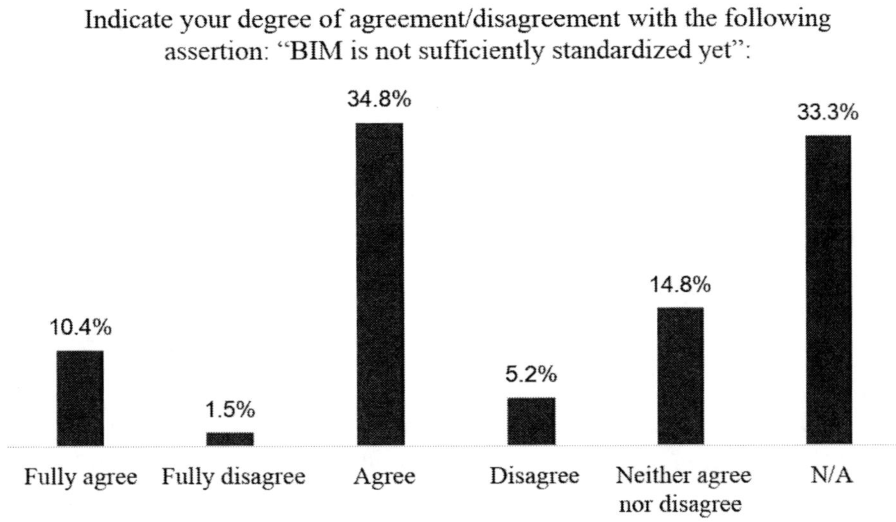

Figure 5. Survey: BIM –standardization

On the Smart City map are Romanian counties like Alba, Timisoara, Sibiu, Cluj, counties whose cities or communes start to develop a sustainable ecosystem with community inclusion, and where the public administrations understood that BIM could generate growth and efficiency in urban planning. The community needs, the big companies' requirements and, more importantly, universities should all contribute by making new curricula, training programmes regarding LCA of construction materials by means of BIM technologies.

4. Conclusions

For higher education organizations, UrbanBIM will contribute by giving them new curricula and material for study and training BIM programs including the LCA and EE of construction materials. Professors will provide new courses for their students and high-technological education in BIM systems and working methods.

The construction sector enterprises need employees with high technological skills and a starting point is UrbanBIM – project grounded on EU Directives and EU communication, p.m. 2015/C 195/06) Resource efficiency opportunities in the building.

The Romanian construction sector has continued a classic design being in the maturity level 2D up to levels 3D and 4D, BIM awareness depending on the size of the company.

The study shows that Romania is at the beginning of establishing rules for the BIM implementation.

The National Strategy for Research, Development and Innovation for 2014 -2020 (*Strategia natională de cercetare, dezvoltare şi inovare 2014-2020*) - supports the construction of new infrastructure and the development of the existing infrastructure as a key priority area, but there is no clear reference to BIM.

The Directive 2014/24/Eu, article 22, para. 4, allows the requirement for the use of BIM provided it does not create discrimination and that alternative means of access are offered to those that do not have access to such technology (this includes free access via a token for instance the day of the publication of the notice to tender).

Governments and public sector organizations can provide leadership to encourage the sector towards the untapped opportunity of digital, and in turn provide better public services and better value for public money. However, governments cannot do this alone: working together with industry at European and national levels is essential to achieve this digital transformation with due consideration given to commercial models, education, skills development and changes to current practices.

Acknowledgement
The work presented in this paper was funded with support from the European Commission. This publication reflects the views only of the authors and the Commission cannot be held responsible for any use which may be made of the information contained therein.

References

[1] National Institute of Building Sciences 2007 *Introduction to the national Building Information Modeling Standard™ Version 1 - Part 1: overview, principles, and methodologies*
[2] Deutsch R, AIA LEED AP 2011 *BIM and Integrated Design,* WILEY U.S.A. p 3, 13
[3] Becerik-Gerber B and Rice S 2010 The perceived value of building information modeling in the U.S. building industry *Electronic Journal of Information Technology in Construction* **15**:1874-4753 p 185
[4] IO2/A1.3 2019 *Report on regulations related to BIM technologies (UrbanBIM - Innovative Educational Integration of Urban Planning based on BIM-GIS technologies and focused on Circular Economy Challenges*
[5] Czmoch I and Pekala A 2014 Traditional Design versus BIM Based Design *Procedia Engineering* **91** 210 – 215 p 211
[6] https://map.romaniansmartcity.ro/#
[7] BIMcon 2019, Annual Conference, Bucharest 23rd of May 2019

IOP Conf. Series: Materials Science and Engineering **789** (2020) 012044 doi:10.1088/1757-899X/789/1/012044

Freeze-thaw testing of stabilized soil samples used for riverbank consolidation

A-Cs Nagy[1*], A-P Cîrcu[1], V-C Ciubotaru[2], B M Crăciunescu[2], G Topan[2], D V Moldovan[1], L-E Moldovan[3] and D M Paşca[1]

[1]Technical University of Cluj-Napoca, Faculty of Civil Engineering, Cluj-Napoca, Romania
[2]Technical University of Petrosani, Faculty of Mines, Petrosani Romania
[3]University of Agricultural Science and Veterinary Medicine Cluj-Napoca, Faculty of Animal Science and Biotechnologies, Cluj-Napoca, Romania

[*] andor.nagy@yahoo.com

Abstract. Riverbank erosion is a common problem of the Romanian freshwater system. Classic riverbank stabilization methods consisted of using different type of concrete, however concrete manufacture consumes large amount of non-renewable raw materials and energy, and it is a carbon-intensive process. Many efforts are, therefore, being undertaken towards the developing "greener" solutions. There are different type of retaining walls resulted from the need to avoid reinforced concrete wall solutions, such as gabion walls, solid masonry walls, rocky embankment prisms, and wooden pile retaining systems. But all of these solutions converge on gathering raw material from an external source, and transporting it on location. In Romania acquiring crushed stone tends to become problematic, especially, in the areas outside the Carpathian mountain range. Therefore a solution of using local soil as a base for a stabilized soil layer is very practical and economical from a technologic point of view. Being a hydrotechnical application, the freeze-thaw resistance of the stabilized layer is crucial for the lifetime of structures. The objective of our study was to determine this parameter, in accordance with the Romanian norms and regulations.

1. Introduction

1.1. Objective of the study

An experimental project was carried out on "Căpuş" river from Cluj county, by building a slope protection system on both sides of the riverbank. The initial state revealed severe erosion problems, that were only aggravated by the seasonal floods, as seen in figure 1.

Riverbank stabilization was made from stabilized soils, using hydraulic binders from our partners the Lafarge Holcim Group Romania, and were selected according to their expertise, and results of the laboratory analysis. On the left bank a small dam was built, stabilizing the local soil with Doroport TB25© [1]. The right side of the river was considerably steeper, so designing a slope protection was more convenient. This system was built using soil-Dorosol C30© [2] admixture. Being in constant contact with water through all seasons raises the question of durability, and stability of formula under various temperature and moisture conditions. Therefore, one of the most important parameters for the present project was to determine the weight loss through freeze thaw testing.

Figure 1. Erosional problems on both riverbanks of the Căpuş river.

1.2. State of the art

There are several studies concerning freeze thaw testing of stabilized materials. Su et al. [3] evaluated the effect of freeze-thaw cycling on the mechanical properties of cementitiously stabilized materials (CSM). Ultrasonic testing was applied to monitor the P-wave velocity (constrained modulus) change during the freeze-thaw cycling. Unconfined compression strength (UCS) test was performed at the end of freeze-thaw cycling. It was found that freeze-thaw cycling could be detrimental to CSM as the constrained modulus decreased to 7% to 96% depending on the CSMs. The residual UCS was 35% to 84% of the initial UCS at the end of the test. The results showed that freeze-thaw cycling is an aggressive test, which result in degradation in the constrained modulus (D) and UCS for cementitiously stabilized materials (CSMs). Binder content has been proven to influence freeze-thaw durability significantly. Resistance against freeze-thaw cycling increases with increasing UCS.

Yin et al. [4] studied frost resistance durability of strain-hardening cement-based composites (SHCC) under combined flexural loading at different levels and under chloride attack. The loss of weight, dynamic elastic modulus, and microstructure characteristics of SHCC specimens were determined, and the influence of loading level on frost resistance was analyzed. In addition, the effect of freeze–thaw action on the flexural performance and diffusion properties of chloride in SHCC under the combined loads were investigated. The results show that the process of degradation was accelerated due to the simultaneous action of flexural loading and freeze–thaw cycles in the chloride environment, and SHCC suffered more serious damage at a higher loading level. However, flexural strength decreased by only 13.87% after 300 freeze–thaw cycles at load level S = 0.36. The diffusion properties of chloride in SHCC under constant flexural loading were affected by the freezing and thawing cycle.

Another study conducted by Jang [5] provided experimental results to investigate the mechanical properties of sustainable strain-hardening cement composite (2SHCC) for infrastructures after freeze-thaw actions. To improve the sustainability of SHCC materials in their study, high energy-consumptive components—silica sand, cement, and polyvinyl alcohol (PVA) fibers—in the conventional SHCC materials were partially replaced with recycled materials such as recycled sand, fly ash, and polyethylene terephthalate (PET) fibers, respectively. To investigate the mechanical properties of green SHCC that contains recycled materials, the cement, PVA fiber and silica sand were replaced with 10% fly ash, 25% PET fiber, and 10% recycled aggregate based on preliminary experimental results for the development of 2SHCC material, respectively. The dynamic modulus of elasticity and weight for 2SHCC material were measured at every 30 cycles of freeze-thaw. The effects of freeze-thaw cycles on the mechanical properties of sustainable SHCC were evaluated by conducting compressive tests, four-point flexural tests, direct tensile tests and prism splitting tests after 90, 180, and 300 cycles of rapid freeze-thaw. Freeze-thaw testing was conducted according to ASTM C 666 Procedure A. Test results show that after 300 cycles of freezing and thawing actions, the dynamic modulus of elasticity and weight loss of damaged 2SHCC were similar to those of virgin

2SHCC, while the freeze-thaw cycles influence mechanical properties of the 2SHCC material except for compressive behavior. Freeze-thaw exposure has little effect on the tensile performance of conventional SHCC materials. However, tensile performance of the sustainable SHCC replaced partially with recycled components is slightly different from that of conventional SHCC tensile specimen after repeated freeze-thaw cycles.

2. Materials and methods

Two different soil types were encountered on the sides of the riverbank. Granulation analysis shown in figures 2 and 3 reveal that the left embankment is composed of a slightly clayey sand, while the right embankment is formed from a clayey soil.

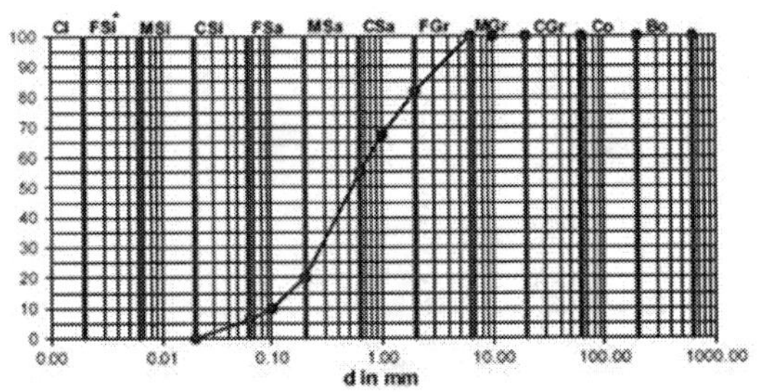

Figure 2. Granulation analysis on left embankment.

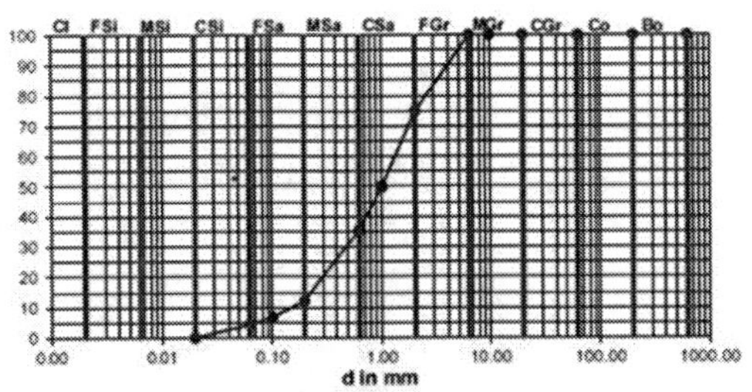

Figure 3. Granulation analysis on right embankment.

This fact led to using two different hydraulic binders, after performing a series of geotechnical analysis. Stabilization effects on shear strength parameters (cohesion, internal friction angle), and oedometric modulus were described in a separate article [6]. Doroport TB25 was used on the left side, and Dorosol C30 was mixed with the soil from the right side. From an economic point of view 3% and 5% hydraulic binder content was considered.

2.1. Romanian regulation

Freeze-thaw testing methodology for Romania is described in STAS 10473-2-86 [7]. It sets a curing time of 13 days for stabilized samples, which then should be subjected to 14 freeze-thaw cycles, each of them consisting in 16 hours of exposure to a temperature of -5°C (±1 °C), followed by 8 hours of immersion in water (the water temperature should be 25 ± 2 °C). Prior to freeze-thaw cycles, samples are to be submerged in water for a day, and weighed before starting the test. The stabilized soil samples should be visually inspected after each cycle and one should note the number of cycles at which cracks and other degradation are observed. After completing the test, samples should be wiped with a wet rag, and brushed to remove all the parts affected by freeze-thaw cycles. Then another set of weighing will determine the weight lost during the test.

The weight loss parameter (P_{id}) is determined from the following formula:

$$P_{id} = \frac{m_{13+1_{im}} - m_{14\ cicl}}{m_{13+1_{im}}} \text{ x } 100 \quad (1)$$

where m_{13+1im} is the weight of the samples at 13 days of age, plus 1 day immersed in water; $m_{14\ cicl}$ is the weight of the samples weighed after 14 freeze-thaw cycles.

2.2. Sample preparation and testing

A number of 12 samples were prepared using combinations of local soils with 3 % and 5 % of Doroport TB25, and Dorosol C30. Extracting unaltered samples from the location of the project with a double core boring device would have been problematic and costly, due to the steepness of the embankments. There would have been a high chance to extract already cracked samples, which would be irrelevant for the present stage of the study, and this determination in particular. Proctor test were carried out on both soil types in laboratory conditions, before building the slope protection system, in order to determine optimal compaction moisture content for all combinations with 3 % and 5 % hydraulic binder [6]. Optimal water content for the left riverbank was around 11%, while on the right side it was near to 13 %. The recipe for stabilization included 5 % of hydraulic binder on both sides of the embankment. On site compaction was made considering these values, and closely monitored by our team and our partners from Lafarge Holcim Group Romania. Therefore, the samples for the present study were prepared and extracted from the Proctor device, according to previously set values, in the Lafarge Holcim factory's laboratory from Turda, as shown in figures 4 and 5.

Figure 4. Sample compaction inside the Proctor device vessel.

Figure 5. Extracted sample

After a 13 day curing time the samples were submerged and weighed, the recorded values are presented in table 1.

Table 1. Weight of the samples at 13 days of age, plus 1 day immersed in water (m_{13+1im}) in grams (g).

	Dorosol C30 3% - right embankment	Dorosol C30 5% - right embankment	Doroport TB25 3% - left embankment	Doroport TB25 5% - left embankment
P1	1992	2142	1985	2080
P2	1971	2150	1970	2101
P3	2001	2155	1965	2092

Samples before testing are presented in figure 6.

Figure 6. Stabilized soil samples, before freeze-thaw testing (from left to right Dorosol 3%, Dorosol 5%, Doroport 3%, Doroport 5%).

After 14 days of testing in the laboratory of the Technical Univeristy of Cluj-Napoca, constant visual inspection concluded that the clayey soils treated with Dorosol C30© presented several cracks, while the sandy soils from the left embankment treated with Doroport TB25 showed no or minor alterations. At the end of the cycles samples were wiped and brushed. Another weighing was conducted, values are presented in table 2.

Table 2. Weight of the samples after 14 freeze thaw cycles ($m_{14\ \mathrm{cicl}}$) in grams (g).

	Dorosol C30 3% - right embankment	Dorosol C30 5% - right embankment	Doroport TB25 3% - left embankment	Doroport TB25 5% - left embankment
P1	1987	2137	1975	2071
P2	1887	2144	1960	2072
P3	1952	2149	1958	2086

Samples after testing are presented in figure 7.

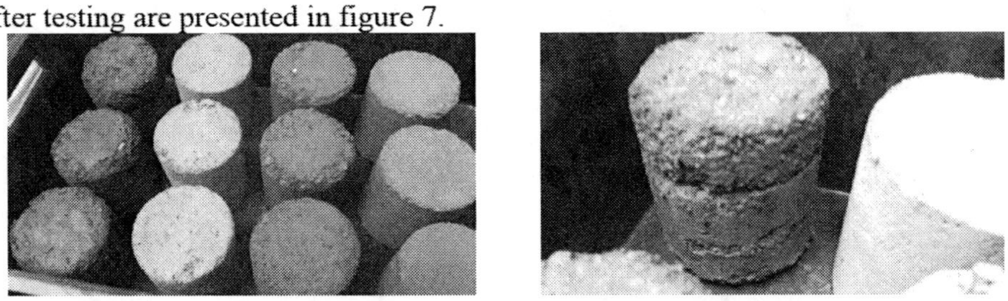

Figure 7. Stabilized soil samples, after freeze-thaw testing (from left to right Dorosol 3%, Dorosol 5%, Doroport 3%, Doroport 5%) Significant cracks appear on the samples treated with Dorosol, as seen on the right picture.

3. Results and discussion

Applying the formula from STAS 10473-2-86 gives the weight loss parameter (P_{id}), presented individually for every sample, and as an average value for the groups of three, in table 3.

Table 3. Weight loss parameter (P_{id}) in percentage (%).

	Dorosol C30 3% - right embankment	Dorosol C30 5% - right embankment	Doroport TB25 3% - left embankment	Doroport TB25 5% - left embankment
P1	0.25	0.23	0.50	0.43
P2	4.26	0.28	0.51	1.38
P3	2.45	0.28	0.36	0.29
Average	2.32	0.26	0.46	0.70

A significant improvement is noted on the average resulted from the samples mixed with Dorosol C30©. The samples with 3% hydraulic binder were the first to present cracking marks after cycles 3 and 4, and by the end of the test, the successive immersion, wiping and brushing left visible marks on the sides. Given the differences between obtained values more tests should be carried out on this type of admixture to draw viable conclusions. The results from the set with 5% Dorosol are remarkably similar, so this can be considered a reference value. The same thing can be said about the samples stabilized with 3% Doroport TB25©. The 5% admixture's average value is influenced by the P2 sample, which might have suffered some alterations independent from freeze-thaw cycles. Otherwise if we eliminate that result we have a slight improvement against the 3% admixture. More tests should be carried out on this recipe too.

4. Conclusions

The experimental protection system on Căpuş river behaves differently from one embankment to the other, largely eight months after constructions. The structures were built in October 2018, and were left exposed to the winter conditions of 2018. A visit in June 2019 revealed erosion problems on the right embankment, where Dorosol was mixed with a clayey type soil, as presented in figures 8 and 9.

Figure 8. Right embankment October 2018

Figure 9. Right embankment June 2019

Having different soils on the embankments, stabilized with two kinds of hydraulic binder raises a sum of uncertainties in interpreting the different behavior of the structures exposed to the same enviroment and conditions.

A freeze thaw sensibility potential revealed by laboratory tests on the right embankment stabilized with Dorosol C30© can be an explanation, however more tests would be required to draw long term conclusions, including swelling index determination.

Figure 10. Left embankment October 2018

Figure 11. Left embankment June 2019

An internal study of the Lafarge Holcim Group revealed that less aggressive cycles of freeze-thaw have much lower effect on freeze-thaw sensibility. At the factory freezing cycles of 7 days were applied, maintaining a temperature of -8°C. Only after this period were the samples sunk in water, and a number of 12 wet-dry cycles performed afterwards. This opens a discussion about the actuality of the Romanian standard. The American standard [8] also sets a different timing for the cycles, having 24 hours of freezing followed by 24 hours of immersing in water. The rapid changes of temperature and humidity in the testing procedure of the Romanian standard may cause more damage to the samples than other acknowledged testing procedures. Stabilized soil admixtures are not designed to be left exposed to the elements, as they are usually included in road bed structures. The damage of the embankment of the experimental sector shows that a protective layer is necessary over the stabilized soil structures. Considering the economy caused by using local soil, and the fact that grassing would be the cheapest and most expedient method of protection, this method of riverbank consolidation is considerably "greener" than conventional structures used today.

Acknowledgments
The authors would like to thank our colleagues Adrian Dârgău, Anca Gheorghe and Daniel Stănică from the Lafarge Holcim Group Romania for all their support during the execution phase, as well as granting access to the factory's laboratory and equipment, and providing documentation.

References

[1] Doroport – technical sheet
[2] Dorosol – technical sheet
[3] Su J, Senoh M, Tinjum J M, Ahmet Gokce and Edil T B 2014 Freeze-Thaw Cycling Effect on the Constrained Modulus and UCS of Cementitiously Stabilized Materials *ASCE* 2014
[4] Yin L, Yan C and Liu S 2018 Freeze–Thaw Durability of Strain-Hardening Cement-Based Composites under Combined Flexural Load and Chloride Environment *Materials* 2018 **11** 1721
[5] Jang J S, Rokungo K, Park W-S and Yun H D 2014 Influence of Rapid Freeze-Thaw Cycling on the Mechanical Properties of Sustainable Strain-Hardening Cement Composite (2SHCC) *Materials* 2014, **7**
[6] Nagy A Cs, Cîrcu A P, Ciubotaru V C, Crăciunescu B M, Topan G, Cherecheş T and Fărcaş V S 2019 Applications of stabilized soil in hydraulic constructions *(Conference paper, in course of publication)*
[7] STAS 10473/2-86 Road structures from natural aggregates and stabilized soils with hydraulic binders. Testing methods.
[8] ASTM D560/D560 M Standard Test Methods for Freezing and Thawing Compacted Soil-Cement Mixtures

Environmental impact evaluation associated to road traffic. A case study for Iasi city

A M Nicuţă[1*] and M Condurat[2]

[1] "Gheorghe Asachi" Technical University of Iasi, Faculty of Civil Engineering and Building Services, Iasi, Romania
[2] Regional Direction of Roads and Bridges (DRDP) Iasi, Romania

[*] alinanicuta@yahoo.com

Abstract. Development means a strong transportation network. Still, it comes associated with negative consequences, among them climate change and environmental disruptions. The pollution effects associated with road transportation are determined by a serial of factors like atmospheric conditions, topography, climate, traffic, fuel type, vehicle age but also improper maintenance of vehicles. All these components lead to atmospheric emissions caused by fuel leakage on rolling surface, dust and noise, accidental leakage of toxic chemical products. The present paper brings its contribution to the issue of sustainable development, approaching the problem of environmental impact associated with road traffic, quantifying pollutant emissions and comparing values. The evaluation is based on statistics for the most crowded road in Iasi City, namely Independence Boulevard, consisting of traffic data for the last 3 years obtained from the local Office of Traffic Management.

1. Introduction to the Environmental Impact of Road Traffic

The pollution effects associated with road transportation are extended and subjected to multiple factors like atmospheric conditions, topography and climate, traffic congestion, fuel type, vehicle age, maintenance and so on. These effects consist of gases emissions in atmosphere, fuel leakage on road surface, accidental spilling of toxic substances, dust and noise.

CO_2e emissions, ecosystems fragmentation and elimination in atmosphere of suspended particles PM_{10} are currently the most urgent problems on mankind [2]. Exhaust gases disorder the air quality, population health, ecosystems equilibrium but also intensify the global warming. There is a wide diversity of toxic gases associated with road traffic. In table 1 below can be found a detailing of the most significant emissions.

Prevention and remission methods of the impacts produced by transportation consist of improving the vehicles production and maintenance techniques, usage of electric, hybrid and " flex fuel" vehicles, where fossil fuel is replaced with bio fuel or electric systems, but also development of public transportation and usage of other way of transport.

Reduction of negative effects associated with road transportation can be realised also by changes of vehicles characteristics [10] like reducing the inertial load, improvement of aerodynamic characteristics, reduction of friction forces inside the engine, transmission improvement, usage of alternative fuels and electric systems.

Table 1. Emissions associated with road vehicles.

Emission	Symbol	Description
Carbon dioxide	CO_2	Results from the chemical reaction of fuel burning, is considered the most important greenhouse gas which contributes to oceans acidification
Carbon monoxide	CO	Results from the incomplete burning of fuel due to low combustion, contributes to the reduction of oxygen necessary to the normal functioning of the human body. Also it increases the ozone content in atmosphere, contributing to global warming.
Nitrogen oxides	NO_x	Resulted from combustion process, contribute to acid rain.
Sulphur oxides	SO_x	Take in atmosphere the form of sulphur dioxide (SO_2), sulphuric acid (H_2SO_4) and sulphate particles. Contribute to acid rain, tropospheric ozone, suspension powders and engine corrosion.
Hydrocarbon	HC	Emitted together with exhaust gases and during the supply process with fuels.
Benzene	C_6H_6	Can be found in small quantities (below 2%) in fuel, emitted by exhaust gases as unburned fuel or during alimentation. It contaminates the groundwater and is highly carcinogenic.
Material particles in suspension	PM	Are fuel components partially burned, characteristic to diesel motors and tire friction with road surface.
Volatile organic compound	VOC	Contribute to urban smog and have a high toxicity level
Smoke	-	The particles contain a carbon nucleus that absorbs burned and unburned hydrocarbon.
Lead emissions	-	Determined by additive fuel with lead tetraethyl. This type of pollution has severe effects on health and environment, its effects maintaining for hundreds of years. [6]

Statistics of Association of Vehicles Producers and Importers (APIA) [1] refer to an increase of 69% in 2019 for electric and hybrid car selling, as compared with 2018. Most important sellers for full electric are Nissan (119 units) and Renault (81 units), for hybrid Toyota (1611 units) and Lexus (82 units). [3]

2. Case Study on Road Traffic for the most crowded road sector in Iasi

Road transportation is currently the leading way of transport for people and merchandize. Due to a significant increase of traffic in urban areas, it is important to reconsider the traffic management in these areas, proving solutions like rerouting, signalizing, increase of road capacity, using alternative transportation vehicles (scooters, bicycles) and so on.

Similar to other large cities, the City of Iasi follows the same trend in road transportation. The Office of Traffic Management in Iaşi provided a serial of information on the local traffic and corresponding statistics. For further analysis of the environmental impact associated with road traffic, will be considered the sector with the highest registered values. This is Independence Boulevard, a 6 lane street, positioned in Iaşi City Centre. This boulevard connects the city entrances with the commercial and business centre, connects the University Centres (Al.I.Cuza University, Medicine and Pharmacy University, Agriculture and Veterinary Sciences University, „Gh. Asachi" Technical University of Iasi) and the road to Iaşi airport.

Figure 1. Independence Boulevard Mapping Overview [7].

Figure 2. Independence Boulevard Street View [8].

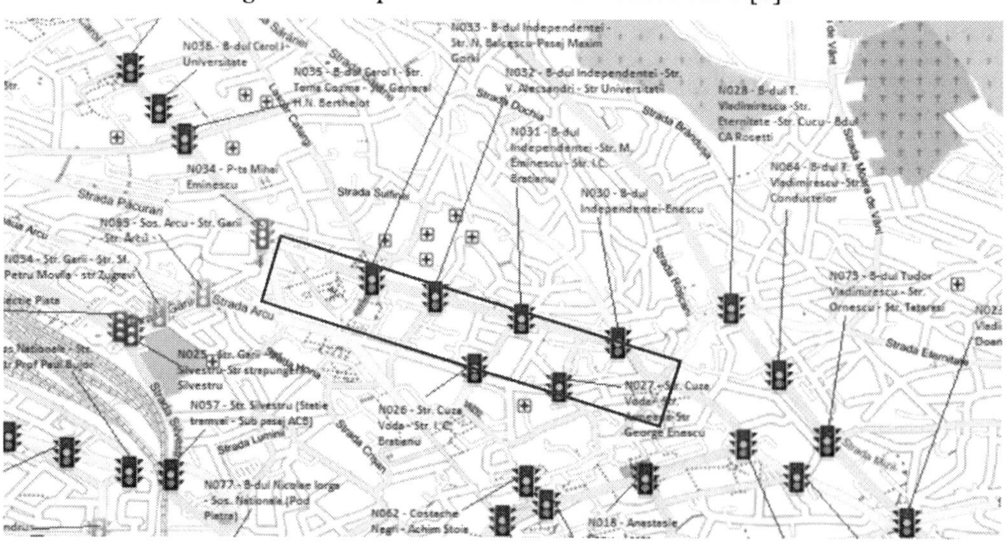

Figure 3. Independence Boulevard Traffic Light System.

Figures 1, 2 and 3 above provide visual information on the most crowded street of Iaşi City. As seen in figure 1, Independence Boulevard is very close to the old city centre, providing connections with main touristic objectives (marked with yellow bullet). Figure 2 details the number of lanes on the Boulevard while figure 3 is an overview of the Traffic Light System on the boulevard. Figure 1 and 2 show images provided by Google Maps, while figure 3 image was put at our disposal by the Office of Traffic Management in Iaşi.

According to the information received from the Office of Traffic Management in Iaşi the traffic for Independence Boulevard scheduled for 2016 – 2018 is presented in table 2 below and associated graph 1, where can be observed the increase with almost 8%, in road traffic, in a very short time period (2 years).

Table 2. Traffic Report on Independence Boulevard.

	Days month of October									**Average traffic**
Years	1	2	3	4	5	6	7	8	9	
2016	43082	36833	55818	478735486354800			56576	40773	38908	47725.1111
2017	43326	54675	56320	566735744058155			42684	38955	53446	51297.1111
2018	54278	50029	54059	572355868343572			42231	46805	56510	51489.1111

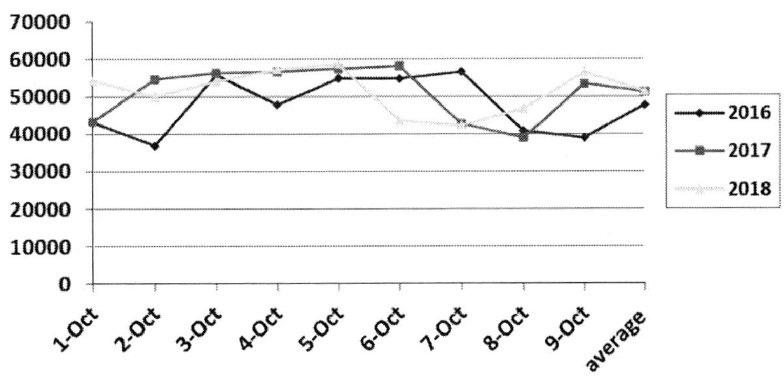

Figure 4. Graphical representation of road traffic for 2016-2018 on Independence Boulevard of Iasi.

3. Environmental Impact associated with Road Traffic for the most crowded road sector in Iasi

For a proper analysis of pollutant emissions associated with road traffic has been used the computer software COPERT 5, provided by European Environmental Agency (EEA). [4] This software was developed in order to determine quantitative emissions characteristics to road traffic [5]. Specific information for the analysis refer to local environment temperature and humidity, average annual fuel consumption, configuration of vehicle fleet and their number, vehicle speed. [9]

The objective of the case study was to determine the environmental impact characteristic for the road with the highest traffic in Iasi, namely Independence Boulevard presented in the previous chapter. The analysis was performed for years 2016 and 2018 in order to highlight the level of emissions and the traffic changes.

For the evaluation, the available number of vehicles was divided into 33 categories like passenger cars, light commercial vehicles, buses and motorcycles, with specific characteristics. COPERT 5 software provides information on 26 types of pollutants. Among these have been considered the value of CO_2 and CO and presented in table 3.

Table 3 below details the values characteristic for passenger cars (petrol, diesel and hybrid) considered for urban off peak and urban peak time intervals. The values are expressed in tones of pollutant.

As observed in the table, the increase in the number of vehicles by 3764 vehicles in 2 years' time brought an increase of approximately 1 tone of pollutant emissions of CO_2, respectively 0.01 tons of CO. If it is considered that this increase is valid only for one road under evaluation, can be concluded on the significant increase of environmental impact associated with road traffic for the entire city of Iasi.

Table 3. Pollutant emissions by type.

Emission	2016		2018	
	urban peak (t)	urban off peak (t)	urban peak (t)	urban off peak (t)
CO_2	11.94	18.42	12.74	19.69
CO	0.039	0.053	0.059	0.0068

The results of this evaluation showing significant increase of pollutant emissions in a short time period, brings us to the idea that it is highly important for the transportation management system to start taking action in this direction by identify methods to prevent and reduce the pollution level, improving the vehicle fleet, replacing fossil fuels cu biofuels or by improving their characteristics.

4. Conclusions

Considering the high level of pollution and the fact that, according to European Commission, the transportation activities are responsible for 32% of Europe energetic consumption and 28% of total CO_2 emissions (Eurostat, 2016), technologies development and sustainable construction processes from road sector becomes more and more important. More than this, the road traffic represents a major factor in greenhouse gas emissions expansion. According to PIARC report, the highest energetic consumption is associated with road traffic and less to the constructive process of road structures. [11]

The evaluation of greenhouse gas emissions associated with road traffic, performed for the road with the most traffic from Iasi City, called Independence Boulevard, confirmed the general trend of increasing the number of vehicles and of the pollutant emissions. If no measures are taken and the trend continues to be the same, the consequences might be more severe than imagined.

COPERT 5 computer software facilitated the environmental impact evaluation process, still there is a wide diversity of programs and methodologies that implement similar types of analysis.

The pollution effects due to road traffic are extended, caused by a significant number of factors like atmospheric, topographic and climatic conditions, traffic congestion, fuel type, vehicles age and wrong vehicle maintenance. Consequences refer to gases emissions in atmosphere, fuel leakage on road surface or accidental leakage of toxic substances in case of accidents. As an answer to all these issues, must be developed and implemented specific measures for prevention, reducing and compensation. Newly appeared technologies, in the area of CO2 emissions reduction, refer mainly to efficient vehicles which use alternative fuel types, hybrid, electric fuel cell cars, or the so called flex – fuel vehicles. The new configuration of vehicles-park in the transition process towards mobility with reduced carbon emissions consists of efficiency in fuel consumption by the use of improved vehicles structure and engines but also the use of alternative fuels. These measures have the purpose of reducing the CO2 and NOx emissions but also suspension dusts emissions. The increase of fuel efficiency can be obtained by improvements in their aerodynamic characteristics, reducing the vehicles total load and use of high performance engines. Also, the development of new polymers, composite materials and the so-called "memory metals", more light and resistant, should determine significant reduction of burned fuel quantities. The use of alternative fuels like liquid hydrogen (H2), bio-fuels, synthetic fuels and liquefied petroleum gases (GPL) as alternative energy sources also leads to a significant reduction in the quantity of emissions released in the atmosphere.

Complementary to the new technologies, materials and processes it`s recommended a significant change of transport policies in road traffic in order to support the city sustainable development. These changes refer to integration, in the current transport network, of cycling infrastructure, promote un-motorized vehicles, intelligent use of resources and joint transportation (like electric buses) or use of eco-efficient vehicles.

References

[1] Association of Vehicles Producers and Importers (APIA) http://www.ziare.com/auto/masini-electrice/romanii-cumpara-tot-mai-multe-masini-electrice-vezi-topul-celor-mai-vandute-marci-1567465
Borken, J., 2003 Assessment of Transport's Impact on Health and Environment for Germany *European Conference of Transport Research Institutes,* Brussels http://www.ectri.org/YRS03/Session-1/Borken.pdf.

[2] Condurat M, 2016 Prevenirea şi reducerea impactului negativ al sistemului de transport rutier asupra mediului, *Doctoral Thesis*, "Gheorghe Asachi" Technical University of Iasi

[3] EEA, 2009, European Environment Agency. *Glossary European Agency,* http://glossary.eea.europa.eu/terminology/concept_html?term=environmental%20indicator

[4] Gkatzoflias D, Kouridis C, Ntziachristos L and Samaras Z, 2012, *COPERT 5.* Computer programme to calculate emissions from road transport. User manual (version 9.0). *European Environment Agency*

[5] Gorham R 2002 Air Pollution from Ground Transportatio *Division for Sustainable Development* Departament of Economic and Social Affairs, United Nations

[6] Google Maps https://www.google.com/maps/place/Bulevardul+Independen%C8%9Bei,+Ia%C8%99i/@47.168563,27.5844459,2845m/data=!3m1!1e3!4m5!3m4!1s0x40cafb7bec2c0d1f:0x610ba55fe0373ab3!8m2!3d47.167272!4d27.5833945

[7] Google Maps https://www.google.com/maps/@47.1666688,27.585568,3a,75y,323.84h,100.13t/data=!3m6!1e1!3m4!1snumoWXO5VTDRpv4AcVtkrg!2e0!7i13312!8i6656

[8] Horobeţ I 2013 Assessment of the impact of traffic and transport infrastructure on the environment (Evaluarea impactului traficului şi infrastructurii de transport asupra mediului), Phd Thesis *Universitatea Tehnică "Gheorghe Asachi"*, Iasi

[9] Kahn R, et.al. 2007 Transport and its Infrastructure. In Climate Change 2007: Mitigation. Contribution of Working Group III to the Fourth Assessment Report of the Intergovermental Panel on climate Change [ed.] B. Metz, B.; Davidson, O.R.; Bosch, P. R.; Dave, R.; Meyer, L.A.. Cambridge, United Kingdom and New York, USA : *Cambridge University Press*, pg. 336-351

[10] PIARC. 2015, Reducing the life cycle carbon footprint of pavement, PIARC TC 4.2.3 report

IOP Conf. Series: Materials Science and Engineering **789** (2020) 012046 doi:10.1088/1757-899X/789/1/012046

Seismic Analysis of Steel Structures Using Friction Dampers

I S Oance[1*] and S Gelmambet[2]

[1] Ovidius University of Constanta, Doctoral School of Applied Sciences, Constanta, Romania
[2] Ovidius University of Constanta, Constanta, Romania

[*] ionut_oance@yahoo.com

Abstract. In modern seismic design, damping devices are used to increase the capacity of structures to dissipate energy. This paper evaluates the efficiency of using a passive friction damper system in a structure compared with typical structures and the influence of the damper's capacity on the structural response. The analysis concludes that dampers with lower capacity slip more times during earthquake than dampers with bigger capacity but the acceleration result increases.

1. Introduction

Recent earthquakes have shown that the ductile design of structures, even in developed countries, leads to substantial degradation of structures as result of seismic action.

If the energy induced in the structure by the earthquake action can be controlled or can be mechanically dissipated by devices independent of the structure, the building's seismic response is improved, thus substantially decreasing the potential damage. [1]. Therefore, there are two methods for the seismic design of structures [2]:

- Conventional method: the traditional method to resist lateral force by increasing the design capacity and stiffness e.g. shear wall, braced frames or moment resisting frames;
- Nonconventional method: based on the reduction of seismic demand instead of increasing capacity e.g. base isolation, dampers.

Passive control devices have been successfully used to reduce the dynamic response of structures subjected to earthquake. Friction devices have been used as a component of these dampers because they present high energy dissipation potential at a low cost and are easy to install and maintain [3]. Several friction devices have been tested experimentally – Pall (1982) [4], Fitzgerald (1989) [5], Aiken (1990) [6], Constantinou (1991) [7], Grigorian, (1993) [8], Nimbus et al (1993) [9], Dorka (1998) [10], Mualla (2000) [3] and some of these have been implemented in buildings around the world.

This article presents the benefits resulting from using frictional dampers to improve the seismic response of structures.

2. Experiment description

A nonlinear dynamic time-history analysis was used for evaluating the behaviour of structures equipped with friction dampers. This type of analysis shows the response of the structure during seismic action. Dampers were modelled given their hysteretic characteristics, as is shown in figure 1.

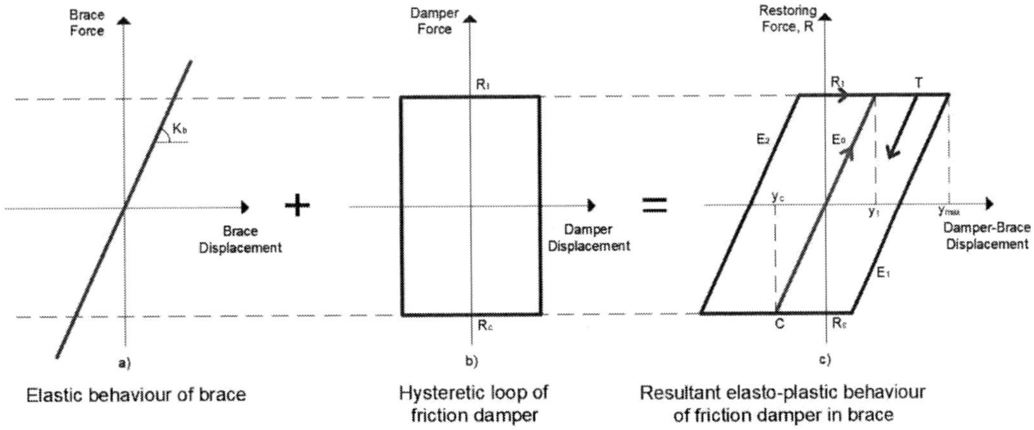

Figure 1. Resultant elasto-plastic behaviour of friction damper in brace [11].

Four design scenarios were taken into consideration: unbraced structure in figure 2, centric brace structure in figure 3, eccentric brace structure in figure 4 and frictionally damped structure in figure 5. Friction dampers from figure 6 were modelled as nonlinear elements – plastic (Wen). The program ETABS 17.0.1 (2018) was used to assess the response of a 2-D structural steel frame.

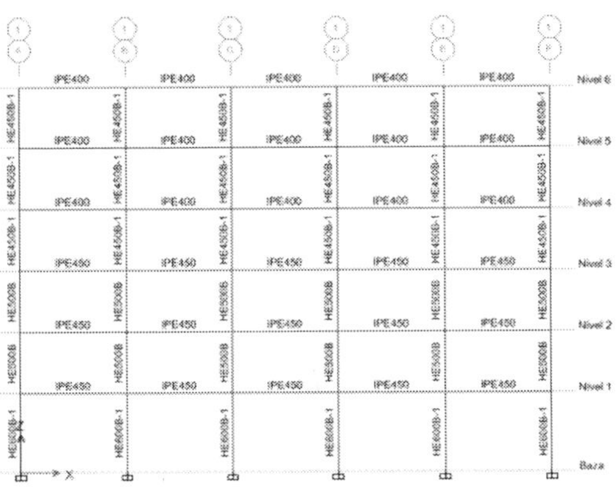

Figure 2. Unbraced structure.

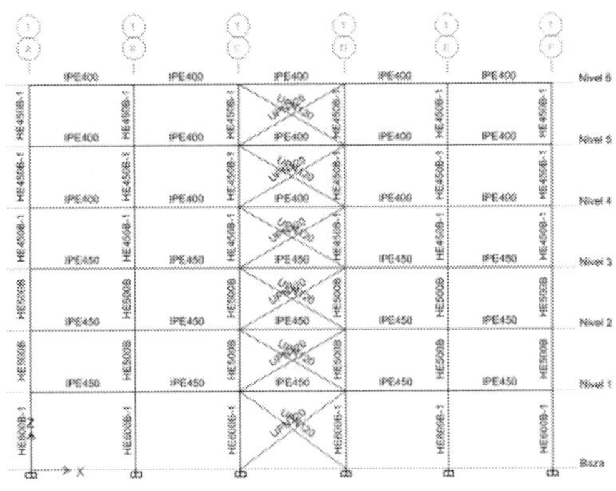

Figure 3. Centric brace structure.

Figure 4. Eccentric brace structure.

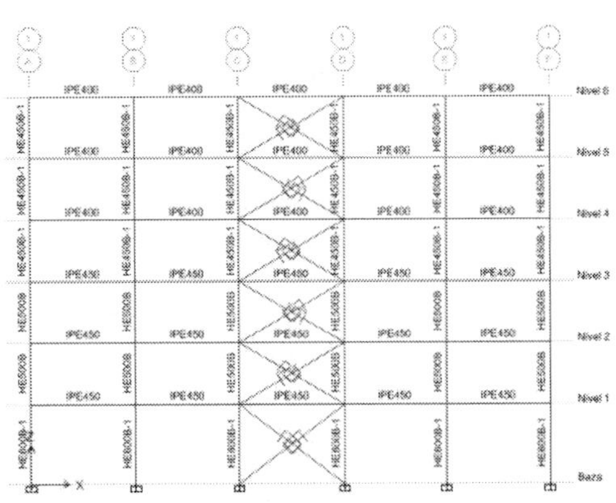

Figure 5. Frictionally damped structure.

Figure 6. Rotational friction damper [3].

The general characteristics of the structures are listed below:

- The distance between columns is 6 m;
- The first level is 4.5 m high, while the rest of the levels are 3.5 m;
- The material is Steel S355;
- The response in all versions of this structure was created using the seismic action of the 1977 Vrancea earthquake (figure 7).

The structure's response is in terms of top roof maximum displacement, story drift, story displacement, story acceleration, variation of dissipated energy.

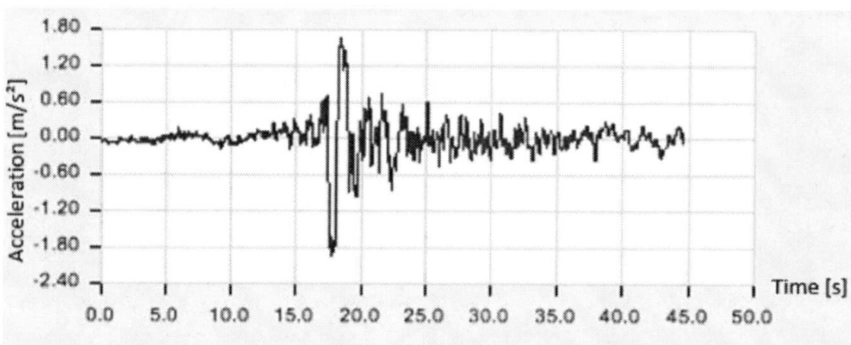

Figure 7. Earthquake Vrancea 1977 accelerogram.

3. Results and significances

Table 1 presents the dynamic characteristics of the structures in terms of periods of vibration and mass participation factors in each mode of vibration.

Table 1. Dynamic Characteristics of Structures

	Unbraced structure	Centric brace structure	Eccentric brace structure	Frictionally damped structure (50 kN)
Periods [s]	1.357	0.932	1.068	0.905
	0.443	0.307	0.358	0.316
	0.229	0.169	0.194	0.183
	0.144	0.118	0.128	0.127
UX	0.7982	0.8184	0.8219	0.8384
	0.1234	0.1207	0.113	0.1055
	0.0444	0.0348	0.0371	0.0313
	0.0213	0	0.0173	0.0156

The fundamental period of the centric brace structure is lower than the unbraced and eccentric structures but bigger than the frictionally damped structure.

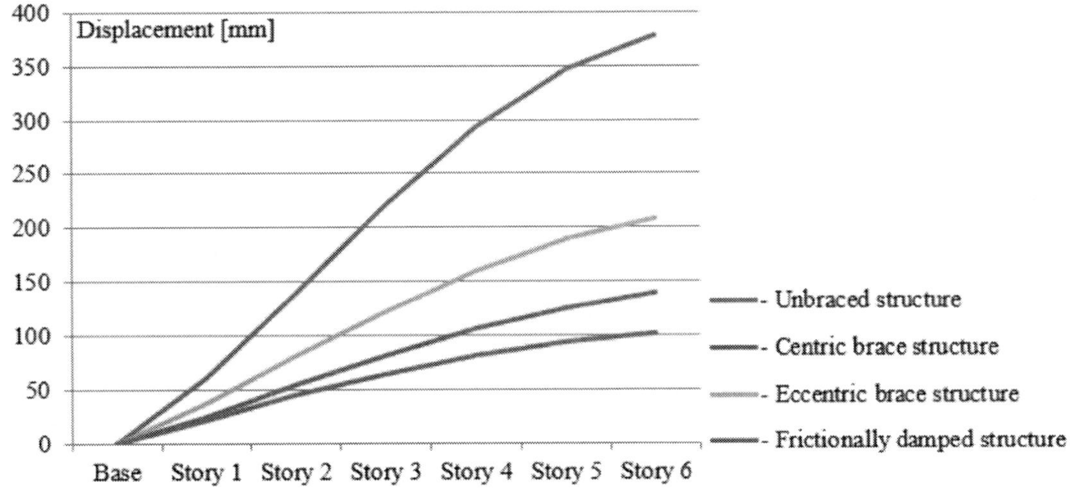

Figure 8. Maximum story displacement.

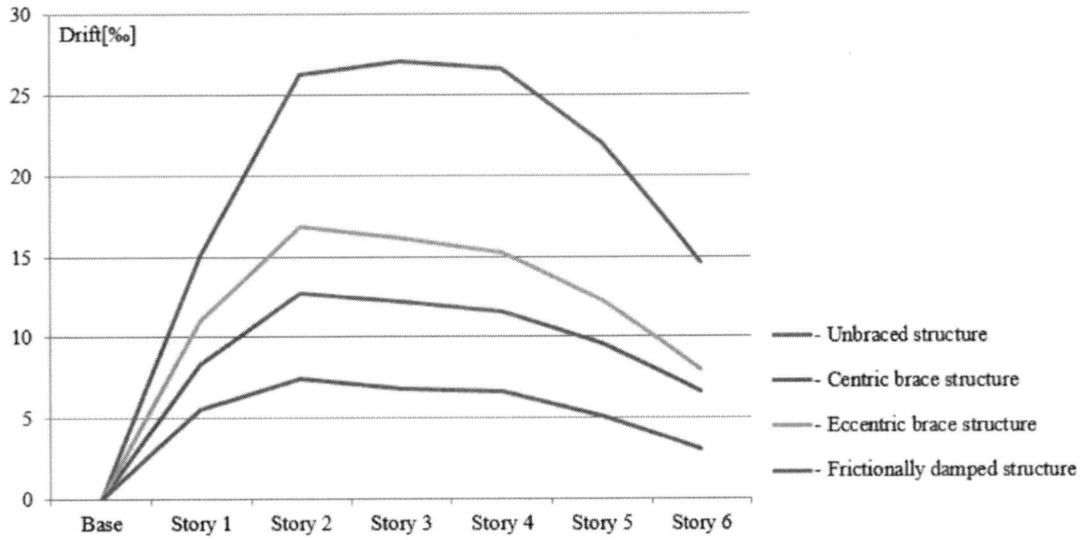

Figure 9. Story drifts.

Story displacement is the absolute value under action of the lateral forces.

The maximum displacement of the damped structure is 3.75 times smaller than the unbraced structure. Even if the structure has a centric brace, the displacement is bigger than frictionally damped structure, as shown in figure 8.

The smallest top displacements are for the structure equipped with 150 kN dampers, figure 10, but the maximum values of acceleration are larger than the structure equipped with 50 kN dampers, figure 11.

Story drift is the displacement difference between two consecutive stories divided by the height of that story.

The story drifts can provide useful information for story damage assessments. The story drifts in the frictionally damped structure are the lowest, as shown in figure 9.

This means that using dampers in structures improves the response in terms of maximum story displacement, story drifts, top displacement and top acceleration.

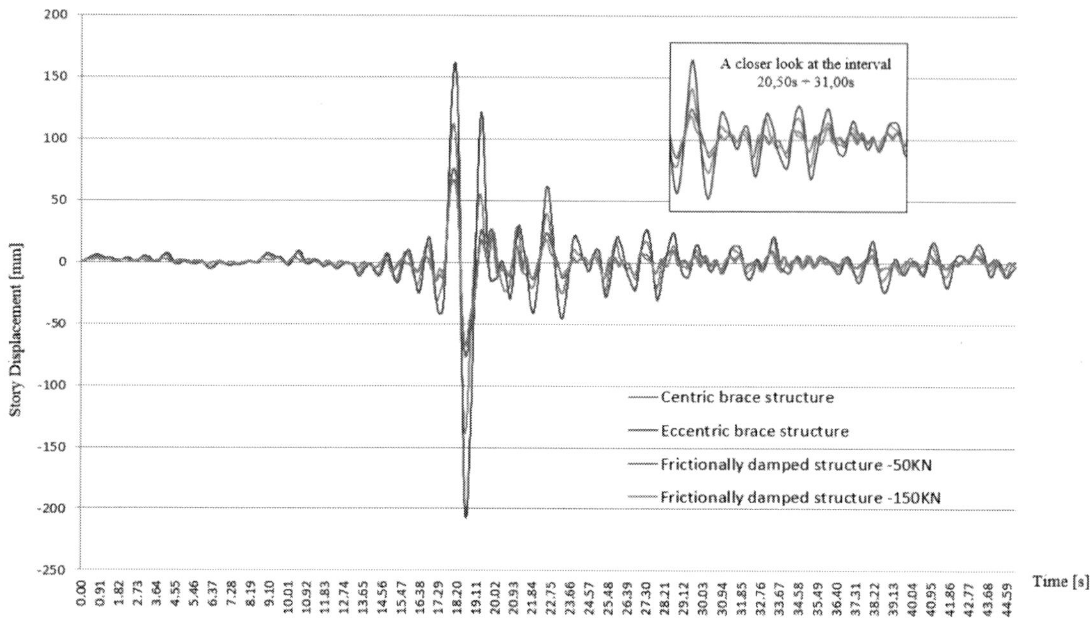

Figure 10. The response of the structures to Vrancea earthquake – top displacement.

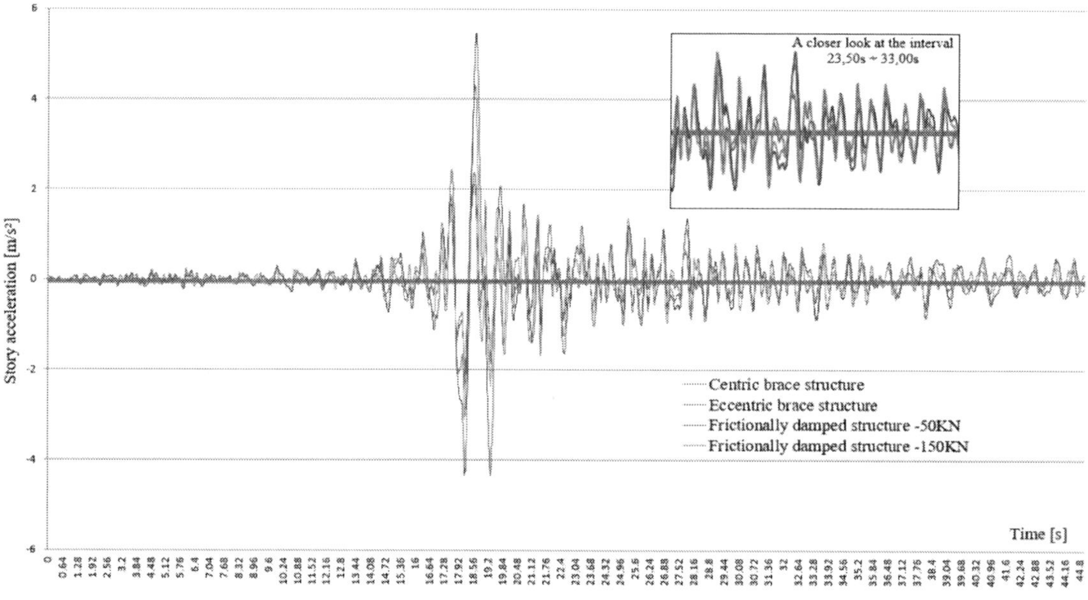

Figure 11. The response of the structures to Vrancea earthquake – top acceleration.

During the seismic action, the damper with a 50 kN capacity, figure 12, slips more than the damper with a 150 kN capacity, figure 13.

Figures 14 and 15 show the energy components in a nonlinear time-history analysis. The input energy is greater in the structure equipped with the 50 kN friction damper than the other structure with a 150 kN damper. The dissipated energy is lower for the structure with 50 kN damper than the structure with 150 kN damper. Even though both structures have the same design, applying a different capacity for dampers can significantly change the result in terms of top displacement, top acceleration and dissipated energy.

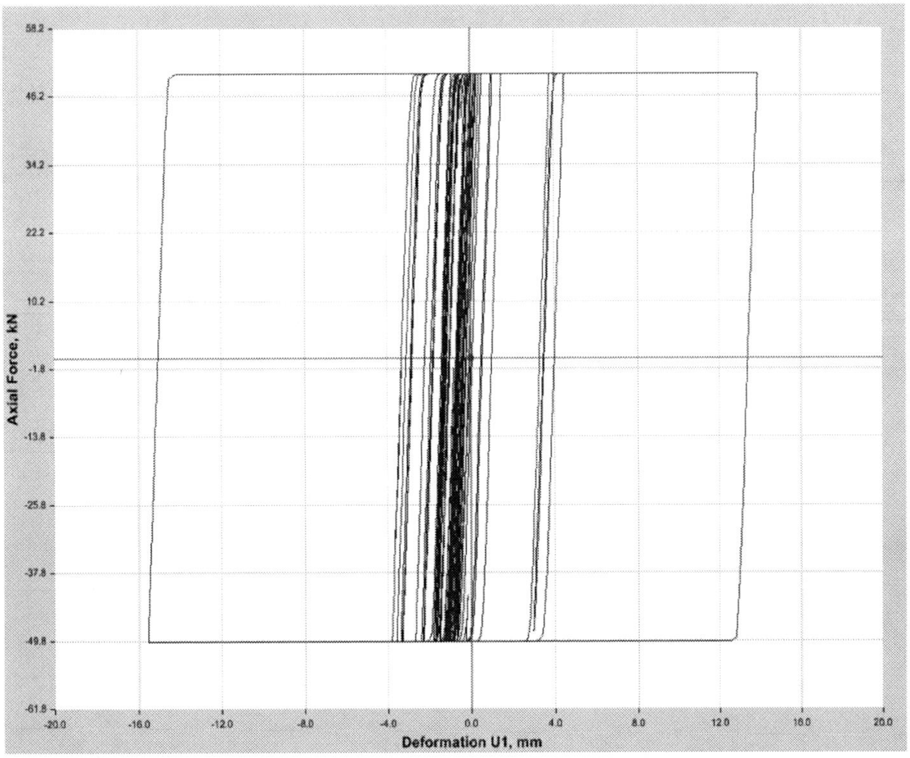

Figure 12. Hysteretic curve for a brace at ground floor - Damper with slip load 50 kN.

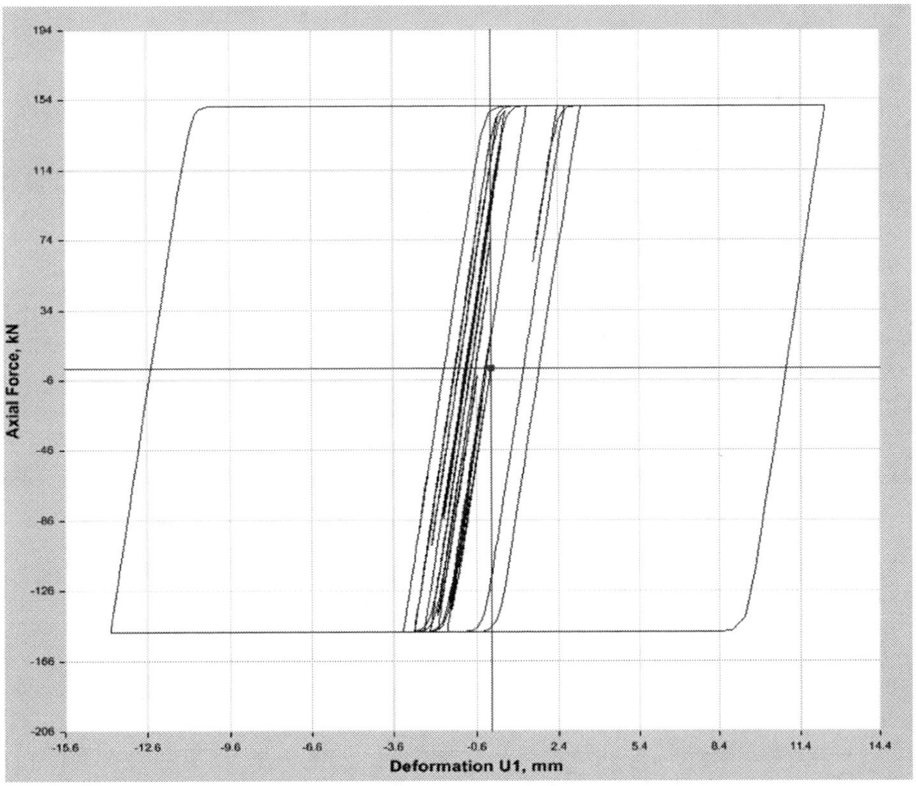

Figure 13. Hysteretic curve for a brace at ground floor - Damper with slip load 150 kN.

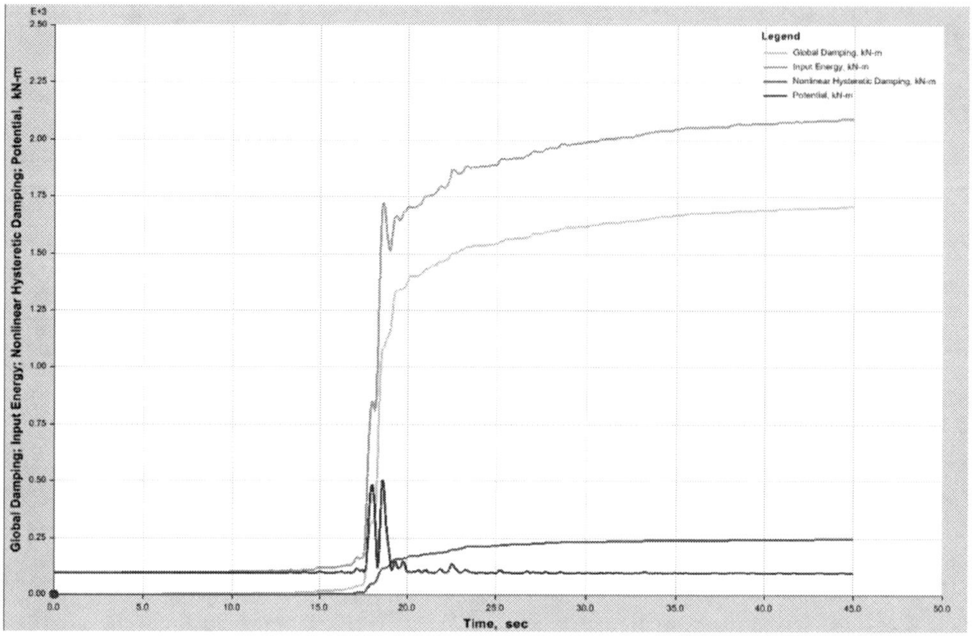

Figure 14. The dissipated energy for a damper with slip load 50 kN.

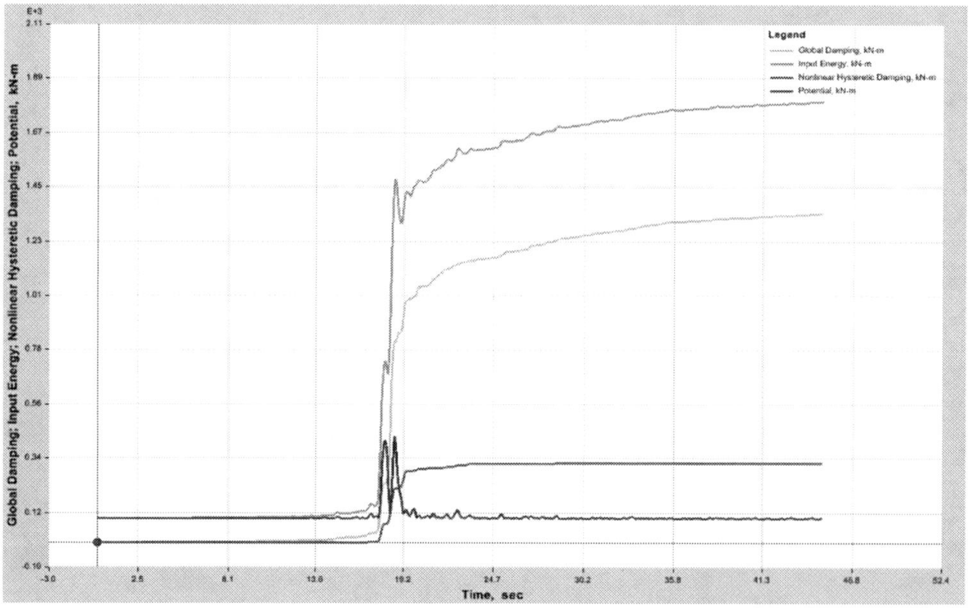

Figure 15. The dissipated energy for a damper with slip load 150 kN.

4. Conclusions

The dynamic nonlinear time-history analysis shows that using dampers in structures improves the response in terms of maximum story displacement, story drifts, top displacement and top acceleration.

It is evident that dampers with lower capacities slip more times during earthquakes than dampers with bigger capacities, but the resulting acceleration increases.

Given their relative independence of the environmental conditions, the friction dampers are a viable alternative to enhance the structure's capacity to dissipate energy.

References

[1] Ştefancu A I, Budescu M and Păduleţ-Crăiniceanu F 2011 Time-history analysis of frictionally damped structures *Bulletin of the Polytechnic Institute of Iasi* Tome LIV (LVIII) Fasc. 4
[2] Khannavar S, Kolhar M H and Algur A 2017 Seismic analysis of RC Structures Using Friction Dampers *International Journal for Research in Applied Science & Engineering Technology* (IJRASET), Volume **5** Issue XII
[3] Muala I H and Belev B 2000 Performance of steel frames with novel friction damper – testing and numeral simulations *Proc. Third Cairo Earthquake Eng. Symp.* CEES 2000 Cairo Egypt
[4] Pall A S and Marsh C 1982 Response of friction damped braced frames *J Struct Div* ASCE 1982;108:1313–23
[5] Fitzgerald T F, Anagnos T, Goodson M and Zsutty T 1989 Slotted bolted connections in a seismic design of concentrically braced connections *Earthquake Spectra* EERI 1989; **5** (2):383–91
[6] Aiken I and Kelly S 1990 *Earthquake simulator testing and analytical studies of two energy absorbing systems for multi-storey structures* Report No. UCB/EERC-90/03 EERC Berkeley
[7] Constantinou M C, Reinhorn A M, Mokha A S and Watson R 1991 Displacement control device for base isolated bridges *Earthquake Spectra* EERI 1991; **7** (2):179–200
[8] Grigorian C E and Popov E P 1993 Slotted bolted connection energy dissipaters *Earthquake Spectra* EERI 1993;9(3):491–504
[9] Nims D K, Richter P J and Bachman R E 1993 The use of the energy dissipation restraint for seismic hazard mitigation *Earthquake Spectra* EERI 1993; **9** (3):467–87
[10] Dorka U, Pradlwarter H J and Schueller U 1998 Reliability of MDOF-system with hysteretic devices *Eng. Struct* 1998; **20** (8):685–91
[11] Nikam S G, Wagholikar S K and Patil G R 2014 Seismic Energy Dissipation of a Building Using Friction Damper *International Journal of Innovative Technology and Exploring Engineering* (IJITEE) Volume-3 Issue-10

IOP Conf. Series: Materials Science and Engineering **789** (2020) 012047 doi:10.1088/1757-899X/789/1/012047

Analysis of factors affecting the performance of reinforced concrete structures of engineering constructions under the influence of aggressive environment

E G Pakhomova[1], **S G Emelyanov**[1], **N V Bredikhina**[1] **and N E Semicheva**[1]

[1] South-West State University, Kursk, Russia

[*] egpakhomova@yandex.ru

Abstract: The work is devoted to the study of the performance of reinforced concrete structures of engineering structures operating under the synergetic influence of aggressive environment with damage to concrete and main reinforcement.

1. Introduction

During health ensuring of reinforced concrete structures engineering structures should be ensured by both technological and constructive requirements. When exposed to aggressive environments, it is necessary to protect building structures from corrosion and it is one of the main and big problems in addressing the issue of ensuring the durability of buildings and structures.

One of the main causes of reinforced concrete structures of engineering constructions corrosion is the impact of aggressive environments of man-made and natural character. This leads to active destructive processes. It is well known that the change of material properties in time depends on the interaction with the environment and is irreversible. Especially clearly destructive processes are observed in areas of variable water level, active chemical and physical effects of the environment. For example, in the construction of industrial hydraulic engineering (hyperbolic tower cooling towers, fan cooling towers, air tanks, filters, settling tanks, dock chambers, gravity embankments). During the operational period of 6-8 years, the depth of corrosion of concrete reaches 8-10 cm, and for the period of 25-30 years can reach 1-1.5 m. In addition, as the aggressive environment penetrates, the bodies of the structure, the protective properties of the concrete decrease with respect to the reinforcement, which begins to corrode. All this affects the bearing capacity of reinforced concrete structures [1].

Thus, when considering the corrosion of reinforced concrete, it is possible to determine synergistic effects, such as the combined action of the external environment (high humidity, temperature, etc.), aggressive environment (various liquids, gases, solid aggressive formations), taking into account the stress-strain state of the reinforced concrete element [1-3].

2. The synergistic effects of environments on the concrete properties

Synergetic effects of environment lead to significant changes in the deformation and strength properties of concrete affected area. The change in material properties over time is irreversible and depends on the deformation conditions and the interaction with the environment. In addition, as the aggressive environment penetrates into the body of the structure, the protective properties of the concrete decrease with respect to the reinforcement, which begins to corrode. As a result of corrosion, the cross-sectional area of the reinforcement decreases and its adhesion to concrete is broken. All this affects the bearing capacity of reinforced concrete structures.

Damages of concrete are noted at action on it of water solutions of acids or acid gases, solutions of salts and even alkalis, some organic compounds. The degree of aggressive action depends not only on the composition of the aggressive environment, but also on the contact conditions, the speed and pressure of the liquid environment, the density of the adjacent soil under the action of groundwater, the temperature of the environment, power loads, the stress state of the material of structures and other factors. Almost all the effects that occur with the participation of aggressive solids and gases can be attributed to synergistic, since the presence of water is necessary for the chemical reaction under normal conditions. All acidic gases acting on the concrete structures, together with the CO_2. In most cases, the advanced process is the carbonation of concrete, which begins with the manufacture of the structure, while specific acidic gases begin to act, as a rule, only after the start of operation of the building. The effect of gases on the concrete causes its neutralization, and the resulting salts penetrate deep into the concrete at a rate depending on their solubility, permeability and humidity [4]. The nature of the main destructive processes is presented in table 1.

Table 1. Typical cases of structural interaction with the environment.

Environment	Environmental exposure conditions	Predominant processes in concrete
Air-wet	Free-flow	Neutralization
	Pressured	Same, accelerated
Air-wet, with the presence of solutions of salts, acids, etc. and with direct periodic humidification	Free-flow	Increase in the amount of introduced aggressive components or products from interaction with cement stone, neutralization, diffusion of aggressive ions
	Pressured	The same processes, accelerated +leaching
Aqueous with the presence of solutions of salts, acids, etc.	Free-flow	Diffusion of aggressive ions
	Pressured	The same leaching

3. Mathematical model of concrete corrosion

To assess the corrosion damage of concrete in reinforced concrete structures of engineering structures, the following indicators can be used: the depth of damage to concrete and service life.

It is obvious that the ultimate goal of constructing a mathematical model of corrosion $L=f(t)$ is to obtain a simple formula, which is convenient to perform engineering calculations. However, almost all researchers in the derivation of the dependence $L=f(t)$ to simplify proceed from the interaction of only two substances.

As a result of the study of the methodical approach to the determination of the service life of concrete and the preparation of proposals for the quantitative assessment of the kinetics of corrosion processes arising from the contact of liquid corrosive environment with concrete, which were based on the analysis of the nature of corrosion processes, the following conclusions were obtained in the works of domestic authors [1, 5]:

- it is established that the intensity of corrosion processes is determined by the intensity of penetration of aggressive components of the environment into the pore structure of concrete;
- the movement of the aggressive environment from the outer surface deep into the concrete is carried out under the influence of hydrostatic pressure, molecular diffusion and capillarity; the pressure of the external environment on the open surface of the concrete accelerates this process:
- such classification of the acting forces inducing movement of the aggressive environment in concrete allows for stationary conditions to carry out quantitative calculations of a stream of

aggressive substance through a concrete surface and to give an assessment of its influence on a condition of concrete in time for simple boundary conditions.

Table 2. Characterization of the synergistic interaction of the two substances.

Examples of the interaction of concrete with a diffusing external environment	Description process's	Differential equation, describing the process [2]
Contact of cement stone with solutions of salts, acids, etc.	The presence of chemical interaction of concrete with the environment, the process is controlled by diffusion and chemical reaction. Dissolution of cement stone minerals, enhanced by the action of acids. The accumulation in the concrete pores of salts that are able to switch into other crystalline hydrate forms with a volume change. Destruction of aggregate contacts with cement stone.	$\begin{cases} \frac{\partial c_A}{\partial t} = D_A \cdot \frac{\partial^2 c_A}{\partial x^2} + RC_i \\ \frac{\partial c_B}{\partial t} = D_B \cdot \frac{\partial^2 c_i}{\partial x^2} + \frac{k_B s_B}{\Pi_B} R(C_i) \end{cases}$ (1)
Contact of cement stone with chlorides	The absence of chemical interaction of concrete with the environment, the process is controlled by diffusion. The formation of gypsum with a significant volume increase.	$\frac{\partial c_A}{\partial t} = D_A \cdot \frac{\partial^2 c_A}{\partial x^2}$ (2)

Corrosion of concrete in the product or structure is caused by the activity of organisms. Biological corrosion is most common in those structures where organic substances come into contact with the surface of reinforced concrete. Corrosion destruction is accompanied by leaching of calcium and magnesium in the form of sulphates from concrete [2].

4. Influence of environment aggressiveness on the loss of bearing capacity of structures

Climatic effects on concrete, which include temperature, humidity, the number of transitions through 0°C, destroy the surface structure of its layers, lead to the formation of closed microcracks, which connecting with each other, form a through porous system, facilitating access to subsequent effects of chlorine ions or carbon dioxide.

The General theory of the processes occurring during the freezing of water-saturated concrete is still in the stage of its development. Features of the destruction process during freezing of water-saturated concrete are reflected in the works of V. M. Moskvin, V. B. Gusev, N. K. Rosenthal and other authors [6-8]. The movement of the freezing front and the increase in volume during the phase transition to ice causes the movement of water. At the same time, the pore pressure increases sharply, which reduces the crystallization temperature. In the presence of salt solutions in the liquid phase of concrete it is affected by the concentration of salt solutions [3]. The practice of surveys shows that one of the main reasons for reducing the bearing capacity of reinforced concrete structures is the influence of aggressive environmental influences. The most unfavorable result of this effect is the chemical corrosion of reinforced concrete. In industrialized countries, corrosion damage is estimated at 3-5 % of gross national income, with 13-19 % of building structures. Currently, the effect of corrosion on the bearing capacity of reinforced concrete structures during the survey is estimated [9-11].

Thus, in the surface layers of concrete in contact with the external environment, under the action of carbonic and hydrochloric acids and magnesia salts, the structural elements of hydrated cement stone and sometimes non-hydrated grains of cement clinker are destroyed. New formations do not possess astringent properties and sufficient density to inhibit further penetration of the aggressive environment. They are washed off; dissolved and deeper concrete layers are exposed.

Aggressive factors causing the destruction of reinforced concrete structures are: the dissolving ability of water, the content of hydrogen ions, the content of salts and sulfates, the alkali content, acids and direct current flow, biodeterioration of concrete by microorganisms. The degree of their aggressive effect on concrete determined in accordance with the requirements of GOST 31384 [12] and SP 28.13330 [13].

The degree of aggressiveness of the medium and the depth of destruction of the surface layer of concrete, affecting the loss of the bearing capacity of concrete structures, are presented in table 3 [13]. With the simultaneous exposure of various aggressive environments, the degree of exposure of the medium to concrete is determined by the more aggressive considering the operating conditions of the structure [13].

Table 3. Loss of bearing capacity in the operation of structure.

The degree of aggressiveness of the environment	Depth of destruction of the surface layer (mm/year)	Average annual loss of bearing capacity in the operation of structures (%)	
		underground	bearing and enclosing
Weak	up to 0.4	3	5
Average	0.4-1.2	5	10
Strong	More than 1.2	8	15

Protective action of concrete in relation to reinforcement determined by the ability of cement stone to passivate steel [14]. The main reasons for the decrease in steel passivity are: carbonization of the protective layer of concrete under the influence of carbon dioxide; accumulation of ions, aggressive towards steel (chlorides, sulfates) under the influence of the operating environment of the structure; the introduction into the concrete mix of chemical additives containing aggressive towards steel ions. Necessity of protection steel embedded parts and the choice of corrosion protection methods are determined by corrosion damage, depending on the degree of aggressiveness of the medium and affecting the decrease in the strength of the reinforced concrete structure in the corrosion zone [11, 12].

The degree of aggressiveness of the environment, affecting the depth of damage to the reinforcement, is presented in table 4 [11].

Table 4. The degree of aggressiveness of the environment.

The degree of aggressiveness of the environment	Corrosion damage, (mm/year)	Points by GOST 13819-68 [11]	Decrease in strength in the area of corrosion (%)
Nonaggressive	0.1	1-3	0
Weak	0.01-0.05	4.5	up to 5
Average	0.05-0.5	6	up to 10
Strong	>0.5	>7	>10

Evaluation of the life of the operating structures is proposed to perform by formula (3):

$$t_{np} = \frac{t_{o6} \cdot \delta^2}{(m_1 \cdot x_{o6})^2} - t_{o6}, \tag{3}$$

where $t_{об}$, $t_{пр}$ is taken accordingly, the service life of the structure at the time of the survey and the projected service life (resource), $x_{об}$ – depth of concrete carbonation at the time of the survey.

5. Summary

Currently there are different approaches various approaches to assessing the corrosion damage of concrete and fittings. In relation to the definition of reinforced concrete structures performance subject to synergistic effects of the media, it is recommended to apply the following criteria for assessing the state of a reinforced concrete element: depth of corrosion damage to concrete; decrease in concrete strength; assessment cumulative valve damage reinforced concrete structures.

References

[1] Moskvin V M, Ivanov F M, Alekseev S N and Guzeev E A 1980 Corrosion of concrete and reinforced concrete. Methods for their protection Stroizdat Moscow p 536

[2] Ivanov F M and Gorshina S N 1984 Biodegradation in construction Stroizdat Moscow p 320

[3] Phonto L M 2004 Durability of concrete structures of engineering constructions (silos, bunkers, reservoirs, water towers, water walls) Monograph Moscow: Publishing house of the DIA p 424

[4] Thomes T, 1992 Durable Concrete Structures vol **182** (*CEB Design Guide*) p 128

[5] Verjuzhskij Ju V 2005 Metody mehaniki zhelezobetona (Reinforced concrete mechanics methods) *Kiev Knizhnoe izd-vo NAU* p 653

[6] Gvozdev A A and Karpenko N I, 1965 Rabota zhelezobetona s trewinami pri ploskom naprjazhennom sostojanii (Reinforced concrete operation with alarms in the case of plane stress) *Stroitel'naja mehanika i raschet sooruzhenij* (Structural mechanics and structural analysis) vol **2** pp 20-2

[7] Gvozdev A A, Petrova K V, 1978 Prochnost', strukturnye izmenenija i deformacii betona (Strength, structural changes and deformation of concrete) Stroizdat Moscow p 299

[8] Oatul A A, Kutin Ju A and Paseshnik V V 1977 Sceplenie armatury s betonom Izvestija vuzov (Clutch reinforcement with concrete) *Stroitel'stvo i arhitektura* (Construction and architecture) vol **5** pp 3-15

[9] Pahomova E G, Gordeev A V 2008 Issledovanija prochnosti izgibaemyh zhelezobetonnyh konstrukcij pri korrozionnyh povrezhdenijah Jeffektivnye stroitel'nye konstrukcii: Teorija i praktika (Strength studies of flexible concrete structures under corrosion damage Effective building structures: Theory and practice) *VII Mezhdunarodnoj nauchno-tehnicheskoj konferencii* (International Scientific and Technical Conference) vol **1** (Volga House of Knowledge) pp 107-109

[10] Pahomova E G, Merkulov D S and Gordeev A V 2008 Prochnost' i deformativnost' izgibaemyh zhelezobetonnyh konstrukcij pri korrozionnom povrezhdenii betona i armatury (Strength and deformability of flexible concrete structures with corrosion damage to concrete and reinforcement) Izvestija Orel Serija Stroitel'stvo. Transport vol **3** p 29-32

[11] GOST 13819-68 Yedinaya sistema zashchity ot korrozii i stareniya. Pokrytiya lakokrasochnyye. Metod ispytaniy na stoykost' v atmosfernykh usloviyakh (GUEST 13819-68 Unified system of protection against corrosion and aging)

[12] GOST 31384-2008 Zashchita betonnykh i zhelezobetonnykh konstruktsiy ot korrozii. Obshchiye tekhnicheskiye trebovaniya (GUEST 31384-2008 Protection of concrete and reinforced concrete structures from corrosion)

[13] SP 28.13330.2017 Zashchita stroitel'nykh konstruktsiy ot korrozii. Aktualizirovannaya redaktsiya SNiP 2.03.11-85 (s Izmeneniyem N 1) (SP 28.13330.2017 Protection of building structures from corrosion. The updated edition of Snip 2.03.11-85 (with Change N 1)

[14] Alekseyev S N, 1968 Korroziya i zashchita armatury v betone (Corrosion and reinforcement protection in concrete) Stroizdat Moscow p 120

IOP Conf. Series: Materials Science and Engineering **789** (2020) 012048 doi:10.1088/1757-899X/789/1/012048

The choice of aversion of the project proposal on restoration of the cultural heritage property on the basis of multicriteria comparative analysis

E G Pakhomova[1*], V A Jezersky[2], P V Monastyrev[3] and N V Kuznetsova[3]

[1] South-West State University, Kursk, Russia
[2] Bialystok University of Technology, Bialystok, Republic of Poland
[3] Tambov State Technical University, Tambov, Russian Federation

[*] egpakhomova@yandex.ru

Abstract. With time a problem of preserving architectural heritage sites becomes more acute, since many unique listed buildings are in an emergency state and require the restoration measures that take into account the specifics of the historical environment, an analysis of the technical condition, evaluation of the monuments' value, the possibility of their functional adaptation to modern needs. The search for solution to preserve such objects is often complicated by the development of a modern urban renovation area around them and by the difficulties of maintaining a complete historical appearance of cities when adapting listed buildings to the functions demanded in modern society. The article proposes an approach to choose the best version of the project proposal for the restoration of a cultural heritage property of regional significance. The building is designed in an eclectic style in the city of Tambov. 11 versions of the author's project proposals for the restoration transformations are considered, which preserve the historical appearance of the building and various transformations of the internal volume and development options of the adjacent territory. 14 criteria are proposed to evaluate the qualities of versions. To select the best restoration version a scientific method of a multicriteria comparative analysis was used, which made it easy to compact a large amount of information about the object to a single quality parameter in the form of a synthetic indicator for each version. Simultaneously the weight factors of the criteria were established on the basis of expert evaluation obtained from the professional groups of various profiles: architects, representatives of the contractor, state supervisory authorities, the owner of the object, the tenant of the object.

1. Introduction

In seeking to resolve development problems, the historical town image is sometimes distorted due to the implementation of programs for a new construction and restoration of the architectural heritage without taking into account the specifics of the historical environment, an analysis of the technical condition, evaluation of the monuments' value, the possibility of their functional adaptation to modern needs [1,2]. The problem of restoration is particularly acute for the listed buildings that have architectural and historical merits but which are in an emergency state and could be demolished during comprehensive development. A mansion, a house number 37 Komsomolskaya street in the city of Tambov is in an ultimate limit state now, it is a cultural heritage property (CHP) of a regional significance "*The house in which the writer S.N. Sergeev-Tsensky lived during his studies at the Tambov gymnasium in 1887-1892*" (Resolution of the Tambov Region Administration of 08/10/1993, No. 280).

The mansion of the middle of the XIX century belonged to the state councilor Ivan Skvortsov. He was a doctor, served as a director in a hospital for mentally defective patients, participated in the work of the city medical and sanitary commission, and he opened a hydropathic in a neighboring house. Skvortsov also let apartments for rent in his house. This house is also associated with a famous writer of the first half of the twentieth century, Sergey Nikolaevich Sergeyev. There was a river Tsne there, and later he added the pseudonym Tsensky to his last name. The future writer lived in this house in 1887-92, when he studied at the Tambov gymnasium. So, this house seems to be a part of the cultural treasure of Tambov and it forms the architectural space of the district [3 - 5].

The two-story building with a rustication of corners and symmetrically arranged windows framed with decorative elements in appearance runs to the architectural stylizations of the turn of the century (large windows, motifs of neoclassicism and oriental ornamentation of facades, etc.). An eclectic architecture of the facade of the building facing the Komsomolskaya street is not typical for the housing development of Tambov of that period. Currently, this building is in an ultimate limit state and needs an urgent conservation and a restoration work (figure 1). The historical territory of the monument (figure 2) is not identified and completely violated, despite the fact that there is an established protection zone and a development control and economic activity zone [6]. According to the approved subject of protection, when planning restoration work on a listed building, the parameters of cultural heritage property are not subject to change that is recorded in the subject of protection (Federal Law N 73 "On cultural heritage property (historical and cultural monuments) of the peoples of the Russian Federation" of June 25, 2002): location and spatial compositional construction of the building; the plane of the southern facade with all the architectural elements located on it (Directive of the culture and archiving department of the Tambov Region No. 260 of 10/15/2012). The restoration of the building with its adaptation to modern use (valorization) involves the consideration of several options to transform the object with possible additions of new volumes to it and options to develop the adjacent territory in accordance with modern approaches to the adaptation of monuments [7 - 9].

Many authors associate a quality assessment of cultural heritage property with categories of authenticity [10], value [11,12], social and semantic role [3, 13], and for the objects used for the practical purposes, safety requirements must be taken into account, [14]. There are methods of numerical evaluation of the significance of the objects, taking into account a number of value characteristics of the monument (historical, architectural, artistic, city planning, construction and technological, etc.) [4,15,16]. In addition to historical and cultural value, vital importance when estimating CHP is given to the cost indicators of the object [17, 18]. The value of the object can be expressed by a balance cost, an estimated value and a value, taking into account the restoration of the objects of protection; the methods to determine are based on the cost, income and comparative (market) approaches [19, 20]. The listed methods for determining the CHP quality indicators relate primarily to existing buildings and they are based on separate accounting of groups of nonmaterial indicators (authenticity, integrity, environmental perception, etc.), or technical (percentage of loss, physical deterioration, burdens, etc.) or cost; they do not imply the price adjustment during an object restoration with an adaptation for modern use. In this regard, to assess the quality of versions of the project proposals for the restoration of a listed building, we need a method considering a set of criteria for assessing the object and making it possible to take up the reasonable choice of the best option for a comprehensive indicator.

The purpose of this study is to develop the versions of the project proposals for a restoration of a mansion of the late XIX century in Tambov, which imply preserving the historical facade and transforming the building and the surrounding area, as well as evaluating the options according to the selected criteria and choosing the best option for the project proposal based on the use of the scientific method of multicriteria comparative analysis.

IOP Conf. Series: Materials Science and Engineering **789** (2020) 012048 doi:10.1088/1757-899X/789/1/012048

Figure 1. General view (current state) Cultural heritage site "The writer Sergeyev-Tsensky lived in this house while studying at the Tambov gymnasium. 1887-1892", Address: Tambov, Komsomolskaya street 37.

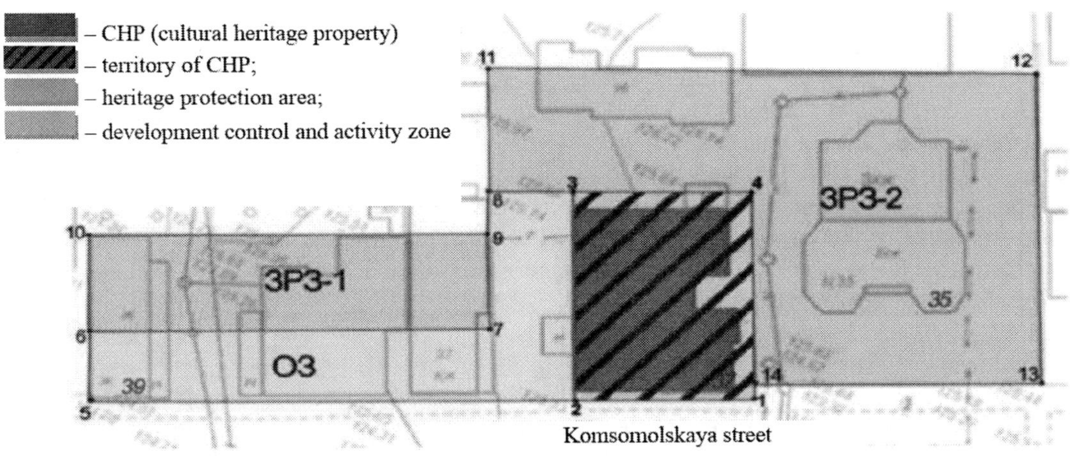

Figure 2. Protection zones scheme Cultural heritage site "The writer Sergeyev-Tsensky lived in this house while studying at the Tambov gymnasium. 1887-1892 Address: Tambov, Komsomolskaya street 37.

2. The choice and justification of a mathematical method of multicriteria comparative analysis

Many factors should be considered when assessing the quality of versions of the project proposals for the restoration of CHP [21,22]. These factors eventually determine the quality of the architectural environment [23-25]. Thus, the consideration of all factors when choosing the final version is very problematic, since a large number of criteria makes it harder to assess them objectively. Problem solving can be facilitated by using computational algorithms and tools that objectively characterize the quality of versions taking into account the preferences included in the project creation stage.

Various methods of multivariate and multicriteria comparative analysis can be used as the basis of a computing tool for assessing the quality of versions of project proposals for the restoration of CHP [26-28].

The authors have decided in favor of the multicriteria comparative analysis, which serves as a computational tool to determine the most suitable options among the many proposed, based on the totality of their characteristics or criteria. This method is based on the construction of a scalar of a synthetic indicator that considers the numerical values of the criterion evaluations and their weight numbers (factors) [28]. We used the adjusted summing indicator J_i as a synthetic indicator, [11], which was calculated using the formula:

$$J_i = \sum_{i=1}^{m} (z_{ij} \cdot v_j) \tag{1}$$

where: z_{ij} is the encoded measure of the j-th criterion for the i-th option; v_j is the weight number of the j-th criterion; m is the number of criteria.

Beforehand when performing an analysis all options must be evaluated in accordance with the selected criteria. The criteria can be quantitative, expressed by numbers, and qualitative, which require the creation of a rating scale and in accordance with the established scale it should be given digital values.

Then, for performing an analysis, the non-dimensional values are assigned to the values of the criterion evaluations expressed in units of measurement. For this purpose, the encoding of the values of the criterion evaluations is performed. In this work, the Neumann-Morgenstern method [28] is used, according to which the encoding for stimulants (criteria when increasing rise the assessment of the quality of the version) is performed using the formula:

$$z_{ij} = (x_{ij} - x_{j\,min})/(x_{j\,max} - x_{j\,min}) \tag{2}$$

as for destimulants (criteria when increasing hold down the assessment of the quality of the version) is performed using the formula:

$$z_{ij} = (x_{jmax} - x_{ij})/(x_{j\,max} - x_{j\,min}) \tag{3}$$

where: z_{ij} is the encoded evaluation value of the i-th option according to the j-th criterion; x_{ij} are the natural values of the evaluations; $x_{j\,max}$, $x_{j\,min}$ - the maximum and minimum value among the evaluations upon the j-th criterion.

In the case of applying methods that consider the criterion weights, the next step should be to determine the values of the weight numbers of the corresponding criteria, taking into account the fact that their sum should be equal to 100%. After establishing all the data, the values of the selected synthetic indicator are calculated for all versions.

3. Characteristics of the versions of the author's project proposals for comparative analysis

Eleven versions of the author's project proposals for the restoration of the building were developed based on the results of the field studies of the selected cultural heritage property and analysis of the city adjacent area. These versions contained proposals for preserving the historical facade of the restoring building and for transforming the building and the surrounding area (table 1). Each version was also considered from the point of view of the potential users of the object.

Each version is interesting and unique in its own way. In order to assess fully the strengths and weaknesses of each of the versions, a special scientific analysis was carried out evaluating each version upon 14 criteria.

Table 1. Versions of the project proposals to renovate the monument.

№ of version	Suggestions for preserving the historical facade and filling the plot up	Description
1	2	3
1	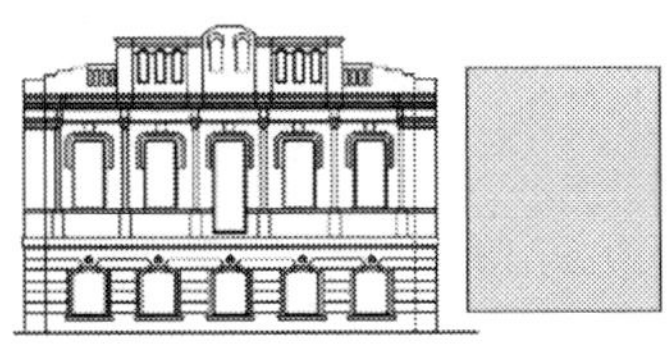	Complete restoration of facades and the building within the old walls. The material of the walls is selected according to the condition of the existing walls. Reconstruction of the historical layout.
2	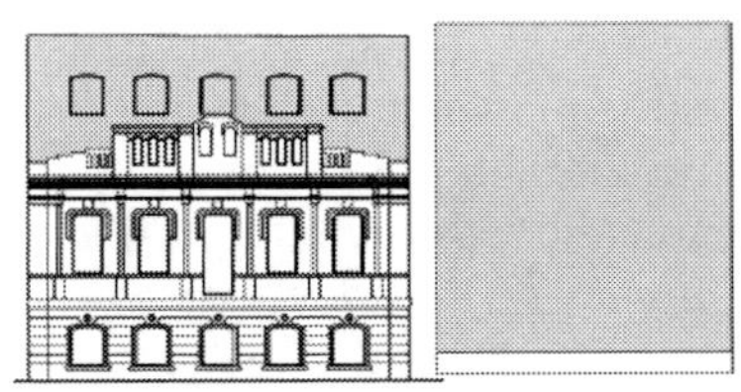	Restoration of the building and the facade, an additional storey of the third floor. Window openings of the additional storey imitate elements of historical window openings. The roof is flat.
3	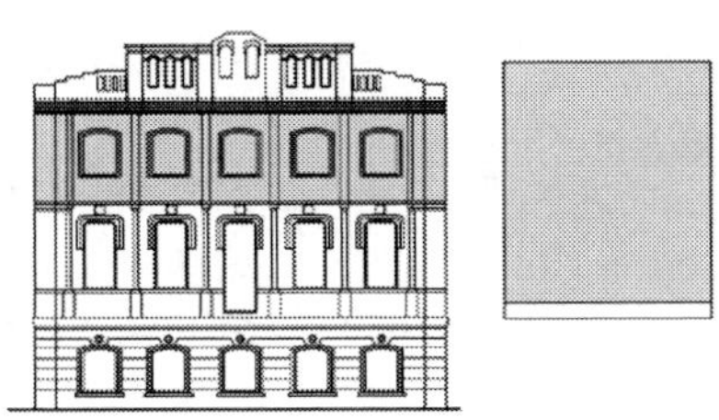	Restoration of the building and the facade, an additional storey of the second floor. Window openings of the additional storey imitate elements of historical window openings, the whole pediment located above the first floor is dismantled and is installed above the additional storey. The gable roof is hidden behind the pediment on the main facade.
4	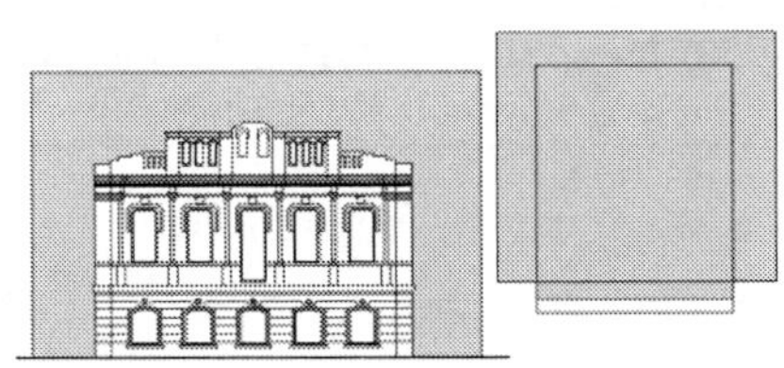	Restoration of the historical facade, adding a new part that goes beyond the boundaries of the historical walls of the building. Number of floors - three floors. The roof is flat.
5	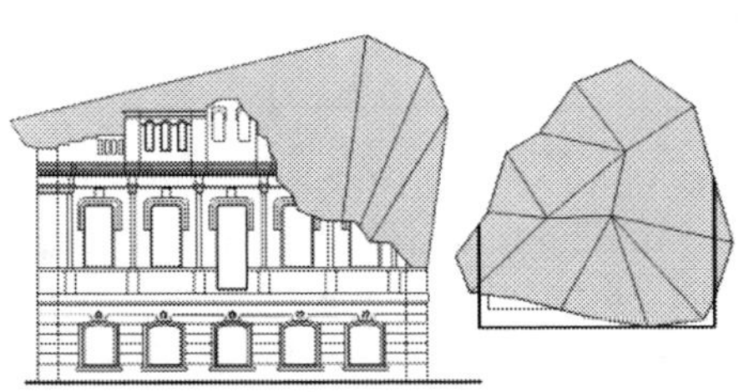	Partial restoration of the historical facade, dismantling of the facing part, adding a new part of the building with a bionic form, which goes beyond the boundaries of the historical walls. The adding floor has a complex roof.

Continuation of table 1

1	2	3
6	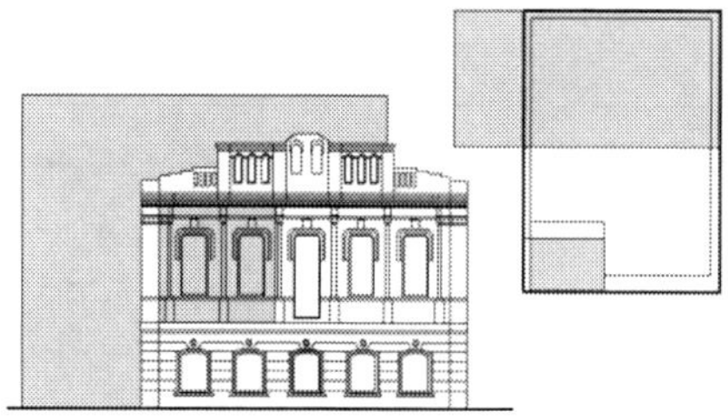	Restoration of the historical facade and the old walls around the perimeter. Dismantling of a part of the main facade and replacement of old parts with identical ones, which are made of modern materials (glass, metal, porcelain tiles), an adding within the old walls and going beyond them on one side, the adding storey is 1 floor higher than the main building. The roof is flat.
7	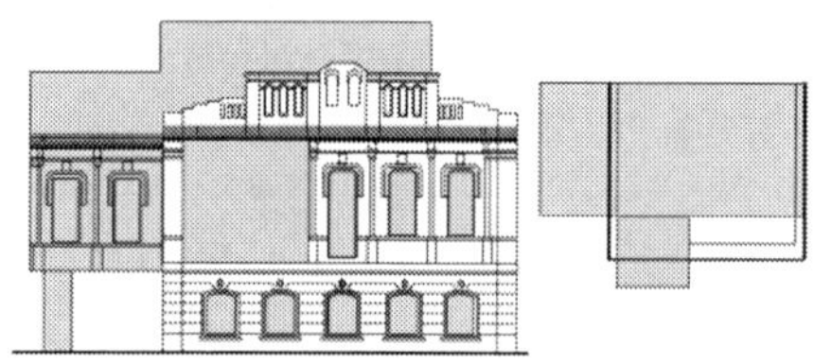	Restoration of the historical facade, partial restoration of the old walls, cutout of a part of the old facade and its use in the new facade of the addition. An identical part of the facade from modern materials (metal, glass, porcelain tile) is reconstructed at the cutout place to emphasize the time frame of the recreation. The additional building extends beyond the boundaries of the old contour of the building. The number of floors of the addition is the same as that of the main facade, additionally there is an attic floor.
8	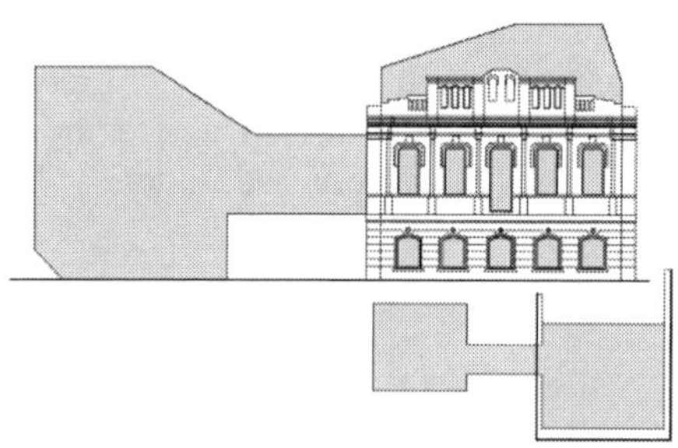	Restoration of the historical facade, partial restoration of the old walls, a building is being erected within the boundaries of these walls, its height is like the historical facade and having an asymmetric attic floor. A passage is built from the new building at the second floor level, leading to a building erected 5 meters from the main building and at the back of the plot. The neighboring building has 2 levels and contrasts with the main historical facade.
9	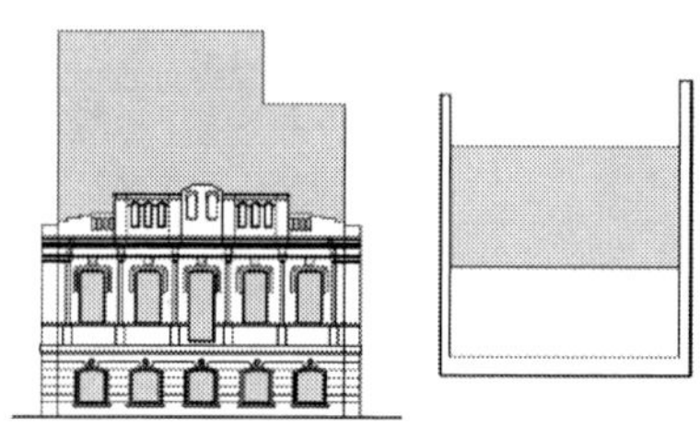	Restoration of the historical facade, partial restoration of the old building, adding to it a new one, with increased number of floors and a different facade.
10	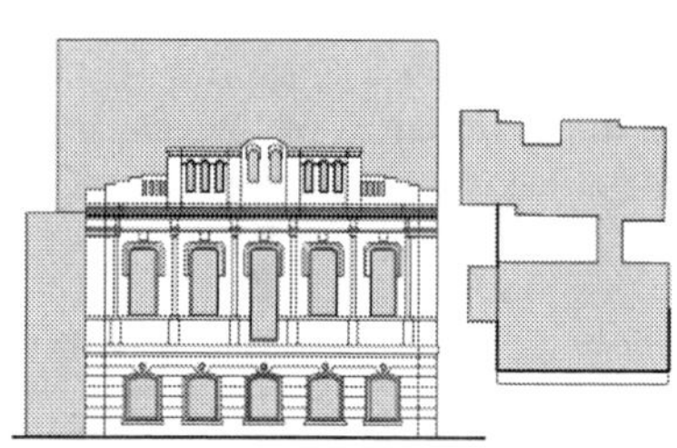	Restoration of the historical facade, extension of a new building to it within the boundaries of the old walls. The new building has two parts connected by a passage. The first part is equal in height to the historical facade, it has an offset (jog) on the left (there, presumably, was the entrance to the old building). The second part has uneven facades and surpasses the first part in height by one floor.

Continuation of table 1

1	2	3
11	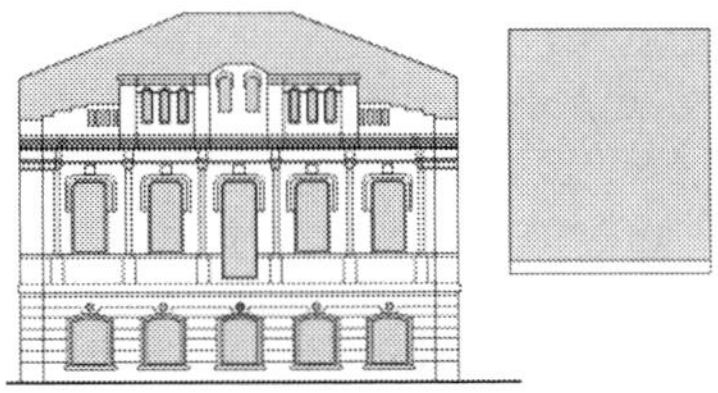	Restoration of the historical facade, the construction of a new building within the boundaries of the old building with the dismantling of the old walls. The new building surpasses the old one in height and has a glazed attic.

4. The choice of quality criteria of versions of project proposals

In order to assess the quality of the versions of the project proposal for the restoration of the monument under consideration, the criteria were selected that reflect various aspects - architectural, construction, social, household, etc. The composition of the criteria was formed taking into account the proposals of the professional groups participating in the discussion. At the same time, the regulations and requirements of design standards for buildings, the planning and development of the city areas, and, first of all, the nature of the architectural composition of the territory adjacent to the monument were taken into account. The selected evaluation criteria were adopted for the restoration of a specific architectural monument, taking into account its adaptation to modern use as an office building (business center). A special condition for choosing options and criteria for their evaluation was the success of the building being restored as a financial product, its investment attractiveness, which is achieved due to the architectural and historical significance of the monument, as well as its location in the central part of the city. Depending on the type of building, the features of its adaptation, the degree of preservation of genuine fragments, restrictions on the use of adjacent territories, etc. assessment criteria are able to change.

The authors do not claim to offer the only correct set of criteria, but as a result of painstaking analytical work for the considered monument, 14 criteria were selected. For each of them the content is formulated, the units of measurement are given, the nature of the contribution to the quality indicator, namely, stimulant (St) or destimulant (Ds), is specified, and for qualitative criteria the rating scales from the possible minimum to the maximum value are developed. The following evaluation criteria have been adopted:

C_1 - the percentage of historical safety of the building. A full restoration is 100%, 0% is the demolition of the building, that is, the loss of all historical elements (stimulant criterion);

C_2 - filling the plot under the historic building. The building in its historical dimensions - 1 (stimulant criterion);

C_3- architectural volume, m^3(stimulant criterion);

C_4 - the versatility of the building. An absolutely universal building is taken per 1, which can have more than 6 functions without any changes inside the building, 0.5 - 3 functions, etc. (stimulant criterion);

C_5 - building perspectives. 0 - demolition, 0.3 - preservation without changes in the existing form, 0.4 - conservation, 0.5 - reconstruction and restoration of individual elements, 0.6 - reconstruction of the entire building for operation, 0.7 - reconstruction of the building with partial ruination , 0.8 - reconstruction of the building with the preservation of some historical elements, 0.9 – addition of new volumes to the existing building with partial preservation of historical elements, 1 - reconstruction and addition with the loss of historical details (destimulant criterion);

C_6 - the height of the building. 1 per a height of 10m., 0.9 - 9m, etc. (destimulant criterion);

C_7 - an aesthetic view in the front of the street. 0.1 – a full identity with historical buildings, 0.3 – a reconstruction using modern materials, 0.5 – a partially preserved historical appearance, but using modern elements, 0.8 – a partially preserved historical appearance, but using many modern elements, 1 – a full contrast with historical buildings (destimulant criterion);

C_8 – an evaluation of innovative solutions while transforming the building. 1 - innovative technologies that have no analogues are applied, 0.8 - a large number of new proposals, 0.5 - the use of new but already applied methods and proposals, 0.3 - a small introduction of innovations into the building reconstruction process, 0 - not used (stimulant criterion);

C_9 - the color index of the external finish: 0- 0.4 - monochrome colors of one scale, 0.5 ... 0.7 - 2 colors with shades, 0.8-1 - more than 3 colors with shades (destimulant criterion);

C_{10} - the ratio of new and historical volumes. A relation between the old part of the building and the new volume is determined graphically (destimulant criterion);

C_{11} - the possibility of modernizing the internal space. 0 - impossible; 0.3 - difficult to redevelop; 0.5 - possible partial redevelopment; 0.7 - a large number of changes, 1 - any changes are possible (stimulant criterion);

C_{12} - the ability to place advertising signboards. 0 - it is impossible to place a signboard, it is possible to place a signboard, but with certain requirements, 1 - it is possible to place any signboard (stimulant criterion);

C_{13} - per square meter price, thousand rubles. It is determined upon aggregated indicators, taking into account the volume of the building, the average market value (destimulant criterion);

C_{14} - per cubic meter price, thousand rubles (destimulant criterion).

Taking into account the selected criteria in the appropriate measurement units and in accordance with the accepted scales for the quality criteria, all 11 versions of project proposals were evaluated.

5. The choice of weight numbers of the criteria

Weight factors of v_j criteria were established on the basis of an expert evaluation obtained from the professional groups of different profiles: architects; representatives of the contractor involved into the restoration; administrative authorities supervising the monuments; owner of the building; tenant of the building. In this respect, six scales of preference were developed. Scale 1 is adopted with the same priorities and equal weight factors for all 14 criteria. Further, the scales of weight factors, corresponding to the priority criteria assigned by specialists, are adopted as follows: scale 2 - priorities of architects; scale 3 - priorities of representatives of the contractor; scale 4 - priorities of representatives of the administration; scale 5 - priorities of the owner of the object; scale 6 - priorities of the tenant of the object (table 2).

Table 2. Weight factors v_j for the selected criteria

Criterion	Scale 1 (J_1)	Scale 2 (J_2)	Scale 3 (J_3)	Scale 4 (J_4)	Scale 5 (J_5)	Scale 6 (J_6)
C_1	0.071	0.10	0.05	0.30	0.03	0.02
C_2	0.071	0.08	0.03	0.01	0.02	0.01
C_3	0.071	0.08	0.15	0.01	0.03	0.01
C_4	0.071	0.08	0.06	0.03	0.02	0.08
C_5	0.071	0.05	0.02	0.10	0.10	0.01
C_6	0.071	0.06	0.15	0.10	0.15	0.15
C_7	0.071	0.08	0.01	0.08	0.04	0.15
C_8	0.071	0.08	0.05	0.10	0.01	0.02
C_9	0.072	0.05	0.05	0.15	0.02	0.01
C_{10}	0.072	0.08	0.15	0.01	0.10	0.11
C_{11}	0.072	0.05	0.05	0.02	0.10	0.10
C_{12}	0.072	0.05	0.03	0.03	0.08	0.08
C_{13}	0.072	0.08	0.05	0.03	0.15	0.20
C_{14}	0.072	0.08	0.15	0.03	0.15	0.05
Σ	1.00	1.00	1.00	1.00	1.00	1.00

The established weight factors v_j were further used when calculating the synthetic quality indicators of versions of the project proposals according to the accepted criteria.

6. The procedure of performing calculations

The first step involved assessment of the versions. The assessed values for individual criteria for each option are given in table 3. These values for qualitative and quantitative criteria were established according to the methods given in the description of each criterion.

Table 3. Assessment of x_{ij} versions according to accepted criteria

Criterion / № of version	C_1	C_2	C_3	C_4	C_5	C_6	C_7	C_8	C_9	C_{10}	C_{11}	C_{12}	C_{13}	C_{14}
	St	St	St	St	Ds	Ds	Ds	St	Ds	Ds	St	St	Ds	Ds
V_1	100	1	1400	0.7	0.3	0.5	0.1	0	0.5	0	0.5	0.2	30	5.0
V_2	20	1	2200	0.7	0.6	0.6	0.3	0.3	0.5	3.11	0.5	0.5	40	8.3
V_3	20	1	2200	0.7	0.5	0.6	0.3	0.5	0.5	3.65	0.5	0.5	40	8.3
V_4	16.6	0.8	3120	0.8	0.5	0.55	0.5	0.3	0.5	1.99	0.5	0.5	45	7.2
V_5	13	0.8	2800	0.4	0.8	0.7	0.8	0.8	0.5	2.18	1.0	0.5	65	10.5
V_6	40	1.2	2000	0.5	0.8	0.6	0.5	0.3	0.7	1.64	0.5	0.5	45	10.1
V_7	30	1.2	900	0.5	0.9	0.6	0.5	0.3	0.7	1.40	0.5	0.5	55	11.6
V_8	43	0.7	1300	0.4	0.8	0.65	0.5	0.5	0.7	1.30	0.5	0.5	45	12.0
V_9	54	0.7	1800	0.7	0.8	0.85	0.5	0.3	0.7	2.20	0.5	0.5	40	13.2
V_{10}	16.6	0.8	1400	0.5	0.8	0.7	0.5	0.3	0.5	1.50	0.5	0.5	38	11.5
V_{11}	16.6	1	1700	0.6	0.5	0.6	0.3	0.3	0.5	1.56	0.5	0.2	40	8.3

The second stage of the study was the encoding of the assessment values from table 3, that is the assessment values of criteria expressed in units of measurement are given non-dimensional values. The encoding was performed separately for stimulant (*St*) and destimulant (*Ds*) in accordance with formulas (2) and (3).

After calculating the coded values of the partial estimates of the particular criteria. they were adjusted taking into account the weight factors of each group of experts. The final stage of the calculation was the summation of the partial corrected coded estimates into a synthetic quality indicator J_j for each i-th version. The calculation of the indicator J_i for each version was carried out six times taking into account the scales of the weight factors of the criteria.

7. The results of the study and their explication

The results of multi criteria comparative analysis, by a hierarchization of versions of project proposals for the restoration of the considered monument in the are presented in table 4 from the highest quality to the lowest.

The distinctions in the obtained values of the synthetic indicator J_i for six scales point to the importance of the correct selection of the weight factors of the criteria, since all the initial data for the calculations, except for the scales of the weight factors developed by the experts, were the same. However these distinctions were small. As the best version for project proposals, all groups of experts have chosen the version №1. The second position, according to architects, the contractor and the administration, was the version № 4. The owner and tenant have chosen № 3 and 2 respectively. The third position according to the architects, the contractor and the administration can be the version №3. The owner and the tenant have settled upon the version № 4.

Pretty close unanimity was also observed in evaluating the worst versions. The most times all groups of experts selected the versions №7, 8, 9 among the three project proposals of the lowest quality. Significantly less often among the weak project proposals the versions № 6. 10 and 11 are identified.

Table 4. The results of multicriteria comparative analysis arranging the versions in the order from the highest quality to the lowest.

Scale 1		Scale 2		Scale 3		Scale 4		Scale5		Scale 6	
J_1	Ver.	J_2	Ver.	J_3	Ver.	J_4	Ver.	J_5	Ver.	J_6	Ver.
0.686	1	0.696	1	0.677	1	0.701	1	0.674	1	0.658	1
0.595	4	0.557	4	0.578	4	0.444	4	0.539	3	0.608	2
0.575	3	0.575	3	0.529	3	0.441	3	0.531	4	0.603	4
0.556	2	0.586	2	0.520	2	0.413	5	0.524	2	0.601	3
0.491	11	0.406	11	0.480	11	0.406	6	0.512	5	0.545	10
0.455	5	0.456	6	0.456	5	0.404	2	0.507	11	0.477	5
0.446	6	0.358	5	0.362	6	0.395	8	0.442	10	0.472	11
0.407	10	0.326	10	0.306	10	0.340	9	0.372	6	0.452	9
0.362	7	0.363	9	0.267	9	0.338	7	0.333	8	0.430	6
0.336	9	0.384	7	0.238	8	0.332	11	0.302	9	0.394	8
0.334	8	0.469	8	0.234	7	0.294	10	0.287	7	0.358	7

Thus the version № 1 was chosen from the author developed versions of project proposals for the restoration of the monument based on the applied mathematical method of comparative analysis and having into consideration the opinions of different expert groups. This version implies a complete restoration of the facades and buildings within the old walls with the selection of wall material based on the condition of existing walls as well as the restoration of the historical layout on the territory adjacent to the object. Six preference scales with different priorities of weight factors have been developed to evaluate specialists' estimates. As a result, among the developed author's versions of the project proposals for the restoration of the monument on the basis of the mathematical method of comparative analysis and having into consideration the opinions of different expert groups. The best version is the one that assumes complete restoration of the facades and the building within the old walls with the selection of the wall material based on the condition of the existing walls as well as a restoration of the historical layout on the territory adjacent to the object.

8. Conclusions

- Using multicriteria comparative analysis for assessing the quality of the project proposals for the restoration of cultural heritage property proved to be very effective. once this made it possible to compact easily a large amount of information about the object up to a single quality parameter in the form of a synthetic indicator J_j for each i-th version of the project proposal.
- The best version of the project proposals for the restoration of the monument by all expert groups was chosen number 1. Versions 4, 3, 2 can be considered as the alternative. similar ones in quality to the version 1.
- Most commonly the experts chose numbers 7, 8, 9 among the lowest quality project proposals for the restoration of the monument. Numbers 6, 10, 11 are identified less often among the weak project proposals.

References

[1] Rubinshtejna A YA 2016 *Aktual'nye problemy ehkonomiki kul'turnogo naslediya (Actual problems of the cultural heritage economy)* Pod red. M.: Gosudarstvennyj institut iskusstvoznaniya (State Institute of Art Studies) p 108

[2] Azatyan K R and Engoyan A R 2014 Problemy integracii starogo i novogo v processe razvitiya gorodskogo prostranstva (Problems of integrating old and new in the process of urban development) *Vestnik MGSU* **6** p 7-16

[3] Bal'zannikova E M 2014 Sohranenie gorodskih ob"ektov istoriko-arhitekturnogo naslediya (Preservation of urban objects of historical and architectural heritage) *Vestnik MGSU* **1** p 15-24

[4] Bondarenko I A 2016 O reabilitacii istoriko-kul'turnyh cennostej na territoriyah novostroek (On the rehabilitation of historical and cultural values in the territories of new buildings) *Academia. Arhitektura i stroitel'stvo* (Academia. Architecture and construction) **1** p 87-93

[5] Dushkina N O 2009 Vnov' o cennosti podlinnosti v arhitekturnom nasledii (Again on the value of authenticity in architectural heritage) *Arhitektura. Stroitel'stvo. Dizajn.* (Architecture. Construction. Design.) **3** p 14-17

[6] Kim O, M'yuller CH U and Klekka U R 1989 Faktornyj. diskriminantnyj i klasternyj analiz (Factorial. discriminant and cluster analysis) *Dzh.— M.: Finansy i statistika.* (Finance and Statistics) p 215

[7] Glavnye komponenty i faktornyj analiz. (Main components and factor analysis) http://statsoft.ru/home/textbook/modules/stfacan.html (data obrashcheniya: 19.12.2018)

[8] Grincevich E A, 2011 Ocenka ehkonomicheskogo potenciala arhitekturnyh ob"ektov kul'turnogo naslediya (Assessment of the economic potential of architectural objects of cultural heritage) *EHkonomika i predprinimatel'stvo* (Economics and Entrepreneurship) **5** (22) p 122-124

[9] Krasheninnikova E S, Norenkov S V and Grauverg T A 2017 Algoritmy taksonomizacii artefaktov arhitekturnogo naslediya v kodah prostranstva (Algorithms for taxonizing architectural heritage artifacts in space codes) *Privolzhskij nauchnyj zhurnal* (Volga Scientific Journal) **1** S. 75-82

[10] Kudryavceva T P 2017 Arhitekturnye modeli sohraneniya ob"ektov kul'turnogo naslediya za poslednie chetvert' veka (Architectural Preservation Models for Cultural Heritage Sites Over the Last Quarter Century) *Vestnik. Zodchij.* 21 vek. **3** (64) p 30-33

[11] Kurashov YU 2017 Kriterii ocenki ob"ektov kul'turnogo naslediya: kul'turno-istoricheskij aspekt i pravovoe reshenie (Criteria for assessing cultural heritage objects: cultural and historical aspect and legal solution) *Akademicheskij vestnik UralNIIproekt RAASN* (Academic Bulletin RAASN) **4** p 40-44

[12] Nazarova M P 2012 Arhitekturnye pamyatniki v strukture kul'turno-istoricheskogo naslediya (Architectural monuments in the structure of cultural and historical heritage) *Voprosy kul'turologii* (Cultural Studies) **6** p 60-64

[13] Prikaz Ministerstva 2018 ehkonomicheskogo razvitiya RF (Order of the Ministry of Economic Development of the Russian Federation 2018) g. N 226 Ob ut-verzhdenii metodicheskih ukazanij o gosudarstvennoj kadastrovoj ocenke (N 226 Ov of the approval of methodological instructions on state cadastral valuation)

[14] Razrabotka proektov zon ohrany ob"ektov kul'turnogo naslediya regional'nogo znacheniya. Tom 2. Materialy obosnovaniya proekta zon ohrany ob"ektov kul'turnogo naslediya (Design of protection zones of cultural heritage objects of regional significance. Substantiation materials of the project of protection zones of cultural heritage objects)

[15] Semencov S V 2009 Sredovoe zonirovanie. gradostroitel'naya reglamentaciya i pravila zastrojki istoricheskih rajonov Sankt-Peterburga (Environmental zoning. urban planning regulations and building rules of the historical districts of St. Petersburg) *Vestnik grazhdanskih inzhenerov* **4** p 13-18

[16] Slabuha A V 2016 Ustanovlenie istoriko-kul'turnoj cennosti ob"ektov arhitekturnogo naslediya (chast' 2): kriterii i metod v sovremennoj ehkspertnoj praktike (Establishment of historical and cultural value of architectural heritage objects (part 2): criteria and method in modern expert practice) *CHelovek i kul'tura.* 2016 **6** S. 9-22

[17] Valegrahov V M 2014 Analiz primenimosti trekh podhodov k stoimostnoj ocenke ob"ektov kul'turnogo nalediya (An analysis of the applicability of the three approaches to valuing cultural heritage sites) *Upravlenie ehkonomicheskimi sistemami* **11** (71) p 1-22

[18] Vavilonskaya T V, 2015 Nerelevantnaya ocenka cennosti arhitekturno-istoricheskoj sredy ehkspertno-kvalimetricheskim metodom (An irrelevant assessment of the value of the architectural and historical environment by the expert qualimetric method) *Vestnik SGASU. Gradostroitel'stvo i arhitektura* (Bulletin of the SSACU. Urban Planning and Architecture) **4** (21) p 6-12

[19] Zelenova S V 2008 Formirovanie ponyatiya «pamyatniki arhitektury». sistema ih ocenki i klassifikaciya v Rossii v period s 1946 po 1964 gg. (The formation of the concept of "architectural monuments." their assessment system and classification in Russia from 1946 to 1964) *Privolzhskij nauchnyj zhurnal* (Volga Scientific Journal) **1** p 77-83

[20] Aigwi I E, Egbelakin T and Ingham J 2018 Efficacy of adaptive reuse for the redevelopment of underutilised historical buildings: Towards the regeneration of New Zealand's provincial town centres *International Journal of Building Pathology and Adaptation.* **36** pp. 385-407

[21] Carbonara G 2012 An Italian contribution to architectural restoration *Frontiers of Architectural Research* **1** (1) p 2-9

[22] Sartorius C. Lévai P. Niederste-Hollenberg J. Nyga I and others 2018 Comparative multi-criteria performance assessment of alternative water infrastructure systems *Water Science and Technology: Water Supply* **18** (6) p 2188-2198

[23] Ferretti V. Bottero M and Mondini G 2014 Decision making and cultural heritage: An application of the Multi-Attribute Value Theory for the reuse of historical buildings *Journal of Cultural Heritage* **15** (6) p 644-655

[24] Gregorová J. Kalesný F. Polomová B and Vojteková E 2017 Architectural Modelling of Alternatives for Verification of New Interventions on the Example of the Romanesque Palace at Spiš Castle in Slovakia *IOP Conference Series: Materials Science and Engineering* **245** (8) art. 082007

[25] Mazzetto S 2018 Heritage Restoration as a Tool to Promote Architectural Identity in the Gulf Regions *Preservation. Digital Technology and Culture* **47** (1) p 3-11

[26] Soliman O A and Aggour M M, 2018 A framework for new architectural additions to heritage buildings *Journal of Engineering and Applied Science* **65** (6) p 423-445

[27] Niang C, Marinica C, Markhoff B, Leboucher E 2017 Supporting semantic interoperability in conservation-restoration domain: The PARCOURS project *Journal on Computing and Cultural Heritage* **10** (3) art. no. 16

[28] Szwabowski J 2001 Metody wielokryterialnej analizy porównawczej: podstawy teoretyczne i przykłady zastosowań w budownictwie. (Methods of multi-criteria comparative analysis: theoretical foundations and examples of applications in construction) PŚ. Gliwice p 129

The dynamics of changes in the physical and mechanical characteristics of the base soil

E G Pakhomova[1] and K O Dubrakova[1]

[1] Southwest State University, Kursk, Russia

[*] egpakhomova@yandex.ru

Abstract. This article discusses the change in the physical and mechanical characteristics of the foundation soil during the operation of construction projects and the possibility of taking these changes into account when designing shallow foundations. Methods Based on the assumption that the weight of the engineering and geological elements of the base remains constant regardless of the level of existing additional vertical stresses, an expression is obtained for determining the coefficient k_γ, taking into account changes in the density and specific gravity of the soil during the operation of buildings and structures. Moreover, based on the assumption that the plot of the additional vertical stresses is triangular, an equation for calculating the coefficient k_c is determined, which allows determining the change in soil adhesion during the operation of construction projects. Using the coefficients k_γ and k_c, an expression is obtained to determine the change in the calculated resistance of the soil of the base during the operation of construction projects. Taking into account the coefficient k_γ with the width of the projected foundation more than two meters with a ratio of settlement of the base to the depth of the compressible thickness equal to 0.3 leads to an increase in the design resistance of the operated base by 5%; by 3% and 1% with S/H_c = 0.2 and with S/H_c=0.1, respectively. In this case, depending on the width of the designed foundation, taking into account the coefficient k_c allows you to exceed R by 25%. Joint accounting of the considered coefficients leads to an increase in the calculated resistance of the soil of the base by an average of 27%. Conclusion The obtained coefficients, taking into account changes in the specific gravity and soil adhesion during the operation of the base, make it easy to estimate the growth of the design resistance during the compression process, which in turn makes it possible to save an average of 10% of the material when designing shallow foundations.

1. Introduction

It is known that in the process of buildings and structures operation there is a change in the physical mechanical characteristics of the soil base. In particular, there is an increase in the specific weight, the angle of internal friction, grip.

Changes in these characteristics lead to a change in the calculated resistance of the base soil [1-21]. Therefore, it is necessary to develop methods of analyzing the dynamics of changing the physical mechanical characteristics of the ground to assess the residual resource of the base during the reconstruction of buildings and structures, as well as to take into account the width of the foundation sole at the design stage [11, 12].

2. Method for assessing changes in the specific gravity of soil

One of the main conditions of the base's strength according to the 2 group of limit states: the average pressure under the sole of the foundation should not exceed the calculated resistance of the base soil [19] $P_{cp} \leq R$. Calculated resistance (R) according to current regulatory documents is determined by the known formula (1):

$$R = \frac{\gamma_{c1} \cdot \gamma_{c2}}{k} \left[\begin{matrix} M_\gamma \cdot k_z \cdot b \cdot \gamma_{II} + M_q \cdot d_1 \cdot \gamma_{II}' + (M_q - 1) \cdot d_b \cdot \gamma_{II}' \\ +M_c \cdot c_2 \end{matrix} \right] \tag{1}$$

The change of soil specific weight is connected with joint deformation of the "building- basis" system [9, 17, 18]: $\gamma_{II,э} = \frac{G}{V_э} = \frac{G}{1\text{кв.м.} \cdot (H_c - S)}$, i.e. change of this physical characteristic can be defined as the soil weight ratio to a difference between compressed thickness depth and draft. To study the dynamics of the change in the calculated resistance when increasing the specific weight, we will determine the k_y coefficient, taking into account the change in the specific weight of the soil during operation: $k_y = H_c/H_c - S = 1/1 - S/H_c$.

3. The results of the assessment of changes in the specific gravity of the soil

The impact analysis of change in the specific weight on the value of calculated resistance is made with application of the offered coefficient and presented graphically (figure 1).

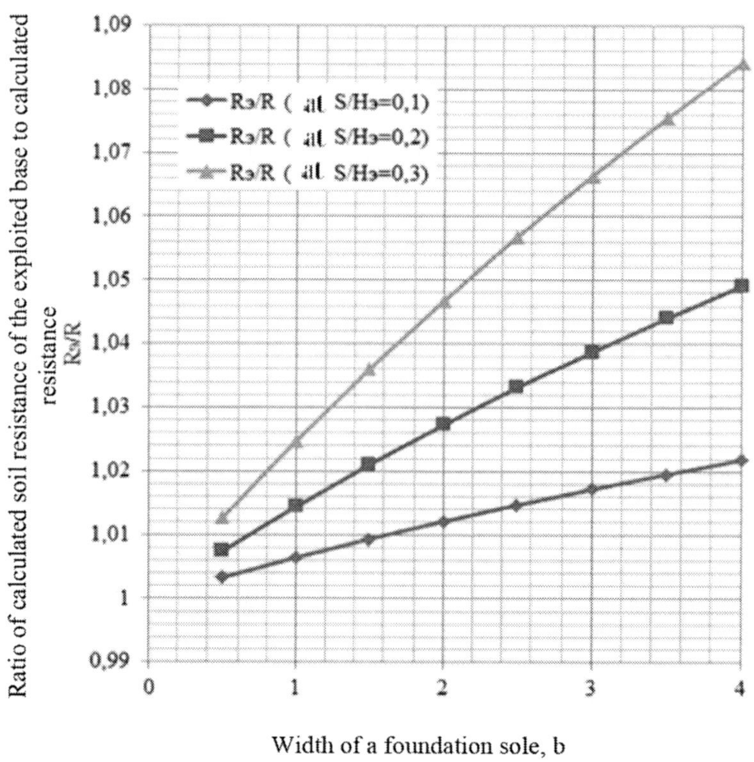

Figure 1. Ratio of calculated soil resistance of the base after R and before R operation in various draft ratio to the depth of the compressed thickness of the S/Hc

From figure 1 it is visible that with an average draft of 4 cm calculated resistance will increase by 4.5%.

4. Method for assessing changes in specific soil adhesion

The change in the specific clutch is connected with soil compaction in the course of joint deformation of the "building base" system. Similarly with k_Y coefficient we will define k_c coefficient which consider these changes (2), where $\frac{S}{H_c}$ is the ratios of the draft to the depth of the compressed thickness:

$$k_Y = \frac{H_c}{H_c - S} = \frac{1}{1 - \frac{S}{H_c}} \quad (2)$$

The analysis of the effect of the clutch on calculated resistance has been fulfilled by means of the offered coefficient and presented graphically in (figure 2). Taking into account the given coefficients, the formula for determining the calculated resistance considering the change in physical and mechanical characteristics of the soil base will take a form, where C is specific adhesion of soil:

$$k_c = 1 + \frac{R \cdot tg\varphi}{2 \cdot C}. \quad (3)$$

5. The results of the assessment of changes in specific adhesion

Calculated resistance of soil base before *R* and after *Re* operation is presented graphically (figure 2).

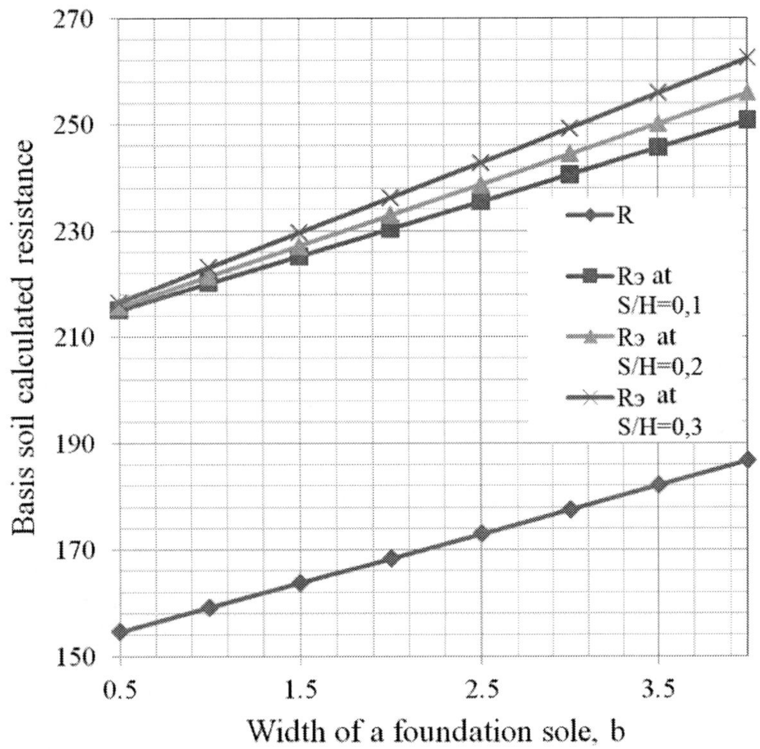

Figure 2. Calculated resistance of soil base before *R* and after *Re* operation at various ratios of the draft to the depth of the compressed thickness *S/Hc*

6. Experimental studies of changes in the physical and mechanical characteristics of soil

From figure 2 it is visible that at the average draft of 6 cm the calculated resistance increases by 35%. In order to check the settlement device for apparatus to assess the dynamics of the change in the calculated soil resistance of the base is developed, the priority of which is protected by the patent for useful model. An experiment essence is the following: the fixed vertical tension is transferred to the soil sample placed in shift the device; further through a polyspast steps put the loading creating horizontal tension, the moment of shift of a sample is registered. Further, according to the well-known formula of Coulomb-Mora, the angle of internal friction and specific coupling of the soil is determined [13-15].

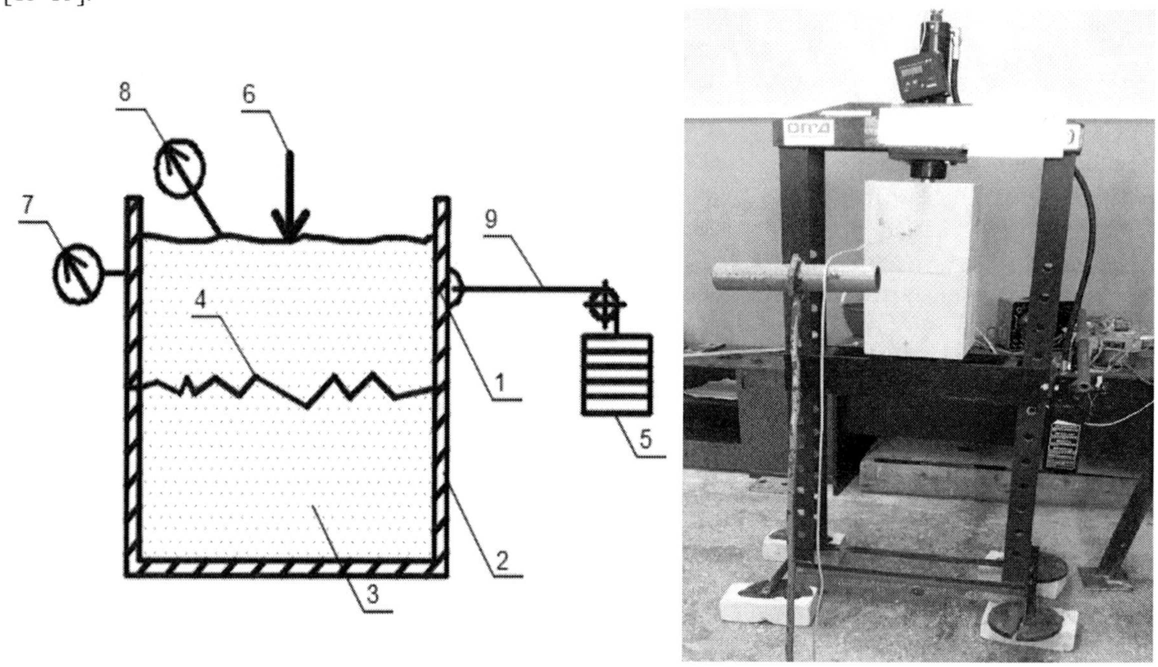

(a) (b)

Figure 3. Scheme (a) and general view (b) of the test unit: 1 – mobile part of the device; 2 - motionless part of the device; 3 - soil sample; 4 - cut and shift line; 5 – freight; 6 – press; 7, 8 – indicators; 9 - steel rope

The model of the operated basis was created in a tray. The foundation is simulated as a rigid square stamp a concentrated load application. During the experiment base draft was fixed until it stabilized. Further soil was withdrawn, and re-testing on shear was carried out.

Figure 4. Scheme (a) and a general view of model of the exploited basis.

7. The results of experimental studies of changes in the physical and mechanical characteristics of the soil

By results of pilot studies specific weight increased by 10%, coupling for 24%, the angle of internal friction for 17%. The matched analysis of experimental and analytical data is presented in Table 1.

Table 1. Results of changes in physical and mechanical characteristics of soil base before and after soil compaction.

№	Before soil compaction	After soil compaction
1	2	3
Angle of internal friction	15,64°	18,8°
Specific cohesion	28 kPa	37 kPa

For further research, the authors plan to obtain additional data that must be collected to refine (confirm) the results.

8. Conclusion

Difference between theoretical values and results of an experiment is an average 10%. As a result of a research it is defined that changes in calculated resistance of soil in operation of buildings and structures process is in direct proportion to the base draft ratio to the depth of the compressed thickness, width of a foundation sole and an average pressure under his soil.

References

[1] Nguyen M D 2013 Behavior of nonwoven-geotextile-reinforced sand and mobilization of reinforcement strain under triaxial compression, *Geosynthetics* **20** (3) p 207–225

[2] Phoon K K, Retief J V 2016 Reliability of geotechnical structures in ISO2394 *Matieland, South Africa* p 249

[3] Akhlustin O E 2013 Regularities of the variability of the physico-mechanical properties of subsidence soils of Anapa district of the Krasnodar Territory *Candidate of Geological Sciences - Yekaterinburg*

[4] Baranovsky A G 2015 Changes in the physico-mechanical properties of eluvial clay soils under the influence of technogenic factors *Collection: Analysis, forecast and management of natural risks in the modern world* p. 92-97

[5] Boldyrev G G 2014 Methods for determining the mechanical properties of soils *State of the issue monograph, Penza PGUAS* 696

[6] Demintseva E A and Weinstein V M 2013 Analysis of changes in the physical and mechanical properties of soils during stabilization by the Penetron modifier, *Modernization and scientific research in the transport complex,* p. 157-162

[7] Dmitrieva K O 2016 Change in the calculated soil resistance of the base of the operated building, *Design and construction. Kursk,* p 19-22

[8] Dmitrieva K O and Dubrakov S V 2017 *Soil mechanics,* Textbook Kursk p 212

[9] Kolchunov V I, Potapov V V and Dmitrieva K O 2017 Calculation analysis of long-term deformation of the building-to-building system of a nuclear waste storage facility No **4** (72) p 27-33

[10] Krasnov A A, Chetverikov A L, Sheina S G and Shumeev V G 2001 Assessing the impact of a multi-storey building being built on the technical condition of nearby buildings. Problems of construction, engineering support and the city ecology. *Collection of materials of the III All-Russian Scientific Conference* Penza Volga House of Knowledge p 3-5

[11] Akbulyakov M A, Ponomarev A B, Sychkina E N and Cherepanov A Y 2013 Comparison of the results of experimental studies of the mechanical properties of mudstones when choosing

the parameters used in the design of buildings and structures, *Bulletin of PNIPU*. Urban Studies. Perm

[12] Mirsayapov I T and Koroleva I V 2010 Studies of the mechanical properties of clay soils under spatial stress conditions, *Izvestiya KGASU* **1** (13) p 170-175

[13] Nikiforov V V, Seliverstov M M and Bereznev V A 2014 Engineering solutions for the reconstruction of the foundation in the conditions of anomalies in the physical and mechanical properties of the soil, *Youth Science* p 282-286

[14] Polishchuk A I 2000 Approach to the assessment of the load of the foundations of the foundations of reconstructed and restored buildings, *Bulletin TGASU* **1** p 313-326

[15] Ponomarev A B 2013 On the issue of changing the physical and mechanical characteristics of soils during the construction and operation of buildings, *Construction and transport* **2** (14) p 104-109

[16] Prostov S M, Smirnov N A and Bakhaeva S P 2015 Forecast of the physicomechanical properties of the alluvial massif according to electrical sounding data, *Physicotechnical problems of mining* **1** p 69-78

[17] Smorchkov A A, Kereb S A, Orlov D A and Baranovskaya K O 2014 Change in the design resistance of the soil of the base, working as a nonlinear inelastic system *Industrial and civil engineering* p. 54 -56

[18] Smorchkov A.A., Kereb S A, Orlov D A and Baranovskaya K O 2012 Calculation of base deformations using a nonlinear inelastic system, *Izv. Southwest. state Univ.* p 182-185

[19] SP 22.13330.2016 Foundations of buildings and structures - Moscow: Institute of OJSC Research Center Construction

[20] Torov V V and Tsymbelman N Y 2014 Change of physical and mechanical properties of soils under seismic impact, *Vologda Readings* **70** p 7-8

[21] Ulitsky V M, Shashkin A G, Shashkin K G and Vasenin V A 2012 Assessment of the mutual influence of buildings and underground structures, *Georeconstruction* **231**

IOP Conf. Series: Materials Science and Engineering **789** (2020) 012050 doi:10.1088/1757-899X/789/1/012050

Current approaches regarding the assessment of quality of the buildings from the perspective of the user's perception

P D Parincu[1*]

[1] Technical University "Gheorghe Asachi" from Iasi, Faculty of Architecture "G. M. Cantacuzino", Iasi, Romania

[*] dorian.parincu@yahoo.com

Abstract. The article focuses on the satisfaction of the users/beneficiaries and on the quality level of the construction industry. Monitoring user satisfaction brings benefits in the construction industry by improving communication between the parties, engaging in mutual agreements, evaluating progress and monitoring the results and changes made. Users have certain claims about how they want to be treated, moreover, the physical product obtained must fit within their internal value system.

1. Introduction

The quality level of the built space represents the extent to which the building can respond by its characteristics to the demands of the users, the specific requirements of the destination and the social order.

Construction companies are engaged in increasingly tough competitions, which require greater attention in the relations with the beneficiaries and the satisfaction of the users. User satisfaction studies provide valuable information to entrepreneurs, who need to understand the needs of users and come to their foreplay. Also, users have expectations regarding the behavior of the service providers, behavior that can positively influence the level of satisfaction when the quality of services exceeds the level of expectations.

The system in which the constructions are integrated, respectively the adjacent external environment, the related socio-economic, urban and temporal environment, together with the interior system defined for the occupants, can be monitored throughout the life of the building and can generate specific actions to respond to the demands made by users.

2. User satisfaction and building quality

2.1. User satisfaction

The satisfaction of a customer is found in the relationship between the perceived quality and the failure to respond to his expectations from the product in question. Customers compare the performance of a product with a standard of performance. Satisfaction occurs when the perceived performance exceeds the standard (in a positive way), while dissatisfaction occurs when the performance falls below the standard (in a negative way).

In the constructions field, the measure of user satisfaction is defined later, when the product is already purchased.

User satisfaction is one of the key elements in the Total Quality Management - TQM building. Analyzing and understanding user requirements is essential in ensuring user satisfaction.

For a fair assessment of user satisfaction, all the actors involved in this process must be identified. Potential clients who may have requirements and expectations that may affect the quality of a construction may be: contractors and their partners, project managers, design team members, contractors and subcontractors, service providers, product and service users and the society in general. User satisfaction as the last element in this decision-making chain is affected by the individual goals and interests of the members of this group.

Determining the quality of a building is a complex task. In general, defining quality can have two approaches: compliance with requirements and user satisfaction. Compliance with the requirements is an approach that follows how well the building responds to the project specifications. The limitations of this approach are that the users cannot know how well the product or services respond to the project conformations.

Addressing user satisfaction defines quality as the extent to which a product or service meets or exceeds user expectations. The advantage of this approach is that it determines what is important to the user, rather than setting standards based on managerial decisions that may be inaccurate. The weakness of this approach lies in the difficulty of monitoring users' expectations and the differences given by the short or long term evaluation.

2.2. The relationship between user satisfaction and quality at the design level

User expectations play an important role in evaluating the performance of the contractor. The satisfaction of the users in the construction industry is related to the degree of response to the formulated demands, and the quality of the project is perceived as meeting these requirements.

There is a distinction between the quality of a product and the quality of a process. The quality of a product refers to the quality of the materials, equipment and technology integrated in the building, while the quality of the process refers to the achievement of quality through the way the project is organized in the three stages: design, construction, use and maintenance [1].

At the design level, the user evaluates the performance of the contractor in relation to three comparisons, with an impact on the degree of satisfaction (figure 1).

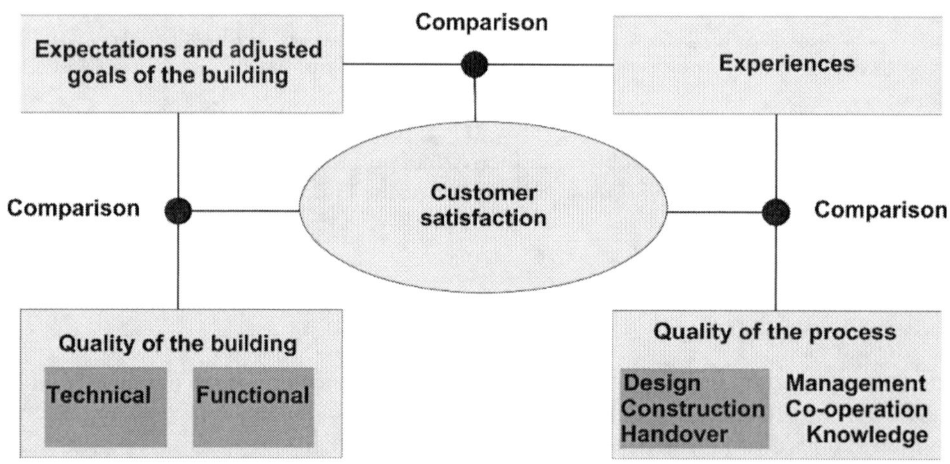

Figure 1. Interrelationships between customer satisfaction and quality at project level [2].

2.3. User perception and quality assessment through the satisfaction level –PMV-PPD

Predicted Mean Vote *PMV* was developed on the principle of thermal balance and experimental data collected in a room with controlled environment, and refers to the performance index of the comfort perceived by the users of a building, on a seven-point scale, from -3 (cold) to +3 (warm), where 0 corresponds to thermal neutrality (ideal value). To find out the satisfaction rate of the users of a building, we use an equation that implies the percentage of dissatisfaction (Predicted Percentage Dissatisfied) *PPD*. This relationship is based on studies in which the interviewed subjects are in a room where the indoor environment conditions are precisely controlled, ignoring the location and adaptation to the thermal environment, maintaining a constant temperature throughout the year. These indicators take into account not only the values of the microclimatic parameters but also the nature of the activity and the type of clothing.

2.3.1 Global indicator PMV. The state of thermal comfort supposes that the temperature of the human body remains constant, near the value of 37 °C. This is done in case of maintaining a balance of the quantities of heat transferred between the human body and the microclimate of the building. The mathematical relation that an ideal thermal balance should respect is as follows:

$$Q_{internal} + Q_{received} = Q_{ceded} \tag{1}$$

where: $Q_{internal}$ – the amount of heat produced by the human body in an arbitrary period of time τ;
$Q_{received}$ – the amount of heat received by the human body over time τ;
Q_{ceded} – the amount of heat lost from the human body during the time τ.

In real conditions, the equality in relation (1) cannot be respected, so that:

$$\Delta Q = Q_{internal} + Q_{received} - Q_{ceded} \neq 0 \tag{2}$$

where ΔQ has the meaning of a thermal residue, the value of which must be as close as 0 to meet the conditions of thermal comfort.

There are several ways to determine the PMV indicator.

A first possibility consists in the use of the relation (3), results from the thermal balance equation of the organism, in which both the internal microclimatic characteristics (objective factors) and the metabolic rate, the energy consumption required for performing a mechanical work and the thermal resistance of the clothing are involved (according to SR ISO 7730/2006) [3].

$$PMV = \left(0.303.e^{-0,036.M} + 0.28\right).\Delta Q \tag{3}$$

where: M - energy metabolism (the amount of heat produced by metabolism, depending on the type of activity carried out, expressed as the average unit thermal flux in W/ m^2);

ΔQ - thermal residue, depending on the temperature of the indoor air and the interior surfaces, the speed of circulation and the humidity of the indoor air, the average temperature of the interior surfaces, but also on the energy metabolism and the thermal resistance of the clothing.

A second way of determining the PMV index is in the annexes of the mentioned standard, the PMV index values are shown for different values of the operative temperature, the air speed and the type of work done and clothing.

The third way to assess the PMV indicator is through direct measurements on a sufficiently large number of subjects, using a special device (integrator capture).

2.3.2. Global indicator PPD. When the thermal residue ΔQ is zero, so the body discharges the amount of heat it produces (and that it eventually receives), according to the relation (1.3) the *PMV* = 0 indicator, so that the thermal sensation of the subjects should be fully comfortable.

Experiments on a large number of people have shown that it is practically impossible to create an environment where absolutely everyone declares themselves in a state of thermal comfort. Even when ΔQ = 0 (so also *PMV* = 0), on average 5% of the subjects feel slightly uncomfortable.

Under these conditions, a new parameter was defined, noted with *PPD* (Predicted Percentage Dissatisfied) which represents the average percentage of people who declare a state of thermal discomfort in relation to a given environment.

The *PPD* indicator can be evaluated according to the *PMV* values based on the relationship:

$$PPD = 100 - 95.e^{-\left(0.03353.PMV^4 + 0.2179.PMV^2\right)} \tag{4}$$

A second way of appreciating the *PPD* indicator, also based on the *PMV* parameter, is using the graph in figure 2.

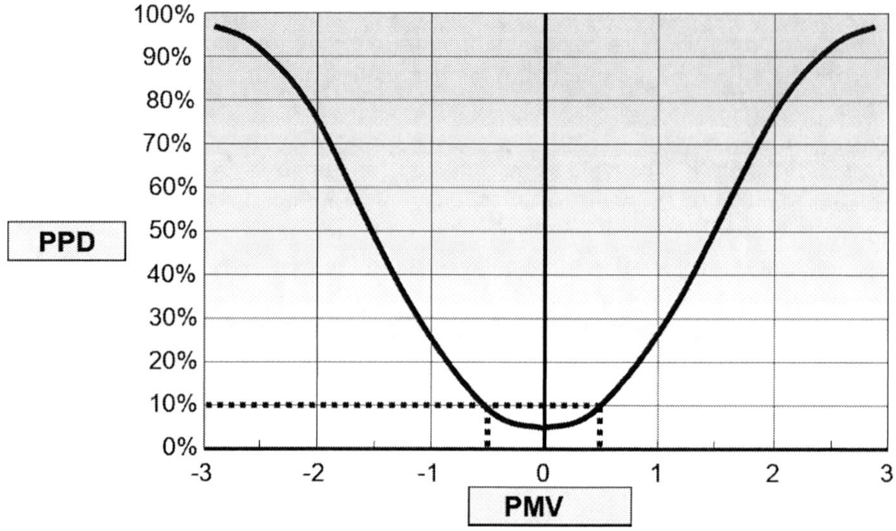

Figure 2. Relationship between *PPD* and *PMV* indicators.

According to the regulations in use, the buildings must be constructed in such a way that the thermal environment in the spaces occupied by people corresponds to the comfort requirements demanded by the activity to be carried out under the conditions of an appropriate clothing. The *PMV* indicator should be within range –0.5...+0.5, and the *PPD* index should be less than 10% (figure 2).

The ASHRAE 55-2010 standard uses the *PMV* model for the regulation of indoor environment conditions, with the requirement that at least 80% of the occupants be satisfied.

2.4. Adaptive comfort

The concept of adaptive comfort emphasizes that the occupants of the buildings do not passively receive the thermal environment suggested by the heat/cold diagrams, but interact with the building actively. The model is particularly applicable to naturally ventilated buildings and hybrid buildings, where the outside climate influences the indoor comfort during different seasons of the year. The questionnaires show that the occupants of the naturally ventilated buildings accept and prefer the higher temperature variations, which are in close relation with the outside temperature, to the detriment of a sealed environment, controlled by the air conditioning systems. The occupants can control the equipment mechanically or non-mechanically, in search of personal comfort, which leads to energy savings. Basically, there are three categories of thermal adaptation, namely behavioral, physiological and psychological. The actions of the residents are related to social, cultural and technical factors [4].

Comparative ASHRAE studies for HVAC (Heating, Ventilation and Air Conditioning) and NV (Naturally Ventilated) buildings in the same climate (Singapore) show that acceptability in the two thermal environments has similar values (78% and 76%) in very different indoor environment conditions. Thermal thresholds in naturally ventilated buildings are larger than those in mechanically ventilated buildings and are accounted for by expectations, physical and behavioral adaptation [5], and to some extent by stronger air movements in naturally ventilated buildings. (on average 0.22 m/s compared to 0.11 m/s in mechanically ventilated buildings) (figure 3).

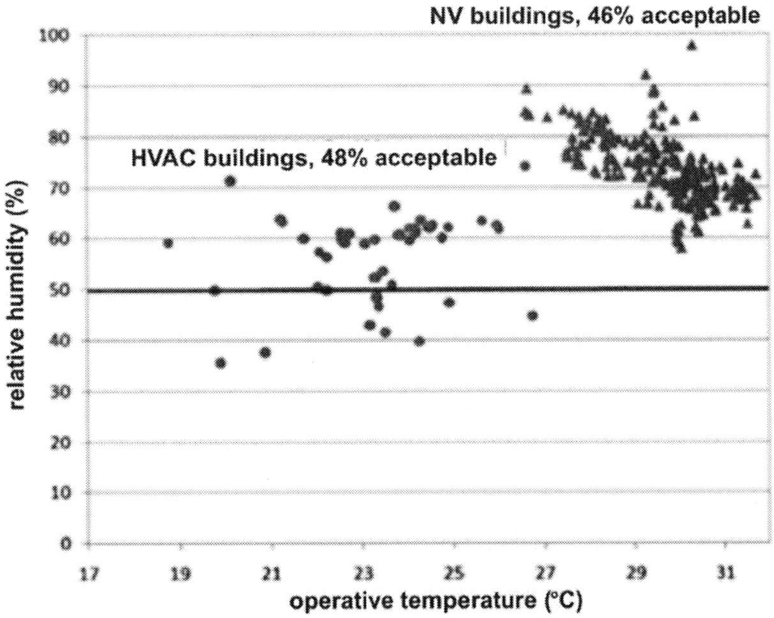

Figure 3. Similar rates of acceptability for different indoor environments - HVAC and NV (Singapore buildings) [5].

The models of adaptive thermal comfort are implemented by the European standard EN 15251 and ISO 7730 in the case of hybrid buildings, and ASHRAE 55 in the case of naturally ventilated buildings.

2.5. Drivers of diversity in human thermal perception

Extensive research on human thermal perception by multiple disciplines has led to extensive knowledge about the general and specific aspects of this topic. Interest in this area has increased in recent years and has led to new approaches in thermal perception modeling.

Global and local challenges related to thermal perception

Climate change in recent years has led to an increase in average temperatures during summer, but also a high frequency of heat waves, precipitation levels and an increase in sea level due to the melting of glaciers. These climate changes affect the energy consumption necessary to obtain the comfort conditions inside the buildings. An increase in the need for cooling in the summer is forecast and a decrease in the need for heating in the winter.

Climate change can be associated with health and productivity - heat waves and cold winds are associated with thermal discomfort, decreased physical performance and increased mortality.

Thermal comfort

Thermal comfort is defined by the American Society of Heating, Refrigerating and Air-Conditioning Engineers (ASHRAE) as "the condition of mind that expresses satisfaction with the thermal environment and is assessed by subjective evaluation" [6].

Physiological drivers of diversity in thermal perception

From the physiological point of view, a great weight in the thermal perception has the distribution of the internal temperature of the body and the state of acclimatization of the skin. Recent research also discusses other physiological factors that may influence an individual's thermal perception, such as physical condition. This can be determined by demographic factors, by their own factors, such as sex and age, and a number of physiological variables, such as body composition and fitness, which may change over time or conform to the situation. Let's review them:

Age
From a physiological point of view, the thermoregulatory capacities change substantially from childhood to old age. These physiological changes include, for example, lower body temperature in both older men and women, as well as structural skin changes and metabolic changes. As the age increases, the thermoregulatory capacities of the human body decrease, especially if the level of fitness is reduced.

Sex
The physiological reasons for the thermoregulatory differences between men and women are manifold. For example, women have a higher surface-to-volume ratio (which allows for greater heat loss through the skin), different body composition (higher fat mass and lower muscle mass) and lower metabolic rate (less active), metabolic free mass in women and thus with up to 20% less heat production [7-8].

Body composition and fitness
Body composition is an important parameter in the thermoregulation of the human body. A thick layer of subcutaneous fat increases insulation and reduces heat exchange through the skin, which helps maintain body core temperature in a cold environment. In contrast, heat loss could theoretically be affected at higher temperatures when the subcutaneous fat layer is thicker. Several previous physiological researches report higher body temperatures in obese people compared to ascetic people when exposed to heat and physical exercise [9–10].

Metabolic rate
Of all physiological parameters, metabolic rate is the only one incorporated in the PMV model as input parameter. The metabolic equivalent of task (MET) is one of the standard values in the PMV model (next to clothing insulation, air temperature, radiant temperature, air speed and humidity). MET is a commonly used physiological concept considered a simple procedure for expressing the energy cost of physical activities as a multiple of the basic metabolic rate [11].

Importantly, absolute metabolic rate is not only determined by activity, but also by other physiological parameters and environmental conditions, such as body composition, diet, adaptation state and temperature.

Physiological adaptation to the thermal environment
The extraordinary ability of the human species to adapt to a wide range of outdoor temperatures is well known. For example, indigenous people who are living in extreme climatic conditions have adapted over thousands of generations. However, the human body, even without any specific genetic predisposition, can adapt to a wide range of thermal conditions, both acute and long-term. The fluctuations of the temperature of the natural environment are counteracted by natural physiological adjustments for the vast majority of the population (the succession of seasons, the day-night cycle, etc.).

Habituation of different climatic zones
Regarding the habit of the different climatic zones and its effect on the thermal perception, two review documents have been identified. If we consider the fact that repeated exposure to a thermal challenge has as a consequence the physiological adaptation, it is obvious that the subjective perception of the thermal environment also changes. The ideal of thermal comfort differs from one climatic region to another, being adjusted to local conditions, but with quite small differences from the average.

Seasonal adaptation
In the climatic regions that have distinct temperature differences in the different seasons, a change of the thermal sensation and of the comfort temperatures throughout the year is perceived, with a variable mainly due to the clothing.

Diurnal rhythm
The temperatures of the human body change during the day and night at a diurnal rate (24 hours), the basic temperature being minimum early in the morning (~ 05: 00 AM) and at the highest in the late afternoon/early evening [12].

Psychological drivers of diversity

Considering the definition of thermal comfort as "that condition of the mind that expresses satisfaction with the thermal environment" [13], we can expect psychological factors to play an important role in thermal perception. Of all the psychological aspects, personal control received the greatest attention in comfort research, followed by the effect of personality, self-efficacy and anticipated costs of an action.

Effects of personal control

Paciuk [14] pointed out that personal control can be one of three things: available control, exercised control, and perceived control. The effect of available but not exercised control can be considered a psychological aspect, as can perceived control. However, exercised control will change environmental conditions and will be discussed in the section on contextual drivers. Hence, only studies were included that allowed assessment of solely the psychological component of personal control, i.e. studies where environmental conditions were identical in the control and no control condition, or where the outcome variable was the neutral temperature calculated based on thermal sensation votes at different operative temperatures.

Contextual drivers of diversity

Besides physiological and psychological aspects of the human body, thermal perception indoors is very much determined by contextual factors of the built environment.

In real life situations, the environmental conditions that people often encounter differ, sometimes easily and gradually, such as increasing the heating point, but sometimes more extreme and sudden, for example when entering an air-conditioned building during a hot summer day. The destination of a room or a building area also determines the different temperature requirements.

Local effects and non-uniform environments

Local effects and uneven environments affect thermal sensation and comfort and are another source of diversity between people even when sharing the same space. From the contextual perspective of the building, this refers to the stratification of air temperature and radiant asymmetry: two aspects that are very determined by the type of air conditioning systems, the thermal insulation level of the building envelope and the level of air conditioning control.

Clothing adaptability

Clothing adaptation is a common and effective strategy used by the occupants of a building. Occupants can tolerate operating temperatures up to 29° C only by adjusting clothing [15]. Therefore, the ASHRAE 55 standard [5] applies to buildings with operable windows, without operation or limited operation of mechanical cooling and heating systems.

In addition, research has shown that the level of isolation of people's clothing is adapted to the external and internal conditions. The clothing of men versus women is different; the level of isolation of men's clothing is less variable than that of women. In addition, the latter show that clothing isolation levels also differ between age groups, with older age groups wearing garments with higher isolation levels.

In conclusion, there is a huge variety of potential drivers of diversity and much more research is required to understand better the underlying mechanism of diversity. This information is required in order to rule out irrelevant factors and in particular to reveal the important and significant drivers of diversity.

Knowing the true drivers of diversity will be helpful in preparation for global challenges because the indoor thermal conditions chosen by an individual do not only affect the energy use – in itself a driver of climate change – but also affect health, wellbeing, and productivity [16].

3. Perception of Transient Thermal Environments: pleasure and alliesthesia

The term "alliesthesia" is defined as the perception of external stimuli as pleasant or unpleasant, depending on the internal state of the human body [5]. The term is used to differentiate thermal pleasure from thermal neutrality or acceptability, or in other words, the perception of neutral thermal conditions as "comfortable", but not "very comfortable". International comfort standards encourage exclusive reliance on HVAC usage by largely restricting thermal asymmetries and transients within indoor environments as potential sources of discomfort. The perception of thermal conditions as "very comfortable" comes from the area of asymmetrical and transient environments, encountered in naturally ventilated or hybrid buildings. Recent research indicates that dynamic thermal environments can potentially deliver higher levels of occupant satisfaction than static, homogenous indoor environments.

Schalter et al. (2010, p. 274) [17] concludes in his work: "... changes in skin temperature mediate the thermo-behavioral response to maintain the level of heat and regulate the temperature. These behaviors are preceded by changes in the thermal comfort index and sensations, contributing to the initiation of a coordinated behavioral effort." [17]. Schalter's emphasis on skin temperature for thermal pleasure can be found in the model proposed by Hardy et al (1971) and Hensel (1981) [18-19] - figure 4. They show that while neural stimuli for cold and heat sensations are primarily skin thermoreceptors, the perception of thermal comfort depends on a more complex suite of signals, including:

- reactions of the thermoreceptor (internal and dermal);
- the neural pathways for temperature regulation, generated in the anterior hypothalamus;
- the effector thematic regulator acts by itself - tremors and vasoconstriction in the cold environment and sweating and vasodilation in the hot environment.

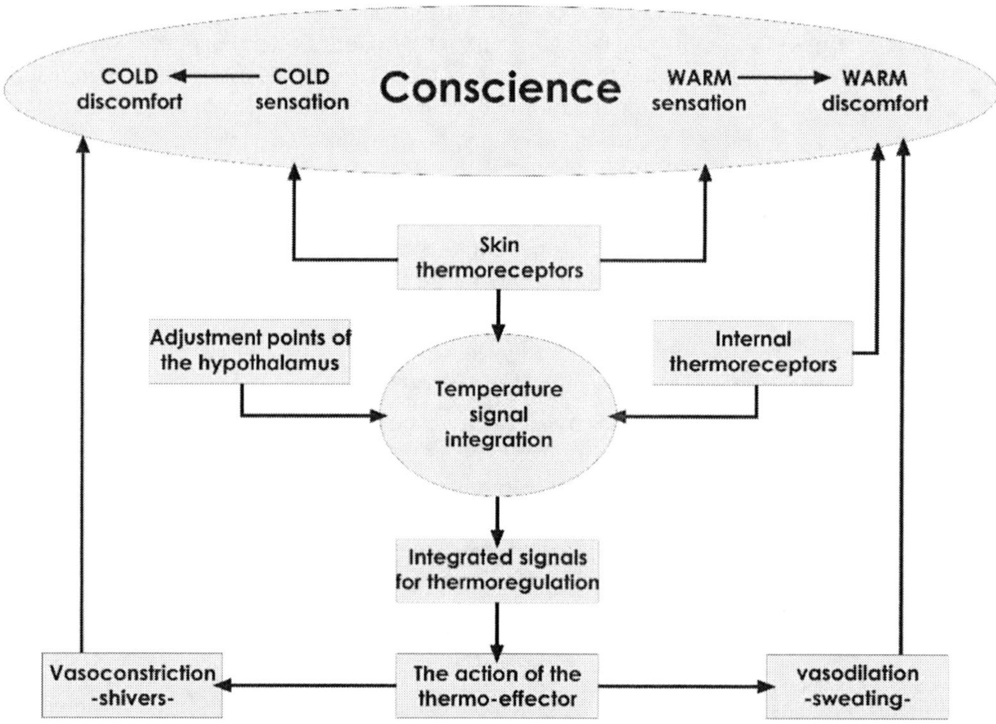

Figure 4. Neural stimuli for thermal comfort [18-19].

Heat balance models are widely acknowledged as inadequate in evaluating the effects of asymmetrical or transient environments on comfort. We don't know *why* certain environments are pleasant sometimes, and distinctly unpleasant other times. The renewed interest in thermal alliesthesia seems timely given the recent advances in modeling and global uptake of adaptive comfort principles.

Thermal pleasure and the built environment

Thermal stressors are ranked according to their impact on homeostasis in order to prioritize behavioural responses. This led Cabanac (1992) [21] to suggest that such a system of negotiation requires a common currency to evaluate each instance of conflicting stimuli. The currency he proposed was pleasure. In his interpretation priority is given to any behavioural response that maximises pleasure (or minimises displeasure) with minimal regulatory strain. Pleasure is deemed greatest immediately after a successful response, continually diminishing until a new equilibrium is reached, and at this point in time pleasure disappears altogether. Connotations of the word 'pleasure' are likely to vary between individuals and between pleasant and unpleasant [20]. The experience of pleasure in our thermal environment is a daily phenomenon that we share and appreciate, but, surprisingly, little is known about the casual process. If either the environment or the subject is static, there is no opportunity for the body to interpret the 'usefulness' of a stimulus for thermoregulation. Thermal pleasure can therefore only be experienced in transient states [21].

For this reason, we are tempted to say that in order to feel pleasure, we must first experience the state of dissatisfaction. It's not entirely true. People are constantly exposed to environments that disturb the thermal balance when they are traveling (between buildings, in traffic to the workplace, etc.). Pleasure is rarer than dissatisfaction simply because any stimulus experienced in excess can become unpleasant.

If we look further afield, a similar discourse is taking place in product development and user experience evaluation where pleasure has emerged as the design goal. Coelho & Dahlman (2000) [22] suggest that concepts of comfort and pleasure overlap, but pleasure holds dimensions that are not included in comfort. In this sense it is not a cause-and-effect relationship; comfort is an aspect of pleasure.

In conclusion, the inherent difficulties of pleasure are recognized and this is a field of research that needs to be further developed. An alliesthesial model of pleasure may help to explain occupant acceptance of dynamic indoor environments. Pleasure depended on the nature of the physiological change caused by transitions, and only lasted for short periods immediately after.

4. Conclusions

The satisfaction of the end users of a building, regardless of the architectural program, is an increasingly prominent factor in the decision-making at the level of designers and contractors.

The purpose of this paper is not to suggest the abandonment of the conventional concept of comfort, but to explore how dynamic indoor environments have a beneficial influence on its occupants, as opposed to the static and uniform ones.

Until the introduction of the concept of alliesthesia by Cabanac in 1971, it was decided to control the temperature of the interior environment of buildings artificially, so that it would be maintained at a preset value, theoretically optimal. With the exploration of the principles of adaptive comfort and alliesthesia, there were developed ways of taking over the concept of dynamic environment by illustrating new methodologies for putting the previous idea into practice, in opposition to the static approach of thermal comfort. The antithesis of the two systems led to the rethinking of the design of buildings, furthering the concepts of natural or hybrid ventilation solutions.

In this new approach, sunlight is an important factor in the variability of indoor temperature, which can become similar to that of the outdoor environment, highlighting an increased level of comfort for the occupants of the building. The latter are assigned an active role in avoiding the extremes of the interior natural lighting, in this case imposing the intelligent control of the illumination, avoiding furthermore the feeling of discomfort. In the areas that cannot fully benefit from the contribution of natural light, one can resort to the use of biodynamic lights, whose intensity varies during the day.

The active role of the occupants of the building can also be extended to ensure thermal comfort, by facilitating access to the ventilation, heat and humidity control devices. Thus, the thermal neutrality and the dissatisfaction associated with it will be avoided. The concept of alliesthesia investigates the thermal pleasure resulting from the transient thermal environments, where they can experience sudden temperature changes, with a beneficial effect in the short term, an effect that ceases with the return to thermal neutrality. The same change in ambient temperature can be both pleasant and unpleasant, depending on several subjective factors specified in chapter 2.5.

Due to the increased attention in the pre-design stage towards the needs of the users it is possible to avoid errors in the design process, which can result in long-term dissatisfaction or costly rectification solutions. Thus, the existing buildings can be monitored, together with the occupants interviewing, resulting in the appropriate completion of the project, in relation to meeting the new comfort requirements generated by the principles of the active and dynamic indoor environment.

References

[1] Andreica H and Dumitras A 1994 *Proiectarea constructiilor*, Editura Universitatea Tehnica, Cluj-Napoca
[2] Karna S (n.a.) Analysing customer satisfaction and quality in construction – the case of public and private customers *Helsinki University of Technology Construction and Management*, P.O. Box 2100, FIN-02015 HUT, Finland
[3] ISO 7730- 2005 *Ergonomics of the thermal environment: analytical determination and interpretation of thermal comfort using calculation of the PMV and PPD indices and local thermal comfort criteria*
[4] Nicol V F and Humphreys M 2002 Adaptive thermal comfort and sustainable thermal standards for buildings *Energy and Buildings* **34** (6) 563–572. doi:10.1016/S0378-7788(02)00006-3
[5] de Dear R and Brager G S 1998 *Developing an adaptive model of thermal comfort and preference.* ASHRAE Transactions, 104(1b), 145-167
[6] ASHRAE Standard 55-2013 *Thermal environmental conditions for human occupancy*. Am. Soc. Heating, Refrig. Air-Conditioning Eng. Atlanta, USA
[7] Kingma B R and van Marken Lichtenbelt W D 2015 *Energy consumption in buildings and female thermal demand. Nat Clim Chang.* **5** 1054–1056
[8] Schmidt-Nielsen K 1984 *Scaling: why is animal size so important?* Cambridge: Cambridge University Press
[9] Havenith G and van Middendorp H 1990 *The relative influence of physical fitness, acclimatization state, anthropometric measures and gender on individual reactions to heat stress*. Eur J Appl Physiol Occup Physiol.; **61**:419–427. Available from: 10.1007/BF00236062
[10] Leites G T, Sehl P L, Cunha G, et al. 2013 *Responses of obese and lean girls exercising under heat and thermoneutral conditions*. J Pediatr.;162:1054–1060
[11] de Dear R J, Brager G S and Cooper D 1997 *Developing an adaptive model of thermal comfort and preference*. Final Rep. ASHRAE Res. Proj. 884. Macquarie Univ. Sydney
[12] Krauchi K 2002 *How is the circadian rhythm of core body temperature regulated?* Clin Auton Res Germany; p. 147–149
[13] *ISO Ergonomics of the thermal environment - assessment of the influence of the thermal environment using subjective judgement scales*, (BS EN ISO 10551:2001). London: BSI
[14] Paciuk M 1990 *The role of personal control of the environment in thermal comfort and satisfaction at the workplace*. Environ Des Res Assoc. **21** 303–312
[15] Fanger P O and Toftum J 2002 *Extension of the PMV model to non-air-conditioned buildings in warm climates*. Energy Build. **34** 533–536
[16] Schweiker M, Huebner G M, Kingma B R M, Kramer R and Pallubinsky H 2018 *Drivers of diversity in human thermal perception – A review for holistic comfort models* Temperature (Austin) **5** (4): 308–342.doi: 10.1080/23328940.2018.1534490, PMCID: PMC6298492
[17] Schlader Y J, Stannard A K and Sekhar C 2010 *Human thermoregulatory behavior during rest and exercise – a prospective review*. Physiology and Behavior
[18] Hardy, J D, Stolwijk, J A J and Gagge, A P 1971 Man, in G C Whitttow (ed.): *Comparative Physiology of Thermoregulation*, Academic Press; New York, NY, Vol. 2.
[19] Hensel H 1981 *Thermoperception and Temperature Regulation*, Academic Press, London;
[20] Cabanac M 1971 *Physiological Role of Pleasure*. Science, Vol. **173**, No. 4002, pp 1103-1107;
[21] Cabanac M 1992 *Pleasure: the common currency*. Journal of Theoretical Biology, Vol. **155**, pp 173-200
[22] Coelho D A and Dahlman S 2000 *Comfort and pleasure. In* Jordan, *P* W & Green B. (Eds.), *Pleasure in Product Use*, Taylor & Francis, London, pp. 321–331

IOP Conf. Series: Materials Science and Engineering **789** (2020) 012051 doi:10.1088/1757-899X/789/1/012051

Road embankment for bridge accessing rural area

D-M Pohrib[1*], D Platică[1], P Turcanu[2] and C Pleşcan[3]

[1,2] "Gheorghe Asachi" Technical University of Iasi, Faculty of Civil Engineering and Building Services, Iasi, Romania
[2] S.C. PROCONRIM S.R.L. Iasi
[3] "Transilvania" University of Brasov, Faculty of Civil Engineering, Brasov, Romania

[*] madalina.pohrib@yahoo.com

Abstract. The article aims to show the construction on soft soil of a road embankment for access to a bridge which passes over a stream. To strengthen, consolidating and draining the embankment near the stream a combined geosynthetic and gabions solutions it was use.

1. Introduction

The construction of a road embankment that gives access to a bridge, which passes over a stream, is to be constructed in order to make the traffic possible again on the only direct road that connects two nearby settlements.

A combined geosynthetic and gabions solutions were used on a soft soil to strengthen, consolidating and draining the embankment near the stream.

2. The causes of bridge access degradation

The studied embankment in which a collapse of the lane has occurred is located on a communal road near the "Ichim" bridge.

Analysing the figure 1, 2 and 4 presented below it can be easily observed that beside the perimeter of the embankment adjacent to the bridge, during periods with high stream flows, the water seeps into the embankment of the bridge road access, washing of fine particles. Thus contributing to the decrease of the shear strength parameters of the constituent soil layers and to the decrease of the stability factor of the bridge road access embankment.

The collapse occurred due the infiltration of water from the stream or rainwater into the embankment adjacent to the bridge during periods with high flows and high amount of precipitation. Is also considered that the improper management of rainwater and the lack of ditches favors the seepage of water into the lane and into the embankment.

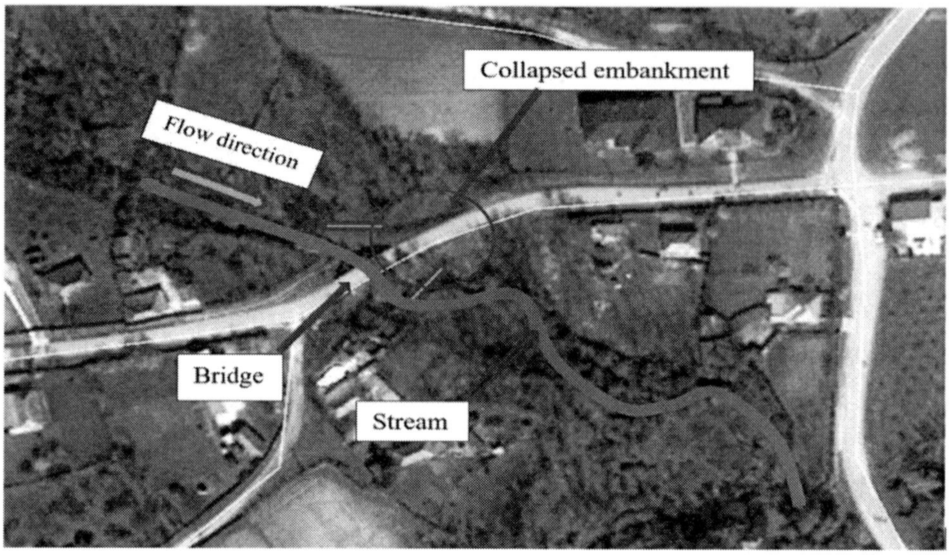

Figure 1. Analysed embankment site.

A cause for the collapsed embankment (see figure 1 above) is the wrong positioning of the bridge with the embankment located on a concave bank (see figure 2 below) with lateral erosion from the stream [2], opposite the stronger right bank that is forested.

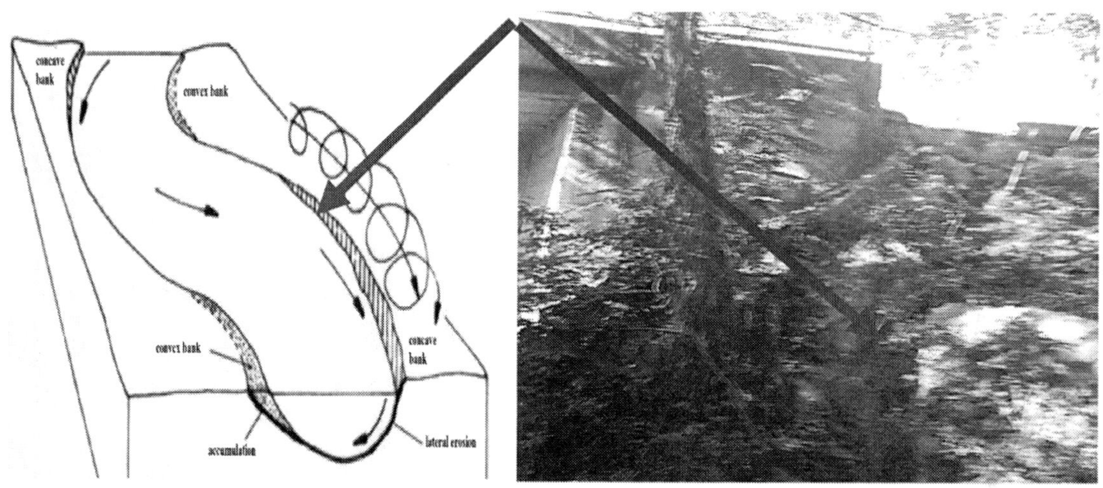

a. Lateral erosion of meandered stream [2] **b.** In situ erosion

Figure 2. Exemplification of erosion.

Another inconvenience of the access to the bridge is the lack of verge and ditches, because of this the discharge of rainwater into the stream takes place at about 2.00 m higher level difference from the stream water level. This is leading to an increase in humidity of the foundation layer of the road and to a continuous erosion on the discharge area, according to figure 5.

The affected embankment bridge access has a length of about 10.00 m with a height of the break step according to in situ measurements in September 2018 of about 0.70 m.

Figure 3. Rotational type slip (collapse).

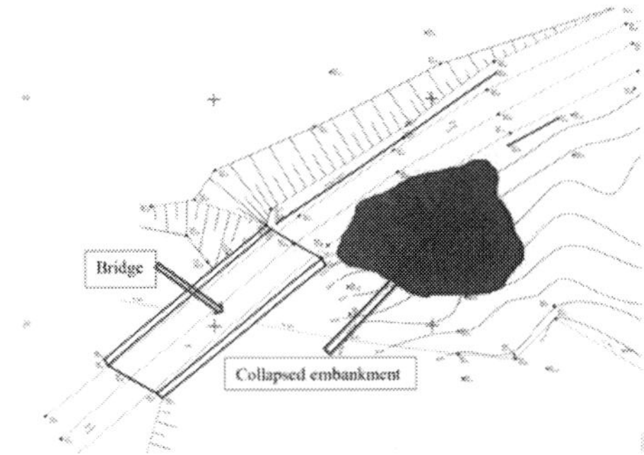

Figure 4. Collapsed embankment map.

a. Collapsed embankment view **b.** Embankment rift

Figure 5. In situ embankment views.

3. Geotechnical context

The soil studies carried out according to the positioning from figure 6 below have revealed that the mechanical characteristics of the soil on site are mediocre. The free ground water level is at a depth of -2.00 m in borehole f01, at a depth of -7.00 m in the dynamic penetration DPSH-B 1and at a depth of -3.00 m in the dynamic penetration DPSH-B 2.

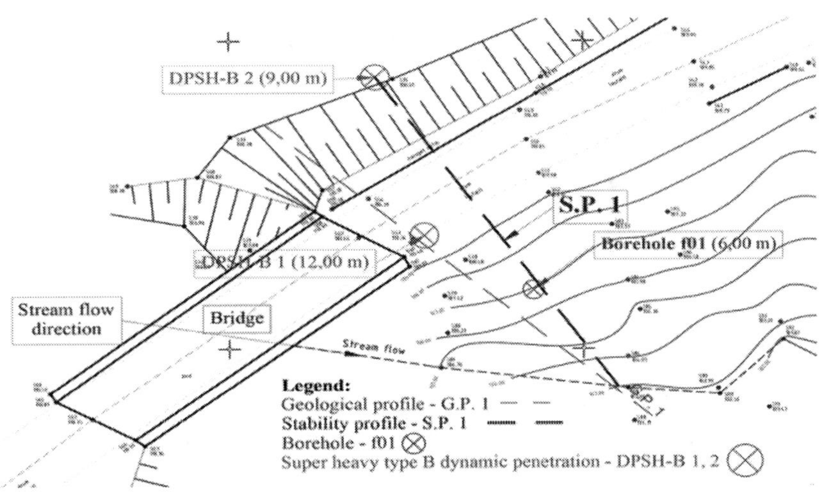

Figure 6. In situ soil investigation positioning.

On average, from –0.20 m to -1.50 m, the soil contains silty clay, from -1.50 to -3.50 m, plastic soft sandy silt with base sand and beyond this depth the soil is composed of sandy clay.

3.1. Characteristic and design calculation values of geotechnical parameters

The determination of the characteristic and design values of the geotechnical parameters was performed according to SR EN 1997:1-2004 [3] and NP 122:2010 [4]. The design values are determined based on euro code SR-EN 1997:1-2004 depending on the Design Approaches used.
Following the calculations made, the Design Approach 1 - Group 2 presents the most unfavorable states of effort, so that in this article only the results of the analyzes performed for this approach are presented.

3.1.1. Characteristic values of geotechnical parameters

Considering the uneven stratification of the soil layers intercepted in the borehole, statistical processing of the geotechnical parameters values was perform as exemplified in figure 7 below, thus resulting the values presented table 1.

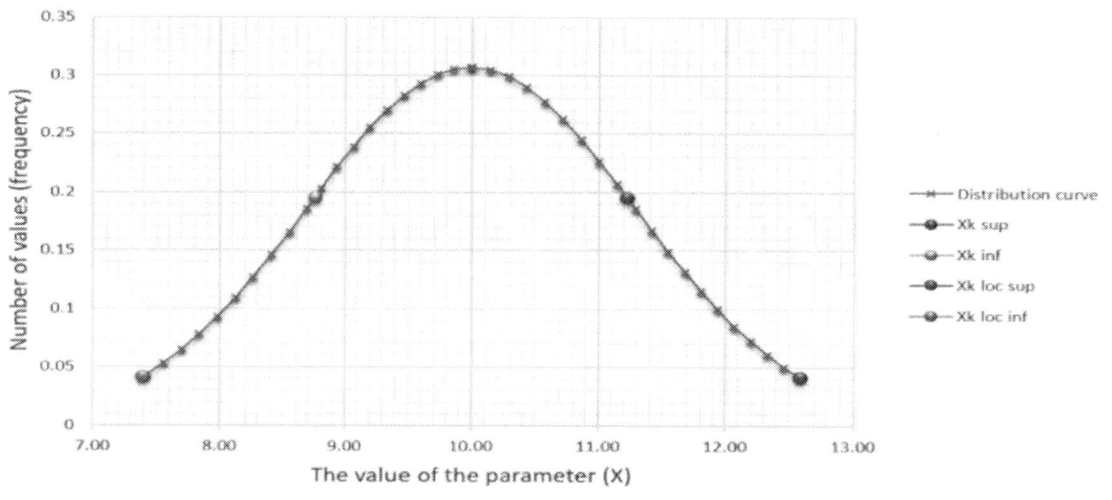

Figure 7. Example of statistical processing of the characteristic values of geotechnical parameters.

Table 1. Characteristic values for geotechnical parameters of the soil layers.

Layer	Upper elevation	Lower elevation	Thickness	Design Approach 1 – Group 2				
				E	φ'_k	c'_k	γ_{nat}	υ
-	m	m	m	kN/m^2	°	kPa	kN/m^3	-
1. Silty clay	0.00	1.50	1.50	1.0E+4	10.00	20.00	19.40	0.35
2. Sandy silt+silt	1.50	3.50	2.00	9.0E+3	32.00	5.00	19.50	0.30
3. Sandy clay	3.50	8.70	5.20	2.2E+4	20.00	25.00	20.85	0.35
4. Hard clay	8.70	13.00	4.30	2.9E+4	12.00	80.00	21.18	0.42

3.1.2. Design values of geotechnical parameters. The values used for the partial factors on Design Approach 1 – Group 2 for permanent and variable unfavourable actions (A2) are$\gamma_{G,unfav}$ = 1.00, $\gamma_{Q,unfav}$ = 1.30, and for soil parameters (M2),γ_φ = 1.25 for internal friction angle, γ_c = 1.25 for cohesion and γ_γ = 1.00 for volumetric weight.

Table 2. Design values for geotechnical parameters of the soil layers.

Layer	Upper elevation	Lower elevation	Thickness	Design Approach 1 – Group 2				
				E	φ'd	c'd	γ_{nat}	υ
-	m	m	m	kN/m²	°	kPa	kN/m³	-
1. Silty clay	0.00	1.50	1.50	1.0E+4	8.03	16.00	19.40	0.35
2. Sandy silt+silt	1.50	3.50	2.00	9.0E+3	26.56	4.00	19.50	0.30
3. Sandy clay	3.50	8.70	5.20	2.2E+4	16.23	20.00	20.85	0.35
4. Hard clay	8.70	13.00	4.30	2.9E+4	9.65	64.00	21.18	0.42

3.2. Methods used in programs for stability analysis

As a synthetic indicator of the equilibrium state of a slope, for a given situation, the stability factor is used, Fs, which in the most general way is defined by comparing the stress states along the potential slip surface, respectively:

$$F_s = \frac{\tau_f}{\tau} \tag{1}$$

where τ_f representing the value of the shear strength of the mobilized ground, and τ the value of the existing tangential stress in the mass for the calculation hypotheses considered; In order to ensure stability, Fs must have super unit values.

In order to assess the stability based on the stability factor Fs, the calculation is based on a method of analysis established in geotechnical practice and based on the concept of limit equilibrium of the Spencer type. The formulation of this method considers the mass of the earth, above the potential slip surface, discredited into bodies, elemental volumes - strips, separated by vertical planes as in figure 8.

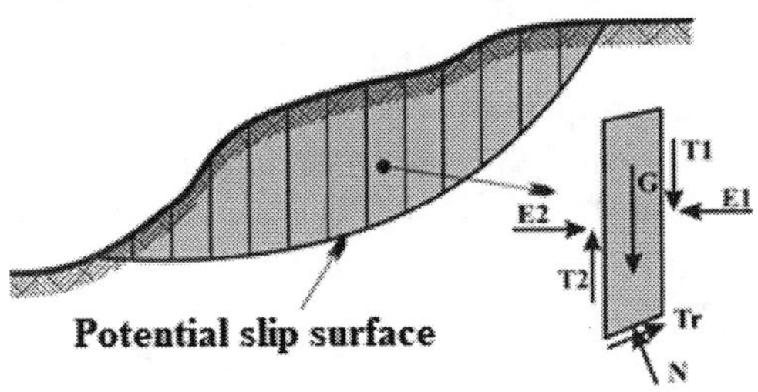

Figure 8. Spencer type limit equilibrium.

In addition, the method admits that the sliding surface is of circular-cylindrical shape with the horizontal axis or of some shape. Figure 8 illustrates a potential landslide divided into strips and, in a simplified way, the forces acting on a strip. The method developed by Spencer considers the complete interaction between the strips (through vertical and lateral components, E modulus and friction, T) and satisfies both the global balance of moments and forces.

The stability conditions are represented by two stability factors, evaluated based on the global balance of moments Fs, M and of forces, according to the horizontal direction Fs, F and which imply the satisfaction of the Mohr-Coulomb yield criterion and the balance of forces according to the vertical direction for each one [5, 6].

As a common feature of the formulations of the methods that take into account the complete interaction between the bands and for which both equilibrium conditions are satisfied (Spencer, Morgenstern-Price, General Limit Balance Method), it is assumed that between the normal (lateral) interaction forces) and shear between the bands there is a relation of the form:

$$T=\lambda\times f(x)\times E \tag{2}$$

where f (x) is a function and λ is a parameter of function adjustment (λ = 0 means that there is no vertical component of the interaction between the bands).

The particularization λ = 0, the expression of the stability conditions through Fs, M and the neglect of the shear interaction forces between the strips ensure the conditions of the simplified Bishop Method; expressing the stability conditions through Fs, F ensures the conditions of the simplified Janbu method. In the Spencer Method f (x) = 1 (the ratio of the interaction forces between the strips becomes constant and the same for each strip) and λ is determined by an iterative calculation until Fs, M and Fs, F are approximately numerically equal (figure 9), the Spencer method being thus the most complete method based on the limit equilibrium.

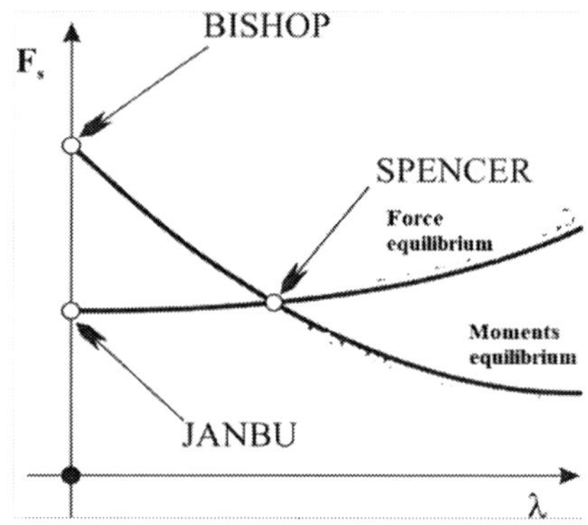

Figure 9. Comparison between used stability methods.

4. Performed calculations

Based on the topographic survey carried out on site, in the analyzed area and the geotechnical design values from anterior chapter, a stability profile "S.P.1" was performed with Geoslope 2D and Plaxis 2D v8.2. This helps to determine the local stability of the embankment and the state of effort on the bridge embankment access in the existing situation and in two proposed situations: first situation with geosyntethics with gabions and the second with interspaced piles.

In the stability, calculations considering the road loads, the forces induced by braking-starting vehicles, analyzing the following situations:

- existing situation - natural state and natural state with earthquake
- existing situation - flooded state and flooded state with earthquake
- proposed situation - flooded state with earthquake

The calculation programs use the calculation approaches according to SR-EN 1997:1-2004 [3], thus a stability factor higher than 1.00 indicates stability.

When carrying out a calculation in the flooded state, the design values of the shear resistance parameters of soils are reduced as presented in table 2 by about 25-60% leading to important horizontal and vertical deformations in the soil.

4.1. Existing situation

Stability analyzes performed for S.P. 1 indicates phenomena of loss of stability in the current situation in both static and dynamic regimes.

In case of the increase of the ground water level (flooded state) as a result of abundant precipitation and the perimetral flooding of the embankment caused by an increase of water level in the stream, the values of the sub unitary stability factors decrease. In this case is resulting in instability both in the seismic situation and in the situation, which considers the earthquake, according to figure 10 and table 3 below.

Figure 10. Allure of the slip surface in existing situation (with Geoslope 2D).

Table 3. Centralizing stability analysis factors (Fs) and total displacements in existing situation.

Stability profile S.P.1		Natural state without earthquake	Natural state with earthquake	Flooded state without earthquake	Flooded state with earthquake
Bridge access road embankment	F.S.	0.95	0.76	0.78	0.52
	Total displacements (mm)	55	115	168	274

The representation of existing displacements and state of efforts in the embankment is shown in figures11 below.

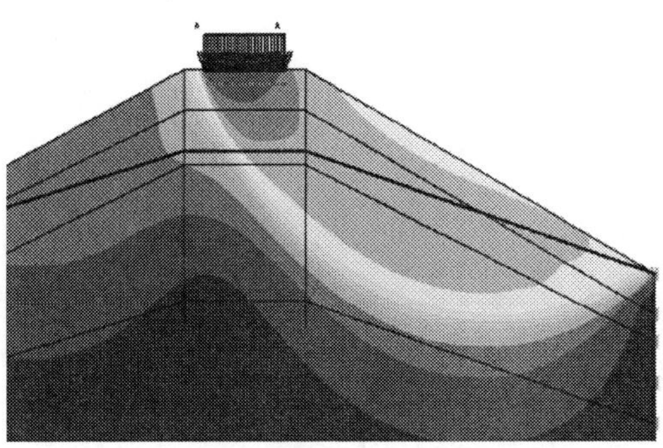

a. Total displacements.

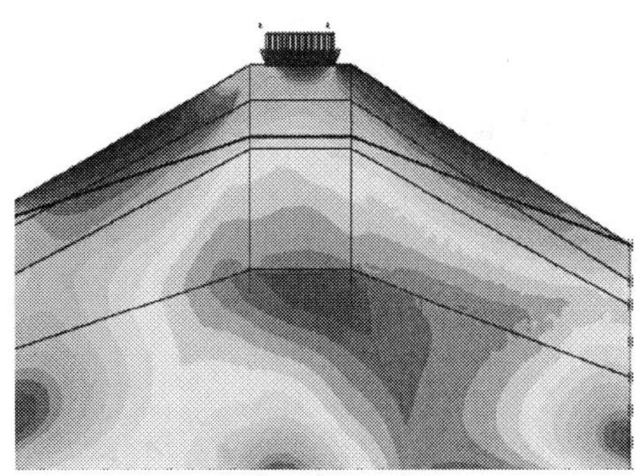

b. Deviatronic stresses.

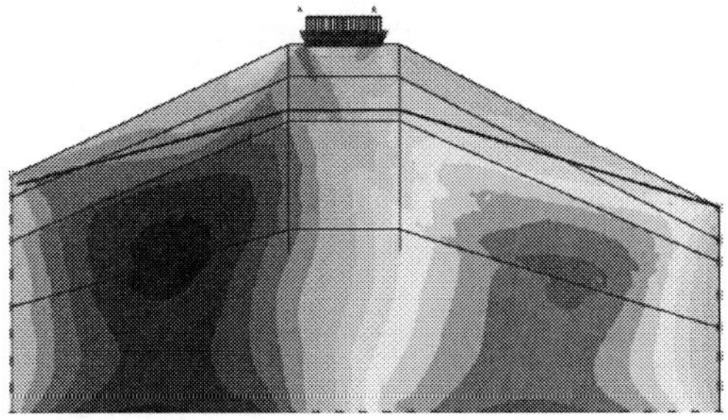

c. Shear stresses.

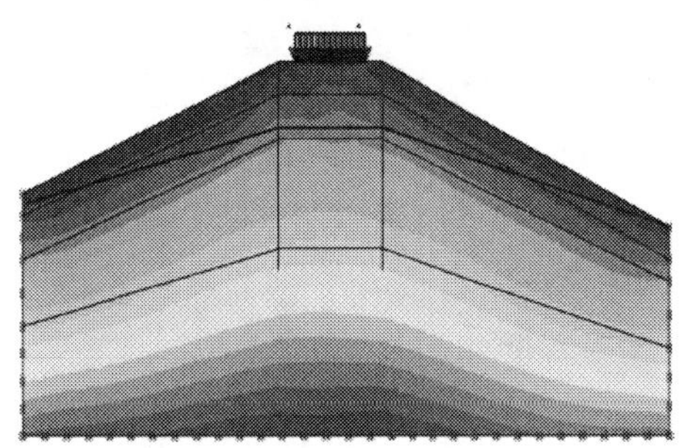

c. Vertical total stresses.

Figure 11. Displacements and efforts in the embankment (withPlaxis 2D).

4.2. Proposed situation

The article shows the construction on soft soil for a road embankment that gives access to a bridge that passes over a stream. To strengthen, consolidating and draining the embankment near the stream it was used combined geosynthetic and gabions solution [1].

In the case of carrying out mixed support works from reinforced soil at the level of the road sub base and protection from gabions on the left-right slope of the road area, the values of the stability factors are higher than 1.00, both situations without earthquake and with earthquake, according to figure 12.a and table 4 below.

In the case of making the bridge road access support from interspaced piles, the values of the stability factors are higher than 1.00, both situations without earthquake and with an earthquake. There will be problems of stability of the slope situated downstream of the piles that will be eroded further by the stream, according to figure 12.b and table 4 below.

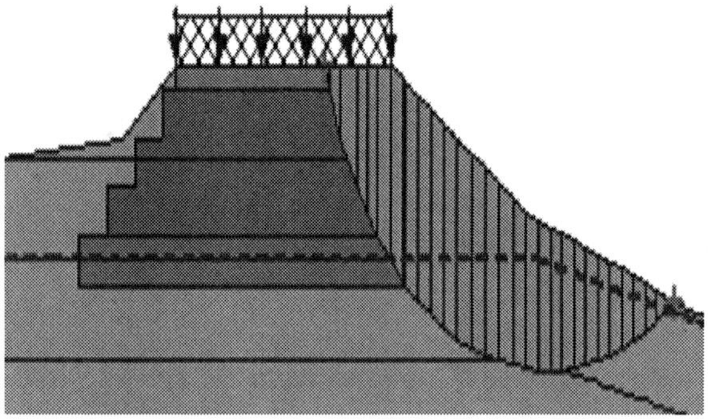

Figure 12.a Allure of the slip surface in proposed situation with geosynthetics and gabions. (with Geoslope 2D).

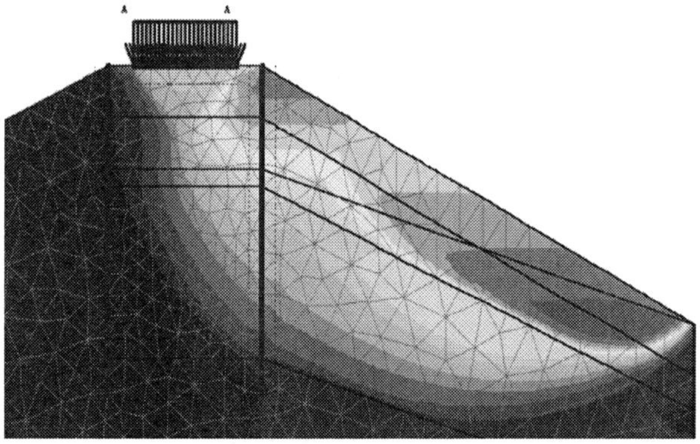

Figure 12.b Allure of the slip surface in proposed situation with piles. (with Plaxis 2D).

Table 4. Centralizing stability analysis factors (Fs) and total displacements in proposed situation.

Bridge accessroadembankment	Stability profile	Flooded state with earthquake	Total displacements (mm)
Solutionwithgeosyntheticsand gabions	S.P. 1	1.41	11
Solutionwithinterspaced piles	S.P. 1	1.46	8

The pile solution is very expensive, time consuming, and does not solve the erosion problem downstream of the road, resulting from a technical and economic point of view, that being more plausible with geosynthetics and gabions.

The representation of displacements and state of efforts in the gosynthetics with gabions reinforced embankment is shown in figures 13 below.

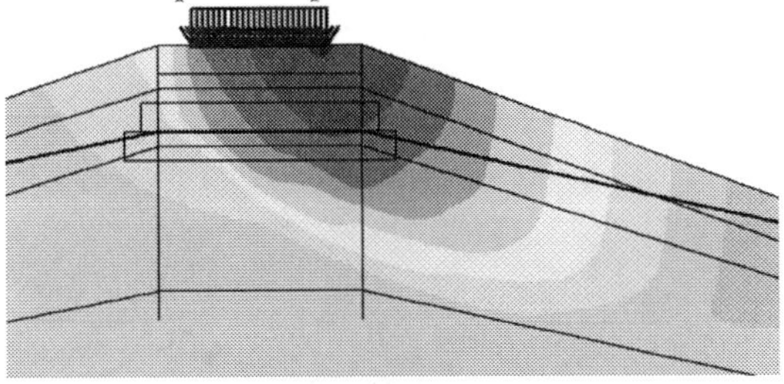

a. Total displacements.

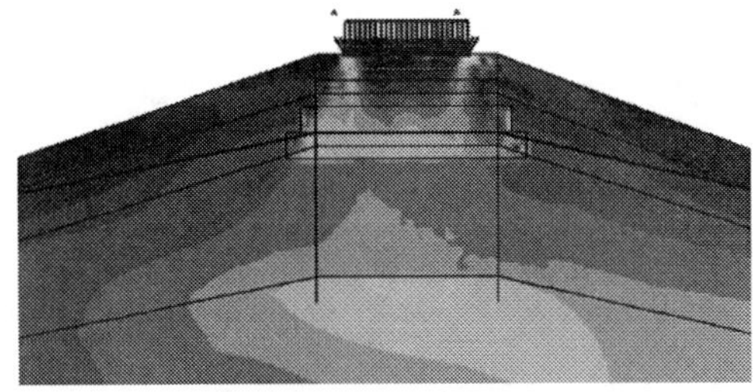

b. Deviatronic stresses.

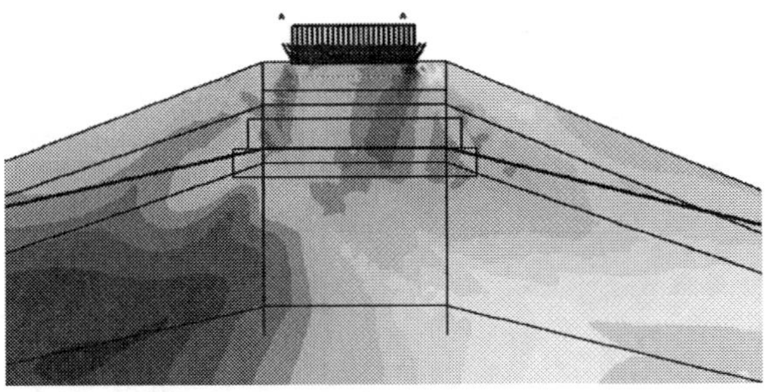

c. Shear stresses.

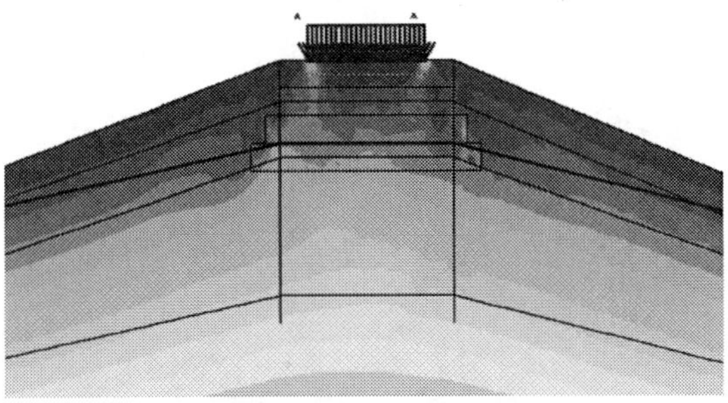

c. Vertical total stresses.

Figure 13. Displacements and efforts in the embankment (with Plaxis 2D).

5. Applied solutions

The article shows the construction on soft soil of an road embankment that gives access to a bridge which passes a stream.

The embankment near the bridge access was excavated on a length of about 10 m from the bridge and was restored in a mixed terramesh system with georgics filled with ballast in elemental layers of maximum 30 cm and a height of 1.00 m and protection on both slopes with gabions. Before making the first layer of ballast with geogrid, a blockage was made of compacted stone until refusal as exemplified in figure 14 and figure 15 with execution phases below.

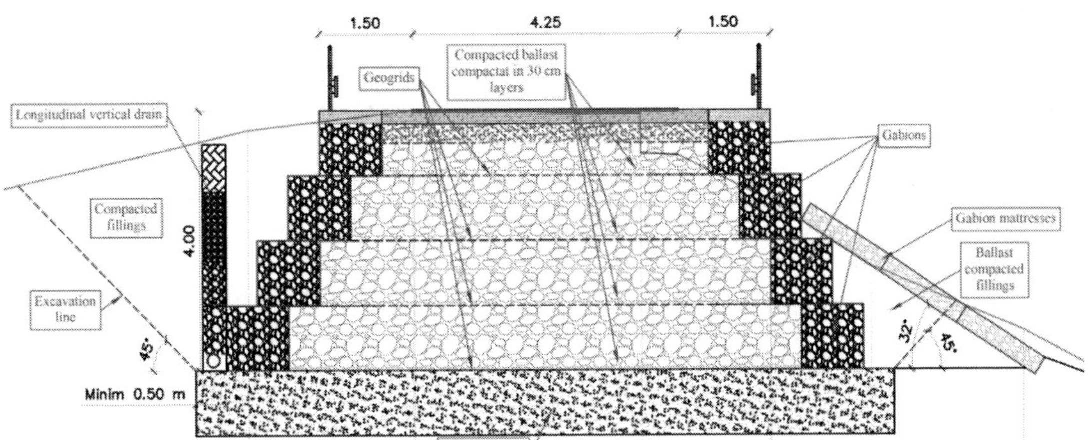

Figure 14. Characteristic cross section for the embankment with the applied solution.

a. Excavation.

b. Blockage with compacted stone.

c. Geosyntheticsand gabions executions.

d. After execution.

Figure 15. Execution period.

On the slope of the south side (to the stream), compacted fillings were made and after that gabions mattresses were placed for protection.

In order to collect the waters on the north side of the road, a longitudinal vertical drain was made up to a depth of -4.00 m measured from the carriageway level with controlled gabion mattress evacuation in the stream near the bridge abutment.

6. Conclusions

The article shows the construction on soft soil of a road embankment that gives access to a bridge, which passes over a stream. To strengthen, consolidating and draining the embankment near the stream it was used a combined geosynthetic and gabions solution.

This solution resulted from a technical and economic point of view, being more plausible than a more expensive solution with drilled reinforced concrete piles along the affected road embankment.

Reinforced earth is somewhat analogous to reinforced concrete. But the direct comparison between the reinforcement functions in the two cases is not valid. The mode of action of the reinforcement in the ground is not one of taking over the tensile stresses developed as in reinforced concrete, but of anisotropic reduction of the normal tension.

A drainage system with protected evacuation areas are very important for the lifetime of an embankment situated near a stream with fluctuating water levels.

When carrying out a calculation in the flooded state, the design values of the shear resistance parameters of soils presented in table 2 above are reduced by about 25-60% leading to important horizontal and vertical deformations in the soil.

References

[1] Arab R, Faure Y H, Lecendre V, Gendrin P and Douillat M 2004 Embankment access road to a civil railway structure on soft soil *Third European Geosynthetics Conference Munich-Germany* 135

[2] Loghin V 2009 *Elements of fluvial geomorphology* Valahia University Press Târgovişte **55** 80

[3] SR EN 1997-1:2004 *Eurocode 7: Geotechnical design. Part 1: General rules*

[4] NP 122:2010 *Normative regarding determination of characteristic and calculation values of geotechnical parameters*

[5] Prashant P, Ipupu M, Shubham G and Yogesh U 2016 Soil Reinforcement Techniques *Prashant Patil.et al. Int. Journal of Engineering Research and Application* **25** 31

[6] Abrudan I and Pleşcan C 2015 Land slide stabilization for a working platform which occurred during excavation in Cluj-Napoca, *Proceedings of The International Scientific Conference CIBv 2015* **179** 177

International Conference CIBv2019 Civil Engineering and Building Services IOP Publishing
IOP Conf. Series: Materials Science and Engineering **789** (2020) 012052 doi:10.1088/1757-899X/789/1/012052

The analysis of influencing factors over the computerised designing process of deep excavations near roads in urban environment

D-M Pohrib[1*] and P Turcanu[2]

[1] Faculty of Civil Engineering and Building Services, Iasi, Romania
[2] S.C. PROCONRIM S.R.L. Iasi, Romania

[*] madalina.pohrib@yahoo.com

Abstract. The article aims to show the errors that a user can make in the use of inappropriate computing programs for the calculation of support structures and / or inadequate use of geotechnical data when designing support for deep excavations near roads in urban areas, and to indicate the software that offers plausible and reality-related results. Comparisons were made between the displacements and the efforts of ground support elements from numerical computational models in which was varied the coefficient of bedding for horizontal actions and between geotechnical calculation models.

1. Introduction

Design solutions for geotechnical road support structures are currently based on simplified numerical experience and calculations. Three-dimensional numerical modeling is not commonly used due to the lack of codes for proper and correct interpretation of the results obtained from the Finite Element Methods calculations.

2. Material and Methods

Retaining walls are inherently retaining soil structures, i.e. providing lateral support with sufficient moment capacity to resist the action of lateral load from soil pushing. In contrast, piles are traditionally single structures with the ability to resist axial loads. In practice, this black or white approach is not feasible and piles are often relied on to resist lateral loads [1].

In the following chapter, we will present the theoretical bases on modelling the interaction of interspaced pile land-support structures near roads in urban areas.

A particular problem is the correct choice of the available and realistic constitutive models to describe as accurately the behavior of the foundation ground as the task.

A mandatory condition for correctly using soil behavior patterns in the Finite Element program is the use of geotechnical parameters (cohesion, internal friction angle, stiffness, etc.) obtained from in-situ or laboratory tests. In order to assess the efforts and the state of tension in the support structures, the numerical modeling of the supporting structure, the foundation ground and the supported ground as well as the cooperation between the support structure and the ground should be as close as possible to the reality.

At present, the spatial computing programs commonly used for calculating road support structures in our country are Plaxis 3D and AxisVM.

Numerical calculations of buildings and structures in the framework of soil-structure interaction do not pertain only to the domain of design of unique and technically sophisticated or hazardous buildings and structures any more.

They become actively involved in everyday design practice which, as a rule, includes geotechnical investigations [2, 3].

2.1. Plaxis 3D Program

The results obtained from spatial calculations with the PLAXIS - 3D Foundation Program are in the form of displacements, efforts, internal forces, moments and, in particular, failure of the support (tension and cracks) and the ground, results that allow the designer to create a close vision the reality of the projected support risk.

Numerical modelling of the program - The mechanical behavior of the terrain used for the calculations in this paper is elastic-plastic behavior Mohr-Coulomb Model.

The Mohr-Coulomb Model - This constitutive model implies 5 characteristic parameters: ground stiffness - E and Poisson coefficient υ for the elasticity part as well as φ, c, dilatandcy angle ψ to model the plasticity of the terrain.

The "embedded pile" used for the calculation of support structures, is a pile made up of bar elements that can be placed in random direction in the soil and which interacts with the perimeter soil through special interface elements.

This interaction may involve a base resistance besides a mantle resistance. Although an embedded pile does not occupy a volume, the program considers around the pile a particular volume depending on the pilot diameter (elastic area) excluding the plastic behavior of the soil. The basic parameters include the rigidity of the pilot, the unit weight of the material in the pile γ, geometric cross-sectional parameters, mantle resistance and base resistance.

2.2. Axis VM Program

The program is a static and sizing program based on the finite element method used predominantly for the constructions superstructure calculation.

For the calculation of an earth support structure in Axis VM, according to the calculation methods from geotechnical norms and specialty literature it can be considered in the calculation the beam Theory on Winkler elastic environment of a pile supposed to transversal loadings.

According to the calculation model in the book "Geotechnical Engineering Applications" [4] there is the following numerical calculation method: the ground on the plug *f* is considered to be on a Winkler elastic environment characterized by the horizontal displacements bed coefficient k_h and by the vertical displacements bed coefficient k_v.

The k_h coefficient increases linearly with the depth after the relation $k_h = mz$, *m* being the proportionality coefficient from table 1.

If a bar is placed vertically in this environment and it is loaded with a horizontal force (figure 1), at a depth *z* it will record a horizontal pressure $\sigma_h(z)$. According to Winkler Hypothesis, this pressure is proportional to the displacement following the relation:

$$\sigma_h(z) = k_h y(z) \tag{1}$$

From a physical point of view, the Winkler environment is made up of elastic arches arranged between the bar and a rigid base.

According to the relation (1), the bed coefficient k_h is the horizontal pressure appearing in a section if the displacement in that section is equal to 1. Dividing the plug *f* embedding section into segments of length a, and the continuous bearing between the bar and the ground is replaced by springs in the centers of these segments.

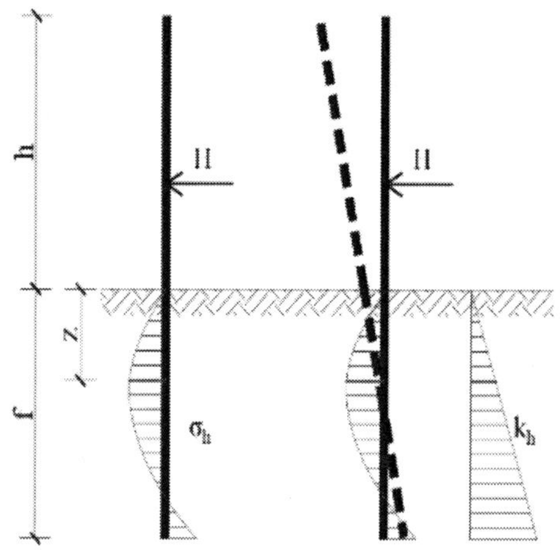

Figure 1. Beam in elastic environment.

The punctual support is achieved by means of elastic arcs between the bar and the rigid base. Since the bar is also supported vertically on the ground, an arc is also inserted in this direction at the bottom end of the bar, figure 2.

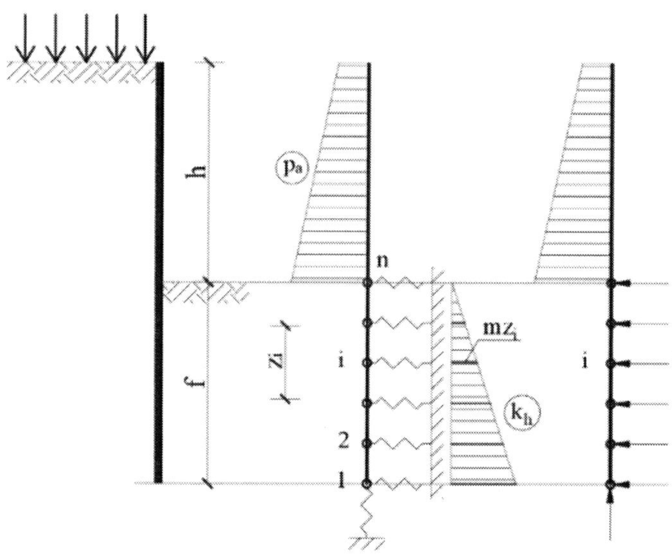

Figure 2. Modelling of bar-soil interaction.

When the element deforms under the action of earth pushing, elastic reactions occur in the arches. The elastic reaction in the horizontal arc *i* is:

$$R_{ei}=mz_i aby_i \tag{2}$$

Considering the terrain bedding coefficient for each vertical displacement k_v equal to the horizontal bed coefficient from the depth f (k_v = mf), the elastic reaction in the vertical spring is:

$$R_{ev}=mfAz_v \tag{3}$$

Considering *A* the area of the cross sections of the earth supporting structure, z_v s the vertical movement of the wall; *b* is the width of the earth-supporting wall element.

The value mz_iab from relation (2) represents the reaction from the point i if the displacement $y_i = 1$, meaning the rigidity of the elastic bearing in *i*. Similarly, mfA from relation (3) is the rigidity of the vertical elastic spring.

These rigidities must be entered into the calculation program along with the other input data. The calculation program will give the values of the displacements, reactions and sectional stresses (N, M si Q) at all points *i* selected on the wall.

From the reactions, the horizontal pressure σ_h between the wall and the earth is obtained: $\sigma_h = R_{ei}/(ab)$, which can be calculated with the relation (1).

The maximum value of this pressure must be less than the horizontal load bearing capacity (CP_h), which can be calculated with the relation:

$$CP_h = 2(\gamma z tg\phi + c) \tag{4}$$

in witch γ, φ , c are the ground characteristics of the plug *f*.

Since the load capacity varies with the z depth, the check $\sigma h\ (z) \leq CP_h$ will be done both for the maximum positive pressure and its negative maximum value. If the check is not satisfied, the plug in the soil *f* is insufficient.

Considering the result of the $\sigma_h(z)$ diagram is equal to the result of applied load, by increasing the plug the ordinates of the diagram will decrease, and the verification will be fulfilled.

Table 1. Coefficient of proportionality *m*– (Table C.1 [8]).

Soil type	Coefficient *m* [kN/m^4]	
	Prefabricated piles	On site executed piles
Clays and silty clays with Ic ≤ 0.25	650-2500	500-2000
Clays and silty clays with Ic = 0.25 – 0.50, sandy silts with Ic ≤ 1.00 and silty sands with e = 0.6-0.8	2500-5000	2000-4000
Clays and silty clays with Ic = 0.50 – 1.00, sandy silts with Ic > 1.00 and fine and medium sands	5000-8000	4000-6000
Clays and silty clays with Ic> 1.00 and big sands	8000-13000	6000-10000
Sand with gravel, gravel and cobble sand	-	10000-20000

2.3. Input data of computing programs used

In this paper were used, for the two programs, the following partial factors for Design Approach 1 – Group 2:

Partial factors for permanent and variable unfavorable actions (A2): $\gamma_{G,unfav} = 1.00$, $\gamma_{Q,unfav} = 1.30$.

Partial factors for soil parameters (M2): $\gamma_\varphi = 1.25$ for internal friction angle, $\gamma_c = 1.25$ for cohesion and $\gamma_\gamma = 1.00$ for volumetric weight.

It is considered that there is no hydrostatic level. There is a 5 kN/m^2 surcharge load from the road. he actions, the effect of the actions as well as the geotechnical properties of the stratification are the same for the two programs used in this paper.

The soil layers considered are shown in figure 3 below.

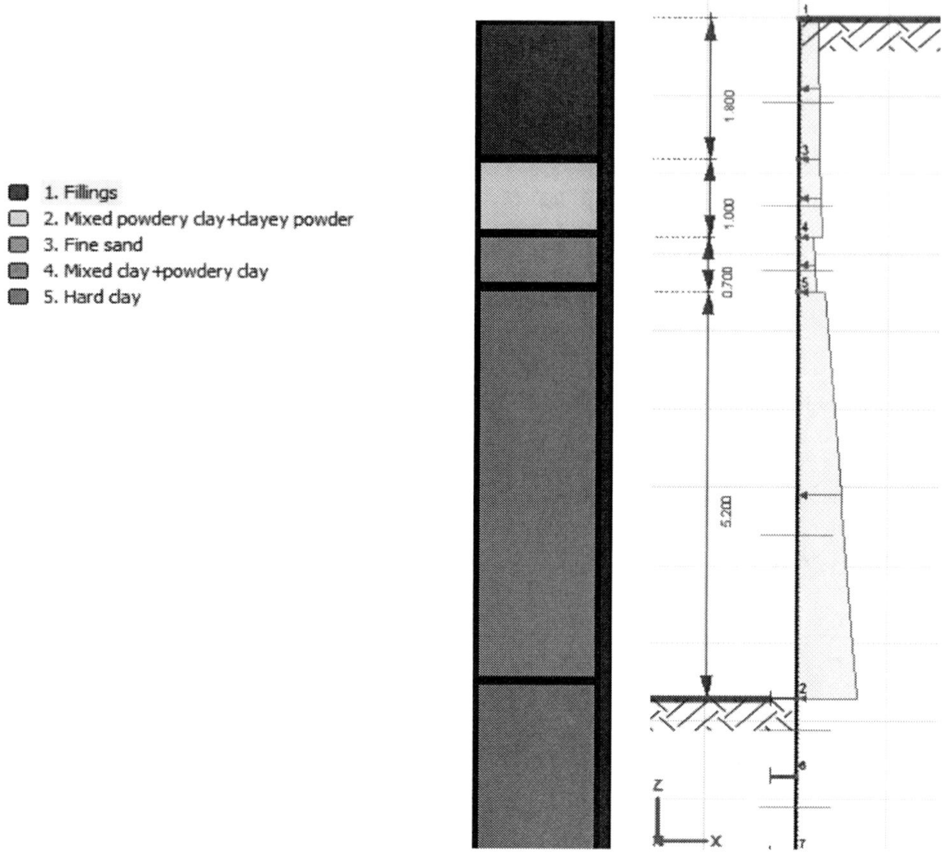

Figure 3. Soil layers and stratigraphic section.

2.3.1. Inputs for Plaxis 3D. Ground characteristics: The design values of soil rigidity by module of elasticity (E), Poisson coefficient (υ), internal friction angle (φ), cohesion (c), are the input data for the soil layers and are presented in the table 2 below.

Table 2. Geotechnical input data for the soil layers.

Layer	Upper elevation	Lower elevation	Thickness	Design approach 1 – Group 2				
				E	φ_d	c_d	γ_{nat}	υ
	m	m	m	kN/m^2	°	kPa	kN/m^3	-
1. Fillings	0.00	1.80	1.80	5.0E+3	34.82	5.00	19.00	0.30
2. Mixed podery clay+clayey powder	1.80	2.80	1.00	1.0E+4	19.18	22.00	19.40	0.35
3. Fine sand	2.80	3.50	0.70	9.0E+3	24.79	5.00	19.50	0.30
4. Mixed clay+powdery clay	3.50	8.70	5.20	2.5E+4	14.57	20.00	20.85	0.35
5. Hard clay	8.70	20.00	3.00	2.9E+4	7.02	50.00	21.18	0.42

Pile characteristics are: elastic modulus for C25/30 concrete E = 3.15E+07 kN/m^2; volumetric mass of the concrete in the pile γ = 25 kN/m^3.

The skin resistance and base resistance are inputs evaluated according NP 123-2010 [5] or according with results from displacement measurements on site.

The active pushing and bed coefficients for the embedded plug are automatically calculated by the program, representing output data.

2.3.2. Inputs for Axis VM. Ground characteristics: The active pushing are calculated according to [9] and the horizontal load bed coefficients are quantified by the proportionality coefficient. The active pushing and the horizontal bed coefficients are the input data, so, if not correctly approximated, the results are not realistic.

The active pushing's design values have been synthesized in the table 3 below for Design Approach 1 - Group 2 because the results are the most unfavorable states of effort [7] for the case treated in the paper.

Table 3. Active pushing's table.

Layer	Upper elevation	Lower elevation	Thickness	Design approach 1 – Group 2				
				k_a	φ_d	c_d	γ_{nat}	p_a
	m	m	m		°	kPa	kN/m^3	kN/m^2
1. Fillings	0.00	1.80	1.80	0.03	34.82	5.00	19.00	1.10
2. Mixed podery clay+clayey powder	1.80	2.80	1.00	0.23	19.18	22.00	19.40	12.72
3. Fine sand	2.80	3.50	0.70	0.14	24.79	5.00	19.50	9.25
4. Mixed clay+powdery clay	3.50	8.70	5.20	0.35	14.57	20.00	20.85	39.09
5. Hard clay	8.70	20.00	3.00	0.61	7.02	50.00	21.18	72.98

Pile characteristics are: elastic modulus for C25/30 concrete E = 3.15E+07 kN/m^2, volumetric mass of the concrete in the pile $\gamma = 25$ kN/m^3.

According to the Winkler Approach, the coefficient of bedding for horizontal displacements increases linearly with the depth after the expression k_h = mz, where m is the proportionality coefficient [kN/m^4], table 1, depending on the nature of the soil (cohesive, non-cohesive), physical and mechanical characteristics of the soil, of the pile type (prefabricated, executed on the spot); z - the depth of excavation

The active pushing and bed coefficients for the embedded plug represents input data.

3. Results of modelling an earth support structure near urban road in non-specialized geotechnical program

In this chapter we present the calculation of a road support structure made of 40 cm diameter piles with interspaces at 70 cm between the axes, with a depth of excavation of 8.70 m and a plug length of 4.00 m in Axis VM, a non-geotechnical program.

With the geotechnical data taken from table 4, several calculation models were made:

- (a) variation of the bed coefficient k_s for vertical loads in the case of a similar length pile embedding with and high consistency index - inadequate modelling; The values of the coefficient of bed k_s will be taken from table 4 below, according to norm NP 112-2014 [6], table K.2, for a clayey soil:

Table 4. Bed coefficient values k_s [4].

The vertical bed coefficient in correlation with consistency index for clayey soils	Plastic flowing soil	Soft plastic soil	Consistent plastic soil	Plastic strong soil
I_C	0÷0.25	0.25÷0.50	0.50÷0.75	0.75÷1.00
k_s	-	7000÷34000	34000÷63000	63000÷100000

- (b) variation of the bed coefficient for horizontal loads k_h by means of the coefficient of proportionality m [kN/m^4] in the case of a pile embedding with a low, medium and high consistency

IOP Conf. Series: Materials Science and Engineering **789** (2020) 012052 doi:10.1088/1757-899X/789/1/012052

index for a clayey soil - accepted modelling for pre-dimensioning. The values of the proportional coefficient m will be taken from table 1 [3, 8].

The calculations models performed whom taken into account the proportionality coefficient m [kN/m^4] with the maximum value for clayey soil will be compared to the calculation model made by calibrating the geotechnical data of the constitutive model in the Plaxis 3D program.

3.1. *Inadequate modeling – the use of the vertical loads bedding coefficient k_s*

According to above point 3(a), in the following figure 4 we present the results consisting in charts of efforts and horizontal displacements for the variation of the coefficient of bed for vertical loads k_s used for horizontal loads from low values of 63000 kN/m^3 to maximum values of 100000 kN/m^3 on the considered high consistency index with similar length clay embedding.

Vertical Bedding coefficient k_s

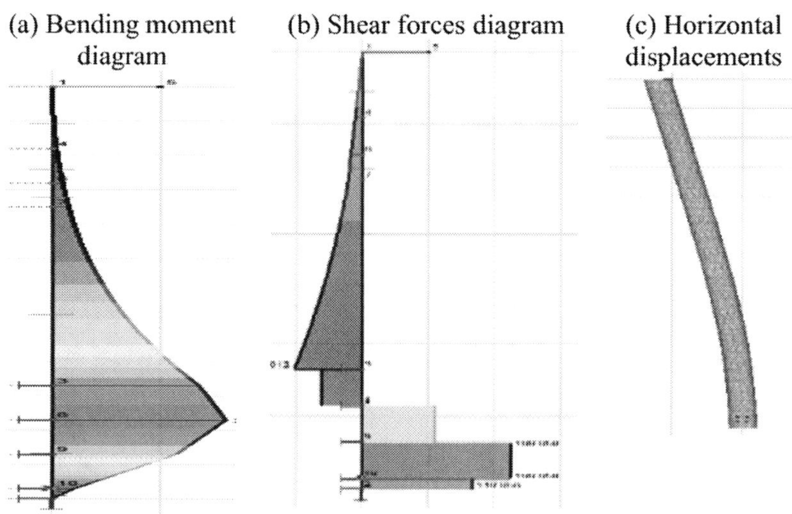

Value	[kN/m^3]	M_{max} [kNm]	Q_{max}[kN]	e_x[mm]
Minim	63000	289	145	189
Middle	81500	289	160	182
Maxim	100000	289	173	178

Figure 4. Charts for: (a) Bending moment; (b) Shear force; (c) Horizontal displacements.

3.2. *Acceptable modeling - Bed coefficient variation for horizontal displacements k_h (horizontal loads)*

According to above point 3. (b), in the following figure 5 are synthesized the results consisting in charts of efforts and horizontal displacements for the variation of the coefficient of horizontal soil reaction k_h from the low values of 6000 kN/m^4 to the maximum values of 10000 kN/m^4 on the considered high consistency index with similar length clay embedding.

Horizontal bedding coefficient		(a) Bending moment diagram	(b) Shear force diagram	(c) Horizontal displacements diagram

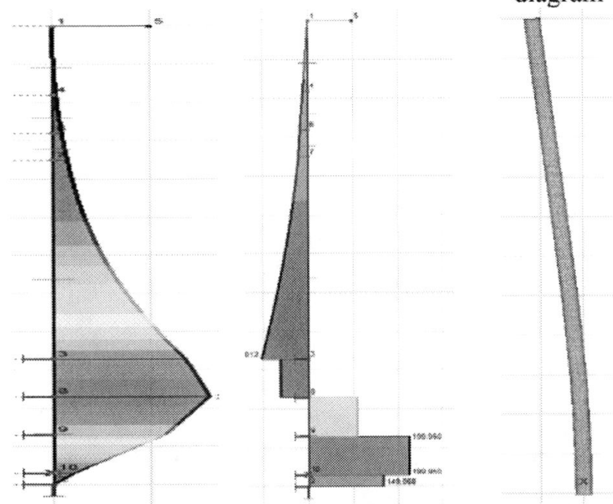

Value k_h	[kN/m^3]	M_{max} [kNm]	Q_{max} [kN]	e_x[mm]
Minimum	6000	343	200	412
Middle	8000	339	196	375
Maximum	10000	336	191	349

Figure 5. Charts for: (a) Bending moment; (b) Shear force; (c) Horizontal displacements.

4. Results of a site executed earth support structure near an urban road

Based on the topographic survey carried out on site in the analysed area and the geotechnical design values

4.1. Accepted modeling - Bed coefficient variation for horizontal displacements k_h (horizontal loads) - modeling performed with Axis VM

The following figure 6 presents the results calculation of a earth retaining wall near urban road made of piles with 40 cm diameter, with interspaces at 70 cm between axes, with a maximum excavation depth of 8.70 m and a ground embedding length of 4.00 m.

The coefficient k_h was used depending on the characteristics of the soil in which the piles are embedded for maximum values of 10000 kN/m^4.

Model Horizontal bed coefficient		(a) Bending moment	(b) Shear force	(c) Horizontal displacement
Value k_h	[kN/m^3]	M_{max} [kNm]	Q_{max} [kN]	e_x[mm]

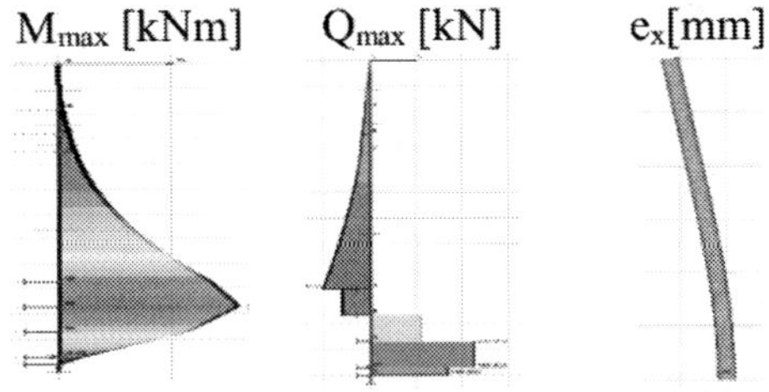

Maximum	10000	336	191	349

Figure 6. Charts of efforts for: (a) Bending moment; (b) Shear force; (c) Horizontal displacements.

4.2. Optimal modeling - geotechnical data calibration with topometric measurements

The following figure 7 presents the results calculation of an earth retaining wall near urban road made of reinforced concrete piles with 40 cm diameter, with interspaces at 70 cm between axes, with a maximum excavation depth of 8.70 m and a ground embedding length of 4.00 m after a model calibration according to in-situ measurements of the piles displacements.

(a) Bending moments diagram [kNm]	(b) Shear forces diagram [kN]	(c) Horizontal displacements diagram [mm]
M_{min} = -107	Q_{min}= -87	u_{xsup} = 253
M_{max} = 41	Q_{max}= +110	u_{xinf}= 22

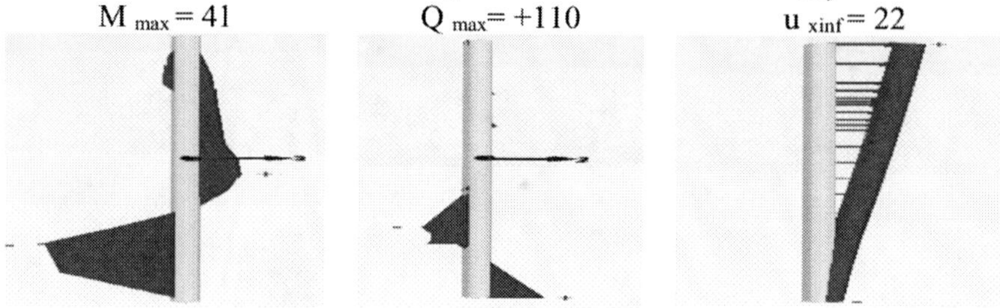

Figure 7. Charts for: (a) Bending moment; (b) Shear force; (c) Horizontal displacements.

4.3. Situation with half pile length embedding

Given that the depth of excavation is more than 8.00 m and the pile diameter is only 40 cm, it presents high rotations and displacements, therefore in the following it will be presented the comparative results of the 40 cm diameter and of 80 cm diameter with half-length embedding calculations for the earth supporting wall, executed near urban road, made with Axis Vm and the calculations performed in Plaxis 3D.

4.3.1. Axis VM – Reinforced concrete piles, L=16.70 m (excavation 8.70 m – embedded 8.00 m), D=40 cm. The following figure 8 presents the results calculation of an earth retaining wall made of reinforced concrete piles with 40 cm diameter, with interspaces at 70 cm between axes, with a maximum excavation depth of 8.70 m and a ground embedding length of 8.00 m after a model calibration according to in-situ measurements of the piles displacements.

It can be noted that the movement recorded at the free end is very high and that on the embedding length, the pile has buckling characteristics.

Model	(a) Bending moment diagram Mmax = 341 kNm	(b) Shear force diagram Qmax = 132.36 kN	(c) Horizontal displacements diagram ex = 345 mm

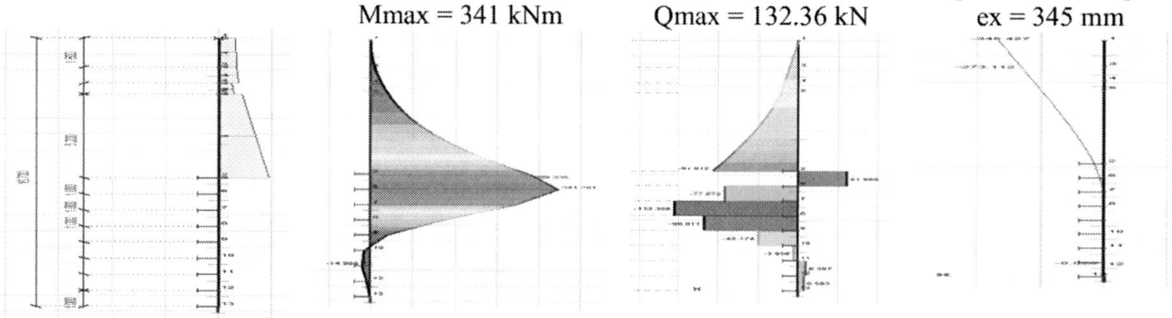

Figure 8. Charts of efforts for: (a) Bending moment; (b) Shear force; (c) Horizontal displacements.

4.3.2. Axis VM - Reinforced concrete piles, L=16.70 m (excavation 8.70 m – embedded 8.00 m), D=80 cm. The following figure 9 presents the results calculation of an earth retaining wall, near an urban road, made of reinforced concrete piles with 80 cm diameter, with interspaces at 110 cm between axes, with a maximum excavation depth of 8.70 m and a ground embedding length of 8.00 m after a model calibration according to in-situ measurements of the piles displacements. It can be noted that the displacement at the free end and the rotation is reduced by about 90% compared to the 40 cm diameter pile showed at previous point and that the moment diagram only appears on the active pushing side.

Model | (a) Bending moment diagram | (b) Shear force diagram | (c) Horizontal displacements diagram

M_{max} = 398 kNm | Q_{max} = 94 kN | e_x = 41.86 mm

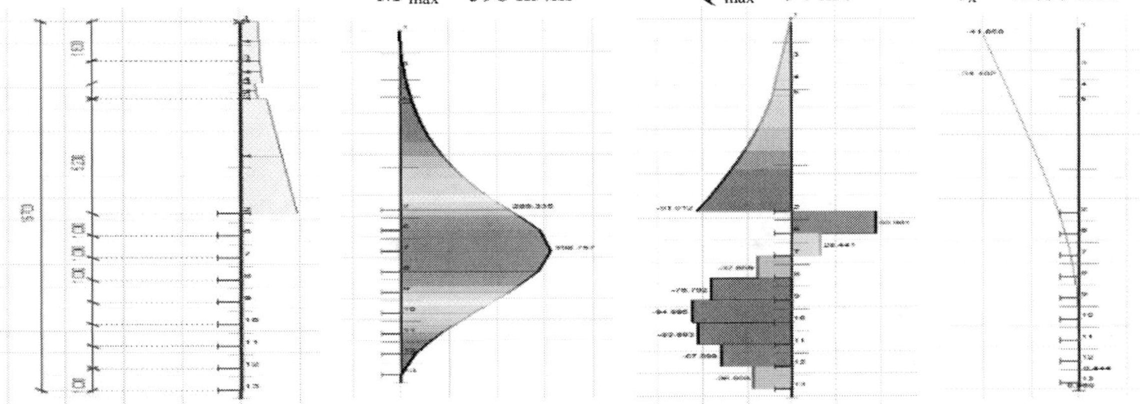

Figure 9. Charts for: (a) Bending moment; (b) Shear force; (c) Horizontal displacements.

4.3.3. Plaxis 3D - Reinforced concrete piles, L=16.70 m (excavation 8.70 m – embedded 8.00 m), D=40 cm. The following figure 10 presents the results calculation of a earth retaining wall, near an urban road, made of reinforced concrete piles with 40 cm diameter, with interspaces at 70 cm between axes, with a maximum excavation depth of 8.70 m and a ground embedding length of 8.00 m after a model calibration according to in-situ measurements of the piles displacements.
It can be noted, compared to Axis calculations for the same situation (see point 4.3.1) that the displacement and the internal forces are much smaller.

Model | (a) Bending moment diagram [kNm] | (b) Shear force diagram[kN] | (c) Horizontal displacements diagram [mm]

M_{min} = -76, M_{max} = 33 | Q_{min} = -78, Q_{max} = +44 | u_{xsup} = 79, u_{xinf} = 19

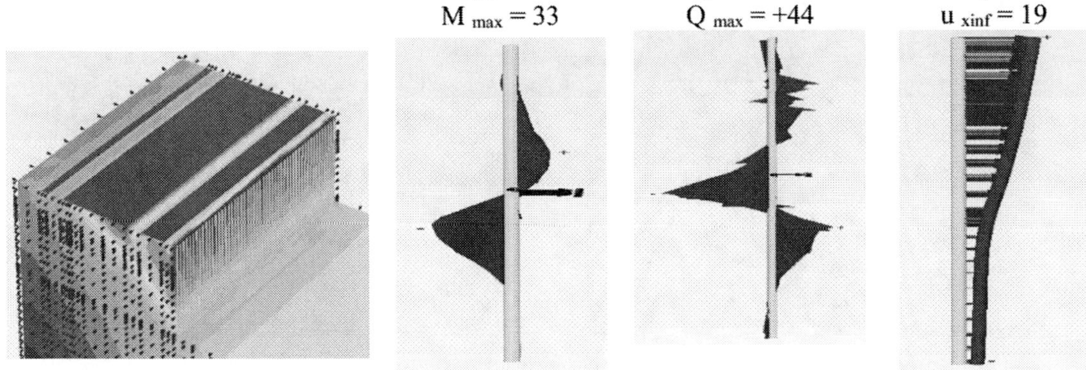

Figure 10. Charts for: (a) Bending moment; (b) Shear force; (c) Horizontal displacements.

4.3.4. Plaxis 3D - Reinforced concrete piles, L=16.70 m (excavation 8.70 m – embedded 8.00 m), D=80 cm. The following figure 11 presents the results calculation of an earth retaining wall, near an urban road, made of reinforced concrete piles with 80 cm diameter, with interspaces at 110 cm between axes, with a maximum excavation depth of 8.70 m and a ground embedding length of 8.00 m after a model calibration according to in-situ measurements of the piles displacements.
It can be noted, compared to Axis calculations for the same situation (see point 4.3.2) that the displacement at the free end is higher, but the value of the moments is reduced by 25 %.

Model	Bending moment [kNm]	Shear force [kN]	Horizontal displacements [mm]
	M_{min} = -248 kNm	Q_{min} = -58 kN	u_{xsup} = 69 mm
	M_{max} = -200 kNm	Q_{max} = +64 kN	u_{xinf} = 21 mm

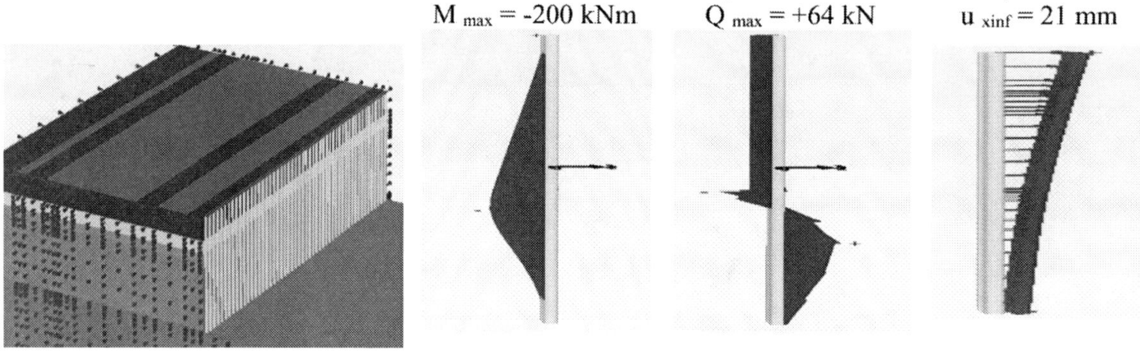

Figure 11. Charts for: (a) Bending moment; (b) Shear force; (c) Horizontal displacements.

5. Discussions – Errors made at data input

Incorrect linear estimation of the bed coefficient, given the spatial behaviour (3D) of the state of effort in the pile surrounding soil. Also the coefficient of bedding for vertical loads in the geotechnical literature is estimated by tests with rectangular and not circular surfaces, as is the case of a circular pile from an earth supporting structure for a deep excavation in an urban area.

Some programs allow the choice of geotechnical parameters but not all users have sufficient knowledge to correctly approximate these geotechnical parameters. The most common mistake is that the calculations of an earth supporting structure near an urban road in computer programs is made by non-specialist engineers in geotechnics. The main error in introducing data into programs, used just for structural calculation, is the inadequate calculation of the springs rigidity for the Winkler approach, by confusing the coefficient of bedding from tabs 8.1 and 8.2 of NP 112-2014 [4] k_s [kN/m^3] for vertical loads on the soil with the coefficient of proportionality m [kN/m^4] [4, 6, 8], for horizontal actions on the soil. This leads to erroneous results (less displacements and less efforts versus reality) and hence to the geotechnical failure of the piles after the excavation, figure 12.

Figure 12. Pile spin and measured top displacement of 26 cm.

Figure 13. Piles who did not yielded D=60cm (left side) and piles who yielded D=40 cm (right side).

There are also cases where the coefficient of bedding was considered constant on the embedded depth, although according to the literature, the bed coefficient transformed into the proportionality coefficient varies linearly with the depth.

The calculations in the two programs are made without considering a piezometric line.

It was considered, as shown in figure 12 above that the soil does not flow among the piles.

6. Conclusions

6.1. Comparison between pilot diameters

The moment diagram changes its character: in the case of the 40 cm diameter pile, the stretched concrete fibre appears also on the active pushing side.

The bending moment value of the D = 40 cm to D = 80 cm is increased by 10 % in the Axis program and by more than 50 % in the Plaxis program.

For the considered situation, in the Axis program the rotation is lower when using piles with diameter of 80 cm; the value is reduced by about 90 %;

The upper displacement is about 88 % lower for the piles with a diameter of 80 cm.

The value of the shear forces in the concrete element decreases with the diameter of the pile.

Table 5. The variation of efforts values with the increase of pile diameter.

Computer program	Effort	40 cm pile diameter	80 cm pile diameter	Variation of efforts value
Axis	Bending moment	341	398	+10 %
Plaxis 3D		76	248	+50 %
Axis	Shear force	132	94	-10 %
Plaxis 3D		78	58	-10%

6.2. Comparison of the computing programs

There are specialized programs for support elements, but for any pre-dimensioning and a primordial cost estimation, any structural calculation program that accepts elastic supports can be used only if the provided input for sensible estimated geotechnical parameters are as possible as accurate.

In subsequent design and execution phases, when it is necessary to determine the reinforcing and the displacement of the earth outside the excavation, only specialized geotechnical engineering programs will be used.

It is imperative to model earth supporting structures for deep excavations in urban areas with an 3D program to take into account corner areas of the supports.

For the estimation of the rigidity of springs for the horizontal reaction of the soil, the tables for the proportionality coefficient m (kN/m^4) [3, 6] shall be used and not the value of the coefficient of bed k_s (kN/m^3) [4], which is approximately 10 times higher, thus leading to less effort and displacements in earth supporting deep excavation walls made from piles cast on site than real displacements.

Although the Winkler model admits the hypothesis of using a bed coefficient for horizontal displacements, in reality the displacement of a point from the support structure - perimeter soil interface has at least two components: horizontal and vertical, so, it is preferably to model a support structure in a special 3D Geotechnical program.

Also, according to NP 113-2004 [7], the use of the finite element method that are at the base of the geotechnical programs is considered to be a method that offers complete theoretical solutions. The necessary incorporation depth of the wall is generally determined from the required slide stability [10].

The slide stability is defined in such a way that the total of the stimulated laying forces at the foot of the wall is compared with the total of the possible passive soil pressure [11].

The existing modern software complexes making numerical calculations do not always consider these particularities. Moreover, commercial software largely used in the global practice of construction design not always can confirm data obtained at real sites concerning actual strain-stress behaviour of buildings and structures. In a number of cases, software becomes a "black box", whereas a geotechnical engineer involved in calculations is not able to be an active participant of simulation. Application of soil models which are not tested and adapted to geotechnical conditions of a certain area is a factor of excessive risk for high-level responsibility of structures [12].

For deep excavations support in urban areas, for obtaining small effective displacements, it is recommended to use piles with a minimum diameter of 80 cm with a sufficient embedding length.

6.3. Influence of increasing pile diameter and depth of embedding

The calculations presented above shows, for the calculation comparison of 40 cm pile to the 80 cm diameter pile, that the horizontal displacements decrease by an approximate 1.00 cm, the shear forces also decrease by an approximate 10% and the character of the bending moments is changed.

Compared with Axis VM modelling, in the 3D Plaxis modelling for larger diameter piles and half-length embedding, it is found that the moments and the shear forces in the piles of an earth support element for a deep excavation in urban areas are smaller, but the displacements are at least 25 % bigger.

Table 6. The variation of efforts values with the increase of embedding length.

Computer program	Embedding length for 80 cm diameter piles	Bending moment [kNm]	Shear force [kN]	Displacements [mm]	Variation of displacement values
Axis	4 m	289	173	178	
Plaxis	4 m	107	110	253	+25%
Axis	8 m	398	94	54	
Plaxis	8 m	248	58	69	+27%

References

[1] Jørgen S S 2017 *Piles as retaining structure* 2017 (COWI, Kongens Lyngby, Denmark Lateral soil support enhanced by piles, Prague Geotechnical Days)
[2] Michael L V U 2015 *Preface of seminars on soil – structure interaction and retaining walls* (Edinburgh – Scotland)
[3] Valaskova V, Papan D and Barros R C 2016 *Assessment of the roadway dynamic response due to the Tatra 815 lorry excitation and experimental verification* (Scientific letters of the University of Žilina) **18** 55
[4] Roman F 2011 *Geotechnical Engineering* (Cluj-Napoca, Romania, Papyrus Print)
[5] * * * NP 123:2010 *Normative on Geotechnical Design of Foundations on Pilots*
[6] * * * NP 112-2014 *Normative regarding the geotechnical design of surface foundations*
[7] * * * NP 113:2004 *Normative for the design, execution, monitoring and reception of buried walls*

[8] * * * NP 120:2010 *Normative requirements for the design and construction of deep enclosures in urban areas*
[9] * * * NP 122:2010 *Normative regarding determination of characteristic and calculation values of geotechnical parameters*
[10] * * * NP 124:2010 *Normative regarding the geotechnical design of the supporting works*
[11] * * * GP 129-2014 *Guidance on Geotechnical Design*
[12] * * * RVS 9.31 2004 *Open Building Method*

IOP Conf. Series: Materials Science and Engineering **789** (2020) 012053 doi:10.1088/1757-899X/789/1/012053

Critical SCADA applications to parallel operation of transformers in railway electrical traction

A Popov[1*] and M Fratu[2]

[1] Dept. of Automatics and Technology of Information, Transilvania University of Brasov, Romania.
[2] Dept. of Building of Services, Faculty of Civil Engineering, Transilvania University of Brasov, Romania.

[*] alexandru.popov@unitbv.ro

Abstract. The article describes the parallel operation of medium voltage (MV) transformers utilized in electric traction substations (ETS), their connection to the national power system (NPS), and the conditions required for the parallel operation of the ETS transformers. Solutions for the optimization of the parallel operation the MV transformers that power the electric railway traction system (ERTS) are discussed. In the case of multiple failure (fault transformers that serve a section) or load increase over the rated power provided by an electric traction substation (ETS), connection to the respective section of the adjacent ETS transformers is required. The parallel coupling of ETS is possible if High Power Transformers ERTS meet all the conditions for parallel operation. This mode of operation ensures the stability of the electricity distribution system for the consumer. SCADA systems can be used for monitoring critical situations and general optimization of electrical power supply of the railway electric traction system.

1. Introduction

Typically, electricity is generated and distributed in the three-phase electrical system, and therefore it is obvious that we would need voltage-lowering triphasic transformers. The main drawback of a three-phase transformer is that, even if a fault occurs in one phase, the whole transformer is removed from service for repairs.

In the railway electrical traction, it is virtually possible to use single-phase transformers instead of triphasic transformers, properly distributed along the railways. This will prevent imbalance in the three-phase distribution network [1].

Along the railways it is necessary to provide two successive transformers of the same phase so that there is the possibility of supplying one of the two successive railway segments, from transformers that feed the adjacent rail segment. It creates the possibility that the single-phase transformers operate in parallel by allowing the power locomotive to be fed from two transformers connected in parallel. Parallel transformers in the electric rail system will feed the locomotive without total interruption of electric power

IOP Conf. Series: Materials Science and Engineering **789** (2020) 012053 doi:10.1088/1757-899X/789/1/012053

2. Electrical traction substation (ETS)

Electrical railway traction system (ERTS) are supplied with electrical power by electric traction substations (ETSs) which in turn are powered by 110kV overhead lines, that are part of the national power system (NPS). A block diagram of the system is shown in figure 1.

The main element of the electric traction substation (ETS) is the high power transformer, which has the following characteristics: Sn = 16 MVA, U1 / U2 = 110 / 27.5 kV, f = 50 Hz, ΔPo = 28 kW, ΔPsc = 10.6%, Io = 2.5%. The transformer is bi-phase, powered by two phases of the 110 kV national power system network. The transformer powers the traction electric system with a voltage of 27,5 kV. One running transformer and one backup transformer are mounted in an ETS [2].

ETSs are located along the railway at distances of approximately 50-60 km. These provide power to railway consumers through the feeders, covering areas of about 30 km. The bounding of the feed zone of an ETS is done through neutral zones [3].

For safety reasons the electric power supply of the railway system, in special situations (damage, load peaks, decommissioning for equipment overhauls or failures) it is necessary to connect the ETS transformers or the adjacent ETS transformers in parallel. For coupling in parallel of the ETSs it is necessary to close the neutral zones and supply the ETS transformers to be made from the same two phases of the NPS [4]. The schematic of parallel connection of the ETSs is shown in figure 2.

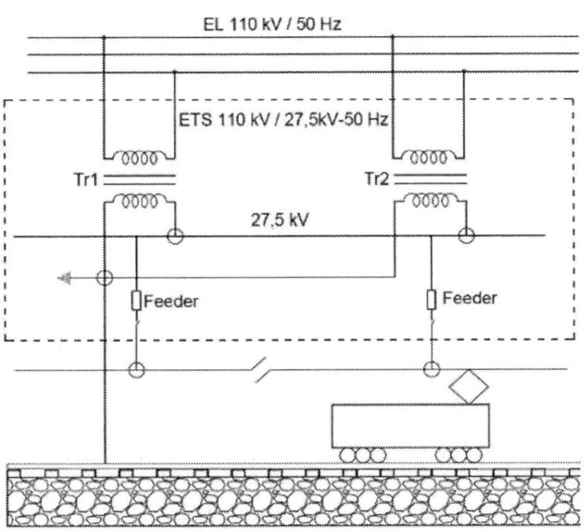

Figure 1. Schematic of the electrical power supply system of the electrical railway traction system. Tr=transformer.

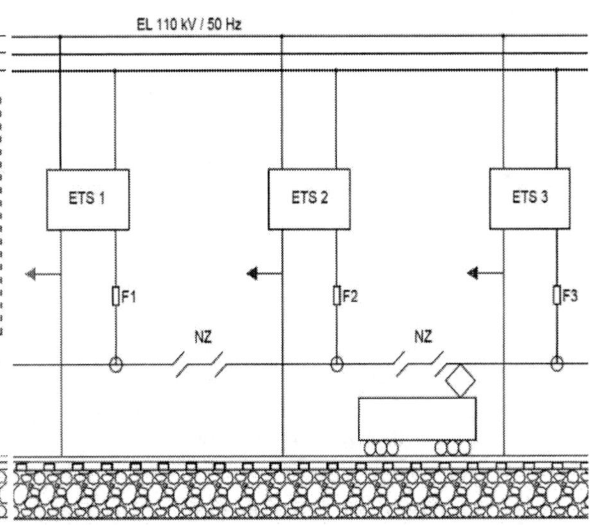

Figure 2. Parallel connection diagram of ETSs by closing the neutral zones a section will be fed from both ends which will limit the longitudinal voltage drops on that area. F=feeder, NZ=neutral zone.

3. Parallel operation of transformers in rail electric traction

Usually power is generated and distributed in a three phase system, therefore it is necessary to use three phase transformers to step up and step down voltages. However, it is practically possible to use three suitably interconnected single phase transformers instead of one three phase transformer [5]. If two or more transformers are connected to the same supply on the primary side and to a same load on the secondary side, then it is called *parallel operation* of transformers. Sometimes, when attempting electric traction on an inclined railway section, it becomes necessary to connect one transformer in parallel in order to avoid excess load of existing active transformer.

In electric traction, another situation in which the transformers must be connected in parallel is the case of simultaneous faults of the both active and passive transformers (allocated to a segment) which are then removed from service for repairs. In this situation the respective section can be supplied from transformers which normally feed the adjacent segments [6].

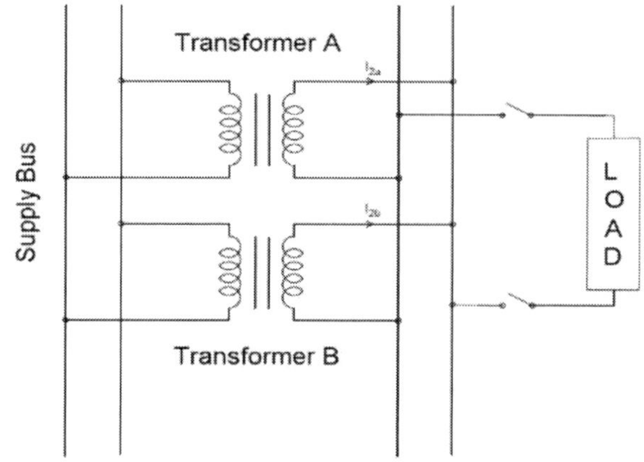

Figure 3. Parallel operation of single phase transformers- scheme of principle.

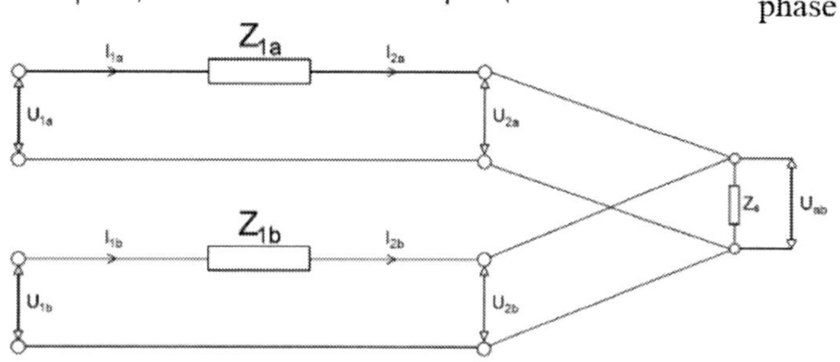

Figure 4. Parallel operation of single phase transformers - electric scheme.

3.1. Necessity of Parallel Operation of Transformers

Parallel operation of transformers (depicted in figures 3 and 4) is needed for:

- increased load: When load is increased and it exceeds the capacity of the existing transformer, another transformer may be connected in parallel with the existing transformer to supply the increased load.
- non-availability of large transformer: If a large transformer which can meet the total requirement of the load is not available, two or more small transformers can be connected in parallel to increase the capacity.
- increased reliability: If multiple transformers are running in parallel, and a fault occurs in one transformer, then the other parallel transformers will continue to supply the load. Furthermore, the faulty transformer can be removed for the maintenance [7].
- transportation is easier for small transformers: If installation site is located far away, then transportation of smaller units is easier and may be economical.

3.2. Conditions for parallel operation

Parallel operation of transformers is required when the necessitated power is greater than the individual transformer's rating. In such cases operating two or more transformers to facilitate the power flow is possible, but certain conditions to be followed while operating transformers in parallel condition [7]. These conditions are:

- the voltage ratio of all connected transformers must be same. If the voltage ratio is not same, then the voltage of the secondary windings will differ, even though the primary windings are connected to same busbar. This results in current circulating between the secondary windings, as well as reflected circulating current on the primary side [8]. In this case, considerable amount of current is drawn by the transformers even without load.

$$K = \frac{U_{1a}}{U_{2a}} = \frac{U_{1b}}{U_{2b}} \tag{1}$$

- the per unit (pu) impedance of each transformer on its own base must be same. Sometimes, transformers of different ratings may be required to operate in parallel. For, proper load sharing, voltage drop across each machine must be same. That is, larger transformer has to draw equivalently large current. That is why per unit impedance of the connected transformers must be same [11].

$$\frac{I_{2a}}{I_{b2}} = \frac{Z_{2a}}{Z_{2b}} = \frac{I_{1a}}{I_{1b}} \tag{2}$$

- the polarity of all connected transformers must be same in order to avoid circulating currents in transformers. Polarity of a transformer means the instantaneous direction of induced EMF (electric magnetic flow) of the secondary winding. If polarity is opposite to each other, the currents of the two secondary windings will take opposite directions.

$$U_{2a} = U_{2b} \tag{3}$$

- the phase sequence must be identical of all parallel transformers. This condition is relevant to poly-phase transformers only. If the phase sequences are not same, then transformers can not be connected in parallel. The currents of the secondary windings will be in phase and will be summed algebraically.

$$\varphi_{2a} = \varphi_{2b} \tag{4}$$

$$I_s = I_{2a} + I_{2b} \tag{5}$$

- the short-circuit impedances should be approximately equal (as it is problematic to achieve identical impedances practically) [9].

$$U_{1a} - U_{1b} = Z_{1a} * I_{1an} - Z_{2a} * I_{2an} \tag{6}$$

4. Choosing the optimal solutions for ETSs in different situations:

4.1. In parallel connection of electric transformers

To increase performance in parallel operation of ETS transformers, the following measures are proposed:

- transformer coupling and decoupling automatization: transformers will be coupled in parallel when the value delivered by a transformer reaches In and switched off when the current per transformer is <40%;
- protection of the transformers with the help of relays;
- protection of transformers against internal faults (such as defects that occur between the transformer's windings);
- protection of transformers on external faults (maximum current, maximum inverse current, homopolar sequences of current or voltage, minimum or maximum voltage, etc.).

The possibility of parallel coupling of the transformers is conditioned by:

- the transformers are on the same plot;
- identical voltage on the 27.5 kV bar;
- the load coupling should be made only if the equipment has this ability;
- the use of complex devices that can measure the active and reactive power, that interfere with a deforming regime.

Total integration of equipment operation in SCADA systems as well as the use of modern relays based on numerical microprocessors bring the following advantages:

- a programmable logic chart that allows changing the configuration of the ETS system;
- a self-supervision function that optimizes maintenance and reduces the likelihood of human errors occurring;
- by integrating the protections, the volume of the secondary circuits in the ETS is reduced;

- improve performance of transformer protection;
- specific operating protections can be configured under special conditions;
- the time for actioning the protection is reduced;
- the operating parameters can be changed by easy reconfiguration whenever needed;
- the ability to record events and defects in a database;
- remote transmission of information and operating parameters.

4.2. For double track railway;

Transformers of ETS that supply dual-track railway paths need to be dimensioned for a much larger (almost double) load compared to transformers that feed a simple railroad.

The supply voltage and the voltage in the contact line must be identical on both runways. If there is a potential difference in the crossing points from one path to the other, the pantograph of the electric locomotives would be equalized. This would produce an electric arc and risk of damaging the pantograph collector brush [6].

For the uniform distribution of currents and reduction of voltage drops in the contact line, on both runways, halfway between two ETSs or halfway between an ETS and a sectioning station (PS-SS), a Parallel Linking (PLP-PL) that must ensure:

- automatic opening of the circuit breaker or separator under load when the voltage drops on a running track,
- automatic switching of the circuit-breaker or the under-load separator when the electrical voltage on both runways.

Achieving these conditions eliminates the possibility that in the event of a fault, on a runway, when the feeder that feeds it is triggered, the PLP will supply the fault area and maintain the short circuit current in the plant.

By installing the PLPs in a current line between the two running paths, the power losses are reduced due to longitudinal and transverse stresses.

4.3. Parallel coupling of electrical group lines.

In order to balance the load and the electrical voltage in the contact line, in stations having several groups of lines, or in the classification yard (where there are many sorting lines), transverse separators are installed. These have the role of coupling in parallel the groups of lines.

Since the transverse separators do not switch on or do not interrupt the load currents, it is not necessary to install separators under load. Instead, separators can be installed that support the current flow from the power supply to the consumer (electric locomotive) through its closed contacts. The transverse separators for remote manipulation by motion sensors or remote controls by zonal dispatchers must be fitted with actuators equipped with electric servo motors. The power supply of the servo-motors must be provided from several sources: one basic and at least one reserve. The basic source can be secured from the national grid or from the contact line transformation station (CLTS). The backup source may be a generator or a stationary battery coupled to a buffer. The necessity of these measures ensures the parallel connection of the groups of lines and their disconnection whenever necessary for maintenance or isolation of defective areas. If the remote control does not work for various reasons, the occurrence of a fault in the contact line can not restrict the under-voltage area. This type of fault would affect the movement of electric locomotives on all the groups of lines that can not be isolated from the parallel connection, with a direct impact on the circulation of the trains. This type of equipment malfunction may affect trains that would otherwise be circulating normally.

5. Critical SCADA Applications to choosing the optimal solutions in ETS-s

Designing a SCADA system for monitoring ETSs must account for all critical situations that can arise, involving equipment or installations either in operation or in standby. Concurrently, when implementing the SCADA system, it is necessary to take into consideration certain optimization solutions for the functioning of equipment. The main goal is solving critical issues which may arise in the functioning of

ETSs, as well as the introduction of certain optimization solutions for the functioning of equipment. Additionally, the safety and performance of ETSs are enhanced and thus made suitable for the increasing demands of the ERTS [10].

5.1. Critical situation that need to be monitored by the SCADA system:

- in order to protect transformers from internal faults, SCADA will monitor the temperature of the electrical insulant oil; if the upper limit of the functioning temperature range is reached, SCADA will start the forced cooling fans, and if the upper temperature limit is exceeded, SCADA will decouple the transformer from the electrical load and from the current source, and will activate adjacent transformers;
- SCADA will commence acoustic and light signals when the electrical insulating oil inside the transformer reaches the lowest allowed level;
- SCADA and the earth fault protection system will withdraw from operation any transformers with damaged winding insulation. The faulty transformer will be replaced by one or more parallel-coupled transformers according to necessity;
- the maximum current relay sends to the SCADA information about the exceeding of the upper limit of current flowing through the windings. SCADA will then supplement the power delivered to the ERTS by connecting the adjacent ETSs in parallel;
- SCADA will withdraw any transformers with short-circuited medium-voltage or high-voltage terminals from operation;
- the maximum current relay monitors current circulation, the normal direction being from source to consumer. In case of reversal, SCADA will disconnect the transformer from the load, in order to protect the equipment or to prevent hazardous voltages on the primary winding of the transformer;
- homopolar protection together with SCADA will disconnect the secondary winding of transformers when supply circuits are interrupted through melting or sectioning of conductors;
- SCADA will monitor increases in absorbed current and decreases in voltage level due to load variations or equipment faults. If the load current increases SCADA will couple the transformers of adjacent ETSs in parallel

The SCADA system designed for ETSs should be capable to prevent these critical situations, to remedy them if at all possible, to alarm the points of surveillance, and to maintain a constant electrical power supply to the electrical traction railway system at the required level of quality.

5.2. ETSs optimization solutions employing SCADA:

- SCADA should be able to constantly and automatically modify the supply schema, ensuring the operation of MV transformers as close to the nominal load as possible and with a high power-factor;
- SCADA should isolate and quickly replace faulty equipment and that which can no longer perform adequately due to different reasons;
- SCADA should ensure the electrical energy is supplied to ERTS within the quality parameters required;
- SCADA should set the load curves by area, equipment and time interval, to prevent failures resulting from overloads;
- SCADA should allow for remote control and remote reading of parameters of the electrical energy supplied to ERTS

The designed and implemented SCADA system should allow for constant development; through the introduction of new monitored equipment and controlling of new parameters. Thus the system will become safer and will perform better, increasing the cost-effectiveness of ERTS.

5.3. For the circulation of high-speed trains, it is necessary for the SCADA system to implement new functions as follows:

- the signal will be set to "go" after the SCADA system has ordered the closing and supply of neutral zones on the route. If these portions are not supplied, the locomotive circuit breakers will trigger through the protection against power failure and will automatically brake the trains
- the SCADA system will display the indication of the next signal on the monitor of the locomotive driver. This ensures adequate decision time, knowing that signals are frequent due to the high-speed at which these trains are moving.
- on high-speed train locomotives, the SCADA system must control the locomotives automatically, reading and observing the significance of the signals in the field.
- the SCADA system will constantly transmit information about the position of the train to the traffic operator. Synchrophases are used to accurately determine the position and speed of the train's movement.

The SCADA system will be synchronized with the ground track equipment with the help of the GPS signal or with the help of the frequency of the electricity distribution network. For the correct interpretation of the information from the field, all the instruments that ensure the movement of the high speed trains will be synchronized to the same time base.

On the railway sections with high-speed traffic, a collector cable will be fitted to ensure the return of the traction return in ETSs, the isolated joints of the track circuits impede the flow of current on the return path.

6. Conclusions

The article presents ways of parallel coupling of MV transformers that supply ERTS. It describes the use of PLPs with automated control system for double track railway and telecontrolled transversal separators, for the parallel coupling of the groups of railway lines. Losses of electricity due to voltage drops and non-uniform distribution of electrical load are reduced.

Parallel connection of the ETS transformers ensures the continuous power supply of the ERTS, at the parameters required by the consumers, in special situations. Loads that can occur at certain time intervals are taken over by the adjacent ETS transformers engage in parallel. It also increases the stability of the power supply system due to the power supply available at all times. It can be considered at one time for an ETS that the adjacent ETSs are equipment in the reserve, which by their parallel coupling can substitute or increase power offered by ERTS.

Parallel coupling of the ETS transformers feeds each railway section at both ends, which leads to the diminution of the longitudinal voltage drops. The voltage level is kept evenly over the entire area fed. If the parallel coupling conditions are met, ETSs will work as close as possible to their nominal load with a high efficiency. Parallel ETSs make ERTS safer, enable the development of electric traction systems, respond promptly to the new demands required by rolling stock that is permanently upgrading. Monitoring of the critical situations for the railway infrastructure as well as the implementation of solutions for optimizing the operation of ETS with the help of SCADA meets the performance requirements of ERTS. The addition of new SCADA system functions that make it possible to switch to high-speed trains will lead ERTS to a superior level of development and performance. Increasing performance and streamlining ERTS remains a topical issue that concerns both power suppliers and rail traffic operator.

References:

[1] Kaudi A, Barisal AK and Sahu S 2017 A study of various traction transformers & active power compensator in co-phase traction systems *Energy Procedia* **117** *252–59*

[2] Milesevic B, Uglesic I and Filipovic-Grcic B 2016 Power quality analysis in electric traction system with three phase induction motors *Electric power systems research* **138** 327–38

[3] Popovici O 2008 *Tractiune Electrica (Electrical Traction)* (Cluj-Napoca, Romania: Mediamira Press) 25-26

[4] Firat G, Yang G and Ali Hussain Al-Ali H 2015 A comparative study of different transformer cannections for railway power supply-mitigation of voltage unbalance *Proc.of 10th International Conf. on Advances in Power System Control (APSCOM 2015)*

[5] Rohilla V 2015 Study of traction transformer and co-phase traction system *AREEE* 31–35

[6] Keenor G 2016 *Overhead Line Electrification for Railways 4th Edition*

[7] Popescu C 2011 Aspecte privind punerea în paralel a transformatoarelor electrice (Aspects of parallel connection of electrical transformers) *Analele UCB Târgu Jiu - Seria Inginerie (Annals of UCB Târgu-Jiu - Engineering Series)* **1** 127–137

[8] Chen TH 1994 Comparison of Scott and Leblanc transformers for supplying unbalanced electric railway demands *Electric power system research* **28** 235–40

[9] Hu S, Zhang Z, Li Y, Luo L, Luo P, Cao Y, Chen Y, Zhou G, Wu B and Rehtanz C 2015 A new railway power flow control system coupled with asymmetric double LC branches *IEEE Transactions on Power Electronics* **30** (10) 5484–98

[10] Scholl MV and Rocha CR 2016 Embeded SCADA for small applications *IFAC-PapersOnLine* **49** (21) 246–53

[11] Fratu M 2013 *Electrotehnica (Electrotechnics)* (Braşov, Romania: Transylvania University Press)

IOP Conf. Series: Materials Science and Engineering **789** (2020) 012054 doi:10.1088/1757-899X/789/1/012054

Protection of turbine halls of power plants from exposure to high temperatures in fire conditions

A I Pykhtin[1*], M V Tomakov[2], V I Tomakov[2], E I Anikina[3] and A V Brezhnev[4]

[1] Department of IT and new entrants, Southwest State University, Kursk, Russia
[2] Department of labour and environment protection, Southwest State University, Kursk, Russia
[3] Department of software engineering, Southwest State University, Kursk, Russia
[4] Department of informatics, Plekhanov Russian University of Economics, Moscow, Russia

[*] aipykhtin@swsu.ru

Abstract. There are numerous cases when fires in power plant turbine halls causes the collapse of roof metal structures. The use of conventional firefighting equipment in high-bay facilities has proved to be ineffective or unacceptable. In this situation fire robots are the most feasible option. Today the best variant is a fire robot that is based on a traditional fire monitor and equipped with infrared fire seat detectors. Some specific advantages of fire robots include their ability to operate under low visibility with high smoke generation conditions that are typical of a fire in a turbine hall. The program of roof structures cooling down is designed and tested in advance so that fire robots can perform their functions in automatic mode. Fire robots can be included into permanent-type fire safety systems that are designed taking into account specific construction features of a facility and temperature conditions in case of a fire. Such composite system is a multifunctional complex that is able to perform versatile fire protection functions. The most appropriate fire robot types to be used in such complex fire protection system are fire robots based on conventional fire monitors, as they have some notable advantages compared with mobile robots or androids. One of them is their ability to protect big areas starting from 5 and up to 15 thousand square meters with fire suppressing media flow rate range just 20-60 l/sec correspondingly. Water or foam can be delivered via air exactly to the fire seat instead of spraying over the total designed area. Moreover, wetting intensity can be maintained at a level that corresponds to fire seat intensity.

1. Introduction

High fire hazard in the turbine halls of conventional and nuclear power plants is related with the presence of such risky factors as insulation of power supplying and control cable; short-circuit events in power cables; oil spillage onto hot equipment areas; human errors during maintenance activities and test runs; the use of flammable materials in electrical devices and instruments. Flammable waste that results from maintenance activities on equipment components or lubrication lines may also present a serious fire hazard [1-4].

Statistically fires at nuclear power plants occurred in the main plant facilities (47% of the total number of fire events), in auxiliary facilities (21%), in storage facilities and open production plants (16% each), i.e. in the areas that have vital importance for a nuclear power plant performance. The majority of fire events were caused by a failure in the production equipment – 48%; lack of responsible care in fire hazard activities – 24%; violation of fire safety rules during hot works – 8%; violation of equipment operation regulations – 15%; and some other causes - 5% [5].

Turbine halls are the areas with the highest fire risks because of flammable oils used in turbo generator oil systems and flammable highly explosive hydrogen used in the turbo generator cooling system.

The structure of fire events at nuclear power plants is as follows, in %: electrical equipment – 7; switch-yard gear – 21; reactor department – 28; turbine halls – 30 [1].

One of the most frequent causes of fire in the turbine hall is a failure in the turbo generator set [6]. For example, about 70% of fires in turbine halls were caused by a mechanical break of a turbo generator component. Such event may occur with a gush of burning oil, hydrogen jet fire or burning of oil spillage covering a significant floor area. About 18% of events had hydrogen leaks with subsequent combustion and explosion that resulted in disastrous consequences and personnel casualties [2].

According to statistical data within 25 years period, 97 accidents including 31 fires, 2 explosions and 2 explosions followed by a fire occurred in turbine halls of nuclear and fossil-fuel power plants that had turbo-generator installations of 50 MW and more capacity in the USSR and abroad. In 22 events the roofs of turbine halls failed, which is 21% of the total accident number in that period. In 2005-2011 136 fires occurred at Russian fossil-fuel power plants including the events with turbine hall ceiling collapse [5].

2. Some fire cases in the turbine halls

On December 31, 1078, during the first decade of Beloyarsk nuclear power plant (NPP) operation (Russia), an accident occurred at the turbo-generator installation No 2 with turbine oil combustion and a fire in power supplying and control cables. The fire was caused by the loss of integrity in the oil line with oil burst and subsequent oil spontaneous ignition due to its spillage onto hot turbine equipment and steam line surfaces. Before the fire brigades arrived at the site, the turbine hall ceiling had collapsed with failed roof area exceeding 960 m^2.

On October 15, 1982 fire occurred at Armyanskaya NPP (Medzamor, USSR). The fire was caused by explosion of the generator at Unit 1. The fire destroyed the cables on the area of about 400 m^2 and caused serious equipment damage in the turbine hall on the area of about 300 m^2.

In 1984 at NPP Rancho Seco (USA) hydrogen exploded in the turbo-generator cooling system, which caused fire and steam generator failure with subsequent radioactive steam release.

In 2017 the French media reported [7] an accident that took place on February 9 at Flamanville NPP. There was an explosion in Unit 3 turbine hall that caused fire. The event was caused by a short-circuit that resulted in the explosion of a fan. Because of the fire turbo-generator 1 was disconnected from the grid and Unit 1 reactor had to be shut down. Flamanville NPP is an operating nuclear power plant that is located in the North-West of France in the North Normandy Region.

On February 9, 2008 in Ulan-Ude (Russia) one of CHPP-1 turbines got on fire. At first an electrical cable caught fire, then the flame extended onto the turbine. The turbine held about 17 tons of oil. Finally, the fire area spread up to 1000 m^2. The event included a partial failure of the turbine hall roof.

On November 5, 2014 at Tomsk fossil-fuel power plant (Tomsk, Russia) a serious fire occurred in the power switchboard room. The fire was caused by oil line rupture in the area of the first turbine support. As a result, oil spilled into the switchboard room, which caused fire on the area of about 50 m^2.

On January 4, 2015 a fire occurred at Surgut power plant-2 (Surgut, Russia) that caused the collapse of two roof spans in the turbine hall of its power Unit 4 with failed area amounting 1300 m^2 (figure 1).

Figure 1. Failed roof of the turbine hall at Surgut power plant unit 4.

The fire was caused by ignited oil in the turbine hall close to feed water pumps. Surgut power generating plant is the biggest fossil-fuel electricity producing plant and the second one in the world. Its installed capacity is 5597.1 MW, the capacity of combined cycle power units is 797.1 MW; installed thermal capacity is 840 Gkal/h. The plant is located in Surgut, a city in Khanty-Mansi Autonomous Area. The plant utilizes associated petroleum gas and natural gas as fuel.

On December 7, 2015 a big fire occurred in the turbine hall of Vasylieostrovskaya thermal power plant (Saint-Petersburg, Russia). The fire area amounted 400 m^2. The fire was caused by a violation of equipment operating regulations, which resulted in the loss of integrity in the oil system of the fourth turbine and a complete rupture of a feeding pipe. As a result, oil under high pressure got onto the turbine hot surface and ignited to cause extensive fire. The fire flame rose at a height of 30 m causing the failure of the turbine hall roof.

3. Specific features of fires in turbine halls

According to fire and explosion hazard classification a fire in a power plant turbine hall belongs to the highest risk category. For example, a failure of the generator shaft oil seal may result in high temperature (over 2000°C) jet flame of burning hydrogen and oil mixture. Spreading flows of burning oil in combination with high temperature of hydrogen flame present serious threat for the integrity and load carrying capacity of metal structures, roof trusses and columns of the turbine hall. In such situation the temperature of unprotected surfaces of metal structures can reach its critical value (500°C) within one minute, which means that within this time metal trusses and roofing will fail [8].

The amount of oil leak from failed turbine control and lubrication systems can actually be about several tons because the capacity of the oil tank may reach 10 m^3 and the system of high-pressure oil pipelines is extremely branching. Within the time period starting from an oil pipe break till emergency disconnection of oil supply the oil leakage volume can become very big with much oil draining into lower elevation areas.

Traditional equipment layout makes fire conditions with burning oil even more complicated because oil tanks, oil pipelines and oil pumps are located at the zero elevation where oil spillage may catch fire and start burning, while generators and turbines with their control and instrumentation systems are located at higher elevations of + 8 … +10 m, i.e. in the zone affected by fire flames. So, there may be cases when fire begins simultaneously at different elevations: at the turbine floor and at mid-floors. In case of an oil system fire, the fire area may depend on the type of equipment damage and its location. Maximal known fire spread rate is 25 m^2/min. [9].

Oil that is burning at mid-floors or at generator floor presents significant hazard for unprotected roof trusses, because flammable products of combustion rise up to accumulate under the ceiling and heat metal trusses up.

Fire risks can be reduced by implementing fire protection measures. Turbine oil tank, turbine oil coolers, oil pumps of the generator shaft seal and hydrostatic rotor lift, oil pumps of the turbine oil supply and control system, generator and turbine surge tanks are protected by water spray systems that can be controlled both remotely and manually. However, the analysis of fire cases has shown that such measures may be either insufficient or inefficient.

4. Performance of metal structures under fire impact

In case of fire impact, the performance of building structures is characterized by the following processes:

- steel resilience and strength are reduced;
- steel becomes ductile with developing creep effects;
- as structures are heated up their inner internal stresses (such as bending and tensile stress) get redistributed;
- because structures are heated up unevenly, significant thermal stresses are developed therein, which can be compared with extra load impacts;
- eventually cross-section areas with developed limit conditions become inoperable while stresses in load-carrying elements are increasing; and
- plastic centroids are formed in the affected cross-sectional areas.

Metal trusses have an essential disadvantage as in case of fire they heat up faster, which causes their extensive destruction under weight loads at early fire stages.

As is shown in [10] theoretical and experimental studies of the performance of metal structures in fires have demonstrated that when metal trusses that are experiencing rated static loads are heated up over 500°C, they get deformed, lose their strength and collapse. That is why the temperature of 500°C is considered to be the limiting heating state. With heat source areas of 80, 50 and 35 m^2 the limiting heating state of a truss (with its subsequent deformation and collapse) is reached within 5, 9 and 16 minutes starting from the fire onset correspondingly.

Load carrying structures that have been damaged by fire may be abnormally deflected or fail completely. Considering multiple connections of building structures even a local failure can result in disastrous consequences for the building on the whole. That is why in case of fire such structures have to be cooled down. Therefore, one of immediate fire fighting measures in a turbine hall is to protect its enclosure and roofing against heat impact by discharging water from hydraulic monitors.

5. Standard methods of protecting metal structures

In engineering practice drencher or sprinkler systems are used to protect metal structures against high temperature impacts. However, they may be inefficient to use in turbine halls due to the following deficiencies [5, 11]:

- sprinkler system pipelines and sprinklers should be fixed directly on the trusses that support roof structures. It increases the loads onto the trusses and columns, especially when the system is filled with water, which may result in complete utilization of the load bearing capacity of these structures;
- sprinkler and drencher systems are usually hard to maintain and test because of their pipelines range and high positioning of piping and sprinklers;
- it is hardly possible to install a fire protection sprinkler or drencher system at existing operating power plants because it may overload the trusses beyond design based values;
- in case of fire the pipelines of sprinkler systems filled with water can be heated intensely with steam formation inside the pipeline, which may result in internal pressure growth and pipe rupture, or cause excessive actuation of sprinklers;

- both vertical and horizontal pipelines located at a significant height have low explosion resistance. In case of explosion one of main feeding or discharging lines can be easily damaged with loss of integrity, and that will disable a total section of the system.

Fire suppression agents applied onto the surface of metal trusses can also increase weight load, because the protected area may be rather large. Moreover, it may be very difficult to apply a fire suppression agent onto the structures of an existing building of an operating facility [11].

Thus, the most feasible method of protecting metal structures in an operating turbine hall is their cooling down by means of water discharged from hydraulic monitors. In this case it is very important that each point on the metal truss should be wet down by two dense water jets. Hydraulic monitors are suitable to focus high-rate water supply on the required area with less consumption of fire extinguishants.

After analysing the possibilities of using fixed manually operated hydraulic monitors in NPP turbine halls it has been found out that they are rather limited due to the following disadvantages:

- manually operated monitors require human presence in the fire area where the operator is exposed to such hazards as open flame, high temperature, toxic combustion products and smoke, as well as to the risks of building structure collapse and even radioactive releases;
- it is rather difficult to aim water jets precisely onto metal structures, because the pointing has to be done visually under heavy smoke formation conditions considering that according to available data the combustion of just 8 kg of turbine oil reduces the reach of sight down to 3 m at the elevations above 7.4 m [11];
- as roof trusses can fail very quickly, water discharge should begin immediately after fire has been detected. However in case of fire the operators in the turbine hall must take prompt actions to protect equipment and may not be able to operate several monitors at the same time; moreover, manual operation of monitors may require additional human effort of fire fighting detachments, which means that people will have to work in extremely hazardous conditions related with open fire, high temperatures, combustion products and smoke with taking risks of enclosure collapse or radiation explosion.

6. Modern fire fighting systems and advanced methods of protecting metal structures against high temperature impacts

Since lately fire fighting robots (FR) have been developing very quickly [12, 13]. They are based on remotely operated hydraulic monitors of permanent-type (figure 2). Their use ensures effective cooling down of metal trusses and guarantees fire resistance of structures within 30 min. This time is sufficient to isolate the fire and suppress it. Practical fire fighting experience has shown that operation time of hydraulic monitors in an NPP turbine hall can range from 8 to 20 hours.

Figure 2. Fire robot (left), discharged water jet (right).

A fire robot generates water or foam jet and discharges it directly into the fire hot spot located at a significant distance. FR arm tool is a branch-pipe with a nozzle; water pressure is controlled automatically at a rate from 8 to 80 l/sec, which ensures the most optimal jet parameters. The robot can generate different jet types starting from a direct continuous jet and finishing with water screen. Water leaves the nozzle in the form of finely dispersed particles, so called Jet Fog (JF), with particle sizes ranging from 50 to 400 mcm. Due to high kinetic energy of big droplets the jet is able to reach a burning surface directly through the fire flame and deliver finely dispersed water particles that have high heat radiation absorption ability into the fire seat. Dosed water supply guarantees desired wetting intensity. Thanks to this performance feature the robot can respond to different fire development scenarios and ensure maximal capacity of feeding fire suppressing agents to the fire hot spot at an early fire stage.

FR is designed to vary the jet direction horizontally in 360° range and vertically in 180° range, thus covering the total space in its operation area. Jet angle can also be changed from 0 to 90°, so that different jet profiles are generated. One robot with water flow rate capacity of 20 to 100 l/sec can protect an area of 5000 - 15000 m^2. For example, a small-size fire robot operating at a water flow rate 20 l/sec with jet supply distance 50 m is able to protected an area over 7500 m^2 ($\pi R^2 = \pi 50^2$). The total amount of fire suppressing agent can be fed into the fire hot spot with wetting rate of about 1.2 l/(sec×m^2) on an area of 12 m^2, which ensures quick fire suppression at an early development stage. Thanks to dosed feed the wetting strength corresponds the heat rate of the fire hot spot.

To supply water to fire equipment only one main pipeline is used; it has no distribution network that is typical of sprinkler and drencher systems.

All parameters of the fire robot meet the requirements to automated firefighting equipment: it has a feature of automatic fire alarm; is able to fix hot spot coordinates and can be actuated automatically to start fire suppression either with water or with foam, if needed; it can also be used to cool down metal structures.

Compared with remotely operated hydraulic monitors, FR has machine vision function that is realized by an IR sensor with a scanner and a TV camera, and machine intelligence that corresponds to the level of performed tasks, including object recognition, target location and pointing, etc.
Fire hot spot detection sensitivity is 0.1 m^2 at a distance of 20 m; response time is several seconds and this time is enough to determine the fire size in 3D coordinate system.

A specific feature of fire robots is their ability to function under heavy smoke conditions with low visibility, which is a typical fire situation in turbine halls. Sophisticated trusted program of cooling the enclosure structures down is developed and tested in advance. Fire robots have wide operation range and high positioning precision. For example, the program can set up complex paths of the monitor nozzle motion and the patterns of water jet travel along the roof trusses of the turbine hall taking into consideration ballistic laws.

As the time period before trusses can fail is rather small, water discharge should start immediately after fire has been detected. Considering this requirement it is advisable to use fire robots united into a robotic firefighting system (RFS) controlled by a common software system [12, 14, 15].

RFS also comprises fire alarm system and TV monitoring system that are used to detect fires and send alarm signal to the security control room (fire point) and generate the signal to actuate fire suppression equipment and on-line monitoring of the situation in the fire hot spot area. Fire robots are connected via mains with network controllers and control devices. All information on firefighting progress is registered by means of TV cameras and the protocol of action development succession. During on-duty period the system operates in self-testing mode and informs operators about any adjustment needs by using a pre-set address, thus maintaining its functional readiness.

RFS is a multifunctional system that is able to accomplish different fire safety and fire protection tasks, such as fire detection and suppression with synchronous sending fire alarm signal and data on the current status of equipment to the plant security station (fire point room). RFS with water cooling unit can be used to cool down load bearing construction elements and protected equipment components that are located close to the fire hot spot.

Fire alarm system and TV monitoring system make a part of RFS and are meant to detect fires at early stage, generate fire alarms and fire fighting equipment actuation signals. They are also used for on-line monitoring of fire development conditions.

In the stand-by mode (alert mode) RFS water pipe system is filled with water up to the disk valves of FRs. The water is supplied either from fire suppression pump station or from utility-firefighting water pipeline under design pressure of 650 kPa. Foam maker pipeline is filled up to solenoid valves.

If addressable fire detectors of flame or addressable loop failure (in case of short-circuit or wire rupture) get actuated, light alarm window will be actuated on the control device with loop number identification and simultaneous actuation of internal audible alarm. Audible fire alarm and loop failure alarms have different sound tones. Multiple interface units will generate fire alarm and forward it to the RFS interface unit to launch fire fighting program. Following a preset program the RFS interface unit selects fire robots to be actuated, specifies fire hot spot angular coordinates and chooses fire suppression media with its discharge mode.

RFS has three operation modes:

- remote;
- self-acting;
- automated.

The first variant can be combined with the second or the third modes, in which case the operator can adjust RFS operation pattern so that to increase its fire fighting efficiency and ensure better cooling of building structures.

The second mode should be used for fire protection purposes in the absence of duty personnel.

As distinct from the second mode in the third one such RFS operations as searching the fire seat, opening disk valves and actuation of RFS interface terminals should be authorized by the operator. This mode is recommended as a regular method of the system operation in duty personnel's presence.

If needed the fire fighting system can also be operated manually.

To cool down the building structures that are located close to the fire seat maximum two fire robots should be activated.

6.1. Remote mode of RFS operation.

The system is operated remotely from the control board (RCB) that is connected either with FR connection box or with RFS interface terminal, or radio control unit (RCU) that is operating within a radio signal range. In this mode the system can perform the following commands:

- select FR(-s);
- open/close disk and solenoid valves;
- point FR (control its horizontal and vertical motion);
- set FR travel speed (8 rates);
- change the jet angle;
- set up program mode (spherical rectangle line-by-line scanning) and register program mode parameters in the FR non-volatile memory – 8 programs;
- start up/stop the program mode; and
- set up FR motion limits in vertical and horizontal planes.

The following information on the current state of the controlled FR is displayed either at RCB or RCU:

- set up FR speed;
- current magnitude of the operating electrical drive;
- water pressure;
- operating data of electrical drives;
- current state of disk valves (open/closed);
- emergency alarms.

6.2. Self-acting mode of RFS operation.

The decision about RFS operation in automatic mode shall be taken by the duty operator at the permanently manned control centre. If within 5 minutes after receiving a fire alarm the operator issues no control signal, fire suppression will start automatically.

Visual control is performed by means of a TV camera that is installed on the fire robot and can be pointed at the fire seat together with the FR to transmit video data on the current situation in the protected area to the operator's display. RFS program ensures the following mode of automatic operation of a FR with fire seat detection function:

- at addressable fire indicator actuation the control-and-indicating unit sends the signal "Fire" to the RFS interface with identifying the number of the actuated indicator;
- upon this signal the RFS interface generates control signals to point minimum two respective fire robots at the identified zone;
- upon entering the identified zone, the fire seat searching program is launched and fire detectors send seat location coordinates to RFS interface;
- upon receiving the signals from two FR the RFS interface determines fire seat coordinates in 3D system and forms fire suppression program;
- at RFS start up for fire suppression the RSF interface forms the following instructions:
  - to disconnect production and electric equipment (if needed), disable ventilation and actuate public fire alarm system;
  - for the pump control cabinet to start up pumps according to a specially designed program;
  - to open disk valves and solenoid valves of the relevant fir robots;
  - to start up FR;
- FR control unit controls the pressure in the RF at the designed level by controlling disk valve position;
- minimum two fire monitors start fire seat suppression.

When the distance to the fire seat is relatively small (up to 15 m) the fire is suppressed by water spraying at a preset angle; at a bigger distance line jets are applied airy.

7. Conclusion

Fire robots have enjoyed extensive use since early 2000-ties. They are one of the most reliable firefighting means, because this equipment is started following actual conditions in the protected area and are able to suppress fire without human involvement. Today fire robots are widely used in different industrial and agricultural spheres. For example, fire robots are used to protect turbine hall of fossil-fuel power plants, for instance, they have been on duty at Petrozavodsk thermal power plant since 1997. Automated stations based on robotic fire suppression systems have already been implemented to protect high-bay building structures at dozens of facilities in different Russian cities and outdoor facilities, examples are sports and concert centres, wood processing factories, refineries and chemical plants, plane docks and major oil storage sites.

8. References

[1] Mikeyev A K 2000 *Fires on radiation-dangerous objects. Evidence. Summary. Recommendations* (Russian research Institute of the Ministry of internal Affairs of Russia, Moscow)
[2] Tomakov V I, Tomakov M V and Petrukhin S L 2012 *News of Southwest State University. Series: Equipment and technologies* **2** 170-176
[3] Ishchenko A D 2016 *Fire and explosion safety* **5** 26-36
[4] Aleshkov M V, Tsarichenko S G, Kholostov A L and Gusev I A 2018 *Fire and explosion safety* **9** 35-49
[5] Tomakov V I, Tomakov M V, Pahomova E G, Andrienko V V and Pashkova M E 2018 *News of Southwest State University. Series: Equipment and technologies* **1** 67-80
[6] Belov V V and Pergamenshchik B K 2013 *Proceedings of the Moscow state University of civil engineering* **4** 61-69
[7] An explosion occurred at a nuclear power plant in France *https://lenta.ru/news/2017/02/09 /franceexplosion/*
[8] Vakhitova L N and Chepovskii V O 2010 *F+S: safety and fire protection technologies* **1** 62-66

[9] Mikeyev A K 1990 *Fire protection of nuclear power plants* (Energoatomizdat, Moscow)
[10] Strakhov V L, Davydkin N F and Kaledin V O 2011 *Fire safety in construction* **2** 20-29
[11] Gorban Y I and Sinelnikova E L 2011 *Security algorithm* **3** 32-36
[12] Gorban Y I 2010 *Security algorithm* **3** 66-71
[13] Fire-fighting robots and cannon fire fighting equipment is the basis of modern and effective technologies extinguishing *http://firerobots.ru/ru/press-center/info/item_9479.html*
[14] Sinelnikova E A, Sleptsova I N, Gorban Yu I and Gorban M Yu 2017 Fire safety **1** 60-66
[15] GorbanY I, Gorban M Y, Tsarichenko S G and Sinelnikova E L 2007 *Fire safety* **1** 105-109

A study into prevention of antibiotics resistance through enhanced building infrastructure design and management strategies

C N Rolfsen[1*], P O Rønning[2], J L Bull[2] and G C Chițonu[3]

[1] Department of Civil Engineering and Energy Technology at The Faculty of Technology, Art and Design (TKD) at Oslo Metropolitan University - OsloMet
[2] Department of Mechanical, Electronic and Chemical Engineering at The Faculty of Technology, Art and Design (TKD) at Oslo Metropolitan University - OsloMet
[3] Department of Civil Engineering, Faculty of Civil Engineering, Transilvania University of Brasov

* crolfsen@oslomet.no

Abstract: This study is part of a larger research project, which looks at how modern hospitals are built in Norway, and what can be done to make hospitals more user friendly and purpose built. It looks at how medical facilities can be made to encourage a quicker improvement in the patient's recovery, and what effect the way modern hospitals are being built and operated, has on the patients, the employees and to the visitors to the health facilities. A key focus of the research is to look at nosocomial infections (hospital-acquired infections) from medical care and to study the prevalence of antibiotic resistance at these facilities. The project aims to find conclusions and areas of improvement along larger parts of the hospital and patient chain. The greater research project aims to look at this scientific field through studies in 3 countries, in Northern Europe, Eastern Europe and Central Europe, to get an idea of the inter-European challenges in a very complicated scientific field, with different financial resources, cultures, focuses, and approaches to the frequency of use of antibiotics.

1. Introduction

This article is related to a larger research project ("Hygiene-Intensive Hospitals"), which looks at how modern hospitals are built in Norway, and what can be done to make hospitals more user friendly and purpose built. It touches on a few aspects which are relevant to the session topics for the International Scientific Conference. The articles main theme is a look at "New materials and technologies in building industry", in this case related to medical facilities. Additionally, it also touches on another important subject related to efficiency and technological improvements for medical facilities, which lies in the ventilation systems.

The greater research project has in its initial phase focused on Norwegian hospitals, but the intention is to look at challenges in Eastern Europe with our Romanian partners, and study the challenges faced in a third Central European country to be selected at a later stage. This approach has been chosen in an effort to get an idea of the inter-European challenges in a very complicated scientific field, with different financial resources, cultures, focuses, and approaches to the frequency in use of antibiotics.

A key focus of the research is in nosocomial or hospital-acquired infections (HAI) from medical care, and to look at the prevalence of antibiotic resistance in medical facilities. This may seem like a broad field to approach, but it is the hypothesis of the research group that large benefits in reducing antibiotic resistance can be found in looking at the problem from a broad perspective. This will allow for identifying areas of improvement along larger parts of the hospital and patient treatment chain. The research project also has a key focus on how the building mass used for medical purposes, and the way modern hospitals are run, affect the patient's recovery, and the chance of survival. It looks at which improvements can be made to reduce infections, how to aid the quick and successful recovery of the patients, to improve the workdays for the staff, and to look at how improvements can benefit the hospitals and their operating economy.

2. The modern Norwegian medical facility building mass

For a greater overview of the building mass of Norwegian hospitals it is necessary to generalize a bit, and for the purposes of this article we will divide the medical facilities into 4 categories:

- general practitioner's facilities
- private and public health clinics
- smaller hospitals
- big hospitals

These facilities serve different purposes and their exposure to germs, viruses, microbes and bacteria, including antibiotic resistant strains, all differ to some degree. However, they all have a common enemy and interest in reducing the frequency of this exposure. In the Norwegian health system, the general practitioners are the front line of the medical treatment. The GP system is alongside the private and public health clinics the first medical professionals that patients turn to for non-emergency room treatment. These GP-facilities experience dealing with the full width of the population, germs and diseases. They serve the function for the health system as very important evaluation stations for the medical treatment chain, to assess the medical situation of the patients, and to determine whether the medical issues can be treated at their facilities by the general practitioners (GPs), or whether the patient needs to see a specialist or go to a hospital. Despite getting the broadest patient groups, and a broad exposure to medical issues and germs, sadly these general practitioner's offices are usually not built for the purpose they serve. They usually have the least specialized systems, assets and methods available to deal with the challenges. The general practitioners are in many cases able to determine whether the specialists or the hospitals need to take special measures during evaluations and treatments of patients, so they form an important line in the fight to protect the health system from extra complications. As for the building mass and equipment, it is difficult to keep non-purpose built buildings and equipment hygienic, and all indications are that there is ample scope for improvement of these medical facilities. Cleaning of them is limited to normal routine hygiene tasks, the ventilation systems are rarely dimensioned for the tasks, the facilities are never disinfected with pressurized droplet systems (aerosols), so that the disinfectant go in to the difficult to clean spots of rooms, technical installations, furniture and so on.

For many patients including immune suppressed patients the general practitioners office is a place of fear of what you may contract while waiting for the doctor, and it is a well-known fact that diseases spread easily at the doctor's office. The consequence is that you may come home with an additional problem to the one you went to the doctor with in the first place. In Norway like in many other countries the precautions one finds in hospital settings are usually only applied to GP offices during pandemic influenzas or when patients have been reporting with highly infectious diseases. In Canada the College of Family Physicians (CFPC) advised the GP or family doctors as they are referred to in the article to enforce tougher infection control guidelines to stop the pandemic outbreak of influenza. The article "Infectious risks in family doctor's offices" by Emily Senger advises that this practice should be kept up, continuously [1]. Infectious diseases of course spread all year round, and they are not limited to when there are very easily transferred and contagious diseases like influenza.

The general practitioner system has limits in which medical cases it is able to treat, and how quickly it can provide access to doctors. Many doctors have 1250 to around 2000 patients on their patient list, which means that there can be significant waiting times. In many cases the patients will either evaluate the situation themselves and see that they need treatment that the GPs cannot provide, and that they need to go to emergency rooms at private or public health clinics. Alternatively, they will they choose to go to a GP or a specialist at a private health clinic instead of their assigned GP.

These health clinics face a wide array of possible infectious cases, and they may often be dealing with open fractures, open wounds, surgeries and other complicated cases where the risk of dangerous infections is considerable. The medical facilities are usually either purpose built, or have had alterations so that they have to a degree been designed for the purpose of being medical facilities. They usually have more advanced systems for hygiene and prevention of infections.

The next chains in the medical treatment line are the smaller and bigger hospitals. They face most of the same challenges, they are to a large degree purpose built, they have good systems in place, and usually knowledgeable staff who works to prevent infections. However, the age of the hospitals, the lack of financial resources to maintain and upgrade them, make individual hospitals very different between the best designed and the older and more average designed hospitals. As the hospitals handle the most seriously ill patients, the most complicated medical procedures, and has the patients under their intensive care for the longest period of time, they face complicated infections issues, and carry great risks for the transfer of secondary infections beyond the initial medical problem. In many cases it is close to impossible to do something about this, and certainly with the resources available. But it is the belief of medical professionals and researchers, that improvements can be made so that resources can be prioritized to minimize the risk of complicated and expensive medical problems that will have to be dealt with later. As it is, HAIs cost the health systems billions of Norwegian Kroner a year according to available research, but the general research data in this field for the Norwegian health system seems to be lacking significant parts of the greater picture. Some studies have been performed on what factors lead to HAIs, and both the building mass, the routines, methods, technical equipment and other factors all come into play. To the research group's knowledge there is room for a substantially improved research basis and data to find the differences, to find best practices, to find general improvements for all hospitals, and to find individual improvements for specific hospitals. Some researches indicate differences in the percentage prevalence of HAIs based on the size of the hospitals, and this is an interesting field that will be looked at, as there are many complicating factors involved [2].

The large numbers of hospital-acquired infections indicate that there should be significant room for reductions throughout the entire health system. There is a desire for this, but good intention, will and discussions must be follows by appropriate actions to make a change. This research project is working to engage all stakeholders in finding practical solutions to how we as a unified group can help each other reduce infection risks. We will make some proposals for improvements and solutions in the next part of this article.

3. Toilets / bathroom facilities

A general public health related weakness with buildings in society in general, are the restroom and bathroom facilities. In hospitals these facilities bring due to the accumulation of bacteria, viruses and germs a considerably greater chance of spreading diseases. Studies have shown that germs in feces can be propelled into the air when one flushes the toilet. Putting the lid down before one flush the toilet, leaving the bathroom stall where you have just flushed as quickly as possible, so that the amount of time you are exposed to germs is reduced [3]. Washing one's hands thoroughly for at least 20 seconds, and being careful with touching the sinks afterwards, are small and important steps that can be taken to reduce infection risk. Studies at the University of Arizona in Tucson found that "sinks are the greatest reservoir of germ colonies in restrooms, thanks in part to accumulations of water that become breeding grounds for tiny organisms" [3].

Research and interviews with medical professionals show that there is a great need for providing better education about bathroom hygiene in general. For hospitals it could be an idea to provide information material on how to take precautions to help the staff in their work, and to help society fight a common enemy.

For hospital wings, specifically for gastro-intestinal diseases, the use of toilets that flush and self-clean automatically, and faucets and showers that are triggered by motion sensors can be good uses of new building technologies. This could have a significantly positive effect in reducing nosocomial infections or hospital-acquired infections (HAI). These are simple and not too expensive improvements that can be utilized in large sectors of the building mass.

Better information material and education of general hygiene and health principles can greatly reduce the spread of infectious diseases. One area where improved education can quickly lead to reduced spreading is in prevention of urinary tract infections (UTI). According to Vyas et al. "Urinary tract infection (UTI) is the most common infection experienced by humans after respiratory and gastro-intestinal infections, and also the most common cause of nosocomial infections for patients admitted to hospitals indeed UTIs are the most frequent bacterial infection in women" [4].

Urinary tract infection (UTI) is one of the most important causes of morbidity in the general population and is the second most common cause of hospital visits [4].

This is an area where preventive measures can be made at any modern health and medical facility, and which in turn can lead to significant reductions in nosocomial infections, and financial expenses.

4. New building technological methods and materials

In addition to use of automatic faucets, toilets that can flush themselves, use of automatic doors, automatic soap and disinfectant dispensers and other simple measures, there are a number of changes that could be made when refitting or building new hospitals. Several hospitals in Norway use pneumatic tube postal systems to make sure blood tests, medicines, blood for operations and critical materials arrive quickly where they are needed. Although not a new technology, having been introduced in the 18th century, this efficient and time saving system is considered as a part of the modern automation of hospitals in Norway in the coming years. It is set to be included in the new hospital being built at Ulandhaug in Stavanger, Norway and is in use in at Ahus and Kalnes Hospital. Increased use of automation and robotics can allow for reduction in man hours required for internal transport, maintenance and operational tasks, freeing up employees to spend more time looking after the patients. Reducing the human handling of food, bed linen, dirty washing and waste can lead to the staff being less exposed to germs which they may bring with them throughout the hospital and into contact with patients.

Another key area for improvement is in the materials used for the various functions of the medical facility building mass. Most hospitals are built with significant weaknesses in the form of porous materials, and one should make more of an effort to reduce the amounts of hiding spaces and possible reservoirs for germs. This is a key area of the research, with the research and data collection in initial phases at the moment. Several research studies over decades have suggested that aesthetics can have positive health effects, and this study will look at the lessons learned and how this perspective can be applied in modern medical facilities. Many GP's offices, health clinics and hospitals have an eye to this field in their focus on art, gardens, plants (or plastic flowers) and so on, but a more holistic approach can probably be taken.

5. HVAC in medical facilities

Another key area of hospital design and operations is in the Heating, Ventilation and Air Conditioning Field (HVAC). As stated by major HVAC designers: "Hospitals are complex environments that require special HVAC system design, maintenance and repair considerations. HVAC systems are responsible for keeping indoor air quality (IAQ) high and providing a safe temperature for patients and staff" [5]. Furthermore: "Hospitals need a functional HVAC system to stay operational. This is especially true for places like a clean room or an operating room (OR)" [5].

The air quality in operating rooms, intensive care units, neonatal wards and isolation rooms etc. must meet several strict requirements. The air must ideally be free of pathogens and have a stable temperature and constant humidity. Furthermore, the air currents supplied by the ventilation system should have a sufficiently low velocity in order to avoid resuspension and recirculation of particles and microbes throughout a room [6]. In addition to the hygienic aspects, effective ventilation will provide a better working environment for the medical staff as well as aiding in the recovery of the patients. It is well documented that good in-door air quality has a positive effect on patient recovery time [7]. Hygiene guidelines for the planning, installation and operation of ventilation and air condition systems in health care facilities are found in the DIN 1946-4-12-2008 standard published by the German Society for Hospital Hygiene. In this guideline, the HVAC recommendations for different classes of rooms in medical facilities are given.

As previously stated, proper ventilation and temperature control is of vital importance in medical facilities. To prevent the spread of airborne microorganisms and pathogens, efficient air filtration systems must be in place. These systems must be able to filter out microscopic particulates and aerosols that may otherwise cause contamination to aseptic areas and/or compromise the recovery of patients. For vulnerable patient groups, like immunosuppressed patients, exposure to airborne viral or bacterial pathogens may result in life threatening complications. High efficiency particulate filters for use in medical facilities and clean rooms are classified according to the EN 1822 standard (EPA, HEPA or UPLA filters) [7]. For medical facilities, choosing the appropriate filter class to ensure enough filtering efficiency and minimal filter penetration and by-pass is essential in preventing the spread of airborne pathogens and other contaminants.

Attention should also be paid to air current patterns and air current velocities within a room. Ventilation systems supplying high velocity air currents may cause settled dust particles and microorganisms on surfaces to be whirled up and resuspended in the air. This will cause a recirculation of pathogens within the room, exposing patients to new infection hazards.

Temperature control is important for the comfort of both the hospital staff and patients. Controlled room temperatures will aid patient recovery and provide a safer working environment for the staff. Furthermore, modern medical equipment and instrumentation require efficient temperature control within certain limits to function optimally. As an alternative to ventilation systems based on conventional HVAC principles, we propose the use of thermal heating and cooling systems. While air condition and other systems using heavy duty fans may produce strong drafts and uneven temperature distributions, thermal systems can produce a more stable and evenly distributed heating/cooling effect in a room. Heating and cooling may be based on radiant heat panels (i.e. heat exchangers), where the heat can be supplied by the circulation of warm water from solar thermal collectors [REF needed]. Cooling is achieved by injection of cold water into the circulating water to give temperatures above 15-16 °C.

There are many ways of incorporating thermal heating and cooling devices structurally into a building. In several countries, like Spain, Australia and the UK, active chilled beams HVAC systems are in use [7]. Other technical solutions may be based on waterborne ceiling, wall or underfloor heating. In general, the use of solar thermal collectors and waterborne heating/cooling devices will provide significant savings in total energy costs compared to conventional HVAC systems. An attractive alternative to systems based on solar thermal collectors is geothermal heating and cooling. In Switzerland and Sweden 75% of new homes have geothermal heating and cooling installed [8]. Geothermal energy can generate energy cost savings up to 72% compared to conventional electric resistance heaters [9]. Local climate factors will of course be a deciding factor in which technical solution that is most suited

Finally, we must stress the importance of properly implemented maintenance and cleaning routines of HVAC systems to prevent system hazards and spreading of pathogens. Changing air filters regularly, cleaning coils, fluid reservoirs and piping when necessary and implementing suitable cooling tower maintenance procedures are all vital parts of the running maintenance. Along with precise air quality monitoring systems, this will ensure that the HVAC system is geared towards achieving optimal hospital hygiene, staff and patient comfort [10].

6. Aims and goals of the research project

The research output aims to come up with proposals for improvements in health care and draw conclusions on how we can reduce the number of nosocomial infections and reduce the risk of antibiotics resistance in our hospitals and medical facilities. The goal is to work with the hospitals, the medical staff, the regional and national health authorities, the patient interest organizations and the patients themselves to come up with concrete proposals for reducing the cases of hospital infections. It is a key hypothesis of the researchers that the current building mass plays a significant role in the problem, and we will aim to make suggestions and concrete proposals for improvement in this. BIM, CAD and computer modeling alongside non-invasive and minimal invasive analysis techniques will play a key part of the research.

7. Conclusion

For this article we will focus on the conclusions to date of the research related to the topic sessions which are:

Improve the access to information materials and improve the health education and increase hygiene awareness in the medical and general communities. This is a simple, highly efficient and cost-effective way of reducing nosocomial infections at medical and health facilities. This method can be utilized at any modern facility and is easily implemented in the prevention chain.

The use for new technologies like motion sensors for faucets, automatic doors, automatic soap dispensers can reduce the transfer of germs. Robotic technology, pneumatic tube systems, more appropriate materials to fight germs, or to deny them hiding spaces are important ways in which modern hospitals can fight nosocomial infections. The use of new building materials with non-porous surfaces may reduce hiding spaces for bacteria, and can help in preventing resistance in hospital settings. The need for appropriate HVAC solutions is important in reducing spreading of germs and bacteria and has potential for considerable reductions in suffering and expenses.

Making sure that the quality of the ventilated air is appropriate, and performing systematic maintenance of the HVAC system is another area where the spread of diseases can be reduced compared to more traditional hospitals and medical facilities.

The authors believe that there are substantial steps that can be made to prevent the spread of antibiotic resistant bacterial strains and other pathogens throughout the hospital facilities. It is the opinion of the authors that several of these improvements can result in both significantly reduced medical complications, suffering of patients and in operating costs for running the medical facilities.

References:

[1] Senger E 2011 Infectious risks in family doctor's offices *CMAJ.* 183(2) p 1-2
[2] Consumer Reports 2015 How Your Hospital Can Make You Sick *America's Antibiotic Crisis* July 29
[3] https://www.webmd.com/balance/features/what-can-you-catch-in-restrooms#2
[4] Vyas S, Varshney D, Juyal R, Nautiyal V and Shrotriya VP 2015 An Overview of the Predictors of Symptomatic Urinary Tract Infection Among Nursing Students *Annals of Medical & Health Sciences Research* **5** (1)
[5] https://donnellymech.com/blog/importance-hospital-hvac-repair-preventative-maintenance/
[6] Balaras C A, Dascalaki E, Gaglia A, 2007 HVAC and indoor thermal conditions in hospital operating Rooms *Energy and Buildings* vol. **39** Issue 4 April 2007 p 454
[7] https://www.trox.no/downloads/e7cf3a7a426ce635/TROX_life_oct2013.pdf
[8] https://www.renewableenergyworld.com/2016/02/24/the-hidden-genius-of-geothermal-hvac-systems/
[9] https://www.energy.gov/energysaver/choosing-and-installing-geothermal-heat-pumps
[10] https://www.speedclean.com/news/how-hvac-keeps-hospitals-healthy/

Tests upon Savonius turbine and its usage in street lighting pole

T Rus[1*]

[1] Technical University of Cluj-Napoca, Faculty of Building Services, Cluj-Napoca, Romania

[*] tania.rus@insta.utcluj.ro

Abstract. The tendency nowadays is to reduce the consumption of conventional energy and to efficiently use renewable energy resources. Thus, the energy independence of the different consumers is tried, along which is the street lighting. There are many applications where street lighting is powered by renewable energy, using photovoltaic panels. This article presents a different solution for powering street lighting, with the help of wind energy, through Savonius turbines. A Savonius wind turbine with the overlap ratio of blades of 0.35 is tested, in order to determine its actual power coefficient and power extracted from the wind. According to the tests, a power coefficient of 0.3 was obtained for wind speeds of 4 and 5 m/s and the maximum power at a wind speed of 7 m/s. With the values obtained, for this type of turbine, different configurations of Savonius turbines have been dimensioned to supply an energy-efficient street lighting pole.

1. Introduction

One of the current trends regarding public lighting is to ensure the independence of its electricity supply from the public electricity grid, this being achieved by using renewable energy (in most cases with solar energy through photovoltaic panels) or hybrid solutions. This is not possible if high-pressure sodium lamps, which are the most widespread, are still used for street lighting, because they have a very low efficiency (low lumen/W ratio). The best solution for the replacement of high-pressure sodium lamps, to ensure the lowest energy consumption, achieved with the condition of preserving or even improving lightning quality are LED lamps. Their low consumption makes the solution of energy supply from renewable sources much easier to implement, thus being able to ensure the independence of these installations.

Wind energy is not as widely spread as solar energy in powering street lighting consumers. The energy contained in wind can be transformed into electrical energy through wind turbines.

Since the beginning of wind energy conversion technologies development, installations of different types and forms have been created and developed. Thus, wind turbines can be classified according to several criteria, but the most important ones are based on rotors operation principle and those that refer to their constructive form [1].

Considering the aerodynamic principles on which the operation of wind turbines is based, they can be divided into two categories, namely, wind turbines whose rotors use exclusively the effect of aerodynamic lift, also called "lift-type rotors", and wind turbines whose rotors use only the aerodynamic drag effect to create a rotational movement, also called "drag-type rotors", category in which Savonius turbines are included [2].

However, in most cases the wind turbines are classified according to their constructive form, and in this situation, the most convenient classification is that according to the position of the rotor axis. Depending on its position, wind turbines are divided into two categories: Horizontal Axis Wind Turbines (HAWT) and Vertical Axis Wind Turbines (VAWT - Vertical Axis Wind Turbines) [3].

The most used type of turbine for electricity production is the turbine with horizontal axis with propeller rotor. The main advantage of this type of turbine is the high-power coefficient, with values between 0.35 and 0.45. Because propeller-type turbines use only the aerodynamic lift effect to create rotational motion, their proper locations are areas with high wind potential, i.e. areas where the average annual wind speed exceeds 5-7 m/s [1,4].

In contrast, wind turbines with vertical axis (VAWT) may represent the future solution for power generation in areas with low air stream speeds (between 2 m/s to 5 m/s) or in areas where the wind has a turbulent character, even if the power coefficient registers values between 0.2 and 0.3 [3,5-8]. In addition to this, turbines with vertical rotors have the main advantage that they can operate and produce energy in severe weather conditions, with low intensity or turbulent wind of nature and frequent changes of direction [1-3].

Due to the simple design and low costs of production and operation, Savonius rotors, have become increasingly interesting for electricity production by converting wind energy into mechanical energy and then into electricity through an electric generator.

Many studies showed that Savonius wind turbines, with different overlap ratios of blades, have a power coefficient between 0.2 and 0.3 [5-8]. Starting from these considerations, in this article, a Savonius wind turbine with overlap ratio of blades of 0.35 is tested. These tests aim to determine the actual power coefficient and power extracted from the wind at different air speeds. Then, based on the values obtained from the tests, a Savonius turbine is dimensioned to power an energy-efficient street lighting pole.

2. Materials and Methods

2.1. Materials

For the test part, a Savonius rotor with semi-cylindrical blades was manufactured [9,10]. The rotor has the overlap ratio of blades (e/D) 0.35. In figure 1 is presented the constructive shape of the turbine and in table 1 are shown the dimensional characteristic of the turbine.

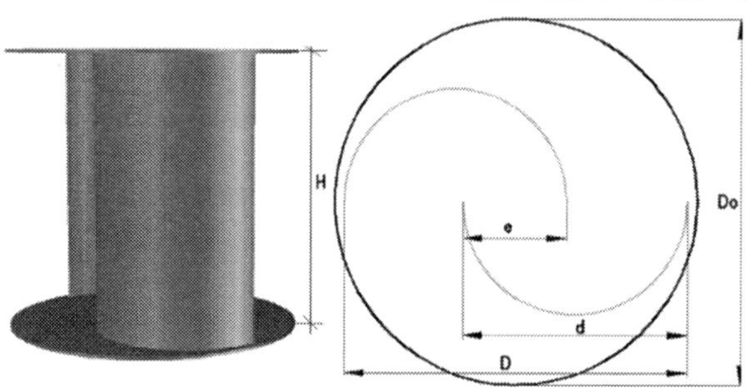

Figure 1. Savonius turbine constructive shape used in experiments.

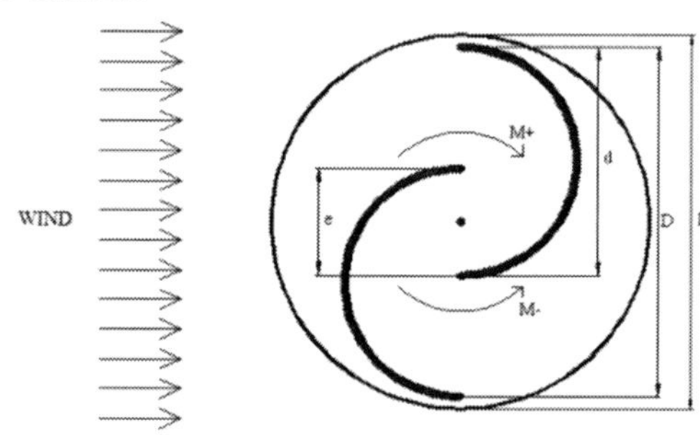

Figure 2. Scheme with operation principle of Savonius turbine.

Table 1. Dimensional characteristics of made-up Savonius turbine.

Rotor diameter, D (cm)	Blade diameter, d (cm)	Rotor height, H (cm)	Endplate diameter, D0 (cm)	Overlap distance, e (cm)	Overlap ratio, e/D (-)
18	12.2	18 cm	19 cm	6.3	0.35

The semi-cylindrical blades placed in "S" shape, as it's shown in figure 1 and 2 (figure 2 - the operating principle of a Savonius turbine), make up the rotor. The concave and the convex side of the rotor are simultaneously under the wind influence [11]. The rotation of the rotor is possible due to different values of the drag coefficient of the two blades during the wind action - the drag coefficient of the concave surface of the blade is greater than that of the convex one. This has a direct influence on the forces acting on the two blades: the force acting on the concave blade is greater than the force acting on the convex blade. Thus, the moment generated by the first force is also greater than the one given by the second force, thereby ensuring the rotation of the vertical axis rotor [2, 9].

Plastic was used for the wind turbines rotor blades. The end plates were made of comatex. A silicone-based adhesive was used for blade assembly with the end plates and with the rotor shaft. Then, the rotor was mounted on a metal frame by means of thrust bearings [12, 13].

The wind turbine was tested in front of an open subsonic wind tunnel (see figure 3). The wind tunnel simulates wind speeds from 0 m/s up to 7 m/s. The wind tunnel has three different sections and all of them are made of galvanized steel sheet. The inlet section of the tunnel is imposed by the axial fan diameter. The second section makes the transition from circular to square section. The outlet section requires a size of 1.2-1.5 times higher than the rotor in order the airflow coming out of the tunnel to cover the entire surface of the turbine [12, 13]. All the tunnel dimensions are presented in the figure below.

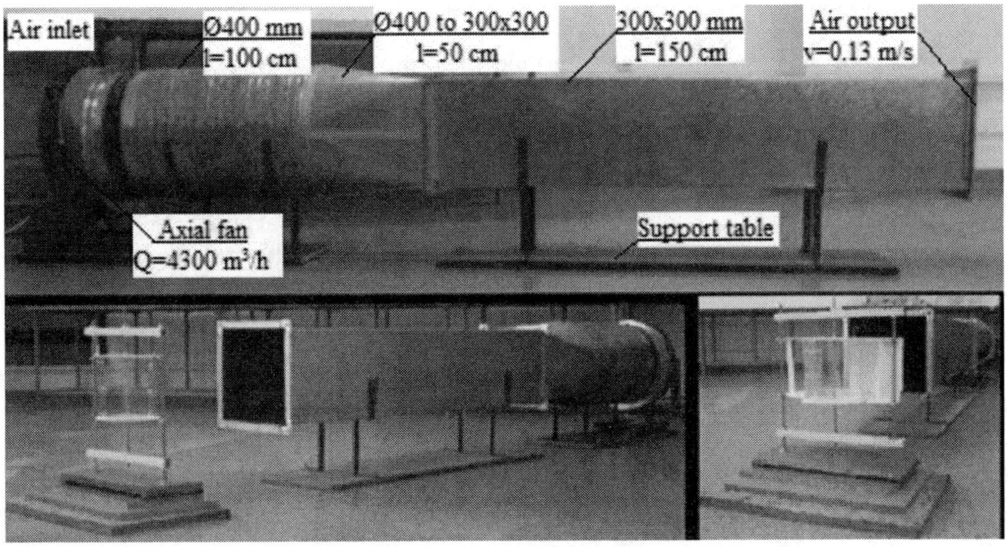

Figure 3. Open-circuit subsonic wind tunnel. [12, 13].

The air stream given by the tunnel was measured with a high precision anemometer, AM 50. The anemometer was placed at the outcome of the aerodynamic tunnel. With an accuracy of ±3% the anemometer can measure airflow speeds between 0 and 45 m/s. The rotation speed of rotor was recorded with a digital tachometer, Voltcraft DT-10L, which was placed at approximately 35 cm. The measuring range of the tachometer is from 2 to 99.999 rotation per minute with an accuracy of ±0.05%. The recommended measurement distance is between 5 and 50 cm. In order to measure the angular velocity a torque meter, PCE-TM 80, was used. This torque meter records values from 0 to 147 (N·cm) and has an accuracy of ±1.5%. The sensor of the torque meter was placed on the rotor shaft.

Preliminary determinations proved that a distance about 3 to 5 equivalent diameters of the wind turbine from the wind tunnel outlet section, was the best solution in obtaining a laminar airflow regime [12].

The operational scheme of the experimental stand is presented in figure 4.

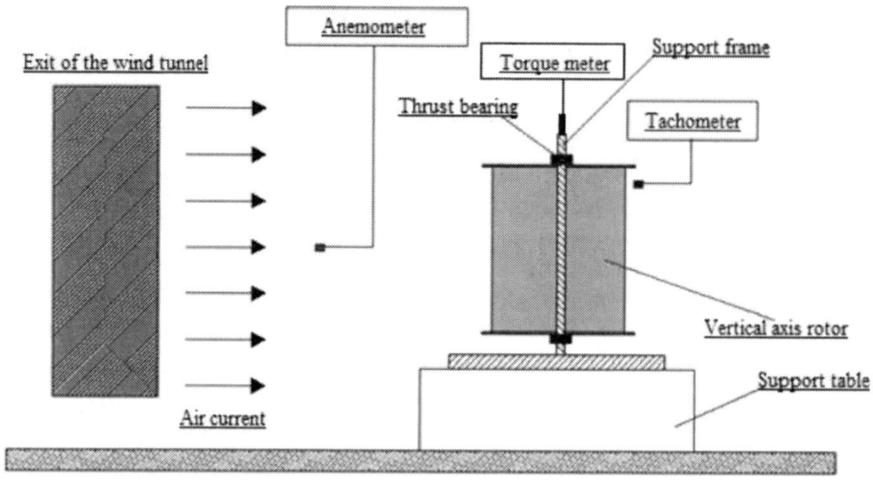

Figure 4. The operation scheme of the experimental set-up.

For the dimensioning part of the Savonius turbine, the principle diagram of an energy-efficient outdoor lighting pole is presented in figure 5.

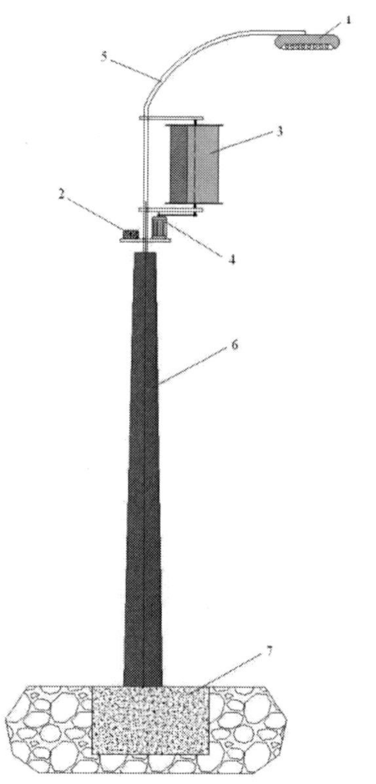

Figure 5. Multifunctional pole for energy-efficient outdoor lighting:
1 - LED street light; 2 - batteries for energy storage; 3 - Savonius wind turbine; 4 - electric generator with permanent magnets; 5 - arm for the installation of the lighting apparatus; 6 - pole; 7 - reinforced concrete foundation.

2.2. Mathematical equations

The mechanical power, P, which may be absorbed from the air streams by means of a wind rotor, can be expressed by the following equation [14]:

$$P = \frac{1}{2}\rho A v^3 C_p \tag{1}$$

where: ρ – air density [kg/m^3]; A – rotor area [m^2]; v – wind speed [m/s]; Cp – power coefficient.

The rotor area of Savonius turbine can be assimilated with the area of a rectangle defined by diameter D and height H of rotor, as it can be seen in the next relationship [3, 14]:

$$A = D \cdot H \quad (2)$$

The power coefficient, Cp, express the amount of energy extracted from wind by using Savonius rotor. It has variable values depending on the tip speed ratio (λ) and coefficient of torque (Ct) [4,15]:

$$C_p = \lambda \cdot C_t \quad (3)$$

The tip speed ratio is determined with the following equation [2, 4]:

$$\lambda = \frac{\omega R}{v} \quad (4)$$

where: ω - angular speed [rad/s]; R – rotor radius [m]; v – wind speed (m/s).

Coefficient of torque is given by the next formula [3, 4]:

$$C_t = \frac{2T}{\rho \cdot v^2 \cdot A \cdot R} \quad (5)$$

where: T - torque (N·m).

3. Results and discussions

The objective of this study was to determine the actual power coefficient and mechanical power extracted from the wind using a Savonius turbine. The wind turbine was tested at airflow speeds between 0 and 7 m/s. These speeds were obtained by modifying the fan speed with the speed regulator which the open subsonic aerodynamic tunnel is equipped. The turbine rotation speed, instantaneous airflow velocity and angular velocity were recorded using a laser tachometer, an anemometer with propellers respectively a torque meter. Regarding the rotor speed, the minimum, average and maximum values corresponding to the different wind speeds in the range 0 to 7 m/s were recorded.

Using average values obtained from test, a graph was drawn that shows the curve of variation of the rotor speed according to the speed of the air flow, graph shown in figure 6.

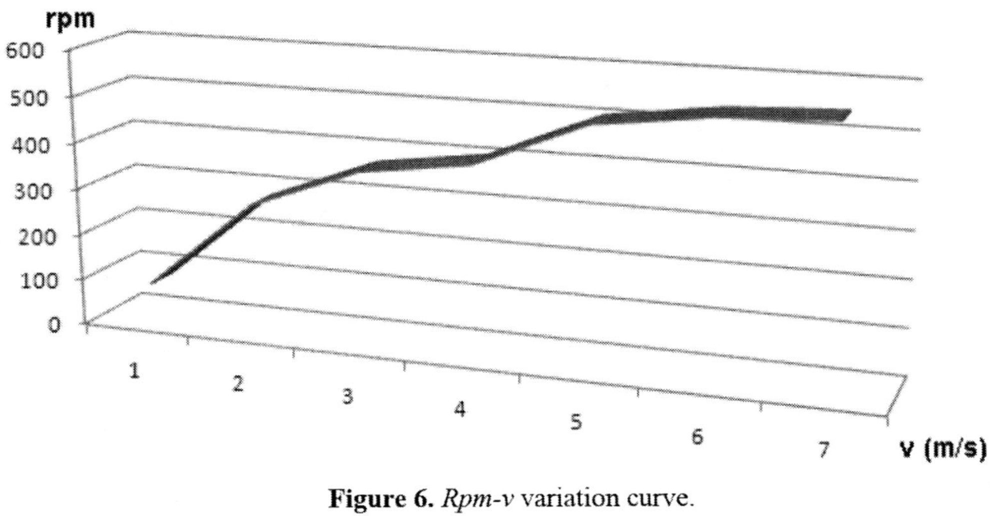

Figure 6. *Rpm-v* variation curve.

Analysing the figure above, it can be observed that rotation per minute increases with air speed. For wind speed between 1 and 3 m/s the rpm values increase rapidly, a slowdown is observed from 3 to 4 m/s, from 4 to 5 m/s rpm values increase rapid again and then again slowdown from 5 to 7 m/s.

In table 2 are presented the medium torque values recorded during tests for *rpm* average values.

Table 2. Torque's medium values.

v (m/s)	1	2	3	4	5	6	7
T (N·cm)	0.06	0.16	0.36	0.95	1.31	2.4	3.12

By comparing the values from the above table with the rpm-v curve evolution, is observed that the torque is inversely proportional to the increase of the rotations per minute curve. Torque values, for wind speeds from 1 to 3 m/s, have a slowly increase compared with rpm values where have a rapid increase, then from 3 m/s the torque values increase rapidly and the rpm values slowdown.

Knowing the curve and the variation equation of the rotor speed for any value of airflow velocity is very useful in the process of choosing the electric generator that will fit the wind turbine. The electric generator can be connected directly to the rotor shaft or coupled to it by a gearbox or speed multiplier [13].

In order to obtain the curve and the variation equation of rotor speed, the experimental data obtained were processed using the CurveExpert calculation program. Figure 7 shows the *rpm-v* variation curve, and in table 3 the equation of the rpm-v characteristic curve of the tested turbine is presented.

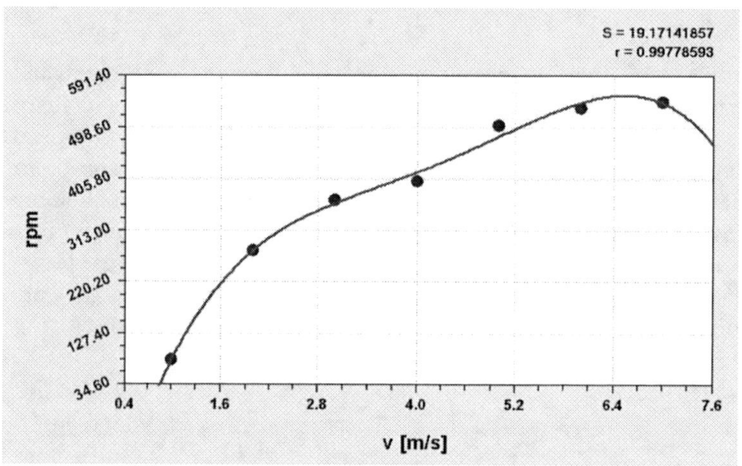

Figure 7. *Rpm-v* characteristic curve for Savonius turbine.

Table 3. Equation of the *rpm-v* characteristic curve.

Function type	Function	Coefficients	Standard error Correlation coefficient
Polynomial degree 4	$rpm = a + bv + cv^2 + dv^3 + ev^4$	a=-358.714 b=617.523 c=-209.25 d=33.813 e=-1.9848	E=11.1714 C=0.99778

From the analysis of the data processed with CurveExpert it is observed that the equation of the characteristic curve of turbine operation is a polynomial of degree 4. The standard error obtained must be as small as possible, ideally the residual values at the reference points are balanced over the whole range, ie the sum of the positive errors should be approximately equal to the sum of the negative errors. The correlation coefficient must have a value as close as possible to the unit, and for this turbine the difference is only 0.2%.

To establish the mechanical power available at the rotor shaft, the power coefficient of the turbine needs to be determined. With equation (3) is determined the power coefficient. The tip speed ratio (λ)

was calculated with relation (4). The power coefficient, Cp, is determined based on tip speed ratio, λ, and on coefficient of torque, equation (5). In figure 8, variation of power coefficient is presented.

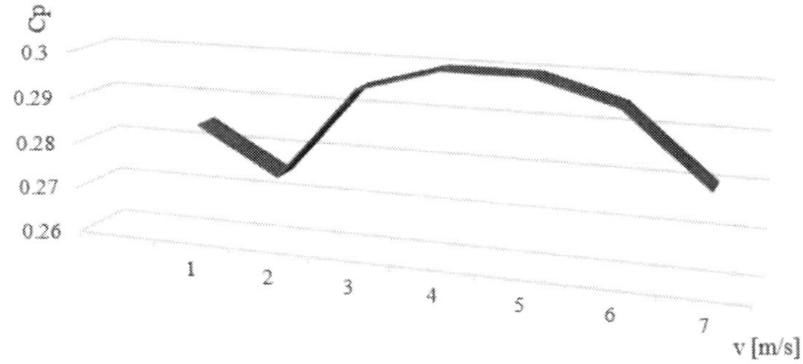

Figure 8. Variation of the power coefficient.

In figure 8 is observed that the power coefficient of Savonius turbine increases from wind speed 3 to 6 m/s. The highest values obtained are for wind speed of 4 and 5 m/s. At 7 m/s wind speed, Cp decreases as a probable consequence of air stream turbulence.

Returning to relation (1), the mechanical power available at the rotor shaft was determined, values presented in table 4.

Table 4. Mechanical power extracted from wind.

v (m/s)	1	2	3	4	5	6	7
P (W)	0.005	0.04	0.16	0.48	0.8	1.24	1.7

From the values presented in table 4 is observed that with the increase of air speed, the mechanical power extracted with Savonius turbine increases.

Following tests performed, a Savonius wind turbine was dimensioned to supply electricity to an energy-efficient street lighting pole. The vertical axis wind turbine must supply energy to a led lamp of 36 W that operates approximately 10 h/day. The energy provided by the turbine, under these conditions is 0.36 kWh/day, which is equivalent to a turbine power of 15 W. Area of the turbine is determined using relation (1). The values taken in the turbine sizing calculation are presented in table 5.

Table 5. Values considered from tests for Savonius turbine dimensioning.

v (m/s)	Cp [-]	ρ [kg/m^3]
4	0.3	1.225

In these conditions, the area of the Savonius wind turbine is A = 0.72 m^2. If it's wanted that the shape ratio of the rotor, H/D to be unitary, meaning the diameter and the height rotor to have equal dimensions, the actual dimensions of the rotor are determined and presented in table 6, along with other possibilities of configuration of the rotor.

Table 6. The calculated dimensions of the Savonius turbine.

Application	Energy consumed (kWh/day)	Turbine's power (W)	Dimensions D x H (m)	Blade diameter, d (m)
Street lighting	0.360	15	0.85 x 0.85	0.51
			0.6 x 1.2	0.36
			0.72 x 1.0	0.43

IOP Conf. Series: Materials Science and Engineering **789** (2020) 012056 doi:10.1088/1757-899X/789/1/012056

4. Conclusions

Following tests performed, it was observed that the power coefficient reaches maximum value 0.30, at wind speeds of 4 and 5 m/s. The power extracted from wind increases proportionally with the air speed.

The values considered for the wind turbine dimensioning were chosen for a lower wind speed, 4 m/s, where were obtained good results for the power coefficient. At an air speed of 4 m/s, a horizontal axis wind turbine is not efficient, the best option to supply energy to a street-light lamp is the vertical axis wind turbine.

For the street-lighting pole was imposed a led lamp of 36 W that operates 10 h/day. The area of the Savonius wind turbine was calculated, 0.72 m^2. For this required area three different configuration of dimensions were presented and each one can assure the energy for the led lamp. The Savonius turbine can be placed directly on the pillar supporting the luminaire of the street lighting pole, either in the axis of the pillar or on a console. This greatly reduces the investment required to build the wind turbine, as it is no longer necessary to build a separate structure for the turbine's location.

If it is necessary to equip the pillar with two or more luminaires, the electricity requirement can still be ensured, either by installing a larger turbine or by installing several identical turbines, each having the capacity necessary to ensure the consumption of a lighting lamp.

Also, the turbines with vertical axis are suitable for electricity supply for the lighting of bus stations, pedestrian crossings, road markings and advertising panels, giving them the possibility of energy independence.

References

[1] Vaughn N 2009 *Wind Energy: Renewable Energy and the Environment*, CRC Press, Taylor & Francis Group, USA

[2] Erich H 2006 *Wind Turbines: Fundamentals, Technologies, Application and Economics*, Springer-Verlag Press, Berlin, Germany

[3] Jha A R, 2010 *Wind Turbine Technology*, CRC Press, Taylor & Francis Group, USA

[4] Hansen M O L, 2008 *Aerodynamics of Wind Turbines*, Earthscan Press, USA

[5] Pope K, Naterer G F , Dincer I et al., 2001 Power correlation for vertical axis wind turbines with varying geometries *International Journal of Energy Research*, **35** p 423–435

[6] Nasef M H, El-Askary W A, et al. 2013 Evaluation of Savonius rotor performance: Static and dynamic studies *Journal of Wind Engineering and Industrial Aerodynamics* **123** pp 1–11

[7] Akwa J V, Vielmo H A and Prisco Petry A 2012 A review on the performance of Savonius wind turbines *Renewable and Sustainable Energy Reviews* **16** p 3054– 3064

[8] Ramos D M, Sauceda C D 2017 CFD study of a vertical axis counter-rotating wind turbine *IEEE 6h International Conference on Renewable Energy Research and Applications*, San Diego, p 240-244

[9] Mussell D, 2006 Build your own wind turbine The Pembina Institute, Ontario

[10] Al-Bahadly 2009 Building a wind turbine for a rural home *Energy for Sustainable Development* **13** 159-165

[11] Savonius S J 1931 The S-rotor and its applications *Mechanical Engineering* **53** pp 5

[12] Rus T and Rus L F et al. 2018 Experimental Investigation on the Influence of Overlap Ratio on Savonius Turbines Performance *International Journal of Renewable Energy Research* Vol. **8** No.3 pp 1791-1799

[13] Rus T, Rus L F et al. 2016 Experimental Tests in Equipping Vertical Axis Wind Turbines with Electric Generator, *International Journal of Renewable Energy Research* vol.**6** No.2 pp 465-471

[14] Mukund P R 2000 Wind and Solar Power Systems, U.S. Merchant Marine Academy Kings Point Press, New York, USA

[15] Devashish A and Thaku 2017 "A Comprehensive Review on Wind Energy System for Electric Power Generation: Current Situation and Improved Technologies to Realize Future Development" *International Journal of Renewable Energy Research* vol.**7** No.4 pp 1786-1805

IOP Conf. Series: Materials Science and Engineering **789** (2020) 012057 doi:10.1088/1757-899X/789/1/012057

Numerical study on the effect of confinement on thermal spallation drilling of hard rock

T Saksala[1*]

[1] Tampere University, Civil Engineering, Tampere, Finland

[*] timo.saksala@tuni.fi

Abstract. This paper presents a numerical study on the effect of confining pressure (down-to-hole pressure) on the performance of the thermal jet drilling on rock. For this end, a numerical method including a viscoplastic-damage model for rock and an explicit scheme for solving the governing thermo-mechanical problem, which is assumed uncoupled. Moreover, rock heterogeneity, which facilitates the thermal spallation phenomenon, is taken into account. Numerical simulations under axisymmetric conditions demonstrate that the efficiency of thermal spallation based drilling is reduces drastically due to confinement.

1. Introduction

Traditional oil well drilling is based on mechanical breakage by rotary or rotary/percussive drilling methods. However, these methods suffer from severe tool wear due to mechanical contact, especially in hard formations. Therefore, under the pressure of ever-increasing demands for efficiency and economy, non-traditional drilling methods using some non-mechanical agent to assist (or to replace) the mechanical breakage have been searched during the last couple of decades.

One particularly attractive non-mechanical method is to utilize the thermal spallation phenomenon illustrated in figure 1. Accordingly, when a rock surface is exposed to an intensive heating the resulting thermal gradient (∇T in figure 1) induces a compressive stress state (σ in figure 1), which leads to crack growth in the rock surface layer. When the cracks reach the critical length, spallation, i.e. ejection of rock chips, occurs. The minimal required temperature for spallation to occur in granitic rocks is about 500-600 °C [1]. This method has been successfully applied in drilling granite rock at the laboratory conditions [1-3]. However, while spalling phenomenon can be applied in thermal drilling at shallow depths, it seems that confining conditions, especially the down-the-hole pressure (p_{conf} in figure 1), impedes rock fracturing thus preventing the ejection of spalls (see figure 1). Under such conditions then, the rate of penetration of drilling expectedly drops drastically.

This paper presents a numerical study on the thermal spallation drilling method under such confining conditions. For this end, a numerical simulation method is developed. Previous numerical studies on thermal spallation drilling are, e.g., by Saksala [4] and Walsh and Lomov [5]. The present model is based on a damage-viscoplastic constitutive model for rock implemented in the finite element method. The rock material is assumed to consist three different minerals, Quartz, Feldspar and Biotite. The underlying thermo-mechanical problem is solved as an uncoupled problem where the only information from the thermal part to the mechanical part is the thermal strain, which leads to the spallation of the rock.

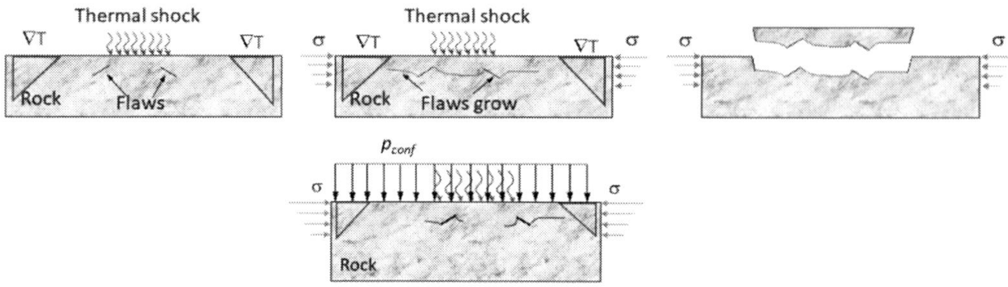

Figure 1. Principle of thermal spallation drilling with and without confinement.

2. Theory of the modelling approach

The modelling approach is briefly described here. First, the constitutive model for rock is explained. Then, the time integration scheme of the uncoupled thermo-mechanical problem is outlined.

2.1. *Viscoplastic-damage model for rock*

The constitutive description of rock chosen here is the damage-viscoplasticity model originally presented by Saksala [6]. In this model, the stress states leading to viscoplastic yield and isotropic damage are indicated by the Drucker-Prager (DP) yield function with the Modified Rankine (MR) criterion as a tension cut-off. Assuming perfectly viscoplastic behaviour, the yield surfaces, along with the rate-dependent cohesion and tensile strength, are written as

$$\begin{aligned} f_{\mathrm{DP}}(\boldsymbol{\sigma},\dot{\kappa}_{\mathrm{DP}}) &= \sqrt{J_2}+\alpha_{\mathrm{DP}}I_1-k_{\mathrm{DP}}c(\dot{\kappa}_{\mathrm{DP}}) \\ f_{\mathrm{MR}}(\boldsymbol{\sigma},\dot{\kappa}_{\mathrm{MR}}) &= \sqrt{\textstyle\sum_{i=1}^{3}\langle\sigma_i\rangle^2}-\sigma_{\mathrm{t}}(\dot{\kappa}_{\mathrm{DP}}) \\ c(\dot{\kappa}_{\mathrm{DP}}) &= c_0+\underbrace{s_{\mathrm{DP}}\dot{\kappa}_{\mathrm{DP}}}_{q_{\mathrm{DP}}}, \quad \sigma_{\mathrm{t}}(\dot{\kappa}_{\mathrm{MR}})=\sigma_{\mathrm{t0}}+\underbrace{s_{\mathrm{MR}}\dot{\kappa}_{\mathrm{MR}}}_{q_{\mathrm{MR}}} \end{aligned} \tag{1}$$

where I_1 ($= \sigma_1 + \sigma_2 + \sigma_3$) and J_2 ($= 0.5((\sigma_1 - \sigma_2)^2 + (\sigma_1 - \sigma_3)^2 + (\sigma_2 - \sigma_3)^2)$) are the first and the second invariants of the stress tensor $\boldsymbol{\sigma}$, σ_i is the ith principal stress, $\langle\rangle$ are the McAuley brackets, α_{DP} and k_{DP} are the DP parameters, c and σ_t are the dynamic cohesion and tensile strength (with their intact values c_0 and σ_{t0}) depending on the rates of the internal variables κ_{DP} and κ_{MR}, respectively. For calibration purposes, the DP parameters are expressed in terms of the friction angle φ: $\alpha_{\mathrm{DP}} = 2\sin\varphi/(3-\sin\varphi)$ and $k_{\mathrm{DP}} = 6\cos\varphi/(3-\sin\varphi)$. By this choice, the uniaxial compressive strength can be matched. Moreover, a plastic potential of the same form as f_{DP} in (1) but with a dilatation angle ψ ($\leq \varphi$) is also employed to mend the poor prediction of dilatancy of the associated flow rule. The internal variable for the DP part in (1) is defined so that the thermodynamic consistency of the model is retained [7]: $\dot{\kappa}_{\mathrm{DP}} = -\dot{\lambda}_{\mathrm{DP}}\partial g_{\mathrm{DP}} / \partial q_{\mathrm{DP}} = \dot{\lambda}_{\mathrm{DP}}k_{\mathrm{DP}}$ with g_{DP} and $\dot{\lambda}_{\mathrm{DP}}$ being the plastic potential and plastic multiplier, respectively, while q_{DP} is a stress like hardening parameter. Similar definition for the MR part leads to $\dot{\kappa}_{\mathrm{MR}} \equiv \dot{\lambda}_{\mathrm{MR}}$. Finally, s_{DP} and s_{MR} are the constant viscosity moduli in compression and in tension, respectively.

In the damage part of the model, the highly asymmetric behavior of rocks in tension and compression is accounted for by separate scalar damage variables in these stress regions. The damage is chosen to be driven by the viscoplastic strain. By this choice and the perfect viscoplasticity assumption made above, both the strength and the stiffness degradation both in tension and in compression are governed by the damage part of the model. The damage part of the model is defined by equations

$$
\begin{aligned}
&\omega_t = A_t\left(1-\exp(-\beta_t \varepsilon^{vp}_{eqvt})\right), \quad \omega_c = A_c\left(1-\exp(-\beta_c \varepsilon^{vp}_{eqvc})\right) \\
&\dot{\varepsilon}^{vp}_{eqvt} = \sqrt{\Sigma_{i=1}^{3}\left\langle \dot{\varepsilon}^{vp}_i \right\rangle^2}, \ \dot{\varepsilon}^{vp}_{eqvc} = \sqrt{\tfrac{2}{3}\dot{\boldsymbol{\varepsilon}}^{vp} : \dot{\boldsymbol{\varepsilon}}^{vp}}, \quad \dot{\boldsymbol{\varepsilon}}^{vp} = \dot{\lambda}_{MR}\frac{\partial f_{MR}}{\partial \boldsymbol{\sigma}} + \dot{\lambda}_{DP}\frac{\partial g_{DP}}{\partial \boldsymbol{\sigma}} \\
&\boldsymbol{\sigma} = (1-\omega_t)\bar{\boldsymbol{\sigma}}_+ + (1-\omega_c)\bar{\boldsymbol{\sigma}}_- \quad (\bar{\boldsymbol{\sigma}} = \bar{\boldsymbol{\sigma}}_+ + \bar{\boldsymbol{\sigma}}_-) \\
&\bar{\boldsymbol{\sigma}}_+ = \max(\bar{\boldsymbol{\sigma}}, 0), \quad \bar{\boldsymbol{\sigma}}_- = \min(\bar{\boldsymbol{\sigma}}, 0) \\
&\bar{\boldsymbol{\sigma}} = \mathbf{E} : (\boldsymbol{\varepsilon}_{tot} - \boldsymbol{\varepsilon}_{vp} - \boldsymbol{\varepsilon}_\theta), \quad \boldsymbol{\varepsilon}_\theta = \alpha \Delta\theta \mathbf{I}
\end{aligned} \tag{2}
$$

where parameters A_t and A_c control the maximum values of the damage variables ω_t and ω_c in tension and in compression, respectively. Parameters β_t and β_c are defined based on the fracture energies G_{Ic} and G_{IIc} by $\beta_t = \sigma_{t0}h_e/G_{Ic}$ and $\beta_c = \sigma_{c0}h_e/G_{IIc}$ with h_e and σ_{c0} being a characteristic length of a finite element and the uniaxial compressive strength of the material, respectively. The equivalent viscoplastic strain in tension and compression, $\dot{\varepsilon}^{vp}_{eqvt}$, $\dot{\varepsilon}^{vp}_{eqvc}$, respectively, are defined by the rate of viscoplastic strain tensor $\dot{\boldsymbol{\varepsilon}}^{vp}$ and its principal values $\dot{\varepsilon}^{vp}_i$. Moreover, the nominal-effective stress relationship is written in (2) with the positive-negative parts decomposition. Finally, the (effective) stress-strain relationship is written with the elasticity tensor $\mathbf{E}$ and the total ($\boldsymbol{\varepsilon}_{tot}$), viscoplastic ($\boldsymbol{\varepsilon}_{vp}$) and thermal ($\boldsymbol{\varepsilon}_\theta$) strain tensors. In the definition of the thermal strain, α and $\Delta\theta$ are the thermal expansion coefficient and the temperature change, and $\mathbf{I}$ is the second order identity tensor.

2.2. Explicit time stepping scheme for solving the discretized thermo-mechanical problem

This problem of thermal jet drilling is governed by the time-dependent heat equation and the equation in motion, which can be written in the finite element discretized form as

$$
\begin{aligned}
&\mathbf{C}\dot{\boldsymbol{\theta}} + \mathbf{K}_\theta \boldsymbol{\theta} = \mathbf{f}_\theta + \mathbf{f}_Q \quad \& \quad \mathbf{M}\ddot{\mathbf{u}} + \mathbf{f}_{int} = \mathbf{f}_{ext} \\
&\mathbf{f}_{int} = \mathbf{A}_{e=1}^{N_e}\int_{\Omega_e} \mathbf{B}_e^T \boldsymbol{\sigma}_e d\Omega \\
&\mathbf{C} = \mathbf{A}_{e=1}^{N_e}\int_{\Omega_e} \rho c \mathbf{N}_\theta^{e,T}\mathbf{N}_\theta^e d\Omega, \quad \mathbf{K}_\theta = \mathbf{A}_{e=1}^{N_e}\int_{\Omega_e} k \mathbf{B}_\theta^{e,T}\mathbf{B}_\theta^e d\Omega \\
&\mathbf{f}_\theta = -\mathbf{A}_{e=1}^{N_e}\int_{\partial\Omega_e} q_n \mathbf{N}_\theta^{e,T} d\partial\Omega, \quad \mathbf{f}_Q = \mathbf{A}_{e=1}^{N_e}\int_{\Omega_e} Q \mathbf{N}_\theta^{e,T} d\Omega
\end{aligned} \tag{3}
$$

where the symbol meanings are as follows: $\ddot{\mathbf{u}}$ is the acceleration vector; $\boldsymbol{\theta}$ **is the** temperature vector; $\mathbf{M}$ is the lumped mass matrix; $\mathbf{f}_{ext}$ is the external force vector; $\mathbf{f}_{int}$ is the internal force vector; $\mathbf{C}$ is the capacitance matrix; $\mathbf{K}_\theta$ is the conductivity matrix; $\mathbf{f}_\theta$ is the external heat loading vector; $\mathbf{f}_Q$ is the mechanical heating term (vector); $\mathbf{A}$ is the standard finite element assembly operator; $\mathbf{B}_e$ is the kinematic matrix (mapping the nodal displacement into element strains); ρ is the density; c is the specific heat capacity; θ is the temperature; $\mathbf{N}_\theta$ is the temperature interpolation matrix; k is the conductivity; q_n is the normal component of the heat flux; $\mathbf{B}_\theta$ is the gradient of $\mathbf{N}_\theta$.

Generally, the specific heat capacity and conductivity depend on temperature but this dependence is neglected in this study. Moreover, the convection at the boundaries of the rock specimen is also ignored as insignificant due to the short heating times. Finally, the term Q in (3) expresses the mechanical heat production through dissipation and strain rate. This term is ignored as insignificant in comparison to the external flux, hence $\mathbf{f}_Q \equiv \mathbf{0}$ from now on. This assumption makes the thermo-mechanical problem uncoupled facilitating its solution significantly.

The uncoupled thermo-mechanical problem governing the thermal spallation is solved with explicit time marching. The forward Euler method is employed here leading to following equation for the new temperature and mechanical response:

$$\begin{aligned}
&\mathbf{C}_t\boldsymbol{\theta}_{t+\Delta t} = \mathbf{C}_t\boldsymbol{\theta}_t + \Delta t(\mathbf{f}_{\theta,t} - \mathbf{K}_{\theta,t}\boldsymbol{\theta}_t) \rightarrow \boldsymbol{\theta}_{t+\Delta t}\\
&\mathbf{M}\ddot{\mathbf{u}}_t + \mathbf{f}_{\text{int},t} = \mathbf{f}_{\text{ext},t} \rightarrow \ddot{\mathbf{u}}_t\\
&\dot{\mathbf{u}}_{t+\Delta t} = \dot{\mathbf{u}}_t + \Delta t\ddot{\mathbf{u}}_t\\
&\mathbf{u}_{t+\Delta t} = \mathbf{u}_t + \Delta t\dot{\mathbf{u}}_{t+\Delta t}
\end{aligned} \tag{4}$$

Now, the solution procedure is as follows. First, the temperature at time $t + \Delta t$ is solved by Equation $(4)_1$. Second, an element level loop is performed for solving the new stress, viscoplastic strain, and other internal variables based on which the new internal force vector is assembled. Third, the acceleration $\ddot{\mathbf{u}}_t$ at t is solved with $(4)_2$. Finally, the mechanical response, i.e. displacement $\mathbf{u}_{t+\Delta t}$ and velocity $\dot{\mathbf{u}}_{t+\Delta t}$, is computed further.

3. Numerical examples

The numerical simulations of the thermal spallation drilling are carried out here. As mentioned, the rock heterogeneity is accounted for by random clusters of finite elements representing three different rock constituent minerals. The material parameters for the minerals used in the simulations are given in table 1. These values, mainly taken from [7-10], do not necessarily represent any particular rock but a granitic rock like behaviour in general.

Table 1. Material and model parameters for simulation.

Parameter/mineral	Quartz	Feldspar	Biotite
E (GPa)	80	60	20
σ_{t0} (MPa)	10	8	7
c_0 (MPa)	50	50	50
ν (Poisson's ratio)	0.17	0.29	0.2
φ (°)	30	30	30
ψ (°)	5	5	5
ρ (density) (kg/m^3)	2703	2703	2703
G_{Ic} (J/m^2)	40	40	28
G_{IIc} (J/m^2)	400	400	280
A_t	0.98	0.98	0.98
A_c	0.95	0.95	0.95
s_{MR} (MPa·s)	0.01	0.01	0.01
s_{DP} (MPa·s)	0.01	0.01	0.01
α (1/K)	1.60E-5	0.75E-5	1.21E-5
k (W/mK)	4.94	2.34	3.14
c (J/kgK)	731	730	770
f (%)	33	50	17

Some of the material parameters are assumed temperature dependent. Namely, the Young's modulus E, tensile strength and the cohesion are assumed to depend linearly on temperature so that their values at 500°C are 50% of the corresponding values at room temperature (293°K). Mathematically, the temperature dependence of a property x is thus of form $x(\theta) = x(\theta_{ref}) + K_x^{500}\Delta\theta$

where K_x^{500} is the modulus of temperature dependence and θ_{ref} is reference temperature. This simplified assumption is sufficient for the present purpose to reveal the effect of temperature dependence of these parameters in general, not its exact quantification for a specific rock. However, the temperature dependence of the thermal expansion coefficient of some concretes [7] and granites [10] is nonlinear. Here, the data in [10] is approximated here by a second order polynomial $\alpha(\theta) = a_1\theta^2 + a_2\theta + a_3$ where a_i are the fitting coefficients.

Due to the vastly differing time scales, and hence the critical time steps of the explicit schemes (4), of the thermal and mechanical problems, mass scaling is used to speed up the solution in time. Moreover, in order for the thermal spallation drilling to be competitive with the mechanical drilling methods, very short heating times and high heat flux intensities must be used.

The boundary conditions and a detail of the mesh used in the simulations are shown in figure 2. The axisymmetric mesh has 17399 standard four-node bilinear 2D solid elements and the average mesh size close to the nodes where the heat flux in applied is 0.25 mm. Different colors indicate different rock constituent minerals.

Finally, the heating time and the flux intensity are, respectively, specified as 0.05 s and q = 7 MW/m^2 as a flux intensity. These values are chosen for demonstrative purposes only and they do not necessarily represent practical values.

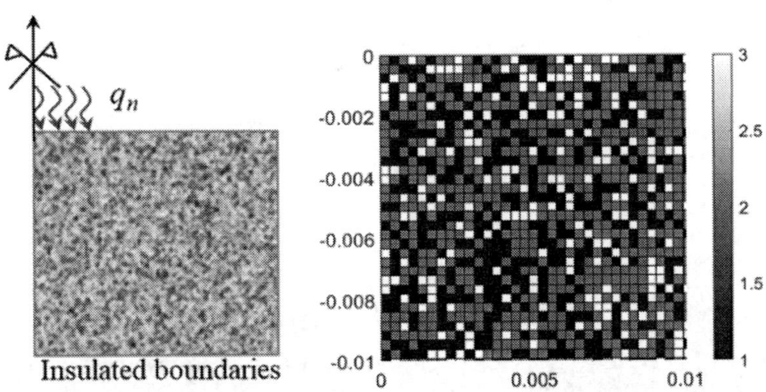

Figure 2. Boundary conditions and a close up of the finite element mesh for thermal spallation simulations.

Simulation results for the boundary conditions and mesh shown in figure 2 are presented in figure 3. According to the results in figure 3, the temperature at some of the flux nodes reaches 800°C, which is well beyond the α - β quartz phase transition temperature (573°C), which is not taken into account in the present modelling. Moreover, due to the short heating time, the temperature rises notably only in a very narrow (< 1 mm) strip. However, the stress state this extremely steep temperature gradient induces in the rock surface area is enough to cause spalling, as attested in figure 3 by the bulging out of the strip of elements adjacent to the rock surface.

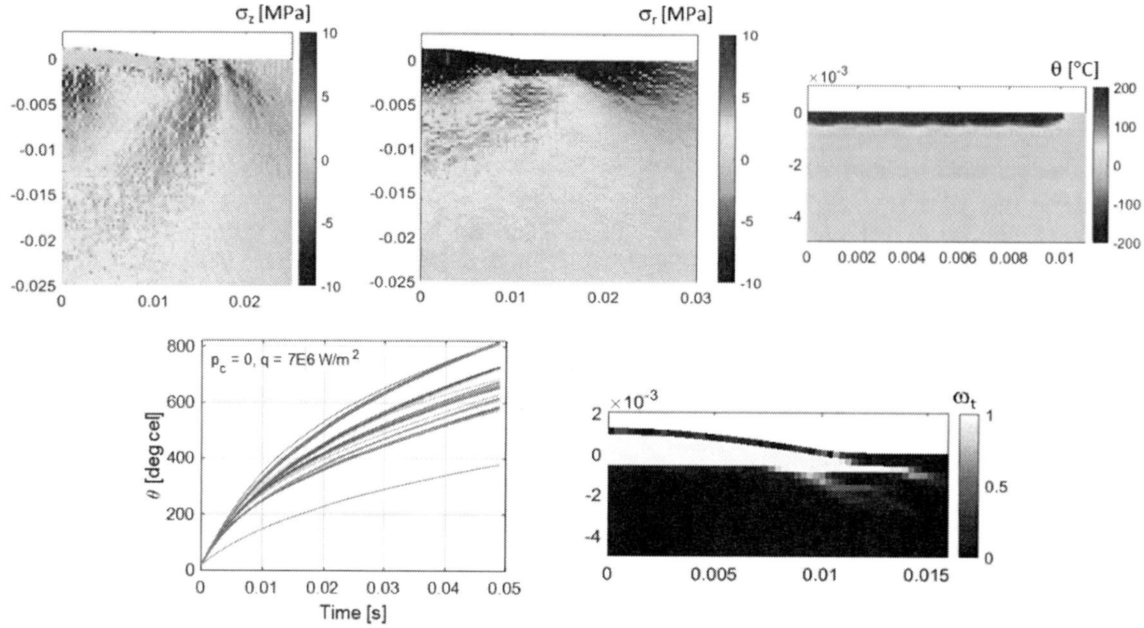

Figure 3. Simulation results for thermal jet drilling, unconfined case.

Expectedly then, this bulging is effectively prevented by confining pressure. The simulation results for the case with 20 MPa of hydrostatic confinement are shown in figure 4.

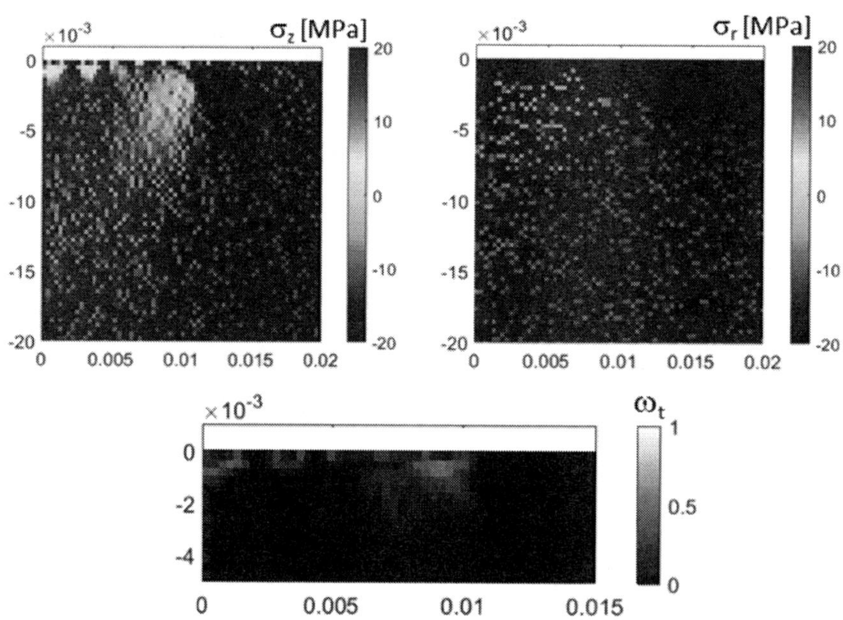

Figure 4. Simulation results for thermal jet drilling, confined case (20 MPa).

Hydrostatic confinement of 20 MPa indeed suppresses the tensile damage, as the tensile stresses induced by the heat shock are not enough to counter the compressive confinement. However, fully hydrostatic confinement may not always be the case in real in-situ drilling.

4. Conclusions

The numerical study on the effect of confinement on the action of thermal jet drilling demonstrates that (hydrostatic) confinement has a detrimental effect on the tensile damage required for the thermal spallation phenomenon to occur. According to the simulations carried out here, confinement of 20 MPa, corresponding roughly to a depth of 1 km, prevents spallation. While fully hydrostatic confining conditions are seldom met in real in-situ drilling, this results means, in the least, that the thermal jet drilling faces serious challenges in deep hole drilling. Finally, it is noted that despite the fact that the simulations were carried on granitic rock, the modelling approach can be applied to other rocks as well.

Acknowledgements
This research was funded by Academy of Finland under grant number 298345.

References

[1] Kant M A and von Rohr PR 2016 Minimal Required Boundary Conditions for the Thermal Spallation Process, *Int J Rock Mech Min Sci*. Vol **84** p 177-186
[2] Rossi E, Kant M A, Madonna C, Saar MO, von Rohr P R 2018 The Effects of High Heating Rate and High Temperature on the Rock Strength: Feasibility Study of a Thermally Assisted Drilling Method. *Rock Mech Rock Eng*. Vol **51** p 2957-2964
[3] Rossi E, Kant M A, Borkeloh O, Saar M O, von Rohr P R 2018 Experiments on Rock-Bit Interaction during a Combined Thermo-Mechanical Drilling Method. *PROCEEDINGS, 43rd Workshop on Geothermal Reservoir Engineering*, Stanford University, Stanford, California, February 12-14
[4] Saksala T, 2018 Numerical modelling of thermal spallation of rock. In: *Proceedings of XII Argentine Congress on Computational Mechanics (MECOM2018)*, 6-9 November, Tucuman, Argentina
[5] Walsh S C, Lomov I N, 2013 Micromechanical modeling of thermal spallation in granitic rock. *Int J Heat Mass Transfer* Vol **65** p 366–373
[6] Saksala T, 2010 Damage–viscoplastic consistency model with a parabolic cap for rocks with brittle and ductile behavior under low-velocity impact loading. *Int J Numer Anal Meth Geomech*. Vol 34 p 1362–1386
[7] Ottosen N S and Ristinmaa M 2005 *The Mechanics of Constitutive Modeling*, Elsevier
[8] Park J-W, Park C, Ryu D and Park E S 2015 Numerical simulation of thermo-mechanical behavior of rock using a grain-based distinct element model. In: Schubert & Kluckner *Proceedings of EUROCK 2015 & 64th Geomechanics Colloquium*, Salzburg, Austria
[9] Tufail M, Shahzada K, Gencturk B and Wei J 2017 Effect of Elevated Temperature on Mechanical Properties of Limestone, Quartzite and Granite Concrete. *Int J Concr Struct M* Vol **11** p 17-28
[10] Heuze F, 1983 High Temperature Mechanical, Physical and Thermal Properties of Granitic Rocks – A Review. *Int J Rock Mech Min Sci & Geomech Abstr* Vol **20** p 3-10

Numerical study of monitoring of early-age concrete strength development using PZT induced stress wave

W Sun[1*], T Y Fu[1], B Li[1] and L Sun[1]

[1] School of Civil Engineering, Shenyang Jianzhu University, Shenyang, China

[*] lgl_315@126.com

Abstract. The purpose of this paper is to verify the validity of concrete strength monitoring method based on piezoelectric wave dynamic technology by the numerical simulation ways. Since the elastic modulus of concrete increases linearly with its strength, the elastic modulus of concrete at different ages is set to simulate its strength growth. The stress wave is transmitted and received by the piezoelectric sensor embedded in concrete, and the stress wave energy value is taken as the strength identification parameter, and its variation trend with the strength is analyzed. The results show that with the increase of concrete strength, the energy value of monitoring signal has a monotonically decreasing trend. In the early stage of concrete, the monitoring signal energy decreases at a large rate and then tends to be flat. In the early stage, the strength growth rate of concrete is faster, and tends to be flat in the later stage. Furthermore, the curve of capability ratio and strength ratio was established, and the two presented good linear lines. The results of this paper show that the concrete strength monitoring method based on pressure wave dynamic technology can be used to monitor the concrete strength.

1. Introduction

Concrete is the most widely used building material in the field of construction. It has been widely applied to the construction of various buildings and structures.

The quality of concrete material is closely related to structural safety. However, in the process of construction, it has been observed that the ultimate strength of the concrete is lower than the expected design strength because of inadequate maintenance and improper construction method, thus leaving hidden danger to the safety of the structure. Therefore, the necessary monitoring of concrete strength is one of the important measures to control the quality of the project.

The methods of spring back value, ultrasonic and compressive strength are widely used in the detection of concrete strength at home and abroad [1, 2].

Although these methods are widely used and mature in technology, there are still some limitations in the practical application process. For example, the spring back value method is susceptible to the influence of concrete aggregate, hence the result is not accurate. The required equipment of ultrasonic method is expensive.

In the compressive strength test method, the test block and the structure cannot be maintained under the same conditions sometime. Therefore, a new method is urgently needed to remedy the deficiency of traditional methods.

In recent years, health monitoring technology based on piezoceramic, represented by lead zirconatetitanate (PZT), has attracted extensive attention in the field of concrete health monitoring, especially in the identification of concrete cracks and damage, and has achieved rich results [3 - 8]. The high sensitivity of piezoceramic materials to concrete state makes it possible for them to be used for concrete strength monitoring [9 - 12].

In this paper, the validity of stress wave method for concrete strength monitoring is verified by numerical simulation. At the same time, the corresponding relationship between concrete strength index and monitoring signal identification parameters is established.

Compared with the traditional concrete detection method, the concrete strength monitoring method based on piezoelectric intelligent materials can realize the in-situ monitoring of concrete strength for the whole age, so as to achieve the purpose of quality control.

This is of great significance to ensure that the actual strength of concrete materials is determined for the state and safety of the structure.

2. Principle of concrete strength monitoring based on piezoelectric wave method

2.1. Piezoelectric smart aggregate

Piezoelectric smart material is one of the main types of smart materials and has piezoelectric effect. Because of the spontaneity and reversibility of piezoelectric effect, piezoelectric smart materials can be made into typical bidirectional sensor elements, and used as signal transmitting actuator and receiving sensor.

Based on this characteristic, piezoelectric sensors have been widely used in the fields of ultrasonic, communication, aerospace, radar and detonating, which are combined with laser and infrared technology, and turn into an important device for the development of new technology and high technology.

The application of piezoelectric ceramic materials in the health monitoring of concrete structures is a new research direction in the academic circle.

The piezoelectric ceramic sensor has many advantages, such as low energy consumption, high sensitivity, fast response, low cost and light weight, so it has a wide application prospect in the field of health monitoring of concrete structures.

However, the enormous structure and complex construction process of the concrete structure make the technology encounter many unavoidable problems. Especially, the material of the piezoceramic is more brittle.

For instance, when directly embedded in the main structure during the construction process, it is easy to damage the piezoelectric ceramic.

Moreover, the service period of concrete structures is longer, which often takes decades or even hundreds of years, and this requires the sensor to have better durability in order to continuous work in a long time.

In addition, piezoelectric ceramic sensors are very sensitive to environmental changes, and the changes in temperature and humidity may affect the measurement results of sensors.

In order to solve the above problems, Song et al. developed the concept of "Smart Aggregate" (SA). The SA is used as the main component of structural health monitoring and embedded in the structure. The steps of the PZT patch cast into the form of SA are shown in figure 1.

The PZT patch welded wire was put into steel mould. Then, epoxy resin was cast into the steel as packaging layer, which can improve the brittleness of PZT patch and isolate the influence of humidity, water and acid-base environment on the performance of PZT patch.

This will not affect the function of PZT, and to a certain extent instead, protect the piezoelectric ceramic sheet, and increase the life of the sensor. The steel mould makes SA with anti-noise properties.

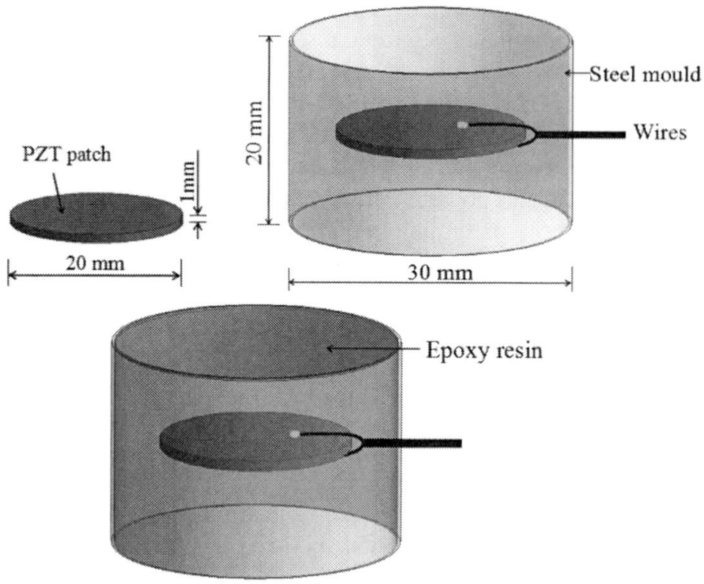

Figure 1. SA structure diagram.

2.2. The basic principle of detecting concrete strength by piezoelectric wave method

A couple of SAs buried in concrete specimen can form a monitoring group. One SA generates stress wave as a signal transmitter, while the other receives stress wave as a signal receiver.

By analyzing the variation of the propagation parameters of the received stress wave such as wave shape, amplitude or frequency, the state change of the medium in the propagation path of the stress wave can be deduced.

The propagation parameters of the stress wave are closely related to the properties of the propagating medium. The generation of the stress waves is the result of the mutual movement between the particles, which is closely related to the elastic connection between the particles, that is, it is related to the elastic density and elastic modulus of the propagating medium. After the concrete is poured and formed, with the maintenance process proceeding, the initially mobile liquid mixture will become the final solid form through hydration reaction.

The most important change of concrete state characteristics is that the strength value has been increasing, and the mechanical parameters such as elastic modulus of concrete will also increase. Combined with the analysis of the stress wave changes and the mechanical properties of concrete, the monitoring signals collected by SAs should be constantly changing before the hydration reaction of concrete is not completed.

By determining and analyzing the changes in the parameters of these acoustic signals, the information about the state of concrete is mastered, so as to realize the monitoring and backstepping of the strength state of the concrete.

3. The establishment of finite element model of concrete beams embedded with PZT patches

3.1. The finite element model of concrete beam embedded with PZT patches

A concrete beam model which size is 150 mm×150 mm×600 mm is used in research word. The size of two built-in PZT patches is Φ20×1 mm. The distance between two PZT patches is 500 mm, which is located at the center line of 50 mm at both ends of concrete.

The model and the locations of PZT patches are shown in figure 2. The meshing of the three-dimensional finite element model of concrete beams with embedded PZT patches is shown in figure 3.

Three strength grades of concrete are studied. The relationship between age and modulus of elasticity of C30 and C40 can be described by equation (1):

$$E(t) = 37.5e^{-0.6/t} \tag{1}$$

The relationship between age and modulus of elasticity of C50 can be described by equation (2)

$$E(t) = 38.2e^{-0.3/t} \tag{2}$$

where $E(t)$ is elastic modulus function of t, and t is the time with a unit of day. Considering medium damping, the concrete material is assumed to be a homogeneous elastic medium with Rayleigh damping to compensate for the effect of microstructures on stress waves.

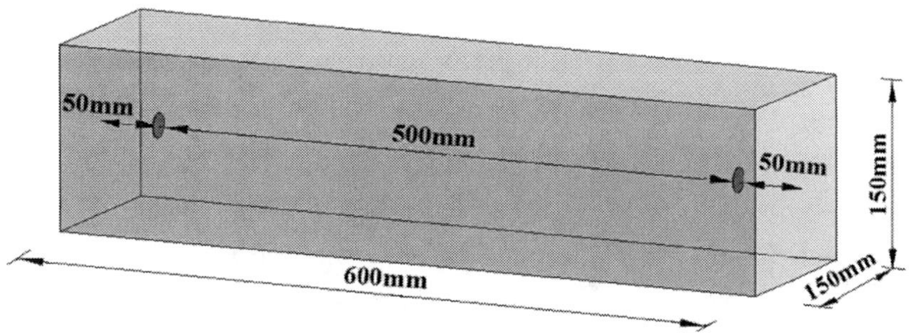

Figure 2. Model size schematic.

3.2. Simulating method of concrete strength increase

The concrete model is simplified to homogeneous elastic material, and the influence of aggregate on the stress wave can be neglected. Some studies indicate that there is a linear relationship between concrete strength and its elastic modulus as shown in figure 4.

This means the growth of concrete strength is bound to increase with the elastic modulus. Based on the relationship between concrete strength and elastic modulus, this numerical simulation will establish a concrete strength growth model, by changing the elastic modulus of concrete. The relationship between age and elastic modulus under different strength grades is described as equation (1) and (2).

The concrete strength models of 3d, 7d, 14d and 28d are established in turn. Therefore, in order to simply simulate the change of concrete strength, only the elastic modulus, Poisson's ratio and density of concrete are considered when setting parameters.

Figure 3. Mesh generation of model.

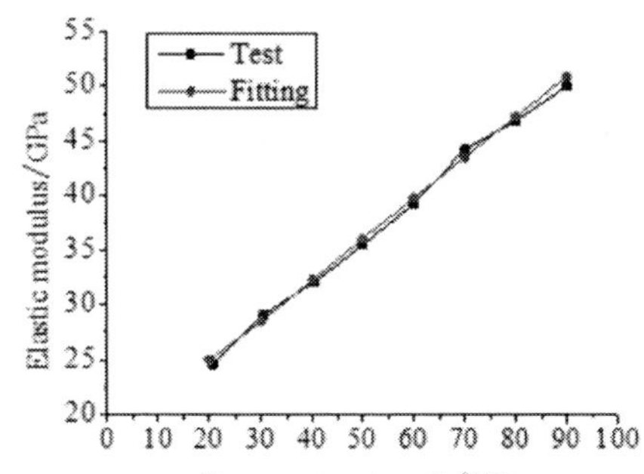

Figure 4. Relationship between compressive strength and modulus of elasticity of concrete.

3.3. Constitutive relationship setting of PZT

The property density of the piezoelectric material is 7600 kg/m^3, and the elastic modulus and dielectric constant are orthogonal. The vibration direction of PZT pieces material is defined as vibration along the thickness direction, that is D_{33} direction. The PZT pieces embedded in concrete is in the second boundary condition, namely mechanical clamping and electrical short circuit.

The constitutive equation can be written in equation (3):

$$\begin{aligned} T &= c^E S\text{-}e^t E \\ D &= eS + \varepsilon^S E \end{aligned} \tag{3}$$

where ***T*** and ***D*** are stress and potential tensor; ***S*** and ***E*** as strain and electric field intensity tensor; $\boldsymbol{c}^E$ and $\boldsymbol{\varepsilon}^S$ are elastic stiffness constants and node constant matrices; ***e*** and $\boldsymbol{e}^t$ are the transposed matrixes of piezoelectric stress constant matrix and piezoelectric stress constant matrix.

The intermediate stiffness matrix is $\boldsymbol{c}^E$, the piezoelectric stress constant matrix ***e*** and the dielectric constant matrix are $\boldsymbol{\varepsilon}^S$, as equations (4) - (6):

$$[c]=\begin{bmatrix} 126 & 77.8 & 74.3 & 0 & 0 & 0 \\ 77.8 & 126 & 74.3 & 0 & 0 & 0 \\ 74.3 & 74.3 & 115 & 0 & 0 & 0 \\ 0 & 0 & 0 & 24.1 & 0 & 0 \\ 0 & 0 & 0 & 0 & 25.6 & 0 \\ 0 & 0 & 0 & 0 & 0 & 25.6 \end{bmatrix} \times 10^9\,\mathrm{N/m^2} \tag{4}$$

$$[e]=\begin{bmatrix} 0 & 0 & -5.2 \\ 0 & 0 & -5.2 \\ 0 & 0 & 15.1 \\ 0 & 0 & 0 \\ 12.7 & 0 & 0 \\ 0 & 12.7 & 0 \end{bmatrix} \mathrm{C/m^2} \tag{5}$$

$$[\varepsilon]=\begin{bmatrix} 6.46 & 0 & 0 \\ 0 & 6.46 & 0 \\ 0 & 0 & 5.62 \end{bmatrix} \times 10^{-9}\,\mathrm{F/m} \tag{6}$$

3.4. Application and acceptance of signals

In the model, one PZT (PZT1) patch is used for signal transmission. The other PZT(PZT2) patch is used to receive the signals. The pulse wave is used as monitoring signal since it is sensitive to crack damage of concrete.

Signal frequency is also an important factor to be considered, which will affect signal sensitivity and propagation distance. Generally, propagates length of the high frequency signal is short, while the propagates length of low frequency signal is long. Because the size of concrete members in buildings is usually large, it is appropriate to select low frequency.

The signal parameters are shown in detail in table 1.

Table 1. Signal types and related parameters table.

Signal type	Amplitude	Frequency
Pulse wave	5v	500 Hz, 1 kHz, 3 kHz, 5 kHz, 10 kHz

IOP Conf. Series: Materials Science and Engineering **789** (2020) 012058 doi:10.1088/1757-899X/789/1/012058

4. Equations and mathematics. Numerical simulation and results analysis

4.1. Monitoring signal and working condition selection

The processes of PZT patches transmitting and receiving signals are simulated. It can be seen from the propagation law of displacement nephogram that: a waveform pulse is generated as soon as PZT1 is stimulated.

With the wave diffuses around, the waveform separates two wave packets. The wave that travels faster is the P-wave and the slower one is S-wave. The displacement nephogram of the wave at different time during its propagation is shown in figure 5. With the wave propagation, there will be reflection wave and other clutters.

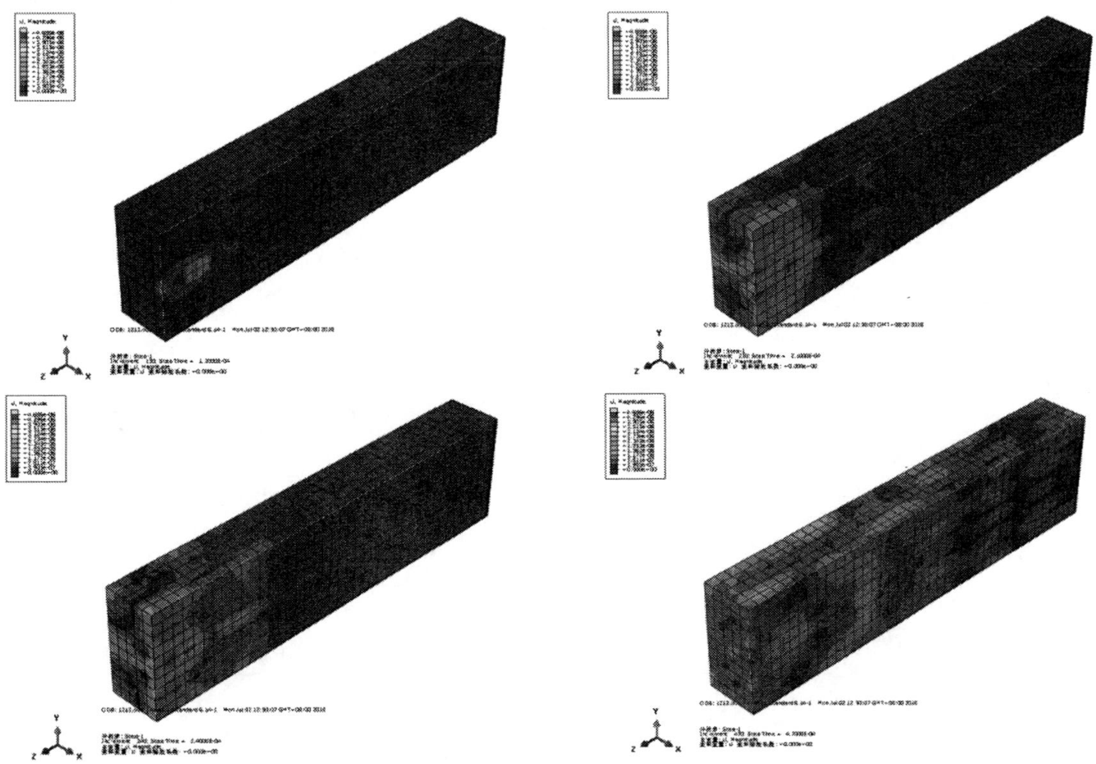

Figure 5. Propagation displacement cloud pattern of wave in the model.

Therefore, it can be judged that the waves PZT2 received were complex which were the superposition of multiple morphological waves. But the head wave received by PZT2 must be the P-wave, because it propagates fastest. Moreover, the P-wave who arrives at PZT2 firstly does not contain clutter. Its change can reflect the state of concrete best. So, the head wave of receipt signals is used to deduce concrete strength. The signal waveform received by PZT2 is shown in figure 6. The head wave in the figure is a P-wave with faster propagation speed and no other waveform superposition.

With the increase of signal time, the superposition of shear wave and clutter appears in the waveform. At this time, the amplitude of the waveform increases obviously. This result is consistent with the output of displacement nephogram.

4.2. Characteristic parameters of concrete strength identification

The signals of different frequencies about different concrete age are obtained and their energy values are shown as in figure 7. It can be seen from the figure that the energy values of signals at each frequency decrease monotonically with the age of concrete. The trend curves of energy values can reflect the change of concrete strength. It can be used as the identification characteristic parameter in concrete strength monitoring.

The corrected determinant coefficient (Adj. R-square) were used as an index for accuracy analysis of fitting curves.

Comparing the fitting curves of different frequency monitoring signals. The closer the coefficient is to 1, the better the accuracy of fitting curve is. Comparing the fitting curves of different frequency monitoring signals, the corrected determinant coefficient (Adj. R-square) statistics are obtained as shown in table 2.

When the signal frequency is 1 kHz, the correction decision coefficient is 0.993, which is the closest to 1. That means the fitting curve is the best when the frequency of the monitoring signal is 1 kHz, and 1 kHz is the optimal monitoring signal frequency.

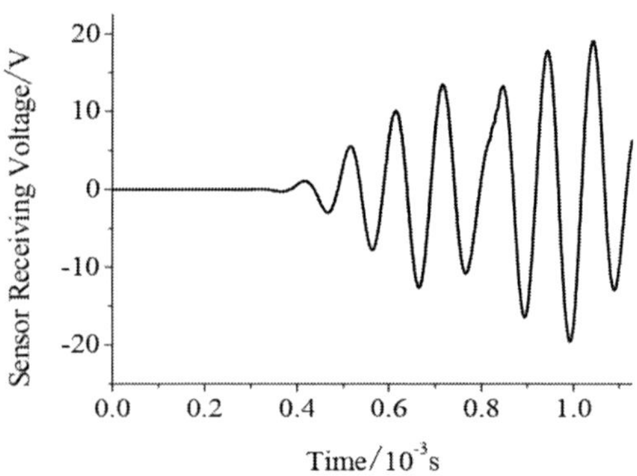

Figure 6. Original waveform received by the sensor.

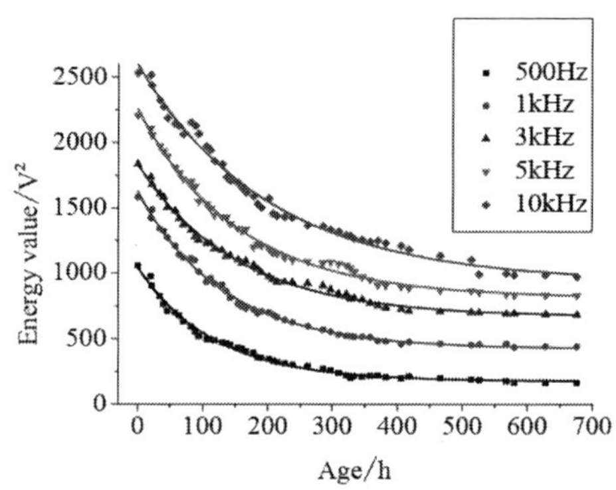

Figure 7. Age-energy relationship curve.

Table 2. Adj. R-square statistical table.

	500 Hz	1 kHz	3 kHz	5 kHz	10 kHz
Adj. R-square	0.899	0.993	0.652	0.503	0.332

4.3. Result analysis

Further, Five-peak pulse with frequency of 1 kHz is used as monitoring signal. The variation of strength of C30, C40 and C50 concrete with age was studied respectively. Relationships between monitoring signal energy, concrete strength and age of those three strength grades concrete were studied by numerical simulation results. Normalizing the date, the relationship curves about energy ratio, strength ratio with age were gained and shown in figure 8.

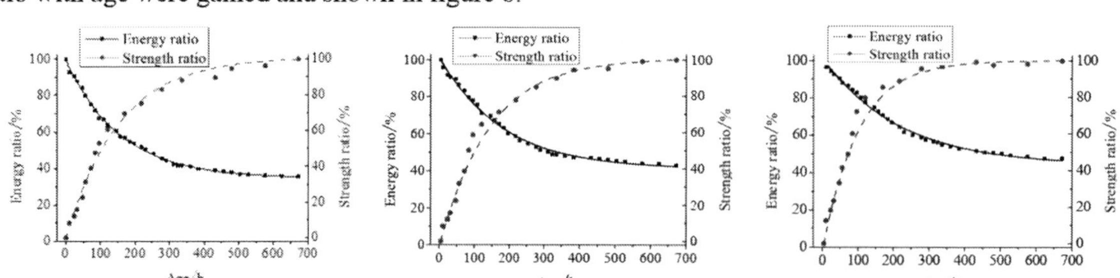

(a) C30 (b) C40 (c) C50

Figure 8. Relationship curves about energy ratio, strength ratio with age of C30, C40 and C50 concrete.

Contrasting two curves in the same coordinate system, they all have monotonic consistency with age. The energy ratio-age curve is monotonically decreasing, and the strength ratio-age curve is monotonically increasing. The development trends of two curves are the same. Before 168 hours (7 days) of age, the two curves all develop sharply. Then they become gently. This proves that there is a corresponding relationship between strength ratio and energy ratio. It's feasible to predict the concrete strength and its development trend by using monitoring signal energy.

The direct relationship between energy ratio and strength ratio can be constructed through the intermediate variable of age.

As shown in figure 9, whatever the strength grade of concrete, the relationship between energy ratio and strength ratio is approximately linear. And they are in inverse proportion, which means with the increase of concrete age, the strength of concrete increases, the strength ratio increases, while the energy ratio decreases.

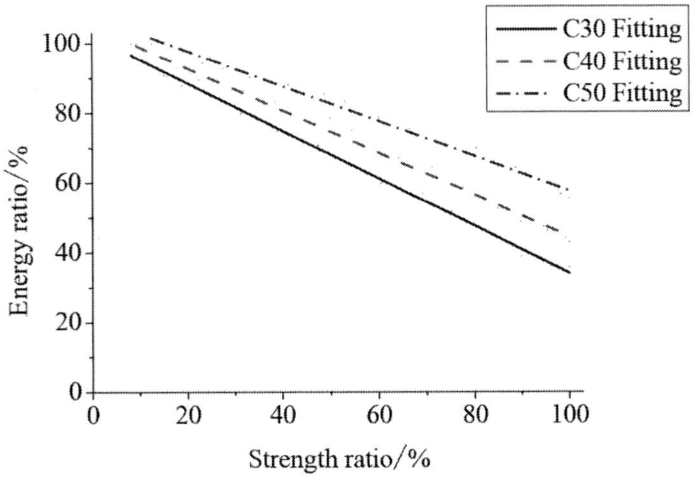

Figure 9. Strength ratio-energy ratio diagram of C30, C40 and C50 concrete.

The three lines in figure 9 represent three different concrete strength grades. It can be judged that: the higher the strength grade of concrete, the upper the line is. And the lower the strength grade of concrete, the lower the line is.

Furthermore, with the increase of concrete strength grade, the slope of the line between energy ratio and strength ratio becomes smaller, the final attenuation rate of the monitoring signal also becomes smaller. In this paper, when the concrete age of 28 days is reached, the final attenuation rate of monitoring signal energy corresponding of C30 concrete is about 65%, which is about 55% of C40 concrete and about 45% of C50 concrete.

5. Conclusions

In this paper, the feasibility of using SA to monitor concrete strength is studied by numerical simulation. The concrete beam model with embedded PZT transducers had been established, and the PZT transducers were stimulated to transmit and receive signals. Wave analysis method is used to study the change of monitoring signal with the increase of concrete strength. The conclusions can be drawn as that:

- it is feasible to use PZT transducers and wave analysis method to monitor concrete strength. The energy of monitoring signal decreases monotonously with the increase of concrete strength. It can reflect the change of concrete strength very well and be suitable as characteristic parameters of concrete strength identification.
- The five-peak pulse signal can be used as the signal type of concrete strength monitoring. When

the signal frequency is 1 kHz, the corrected determinant coefficients of the fitting curve between the monitoring signal energy and the age of concrete are the best. So, 1 kHz can be used as the optimal frequency parameters of monitoring signal.

- The relationship between the energy ratio of monitoring signal and the strength ratio of concrete is established. They approximate satisfies the linear relationship. By comparing the energy ratio-strength ratio lines of three strength grades of concrete, it is found that the higher the strength grade of concrete, the upper the energy ratio-strength ratio relationship line is, the smaller the inclination of line is, the smaller the final energy decay rate is.

Acknowledgments
The research described in this paper was financially supported by Liaoning Province Natural Science Foundation (20170540753); Shenyang Jianzhu University Science Foundation (2014081).

References
[1] Xiao L L and Zheng Y W 2004 J. *Foreign Building Materials Technology* 339-341
[2] Liu Y M, Wei Y D et al 2005 J. *Industrial Building* **6** pp 64-65
[3] Soh CK, Tseng KK, Bhalla S and Gupta A 2000 J. *Smart Materials and Structures* **9** pp 533-542
[4] Tseng K K and Wang L S 2004 J. *Smart Materials and Structures* **13** pp 1017-1024
[5] Bhalla S and Soh C K 2003 J. *Earthquake Eng. Struct.* Dyn. **32** pp 1897-1916
[6] Bhalla S and Soh C K 2004 J. *Journal of Aerospace Engineering* **17** pp 154-165
[7] Song G, Gu H, Mo YL and Hsu T 2007 J. *Smart Materials and Structures* **16** pp 959-968
[8] Yan S and Sun W 2009 J. *Smart Materials and Structures* **18** pp 047 001
[9] Gu H, Song G, Dhonde H, Mo Y and Yan S 2006 J. *Smart materials and structures* **15** pp 1837–1845
[10] Gu H, Mo T and Sanders D 2010 J. *Smart Materials and Structures* **19** pp 065 026
[11] Y Mo, H Gu and A Be 2010 J. *Smart Materials and Structures* **19** pp 065 021
[12] Qiao P Z and Chen F L 2013 J. *Journal of Material in Civil Engineering* **25** pp 836-845

IOP Conf. Series: Materials Science and Engineering **789** (2020) 012059 doi:10.1088/1757-899X/789/1/012059

The influence of waterproofing additives introduced into the mass of fresh concrete on the on the durability feature of the hardened concrete

I Tuns[1*], D N Isopescu[2], F L Tamas[1] and I Boian[1]

[1] Transilvania University of Brasov, Faculty of Civil Engineering, Brasov, Romania
[2] "Gheorghe Asachi" Technical University of Iasi, Faculty of Civil Engineering and Building Services, Iasi, Romania

[*] ioan.tuns@unitbv.ro

Abstract. The constructions with fluid retention and storage in general, between which the water holds a significant weight, are permanently exposed of physical and mechanical actions of deposited agent, with the continuous manifestation of the exfiltration tendency, herby the requirement of performance at tightness becomes essential. If the constructions that fulfill this role are realized of reinforced concrete, the tightness at fluid in general and at water in particular of the elements which compose the storage space is insured in a significant proportion by the waterproofing qualities of the structural concrete put in place. When the process of waterproofing concrete is clean and environmentally friendly and leads to improvement of initial physical- mechanical characteristics, represents an essential element in the combined approach of the performance requirements regarding the tightness and resistance condition. In this paper is presented a research study concerning the positive influence of waterproofing additive "Penetron" introduced into the fresh concrete mass used to realization of structural elements, with the role of storage and treatment of waste water, to mechanical characteristics of concrete after hardening.

1. Introduction

1.1. General aspects regarding the durability of the constructions

The period of time in which the construction provides normal operating conditions, in accordance with the function assigned to the design, represents the "anticipated service life" or the "projected service life", which generically expresses the notion of durability.

By defining the notion of sustainability, this feature is attributed to the construction system, following the assumed development of the stages of conception, design and execution. In this approach, sustainability is the characteristic of a construction based on the performance requirements designed, materialized by an adequate execution and maintenance-repair works carried out properly and on time during the projected service life. Given the complexity of the construction of buildings as well as the diversity of operating or external environmental factors and their mode of action, the assessment of sustainability is a laborious problem, with a variety of variable elements, of limited knowledge, which influence the normal life span.

In this context, the evaluation of the lifetime of constructions is the result of a probabilistic approach, based on interpretations and processing of statistical data, depending on the multitude and variability of the considered characteristics.

Starting from the difficulty of directly determining the anticipated (projected) lifespan, this is evaluated in modern practice based on the performance characteristics attributed to the constructions to the design, in direct correlation with the predictable ones for the existing constructions, similar in terms of design and construction. The operating environment conditions, for which "in situ" checks and tests can be carried out, following the behavior in time and meeting the performance characteristics initially planned [1].

The durability of a construction is determined by a complex of factors, regarding:

- the design activity;
- the execution activity;
- the repair-maintenance-operation activity;
- the characteristics of the materials used and their evolution over time;
- operating environment conditions
- the action of external factors
- other aggressive actions

In this presentation, the characteristics of the materials used in the construction elements are an important factor in ensuring the durability characteristics under the conditions of the function to be performed.

Therefore, in the case of concrete constructions with a liquid retention and storage role, the performance requirement regarding the sealing is directly reflected by the characteristics of the materials used, between which the waterproofing solution occupies a special place.

1.2. General considerations regarding the action of waterproofing additives

The permeability of the concrete is a physical characteristic that expresses the ability to penetrate the gases or liquids in its mass in the presence of a certain pressure difference.

Water penetration in concrete is determined by the hydrostatic pressure gradient, capillary suction, diffusion of water vapor and osmotic pressure.

Contrary to the physical characteristics regarding the permeability of the concrete, the water impermeability of the hardened concrete reflects its ability to prevent the ingress of pressurized water into its mass.

The requirement of impermeability is absolutely necessary in the case of the concrete used in the construction of water retention and storage constructions, specific to the constructions in the water supply and sewage systems of the waste water treatment stations, of the hydrotechnical, maritime constructions etc.

The reduced water permeability of the concrete is influenced by the compactness and existence of the pores, reflected by the composition (dosages, basic materials, and additives), the pouring technology and the protection after casting [2-4].

Additives, in general, are technical substances or products that are introduced into the concrete mass when preparing, for the purpose of the desired modification of the new characteristics, by specific physico-chemical mechanisms.

In the case of waterproofing additives, their introduction into the concrete composition, in preparation, favors the reduction or cessation of the phenomena of physical nature at the level of the matrix under the action of the liquid phase contained in the operating environment.

The physico-chemical mechanisms by which the impermeability of the concrete is ensured, without negatively influencing the mechanical and volume characteristics, are multiple, and can be synthesized according to the action of the additives as follows [5]:

- by forming colloidal compounds that produce clogging of pores and voids in the concrete mass;
- by reducing the A/C ratio, by changing the pore size and distribution;

- by the hydrophobization action on the cement granules, respectively by reducing the capillary ascension of the water;
- by training and favorable distribution of air in the concrete mass.

In this paper, the experimental studies performed are based on the action of the waterproofing additive Penetron admix, introduced into the mass of fresh concrete in a percentage of (0.8-1)% of the amount of cement, before casting, in the concrete mixer, and energy mixing for 15 min., according to figure 1.

Figure 1. Introduction of the Penetron admix additive in the fresh concrete mass and homogenization of the mixture.

The waterproofing action of the Penetron admix additive is due to the multiplication of the crystalline component in the matrix and the filling of the pores in the concrete mass.

The Penetron admix waterproofing additive is supplied in powder form, having the mineralogical composition compatible with that of Portland cement, and in contact with water, the accentuated development of chemical reactions between the components of the additive and the hydration products of the cement with formation and multiplication of the chain is stimulated crystalline compounds.

The chemical reaction that occurs when the Penetron admix waterproofing additive is introduced into the fresh concrete mass, occurs according to the following generic mechanism [6]:

$$Ca(OH)_2 + RCOOH \rightarrow Ca^{+}COOR^{-} + H_2O \text{ [1]} \quad (1)$$

where $Ca(OH)_2$ – calcium hydroxide contained in the mass of fresh concrete; RCOOH – stearate mixture contained in the Penetron admix waterproofing powder; $Ca^{+}COOR^{-}$ – insoluble calcium stearate, the resultant component of the chemical reaction; H_2O – the water released from the chemical reaction.

The increase of the volume of the crystalline component of the concrete nature, after the introduction of the additive Penetron admix, gives to the hardened mixture waterproofing properties, which stop the penetration of water into its mass and recommend the use of the additive in the case of constructions with fluid retention and storage.

2. Description of experimental research

The experimental study aimed to highlight the influence of the Penetron admix waterproofing additive, introduced in the fresh concrete mass as a waterproofing solution, of two decanters with the role of retention and treatment of waste water resulting from the process of processing of animal waste, on the characteristics of durability of hardened concrete.

2.1. Analysis of the factors of influence of the sustainability characteristics

The experimental researches aimed to perform a comparative study on the physico-mechanical characteristics determined in the laboratory on ordinary concrete samples and concrete samples in the composition of which Penetron admix waterproofing additive was introduced and "in situ" on two reinforced concrete decanters. The experimental studies carried out in the laboratory on concrete samples of classes C20/25, C25/30, C30/37 and C35/45 aimed at:

- determination of compressive strength at 28 days;
- determination of the degree of impermeability to water and the penetration depth;
- analysis of the microstructure of the hardened concrete by optical microscopy.

The "in situ" experimental study aimed to determine the time of self-repair of the defects in the walls of the reinforced concrete decanters related to a wastewater treatment plant.

2.1.1. Determination of compressive strength. For having a clear image on the influence of the waterproofing additive Penetron admix introduced into the fresh concrete mass on the compressive strength of the hardened concrete, determinations were made on cubic-shaped test pieces, with a side of 150 mm, 28 days after making.

Laboratory determinations were performed on a number of 24 samples, made of concrete of the classes: C20/25, C25/30, C30/37, C35/45, of which 12 samples without a waterproofing additive and 12 tests with a Penetron admix waterproofing additive, introduced into the mass of fresh concrete at preparation, in the percentage of 0.85% of the quantity of cement.

The results obtained from the determinations made are presented in table 1.

Table 1. The values of compressive strength on concrete grades.

Concrete class			Compressive strength (MPa)			Average compressive strength (MPa)
			Sample 1	Sample 2	Sample 3	
1	Without the addition of waterproofing Penetron admix	C20/25	26.1	25.8	27.6	26.5
2		C25/30	32.7	32.5	33.7	32.96
3		C30/37	38.7	38.1	38.9	38.56
4		C35/45	46.8	47.5	46.7	47
5	With the addition of waterproofing Penetron admix	C20/25	26.4	26	27.8	26.7
6		C25/30	32.9	32.8	33.5	33.07
7		C30/37	38.5	38.5	38.9	38.63
8		C35/45	46.9	48.1	47.2	47.4

The interpretation of the table data shows close values of the compressive strength on concrete classes in the two variants of realization, with a slight increase, but not mechanically significant, for the samples with Penetron admix waterproofing additive.

This slight increase of the compressive strength for the samples made with concrete in which Penetron admix waterproofing additive was introduced in 0.85% of the cement quantity, is explained by the strengthening of the additional crystalline formations favored by the additive.

Therefore, the Penetron admix waterproofing additive introduced into the fresh concrete mass does not influence the class of the concrete, but it increases the compactness of its microstructure, by filling the pores with microcrystals generated by the Penetron admix waterproofing additive, thus reducing the transport phenomena in the concrete mass, respectively the improvement of the service life.

2.1.2. Determining the degree of waterproofness and the penetration depth. The water penetration resistance was determined on standardized concrete samples, classes C20/25, C25/30, C30/37, C35/45 under the specific working conditions for this type of test. The water penetration determination was performed at a pressure of 0.8 MPa, the results being presented in table 2.

Table 2. The depth of water penetration in the concrete sample.

		Concrete class	Depth of water penetration (cm) at working pressure of water 0.8 MPa			Average depth of water penetration (cm)
			Sample 1	Sample 2	Sample 3	
1	Without the addition of waterproofing Penetron admix	C20/25	2.2	2.0	2.0	2.07
2		C25/30	2.0	1.7	1.7	1.8
3		C30/37	1.8	1.5	1.5	1.6
4		C35/45	1.4	1.4	1.3	1.37
5	With the addition of waterproofing Penetron admix	C20/25	1.7	1.8	1.7	1.73
6		C25/30	1.5	1.4	1.4	1.43
7		C30/37	1.2	1.2	1.2	1.2
8		C35/45	1.0	0.9	1.1	1

Following the table data, the following aspects are distinguished:

- the average values of the water penetration depth are less than 1.3 cm, in the case of concrete samples without the addition of waterproofing Penetron admix;
- the average values of the water penetration depth are less than 1 cm, in the case of concrete samples with Penetron admix waterproofing addition;
- the water penetration depth is dependent on the concrete class;
- the Penetron admix waterproofing additive introduced into the concrete mass by 0.85% of the amount of cement, reduces the water penetration depth, in the following percentages: 16.5% for C20/25; 21% for C25/30; 25% for C30/37; 27% for C35/45.

The beneficial aspects of the microstructure of the concrete generated by the increase of the crystalline formations in the matrix as a result of the introduction of the waterproofing additive Penetron admix, which fills the pores of gel and capillaries, reducing the porosity of the concrete and the capillary suction of the water, clearly reflects the improvement of the resistance to water penetration and the durability characteristics of the hardened concrete.

2.1.3. Analysis of the microstructure of the hardened concrete by optical microscopy. The compression-tested concrete specimens, shown in figure 2, were prepared by splitting them into smaller fragments for analysis of microstructure by optical microscopy.

Figure 2. Compression-tested samples, waiting for optical microscopy.

The optical analysis of the hardened concrete fragments was carried out within the Research Institute for Sustainable Development at the Transilvania University of Brasov, using the Keyence VHX600 digital microscope, with a 54 MP video camera and three CCD sensors, which provide internal focusing and magnification from 20x to 200x.

The hardened concrete fragments were extracted from the compression test samples, classes C20/25, C25/30, C30/37, C35/45, respectively 12 pieces without a waterproofing additive and 12 pieces with a Penetron admix waterproofing additive.

The optical analysis carried out aimed at highlighting, in comparative aspect, the microstructure of the concrete, on concrete classes, without and with Penetron admix waterproofing additive, introduced into the fresh concrete mass.

Representative aspects regarding the concrete texture of the analyzed samples are presented in figures 3, 4, 5 and 6, where the microscopic images are focused and magnified by 200x.

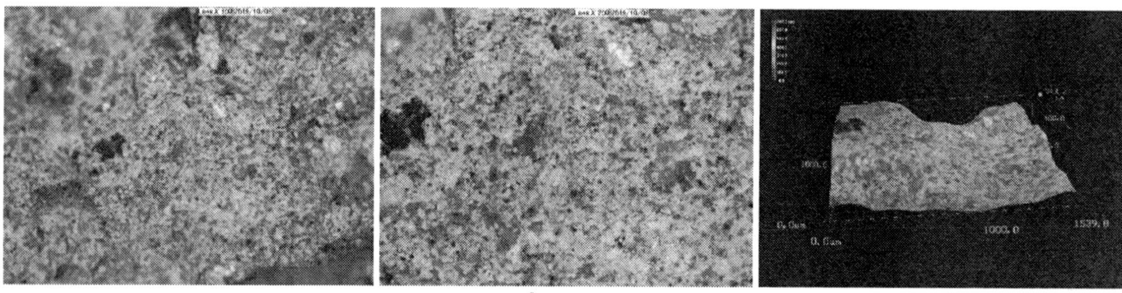

a b c

Figure 3. Microscopic images of the C20/25 concrete test specimen: a – without waterproofing additive; b – with Penetron admix additive; c – 3D representation.

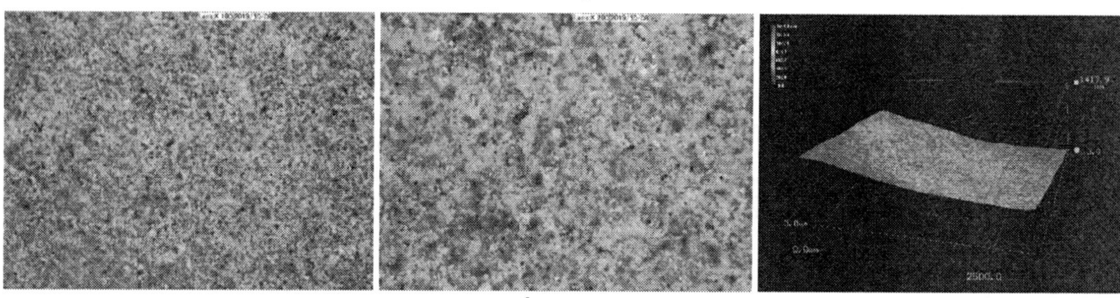

a b c

Figure 4. Microscopic images of the C25/30 concrete test specimen: a – without waterproofing additive; b – with Penetron admix additive; c – 3D representation.

a b c

Figure 5. Microscopic images of the C30/37 concrete test specimen: a – without waterproofing additive; b – with Penetron admix additive; c – 3D representation.

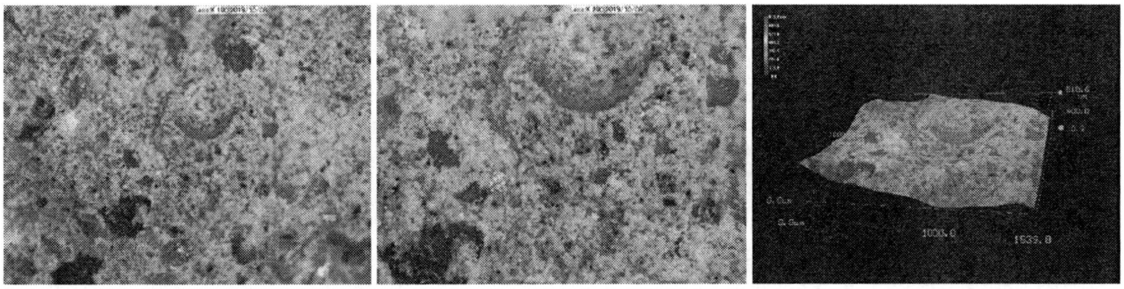

a b c

Figure 6. Microscopic images of the C35/45 concrete test specimen: a – without waterproofing additive; b – with Penetron admix additive; c – 3D representation.

The microscopic images made on the hardened concrete specimens highlight the following aspects:

- the texture of the concrete prepared with waterproofing additive Penetron admix is more uniform and compact for all classes of concrete, compared to that of the concrete of the same class, without waterproofing additive;
- concrete samples with a waterproofing additive highlight a fine particle agglomeration in the matrix of the hardened concrete, compared to the internal structure of the samples without a waterproofing additive;
- concrete samples with a waterproofing additive show a lower internal porosity microstructure, compared to the samples made of concrete without a waterproofing additive.

All the illustrative aspects indicated highlight the beneficial role of the Penetron admix waterproofing additive, for all classes of concrete, on the durability characteristics of the hardened concrete.

2.1.4. The self-repair time, in situ, of the defects in the hardened concrete structure. Starting from the use of the concrete with waterproofing additive Penetron admix to the realization of the elements that are part of the structural composition of the constructions with the role of retention and storage of the liquids, the aim was to highlight the time of self-repair of the cracks due to concrete contraction in the first days after the casting.

The phenomenon of self-repair refers to contraction cracks, which occur frequently in the case of concrete elements.

The observations were made in situ on the walls of two decanters of reinforced concrete with the role of storage and treatment of waste water, in a unit for processing animal waste.

The results showed that after 28 days the process of water infiltration through cracks is reduced by about 50 % compared to the initial situation [7].

3. Conclusions

Experimental researches regarding the influence of the waterproofing additive Penetron admix from the mass of the fresh concrete on the durability characteristic on the reinforced concrete, have led to the following conclusions:

- the Penetron admix waterproofing additive introduced into the fresh concrete mass does not influence the concrete class. The increase of the compressive strength is insignificant, below 1%, but it increases the compactness of the microstructure by filling the gel and capillary pores with newly formed crystals, reducing the phenomena of transport in the concrete mass of the degrading agents.
- the depth of water penetration in the concrete mass is reduced, according to its class, as follows: 16.5% for C20/25, 21% for C25/30, 25% for C30/37 and 27% for C35/45;
- the analysis of optical microscopy shows a microstructure of the concrete in which Penetron admix waterproofing additive was introduced more uniform and compact, compared to the one of the concrete without waterproofing additive;

- the phenomenon of multiplying the crystalline formation following the hydration reactions that take place after mixing the additive Penetron admix with water ensures, in time, the self-repair of the internal cracks due, in general, to the contraction of the concrete.

The results of the experimental researches highlight by the analyzed factors the increase of the durability characteristic of the reinforced concrete in which was introduced waterproofing additive Penetron admix, with more than 10%, compared to the same class of concrete without additional waterproofing.

The solution of waterproofing of concrete by introducing in its mass the waterproofing additive Penetron admix is practicable and friendlier with the environment compared to the classic version of waterproofing, a relevant image in this sense is that of figure 7, which represents a treatment station of waste water from an animal waste processing unit.

Figure 7. Wastewater treatment station in operation.

References

[1] Georgescu D P 2001 *Îndrumător de proiectare a durabilității betonului în conformitate cu anexa națională de aplicare a SR EN 206-1. Clase de durabilitate (Guidance for the design of concrete durability according to the national annex of the application of SR EN 206-1. Durability classes)* Everest typography Bucharest

[2] Burg R G 2001 *Chemical admixture for concrete structure*

[3] Paillare A M, Ben M and Akman S 1992 *Guide for use of admixtures in concrete* Materials and structures vol. **25** nr. 145

[4] Thomas M and Wilson M 2002 *Admixtures for use in concrete* Portland Cement Association

[5] Boicu G G 2011 Contribuții privind realizarea calității betoanelor pe traseul stație-construcție, pe timp friguros (Contributions regarding the realization of the quality of the concrete on the route station-construction, in cold time) PhD Thesys Technical University "Gh. Asachi" Hydrotechnics Facult, Geodesy and Environmental Engineering Iasi

[6] ACI 212.3R-10: *Report on Chemical Admixtures for Concrete* ISBN 978-0-87031-402-5

[7] Tuns I and F-L Tamas 2018 Comparative study of practical solutions concerning water tightness of two reinforced concrete decanters *Bulletin of the Transilvania University of Brașov* Vol. **11** (60) No. 1 Series I: Engineering Sciences

IOP Conf. Series: Materials Science and Engineering **789** (2020) 012060 doi:10.1088/1757-899X/789/1/012060

Comparative application of two methods in the second order elastic analysis of the steel structures

F Ţepeş-Onea[1*] and R Dimofte[1]

[1] University Ovidius of Constanta, Faculty of Civil Engineering, Constanta, Romania

[*] tflorin@univ-ovidius.ro

Abstract. In most situations, in the case of metal structures, the loss of stability occurs before plastic elements enter the structure. The present paper aims to compare the response of a metallic structure obtained by an elastic analysis of the 2nd order developed with the two methods provided in SR-EN 1993-1-1:1) global analysis that takes full account of geometrical and material imperfections and second-order effects; 2)-individual stability check of the equivalent bars using buckling lengths corresponding to the overall instability mode. Obviously the first method is the most complex because it takes into account both global and local imperfections. The second method is the simplest and most used method. In this case, the verification relationships take account of the imperfections of the elements. For global analysis, the global imperfections and local bar imperfections will be considered. These imperfections will be taken into account in the worst form and sense. The check for global analysis will be done using finite element meshing, while using the geometric nonlinearity option so that the stiffness matrix is calculated at each loading step. The individual check of the elements shall be performed according to SR EN1993-1-1 according to the type of request.

1. Introduction

By this comparative analysis it is obtained an imagine of a result of the verification established by the two methods provided by the normative, thus facilitating a choice of the solution chosen in the design. The high slenderness of the metal structures is important compared to the structures in reinforced concrete frames. Consideration of the effects of the second order is mandatory because the deformations have a significant influence on the stress.

This condition is expressed by:

$$\alpha_{cr} = F_{cr} \cdot F_{Ed}^{-1} \quad (1)$$

where α_{cr} the coefficient of multiplication for the structure's elastic instability load; F_{cr} lateral buckling critical elastic load; F_{Ed} the total vertical load applied to the structure.

SR EN 1993-1-1 presents three methods for taking into account the effects of the second order and the imperfections [5]:

- through the global analysis, the imperfections (geometric and material) and the effects of the second order are taken into account;
- the global analysis takes into account the global imperfections and the effects of the second order global (the force-displacement effect) and the stability checks at the element level take into account the local imperfections and the effects of the second order.

- for the basic cases, through individual stability checks of the equivalent bars, using buckling lengths corresponding to the global instability mode.

In the present study, the results obtained from the first method, which is the most complex and, will be compared with the third method which is the most used in the design activity. Stability checks for individual elements for center compression, bending or axial force bending are currently performed in the design activity.

In order to analyze the structure of figure 1 with method 1, both global and local imperfections, curves of elements were taken into account. Figure 2 shows the global imperfections for the frames and the local ones for the columns. As with the columns, local imperfections were also used for the beams.

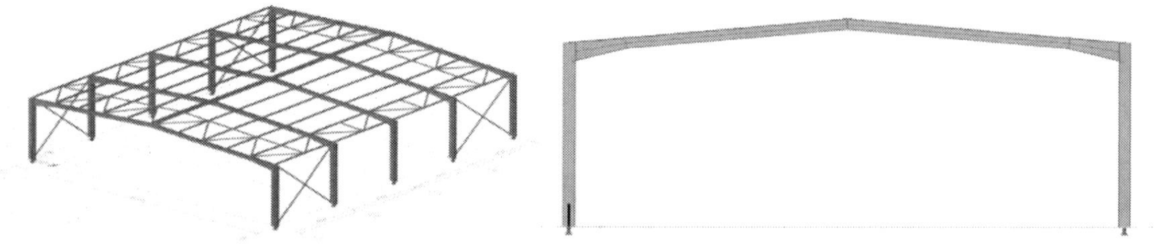

Figure 1. Initial structure. Transversal framework studied.

2. Imperfection

The first type of imperfection used is the initial global deviation from the vertical axis, figure 2.

This was calculated with the relation [4]:$\varphi = \varphi_0 \cdot \alpha_h \cdot \alpha_m$, in which φ_0 have the basic value [3] $\varphi_0 = 200^{-1}$, α_h is the reduction coefficient applicable to the height h of the columns:

$$\alpha_h = 2 \cdot h^{-0.5} = 2 \cdot 7^{-0.5} = 0.7559 \quad (2)$$

where h is the height of the structure in meters, α_m is a reduction factor for the number of columns in a row:

$$\alpha_m = [0.5 \cdot (1 + m^{-1})]^{0.5} = [0.5 \cdot (1 + 2^{-1})]^{0.5} = 0.866 \quad (3)$$

where m *is* the number of columns in a row and:

$$\varphi = 200^{-1} \cdot 0.755 \cdot 0.866 = 3.2 \cdot 10^{-3} \quad (4)$$

For this φ calculated, taking into account the length of the columns, results in a deviation from the vertical of 2.2 cm presented in figure 2.

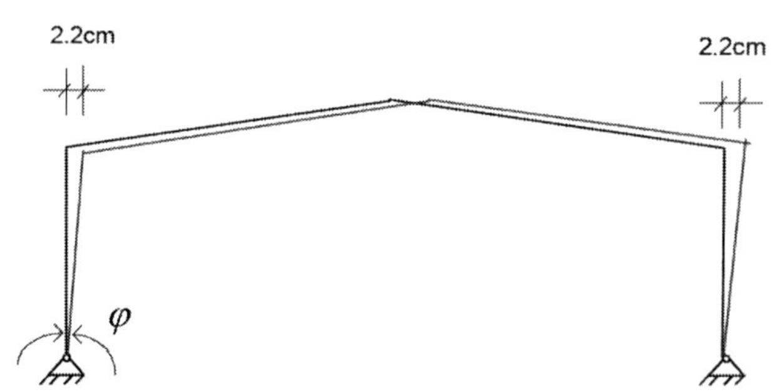

Figure 2. Global imperfections for frameworks used in analysis.

The second type of imperfections used in the analysis are the local imperfections in the arch e_0 of the bends (initial curves) for bending buckling.

The value used for the columns is $\frac{e_0}{L} = 250^{-1}$ [4] and was used $e_0 = 2.2\ cm$ and for beams 4 cm. The table in the National Annex of SR EN 1993-1-1 (the calculation values of the imperfections in the arc) was used [2].

3. Case study

3.1 Linear analysis of individual buckling checks

The case study was performed on the transversally frame of a hall, figure 1, made in Constanta with a snow load of 0.8·1.5 kPa = 1.2 kPa, and a ground acceleration of 0.2 g. The wind load on the structure was also taken into account. It was considered a behavioral factor. The transversal framework is not constrained in the respective direction, thus, the specific provisions P100-1/2013 was used in the design activity. Thus, the design based on the over resistance Ω of Annex F (P100-1) was also taken into account [1].

The linear static calculation was performed with the SCIA Engineer program. Initially the calculation was performed spatially on a hall with four openings, the beam of 6m and a layout of panels of 10 panels / opening.

The columns have 7 m and the opening of the hall is 20.7 m. The beam resulted from the IPE450 stability checks and for the HEB300 columns.

Steel mark S355. The following efforts were used for checking the strength and stability of the beam:

- for the constant section: $N_{Ed} = 83.28\ kN; V_{Ed} = 138.36\ kN; M_{Ed} = 268.2\ kNm$
- for the haunch section: $N_{Ed} = 83.28\ kN; V_{Ed} = 138.36\ kN; M_{Ed} = 445.61\ kNm$

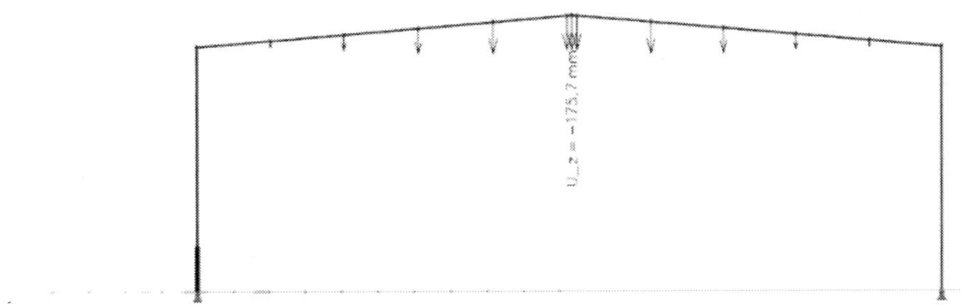

Figure 3. Vertical displacement 175.7 mm for linear calculation.

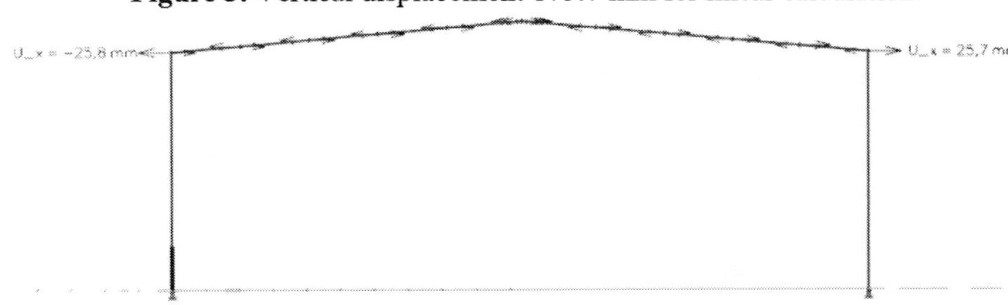

Figure 4. Horizontal displacement 25.8 mm for linear calculation.

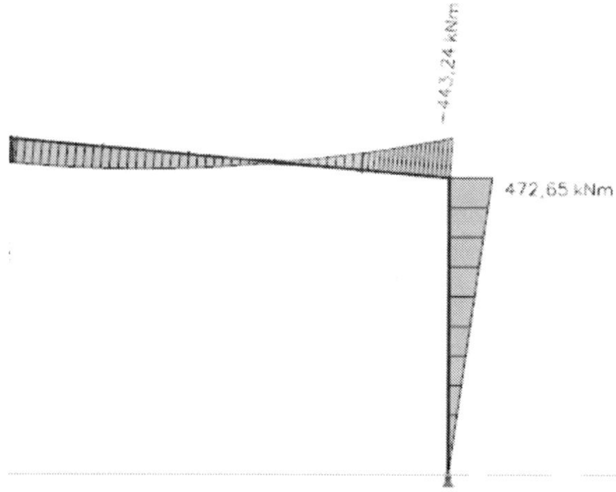

Figure 5. Bending diagram, on the semi-structure, linear analysis (first case).

For the strength and stability checks at the columns in linear analysis, the following values were used for the M-N interaction: $N_{Ed} = 160.16\ KN; V_{Ed} = 67.6\ KN; M_{Ed} = 473.17\ KNm$.

In the case of the linear elastic calculation, the crossed scales are: a) classification in the class of the section; b) resistance checks; c) buckling checks. The calculation of the interaction factors kyy, kzy was performed with method 1.

In the linear analysis, M-N interaction checks were performed for both the column and the beam in the haunch and without a haunch section.

In the case of the columns, the verification of the interaction formulas has the form [2], [3]: $M_{yRk} = w_{pl,y} \cdot fy = 663.4\ KNm$ with:

$$N_{ED} \cdot \gamma_{M1} \cdot (\chi_y \cdot N_{Rk})^{-1} + k_{yy} \cdot M_{Ed} \cdot \gamma_{M1} \cdot (\chi_{LT} \cdot M_{y,Rk})^{-1} = 0.86 < 1 \quad (5)$$

$$N_{ED} \cdot \gamma_{M1} \cdot (\chi_z \cdot N_{Rk})^{-1} + k_{zy} \cdot M_{Ed} \cdot \gamma_{M1} \cdot (\chi_{LT} \cdot M_{y,Rk})^{-1} = 0.491 < 1 \quad (6)$$

In the case of beams (without a haunch section) analytical relation (5) and (6) equal with: 0.468 respectively 0.26.

3.2 Geometrically nonlinear analysis

The following displacement values were obtained in the geometrically nonlinear analysis with global and local imperfections: Dz=182.9 mm; Dx=26 mm.

In the case of the column the values of the efforts are: $N_{Ed} = 159.86\ KN; V_{Ed} = 73.01\ KN; M_{Ed} = 484.7\ KNm$.

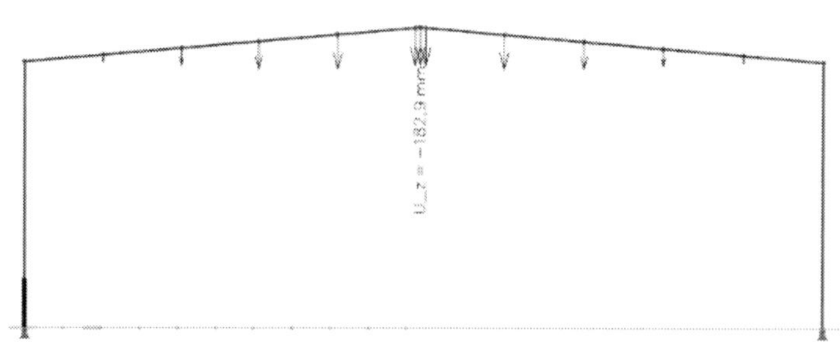

Figure 6. Vertical displacement diagram geometrically non-linear analysis with global and local imperfections in the arc Dz = 182.9 mm.

4. Conclusion

The use of modern calculation software that allows the activation of geometrical nonlinearity modules (Timoshenko) and initial imperfections simplify several calculations for the first method. The present study tries to show whether or not this stable analysis solution is justified.

The existence of global and local imperfections in the arc overlay with a non-linear geometric calculation does not bring major changes in terms of the efforts and displacements to which the design is made.

The realization of a classic design with resistance and stability checks for the individual elements is thus shown to be enough. I think it is more important that the optimization with the computing programs is doubled by the strength and stability checks performed manually. At the same time, the design of a seismically well structure, with correct knot connections introduced in the program and with the use of the super-resistance according to P100-1 are aspects that most often have a decisive influence on the structure's response.

Finally, a non-linear geometrical calculation was made with the amplification of the snow loads, with increasing tracking according to the load (table 1). Obtaining a single matrix resulted in an increase of more than 4 snow loads. On the other hand, the design was achieved with a coefficient of interaction M-N of 0.491.

Table 1. Displacements obtained for the amplification of the load.

Load coefficient	Dy [mm]	Dz [mm]
1.0	27.4	182.9
1.5	29.1	252.2
2.0	30.7	323.1
2.5	32.3	395.8
3.0	38.5	470.5
3.5	44.8	547.8
4.0	50.7	628.0

References

[1] * * * P100-1/2013 *Design of structures for earthquake resistence, part 1*
[2] * * * SR EN 1993-1-1/2006 *Eurocod 3 Design of steel structures. General rules and rules for buildings*
[3] Dubina D and Ungureanu V 2010 *Checking the stability of a steel element in accordance with SR EN 1993-1.1, Recommendations for calculating comments and application examples*
[4] Dubina D and Dinu F 2010 *The global structural design of the steel structures. Recommendations, comments and examples of application according to SR EN 1993-1-1 and SR EN 1998-1*

IOP Conf. Series: Materials Science and Engineering **789** (2020) 012061 doi:10.1088/1757-899X/789/1/012061

Laboratory testing of static and dynamic deformation characteristics of small scale model of pavement subgrade

V Valaskova[1*] and J Vlcek[2]

[1] University of Zilina, Faculty of Civil Engineering, Department of Structural Mechanics and Applied Mathematics, Zilina, Slovak Republic
[2] University of Zilina, Faculty of Civil Engineering, Department of Geotechnics, Zilina, Slovak Republic

* verronika.valaskova@fstav.uniza.sk

Abstract. Interaction system vehicle-pavement represents actual problem in terms of structural, life service and environmental requirements. Experimental measurements with real vehicles and pavements are time consuming and bring high complexity of the inputs, relations and other boundary conditions which need to be identified to achieve reliable outputs. Inevitable uncertainties related to the real structure can be controlled when scale models with identified boundary conditions are prepared. Determination of the characteristics of the individual components of the vehicle-pavement interaction system at controlled conditions is then possible. The subgrade is an important part of the pavement structure. A laboratory and numerical simulations based on the static and dynamic testing methods were used to determine the deformation parameters of the subgrade. A viscoelastic mass was selected as a subgrade for the small scale model of the pavement. Its characteristic allows the simulation of the dynamic effects of the vehicle on pavement structure. A short-time dynamic loading of the subgrade occurs when the vehicle moves on the pavement. Therefore, the response of the subgrade should be also characterized with the dynamic parameters. The results obtained by the experimental observations in-lab will be used for the evaluation of the traffic effect on the pavement structure and surrounding objects such as buildings.

1. Introduction

Observation of the vehicle-pavement system with the real vehicle and pavement is affected by the numerous factors, which many of them cannot be fully controlled within the analysis because of its stochastic nature and complexity also based on mutual influence [1-3]. Small scale model seems to be reliable in terms of controlled observation at given or known boundary conditions. Of course, detailed comparison of model results and real system outputs should be realized to verify the reliability of the small scale model.

Because of small thickness of the pavement, subgrade represents very important part of the structure. Pavement loading is mainly spread in stiffer upper layers but some portion of the loading is transferred to the subgrade and then further to the environment.

Considering this, we selected the material based on the gelatine which should represent the behaviour of the subgrade under dynamic conditions.

In engineering practice is gelatin commonly used to model seismic activities such as earthquakes. One of the important parameters is the properly model scale ratio. Here it is necessary to correctly determine the gelatin material parameters and after that these values can be interpolated to real situations. In this article we focus on the possibility of using a small scale gelatin model for dynamic vehicle-pavement interaction. The tested gelatin consists of animal origin obtained from collagen using hydrolization. Based on the measured experimental results we want to determine the suitability of use for vehicle - pavement interaction [4-6].

This paper presents the results of the static and dynamic loading tests of the subgrade material to estimate its deformation characteristics.

2. Subgrade material

A gelatine like material was selected as a subgrade material. Its viscoelasticity allows to simulate the behaviour under dynamic loading and can be related to the softer subgrade soils or small scale simulations.

Gelatine is protein of animal origin obtained from collagen using hydrolization [7, 8]. In this case, the collagen was prepared from ossein of beef-cattle. Solution containing glycerine was boiled until became liquid and poured into the steel tank. After cool down, a compact mass with flat surface was created.

Poisson's ratio $\mu = 0.5$ corresponding to the soft materials is assumed for the testing at a compound and air temperature of 22°C.

3. Subgrade testing

In real scale conditions, various methods were developed to determine the deformation characteristics of the pavement layers [9]. Despite the small scale, we decided to test the subgrade via conventional methods such as plate load test (PLT) and hammer drop test (HT) which are common in engineering practice.

Using the same methods for the real structures and model allows us to determine the scale relation between calculated quantities.

3.1. Testing stand

The scheme of the tank and PLT and hammer drop tests is plotted in figure 1.

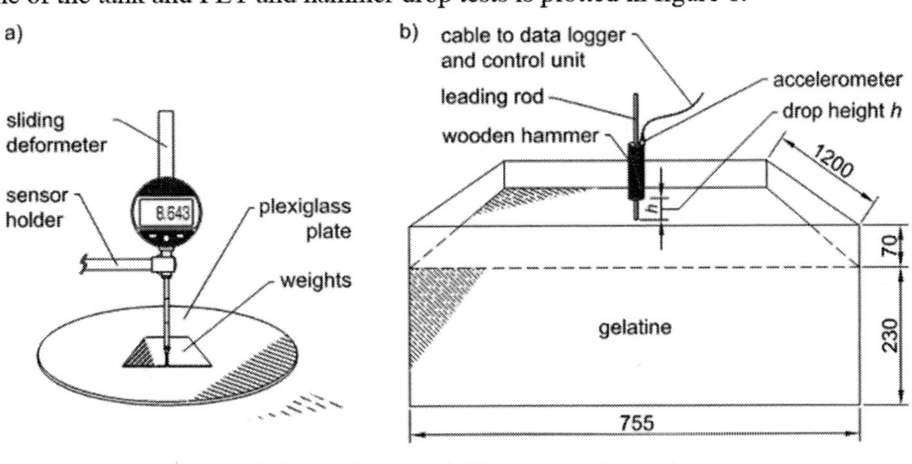

Figure 1. Scheme of the testing stand (dimensions in mm),

a) detail of the PLT test, b) hammer drop test.

3.2. Plate load testing

Plate load testing consists of measurement of the plate settlement at given load acting on the plate. At known plate radius r and Poisson's ratio μ of tested material, we can calculate the modulus of elasticity E for the given pressure interval under the plate Δp and corresponding elastic plate settlement y_e [10].

$$E = [0.5 \cdot \pi \cdot (1 - \mu^2) \cdot \Delta p \cdot r]\, y_e^{-1} \tag{1}$$

A plexiglass plate of thickness of 2.5 mm with the diameter of 215 mm and the weight of 131 g was used for the loading. The plate was rigid and light-weight as well. Additionally, sliding deformeter acts with the weight of 102 g on the plate as a result of pressure of the measurement tip. On each loading stage, 44 or 46 g weight was placed in the middle of the plate, figure 1a. After 30 minutes, all weights were removed and actual settlement was read to calculate the elastic part of the settlement.

This elastic part only corresponds to the effect of the weights because removing of the deformeter and plate is impossible during the test. A 30-minute restriction was selected due to viscous nature of the gelatine when the settlement grows up within the time. Moreover, dynamic load of the vehicle wheel acts only for the short period of time and longer loading stage gives underestimated values of modulus of elasticity.

A total of 5 loading/unloading cycles were realized during each of the 3 PLT tests.

3.3. Hammer drop testing

When a loading plate (typically during lightweight deflectometer test – LWD) is placed on the gelatine, it sticks with the surface and removing is very complicated and cannot be done without some damage of the gelatine mass surface. Stiffness of the plate and the damper influence the recorded values, when the weight impacts the damper on the plate.

To exclude these uncertainties, a hammer drop test based on the Clegg Impact Soil Tester (CIST) was adopted [11]. Wooden cylinder of diameter d = 25 mm with weight m = 39 g slides along the leading aluminium rod from the height h and hits the gelatine surface. Accelerometer attached to the cylinder records the deceleration a during the impact, figure 1b. When the peak deceleration is known, modulus of elasticity E is calculated as follows [12]:

$$E = m \cdot a^2 \cdot (1 - \mu^2)\, (d \cdot h \cdot g)^{-1} \tag{2}$$

Gravitational acceleration is $g = 9.81\ \mathrm{m \cdot s^{-2}}$.

4. Results and discussion

Typical propagations of the load/settlement for the PLT test and the time/acceleration for the hammer drop test are plotted in figures 2 and 3.

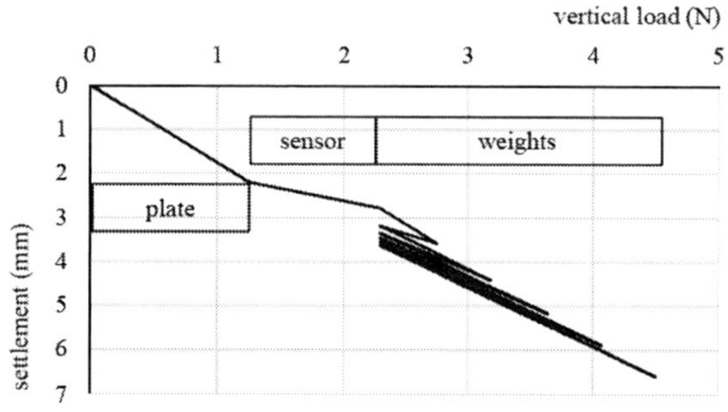

Figure 2. Load/settlement curve for PLT test.

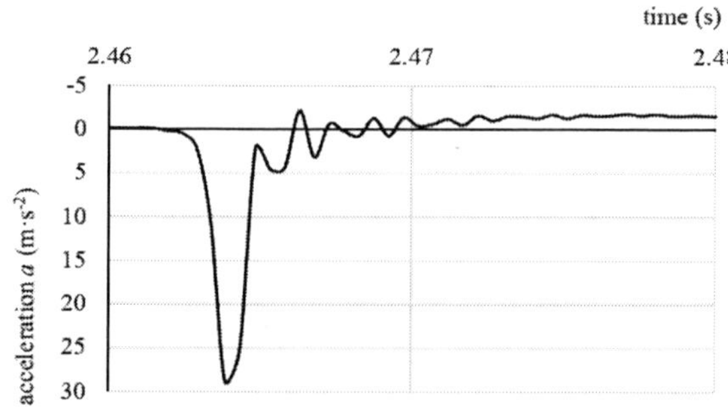

Figure 3. Time/acceleration curve for hammer drop test.

IOP Conf. Series: Materials Science and Engineering **789** (2020) 012061 doi:10.1088/1757-899X/789/1/012061

Calculated values of modulus of elasticity according to equation (1) varies in interval from 5.7 to 2.6 kPa with decreasing of its values with increasing vertical load. Significant drop in E values was observed in the first loading/unloading stage when first 46 g weight was applied ("weights" part of the curve in figure 2). After second cycle, E values oscillate around 2.6 to 3.1 kPa in 3rd to 5th cycle. Final value of the PLT E modulus lies within this interval.

Relation between calculated modulus E and corresponding hammer drop h is plotted in figure 4.

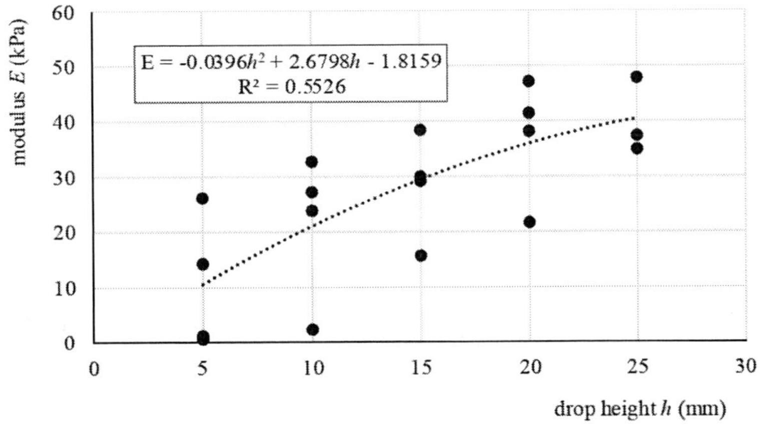

Figure 4. Values of modulus E depending on the hammer drop height.

As shown in figure 4, overall trend of the E values is clearly visible when with increasing hammer drop height E values also increase. Unfortunately, the spread of the data points is very large. Reliability coefficient is $R = 0.74$ for the second degree polynomial relation of the trend line. Despite this, we assume that sufficient large set of test results allows to determine reliable value of E modulus from the hammer drop test.

It's also obvious that E modulus from hammer drop test is significantly larger than E modulus from PLT test. Considering the trend line, the E value lies within the interval from 10.6 kPa to 40.4 kPa which gives the dynamic multiplier between PLT and hammer drop values of modulus E from 4.1 to 13.0. The E values from hammer drop test appear to be a "dynamic values". This statement should be verified by the physical and numerical modelling when the subgrade response to the dynamic load of the pavement will be investigated.

Hammer drop testing method was originally developed for determination of the static values but considering the dynamic nature of the pavement loading and the dynamic effect of the hammer, this method is still suitable for the testing in the model. Determination of the pressure interval on the subgrade is necessary to calculate the E modulus related to the deformation response of the gelatine. The surface tension of the gelatine when the stiffness grows up with increasing impact energy represents problem that can be solved when actual pressure acting on the model subgrade during the vehicle wheel pass will be known. Drop height and hammer weight can be then calibrated to obtain required response of the gelatine.

5. Conclusions

Testing of the gelatine materials is complicated because of its viscoelastic nonlinear nature. Static tests proved that gelatine can swallow up objects after few hours or days depending on the weight and shape of the objects. Time restriction for the static tests was then set to achieve as realistic values of E modulus as possible. Dynamic characteristics of the gelatine strongly depend on the stress state on the subgrade surface. Only exact knowledge of this quantity allows reliable estimation of the E modulus related to the dynamic behaviour of the subgrade.

A large dispersion of the *E* modulus values during hammer drop test should be minimized when a larger set of data will be available. Other methods, such as resonant testing or dynamic excitation, can be combined to describe the behaviour of the gelatine under dynamic conditions.

Based on measured experimental results, we can conclude the possibility of re-evaluating the mixing ratios of gelatin. In this case, we can change the material parameters of the small scale model. However, it is necessary to state that correct identification of gelatin properties is very complicated and complex [13].

Together with the above mentioned statements, dynamic deformation characteristic will be a background for further small scale physical and numerical modelling of the interaction system vehicle/pavement and surrounding objects.

Acknowledgement
The project was performed with the financial support of the Slovak Grant National Agency VEGA 1/0006/20.

References

[1] Park D, Papagiannakis A T and Kim I 2014 *KSCE J. of Civ. Eng.* **18** 2085
[2] Valaskova V and Vlcek J 2017 *CEE* **13** 99
[3] Czerner M, Fellay L S, Suárez M P, Frontini P M and Fasce L A 2015 *Procedia Materials Science* vol **8** (Amsterdam: Elsevier) p 287
[4] Zdao D, Li-Ming P, Chen-hua S and Hong-yuan F 2013 Experimental study on dynamic deformation characteristics of weathered conglomerate under cyclic loading *Advances in Structural Engineering*
[5] Mindaugas M M, Ustinovičius D, Slizyte D and Slizyte T Z 2016 Analysis of static and dynamic deformation modulus *Engineering Structures and Technologies*
[6] Herrmann H and Bucksch H 2014 Small-scale model test *Dictionary Geotechnical Engineering* Wörterbuch GeoTechnik
[7] Raileanu P, Andrei R, Puslau E and Plescan C 2010 Seasonal variation of moisture content in pavement analysis and design, *National Congres of Road and Bridge*
[8] Laurent J-L, Janmey P A and Ferry J D 1980 *J. of Rheol.* **24** 87
[9] Decky M, Drusa M, Pepucha L and Zgutova K 2013 *Earth Structures of Transport Constructions* (Harlow: Pearson Education)
[10] * * * DIN 18134 2012 Soil – Testing procedures and testing equipment – Plate load test
[11] Clegg B 1994 *Clegg Impact Soil Tester Technical Note. Calculation of Penetration and Elastic Modulus from CIV* (Jolimont)
[12] Valaskova V and Vlcek J 2018 *Matec Web of Conferences* Vol 196 (Les Ulis Cedex A: EDP Sciences)
[13] Kiptoo D, Aschrafi J, Kalumba D, Lehn J, Moormann C and Zannoni E 2017 Laboratory Investigation of a Geosynthetic Reinforced Pavement Under Static and Dynamic Loading *ASTM Journal of testing and evaluation* **45** 1

IOP Conf. Series: Materials Science and Engineering **789** (2020) 012062 doi:10.1088/1757-899X/789/1/012062

Review of damage-tolerant solutions for improved seismic performance of buildings

T F Voica[1,2*] and A Stratan[1]

[1] Politehnica University of Timisoara, Faculty of Civil Engineering, Department of Steel Structures and Structural Mechanics, Timisoara, Romania
[2] Popp & Asociații SRL, Research & Development Department, Bucuresti, Romania

[*] toma-florin.voica@student.upt.ro

Abstract. Building structures have been designed for more than 40 years to protect life in the event of a major seismic event. In order to achieve this goal with a minimal initial investment, structures are designed, using the ductility principle, as sacrificial systems, undergoing significant plastic deformations in the main structural elements, which progressively deteriorate. If in the design codes the structural damage is to some extent explicitly accepted and quantifiable, problems related to non-structural damage are addressed much more vaguely. This no longer meets market and society requirements. New structural systems are needed, which are not only efficient in reducing structural and non-structural damage, but also offer good re-centring capability, durability, reparability and affordability, accompanied by accessible design methods. This paper reviews the principal solutions that can be used to achieve these goals. Their advantages and limitations are analysed, as well as their potential as damage tolerant structural systems.

1. Introduction

In performance-based design, designers are supposed to demonstrate that a structure meets specific performance criteria set in advance. Generally, these criteria, as set forth in the 1967 SEAOC "Blue Book" [1], state that the structure must: (1) resist minor earthquakes without damage; (2) resist moderate earthquakes without structural damage but some damage to non-structural components; (3) resist major earthquakes with substantial structural and non-structural damage; and (4) resist the most severe earthquakes without structural collapse. Yet, most modern seismic design codes address these principles only partially and in a greatly simplified manner.

The main objective of modern seismic design codes, like Eurocode 8 [2], is to safeguard life in the event of a major earthquake through dissipation of energy by the development of inelastic deformations in dissipative components. Limitation of damage to non-dissipative (brittle) components, as well as enhancement of total energy dissipation by a global plastic mechanism is promoted using capacity based seismic design of non-dissipative components. Thus, the protection of life at the Ultimate Limit State (ULS) is ensured with the cost of sacrificing the structure in the case of rare, high intensity earthquakes. In the case of more frequent, lower intensity earthquakes (Serviceability Limit State – SLS), the code limits inter-storey drifts, which is intended to control the damage to non-structural elements. However, limitation of structural damage is not addressed at SLS. This design approach has the following disadvantages: (1) the repair of structural damage experienced under a major (ULS) earthquake is not addressed; (2) at the SLS only the non-structural damage is addressed (in a simplified manner).

The need to change the design philosophy towards solutions that are damage tolerant has been highlighted by recent earthquakes such as Christchurch, NZ [3-6] and L'Aquila, IT [7]. The experience of these earthquakes has highlighted that although several structures behaved as designed and damage mechanisms have protected human lives, the associated costs were significant, many buildings being demolished because the replacement cost was lower than the repair cost.

In this context, one very effective way to achieve the desired performance while protecting the structural and non-structural components is to reduce the seismic response of the structure using base isolation and / or viscous dampers. Although this method is very effective and it was successfully applied in the past for some projects, it has a major drawback: it is very costly. This high cost starts with the design stage and ends with the maintenance of the building. First, this kind of solution usually requires very advanced and complex calculation, often including dynamic nonlinear analyses, which are not yet accessible for current design practice. Second, the devices used are very expensive as they are in general subject to patents, and they require testing for prequalification. Third, these systems generally require constant maintenance and some of them require also external power and other data acquisition and interpretation equipment. Therefore, this approach has become more of a niche solution, adopted in general for high profile investments, or where other solutions are not providing the required performance.

Alternatively, a simpler way of reducing the repair costs and minimize the damage to the buildings structural and non-structural components would be to localize the inelastic damage within bolted fuse elements [8, 9], capable to dissipate the seismic energy, that can be replaced after a strong earthquake and which can also afford the structure re-centring capabilities when properly designed. Although this idea is not new, today in Europe there are very few buildings designed on this principle. Following the international trends, extensive research was carried out in Europe during the last decade, leading to the introduction of several innovative steel-based systems [10]. However, only a few designers are confident enough to employ these systems, as provisions for their design have not been included in the Eurocode or other national design codes.

Another important factor limiting the adoption of damage-tolerant structural systems is the insufficient explanation of the basic seismic design principles to the owners and to the general public. Because they are unaware of the potential for damage in a building that complies with the present minimum code requirements, they are less interested in the use of such systems, even though the initial cost difference between the two solutions would be minor compared to any repair costs in the event of a significant earthquake.

2. Main criteria for damage tolerant structural systems

Damage-tolerant structural systems must be able, on the one hand, to reduce the seismic response of the structure in order to limit both structural and non-structural damage, and on the other hand they must be designed to allow the "reset" of structures with minimal cost. This "reset" consists in maintaining or restoring the stiffness and resistance of the structure, the energy dissipation capacity, and finally, in the re-centring of the structure. In other words, the main objective of the damage-tolerant systems is to limit the damage to structural and non-structural components, and to allow the quick and cost-effective return of the structure to its pre-earthquake state.

This general objective must be translated into several clear assessment criteria, which will indicate the optimum solution for a given project. We propose the following five criteria:

- efficiency of damage reduction;
- re-centring capability
- reparability
- durability
- affordability

The first criterion relates to the ability to prevent or reduce the damage to structural and non-structural components of the building. To limit both the direct and the indirect effects of the earthquake on non-structural components (NSC), the system must limit both the maximum inter-storey drifts and maximum story accelerations. As far as the structural components are concern, the system needs to be assessed both locally and globally. For example, systems with dissipative elements placed at the ends of the frame beams allow replacement of parts that have exceeded the elastic stress level (an advantage over conventional systems). However, in the condition in which a floor damage occurs, requiring difficult and expensive repairs, the efficiency of these systems at the global level is greatly diminished.

The re-centring capability is also very important and refers to the ability of the structure to return to the initial undeformed shape after the removal of the damaged dissipative components., Buildings with residual displacements may no longer be acceptable, not only for safety reasons, but also for functionality and comfort reasons. More often, the necessary works to bring the building back to a non-deformed shape may be impractical.

The reparability refers to the ease of bringing the system back to its initial state by repairing or replacing damaged components and the associated cost. In general, the use of less encapsulated systems should be favoured, where state assessment, repair or replacement can be made as simple and as cost effective as possible.

The durability refers to the lifetime, maintenance and possible protective measures required, depending on the sensitivity of the system to environmental actions. If a building has a standard life span of 50 years, components of damage-tolerant systems should also have a similar lifetime. Some devices, such as friction dampers, are particularly sensitive to potential problems with water infiltration. Polymer-based systems have potential problems with the aging of component materials. Therefore, it is essential to provide for a maintenance and monitoring system. It is also advisable to design the structure so that, even in the event of a failure in some of the integrated dissipative systems, it will be able to resist a certain level of lateral load.

The affordability is perhaps the most important factor from a market perspective. It relates to both the initial cost and the maintenance cost. These costs must be balanced against the benefits of the system in terms of reducing the potential repair costs and the downtime following seismic events during the life span of the building. The analysis should therefore cover the entire life cycle. Clients must understand that such an analysis involves much greater effort in the initial concept phase, with higher design costs associated with this phase. An informed decision about the optimal solution can result only based on an accurate assessment of the costs and benefits.

3. Seismic response control systems

Damage limitation during an earthquake is not a new concern. Although, in theory, the damage limitation can be achieved for structural components by increasing the resistance of the structure and designing an elastic seismic response, this approach is not an economical solution, nor does it ensure the limitation of the damage to non-structural components. In general, this goal is achieved by reducing the seismic response of the structure. This is obtained by using devices specifically designed for this purpose, that can be classified, based on the type of control, in passive, semi-active, active and hybrid systems.

Passive control systems consist of mechanical devices to dissipate and reduce the induced seismic energy in the structure, thus reducing structure response and possible damage [11]. They cannot, however, adapt to the overall response of the structure. They are optimized to protect the structure from a specific dynamic load but have limited efficiency for other cases and other types of dynamic loadings [12].

Semi-active control systems are adaptive and can adjust their behaviour based on the information collected about excitation and structural response [13]. The components of this system include sensors, a control computer, a control actuator and a passive damping device. The sensors measure the excitation and / or structural response. The control computer processes the measurements and generates control signal for the actuator, which then acts to adjust the passive device behaviour [11]. Their efficiency is limited by the maximum capacity of the passive devices on which they are based [14].

Depending on the measurement of the global system response, active control systems can have optimal efficiency compared with passive control systems, which depends on the local responses only Active control systems require significant energy to counteract the dynamics loadings, which cannot be ensured during severe natural hazards due to failure of the energy supply during such events. Moreover, active control systems are complicated; they require sensors and controller equipment and give a shift in the dynamic behaviour of the structure by adding or removing energy from it; this may result in an unwanted or even unstable condition [15].

Hybrid control systems introduced since the 1990s represent the parallel grouping of passive systems with semi-active or active systems to achieve the benefits of both hybridized techniques and to greatly mitigate the limitations of each technique. Hybrid devices have a higher efficiency than a passive system and are more reliable and require less energy than active devices because there is no need for large control forces [12].

The extent to which structures containing one or more such systems qualify as damage-tolerant structural systems should be assessed in terms of meeting the five previously proposed assessment criteria.

Passive control systems emerge as the most likely candidate for the role of energy-dissipation in a damage-tolerant structural system. They also offer several advantages that make them particularly interesting: they present intrinsic stability, do not require an external source of energy, do not require structural response measurement to function, have low design, deployment, and maintenance costs [11-13], [16].

Passive control systems can also be roughly classified into two main categories: (1) base-isolation; (2) energy dissipation devices.

4. Base isolation

Base isolation is a passive structural control technique that uses a series of elements interposed between the foundation and the structure to isolate it from the vibrations induced to the foundation by the seismic action. The effect is to increase the structure's fundamental periods, possibly combined with energy dissipation. Base isolation is suitable for both new structures and the protection of existing buildings [3-7]. Such systems have been used for over 50 years both in buildings and bridges [17].

One of the main limitations of this technique comes from the way it works. Such systems are less effective for flexible buildings, whose base un-isolated period is already quite large. Also, in seismic areas with high amplification in the area of large periods, the efficiency of these systems is reduced. Another limitation is related to the fact that isolators are efficient in reducing the horizontal components of the seismic action but are not effective for the vertical component of the seismic action. Some of the most commonly used isolation systems devices can be classified in two categories: isolators who rely on elastomeric materials to provide the recentering force (elastomeric bearings, lead-rubber bearings, combined elastomeric and sliding bearing) and those that do not (sliding friction pendulum systems).

4.1. Isolators based on elastomeric materials

Initially, elastomeric bearings were made up of large blocks of natural rubber without additional reinforcement. Subsequently, horizontal steel plates were added to improve the behaviour and vertical stiffness, and synthetic rubber started to replace the natural rubber. These isolators offer a low damping rate and are generally used with additional damping devices.

Lead-rubber bearings are similar to the elastomeric ones, having in addition a lead core pressed into a vertical bore. This core undergoes plastic shear deformations, dissipating energy with a very stable hysteresis. The composition of the rubber has also undergone changes over time, with the goal of increasing durability and damping, which has led to the elimination of the need for additional damping devices.

In a combined elastomeric and sliding bearing setup, friction-based isolation systems account for sliding at the interface between two materials. One of the materials is usually stainless steel while the other material is usually a polymer. Initially, the latter was Teflon, and more recently synthetic materials specifically developed for this purpose, of the high-density polyethylene type. The efficiency of these devices is influenced by factors such as temperature, wear, slip speed, surface pressure, and degree of contamination of the friction surface. In these systems, the entire vertical force is supported by the friction interface, which also provides hysteretic energy dissipation, while the re-centring force is provided by neoprene elements that do not take up vertical loads. The main advantage of these systems is the low cost.

In general, the primary drawbacks of this category of isolators relate to the dependence on the properties of elastomeric materials, which limits their lifetime and the applicability under severe environmental conditions (especially low temperatures).

4.2. Sliding friction pendulum systems

The friction pendulum isolators combine the horizontal sliding motion with a re-centring force provided by the sliding surface geometry. They consist of an articulated element sliding on a concave spherical surface. The lifting of the building mass, which takes place during lateral movement, ensures the re-centring force. These devices have several advantages: good energy dissipation ability, stable behaviour, reduced torsional effects in the structure, simplicity of calculation, very good durability and very low maintenance costs. They also don't require fire protection, and the sliding surface can be protected against contamination by relatively simple means. Among the drawbacks are the high manufacturing cost, the impossibility of taking up vertical tension forces and the lifting effect of the structure.

5. Energy dissipation devices

There are many devices capable of absorbing energy and increasing the damping of buildings to reduce the seismic response. These additional dampers are used especially in high-rise buildings that cannot be effectively isolated. In these flexible buildings, it is necessary to control lateral displacements, which can be achieved by using damping devices that absorb much of the seismic energy. These systems can also be used in combination with base isolation systems to supplement their energy dissipation capability. The most used types of systems are hysteretic devices, viscoelastic devices, self-centring devices and dynamic vibration absorbers.

5.1. Hysteretic devices

As can be seen from their name, these devices dissipate energy through a mechanism that is independent of the load rate. They can also be divided in two groups: metallic dampers, using metal plastic deformation to dissipate energy, and friction dampers, which use dry sliding friction between two surfaces to dissipate energy in the form of heat [18].

Metallic dampers provide a very efficient energy dissipation mechanism. They have very good durability, very stable properties over time, are relatively insensitive to temperature variations and have low cost. Their main disadvantage is the limited number of load cycles and their nonlinear response.

Friction dampers have good performance, being able to dissipate a large amount of energy. They are less affected by the load frequency, the number of load cycles and temperature variations and exhibit rigid-plastic behaviour. Among the disadvantages are the variability in time of the friction coefficient, mainly influenced by the condition of the friction surface, which is affected by the environmental conditions, and the fact that these devices do not dissipate energy at forces less than the fiction force. Such dampers are generally installed in vertical bracing.

One of the devices tested at the University of Canterbury on steel frames is the high-force-to-volume (HF2V) lead extrusion dissipater [19]. Due to their small size and high capacity these devices can be used as dissipaters in concentrically braced structures, or in beam-column moment connections.

5.2. *Viscoelastic devices*

This category includes a wide range of devices with an energy-dissipating mechanism dependent on the rate of load. The damping force in these devices is proportional to velocity, and the behaviour is viscous. These can be classified into two main categories: solid viscoelastic dampers and liquid viscoelastic dampers.

Solid viscoelastic dampers are usually made from copolymers or glassy substances that dissipate energy through shear deformation in the viscoelastic material. During these deformations, the viscoelastic material manifest features of both elastic solid and viscous fluid, returning to its original shape while dissipating energy as heat.

Viscoelastic fluid dampers are based on the dissipation of energy using the properties of a viscous liquid, which can be contained in an open or closed container. Examples of these dampers include [18]: (1) the cylindrical pot fluid damper, (2) the viscous damping wall system (VDW) and (3) the fluid viscous damper.

In examples (1) and (2), such a device uses an element attached to the top of a structure level which is forced to travel through a high viscosity material contained in an open container attached to the lower side of the structural level. The shape of this container may be cylindrical or parallelepipedal. The advantage of such a device is the very high energy dissipation capacity and relative constructional simplicity, while the main drawback lies in the need to use a very high viscosity material, and such materials have both frequency and temperature dependent properties.

In example (3), the damper is based on the flow of a fluid into a closed container. The movement of a piston forces the fluid through small holes instead of deforming the fluid locally, resulting in high levels of energy dissipation. The main advantage of this type of damper is its ability to reduce both the deflection and stress at the same time, since the damper force is totally out of phase with the stresses resulting from the structure flexing. Moreover, these dampers are practically insensitive to temperature variations. A main disadvantage is the difficulty to reduce the peak structural response in the initial stages of loading due to the dependence of the damper's resisting force on the velocity.

5.3. *Dynamic vibration absorbers*

This type of device, also known as tuned mass damper, is used to reduce energy dissipation demand on the structure under dynamic loading [18]. This is done indirectly by transferring a part of the vibration energy to an absorption device. The main characteristics of the system are mass, rigidity and damping. Its dynamic properties must be tuned to those of the main structure. The applicability of this type of device for seismic actions is limited due to detuning that may occur by changes to the dynamic properties of the structure caused by plastic deformations.

6. Self-centring systems

This type of systems possesses a native re-centring capability due to small residual deformations after load removal [18]. In general, they are based on some type of pre-stressing obtained either with a pressurized fluid or with a pretensioned spring or cable.

Recent development of performance-based design recommendations for self-centring moment resisting frames (SC-MRF) [20] was motivated by results obtained in the last decade in experimental and analytical investigations of these systems for seismic applications. Research shows that these ductile systems resist structural damage after repeated inelastic response cycles under design-level earthquakes. Due to posttensioning that enables self-centring, residual drift after an earthquake is eliminated. Supplemental connection elements are designed to minimize structural damage in the main frame elements at inelastic levels of cyclic loading, and to provide stable energy dissipation through yielding, or through friction-based damping.

The same principle can be applied not only in conjunction with post-tensioning to MRFs, but also to columns base [21] and centrically braced frames (CBFs) [22] using gravity as the restoring force.

Shape memory alloys (SMAs) have been employed to develop various types of SMA-based braces [23-25] and beam-column dissipative connections [26].

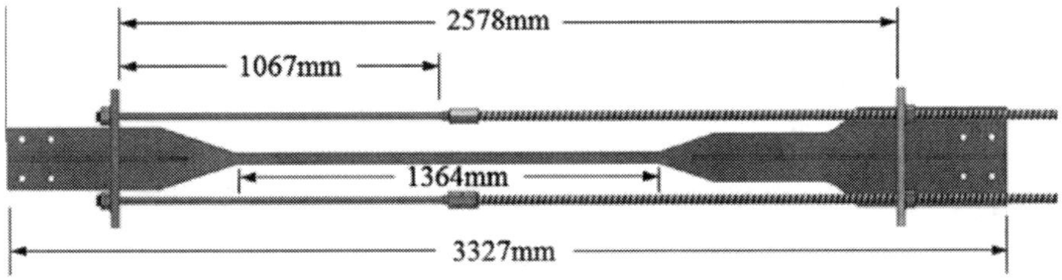

Figure 1. Prototype self-centring BRB dimensions (tubes not shown) [23]

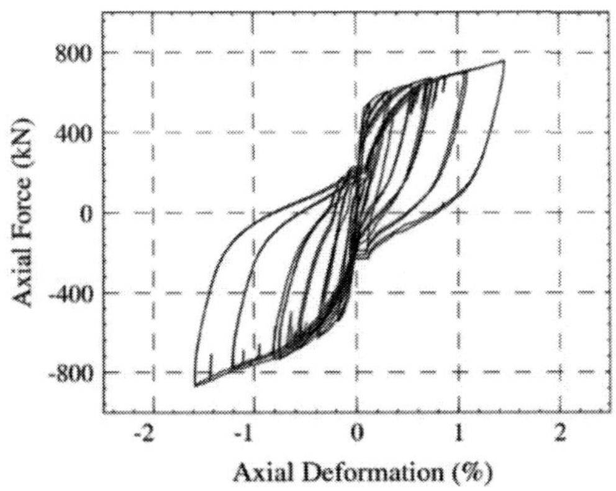

Figure 2. Overall response [23].

Shape memory alloys are a class of materials that can recover from large strains through the application of heat (known as the shape memory effect) or removal of stress (known as the superelastic effect). This results in several unique characteristics, including Young's modulus-temperature relations, shape-memory effects, superelastic effects, high damping characteristics, and re-centring capabilities. SMAs have demonstrated energy dissipation capabilities, large elastic strain capacity, hysteretic damping, excellent high/low-cycle fatigue resistance, re-centring capabilities and excellent corrosion resistance, as in figure 3. All of these characteristics give SMAs great potential for use within seismic resistant design and retrofit applications [27].

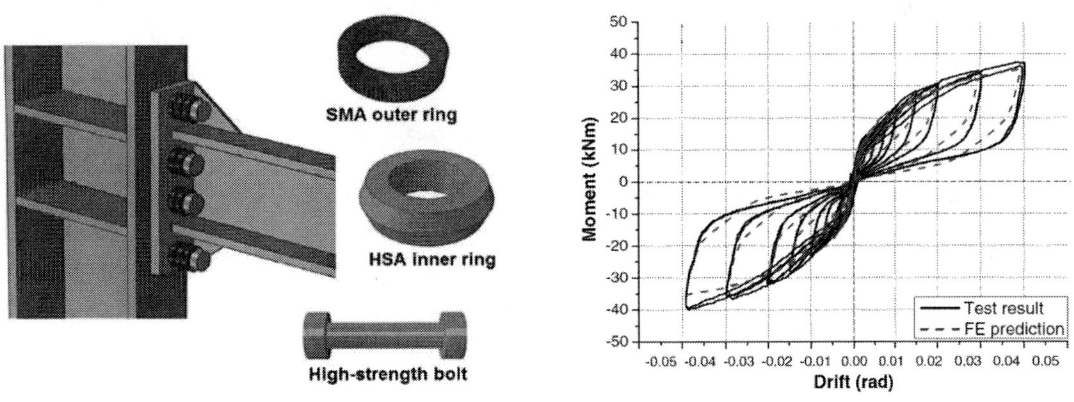

Figure 3. Connections with SMA ring springs [26] and moment-drift response of tested specimen.

In general, the disadvantages of SMA devices are the high cost given by the complexity and the low energy dissipation capacity. Also, varying environmental temperatures may have certain influence on the SMA behavior, and, the transformation temperature must be lower than the working temperature to ensure super elasticity.

7. Damage-tolerant structural systems

The basic concept underpinning these structural systems is the use of replaceable dissipative elements combined with a re-centring solution for the structure [28].

They are generally made up of two structural subsystems: the vertical (gravitational) load resisting system, and the seismic force resisting system. The gravitational system is designed to remain elastic and also to provide the elastic re-centring force, while the lateral resisting system utilizes affordable and easily replaceable or reparable structural devices or elements that ensure both the seismic energy dissipation by concentrating the plastic deformations as well as supplementing the rigidity of the gravitational structure to lateral displacements.

In general, both base-isolation and specially designed energy dissipation devices are represented by complex integrated systems developed by a few specific manufacturers, often protected by patents, which also require testing and validation, and which introduce significant costs both in the design phase as well as the construction stage of the structures that incorporates them.

Therefore, innovative steel-based devices, consisting of engineered solution made from simple, design-specific, non-patent-protected structural elements, that generally do not require additional approval, designed as an integral part of the structure, represent the ideal solution for damage-tolerant structures.

Depending on the general setup of these devices, relative to the main gravity structure, their stiffness and the stiffness of the main structure can be described in a simplified way as serial or parallel springs. Considering the need for an elastic re-centring force in order to ensure the re-centring capability of the structure, only the parallel setup present a viable solution, as it is the only one that retain the stiffness of one system while the other is removed.

Also, the total stiffness of the structure is increased by the addition of the dissipative elements in the parallel setup, while in the serial setup this is decreased.

Such a structural setup is called a dual system, due to its two subsystems: one that is flexible, behaving elastically and a second one that is rigid, behaving inelastically.

The elastic, flexible subsystem can be implemented as a moment frames, that can undergo large elastic deformations.

The rigid, inelastic subsystem can be implemented as a braced frame incorporating bolted, easily replaceable fuses.

Some authors [9, 28] proposed a damage-tolerant structural system based on EBFs with bolted replaceable links, figure 4, and re-centring capability.

The weaker, more flexible subsystem (MRF) should provide a minimum strength (25% of the total seismic force [29, 30]), using the equation below.

$$F_y^{MRF} \geq 0.25 \cdot (F_y^{MRF} + F_y^{EBF}) \tag{1}$$

The re-centring capability should be validated by keeping MRFs in the elastic range and constraining plastic deformations to replaceable dissipative members until reaching the ultimate shear deformation in links. This can be done analytically (predesigned using formulas below [31], level by level.

$$\delta_u^{EBF} = \delta_y^{EBF} + \delta_{pl}^{EBF} = \frac{F_y^{EBF}}{K^{EBF}} + \frac{e}{L-e} \cdot H \cdot \gamma_{pl,u} < \delta_y^{MRF} = \frac{F_y^{MRF}}{K^{MRF}} \tag{2}$$

Where δ_y^{EBF} is the yield storey displacement of the EBF, δ_{pl}^{EBF} – the plastic storey displacement of the EBF, K^{EBF} - the EBF stiffness, $\gamma_{pl,u}$ – the plastic deformation capacity of the link, K^{MRF} – the MRFs stiffness, e – length of the link, H – the height of the frame and L – the span of the frame.

Considering that the analytical approach is a simplified one, nonlinear static and/or dynamic analyses are recommended in order to confirm the re-centring capability.

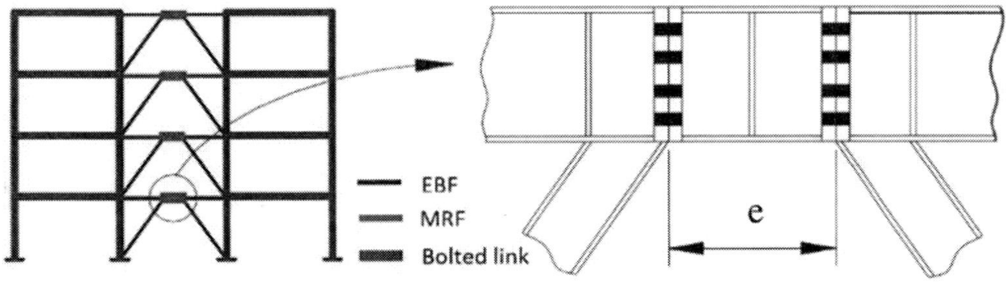

Figure 4. Replaceable link concept [9, 28].

Another damage-tolerant system was based on dual steel frames with replaceable steel shear panels, figure 5 and re-centring capability. In this case the bracing system was represented by the steel shear panels [32].

These can be also made replaceable while the re-centring force is provided by the parallel elastic moment frame.

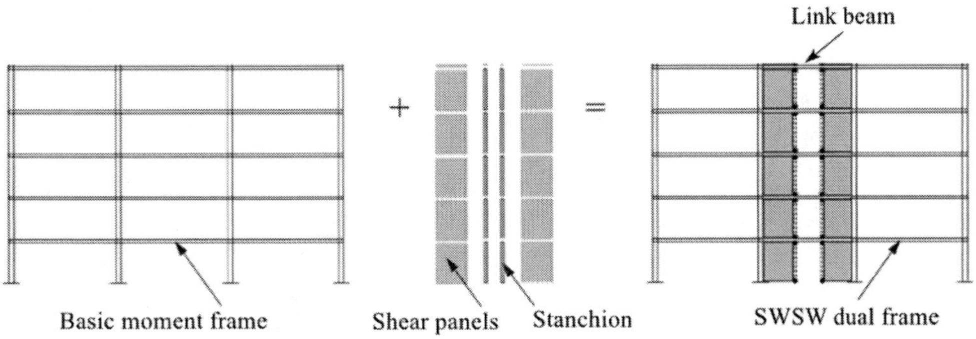

Figure 5. SPSW frame systems: dual system with link beam [32].

Replaceable moment resisting friction connections are conceived to develop the dissipation mechanism by means of the relative slip into ad-hoc devices located between the lower beam flange and the outer cap plate connected to the column flange, while the upper flange of the beam is connected to a plate either bolted or welded to the column, figure 6 [33].

The randomness of the friction properties has to be as much as possible mitigated and accounted for in the design phase, because this variability can inflict in the joint response and, consequently, the global behaviour of the structure. This type of approach is more difficult to implement is high seismic areas because in order to provide re-centring capability this system must be coupled in parallel with an elastic, more flexible subsystem.

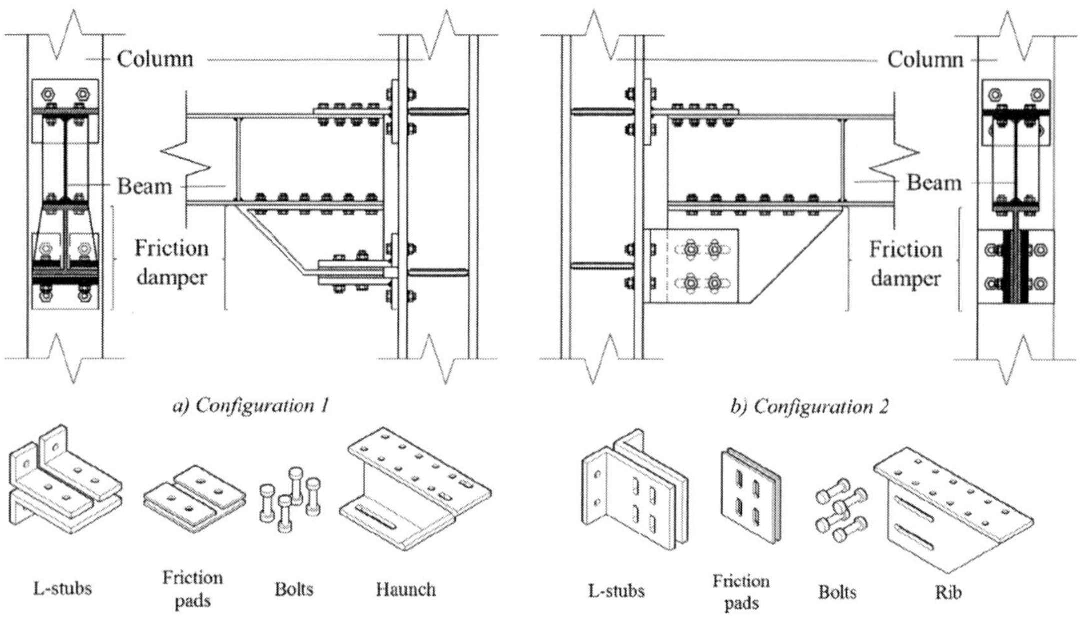

Figure 6. Moment resisting friction connection [33].

8. Conclusions

The state of art for damage-tolerant structures system was reviewed by briefly summarizing the principles of such systems and principal qualifying criteria. Also, the principal types of structural control systems and devices were reviewed to assess the potential use in damage-tolerant structures; these include: (a) passive; (b) semi-active; (c) active; and (d) hybrid systems. To demonstrate the damage-tolerant structural systems potential and future directions in civil engineering, an overview of some innovative practical implementations of these systems was provided. The state of art clearly indicates the huge capability for these systems and their importance in modern buildings. These systems can be used to reduce the damage induced by earthquakes and to allow for easy and inexpensive re-centring of the structures. Further research will address a parametric analysis of efficiency of different damage-tolerant solutions in reducing both structural and non-structural damage.

References

[1] Seismology Committee, Structural Engineers Association of California 1967 *Recommended Lateral Force Requirements and Commentary*, San Francisco, CA

[2] Comite Europeen de Normalisation 2004 *Eurocode 8: Design of Structures for Earthquake Resistance. Part 1: General Rules, Seismic Actions and Rules for Buildings*, CEN, Brussels, EN 1998-1

[3] Buchanan A H, Bull D, Dhakal R, MacRae G, Palermo A and Pampanin S 2011 Base isolation and damage-resistant technologies for improved seismic performance of buildings, *University of Canterbury Research Report No. 2011-02.* Christchurch

[4] Clifton C, Bruneau M, MacRae G, Leon R. and Fussell A 2011 Steel structures damage from the Christchurch earthquake series of 2010 and 2011, *Bulletin of the New Zealand Society for Earthquake Engineering* **44** (4) 297-318

[5] Hare J, Oliver S and Galloway B 2012 Performance Objectives for Low Damage Seismic Design of Buildings, *NZSEE Conference*, Christchurch

[6] Takagi J and Wada A 2019 Recent earthquakes and the need for a new philosophy for earthquake-resistant design *Soil Dynamics and Earthquake Engineering* **119** 499-507

[7] Clemente P and Martelli A 2019 Seismically isolated buildings in Italy: state-of-the-art review

and applications *Soil Dynamics and Earthquake Engineering* **119** 471-487

[8] Balut N and Gioncu V 2003 Suggestion for an improved "dog-bone" solution *Proc. of the Conference on Behaviour of Steel Structures in Seismic Areas* **129-34** Naples: Balkema

[9] Stratan A and Dubina D 2004 Bolted links for eccentrically braced steel frames. Connections in Steel Structures V. Behaviour, Strength & Design; *Proc. of the Fifth AISC / ECCS intern. Workshop*, June 3-5 Delft University of Technology 223-232

[10] Vayas I (ed.) 2017 Innovative anti-seismic devices and systems. Sersilito: ECCS

[11] Elias S and Matsagar V 2017 Research developments in vibration control of structures using passive tuned mass dampers *Annual Reviews in Control* **44** 129-156

[12] Saaed T E, Nikolakopoulos G, Jonasson, J E and Hedlund H 2015 A state-of-the-art review of structural control systems *Journal of Vibration and Control* **21** (5) 919-937

[13] Soong T T and Spencer Jr. B F 2002 Supplemental energy dissipation: state-of-the-art and state-of-the-practice *Engineering structures* **24** (3) 243-259

[14] Spencer Jr. B F and Nagarajaiah S 2003 State of the art of structural control *Journal of structural engineering* **129** (7) 845-856

[15] Christenson R E 2001 Semi active control of civil structures for natural hazard mitigation: analytical and experimental studies *PhD Thesis* University of Notre Dame

[16] Braz-César M T and Barros R 2013 Passive control of civil engineering structures, Proc. Of 4th International Conference on Integrity, Reliability and Failure, IRF'2013. Paper ref: **4756**. Funchal

[17] Calvi P M and Calvi G M 2018 Historical development of friction-based seismic isolation systems *Soil Dynamics and Earthquake Engineering* **106** 14-30

[18] Constantinou M C, Soong T T and Dargush G F 1998 *Passive energy dissipation systems for structural design and retrofit* MCEER Monograph No. 1. Multidisciplinary Center for Earthquake Engineering Research, State University of New York at Buffalo

[19] Rodgers, G.W., Chase, J.G., macRae, G.A., Bacht, T., Dhakal, R.P. and Desomre, J. 2010 *Influence of HF2V Damping Devices On The Performance Of The SAC3 Building Subjected To The SAC Ground Motion Suites*, 9USN-10CCEE, Toronto, July 25-29

[20] Herning G, Garlock MEM, Vanmarcke E. 2011 *Reliability-based evaluation of design and performance of steel self-centring moment frames*. Journal of Constructional Steel Research, **67** (10) 1495–1505

[21] Chi H and Liu J 2012 Seismic behaviour of post-tensioned column base for steel self-centering moment resisting frame *Journal of Constructional Steel Research* 78117–130

[22] Roke D and Jeffers B 2011 Parametric study of self-centering concentrically-braced frame systems with friction-based energy dissipation *Behaviour of Steel Structures in Seismic Areas* CRC Press, p. 691–696

[23] Miller D J, Fahnestock L A and Eatherton M R 2012 Development and experimental validation of a nickel-titanium shape memory alloy self-centering buckling-restrained brace *Engineering Structures*, vol. **40**, pp. 288–298

[24] Kari A, Ghassemieh M and Badarloo B 2019 Development and design of a new self-centering energy-dissipative brace for steel structures *Journal of Intelligent Material Systems and Structures* vol. **30** no. 6 pp. 924–938

[25] Qiu C and Zhu S 2017 Shake table test and numerical study of self-centering steel frame with SMA braces *Earthquake Engineering & Structural Dynamics* vol. **46** no. 1 pp. 117–137

[26] Fang C, Wang W, Ricles J, Yang X, Zhong Q, Sause R, et al. 2018 Application of an Innovative SMA Ring Spring System for Self-Centering Steel Frames Subject to Seismic Conditions *Journal of Structural Engineering* Aug. **144** (8) 04018114

[27] Desroches R and Smith B 2004 Shape memory alloys in seismic resistant design and retrofit: a critical review of their potential and limitations *Journal of Earthquake Engineering* May **8** (3) 415–29

[28] Dubina D, Stratan A and Dinu F 2008 Dual high-strength steel eccentrically braced frames with

removable links *Earthquake Engineering & Structural Dynamics* **37** (15) 1703–1720

[29] NEHRP. 2003 *Recommended provisions for new buildings and other structures (FEMA 450). Part 1: Provisions and Part 2: Commentary*. Building Seismic Safety Council, National Institute of Building Sciences, Washington, D.C.

[30] P100-1 2013 *Seismic Design Code – Part I: Design provisions for buildings*

[31] Stratan A, Dinu F and Dubină D 2010 *Replacement of bolted links in dual eccentrically braced frames*, Proceedings of the 14th European conference on earthquake engineering, August 30 – September 3, Ohrid, Republic of Macedonia

[32] Neagu C, Dinu F and Dubina D 2017 Dual steel frames with re-centring capacity and replaceable steel shear panels *Proceedings of the COMPDYN2017 Conference on Computational Methods in Structural Dynamics and Earthquake Engineering*, Rhodes

[33] Latour M, D'Aniello M, Zimbru M, Rizzano G, Piluso V and Landolfo R 2018 *Removable friction dampers for low-damage steel beam-to-column joints* Soil Dynamics and Earthquake Engineering Dec. **115** 66–81

IOP Conf. Series: Materials Science and Engineering **789** (2020) 012063 doi:10.1088/1757-899X/789/1/012063

Analysis of passenger traffic in Bohai Strait Tunnel

B Wang[1*], Y P Li[1] and F Liu[2]

[1] School of Civil Engineering, Shenyang Jianzhu University, Shenyang, China
[2] Shenyang Highway Plan & Design Institute, Shenyang, China

[*] bingwer@126.com

Abstract. China envisages building a new undersea tunnel called Bohai Strait Tunnel to connect Dalian to Yantai, for the purpose of solving the detour problem in transportation. This paper analyses the passenger traffic in Bohai Strait Tunnel. A modified negative exponent network flow assignment model and an increasing rate method are used to predict the network flow assignment and passenger traffic volume after the completion of Bohai Strait Tunnel in 2050. Two formulas of economic benefits are employed to evaluate the effect of undersea tunnel on passenger traffic. The result indicates that the new undersea tunnel can shorten the distance, reduce the transport cost, save travel time, and produce good economic benefits. Tunnel transportation will attract a large number of passengers and become main transportation way. The analysis result can provide technical storage and references for the future undersea tunnel construction.

1. Introduction

Bohai Strait is located between the Bohai Sea and Yellow Sea, as shown in the figure 1. Bohai Strait is a major route of transportation in China, and is situated in a very important geographical position of coastal defense. The western area of Bohai Strait includes the capital of China, Beijing and an important industrial center, the city of Tianjin. The northern region is the old industrial base of Northeast China. Southern district is the biggest agriculture province of Shandong. Bohai Sea and the three regions constitute Bohai economic circle, which has become the center of economy, politics and culture of Northern China. But there is a defect for Bohai Strait, which hinders Northeast China from connecting to Shandong Peninsula. The straight-line distance from Dalian to Yantai is only 165 km. However, due to the Bohai Strait, the passengers and cargos have to take a detour by land and the distance becomes 1805 km. Even by sea, the sea transportation is inefficient, small scale and slow speed. So, the long distance and slow speed transportation have become the "bottle-neck" of Bohai economic circle development. In order to change this situation, China envisages building a new undersea tunnel to connect Dalian to Yantai. The new undersea tunnel can reduce the transportation distance, save travel time for passengers and freight transportation, decrease traffic pressure of Beijing-Tianjin-Hebei region, strengthen the economic ties between Northeast China and central plain area, boost tourism, push port development, promote the economic developments of Bohai economic circle and eastern coastal area. The construction of the Bohai Strait Tunnel project is a complicated system engineering. The project is in engineering feasibility study period and the preliminary research work of this project is particularly crucial [1]. Future transport capacity and economic evaluation of the undersea tunnel are most important contents for preliminary research work. This paper predicts passenger traffic volume, network flow assignment and economic benefits after the completion of Bohai Strait Tunnel in 2050, and evaluate the effect of undersea tunnel on passenger traffic.

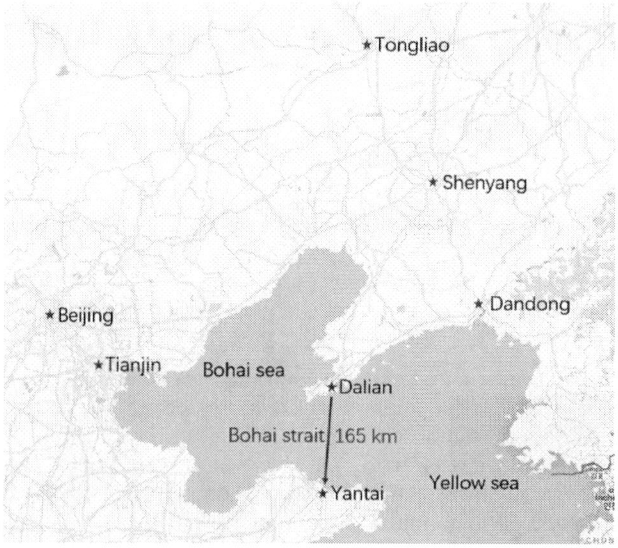

Figure 1. The geographical position of Bohai Strait.

2. Traffic assignment model

Traffic assignment problem is that the traffic volume is distributed into a set of paths based on some plans and models, and people can estimate the traffic demand of each path. According to Wardrop's first principle and second principle [2], traffic assignment models are divided into equilibrium model and disequilibrium model. These models include shortest path algorithm [3], logit model [4], probit model [5] etc. In this paper, a modified negative exponent network flow assignment model (NENFA model) [6], which is one of the disequilibrium model, was used to predict the network flow assignment. This model means that when multimodal transports occur, people tend to choose the minimum-cost path; besides, people also consider other selection factors [7], such as travel time, safety, traffic congestion situation, custom, weather condition, road maintenance, etc. The NENFA model can be evaluated as follows:

$$K_i = A \cdot B_i \cdot e^{\frac{-C_0 \cdot (D_i - D_j)}{D_j}} \tag{1}$$

$$A = \frac{1}{\sum B_i \cdot e^{\frac{-C_0 (D_i - D_j)}{D_j}}} \tag{2}$$

where:

K_i is the ratio of passenger traffic volume in one path to the total passenger traffic volume,

A is influencing factors that affect passenger's preferences, such as travel time, safety, transport distance, traffic congestion situation, custom, weather condition, road maintenance etc,

e is exponent,

C_0 is the cost parameter, which reflects passenger's sensitivity to transport cost,

D_i is the transport expense in path i,

D_j is the minimum transport expense in all paths,

B_i is preference parameter, which affects the choice of path i.

The preference parameter B_i is influenced by many factors, including passenger capacity, transportation cost, transport speed, continuity, flexibility and security, etc. B_i can be obtained by comprehensive assessment method, as listed in reference [8].

3. Economic benefits of passenger transportation

Economic evaluation of transportation project is an essential part of project feasibility study and determines whether the construction scale and time are reasonable, whether the project has good economic benefit, and whether the project is feasible[9].

Economic evaluation of transportation project can be obtained by analyzing its economic benefits by cost-benefit analysis principle. Economic benefits of passenger transportation include economic benefit from shortening distance and reducing transport cost B_{kd} and economic benefit from saving travel time B_{ks}.

$$B_{kd} = C_{k0} Q_k \left(L_0 - L_n \right) \tag{3}$$

where:

B_{kd} is the economic benefit from shortening distance and reducing transportation cost after the completion of Bohai Strait Tunnel (thousands of yuan),

C_{k0} is the unit transport cost of passenger traffic before the implementation of undersea tunnel (yuan/people×kilometer),

Q_k is the predicted passenger traffic volume (thousands of people),

L_0 is the transport distance before the implementation of the undersea tunnel (kilometers),

L_n is the transport distance after the completion of the undersea tunnel (kilometers).

$$B_{ks} = \alpha I_k Q_k T_{ks} \tag{4}$$

where:

B_{ks} is the economic benefit from saving travel time (thousands of yuan),

α is the ratio of passenger travel time for work to the total travel time,

I_k is the national income per head and per hour in planning year (yuan/hour),

Q_k is the passenger traffic volume (thousands of people),

T_{ks} is the total saving travel time of passengers after the completion of Bohai Strait Tunnel (hours).

4. Transport paths

Three logistic transport paths were used to predict the network flow assignment and economic benefits in the year 2050, including Tongliao-Yantai, Dandong-Yantai, Shenyang-Yantai. Analysis and comparison between "before the implementation of undersea tunnel" and "after the completion of the undersea tunnel" were performed.

Multiple transportation ways were considered including road, rail, air, sea and their intermodal transportation. Among them, high-speed rail was used in rail transportation. The details were listed in table 1 and table 2.

The passenger traffic volume of each origin-destination pair (OD pair) was predicted using an increasing rate method, the result was presented in table 3.

$$Q_i^n = Q_i^0 \left(1 + R_i \right)^n \tag{5}$$

where:

Q_i^n is the predicted passenger traffic volume in year *i;*

Q_i^0 is the passenger traffic volume in base year;

R_i is rate of growth in year *i*.

Table 1. The transport distance before the implementation of undersea tunnel.

OD pair	Path	Transport distance (km)					
Tongliao-Yantai	Tongliao-Shenyang-Dalian-Yantai	way	HS	RS	HA	RA	
		distance	824	938	899	1013	
	Tongliao-Chifeng-Beijing-Yantai	way	H	R	A		
		distance	1629	1816	1221		
Dandong-Yantai	Dandong-Dalian-Yantai	way	HS	RS	HA	RA	
		distance	485	457	560	532	
	Dandong-Shenyang-Beijing-Yantai	way	H	R	A		
		distance	1670	1926	1424		
Shenyang-Yantai	Shenyang-Dalian-Yantai	way	HS	RS	HA	RA	A
		distance	561	560	636	635	593
	Shenyang-Beijing-Yantai	way	H	R			
		distance	1430	1719			

Notes: H= highway transport, R= rail transport, A= air transport, HS= highway-sea intermodal transport, RS= rail-sea intermodal transport, HA= highway-air intermodal transport, RA= rail-air intermodal transport.

Table 2. The transport distance after the completion of the undersea tunnel.

OD pair	Path	Transport distance (km)							
Tongliao-Yantai	Tongliao-Shenyang-Dalian-Yantai	way	HT	RT	HS	RS	HA	RA	
		distance	820	934	824	938	899	1013	
	Tongliao-Chifeng-Beijing-Yantai	way	H	R	A				
		distance	1629	1816	1221				
Dandong-Yantai	Dandong-Dalian-Yantai	way	HT	RT	HS	RS	HA	RA	
		distance	481	453	485	457	560	532	
	Dandong-Shenyang-Beijing-Yantai	way	H	R	A				
		distance	1670	1926	1424				
Shenyang-Yantai	Shenyang-Dalian-Yantai	way	HT	RT	HS	RS	HA	RA	A
		distance	557	556	561	560	636	635	593
	Shenyang-Beijing-Yantai	way	H	R					
		distance	1430	1719					

Notes: HT= highway-undersea tunnel intermodal transport, RT= rail-undersea tunnel intermodal transport

Table 3. The predicted passenger traffic volume of OD pair in 2050.

OD pair	Passenger traffic volume in 2050 (thousands of people)
Tongliao -Yantai	2139.9
Dandong -Yantai	716.1
Shenyang -Yantai	4205.9

5. Prediction results

5.1. The results of traffic assignment model

The results of network flow assignment are shown in figures 2 - 4. For the OD pair of Tongliao-Yantai, before the implementation of undersea tunnel, the transportation ways mainly are highway-sea intermodal transport (HS) and rail-sea intermodal transport (RS), the ratios of traffic volume K_i are 40.7% and 39.5%, respectively. Other transportation ways, highway transportation (H) and high speed rail transportation (R), have a certain portion of 8.1% and 6.3%. The sum of K_i by air transportation (HA, RA and AT) is only 5.3%. After the completion of the undersea tunnel, highway-undersea tunnel intermodal transport (HT) and rail-undersea tunnel intermodal transport (RT) become most dominating transportation ways, the sum of K_i by tunnel transportation (HT and RT) reaches up to 90.2%. Meanwhile the sum of K_i by highway-sea intermodal transport (HS) and rail-sea intermodal transport (RS) reduces to 3.1%. The ratio of traffic volume in other transportation ways, such as HA, RA, H and R, and A, is very small, only 1.2%. It is clearly seen that the undersea tunnel has great influence on transportation network and the undersea tunnel is passenger's top choice.

There are similar trends for the OD pairs of Dandong-Yantai and Shenyang-Yantai. After the completion of the undersea tunnel, the ratios of traffic volume in highway-sea intermodal transport (HS) and rail-sea intermodal transport (RS) are greatly decreased. And the intermodal transport of HT and RT become main transport ways. The ratios of traffic volume by tunnel transportation for the two OD pairs are more than 90%. Therefore, shortening the transport distance and saving travel time are greatest advantages for Bohai Strait Tunnel.

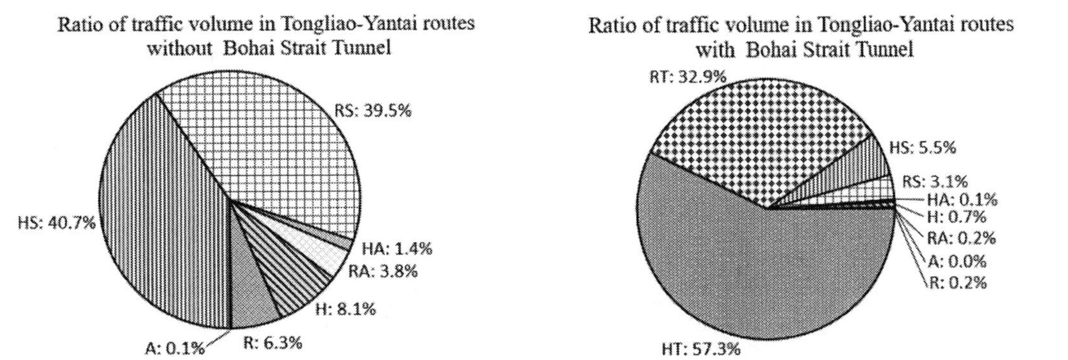

(a) Before the implementation of undersea tunnel (b) After the completion of the undersea tunnel

Figure 2. The result of network flow assignment in the path of Tongliao-Yantai.

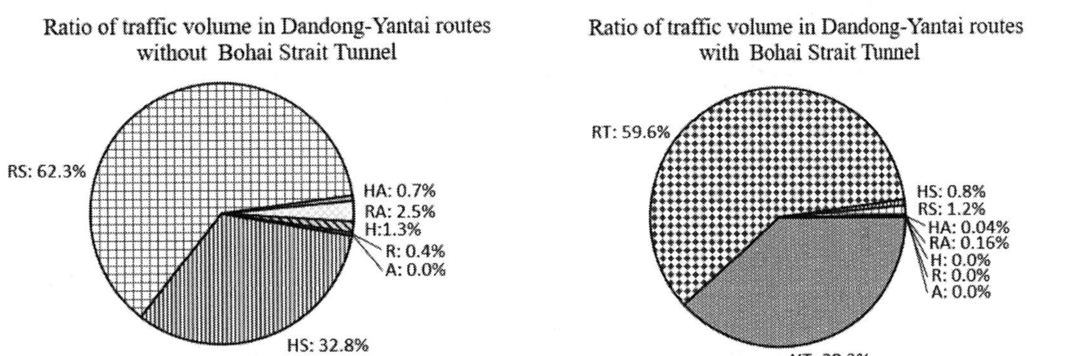

(a) Before the implementation of undersea tunnel (b) After the completion of the undersea tunnel

Figure 3. The result of network flow assignment in the path of Dandong-Yantai.

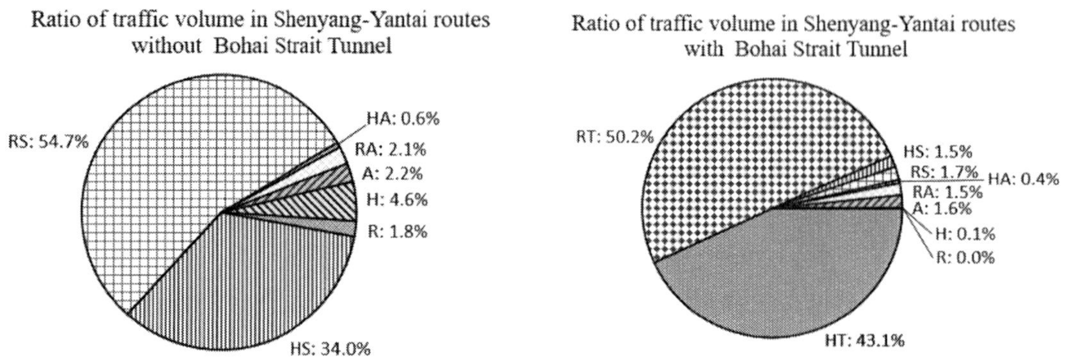

(a) Before the implementation of undersea tunnel (b) After the completion of the undersea tunnel

Figure 4. The result of network flow assignment in the path of Shenyang-Yantai.

5.2. The results of economic benefits of passenger transportation

From the tables 4 - 5, for the three OD pairs of Tongliao-Yantai, Dadong-Yantai and Shenyang-Yantai, economic benefits from shortening distance and reducing transportation cost B_{kd} are 1003.27 million yuan, 633.53 million yuan, and 2649.36 million yuan, respectively. Economic benefits from saving travel time B_{ks} are 35.93 million yuan, 16.67 million yuan, and 74.01 million yuan, respectively.

So, it can be concluded that undersea tunnel can achieve great economic benefits in passenger transportation. Especially for long distance or large traffic volume cites, such as Shenyang-Yantai, undersea tunnel will bring them greater economic benefits.

Table 4. Economic benefit from shortening distance and reducing transportation cost B_{kd} in 2050.

OD pair	Transport way	Q_k (thousands of people)	L_O (kilometers)	L_n (kilometers)	B_{kd} (millions of yuan)	Total (millions of yuan)
Tongliao-Yantai	HT	1227.1	1629.00	820.00	557.89	1003.27
	RT	704.3	1816.00	934.00	445.38	
Dandong-Yantai	HT	274.1	1670.00	481.00	183.14	633.53
	RT	426.5	1926.00	453.00	450.39	
Shenyang-Yantai	HT	1813.3	1430.00	557.00	889.65	2649.36
	RT	2110.5	1719.00	556.00	1759.71	
Sum (millions of yuan)						4286.16

Table 5. Economic benefit from saving travel time B_{ks} in 2050.

OD pair	Transport way	Q_k (thousands of people)	T_{ks} (hour)	B_{ks} (millions of yuan)	Total (millions of yuan)
Tongliao-Yantai	HT	1227.1	6.74	28.95	35.93
	RT	704.3	2.83	6.98	
Dandong-Yantai	HT	274.1	9.91	9.50	16.67
	RT	426.5	4.80	7.17	
Shenyang-Yantai	HT	1813.3	7.28	46.17	74.01
	RT	2110.5	3.77	27.84	
Sum (millions of yuan)					126.61

6. Conclusion

This paper predicts passenger traffic volume, network flow assignment, and economic benefits in the year 2050.

After the completion of Bohai Strait Tunnel, the intermodal transport of HT and RT become main transport ways. The ratios of traffic volume by tunnel transportation for the three OD pairs are more than 90%. The results indicate that the new undersea tunnel can change the transportation network, attract passengers, mitigate traffic pressures of Beijing area and Tianjin area, and become a main transportation way.

Economic benefits from shortening distance and saving travel time of three OD pairs in 2050 are 4286.16 million of yuan and 126.61 million of yuan, respectively. So, the tunnel can produce good economic benefits.

Acknowledgements

The authors would like to acknowledge financial support provided by Liaoning Provincial Natural Science Foundation with Grant No. 20170540496 and Liaoning Provincial Key Research and Development Project 2019JH2/10100014.

References

[1] Wang M S 2013 *Engineering Sciences* **15**(12) pp 4-9
[2] Wardrop J G 1952 *Proceedings of the Institute of Civil Engineers* V1 part 2 pp 325-378
[3] Xiao C Z 2015 Method of traffic flow analysis and simulation on ring road system of urban expressway based on shortest path algorithm (in Chinese) *Beijing Jiaotong University*
[4] Lim Y and Heydecker B 2005 *Transportation Research Part B: Methodological* **39**(2) 97-118
[5] Yi F J, Deng W, Chen Q and Cheng X 2009 *J. of Transportation Engineering and Information* **7**(1) pp 51-56
[6] Chen L F 2008 Study on container transportation modes around Bohai based on negative exponent network flow assignment (in Chinese) *Dalian Maritime University*
[7] Lo H K, Luo X W and Siu B W Y 2006. *Transportation Research Part B* **40**(9) pp 792-806
[8] Wang B, Deng C N and Liu F 2014 *Applied Mechanics and Materials* **587-589** pp 1944-1949
[9] Wu Z L 2009 Comprehensive Transportation Planning (in Chinese) *Tsinghua University Press*

Numerical simulation on heat transfer characteristics of gas pipeline in urban utility tunnel

G Q Wang[1*], H M Wang[1], Yan Cai[2], G H Feng[1] and Z Q Kang[1]

[1] School of Municipal and Environmental Engineering, Shenyang Jianzhu University, Shenyang, China
[2] CCCC Gas and Heat Research and Design Institute

[*] guiq.wang@sjzu.edu.cn

Abstract. Recent years, more and more utility tunnels are adopted in China with the development of urbanization. Compared to the conventional method for utility installation, the utility tunnel that co-locates more than one utility underground eliminates the need for frequent excavation and reinstatement of roads. Ventilation systems are essential in utility tunnel for safety reasons, especially for gas compartments. This paper aims to evaluate effect of the longitudinal ventilation on gas temperature field by investigating the heat transfer process on gas pipeline. Numerical models including convection, conduction and radiation were established and then used to simulate the heat transfer process on gas pipeline. Results show that the ventilation has a great impact on gas temperature. In case of low inlet air temperature, the gas temperature may drop below -10°C, which is too low to guarantee a normal operation.

1. Introduction

A utility tunnel is defined as an underground structure containing one or more utilities, permitting the installation, maintenance and removal of the systems without the necessity of making road cuts or excavations [1]. The utility tunnel has a long history of application around the world, which considered to be more effective on facilitating subsequent repair and renewal. Unfortunately, higher initial costs remain a significant barrier to its adoption [2]. In China, direct buried installation is still the most widely adopted solution for urban utility placement due to its lower short-term construction costs. Recent years, with the promotion of a series of policies, more utility tunnels are invested and built in major cities of China [3].

The utility tunnel is a relative enclosed space with poor ventilation. In order to ensure the normal operation of various municipal pipelines and to provide the maintenance personnel with a safe and sanitary environment, it is necessary to ventilate the utility tunnel to eliminate internal exhaust gas and heat [4-6]. Especially when a fire occurs in the tunnel, the ventilation system should be able to help control the spread of the fire and remove the toxic fumes accumulated in time [7]. However, introducing untreated air may result in changes of ambient temperature in utility tunnel, especially during the coldest weather in winter. Lower temperature may lead to excessive heat exchange between gas pipeline and indoor air, and thus cause the gas temperature to drop.

In this paper, a complete set of numerical models governing the heat transfer process was established, which considers heat convection, conduction and radiation. The simulation on a actual project was conducted to investigate the influence of various boundary conditions.

2. Method

The utility tunnel studied here locates in Shenyang of China with a climate of severe cold winter. The cross section and geometries of the tunnel are shown in figure 1. The upper part of the tunnel is the gas cabin, and the lower part is used for heating pipe installation.

The gas pipeline is made of steel, coated with polystyrene film. The tunnel is buried 5.3 m to 22.2 m underground.

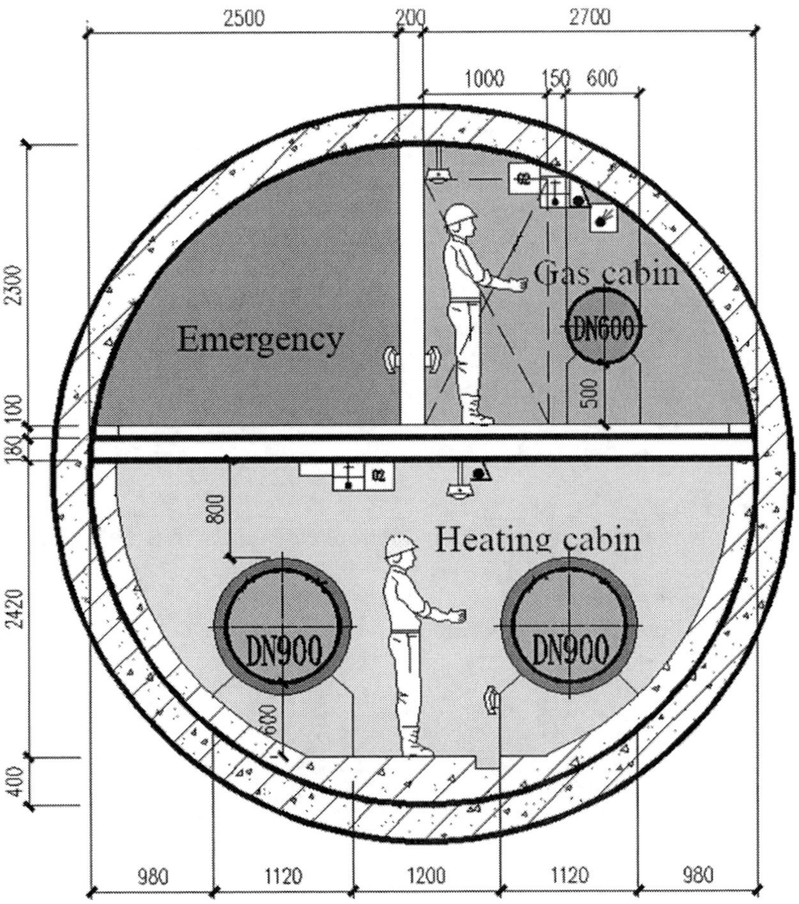

Figure 1. Cross section of utility tunnel.

The full length of the utility tunnel is 10 km, which is segmented every 0.5 km. The entire utility tunnel is divided into 20 segments. For each segment, two ventilation fans were used to circulate the air in tunnel at the start and end positions of the segment. The velocities of gas and air are 9 m/s and 0.5 m/s, respectively.

2.1. Numerical model

As illustrated in figure 2, the heat transfer process between the gas pipeline and the utility tunnel consists of four parts: convection between gas and inner surface of pipeline (Q_g), conduction through the pipeline (Q_w), convection between ambient air and outer surface of pipeline (Q_a), radiation between outer surface of pipeline and the tunnel structure (Q_r).

As heat exchange reaches a steady state, the thermal equilibrium was established and used to simulate the heat transfer between components.

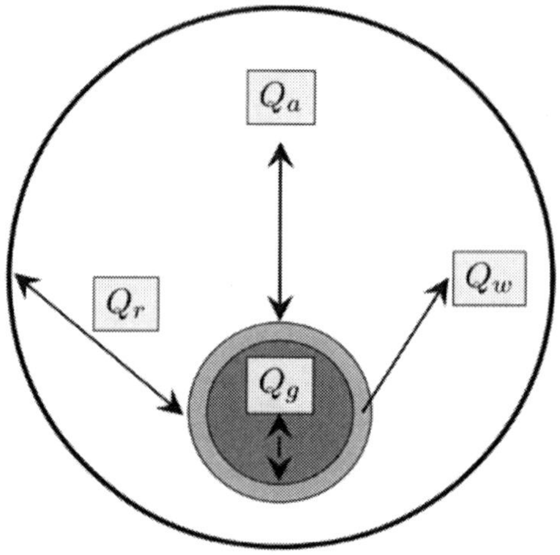

Figure 2. Heat transfer process in utility tunnel.

2.2. Convection between gas and inner surface of pipeline (Q_g)

Gas flow inside the pipeline is pressurized flow, which ensures the heat transfer between gas and pipeline to be forced convection. Dittus-Boelter Equation was used to calculate the convective heat transfer:

$$Nu_f = 0.023 Re_f^{0.8} Pr_f^n \tag{1}$$

Gnielinski equation was also used to calculate the convective heat transfer for comparison.

2.3. Conduction through the pipeline (Q_w)

A steady model based on Fourier's law was employed to simulate the heat conduction through the pipe wall:

$$q = \frac{t_i - t_o}{\sum_{i=1}^{n} \frac{\delta_i}{\lambda_i}} \tag{2}$$

2.4. Convection between ambient air and outer surface of pipeline (Q_a)

The ventilated air in the tunnel was driven by the ventilation fans, which circulate the air at a certain velocity to exhaust excess heat. A forced convection heat transfer correlation for non-circular cross-section channel was used to the heat convection between pipeline and ambient air:

$$Nu_f = \frac{(f/8)(Re-1000)Pr_f}{1+12.7\sqrt{f/8}(Pr_f^{2/3}-1)}[1+(\frac{d}{l})^{2/3}]c_t \tag{3}$$

For liquid:

$$c_t = (\frac{Pr_f}{Pr_w})^{0.11} \qquad (\frac{Pr_f}{Pr_w} = 0.05 \sim 20) \tag{4}$$

For gas:

$$c_t = (\frac{T_f}{T_w})^{0.45} \qquad (\frac{T_f}{T_w} = 0.5 \sim 1.5) \tag{5}$$

2.5. Radiation between outer surface of pipeline and the tunnel structure (Q_r)

Radiation happens between outer surface of pipe and the inner surface of tunnel, in which the heat exchange was calculated by Stephen Boltzmann's law:

$$\Phi = \frac{E_{b1} - E_{b2}}{\frac{1-\varepsilon_1}{\varepsilon_1 A_1} + \frac{1}{A_1 X_{1,2}} + \frac{1-\varepsilon_2}{\varepsilon_2 A_2}} \tag{6}$$

2.6. Governing equations

The governing equation set includes the heat balance for gas and air as follows:

$$(C_{pg}\rho_g A_g)\frac{\partial T_g}{\partial t} + (C_{pg}\rho_g A_g v_g)\frac{\partial T_g}{\partial x} - \alpha_g \pi D_i (T_i - T_g) = 0 \tag{7}$$

$$(C_{pa}\rho_a A_a)\frac{\partial T_a}{\partial t} + (C_{pa}\rho_a A_a v_a)\frac{\partial T_a}{\partial x} - \alpha_a \pi D_o (T_o - T_a) - \alpha_a S_w (T_w - T_a) = 0 \tag{8}$$

$$\alpha_g \pi D_i (T_i - T_g) = \pi \frac{D_i + D_o}{2} \bullet \frac{T_o - T_i}{\sum_{i=1}^{n} \delta_n / \lambda_n} \tag{9}$$

$$\alpha_a \pi D_o (T_o - T_a) + \frac{\pi D_o \sigma [(\frac{T_o + 273.15}{100})^4 - (\frac{T_w + 273.15}{100})^4]}{\frac{1}{\varepsilon_o} + \frac{\pi D_o}{S_w}(\frac{1}{\varepsilon_w} - 1)} = \pi \frac{D_i + D_o}{2} \bullet \frac{T_i - T_o}{\sum_{i=1}^{n} \delta_n / \lambda_n} \tag{10}$$

The transient terms in equations were neglected and a steady simulation was conducted. Finite volume discretization was applied for governing equations, and 4 sets of nonlinear equations including 4 unknown variables (T_g, T_a, T_o, T_i) were obtained at each control volume as follows.

Python was used to implement the algorithm and solve the final equation set.

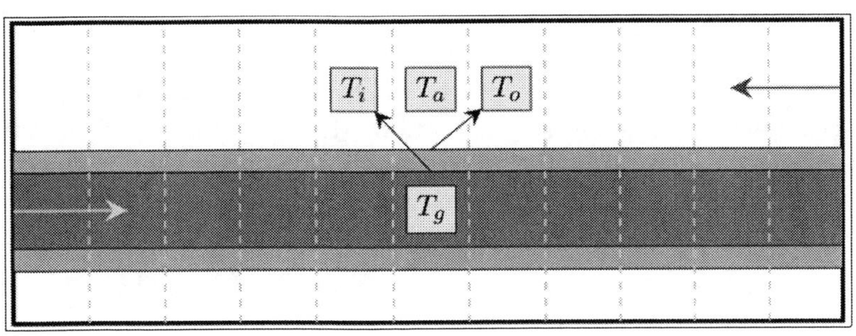

Figure 3. Finite volume discretization of utility tunnel.

2.7. Boundary condition

The utility tunnel studied in this paper was located in Shenyang City. The annual outdoor air temperature was shown in figure 4. It can be seen that the outdoor temperature in Shenyang can be as low as -20°C.

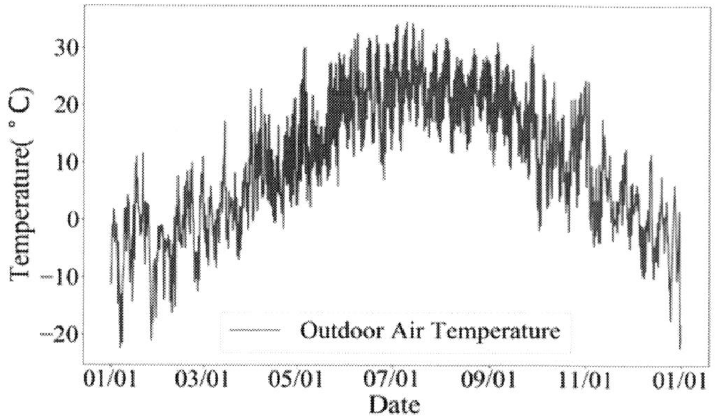

Figure 4. Outdoor air temperature in Shenyang.

The underground temperature at different depth varies greatly in fluctuations, as shown in figure 5.

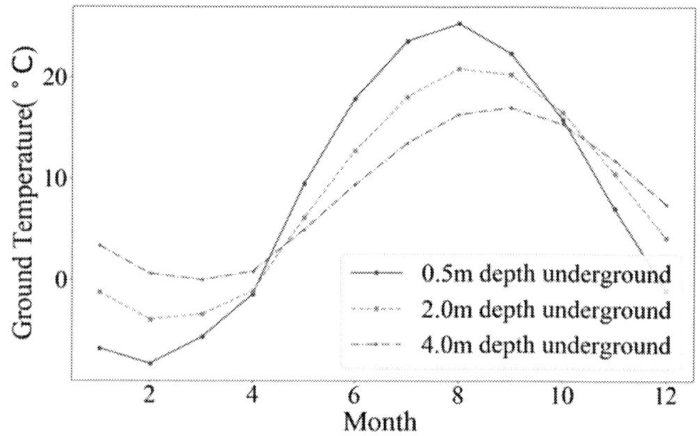

Figure 5. Underground temperature.

3. Results and discussion

The gas temperature profiles with different correlations is shown in figure 6. The correlation equations were used to calculate heat transfer inside gas pipes. Both the temperature differences and surface roughness meet the application condition of correlations. There are not significant differences between temperature profiles calculated with Gnielinski equation and Dittus Boelter equation.

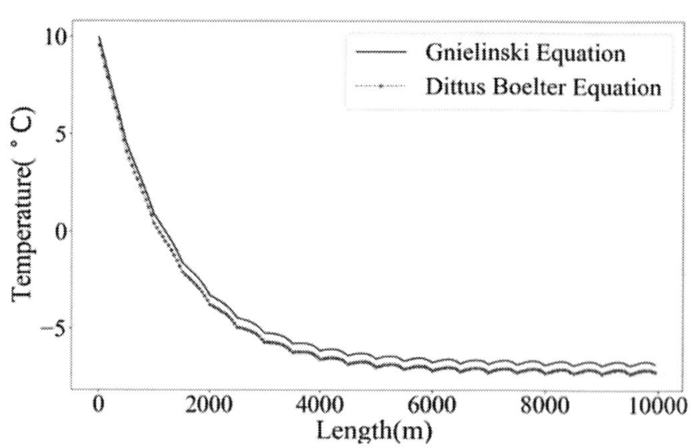

Figure 6. Underground temperature.

The temperature profile for the first segment is illustrated in figure 7. As the gas flows in the forward direction, the ventilation air flows inversely.

Gas transfers heat to ambient air through conduction and convection, accompanied by a slight drop in temperature. The ventilation air receives not only the heat from gas, but also the heat from the utility tunnel. Therefore, the air temperature has increased greatly.

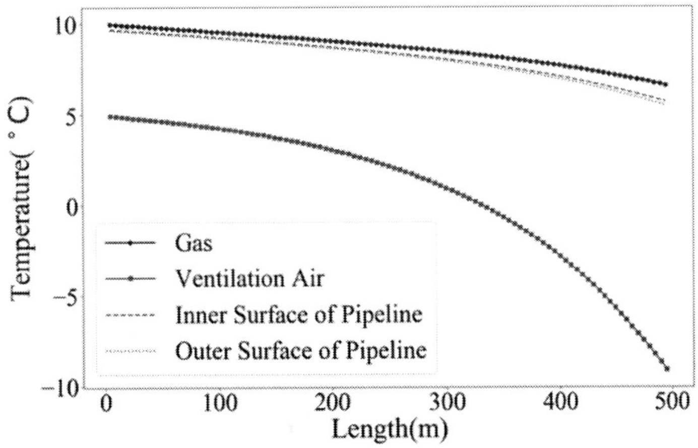

Figure 7. Temperature profile at the first segment.

Figure 8 shows the temperature profiles when the inlet air temperature is -20°C. The gas temperature changes continuously and air temperature change are discontinuous due to the segmentation for ventilation. At each segment of tunnel, the air temperature gradually approaches the temperature of tunnel wall due to the large contact area between the air and the tunnel wall.

As the gas flows, the gas loses heat to air and the temperature decreases to a steady state. Figure 9 shows the gas temperature profiles with different inlet air temperature. The outlet temperature of gas decreases as air inlet temperature decreases. For lowest inlet air temperature, the gas temperature at the exit can be reduced to below -10°C, which is too low to guarantee a normal operation.

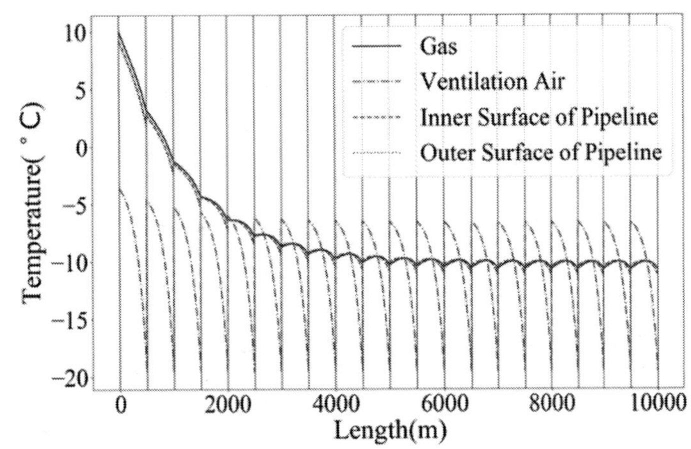

Figure 8. Temperature profiles when inlet air temperature is -20°C.

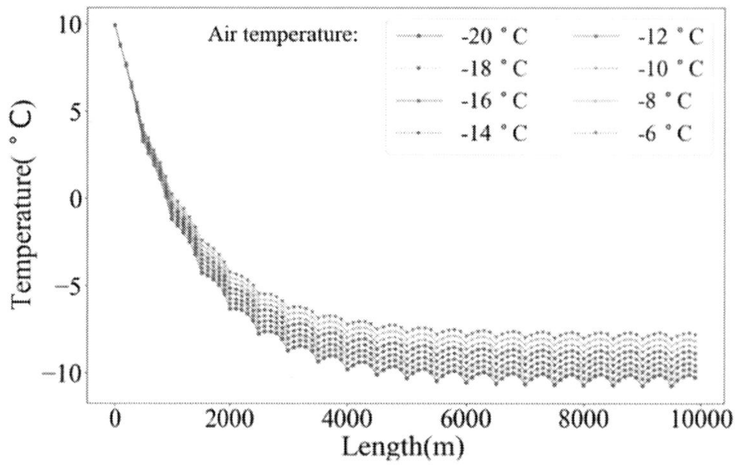

Figure 9. Temperature profiles with different inlet air temperature.

4. Conclusion

Ventilation is an effective way to ensure a safe operating environment in utility tunnel, which also has impact on gas temperature. The studied utility tunnel was segmented for ventilation. At each segment, the temperature of ventilated air increases greatly as the air absorbs the heat from pipeline and tunnel wall. The air temperature depends mainly on the temperature of tunnel wall due to the much large contact area. Different from the discontinuous change of air temperature, the gas temperature is continuously reduced to a relatively stable value along the flow direction. For lower inlet air temperature, the outlet gas temperature may be lower than -10°C, which cannot guarantee a normal operation. More attention needs to be paid on inlet air temperature when operating ventilation.

Acknowledgments

The authors wish to acknowledge the support of Natural Science Foundation of Liaoning Province of China (Grant No. 2019-ZD-0300) and Fundamental Research Project for Higher Education Institution of Liaoning Province (Grant No. LJZ2017032).

References

[1] Sun F, Liu C and Zhou X 2017 Utilities tunnel's finance design for the process of construction and operation *Tunnelling and Underground Space Technology* **69** 182–186
[2] Hunt D V L, Nash D and Rogers C D F 2014 Sustainable utility placement via Multi-Utility Tunnels *Tunnelling and Underground Space Technology* **39** 15–26
[3] Yang C and Peng F-L 2016 Discussion on the development of underground utility tunnels in China *Procedia Engineering* **165** 540–548
[4] Wu D, Zhang Y, Li A, Kong Q, Li Y, Geng S, Dong X, Liu Y and Chen P 2019 Indoor airborne fungal levels in selected comprehensive compartments of the urban utility tunnel in Nanjing, Southeast China *Sustainable Cities and Society* **51** 101723
[5] Yang C, Peng F-L, Xu K and Zheng L-N 2019 Feasibility study on the geothermal utility tunnel system *Sustainable Cities and Society* **46** 101445
[6] Li S, Liu X, Wang J, Fang G, Chen W and Deng S 2019 Reduced scale experimental study and CFD analysis on the resistance characteristic of utility tunnel's ventilation system *Energy Procedia* **158** 2756–2761
[7] Liu H, Zhu G, Pan R, Yu M and Liang Z 2019 Experimental investigation of fire temperature distribution and ceiling temperature prediction in closed utility tunnel *Case Studies in Thermal Engineering* **14** 100493

IOP Conf. Series: Materials Science and Engineering **789** (2020) 012065 doi:10.1088/1757-899X/789/1/012065

Study on concrete damaged plasticity model for simulating the hysteretic behavior of RC shear wall

Q Wang[1*], K K Hou[1], J Lu[1], Q H Dong[2], D P Yao[3] and Z Lu[3]

[1] School of Civil Engineering, Shenyang Jianzhu University, Shenyang, China
[2] Shenyang Newworld Architectural Design Co. Ltd., Shenyang, China
[3] Architectural Design Institution, Shenyang Jianzhu University, Shenyang, China

[*] wangqiangsy@163.com

Abstract. Abaqus is often used to analyze the dynamic elasto-plastic response of tall buildings under strong earthquake. However, the analysis results of the hysteretic behavior of RC shear wall are not very satisfactory. In this paper, the residual strain is proposed to determine the damage parameters in the concrete damaged plasticity model in Abaqus, which include the damage factor and compression stiffness recovery factor. The method can greatly reduce the lagging or exceeding of strain in the compression reloading path. The confinement effect of distributed bars on concrete of RC shear walls is also considered. In addition, a secondary-developed uniaxial constitutive model of steel is adopted to present the Baushinger effect under the reciprocating loading. Simulation results are in good agreement with the test results. It shows that the methods purposed in this paper can effectively realize the simulation of the hysteretic behavior of RC shear wall under the low cyclic reciprocating loading.

1. Introduction

RC shear wall member is the main lateral resistance member in the high-rise buildings because of its characteristics of high stiffness and high bearing capacity. In order to accurately analyze the non-linear response of high-rise buildings under strong earthquake, it is necessary to establish a reasonable mechanical model for RC shear wall member, which means high calculation accuracy and efficiency. In recent years, the layered shell element model of FEM, especially Abaqus, has gradually been regarded as an ideal model for RC shear wall [1].

Although the layered shell element model with the concrete damage plastic model of Abaqus has been used to analyze the nonlinear behavior of RC shear wall members and structures [2, 3], most researchers only study the skeleton curve under reciprocating loading [4].

The existing simulation results of hysteretic behavior of shear wall members are obviously fuller than the test results, which overestimates the energy dissipation capacity of members [5] and will greatly affect the accuracy of dynamic elastic-plastic analysis of the whole structure.

In order to accurately simulate the hysteretic behavior of RC shear wall members based on the layered shell element, the concrete damage plastic model of Abaqus as well as the steel constitutive model is studied in this paper.

IOP Conf. Series: Materials Science and Engineering **789** (2020) 012065 doi:10.1088/1757-899X/789/1/012065

2. Determination on parameters of concrete damage plastic model

2.1. Brief introduction of CDP model

In Abaqus, there are three concrete constitutive models, including the smeared cracking model, damage plasticity model (CDP model) and brittle cracking model. Among those models, the CDP model is recognized as the most effective model to simulate the inelastic behavior of concrete under reciprocating loading.

In the CDP model, a complete constitutive model of concrete material is formed by defining the material parameters of elastic stage, tension and compression plastic damage stage [6]. In the elastic stage, material parameters include elasticity modulus and Poisson's ratio, and the unloading-reloading modulus under reciprocating loading is kept as the initial elasticity modulus. In the plastic damage stage, the CDP model needs to define the stress-strain skeleton curve, plastic damage factor and stiffness recovery factor under compression and tension.

Plastic damage factor and stiffness recovery factor are used to describe the damage behavior of concrete under reciprocating loading, such as crushing, tensile cracking, crack closure and stiffness recovery. The relationship between unloading-reloading modulus E_r and damage factor under reciprocating loading is shown in equation (1).

$$E_r = (1 - d)E_0 \tag{1}$$

where, d is the damage factor in tension or compression, which ranges from 0 to 1. With the increase of the nonlinear behavior of concrete, the damage factor increases, so the unloading-reloading modulus decreases.

Equation (1) shows that the plastic damage factor d is essentially used to characterize the attenuation of the unloading-reloading modulus under reciprocating loading. The rationality of d determines whether the CDP model can accurately simulate the hysteretic behavior of concrete members. According to different damage theories, the relationship between the plastic damage factors and concrete strain varies greatly, which makes the unloading-reloading modulus determined by equation (1) is hard to correctly describe the loading-unloading law of actual concrete materials, and then causes the unreasonable results of hysteretic simulation of RC members. The Sidoroff energy method is usually used to determine the plastic damage factor [7], but its simulation results of hysteretic performance are usually poor.

At present, the Concrete02 model in OpenSees software can simply and reasonably describe the hysteretic behavior of concrete, and good results have been obtained in simulating the hysteretic behavior of shear walls [8, 9].

Therefore, in this paper, the Concrete02 model in OpenSees is referred to determine the material parameters of CDP model in Abaqus.

2.2. Determination of compression skeleton curve

In the Concrete02 model, the constitutive model proposed by Kent and Park [10] and then modified by Scott [11] are adopted, as shown in figure 1.

The Concrete02 model is mainly applied for the beam-column member, and its compression skeleton curve is divided into ascending stage, descending stage and platform stage. The skeleton curve equation is as follows:

$$\varepsilon \le \varepsilon_{cc}, f = f_{cc}[2(\varepsilon / \varepsilon_{cc}) - (\varepsilon / \varepsilon_{cc})^2] \tag{2}$$

$$\varepsilon_{cc} < \varepsilon \le \varepsilon_{cu}, f = f_{cc}[1 - Z(\varepsilon - \varepsilon_{cc})] \tag{3}$$

$$\varepsilon > \varepsilon_{cu}, f = 0.2 f_{cc} \tag{4}$$

where, $\varepsilon_{cc} = 0.002K$, $f_{cc} = Kf_{co}$, $K = 1 + \dfrac{\rho_V f_{yh}}{f_{co}}$, $Z = \dfrac{0.5}{\dfrac{3+0.29f_{co}}{145f_{co}-1000} + 0.75\rho_V(\dfrac{h'}{s_h})^{1/2} - 0.002K}$.

Here, ε_{cc} is the strain corresponding to the peak stress, K is the increase factor of strength and peak strain caused by stirrups restraint, Z is the slope of decreasing strain, f_{co} is the compressive strength, f_{cc} is the restrained compressive strength of concrete, f_{yh} is the yield strength of stirrup, ρ_V is the volume ratio of stirrups in beam-column members, h' is the width of concrete core, s_h is the spacing of stirrups.

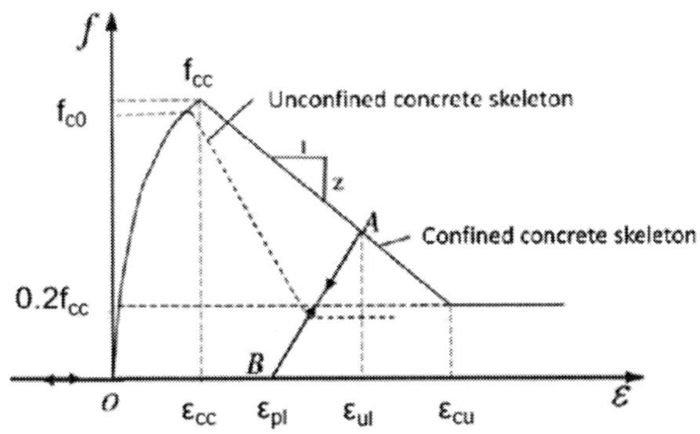

Figure 1. Compression skeleton curve of the modified Kent-Park model.

As shown in figure 1, the strength and peak strain of concrete will be slightly increased when the confinement effect of stirrups is taken into account. At the same time, the ductility of concrete increases significantly, that is, the slope of the descending stage of skeleton curve will be reduced.

In RC shear wall members, the existence of vertical and horizontal distributed bars will also confine concrete, and restrain the development of concrete cracks. When RC shear wall member is simulated by using of the layered shell element, the confinement effect of distributed bars cannot be taken into account in the element configuration, and the bearing capacity and ductility will be underestimated. The confinement effect of distributed bars on concrete in shear wall members should be considered in CDP model. In this paper, the compression skeleton curve of concrete in RC shear wall is also determined by equation (2), (3) and (4). However, ρ_V in these equations is adjusted to the volume ratio of distributed bars in shear wall members. ρ_V can be determined by equation (5).

$$\rho_V = \frac{n \cdot s_v \cdot A_{sh} + n \cdot s_h \cdot A_{sv}}{s_v \cdot s_h \cdot h'} \tag{5}$$

In which, S_v is the spacing of vertical bars, S_h is the spacing of horizontal bars, A_{sv} is the area of vertical bars, A_{sh} is the area of horizontal reinforcement and n is the number of bar layers.

2.3. Determination of compression damage factor

As mentioned above, the compressive damage of concrete in CDP model under reciprocating loading is described by the compressive unloading path of concrete, which is represented by the damage factor d_c in compressive zone. In CDP model, the unloading path is along a straight line from unloading point F to residual strain point A, as shown in figure 2.

However, the actual unloading path of concrete is along the curve with decreasing unloading modulus. Therefore, the CDP model characterizes the unloading behavior of concrete under compression by a simple path. The damage factor d_c determined by the Sidoroff energy method cannot reflect the decreasing of the unloading modulus. In this paper, the damage factor d_c is determined by the residual strain method. The compression hysteresis rule of the modified Kent-Park model is used to determine the compression residual strain point in the CDP model. Residual strain point A corresponding to the unloading point F is firstly determined, and then the damage factor d_c is calculated, as shown in figure 2.

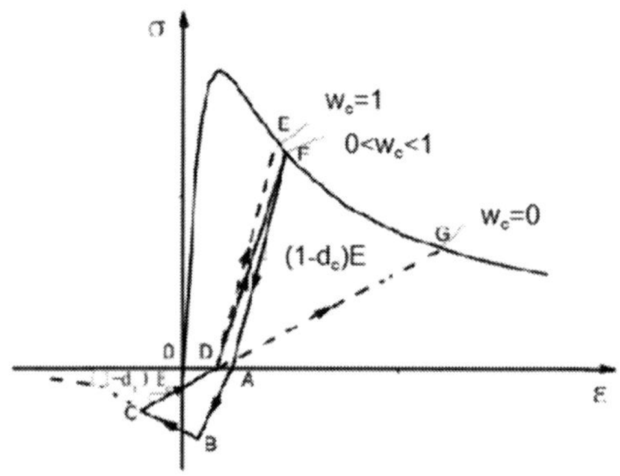

Figure 2. Hysteresis law of CDP model.

In the modified Kent-Park model, the unloading path is a straight line from D point to E point to H point, and the reloading path is a straight line from H point to D point, where H point is the residual strain point in compression zone, as shown in figure 3. The compressive residual strain at H point can be obtained as in equation (6).

$$\varepsilon_{c,pl} = \frac{f_{cc} \cdot (\varepsilon_{ul} \cdot E_c - \sigma_{ul})}{E_c \cdot (\sigma_{ul} + f_{cc})} \tag{6}$$

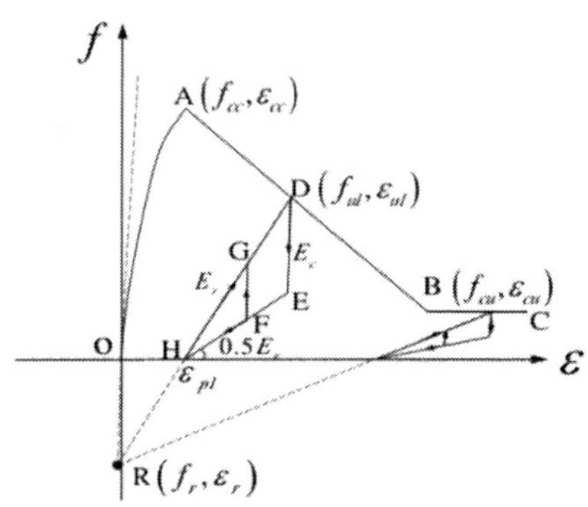

Figure 3. Hysteretic rule of compression zone in modified Kent-Park model.

For the unloading point F of the CDP model in figure 2, its coordinates are ($\varepsilon_{ul}, \sigma_{ul}$). Residual strain point A can also be obtained as in equation (6) accordingly. Thus, the unloading modulus is $E_r = \dfrac{\sigma_{ul}}{\varepsilon_{ul} - \varepsilon_{c,pl}}$, and the compression damage factor $d_c = 1 - \dfrac{E_r}{E_c} = 1 - \dfrac{\sigma_{un}}{(\varepsilon_{un} - \varepsilon_{c,pl})E_c}$. Therefore, the material loading-unloading path in CDP model will be a simplified form of the loading-unloading path of the modified Kent-Park model.

2.4. Tension skeleton curve and tension damage factor

In this paper, the tension skeleton curve is divided into three straight lines, in which the post-failure stress is $0.01f_t$ and ε_t is the peak strain, as shown in figure 4. The tension skeleton curve of concrete ignores the confinement effect of distributed bar.

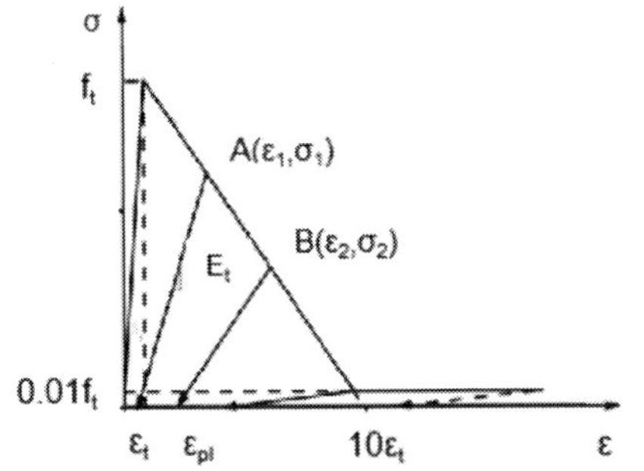

Figure 4. Tension skeleton curve.

It is generally believed that the tensile properties of concrete have little influence on the bearing capacity of members. However, there is a process of tensile cracking and closing repeatedly in concrete under reciprocating loading. The crack closure has an important influence on the hysteresis behavior of members under reciprocating loading. The "pinching" of macro load-displacement hysteresis curve of concrete members is largely affected by the crack closure behavior [12]. Therefore, the tension hysteretic relation representing the crack closing behavior in the CDP model is important for the rationality of the simulation results.

Similar to the unloading rules in the compression zone, the tensile unloading path in the CDP model is a straight line from the unloading point (point A or B) to the tensile residual strain point, as shown in figure 4. Under reciprocating loading, when the concrete in figure 2 is tensioned and unloaded from point C (once compressed, i.e. obtained from point F-A-B-C), the unloading path is a straight line from unloading point C to residual strain point D, and the crack closes at point D. If compressive loading occurs subsequently, by default in the CDP model (compressive stiffness recovery factor $W_c = 1$, i.e. compressive stiffness recovery completely), the compressive reloading will return to point E on the compressive curve along the DE route (DE∥ FA), instead of the previous compressive unloading point F. As a result, it exists that the reloading path lags behind the previous unloading path in the CDP model.

The strain difference between point D and point A is an important index to describe the crack closure behavior. When the tension unloading point is fixed, if the strain difference is larger, it means that the closure of cracks and the entry of compression state will occur prematurely, and the lagging of the reloading route in the compression zone will be more serious. The lagging of the reloading path makes the stiffness of the macro load-displacement hysteresis curve from unloading to reverse loading large, so the "pinching" with lower stiffness con not be represented, and the hysteretic curve tends to be chubby.

In the Concrete02 model, the origin-pointing rule is used to represent the tension unloading behavior of concrete. So, there is no lagging of the reloading path. In order to eliminate the lagging of reloading path in the compression zone of CDP model, the strain difference should be small as far as possible to rationally describe the crack closure. However, the strain difference, i.e. the tension residual strain $\varepsilon_{t,pl}$ and its increment must be greater than 0 in the CDP model. For the sake of simplicity, the tension residual strain is determined by equation (7).

$$\varepsilon_{t,pl} = \left(\frac{\varepsilon_{t,un}}{\varepsilon_t}\right)^{1/2} \varepsilon_t \tag{7}$$

In which, $\varepsilon_{t,un}$ is the unloading point strain on the tension skeleton curve.

Similarity to the compressive damage factor, the tensile damage factor d_t can be determined by equation (8) and (9).

$$E_{r,t} = \frac{\sigma_{t,un}}{\varepsilon_{t,un} - \varepsilon_{t,pl}} \tag{8}$$

$$d_t = 1 - \frac{E_{r,t}}{E_c} = 1 - \frac{\sigma_{t,un}}{(\varepsilon_{t,un} - \varepsilon_{t,pl})E_c} \tag{9}$$

In which, $\sigma_{t,un}$ is the stress corresponding to the unloading strain $\varepsilon_{t,un}$ on the tensile skeleton curve, $E_{r,t}$ is the unloading modulus in tension.

2.5. Determination of compressive stiffness recovery factor.
By the tension residual strain determined by equation (7), the lagging of the reloading route in the compression zone still exists when the tensile strain is large. Therefore, the compressive stiffness recovery factor W_c must be adjusted to further eliminate the lagging of the reloading path.

It can be seen from figure 2 that different W_c values will cause different reloading routes. When $W_c = 1$, the reloading path is along DE, and it exists the lagging of the reloading route; when $W_c = 0$, the reloading path is along DG, and it exists the exceeding of the reloading route. Reasonable W_c should make the reloading path is DF.

Because W_c in CDP model is constant, it is impossible to avoid the lagging or exceeding in different loading-unloading cycles. In this paper, the following methods are used: assuming that the compression unloading point is at the peak point of compression, the tension unloading point under is at $10\varepsilon_t$, and the reloading route under compression can come back to the peak point of compression. So, W_c is calculated by equation (10).

$$W_c = \frac{\varepsilon_{in,cc}}{\varepsilon_{in,cc} + \varepsilon_{t,pl10}} \tag{10}$$

Here, $\varepsilon_{in,cc}$ is the inelastic strain at the peak point of compression, and $\varepsilon_{t,pl10}$ is the residual strain corresponding to the tensile strain ε_t.

3. Constitutive model of rebar

In order to adequately consider the Bauschinger effect of steel under reciprocating loading, a secondary-developed uniaxial constitutive model of steel material is adopted for rebar in shear wall. The skeleton curve of the constitutive model is based on Esmaeily-Xiao model [13], and the loading and unloading curve is based on Légeron model [14], as shown in figure 5, the accuracy verification of the constitutive model and the parameters of the model can be seen in reference [15].

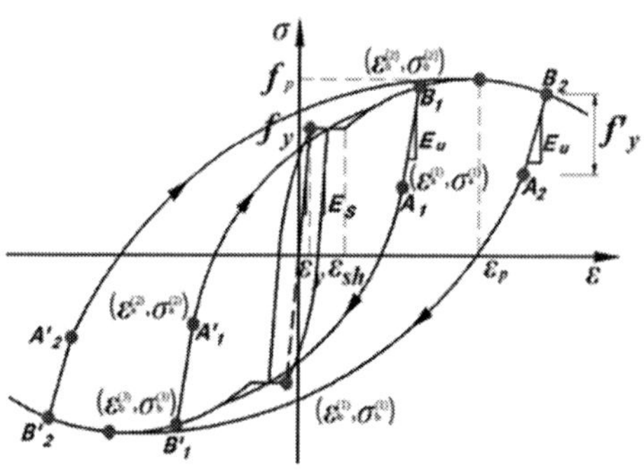

Figure 5. Stress-strain curve of steel.

4. Results and discussions

In order to verify the correctness of the concrete constitutive model, the test of SW1-1 (axial compression ratio is 0.14) and SW1-4 (axial compression ratio is 0.57) in reference [16] are simulated, as shown in figure 6. The layered shell element S4R in Abaqus is adopted in the model. The strength grade of concrete is C30. Yield strength of Φ6 and Φ10 bar are 392 MPa and 352 MPa. The material parameters in simulation are determined strictly according to the experimental conditions, but the number of loading cycles is simplified.

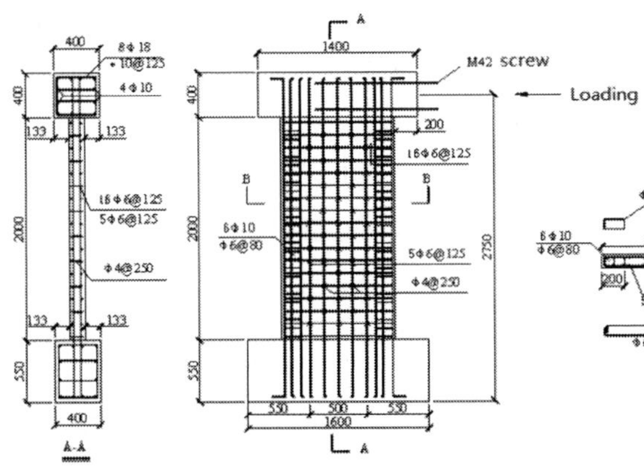

Figure 6. Reinforcement drawing of specimens set SW1.

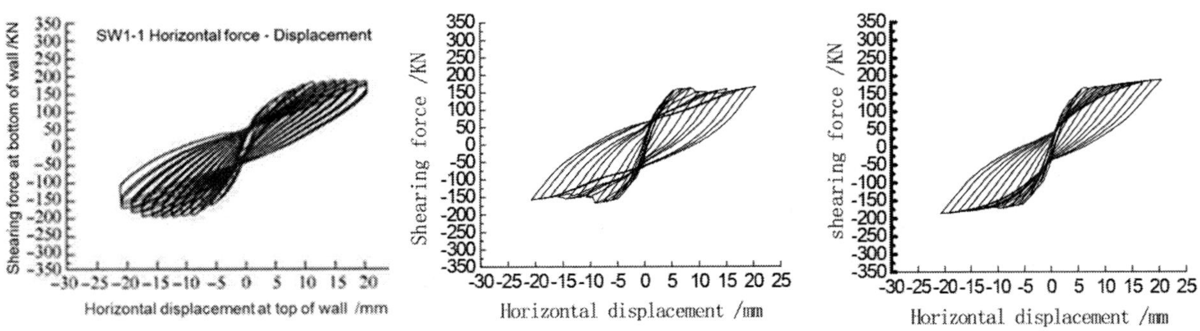

(a) test result (b) result A (c) result B

Figure 7. Comparison of test and simulation results of SW1-1.

The results of testing and simulation of SW1-1 are shown in figure 7. It can be seen that the results of simulation without considering the confinement effect of distributed bars (hereinafter referred to as result A) and with considering the confinement effect of distributed bars (hereinafter referred to as result B) are close to the test result. Figure 8 is the skeleton curve comparison between test and simulation results of SW1-1. It can be seen that the calculated skeleton curves are basically in agreement with the test curve and the stiffness is similar when the loading amplitude is smaller than +5 mm. With the increase of the loading displacement amplitude, the deviation between the calculated skeleton curve and the test increases. In the positive loading direction, the skeleton curve of result A reaches the peak value at the displacement amplitude of 8.5 mm, and then decreases. The bearing capacity deviation between simulation and testing is about 15%. The skeleton curve of result B better coincides with the test curve, and the bearing capacity deviation between simulation and testing is about 4.4%. In the negative loading direction, the deviation between the calculated skeleton curve and the test curve is greater than that of the positive loading. The maximum deviation between the peak bearing capacity of result A and result B is about 23% and 12%, respectively.

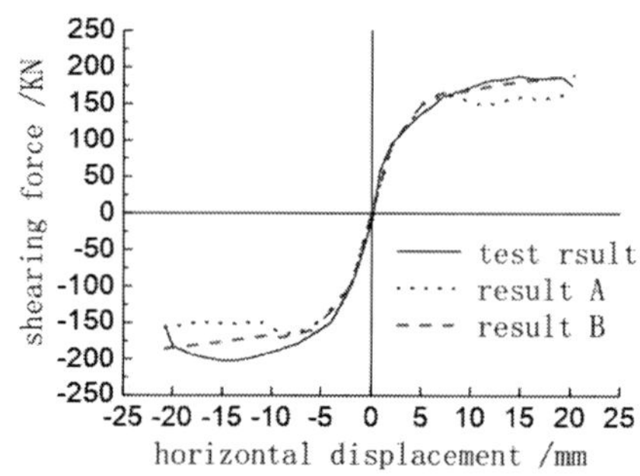

Figure 8. Comparison of skeleton curve of SW1-1.

Figure 9 is the comparison of single-loop hysteresis curves of SW1-1. Table 1 shows the error of single-loop calculation results compared with test results. It can be seen that the peak values of positive and negative loads of results A are lower than those of test results, while the peak values of positive and negative loads of results B are close to those of test results. The energy dissipation area of result A is significantly larger than that of test result at 11.5 mm and 20 mm loading amplitude, while that of result B is only slightly higher than that of test result.

The comparison results show that the method of determining damage parameters by residual strain can better simulate the hysteretic behavior of shear walls, and the simulation results with considering the confinement effect of distributed bars are more accurate and reasonable.

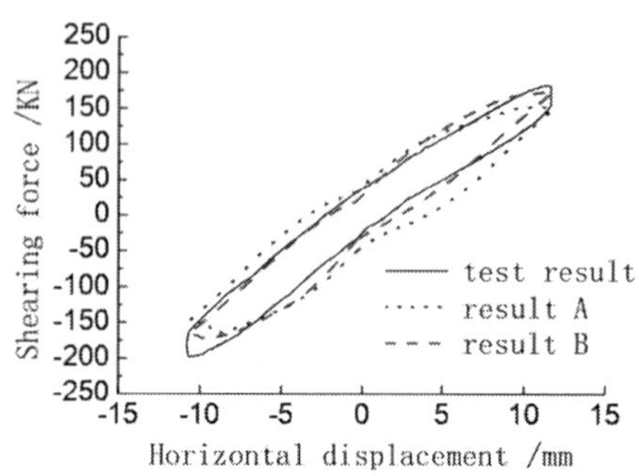

(a) displacement amplitude 10mm

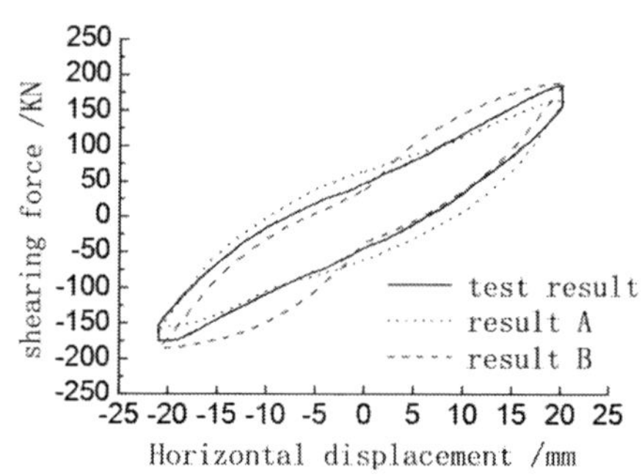

(b) displacement amplitude 20 mm.

Figure 9. Comparison of Single-loop hysteretic curve of SW1-1.

Table 1. Single-loop Error between test results and simulation results of SW1-1.

Displacement amplitude (mm)	result A			result B		
	Positive load capacity	Negative load capacity	Energy dissipation area	Positive load capacity	Negative load capacity	Energy dissipation area
11.5	-16.9%	-15.8%	+33.1%	-4.9%	-12.2%	+2.0%
20	-10.9%	-10.3%	+15.7%	+1.9%	+5.9%	+8.6%

The test results and simulation results of SW1-4 are shown in figure 10. Through the comparison of figure 10(a) and 10(b), it is found that the bearing capacity and hysteretic cycle fullness of the simulation result are close to the test curve. Combined with the analysis results of SW1-1 specimen with low axial compression

IOP Conf. Series: Materials Science and Engineering **789** (2020) 012065 doi:10.1088/1757-899X/789/1/012065

ratio shown in figure 6, it is shown that the model in this paper can reflect the influence of different axial compression ratio on the hysteretic behavior of shear walls.

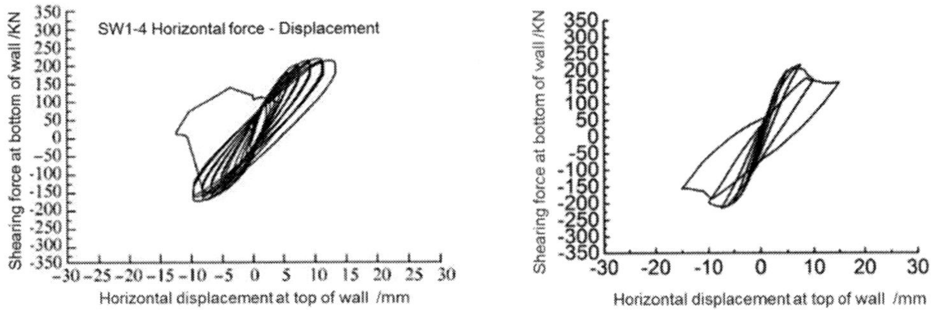

(a) test result (b) simulation result.

Figure 10. Comparison of test and simulation results of SW1-4.

In order to further verify the constitutive model of concrete, test JLQ-3 (Axial Compression Ratio 0.4) in reference [17] is also simulated, in which the strength grade of concrete is C40.

The results of testing and simulation are compared as shown in figure 11. It can be seen that the constitutive model of concrete and reinforcement adopted in this paper can well simulate the pinching of test curve, and the calculated bearing capacity and strength degradation obtained by considering the effect of distributed bars are more consistent with the test results, which further shows that the constitutive model of concrete and reinforcement adopted in this paper is reasonable.

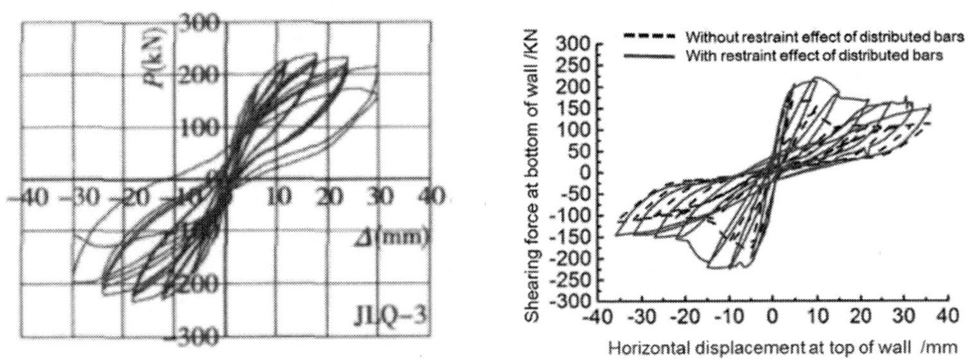

(a) test result (b) simulation result.

Figure 11. Comparison of test and simulation result of JLQ-3.

5. Conclusions

In this paper, the material constitutive model for simulating hysteretic behavior of RC shear wall members under low cyclic reciprocating loading is studied, and the following conclusions are obtained:

- the confinement effect of distributed bars on concrete in RC shear wall can be considered in the constitutive model of concrete, when the layered shell element is used to simulate RC shear wall member;
- the damage factors of CDP model can be determined based on the residual strain method. The residual strain method can reasonably describe the damage and crack closure behavior of concrete under reciprocating loading, and basically overcome the lagging of the reloading path in CDP model.
- By adjusting the compressive stiffness recovery factor W_c, the lagging of the reloading path in CDP model can be further eliminated.
- A secondary-developed uniaxial constitutive model of steel is adopted in the layered shell element, and helpful to simulate the pinching of hysteretic curve under reciprocating loading.

According to the simulation methods proposed in this paper, the hysteretic behavior of RC shear wall members under low cycle reciprocating loading can be simulated accurately and efficiently. These methods can also be applied for dynamic elastic-plastic analysis of RC shear wall structures.

Acknowledgments
This paper is Sponsored by National Natural Science Foundation of China (51178279) and Science and Technology Projects of Shenyang(F16-175-9-00).

Reference

[1] Lu X Z, Jiang Q, Liao Z W and Pan P 2015 *Elasto-plastic analysis of buildings against earthquake* (Beijing: China Construction Industry Press)
[2] Li Z S and Rong B S 2006 *Building struc*ture **36**(s) pp142-149
[3] Pan P, Qian J R and Zhou Y B 2009 *Building struc*ture **39** (10) pp15-18
[4] Zhang J, Wang Q Y and Hu S Y 2008 *Building struc*ture **38**(8) pp 127-130
[5] Wang D S 2013 *Application of dynamic elasto-plastic time history analysis technology in seismic design of building structures (*Shanghai: Shanghai science and Technology Press)
[6] ABAQUS Inc. 2014 *Abaqus Analysis User's Manual*
[7] Cao M 2012 *Road Engineering* **2** pp51-54
[8] Han X L, Chen X W and Dai J H 2008 *J. South China Univ.Tech.***36** (12) pp8-12
[9] Tong S M, Chen L and Zhou Y 2014 *Southern China earthquake* **34** pp102-106
[10] Kent D C and Park R 1971 *J. Struct. Division* **97**(7) pp1969-1990
[11] Scott B D, Park R. and Priestley M J N 1982 *ACI Journal* **79** pp13-27
[12] Nie J G and Wang Y H 2013 *Engineering Mechanics* **30** (4) pp59-67
[13] Esmaeily A and Xiao Y 2005 *ACI Struct. Journal* **102**(5) pp736-744
[14] Légeron F, Paultre P and Mazar J 2005 *J. Struct. Eng.* **131**(6) pp946-954
[15] Wang Q, Zhu L L and Li Z 2013 *China Civil Eng. Journal* **46**(S2) pp100-105
[16] Zhang H M and Zeng S 2014 *Structural engineers* **30** (5) pp165-173
[17] Zheng S S, Hou P J and Li 2012 *China Civil Eng. Journal* **45** (2) pp 51-59

Spindle vibration signal extraction method based on improved all phase Fast Fourier Transform

Z Wang [1], SY Du [1*] and LC ROŞCA[2]
[1] School of Mechanical Engineering, Shenyang Jianzhu University, China
[2] Faculty of Mechanical Engineering, Transilvania University of Brasov, Romania

[*] ttdusiyuan@gmail.com

Abstract. In the modern processing technology, the machine tool plays an irreplaceable role as the main processing tool, and its machining accuracy directly affects the processing quality of the part. Because the spindle vibration caused by the rotor mass imbalance will seriously affect the machining accuracy of the machine tool, it is necessary to dynamically balance the spindle. The key step of dynamic balancing is to accurately extract the characteristics of the vibration signal, the phase of the vibration signal of the spindle was extracted by the all phase fast Fourier transform, the amplitude of vibration signal was extracted by cross correlation analysis, the results are applied to the influence coefficient method to realize the dynamic balance of spindle vibration The results show that the vibration signal feature extraction method combined with the results of the two extraction methods has high precision and stability.

1. Introduction

Machine tools as the most basic equipment in the modern processing process, the machining accuracy plays a decisive role in the quality of the product. Among the many reasons that affect the machining accuracy, the machining accuracy is reduced due to the vibration caused by the imbalance of the machine spindle. When the spindle has a large unbalanced mass during operation, unbalanced centrifugal force will be generated. When the unbalanced centrifugal force due to unbalanced mass is not balanced by the outside, the spindle will generate strong vibration. With the increase of the rotational speed, the centrifugal force will become larger and larger, which seriously affects the machining accuracy of the machine tool and will have a great impact on the processing of the parts. Moreover, the machine tool spindle can`t avoid vibration during operation, so it is necessary to balance the centrifugal force generated by the unbalanced mass of the spindle. The key step affecting the balance accuracy is the extraction of the amplitude phase information of the unbalanced vibration signal.

At present, how to extract the unbalanced signal of the main shaft is the primary problem that needs to be solved to realize the on-line dynamic balance of the main shaft. The research on the extraction and processing of the rotor vibration signal has always been a hot topic in the research field at present, and has achieved certain theoretical results. Guo Junhua [1] have found that the signal is truncated by the whole cycle and processed by DFT method with high extraction precision, but the spectrum leakage and fence effect will occur when DFT is processed. Jiang Zhinong [2] and Xu Juan [3] eliminated the influence of phase lag on phase extraction accuracy by cross-correlation processing the signal, and the extracted amplitude and phase were significantly improved compared with the traditional cross-correlation method. Shang Yiqi [4] proved that the all-phase time-shift phase difference method has higher precision. Zhang Shihai et al. [5] used the least squares algorithm to fit the fundamental frequency signal, which has certain advantages compared with the traditional fast Fourier transform and discrete Fourier transform algorithm [6]. Wang Xu et al. [7] used the Prony method to extract the amplitude and phase information of the vibration signal. Compared with the fast Fourier transform method, the Prony algorithm overcomes the shortcomings of spectrum leakage and low frequency resolution. Mou Yu et al. [8] obtained experimental results that the all-phase FFT algorithm has higher phase extraction accuracy. Wang Zhaohua [9] proved that the all-phase FFT method has the characteristics of phase invariance. Guo Fan et al. [10] used the method of correlation analysis to realize the amplitude and phase extraction of vibration signals. By comparing the vibration response of the spindle system under experimental and simulation conditions, Wang Zhan et al. [11] verified the extraction accuracy of the amplitude and phase of the cross-correlation algorithm. Gao et al. [12] used inverse wavelet transform to reconstruct the signal and extract the vibration signal characteristics of the motor spindle. Dong [13] established the spindle vibration detection and analysis system based on LabVIEW to realize the acquisition, processing and analysis of vibration signals and reduce the hardware cost. Wang [14] used virtual instrument to design the vibration measurement system and could effectively track the vibration of the spindle. Xie [15] proposed an improved wavelet threshold de-noising method to process vibration signals. From the above research results, the all-phase FFT has higher precision in phase extraction and the cross-correlation method has great advantages in amplitude extraction. Therefore, the amplitude and phase of the spindle vibration signal are extracted by the combination of the full phase FFT method and the cross-correlation method. Experimental results show that the method is effective in extracting vibration signal of spindle.

2. Cross correlation method

In the process of signal acquisition, the detection and extraction of vibration signals is very complicated. There is no strict and accurate formula to accurately represent them. In most cases, vibration signal contains fundamental frequency, multiple frequency, sub-multiple frequency, random vibration signal, etc. The expression of the unbalanced vibration signal can be expressed as follows:

$$x(t) = e_0 + e\sin(\omega t + \varphi) + \sum_{i=2}^{n} e_i \sin(i\omega t + \varphi_i) + s(t) \tag{1}$$

In the equation: e_0 is a DC component; $i\omega$ is for each different signal frequency; φ_i is the phase value of each frequency signal; $s(t)$ is interference signals such as noise; where the fundamental component signals: $e\sin(\omega t+\varphi)$ is the vibration signal that needs to be obtained.

Set the standard sine signal and cosine signal with frequency ω and phase 0 as respectively

$$g(t) = \begin{cases} \sin \omega t, 0 < t < 1 \\ 0, other \end{cases} \quad h(t) = \begin{cases} \cos \omega t, 0 < t < T \\ 0, other \end{cases} \tag{2}$$

Two standard sine and cosine signals $g(t)$ and $h(t)$ are cross-correlated with vibration signal $x(t)$ respectively:

$$R_{yg}(0) = \int_{-\infty}^{+\infty} x(t)g(t)dt = \frac{Te}{2}\sin\varphi, \quad t \in [-\infty, +\infty] \tag{3}$$

$$R_{yh}(0) = \int_{-\infty}^{+\infty} x(t)h(t)dt = \frac{Te}{2}\cos\varphi, \quad t \in [-\infty, +\infty] \tag{4}$$

The amplitude and phase of the obtained fundamental frequency signal are as follows:

$$e = \frac{2\sqrt{R_{yg}(0)^2 + R_{yh}(0)^2}}{\mathrm{T}} \tag{5}$$

$$\varphi = \arctan\frac{R_{yg}(0)}{R_{yh}(0)}, \quad \varphi \in [0, 2\pi] \tag{6}$$

The process of extracting the vibration characteristic signal by using the correlation algorithm is actually calculating the average value, and the cross-correlation algorithm can suppress and eliminate the interference component such as the DC component and the noise contained in the vibration signal. The cross-correlation algorithm can ensure the accuracy of extracting information, and does not need to perform all-cycle sampling on the initial vibration signal, and the calculation amount is not large.

3. All phase FFT

The all-phase fast Fourier transform method was first proposed by Wang Zhaohua and Hou Zhengxin of Tianjin University in 2007. After continuous development, it has become an important tool in digital signal processing. The all phase FFT is based on the improvement of FFT, the only difference is that they differ in the pre-processing of the signal.

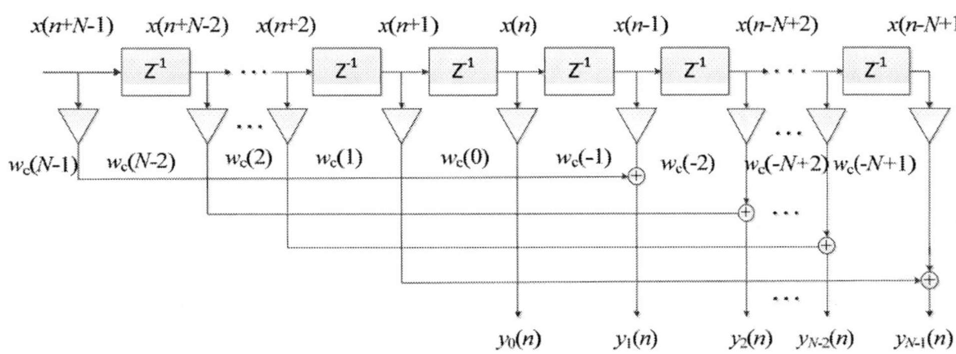

Figure 1. Flow chart of all-phase data preprocessing.

Figure 1 is a unified flow of all-phase data preprocessing. First, input the (2N-1) length of the discrete signal, and then use the convolution window w_c to weight the signals of these lengths, and then separate each data weighting result by N delay units for superposition, and output N. Data. After the pre-processing of the input sequence is performed, the signal needs to be spectrally analyzed, that is, the vibration signal is processed. Spectral analysis yields the amplitude and phase of the vibration signal. Now list the N-point N-dimensional vectors of the time series x(0):

$$\begin{aligned} \boldsymbol{x}_0 &= [x(0), x(1), \cdots, x(N-1)]^{\mathrm{T}} \\ \boldsymbol{x}_1 &= [x(-1), x(0), \cdots, x(N-2)]^{\mathrm{T}} \\ &\cdots\cdots\cdots \\ \boldsymbol{x}_{N-1} &= [x(-N+1), x(-N+2), \cdots, x(0)]^{\mathrm{T}} \end{aligned} \tag{7}$$

Then move each x(0) in x_0 to x_{n-1} to the first place to get another N N-dimensional vectors:

$$\begin{aligned} \boldsymbol{x}'_0 &= [x(0), x(1), \cdots, x(N-1)]^{\mathrm{T}} \\ \boldsymbol{x}'_1 &= [x(0), x(1), \cdots, x(-1)]^{\mathrm{T}} \\ &\cdots\cdots\cdots \\ \boldsymbol{x}'_{N-1} &= [x(0), x(-N+1), \cdots, x(-1)]^{\mathrm{T}} \end{aligned} \tag{8}$$

By aligning x(0) and taking the average, we can get the all-phase data vector:

$$\boldsymbol{x}_{\text{ap}} = \frac{1}{N}\left[Nx(0),(N-1)x(1)+x(-N+1),\cdots,x(N-1)+(N-1)x(-1)\right]^{\text{T}} \tag{9}$$

According to the shifting property of DFT, the pair of the formula $x'_i(n)$ discrete Fourier transform $X'_i(k)$. And $x_i(n)$ discrete fourier transform $X'_i(k)$. There is a very clear relationship between:

$$X'_i(k) = X_i(k)e^{j\frac{2\pi}{N}ik} \qquad i,k = 0,1,\cdots,N-1 \tag{10}$$

For equation (10) $X'_i(k)$ the summation average is the output of the all phase FFT:

$$X_{\text{ap}}(k) = \frac{1}{N}\sum_{i=0}^{N-1} X'_i(k) = \frac{1}{N}\sum_{i=0}^{N-1} X_i(k)e^{j\frac{2\pi}{N}ik} = \frac{e^{j\phi_0}}{N^2}\cdot\frac{\sin^2\left[\pi(\beta-k)\right]}{\sin^2\left[\pi(\beta-k)/N\right]} \tag{11}$$

From equation (11), it is known that φ_0 is the phase value of full-phase FFT, and $X_{ap}(k)$ is the amplitude of full-phase FFT.

4. Simulation and experiment

Use LabVIEW to write the cross-correlation method, and use the formula node to write the MATLAB all-phase FFT program. The two programs were connected together. The cross-correlation method outputs the vibration signal amplitude and the all-phase FFT output phase. We call this method the improved full phase FFT method.

The vibration signal of the main shaft is mainly composed of the fundamental frequency signal of the main shaft and environmental noise. Ambient noise can be divided into multiplier signals and Gaussian white noise. There is also a DC component of the sensor. Concrete equation (12):

$$x(t) = a_0 + \sum_{i=1}^{n} a_i \sin(i\omega t + \phi_i) + s(t) \tag{12}$$

In the equation, a_0 is a DC component; $\sum_{i=1}^{n} a_i \sin(i\omega t + \varphi_i)$ is the fundamental frequency signal and the multiplied signal, which are expressed as the vibration of the main shaft, are the fundamental frequency signal when i=1, and are the multiplied signals when i>1. a_i for amplitude, φ_i for phase $s(t)$ it is a noise signal, including uniform white noise and white Gaussian noise.

The above is a basic mode of the spindle vibration signal, which is included in the signal collected by the sensor. Therefore, it is necessary to filter to obtain a fundamental frequency signal. For the spindle of experiment, the time domain waveform of the simulated vibration signal can be set as: the uniform white noise is the software self-contained, the signal standard deviation of Gaussian white noise is 1, the DC component is 2 μm, the frequency conversion signal is 25 Hz, the corresponding vibration amplitude value is 5 μm, the double-frequency vibration signal is 3 μm, the triple-frequency vibration signal is 2 μm, and the corresponding initial phases are 30°, 20° and 10°, respectively. The sampling point is 1000, and the sampling frequency is 1000 Hz. The corresponding function is as follows:

$$x(t) = 2 + 5\sin(50\pi t + \pi/6) + 3\sin(100\pi t + \pi/9) + 2\sin(150\pi t + \pi/18) + s(t) \tag{13}$$

Vibration signal extraction simulation is performed by two methods: cross-correlation method and improved all-phase FFT. In the measurement, the fundamental frequency of the spindle is 25 Hz, but due to sensor accuracy and system defects, the extracted vibration signal generally does not reach 25 Hz.

Table 1. All-cycle comparison.

	Cross correlation method	Improved all phase FFT
Amplitude / μm	4.78	4.78
Phase maximum / °	30.71	31.2
Phase minimum / °	27.74	29.94
Difference / °	2.97	1.26

Table 2. Non-all-cycle comparison.

	Cross correlation method	Improved all phase FFT
Amplitude / μm	4.70	4.70
Phase maximum / °	33.97	31.2
Phase minimum / °	28.01	29.77
Difference / °	5.96	1.43

The extraction results are shown in table 1 and table 2. Shown, it can be seen that the all cycle frequency and the non-all-cycle frequency have a great influence on the signal processing results. It can be seen from the data in the table that the improved all-phase FFT is more accurate than the cross-correlation method in terms of amplitude extraction and phase extraction, and the jitter amplitude is small. Therefore, the following experimental verification can be performed.

5. Experimental verification

The improved all-phase FFT is applied as an input to the dynamic balance system. The dynamic balance system uses the influence coefficient method to perform dynamic balancing, and compares the method with the cross-correlation method as an input. The experimental platform introduces a mechanical spindle driven by a motor, and a balance head is installed inside, and the unbalanced mass is added by a manual method. Use National Instruments' (NI) data acquisition system with built-in cross-correlation algorithm for dynamic balancing. The experiment uses the NI 9239 to simultaneously acquire the signals of the three Hall elements. The data calculated by the influence coefficient method is dynamically balanced by the controller driving the balance head. The experimental platform is shown in figure 2.

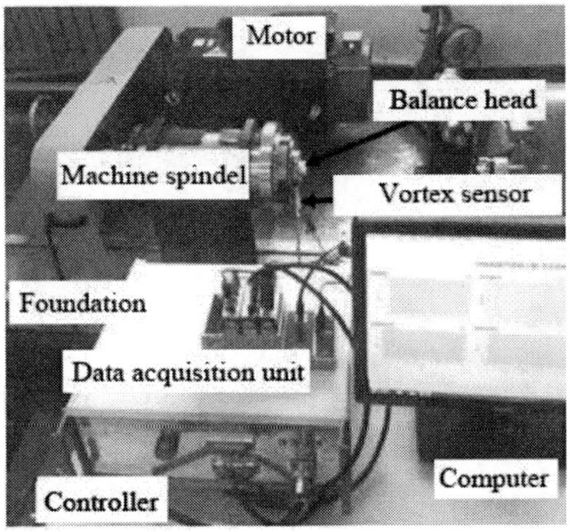

Figure 2. Experimental platform.

Below is the relevant data of the experiment. In the case of a spindle rotation speed of 3000 r/min, the amplitude and phase obtained by massing 8.5 g·mm were obtained, as shown in figures 3 to 4. It can be seen from figure 3 that the amplitudes of the spindle vibration signals extracted by the improved all phase FFT method and the cross-correlation method, although having large fluctuations, are all in the same range. That is, the improved all phase FFT method is comparable to the cross-correlation method in amplitude processing.

However, it is known from figure 4 that the improved all phase FFT method has the advantage of stability in the phase spectrum analysis of the spindle vibration signal, and the phase distribution of the cross-correlation method is about 35°, and there is a large fluctuation. The phase spectrum distribution of the improved all phase FFT method is very stable, showing a straight line within almost 60s, which shows the advantage in phase extraction.

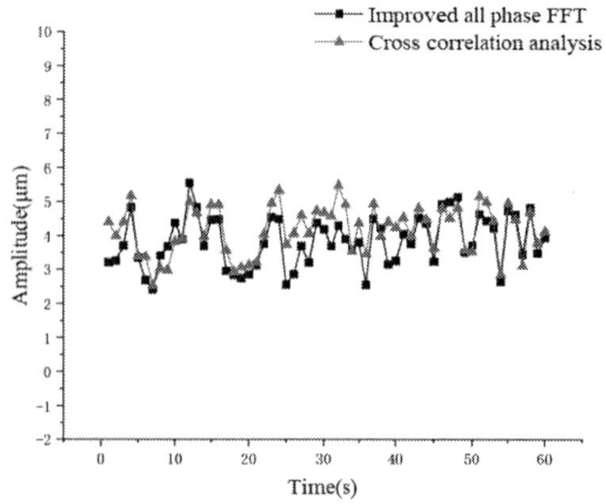

Figure 3. 3000 r/min amplitude contrast.

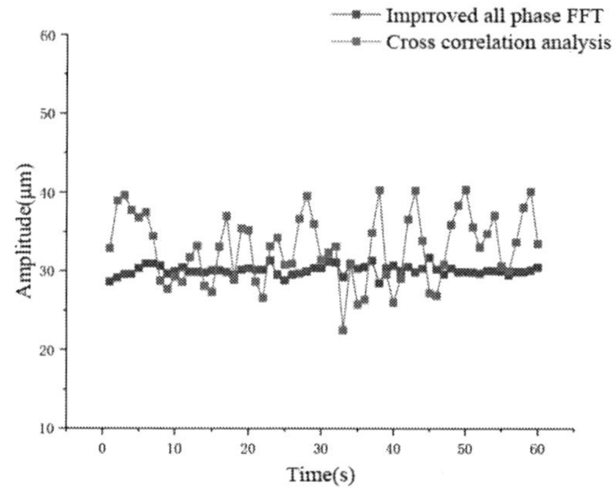

Figure 4. 3000 r/min phase contrast.

The vibration signal extracted by the improved all phase FFT method and the vibration signal extracted by the cross-correlation method are used as the influence coefficient method for online dynamic balance under the experimental conditions of the speed of 3000 r/min, 4000 r/min and the unbalanced mass of 8.5 g·mm.

The vibration characteristics extracted by the two methods were used as input parameters to test the dynamic balance effect, the experimental data are shown in figure 5 and figure 6.

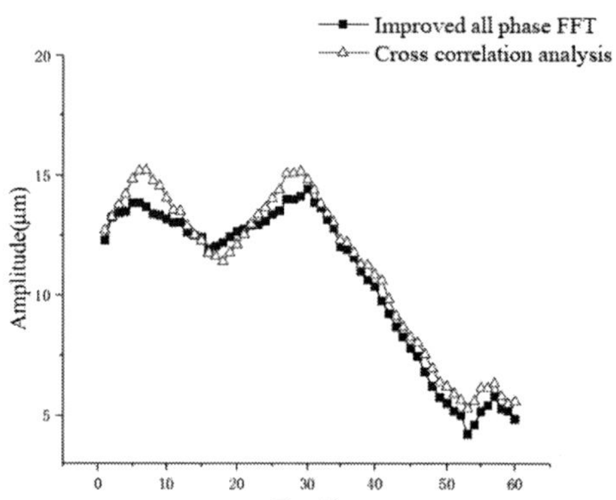

Figure 5. 3000 r/min contrast.

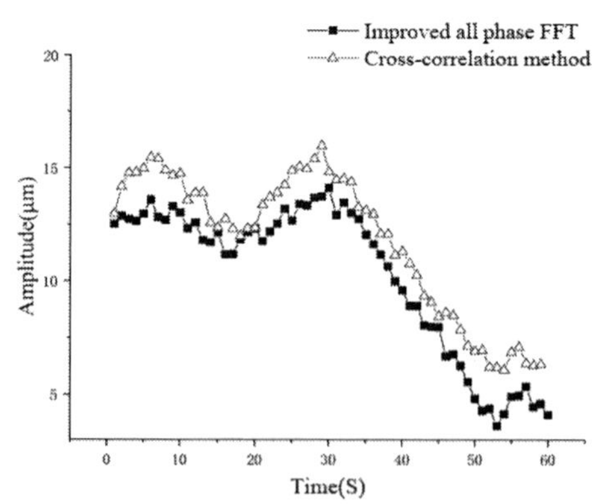

Figure 6. 4000 r/min contrast.

The online dynamic balance experiment shows that the vibration signal extracted by improved full phase FFT method is used as the input of the influence coefficient method, the amplitude of the vibration is obviously reduced, the balance rate is close to 70%, and the balance effect is better than the cross-correlation method built into the existing device. The dynamic balance is ideal.

6. Conclusion

Compared with the cross-correlation method, the improved all phase FFT method draws the advantages of the all-phase FFT and cross-correlation method. There is a great advantage in amplitude and phase analysis. In the phase spectrum analysis, there is a straight line within almost 60s, and the average phase value is 29.71.

The vibration signal extracted by the method is used as the input of the influence coefficient method. The amplitude of the vibration is obviously reduced, the balance rate is close to 70%, the residual balance is lower than the residual balance of the cross-correlation method, and the unbalanced vibration is effectively suppressed. It can be seen that the improved all phase FFT method is applied to the dynamic balance software, and the performance parameters meet the design requirements.

The single-plane dynamic balance process is ideal, and the test system can be applied to the dynamic balance of other similar working conditions, and is extended to high speed spindle double-sided dynamic balance test.

Acknowledgments

This research is supported by the National Natural Science Foundation of China (No.51805337; No.51 675353; No.51705340), National Natural Science Foundation of Liaoning Provincial (No.2017054074 6).

References

[1] Junhua G, Xing W, Xiaoqin L and Tianyan Y 2011 *J. Machinery & electronics* **10** pp 6-10
[2] Zhinong J, Kun F and Jinji G 2006 *J. journal of vibration measurement & diagnosis* **3** pp 234-39
[3] Juan X, Yichao L, Li Z and Meng G 2012 *J. Electr. Measur. Instr.* **44** pp 972-976
[4] Yiqi S, Yiqing C, Yanli C, Ningning Z and Lifen L 2010 *J. Metrology & Measurement Technology* **4** pp 30-32
[5] Shihai Z and Yujun C 2016 *J. Journal of Vibration and Control* **4** pp 1037-1048
[6] Duhamael P and Vetterli M 1990 *J. Signal Processing* **19** pp 259-299
[7] Xu W 2014 *D. Hefei university of technology* pp 44-50.
[8] Yu M, Xin P, Jinji G and Haiqi W 2015 *J. Industrial Instrumentation & Automation* **5** pp 17-20
[9] Zhaohua W, Xiangdong H and Wei Y 2007 *J. World Sci-Tech R & D* **29** pp 28-32
[10] Fan G, Xuming M and Huajie X 2015 *J. Information technology* **44** pp 138-140
[11] Zhan W, Fenglong Z and Wei T 2018 *J. Modul. Mach. Tool Autom. Manuf. Techn.* **2** pp 87-93
[12] Rong G, Peiqing Y, Kerong J and Wen L 2010 *J. Journal of Jilin University* **40** pp 1025-1028
[13] Fang D, Jiancheng Y and Tianzhu G 2014 *J. Applied Mechanics and Materials* pp 556-562
[14] Shunli W, Yufeng D, Lin W, Buyun S and Qibing L 2014 *J. Modular Machine Tool & Automatic Manufacturing Technique*
[15] Zhijie X, Baoyu S, Yang Z and Feng Z 2014 *J. Advanced Materials Research* pp 799-806

IOP Conf. Series: Materials Science and Engineering **789** (2020) 012067 doi:10.1088/1757-899X/789/1/012067

Preparation of catalyst CuO/CeO_2 in multiphase Fenton technology

N N Wu[1*], Y He[1], Y C Tan[1], L C Zhang[2], N Huang[3] and W T Fang[1]

[1] School of Municipal and Environmental Engineering, Shenyang Jianzhu University, Shenyang, China

[2] Shenyang Jianzhu University Urban Planning and Architectural design institute, Shenyang, China

[3] College of Architecture and Urban Planning, Shenyang Jianzhu University, Shenyang, China

[*] nanawu0816@xmu.edu.cn

Abstract. The performance of the catalyst in the multiphase Fenton system is particularly important. It will directly affect the rate of hydroxyl radical generation during the multiphase Fenton reaction which is the main material for oxidative degradation of organic matter in wastewater. Therefore, it is necessary to select a suitable catalyst to construct a multiphase Fenton system. In this study, the preparation of CuO/CeO_2 catalyst in multiphase Fenton technology was discussed. The effects of CuO loading, calcination time and calcination temperature on CuO/CeO_2 were investigated. The best preparation conditions for the catalytic performance of the catalyst were as follows: the loading of CuO was 7%, the calcination time was 3h and the calcination temperature was 400°C. The catalyst prepared under this condition can achieve a COD_{Cr} removal rate of 34% for a plurality of Fenton processes.

1. Introduction

The supported catalyst consists of an active metal or a metal oxide and a carrier. The active metal is the basis of the performance of the catalyst, which determines the reaction rate and reaction direction. At present, domestic and foreign research on catalysts has made rapid progress in transition metals and their oxides, and is divided into catalysts containing noble metals and catalysts containing non-precious metals.

Compared with other metal catalysts, precious metal catalysts have many advantages, but at the same time, they have problems of poor stability, high price, and easy sintering into blocks at high temperatures, so the application of precious metal catalysts is limited. However, the price of the transition metal and its oxide supported on the catalyst carrier is much lower (copper, zinc, iron, nickel, etc.), and it is easier to obtain and has better catalytic performance. At the same time, it also has the advantages of precious metals and avoids its shortcomings. It is currently a hot spot in the field of multiphase Fenton catalysts.

Zhou Lingyun [1] used a multiphase Fenton system composed of CuO and H_2O_2 under low temperature conditions to catalyze the oxidation of dye wastewater and degrade the reactive brilliant red X-3B in water.

The results show that the multiphase Fenton system composed of CuO and H_2O_2 can oxidize reactive brilliant red X-3B quickly and efficiently, and the oxidative degradation rate can reach 99%. Qiu et al. [2] used transition metal oxides as catalysts to form a heterogeneous Fenton system with H_2O_2, oxidized wastewater containing acid red dye, and discussed the performance, catalytic effect and preparation conditions of the catalyst.

The results show that CuO has the strongest catalytic performance in the system compared with the catalytic ability of metal oxides of Ni, Mn and Zn. Therefore, the transition metal oxide CuO was selected as the active component of the heterogeneous Fenton catalyst.

2. Preparation method of CuO/CeO_2

CeO_2 has good oxygen storage and oxygen release capacity, but it is generally not used as a catalyst alone, but as a carrier in combination with other highly stable oxides. Among them, the supported catalyst CuO/CeO_2 is widely used because of its low cost and high activity. Here are a few common methods for preparing CuO/CeO_2.

2.1. Surfactant synthesis method

Surfactants are often used as templating agents because of their good structural distribution. The templating agent-assisted synthesis method means that the added templating agent can modulate the pore structure to obtain a catalyst having a high specific surface area. After 2005, surfactants were applied to prepare catalyst supports and non-silicon aluminium material catalysts with high specific surface area.

For the preparation of multi-component non-silicon-aluminium material catalysts, it is difficult to obtain a regular and uniform pore structure in normal preparation, because in the calcination stage, the valence state of the transition metal changes, resulting in collapse of the pore walls, failing to obtain an ideal mesoporous material.

However, by adding a surfactant during the production process, a catalyst having a large specific surface area and a regular mesoporous structure can be obtained.

2.2. Coprecipitation method

The coprecipitation method refers to precipitating a solution of cerium nitrate and copper nitrate with a precipitant (Na_2CO_3) in an aqueous solution to obtain a precipitate, followed by centrifugation, washing, drying, and finally calcination to obtain a CuO/CeO_2 catalyst. Further, a mixed solution of potassium carbonate and potassium hydroxide can also be used as a precipitating agent to prepare a CuO/CeO_2 catalyst.

2.3. Deposition precipitation method

The deposition precipitation method refers to putting different metal oxides and carriers into a solution containing an active ingredient, adding an alkaline solution for neutralization, and selecting certain reaction conditions to deposit a metal oxide on the surface of the carrier to obtain a precipitate.

After the precipitate is obtained, it is centrifuged, washed, dried, etc., and finally calcined to obtain a CuO/CeO_2 catalyst.Richter et al. [3, 4] precipitated $Cu(OH)_2$ onto Y-type molecular sieves and activated it at high temperature to obtain a chlorine-free Cu/Y catalyst with better catalytic activity.

2.4. Impregnation method

The impregnation method refers to immersing CeO_2 as a carrier in a $Cu(NO_3)_2$ solution. After a period of time, the precipitate is separated for centrifugation, washing, drying, etc., and finally calcined to obtain a CuO/CeO_2 catalyst.

Zhang Haitao et al. [5] prepared CuO supported catalyst by immersing $Cu(NO_3)_2$ and then decomposing $Cu(NO_3)_2$. It was found that the catalyst prepared by activated carbon was catalytically active, and the surface of CuO and Cu_2O was the main catalyst.

2.5. Citric acid hydrothermal synthesis
Citric acid hydrothermal synthesis means adding citric acid to a mixed solution of CuC_2O_4 and $Ce_2(C_2O_4)_3$, pouring the solution into a hydrothermal reaction kettle, and hydrothermal treatment at 150°C for 1 day. Then, the supernatant is removed and the precipitate is separated, followed by centrifugation, washing, drying, etc., and finally calcination to obtain a CuO/CeO_2 catalyst.

Shaikh et al. [6] prepared a CuO/CeO_2 catalyst with CuO attached to the surface of CeO_2 nanorods by hydrothermal synthesis to study the distribution effect of copper species on low temperature CO oxidation. Liu et al. [7] prepared a CuO/CeO_2 composite oxide catalyst by hydrothermal method and studied its catalytic decomposition of N_2O. The CuO/CeO_2 mixed oxide catalyst was found to have higher oxidation activity than pure CuO and CeO_2.

2.6. Sol-gel method
The sol-gel method is a relatively common production method, which is divided into a hydrogen peroxide sol gel method and a urea sol gel method. Hydrogen peroxide hydrogel method refers to separately adding H_2O_2 in a solution of cerium nitrate and copper nitrate, and then mixing the two solutions. After standing for a while, the supernatant is removed and the precipitate is separated, followed by centrifugation, washing, drying, and finally calcination to obtain a catalyst CuO/CeO2.

Ren Jun et al. [8] used copper nitrate and soluble starch as raw materials, and obtained CuO/AC catalyst through sol-gelation, high temperature carbonization and KOH activation. When the carbonization and activation temperatures were 500°C and 850°C respectively, and KOH: C was 1, the specific surface area of the catalyst reached 1690 m^2/g, and the microporous pore volume was 72.4%, which was the highest.

3. Experiment
The supported catalyst CuO/CeO_2 supported by CeO_2 was prepared by sedimentation precipitation method. The reagents used were analytically pure.

The sodium lauryl sulphate solution, the $CeCl_3$ solution and the NaOH solution were prepared for use. Stir the sodium lauryl sulphate solution for more than 40 minutes until the sodium lauryl sulphate is completely dissolved.

The fully dissolved $CeCl_3$ solution was slowly added to the sodium dodecyl sulphate solution using a peristaltic pump and stirring was continued for 30 to 40 minutes. Then, the prepared NaOH solution was slowly added to the solution by a peristaltic pump in the same manner. After the addition, the container was sealed with a plastic wrap and stirring was continued for 24 hours. Subsequently, the mixture is placed in an oven and crystallized at 90°C for 24h, then subjected to 3 to 5 centrifugation washing, 70°C vacuum drying for 12h, 400°C calcination for 4h (under dry air conditions) and natural cooling to normal temperature, etc., a CeO_2 carrier having a high specific surface area can be obtained.

A certain amount of CeO_2 carrier was dissolved in deionized water and stirred for 60 min until the CeO_2 support was completely dissolved in deionized water. he weighed $Cu(NO_3)_2$ was added to the solution, and anhydrous Na_2CO_3 was added to adjust the pH to maintain the pH at about 10. The container is sealed with plastic wrap, heated in a continuous water bath at 60°C for 3h, then washed by 3-5 times, vacuum dried at 100°C for 3h, calcined at 400°C for 3h (under dry air) and naturally cooled to room temperature, a CuO/CeO_2 catalyst having a high specific surface area can be obtained.

4. Results and discussion
At room temperature 25°C, initial pH of 8, the addition of CuO loading of 5%, catalyst CuO/CeO_2 dosage of 1 g/L, H_2O_2 dosage of 0.3 mol/L, reaction time of 120 min, COD_{Cr} removal rate can reach 30%.

4.1 Effect of CuO loading
As shown in figure 1, the removal rate of COD_{Cr} increases first and then decreases with the increase of CuO loading. When CuO loading is 7%, the removal rate of COD_{Cr} is the highest, which can reach 34%.

When the CuO loading is 1%, the COD_{Cr} removal effect is the worst, only 18%. This is because the CuO loading is too low, the active material which plays a major role in the catalysis is reduced, and the catalytic activity is lowered.

When the CuO loading is too high, the pores of the catalyst carrier are blocked, resulting in incomplete contact and reduced catalytic performance.

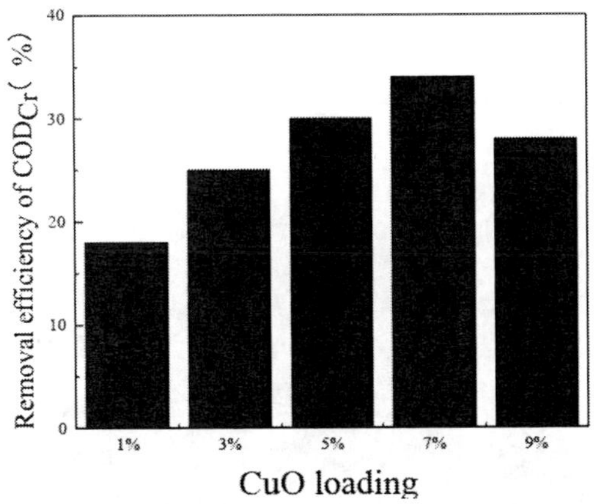

Figure 1. Effect of different CuO loading on COD_{Cr} removal efficiency.

Table 1. Results of BET characterization of catalysts with different CuO loading.

Sample	Specific surface area (m^2/g)	Pore volume (cm^3/g)	Pore size (nm)
1% CuO	156	0.22	4.31
3% CuO	152	0.20	4.33
5% CuO	150	0.18	4.40
7% CuO	150	0.18	4.56
9% CuO	148	0.16	4.78

According to table 1, as the CuO loading increases (1% CuO increases to 9% CuO), the specific surface area gradually decreases from 156 m^2/g to 148 m^2/g, and the pore volume gradually decreases from 0.22 cm^3/g to 0.16 cm^3/g, but the pore size gradually increases from 14.31 nm to 4.78 nm. This is because as the CuO loading increases, the CuO content in the pores of the catalyst carrier increases gradually, resulting in a decrease in pore volume and specific surface area, and a slight change in pore size [9].

It can be seen that too much or too low CuO loading will affect the catalytic performance of the catalyst. There are two possible reasons for this: On the one hand, as the CuO loading decreases, the amount of active component per unit mass of the catalyst is reduced, reducing the catalytic effect.

On the other hand, the structure of the catalyst changes as the CuO loading increases. When the CuO loading is too low, the catalyst has a high specific surface area and pore volume, but the active component is small, and the advantage of large specific surface area of the catalyst is not fully utilized, and the removal effect of COD_{Cr} is reduced.

Therefore, the CuO loading is too low to achieve the best removal effect. When the CuO loading is too high, the specific surface area and pore volume are low, and the effective contact area of CuO and wastewater supported on the catalyst surface is reduced, so the removal effect of COD_{Cr} is decreased.

4.2 Effect of calcination temperature

It can be seen that the calcination temperature has a certain influence on the catalyst CuO/CeO_2, and the overall COD_{Cr} removal rate is between 20% and 35%. At a temperature of 400°C, the COD_{Cr} removal rate is the highest, reaching 34%, while at temperatures below 400°C or above 400°C, the COD_{Cr} removal rate is reduced. This is because, if the temperature is too low, the active material cannot be completely thermally decomposed, the impurities cannot be completely removed, and if the temperature is too high, the active material is sintered [10], thereby reducing the catalytic activity.

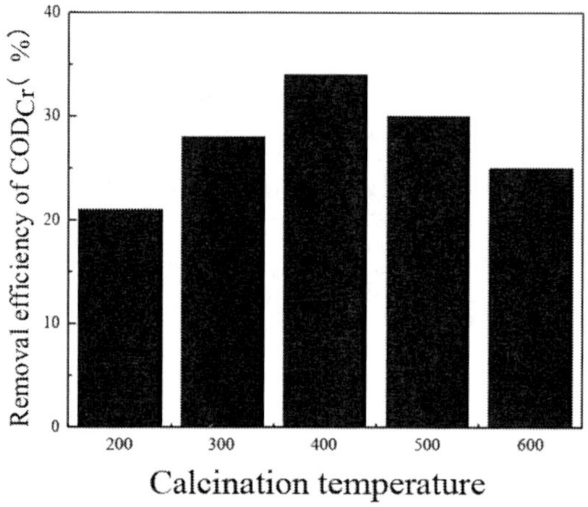

Figure 2. Effect of different calcination temperature on COD_{Cr} removal efficiency.

Table 2. BET characterization of catalysts at different calcination temperatures.

Sample	Specific surface area (m^2/g)	Pore volume (cm^3/g)	Pore size (nm)
200 °C roasting	190	0.22	4.33
300 °C roasting	166	0.21	4.48
400 °C roasting	150	0.18	4.56
500 °C roasting	133	0.17	4.79
600 °C roasting	122	0.17	4.99

It can be seen from table 2 that as the calcination temperature increases, the specific surface area and pore volume gradually decrease, and the pore diameter gradually increases. The specific surface area is reduced from 190 m^2/g at 200°C to 122 m^2/g at 600°C. The difference is large, but the decrease and increase of pore volume and pore size are small, and the pore volume is reduced by 0.05 m^3/g, while the pore size is increased by 0.66 nm. This indicates that the change in the calcination temperature mainly affects the specific surface area of the catalyst, and has little effect on the pore volume.

Combined with figure 2 and table 2, it can be concluded that the specific surface area and pore volume are the highest at 200°C, but the COD_{Cr} removal rate is the lowest, only 21%. At 400°C, the specific surface area and pore volume are compared with those at 200°C having a decrease, but the COD_{Cr} removal rate is the highest.

This is because when the calcination temperature is low, the active material is incompletely loaded, although the specific surface area is large, but the effective catalytically active substance per unit mass is small, and some residual substances also adhere to the catalyst surface, which restricts the catalytic performance of the catalyst. When the temperature is higher, the calcination temperature is higher than the heat resistance limit of the catalyst, causing the catalyst to be sintered, the pore structure is destroyed, and the specific surface area and pore volume are reduced, the catalytic activity is lowered, and the COD_{Cr} removal rate is decreased.

4.3 Effect of calcination time

It can be seen from figure 3 that as the calcination time increases, the removal rate of COD_{Cr} increases first and then decreases. The COD_{Cr} removal rate was the lowest when there was no roasting, only 18%; when the roasting time was 3h, the COD_{Cr} removal rate was the highest, reaching 33%, and the removal effect was better; after the calcination time is more than 3h, the COD_{Cr} removal rate shows a downward trend. This is because the calcination time is too short, the crystal size of the prepared catalyst surface is incomplete, and if the calcination time is too long, the pore structure of the catalyst is destroyed, and finally the COD_{Cr} removal rate can be changed.

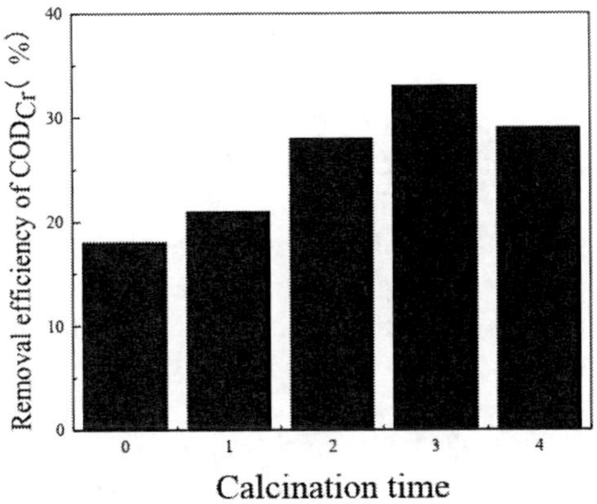

Figure 3. Effect of different calcination time on COD_{Cr} removal efficiency.

Table 3. Results of BET characterization of catalysts with different calcination time.

Sample	Specific surface area (m^2/g)	Pore volume (cm^3/g)	Pore size (nm)
not roasting	178	0.26	4.31
roasting 1h	173	0.23	4.46
roasting 2h	169	0.20	4.51
roasting 3h	150	0.18	4.56
roasting 4h	146	0.17	4.89

It can be seen from table 3 that as the calcination time increases, the specific surface area and pore volume are gradually decreasing, but the pore diameter is gradually increasing. When it is not calcined, the specific surface area of the catalyst is the largest, which can be 178 m^2/g. However, the specific surface area has been decreasing with the increase of calcination time. When the calcination time reaches 4h, the specific surface area has been reduced to 146 m^2/g, and the reduction amount is about 30 m^2/g, and the pore size and pore volume are only changed by 0.58 nm and 0.09 cm^3/g, which indicates that the calcination time mainly affects the specific surface area of the catalyst and has a weak influence on the pore volume.

Combined with figure 3 and table 3, it can be seen that the calcination time has a certain influence on the specific surface area of the catalyst, which in turn affects the COD_{Cr} removal rate. When the catalyst calcination time is 3h, the specific surface area of the catalyst is 150 m^2/g, which is not the maximum nor the minimum, but the COD_{Cr} removal rate is the highest at this time, which can reach 33%, which is much higher than the COD_{Cr} removal rate at the maximum specific surface area. This is because the non-calcination or the calcination time is too short, the active material is insufficiently supported on the surface of the catalyst, and the effective active material content per unit mass of the catalyst is small, and the catalyst specific surface area is large, but the COD_{Cr} removal rate is low.

If the calcination time is too long, the already formed catalyst structure is sintered and damaged due to the calcination time exceeding the limit, resulting in a decrease in the specific surface area of the catalyst, and thus an effective contact with the contaminant in a limited time is reduced, resulting in a decrease in the COD_{Cr} removal rate.

These results indicate that the calcination time of the catalyst CuO/CeO_2 is not shorter or longer, and different adjustments need to be made according to actual conditions. Based on the above discussion, it was finally chosen to prepare the catalyst CuO/CeO_2 with a calcination time of 3h.

5. Conclusion

In the experiment, CuO/CeO_2 supported by CeO_2 was prepared by sedimentation precipitation method. During the preparation process, the content of CuO will affect the content of CuO and the amount of Cu^{2+} dissolved and dispersed in the unit mass catalyst.

The temperature and calcination time can change the appearance and structure of the catalyst, which in turn affects the catalytic activity of the catalyst. Therefore, after investigating the influence of CuO loading, calcination time and calcination temperature on the CuO/CeO_2 catalyst, the best preparation conditions of the catalyst were as follows: CuO loading was 7%, calcination time was 3h, calcination temperature is 400°C, the catalyst prepared under these conditions can achieve a COD_{Cr} removal rate of 34%.

References

[1] Zhou L Y and Guo W 2013 *J. Industrial Water Treatment* **33** pp 61-64
[2] Qiu Z M and Liu X C 2003 *J. Journal of Safety and Environment* **03** pp 50-53
[3] Richter M, Fait M J G, Eckelt R, Schreier E, Schneider M, Pohl M M and Fricke R 2007 *J. Appl Catal B: Environ.* **73** pp 269-28
[4] Richter M, Fait M J G, Eckelt R, Schneider M, Radnik J, Heidemann D and Fricke R 2007 *J Catal.* **245** pp 11-24
[5] Zhang H T, Wang S F, Zhao X Q and Wang Y J 2004 *J. Hebei Univ Technol.* **33** pp 36-40
[6] Shaikh T H, Elizaveta A, Kefu Z et al. 2018 *J. Applied Surface Science* **455** pp 132-143
[7] Liu Z M, He C X, Chen B H, et al. 2017 *J. Catalysis Today* **297** pp 78-83
[8] Ren J, Guo C J, Li L L and Li Z 2013 *J. Chin J Catal.* **34** pp 1734-1744
[9] Ranga G R, Ranjan H S and Braja G M 2003 *J. Colloids and Surfaces A: Physicochemical and Engineering Aspects* **220** pp 261-269
[10] Chen Y N and Meng M 2015 *J. Chemical Industry and Engineering.* **32** pp 1-5+12

IOP Conf. Series: Materials Science and Engineering **789** (2020) 012068 doi:10.1088/1757-899X/789/1/012068

Experimental study on static behavior for assembled monolithic reverse-rib slab

X Yan[1*], J X Li[1], B Zhou[1] and M Liu[1]

[1] School of Civil Engineering, Shenyang Jianzhu University, Shenyang, China

[*] cexyan@sjzu.edu.cn

Abstract. In this paper, the static loading test of two assembled monolithic reverse-rib slabs is carried out, and the cracking process, mechanical mechanism and performance under different load combinations are emphatically studied. The test results show that the development of cracks of the assembled monolithic reverse-rib slab is fitted with the characteristics of the two-way plate. The rib beam and the floor slab between the rib beams can work well together, which can meet the requirements of normal use limit and bearing capacity limit.

1. Introduction

The large-scale promotion of prefabricated buildings can effectively improve construction efficiency, shorten construction period, and reduce environmental pollution, which is of great significance for promoting urbanization in China [1-2]. As an important component of the whole building structure system, the floor slab in the building structure mainly bears the vertical load and is an important part of the assembled building [3]. At present, the most commonly used assembled floor slabs are composite slabs, but some problems have been exposed with the large-scale use of composite slabs, such as low prefabrication rate of the slabs, excessive joints between the slabs, and a large number of reserved reinforcements around the prefabricated slab [4-8].

These problems are contrary to the concept of industrialization of buildings, and they also restrict the development of fabricated building [9]. Based on this, this paper presents a kind of the assembled monolithic reverse-rib slabs which is composed of ribbed beam and multi-ribbed hollow slab composed of rib beam grid. The rib beam is on the upper part of the district slab, and the reinforced bar is arranged in the compression zone to reduce the area of the concrete. It can effectively reduce the quality of the floor, because only the rib beam is connected with the surrounding components. It can directly reduce the floor reinforcement rejection and improve the connection efficiency.

In order to study the load-bearing performance test of the four-sided fixed support of the assembled reverse rib floor slab, the full-scale static load test of the two assembled reverse rib slabs was carried out to study the deflection, steel strain and crack development under static load, and to analyze the force performance under the combination of load standard and basic load.

2. Overview

2.1. Specimen design

The dimensions of the assembled reverse rib floor slab specimens in this study are based on the design of an open room in an ordinary family home. The dimensions are 3400 mm × 4300 mm, the rib height is designed to be 180 mm, the rib width is designed to be 150 mm, and the rib beam is square. The rib beam spacing (the distance between the central axes of the rib beams) is 860 mm, and the thickness of the rib beam panel is 50 mm. See table 1 for the specimen's parameters.

Table 1. Specific parameters of the specimens.

Specimens number	Prefabricated reverse rib floor slab size (mm)	Post-casting strip width (mm)	Floor name size (mm)	Rib form	Boundary conditions
FL-1	4300×3400	500	4500×3600	prefabricated part	The prefabricated part is free of stirrup, and the rib beams and plates are prefabricated by the additional steel bars and the surrounding beams.
FL-2	4300×3400	500	4500×3600	prefabricated Rib beam	Only the rib beam is cast, and the edges of the four corners of the slab are integrally prefabricated by the additional steel bars and the surrounding beams.

In order to ensure that the assembled anti-ribbed floor can be reliably anchored into the support, and it is convenient to simulate the boundary conditions of the four-sided fixed support, after the prefabricated component is completed, the member is poured into the 300 mm support side beam, and at the same time, in order to prevent the four corners of the floor early cracking, pouring concrete on the four corners of the floor and arranging the steel mesh at the four corners, as shown in figures 1 and 2.

2.2. Material mechanical properties

The type of reinforcement used for the specimen is HRB400, and the measurement method is based on GB228.1-2010 "Metal material tensile test method at room temperature". The measured values of steel bar strength are shown in table 2.

Table 2. Measured values of steel bar strength.

D (mm)	f_y (MPa)	f_u (MPa)	Elongation (%)
8	477	600	23
10	443	616	23
14	462	631	24

The concrete strength grade of the assembled reverse rib floor slab specimens is C30 (this strength is measured according to Chinese specifications), and the measured values of concrete strength are shown in table 3.

Table 3. Measured values of concrete strength.

Compressive strength (MPa)	Poisson's ratio	Elastic Modulus ($\times10^4$ MPa)
32.1	0.3	3.0

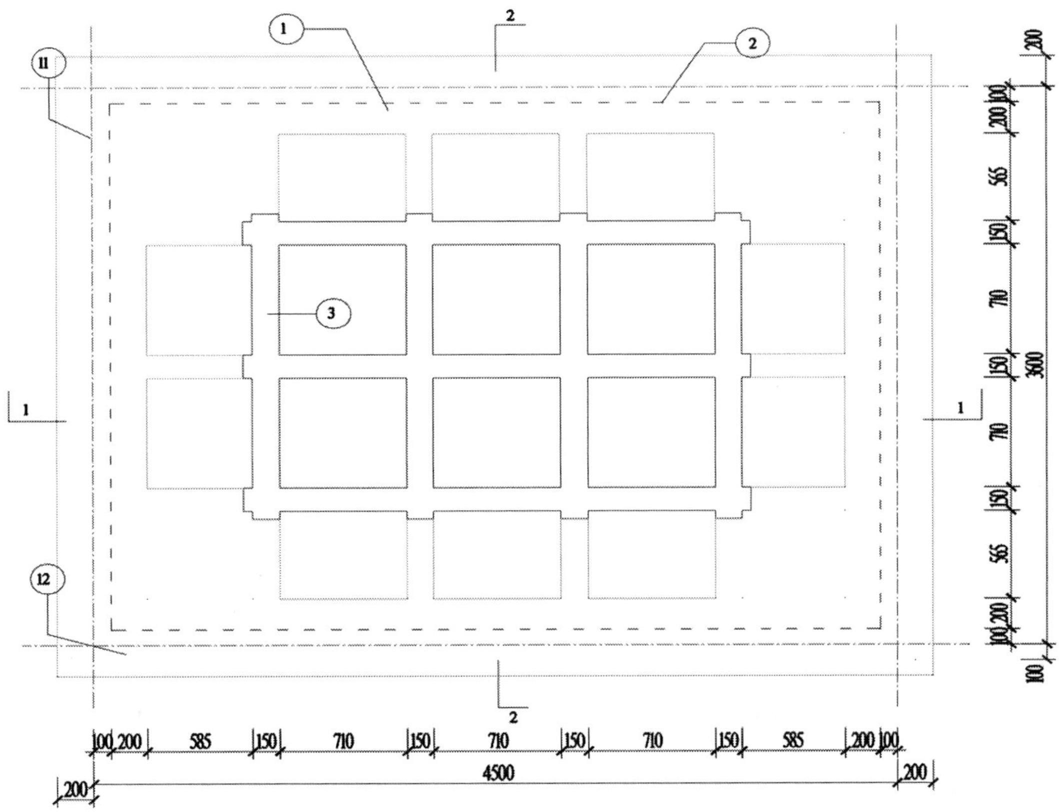

1- Prefabricated part of floor slab, 2- rib beam area panel, 3-rib beam of the assembled reverse rib floor slab 11- Extension support post post-poured strip.
(a)

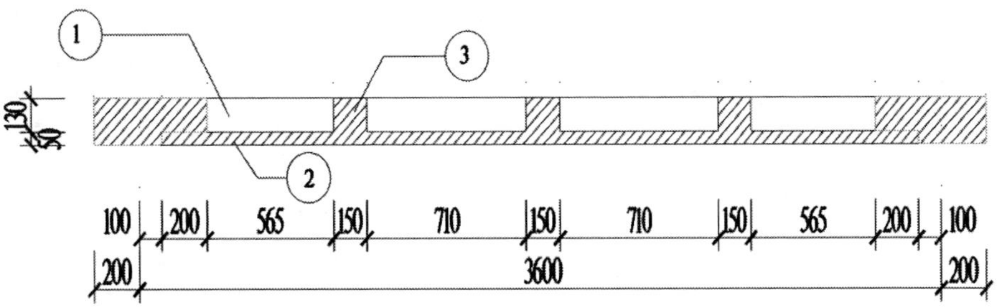

(b)
Figure 1. Specimen FL-1 plan: (a) plan view, (b) 1-1 section.

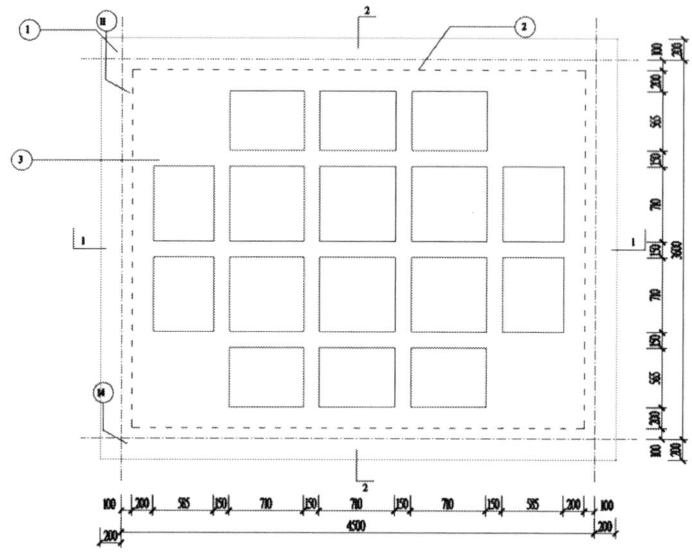

1- Prefabricated part of floor slab, 2- rib beam area panel, 3-rib beam of the assembled reverse rib floor slab 11- Extension support post post-poured strip.
(a)

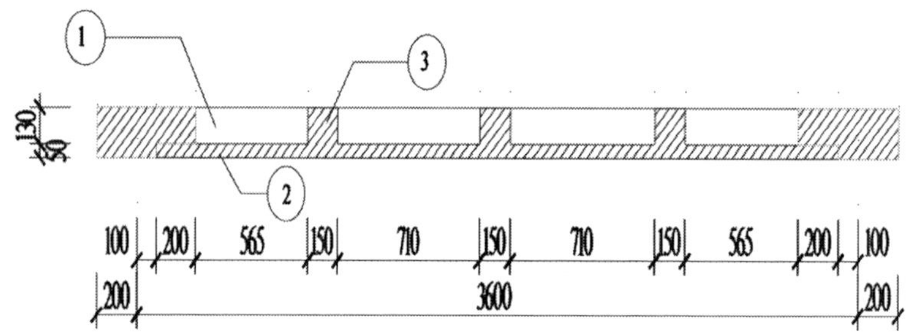

(b)
Figure 2. Specimen FL-2 plan: (a) plan view, (b) 1-1 section.

2.3 Test equipment

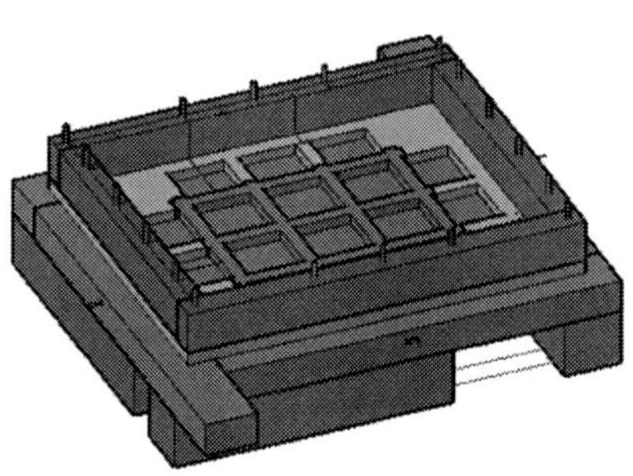

(a)

(b)

Figure 3. Test equipment.

The boundary condition of this test is four-sided fixed support. In order to simulate the force of the slab in actual engineering, the boundary condition of four-sided fixed support is realized by the bottom beam-floor-pressure beam mode.

The bottom beam is used, the pressure beam is prefabricated, and the screw is clamped. The form ensures that the bottom beam, floor slab and pressure beam are not closely separated, and the loading device is shown in figure 3. In figure 3, the first diagram is the schematic diagram, and the second diagram is the actual loading diagram

3. Static test

3.1. Loading test plan

In this test, the weights were piled up. The loaded test blocks were iron and brick, of which the weight of each iron block is 20 kg, and the weight of each brick is 4.2 kg. Since the reverse rib floor slab is applied in practical engineering, the inter-ribbed zone will be filled with lightweight materials. In order to simulate the working state of the actual use of the slab, sand is pre-filled and compacted in the rib beam zone before loading [10]. This is the first stage load and record the data, as shown in figure 4. The number of weight iron blocks arranged in each layer is 98, which is equivalent to a uniform load of 1.34 kN/m^2 applied to each layer. The loading layout is shown in figure 5.

Figure 4. Filled the hole with sands.

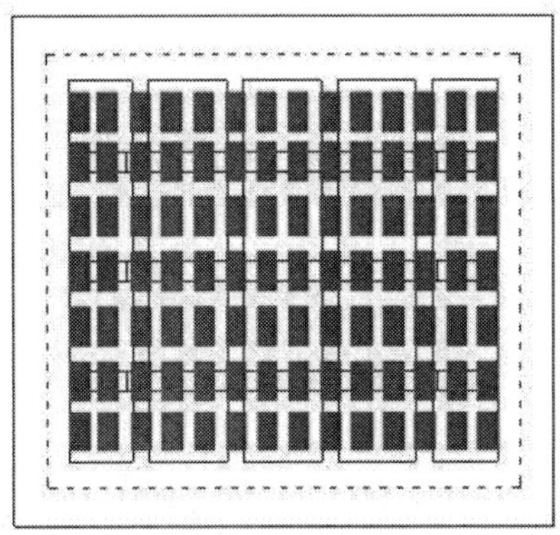

Figure 5. The load layout.

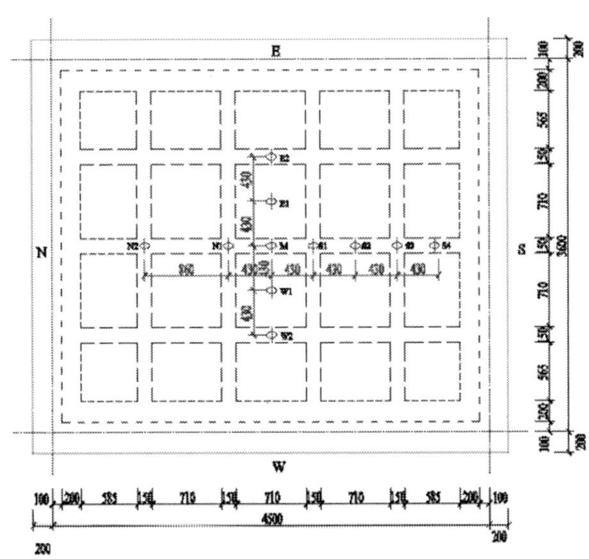

Figure 6. Displacement meter layout.

After each stage loading is completed, it should last for 10 min - 15 min; under standard loads, the duration shall be 30 minutes. During the duration, the occurrence and development of cracks should be observed, as well as the slippage of reinforcement; at the end of the duration, each reading should be observed and recorded.

In order to study whether the rib beam and the rib beam area grid work together, the displacement meter arrangement is shown in figure 6. At the same time, in order to observe the stress of the steel bar in the plate, the strain gauge is attached at the key part, and the arrangement is as shown in figure 7.

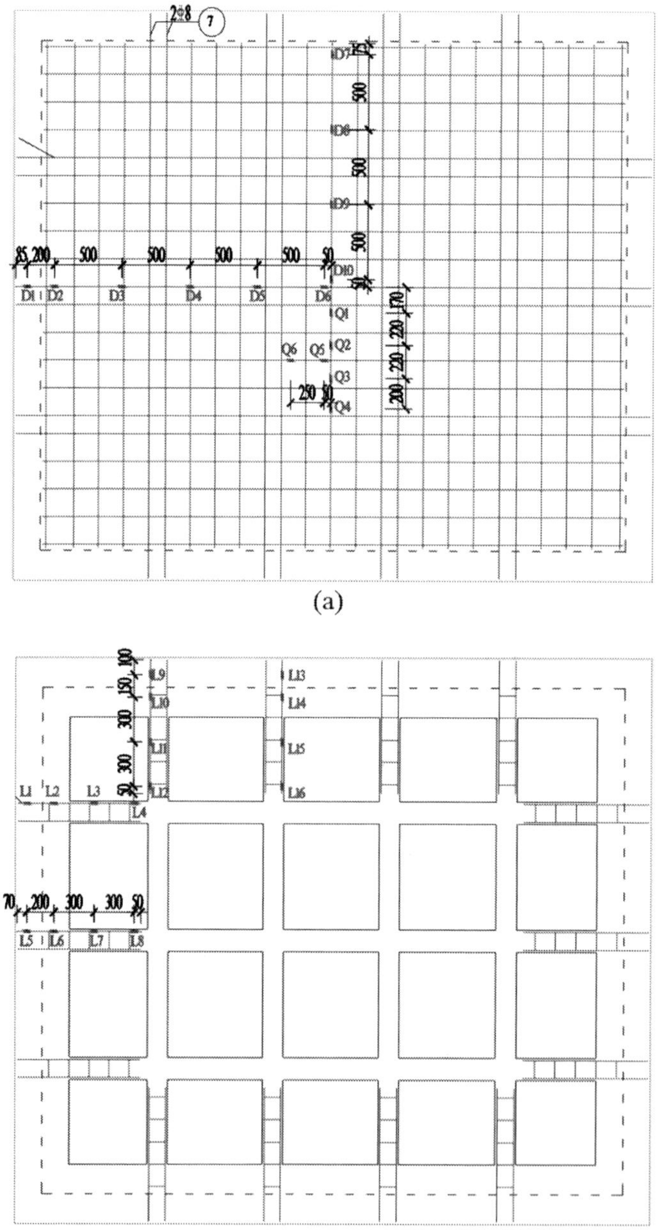

Figure 7. Rebar strain gauge layout: (a) slab reinforcement sheet layout, (b) rib beam reinforcement strain gauge layout.

3.2. Failure criteria
According to the research purpose, the signs of normal use of ultimate load and ultimate capacity of bearing capacity are as follows:

One of the following two conditions is met: the normal use limit load is reached:

- the mid-span deflection reaches $L_0/200$ (L_0 is the calculated span of the board, and the two-way board is calculated by short span);
- the maximum crack width is 0.2 mm.

One of the following four conditions is met: the bearing capacity limit state is reached:

- the main rib is pulled or strained to 0.01;
- the concrete in the compression zone is crushed or strained to 0.0033;
- the mid-span deflection reaches $L_0/50$;
- the maximum crack width at the main rib is 1.5 mm.

3.3. Data collection
The uniform load value of this test can be directly converted by the weight of the iron block without collecting.

The deflection of the critical point of the test piece and the strain of each measuring point are obtained by connecting the displacement meter and the resistance strain gauge to the UCOM-70A data acquisition analyzer.

4. Test phenomena and results analysis

4.1. Test phenomena
When the test blocks are stacked to 4 floors, the uniform load value is 6.3 kN/m^2. When the quasi-permanent combination of loads is reached, the specimen FL-1 is in the elastic stage, but the first crack occurs in the mid-span position of the specimen FL-2, and the crack direction develops along the long-span direction; when the load is added to the fifth layer, the uniform load value is 7.64 kN/m^2, and the first long-side crack occurred in the long span of the FL-1 specimen, and the crack was 1500 mm away from the east bearing.

When the load is added to the sixth layer, the uniform load value is 8.98 kN/m^2 at this time, and the standard combination of the loads is reached.

The mid-span deflection of the two floor specimens is less than 17 mm ($L_0/200$), and the crack width at the bottom plate is also it has not reached 0.2 mm, indicating that the two slabs meet the normal use requirements under the load of this stage.

When the load continues to be loaded to the 7th floor, the uniform load value of the floor is 10.32 kN/m^2 and the basic combination of loads is reached. At this time, the first crack in the FL-1 specimen crosses further and small cracks appear in other parts of the bottom plate, but the crack width is small and does not start to communicate.

The mid-span deflection is 3.86 mm and less than 68 mm ($L_0/50$). The strain of the main rib is 162, and the maximum crack width at the bottom of the plate has not reached 0.2 mm, and the load capacity limit state has not been reached.

FL-2 floor cracks begin to communicate and develop from the middle crack to the four corners, cracks appear in the short span direction, the maximum width of the crack reaches 0.2 mm, reaching the normal use limit state; when the load is added to the 11th layer, the load is applied to 229.3 kN (limit when the bearing capacity calculation value is 55%), the first crack width of the FL-1 specimen reaches 0.2 mm, reaching the normal use limit state.

When the load is fully applied, the load is applied to 342.5 kN (82% of the calculated ultimate bearing capacity), the mid-span deflection of the FL-1 floor specimen reaches 16.85 mm, and the maximum crack width is 1.0 mm; the mid-span deflection of the FL-2 specimen reached 21.8 mm and the maximum crack width was 1.2 mm, which did not reach the limit of bearing capacity.

4.2. Test results

The load-deflection curve of FL-1 specimen and FL-2 specimen is shown in figure 8. It can be seen from the curve that the deflection of the two test pieces is almost the same, indicating that the post-cast concrete rib beam has little effect on the deflection of the floor.

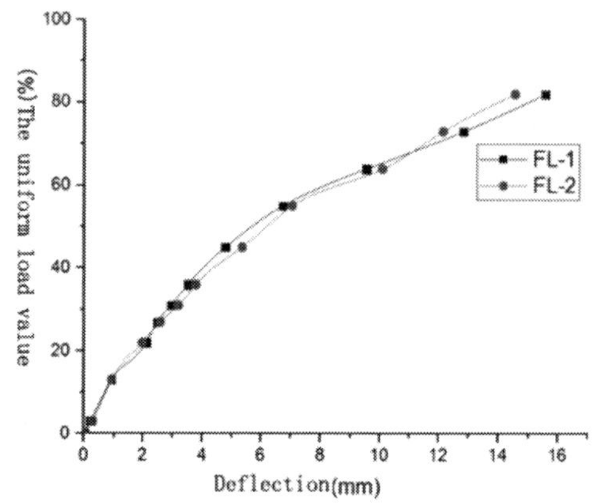

Figure 8. Load-deflection curve.

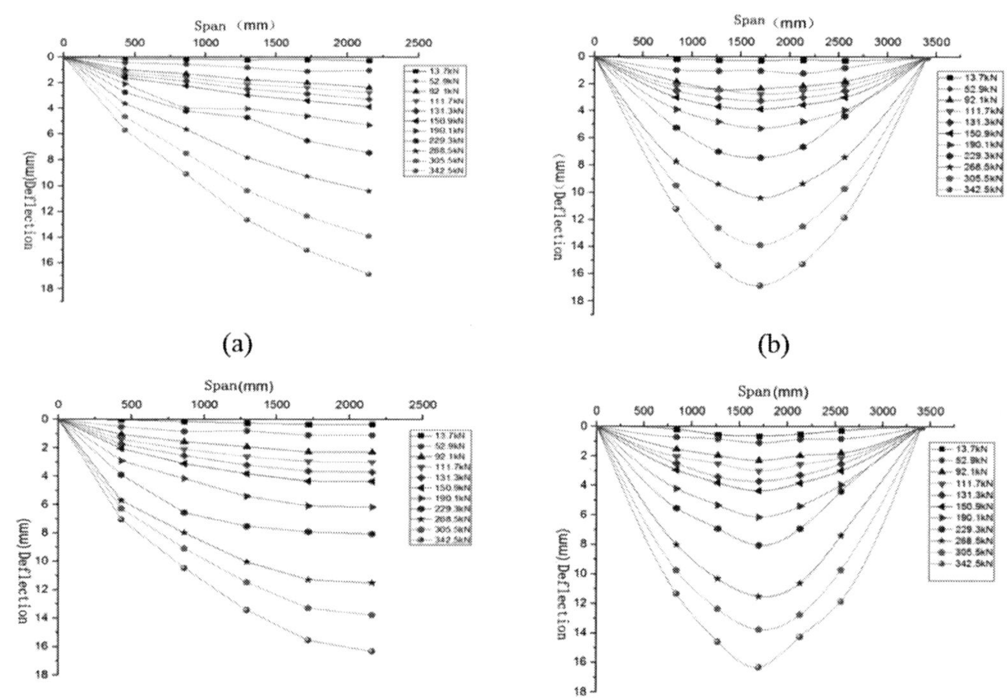

(a) (b) (c) (d)

Figure 9. Specimen span - deflection curve: (a) specimen FL-1 span-deflection curve (long span), (b) specimen FL-1 span-deflection curve (short span), (c) test piece FL-2 span-deflection curve (long span), (d) specimen FL-2 span-deflection curve (short span).

Figure 9 is a mid-span deflection curve of the FL-1 and FL-2 specimens. As shown in the figure, under the same load, the deflection of the two specimens increases smoothly along the span direction, and there is no significant change at the intersection between the rib beams and the rib beams, indicating that there is not rib joint between the rib beam and the rib beam.

Due to the intersection of the transverse longitudinal beams, the joint work is affected; the deflection curve develops in a parabolic shape, and the deflection of each measuring point does not change greatly, indicating that the deflection between the beams is not affected by the influence of the surrounding rib beams. The bottom plate between the beam and the rib can work together.

As shown in figure 10 and figure 11, the cracks of the specimen FL-1 and FL2 are not developed in the long-span direction after the first-span mid-span crack develops to a certain extent in the long-span direction, but in the 45° direction.

Extending to the four corners of the floor, all the cracks penetrate when the load is fully applied. However, the cracks in the bottom plate of the FL-1 specimen show that the crack spacing is larger and the number is smaller, while the crack in the bottom plate of the FL-2 specimen shows that the crack spacing is smaller and the number is larger.

It can also be seen from figure 10 and figure 11 that the crack development of the integrated integral anti-rib floor slab is that the first crack appears parallel to the long side direction in the span, and extends from the mid-span crack to the four corners along the 45° direction, and the crack develops. The law conforms to the characteristics of the two-way floor.

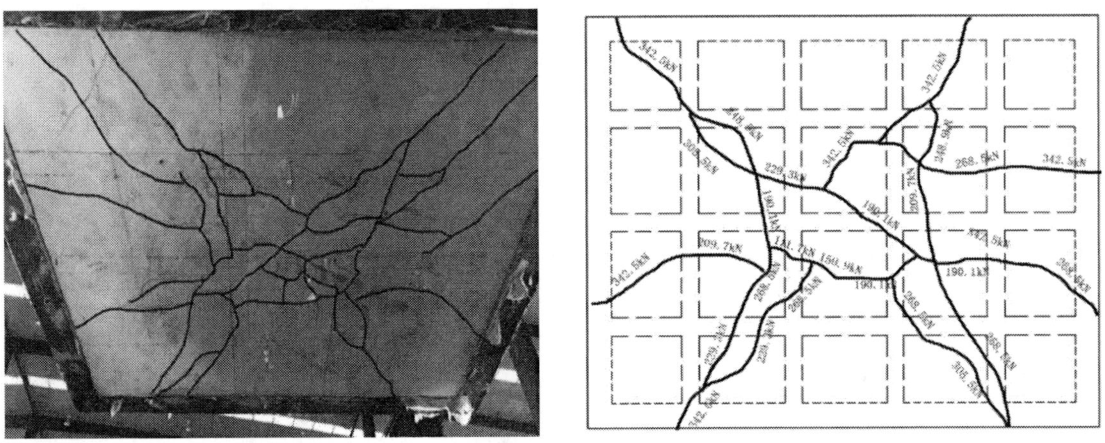

(a) (b)

Figure 10. Crack development of specimen FL-1: (a) crack real view, (b) crack detail.

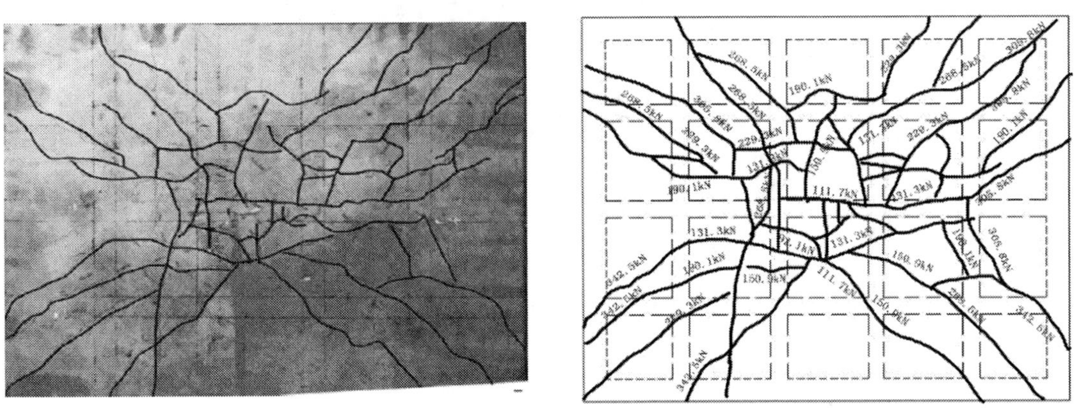

(a) (b)

Figure 11. Crack development of specimen FL-2: (a) crack real view, (b) crack detail.

The figure below shows the load-strain curve of the steel bar at the typical position of the two specimens.

As shown in figure 12, the strain of D6 of the FL-1 specimen is not much increased before the crack of the slab. This is because the concrete has not cracked the steel and the concrete works together.

When the slab cracks, the load-strain curve slope changes suddenly, indicating that this when the concrete at D6 exits the work, the stress at this time is all borne by the steel bars.

Under the same load, the strain of the steel increases with the distance of the bearing, which indicates that the span of the fabricated concrete anti-ribbed slab is the largest.

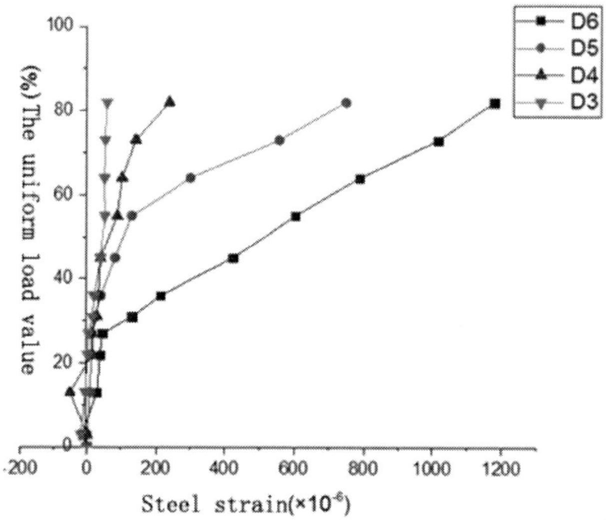

Figure 12. The load-strain curve of the FL-1 specimen (long span).

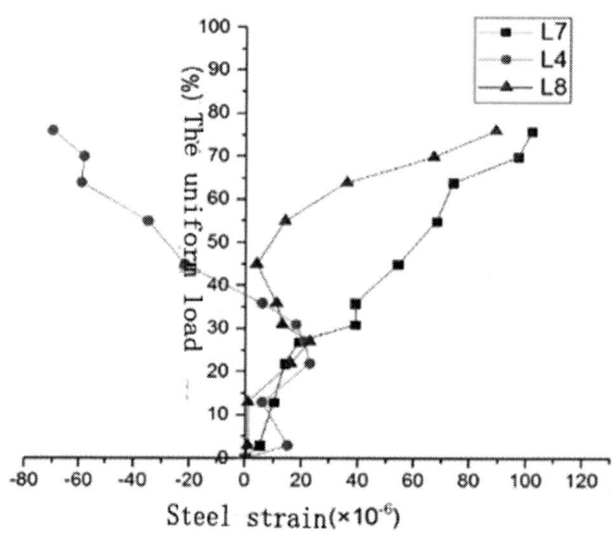

Figure 13. Load-strain curve of the rib beam of FL-2 specimen.

It can be seen from the load-strain curve of the rib beam of the FL-2 specimen in figure 13 that the measuring point L7 is the closest point to the bearing, where the steel strain is the largest and the steel strain is positive, indicating that the rib beam of the floor is close to the branch.

The seat is pulled and the closer to the support, the greater the bending moment. The strain of the steel bar at the measuring point L7 is positive, and the strain of the steel bar at the measuring point L8 is negative, indicating that the rib beam is pulled at L7 without being pressed at L8, and it can be judged that the zero bending moment of the rib beam is located between L7 and L8. Therefore, the negative bending moment area of the slab is taken as l0/5 (l0 is the short span), and it is also the lowest value of the negative bending moment steel of the rib beam prefabricated anti-ribbed floor; the measuring point L8 is compared with the measuring point L4 steel strain. Positive value, the measuring point L8 is pulled here, and the negative bending moment area of the upper corner of the floor is larger than the middle position.

Therefore, the negative bending moment steel at the four corners should be appropriately lengthened in the design of the floor.

5. Conclusions

In this paper, the static performance test of two assembled monolithic slabs is carried out, and the following main conclusions are obtained:

- under the uniform load, the cracks at the bottom of the slab appear in the span, parallel to the long side direction, and the cracks clearly show the crack characteristics of the two-way plate;
- it can be seen from the deflection curve that the post-cast concrete rib beam has little effect on the deflection of the slab, and the floor between the ribbed beam and the ribbed beam can work well together;
- under the action of the basic combination load, the assembled monolithic reverse-rib slab is assembled to meet the requirements of the normal use limit state and the ultimate bearing capacity state, and can be used for the prefabricated building.

References

[1] Wu L W, Chen H B and Liu Y B 2018 *J Build Struc* **39**(S2) pp36-42
[2] Yan F and Gao J 2016 *Building Structure* **46** pp56-60
[3] Ding K W, Chen D, Liu Y L and Xia S 2015 *Chin Civil Eng J* **48** pp64-69
[4] Zhou X H, Deng L B and Wu F B 2015 *J Build Struc* **36** pp82-90
[5] Wu F B, Liu B, Deng L B and Li J 2014 *J Build Struc* **35** pp10-19
[6] Li Y Z, Jiang Q Q and Huang S C 2003 *J Central South Univ (Sci Technol)* **34** pp695-698
[7] Xu T S and Xu Y L 2003 *Building Science* **19** pp11-14
[8] Wang L M, Cai R Z and Zhao J 2001 *Concrete* **5** pp54-56
[9] Shen P S 2003 *Floor structure design principle* (Beijing: Science press)
[10] Elghazouli A Y 2004 *Journal of Structural Engineering* **130** pp1985-1996

Static test study on precast concrete frame beams with large-diameter and high-strength steel bars

Y Zhang[1], B Shi[1], L Jin[1*] and M Liu[1]

[1] School of Civil Engineering, Shenyang Jianzhu University, Shenyang, China

[*] jinlu@sjzu.edu.cn

Abstract. The paper is focused on the mechanical properties of precast concrete frame beams with large-diameter and high-strength steel bars. The full-scale static tests and nonlinear finite element analysis were carried out on the two proposed beams with steel bars of HRB500 Ø32-Ø25 and Ø40-Ø32 to study the load-carrying capacity, flexural capacity and plastic rotational capacity. The influence of steel bar diameter on mechanical properties were explored. On the basis of equal-strength principle of flexural capacity at the same time, the finite element model of the cast-in-place concrete frame beam with large-diameter of HRB500 and ordinary steel bars of HRB400 were compared. Results indicated the properties of the precast concrete frame beams with large-diameter and high-strength steel bar. The load-carrying capacity of the prefabricated concrete beam with steel bars of Ø40-Ø32 was higher. The coefficient of plastic rotational capacity was 3.41 and 2.88 of specimens with steel bars of HRB500 Ø32-Ø25 and Ø40-Ø32, respectively. The positive section yield bending moment could be calculated by the current concrete codes. The properties such as yield load, flexural bearing capacity and the plastic rotational capacity were similar to the cast-in-place concrete frame beam, and could be verified by the current standard.

1. Introduction

The precast concrete structures with small-diameter and low-strength steel bars and various joint of the on-site consolidation leads to the cumbersome installation procedures and the lack of compactness of concrete pouring, which are hard to ensure the quality of concrete pouring and construction efficiency [1, 2]. To address these shortcomings, the paper proposes a precast concrete member with large-diameter and high-strength steel bars. Yin Xiaoli et al., proposed a series of design methods and structural suggestions of precast composite beams and precast frame beam-column joints experimentally and numerically, which can be found in [3 - 5]. The method of at the end of proposed beam in the paper supplement the content of the above-mentioned assemble method. The full-scale static tests and numerical simulations monotonic were carried out on two full-scale precast concrete frame beams with steel bars of HRB500 Ø32 and Ø40, which provide a basis for. Based on the results, the application gist of large-diameter and high-strength steel bars in prefabricated construction have been proposed.

2. Test survey

2.1. Specimen design

Figure 1 shows the detail on the experimental specimens had a net span of 3200 mm, the section was

$h \times b = 350 \times 700$ mm (table 1). The concrete strength grade of precast part and post-pouring part was C40 and C45, respectively. The thickness of concrete cover was consistent with the diameter of stressed steel bar. In order to ensure the quality of connection, shear keys for surface roughness treatment were reserved at the end of prefabricated frame beam, and grouting sleeve was used for connection of stressed longitudinal bars [6]. The end of double-sided welded anchor bars was used for anchorage of stressed longitudinal bars in joint area.

Table 1. Experimental parameters of specimens.

Specimens	Tensile reinforced	Compressed reinforced	Middle reinforced bars	Stirrups	Additional reinforcement
KL-32-25	2 Ø 32	2 Ø 25	3 Ø 12	Ø 10@100	No
KL-40-32F	2 Ø 40	2 Ø 32	3 Ø 12	Ø 10@100	Lay out

Note: The additional reinforcement was Ø 12, and the strength level of stirrups, middle reinforced bars and additional reinforcement were HRB400.

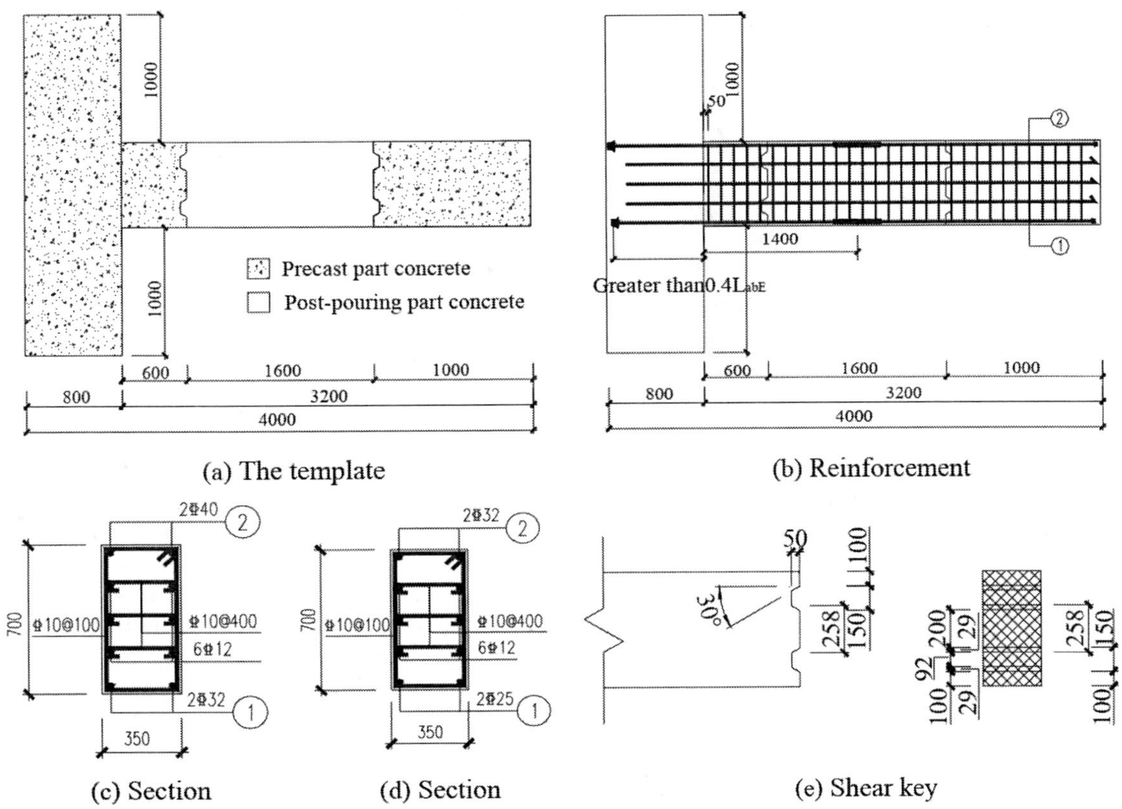

Figure 1. Detail of specimens (units: mm).

2.2 Material Properties

The average measured mechanical properties of the HRB500-grade steel bar and concrete are listed in table 2 and table 3, respectively. More detail about the material properties can be found in [7 - 9].

2.3. Test loading device and loading system

The support end of the specimen beam was fixed on the reaction wall through a steel beam, and the monotonic load was exerted on the beam end by an actuator. The experimental loading device is shown in figure 2.

Table 2. Mechanical properties of steel bars.

D/(mm)	σ_s/(MPa)	σ_b/(MPa)	Grain fineness	A_s/%	A_{gt}/%	E_S (MPa)
25	543	725	1.34	12.4	22.0	1.83×10^5
32	548	735	1.34	13.4	20.0	1.82×10^5
40	542	704	1.30	13.2	23.0	1.81×10^5

Note: σ_s is the yield strength; σ_b is the tensile strength; A_s is the total elongation rate of maximum load; A_{gt} is the elongation after fracture.

Table 3. Structural parameters of concrete.

Batchs	Design strength	f_{cu}^{0}	f_c^{0}	f_t^{0}	E_S (MPa)
First	C40	47	35.7	3.28	34033
Second	C45	43	32.7	3.13	33256

Note: f_{cu}^{0} is the cubic compressive strength of measured strength; f_c^{0} is the axial compressive strength of estimating value; f_t^{0} is the axial compressive strength of standard value.

Figure 2. Test loading device.

3. Experimental phenomena and failure modes

The figure 3 shows the ultimate failure pattern. The experimental phenomena and failure modes of the two specimens were very similar. Bending failure in the end of the beams. The first crack appeared about 350mm away from the column edge. With the increased of load, more flexural cracks formed and propagated, and the original vertical cracks developed into shear diagonal cracks. The number and length of hair cracks developed rapidly, the crack range gradually extended from beam root to beam end. After the longitudinal bars yielded, the displacement loading was adopted. At this stage, the steel bars were mainly stressed, and concrete gradually drops out of bearing capacity. Therefore, the number and length of cracks in beam body tend to be stable, and crack width starts to develop after the longitudinal reinforcement is yielded, the displacement is applied. Ultimately, the beam deflection exceeded 1/25 of the beam span and reached 1/16 [10], and the concrete at the lower end of the beam is crushed with large pieces falling off, and the concrete at the upper end is slightly damaged.

(a) KL-32-25 (b) KL-40-32F

Figure 3. The most damaging pattern of specimens.

4. Test results and analysis

4.1. Displacement -load curve
Figure 4 displays the load-displacement curve. Similar development trend is observed among the two specimens, and both had obvious yield steps. After yielding, the growth rate of bearing capacity gradually decreased. Ultimately, the beam deflection exceeded 1/25 of the beam span and reached 1/16, and the bearing capacity had no obvious decline trend. The ductility of the specimen was good and the overall reinforcement structure was reasonable.

It can be seen the initial stiffness of specimens KL-40-32F was higher than specimens KL-32-25 and the bearing capacity of specimens KL-40-32F also shows obvious advantages, which increases by 22%. It shows that the bearing capacity of the structure can be obviously increased by increasing the diameter of steel bar appropriately.

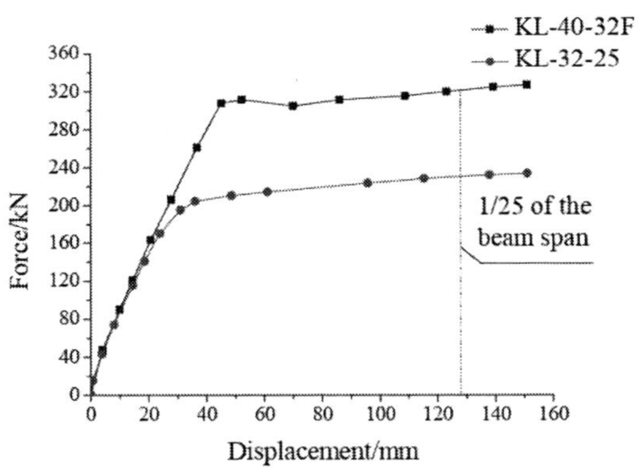

Figure 4. Displacement-load curve.

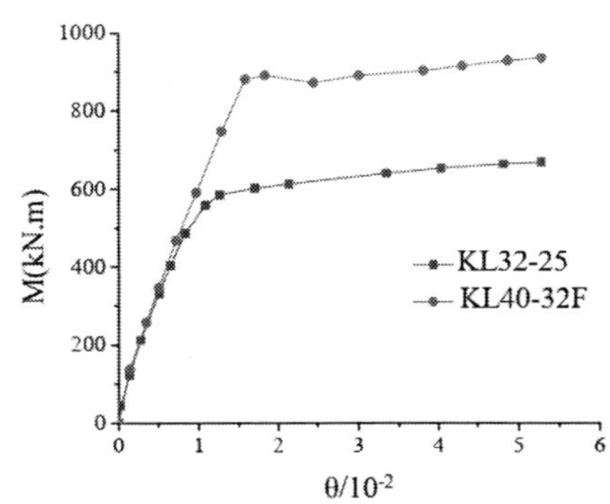

Figure 5. Moment-rotation curve.

4.2. Crack load calculation and analysis
The cracking bending of beam is shown in table 4, more detail about the calculation method can be found in [11-13]. It can be seen that the measured cracking moment of specimens was smaller than the calculated by the standard due to the concrete cover thickness was consistent with the diameter of tensile steel bar (32/40mm). Therefore, the anti-crack performance of specimens decreases with the increase of the concrete cover thickness.

Table 4. Comparison of actual cracking load of test beam and theoretical calculation results.

Specimens	Experimental value	Calculated value of Gu Xianglin method		Calculate value of Hydraulic code method	
	M_{cr}^{t} (kN.m)	M_{cr}^{c1} (kN.m)	M_{cr}^{c2}(kN.m)	M_{cr}^{c3}(kN.m)	M_{cr}^{c4}(kN.m)
KL-32-25	92	193	140	137	104
KL-40-32F	137	238	153	154	124

Note: Superscript t is the test value; superscript c is the calculated value; M_{cr}^{c1}, M_{cr}^{c3} are the yield moment calculated by measured value; M_{cr}^{c2}, M_{cr}^{c4} are the yield moment calculated by standard value.

4.3. Crack width under service load
It can be seen that the maximum crack width under the service load, the measured value is more than 300% of the calculated value (table 5). The results show that anti-crack performance of specimens decreases with the increase of the concrete cover thickness. It is suggested that add steel mesh or reduce the stirrups spacing in order to enhance their anti-crack performance in practical engineering applications.

Table 5. Maximum crack width of test piece under load.

Specimens	M_k (mm)	ω_s (mm)	ω_l (mm)	ω_l^c (mm)	ω_l/ω_l^c
KL-32-25	435	0.45	0.675	0.22	3.07
KL-40-32F	605	0.35	0.525	0.15	3.50

Note: ω_s, ω_l are the short-term and long-term maximum crack width test values of test beams under service loads, respectively; ω_l^c is the calculation value of maximum crack width.

4.4. Analysis of normal section bending bearing capacity

Table 6 summaries the comparison between the yield moment (M_y) and the failure moment (M_u) in the measurement and calculation, respectively.

The average value of M_y^t/M_y^{c1} is 1.05, so the yield moment can be calculated by the current concrete codes. It is not significant to calculate ultimate moment by the current codes due to the bearing capacity of specimens decrease unconspicuous during the test. The mean value of M_y^t/M_y^{c2} is 1.24, which indicates that the flexural capacity calculated by current concrete codes has sufficient safety reserve.

Table 6. Flexural bearing capacity of test beam.

Specimens	Yield (kN.m)			$\frac{M_y^t}{M_y^{c1}}$	$\frac{M_y^t}{M_y^{c2}}$	Failure (kN.m)		$\frac{M_u^t}{M_u^c}$
	M_y^t	M_y^{c1}	M_y^{c2}			M_u^t	M_u^c	
KL-32-25	563	563	455	1.00	1.23	661	751	0.88
KL-40-32F	883	835	672	1.09	1.24	920	1072	0.86

Note: M_y^{c1} and M_y^{c2} is the yield moment calculated by measured value and standard value, respectively.

4.5. Calculation and Analysis of Plastic Rotation Capability

Figure 5 shows the moment-rotation curve of test beam. Table 7 displays that the plastic rotation capacity of the steel reinforcement diameter 32 was better than diameter 40.

Table 7. Calculation of plastic turning ability of test beam (unit: rad^{-1}).

Specimens	θ_y (10^{-2})	θ_u (10^{-2})	θ_p (10^{-2})
KL-32-25	1.06	4.47	3.41
KL-40-32F	1.59		2.88

Note: θ_y is the yielded rotation angle; θ_u is the failure rotation angle; θ_p is the plastic rotation angle.

5. Experimental finite element verification

5.1. Moment-turn curve comparison

It can be illustrated that the development trend of test curve is similar to the simulation curve in figure 6 and table 8 show that the simulation results were in good agreement with the test results. Therefore, the finite element model can simulate the bearing capacity of the test specimens better.

Table 8. Comparison of measured eigenvalues and simulated analytical eigenvalues.

Specimens	Yield load		$\frac{M_y^t}{M_y^c}$	Relative Error (%)	Ultimate load		$\frac{M_u^t}{M_u^c}$	Relative Error (%)
	M_y^t (kN.m)	M_y^c (kN.m)			M_u^t (kN.m)	M_u^c (kN.m)		
KL-32-25	563	572	0.98	1.60	661	660	1	0.10
KL-40-32F	883	844	1.05	4.42	920	940	0.98	2.17

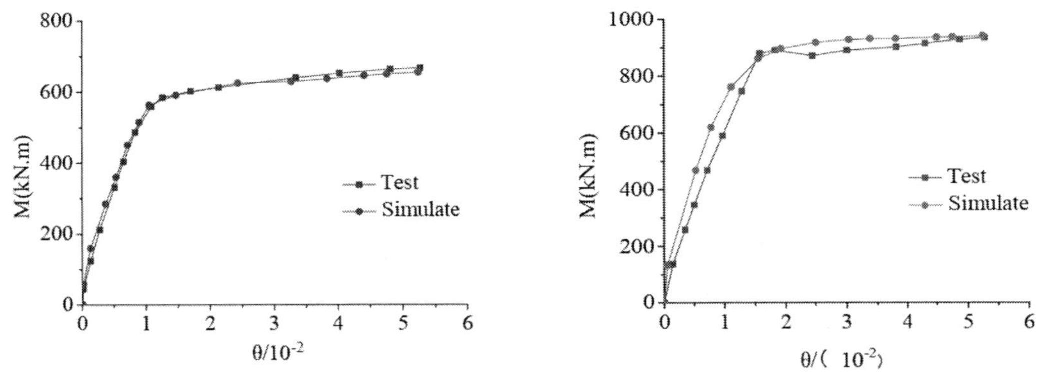

(a) KL-32-25 (b) KL-40-32F

Figure 6. Comparison of specimens bending moment-turn curve.

5.2. Stress and strain cloud analysis

The stress cloud of specimens is shown in figure 7 and the maximum stress of the component can be seen at the root of beam. Tracking changes of cloud shows that the tension zone of concrete specimen quits work before the shear-compression zone, the bearing capacity was mainly provided by the tensile reinforcement. But the shear-compression zone was still supported by concrete and steel. Until the concrete was crushed, and the neutral axis of the beam moved down to 100 mm from the lower edge of the beam, the bearing capacity of the shear-compression zone was provided by the steel bar and the concrete in the hear-compression zone. The above process was consistent with the test phenomena.

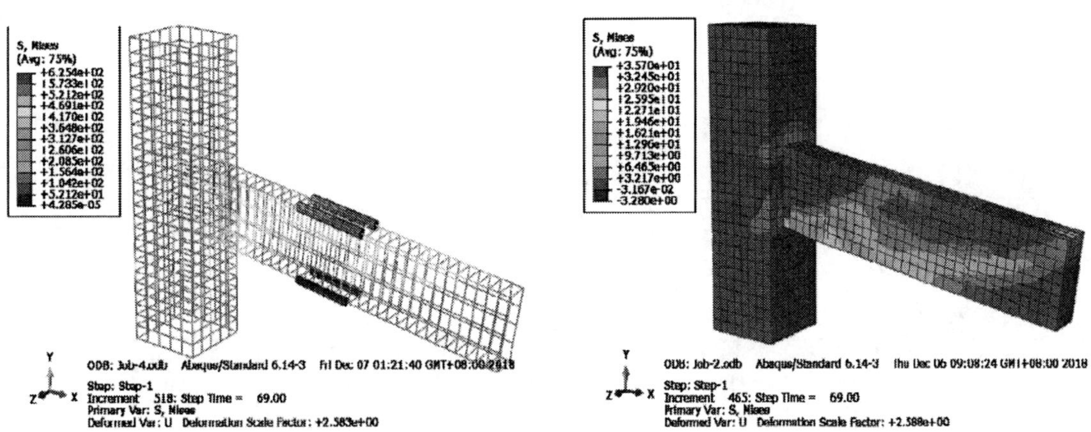

(a) Stress cloud of reinforcement cage (b) Stress cloud of concrete

Figure 7. KL-40-32F stress cloud.

5.3. Fracture contrast

The failure crack state of specimens is depicted in figure 8. The cracking damage of the concrete on the upper side of the beam root is almost complete.

The damage area of the beam body develops from the beam root to the end of the beam with a slope-like reduction. which was basically consistent with the phenomena observed in the experiment.

According to the simulation result, it is shown that the finite element model can accurately simulate the static test study on precast concrete frame beams of large-diameter and high-strength steel bars, and the finite element model of relevant parameters can be used to analyze the model.

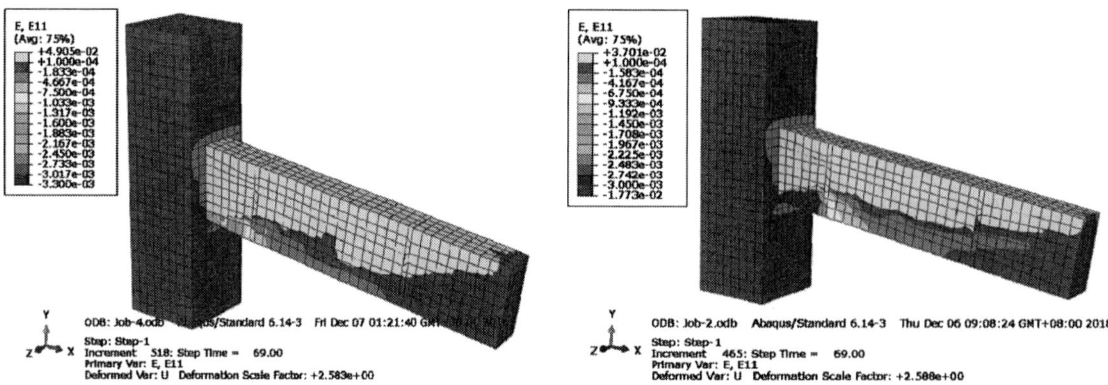

(a) KL-32-25 (b) KL-40-32F

Figure 8. Concrete crack development range of specimens.

6. Finite element comparison analysis

6.1. Basic model parameters

Based on the principle of strength of flexural capacity, the model of the cast-in-place concrete beam with small-diameter steel bars of HRB400 was compared with the experimental model.

The stirrups were Ø10@100 and the basic parameters are shown in figure 9 and table 9.

Table 9. Comparison of measured eigenvalues and simulated analytical eigenvalues.

Group	Specimens	Sectional (mm)	Tensile reinforced	A_s (mm^2)	Compressed reinforced	A_s' (mm^2)
Frist group	KL-32-25	350×700	2 Ø 32	1607.68	2 Ø 25	981.25
	XKL-32	350×700	2 Ø 32	1607.68	2 Ø 25	981.25
	XKL-20(25)	350×700	2 Ø 25+2 Ø 20	1609.25	2 Ø 25	981.25
Second group	KL-40-32F	350×700	2 Ø 40	2512.00	2 Ø 32	1607.68
	XKL-40	350×700	2 Ø 40	2512.00	2 Ø 32	1607.68
	XKL-22(25)	350×700	2 Ø 25+4 Ø 22	2501.01	2 Ø 25+2 Ø 20	1609.25

Note: XKL-32 and XKL-40 are integral casting models with small-diameter reinforcement as test specimens KL-32-25 and KL-40-32F; XKL-20 (25) and XKL-22 (25) are the models of integral pouring of small-diameter reinforcing bars with strong bending capacity of test specimens KL-32-25 and KL-40-32F.

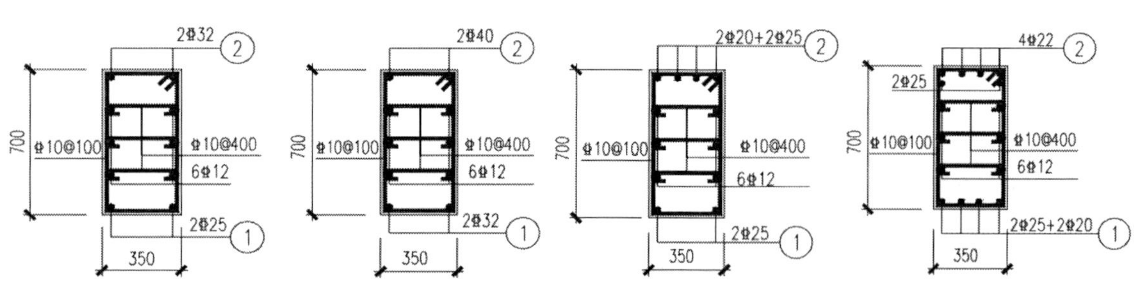

(a) XKL-32 (b) XKL-20 (c) XKL-40 (d) KL-22 (25)

Figure 9. Reinforcement diagram of each model.

6.2. Moment-turn curve comparison

Figure 10 shows that the stiffness of specimens and the development trend of the moment-corner curve was approximately the same. It shows that the stiffness and bearing capacity of the large-diameter steel precast concrete frame beam can meet the requirements of the overall pouring concrete frame beam.

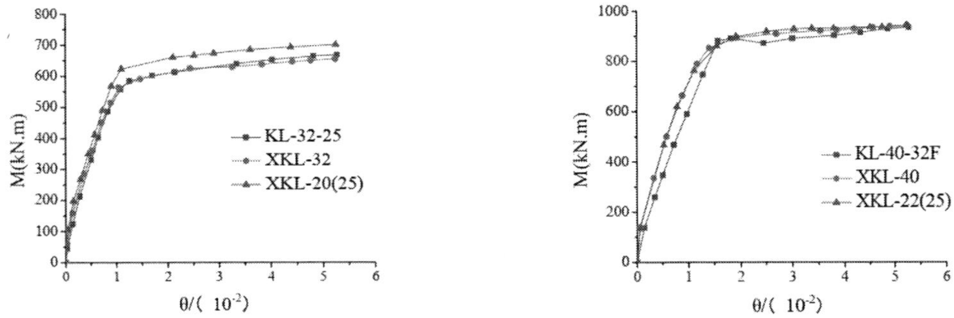

(a) Frist group (b) Second group

Figure 10. Comparison of the second set of load-displacement curves.

6.3. Comparative analysis of bearing capacity

It can be seen that the yield load and ultimate load calculated by the finite element model are compared with the test results in table 10. The median ratio of each group is approximate to 1, which indicates that the bearing capacity of large-diameter precast reinforced concrete frame beams was basically consistent with monolithic concrete frame beams.

Table 10. Comparison of load eigenvalues between test beam and each finite element model.

Group	Specimens	M_y (kN.m)	M_y^t/M_y	M_u (kN.m)	M_u^t/M_u
Frist group	KL-32-25	563	1.00	661	1.00
	XKL-32	563	1.00	655	1.01
	XKL-20(25)	569	0.99	701	0.94
Second group	KL-40-32F	883	1.00	920	1.00
	XKL-40	852	1.04	944	0.97
	XKL-22(25)	861	1.03	941	0.98

Note: M_y is the yield load; M_u is the failure load.

6.4. Comparative analysis of plastic rotation capacity

The compare result between the finite element model and the test are listed in table 11.

Table 11. Calculation of plastic rotation capability of each beam (unit: rad^{-1}).

Group	Specimens	θ_y (10^{-2})	θ_u (10^{-2})	θ_p (10^{-2})
Frist group	KL-32-25	1.06		3.41
	XKL-32	1.05		3.42
	XKL-20(25)	1.12	4.47	3.35
Second group	KL-40-32F	1.59		2.88
	XKL-40	1.40		3.07
	XKL-22(25)	1.55		2.92

The plastic rotation angle in each group is close, which indicates that the plastic rotation capacity of precast concrete frame beams large-diameter meets the requirements of current codes.

7. Conclusion

In this paper, the following conclusions are provided:

- the bearing capacity and plastic rotation capacity of large-diameter and high-strength precast concrete frame beams with steel bar Ø40-Ø32 are superior;
- after the beam-end deflection exceeds 1/25 of the beam span and reaches 1/16, the trend of bearing capacity without obvious falling, which indicates that the ductility of precast concrete members with large-diameter reinforced was excellent;
- for large-diameter reinforced precast concrete frame beams, the yield moment of normal section can be calculated by current concrete codes. The cracking moment and the ultimate moment are quite different from those calculated by the codes. The flexural capacity calculated by current concrete codes has sufficient safety reserve. The plastic rotation capacity coefficients of the specimens with steel bar of Ø32-Ø25 and Ø40-Ø32 are 3.41 and 2.88, respectively.
- There is no crack and mutual dislocation between the prefabricated and cast-in-place joint interface concrete of the test beam, which shown that it was feasible to set up rough surface and shear keys at the assembly interface. The effective transmission of force can be ensured by using double-sided bonded anchor bars and sleeve splicing in the beam-column joint area.
- The anti-crack performance of the large-diameter steel bar specimens decreases with the increase of the concrete cover thickness. It is suggested that add steel mesh or increase the density of transverse stirrups in order to enhance their anti-crack performance in practical engineering applications.

Reference

[1] Yin X W Huang X K Tian C Y Li Y Liu Q and Liu L 2016 Experimental study on flexural performance of RC composite beams with large spacing HRB500 high-strength longitudinal rebar *Journal of Building Structures* **37** pp 291-296

[2] Yin X W Huang X K Tian C Y Li Y Liu L and Liu Q 2016 Experimental study on seismic behavior of RC columns with large spacing and high strength longitudinal reinforcements *Journal of Building Structures* **37** pp 41-47

[3] Wang J Tian C Y Yan F Gao J and Tan Y A 2015 Experimental Study on Bending Performance of Full-tale Superposed RC Beam with Cast-in-place Joint *Building Technology* **31** pp 57-61

[4] Li G Q Huang X K Tian C Y Yin X W and Li Y 2015 Experimental study on shear performance of reinforced concrete superposed beams with overlapping hoops *Journal of Building Structures* **36** pp 286-291

[5] Gao J Tian C Y Hao W and Tan Y A 2015 Experimental study on seismic behavior of precast concrete layered slab and beam to column interior joints *Journal of Building Structures* **36** pp 196-202

[6] Liu L 2015 *Experimental study on seismic performances of precast concrete frame connections with rebars of large diameter and enlarged spacing* (Beijing: Chinese Academy of Sciences)

[7] Ertas O Sevket O and Turan O 2006 Ductile Connections in Precast Concrete Moment Resisting Frames *PCI Journal* **51** pp 66-76

[8] Gao Y F Liang X B and Deng X L 2010 *Metallic materials - Tensile testing - Part 1: Method of test at room temperature* (Beijing: China Architecture & Building Press)

[9] MOHURD 2014 *Standard for test method of concrete structures* (Beijing: China Architecture & Building Press)

[10] Liu Y L Chu D H Liu J J and Zhong X 2015 Comparative study on calculation formula of crack load for reinforced concrete beam *Building Structure* **45** pp 14-17

[11] Gu X L 2011 *Basic Principles of Concrete Structures* (*second edition*) (Shanghai: TongJi Univ. press)

IOP Conf. Series: Materials Science and Engineering **789** (2020) 012069 doi:10.1088/1757-899X/789/1/012069

[12] National Energy Administration 2009 *Design specification for hydraulic concrete structures* (Beijing: China Electric Power Press)
[13] MOHURD 2010 *Code for design of concrete structures* (Beijing: China Architecture & Building Press)

The weaken technique of bending moment for support beam of internal climb tower

C X Zhu[1], N Wang[1*], S Q Gu[1] and W B Ma[1]

[1] School of Mechanical Engineering, Shenyang Jianzhu University, Shenyang, China

[*] i492166613y@gmail.com

Abstract. This paper develops a new analysis method to weaken bending moment at the end of support beam of the inner climbing tower crane. Firstly, the equivalent length of the variable cross-section beam is calculated according to the principle which the angular displacement of the variable cross-section beam is equal to that of the equal cross-section beam under the action of the unit bending moment. Then, the deformation of the variable cross-section can be calculated by using the calculation formula of the equal cross-section beam. Finally, Graphic multiplication is used to solve the stepped variable cross-section statically determinate beam and the function relation between bending moment and inertia moment ratio and variable cross-section length ratio is obtained by MATLAB. It is of great significance to the thin-walled supporting and protection by using the function between the bending moment and the parameter to determine the optimal supporting main beam.

1. Introduction

With the rapid development of science and technology, advanced structural theory, efficient computing technology, new construction techniques and high quality, lightweight construction materials have created favorable conditions for the further breakthrough of building height, and the fierce competition for building height has led to the upgrading of world architecture height ranking. Simultaneously, the requirements for crane support beams have also increased [1].

In recent years, there has been a lot of research on the bearing capacity of the beam itself. Cristian Vulcu [2] taking the joint between the support beam and the load-bearing column as the research object, the joint model was established and the numerical simulation of the beam-column joint strength was carried out by finite element simulation, the influence of factors such as concrete core on the joint strength of the supporting beam is also obtained.

Ayman El-Zohairy [3] tested the fatigue strength of four steel-concrete composite beams under different shear connection degrees and revealed that the range of shear stress is determined by the shear joint between concrete panels. This conclusion is playing an important role in controlling fatigue cracks. Hamdolah Behnam [4] did a set of comparative experiments which carried out on four large external wide beam-column connections of different sizes and different reinforcement ratios. The results showed that the well-reinforced wide-span beam can significantly enhance the shear resistance of the beam-column link surface and the overall connection beam. The torsional strength. Sweedan AMI, Rojob HN [5] discussed the numerical analysis of some composite steel-FRP beams and its performance. Through the combination of finite element analysis and experiment, an analytical solution for the closed moment form of the nominal torque capability and related deflection of the steel-FRP beam system is obtained.

Xiaoqin Liu [6] taking the metal-plated wood truss as the research object, proposed the reliability research of the metal (MP) wooden truss web supported by single or double support and provides a reference for modifying the relevant regulations in the truss design specification. Many scholars are also working on the bearing capacity of supporting beams under different special circumstances.

The fatigue strength of cracked steel beams under different configurations and materials was investigated by Qian-Qian Yu [7]. Mohammad M. Kashani [8] considering the existence of a large number of old bridges in real life, and studied the residual bearing capacity of reinforced concrete bridges under corroded conditions. Through the simulation, Salem O. [9] studied the parameters on load ratio effect on the flexural bending behavior of axially-restrained HSS steel beams under fire condition.

In the optimization and improvement of the beam-based support system, E. Kaya, C [10] studied the effectiveness of two different modification technologies (NSM + NSM; NSM + ETS) in improving the bending strength of the beam support region with experiment. Both of the improvement can make bending resistance of the support beam better. Xie Shengrong [11] considering the stability of the roadway roof as a deep beam, the stability control of the roof of the roadway under the action of monorail is realized, and provides theoretical and technical references for the surrounding rock control under similar conditions. Sabahattin Aykac [12] studied the performance of reinforced concrete beams under monotonic lateral loads, and proposed a post-manufacturing method of RC beams for design and construction errors, as well as overcoming the strength and maintainability of structural systems under changing conditions. Nestor Iwankiw [13] improving steel beam torque connection details and related design criteria to Improved seismic performance. In order to calculate the dynamic resistance of reinforced concrete (RC) components, Foad Mohajeri Nav [14] developed a theoretical method and evaluated in three different stages, the theoretical method can provide a reliable framework for the gradual collapse analysis of reinforced concrete frames.

Fei Peng [15] studied a method of prestressed bending capacity which is based on concrete beams with external ribs. Compared with existing guidelines or models, the method has greater consistency and accuracy. Since the internal climbing tower crane can economize the tower body and reduce the vertical load, etc., it is an important equipment in the construction of high-rise and super high-rise buildings. Because of the climb of the inner climbing tower depends on its support system, it's very practical significance to create a research method of improving the high level support technology for the inner climbing tower crane and becomes a new subject. Internal climbing tower crane generally adopts two support, each is supported with two main girders to the elevator or to the reserved interval on the wall of the room. The tower crane transfers the vertical load to the two main girders via the connection frame, and the horizontal force of the tower at the supporting points is transferred to the side wall with horizontal strut which is supported on the connection points of the connecting frame, the main girder and the side wall. Both ends of the support girder are fixed to the wall mount by the flange seat, bending moment will generate simultaneous in body and the end of the beam, which is transferred to the wall and form a disadvantageous factor. So, it's very realistic to study the weaken technique of bending moment and reduce bending moment at the end of the beam to reduce the stress state of the wall.

2. Research on structure form of ZSL2700 tower support structure

2.1. The form of supporting structures

The internal climbing tower crane supporting system studied in this paper are composed of the following main components. As shown in figure 1: 1. main beam 1, 2. main beam 2, 3. first inclined supporting rod, 4. second inclined supporting rod, 5. third inclined supporting rod, 6. first ear plate, 7. second ear plate, 8. connection base, 9. wall, 10. embedded parts, 11. steel bone column, 12. pin axis, 13. connect frame.

The specific implementation plan is: when the tower crane is in working state, it is required to adopt a supporting structure with two sets of the same structure which above and below, and a spare supporting structure for climbing. The supporting structure comprises a connection frame 13, a main beam 1, a main beam 2, and first, second and third inclined supporting rod 3, 4, 5 which are connected to the main beam 1 in the horizontal direction.

One end of the main beam 1 is connected to the embedded parts 10 in the wall by the connecting base welded on the wall 9, and the other end is connected by connecting parts at the first, second and third inclined supporting rod, 3, 4, and 5, respectively.

The first inclined supporting rod 3 is connected with a pin shaft through the first ear plate 6 and the embedded parts in the wall 9, and the second and third inclined rods 4 and 5 are connected with the pin axis 12 through the second ear plate and the embedded parts in the wall 9. The main beam 2 is connected to the embedded parts 10 in the wall by the connecting base welded on the wall 9.

The side view of climbing tower crane is shown in figure 2 and the installation diagram of the supporting system showed as figure 3 [16].

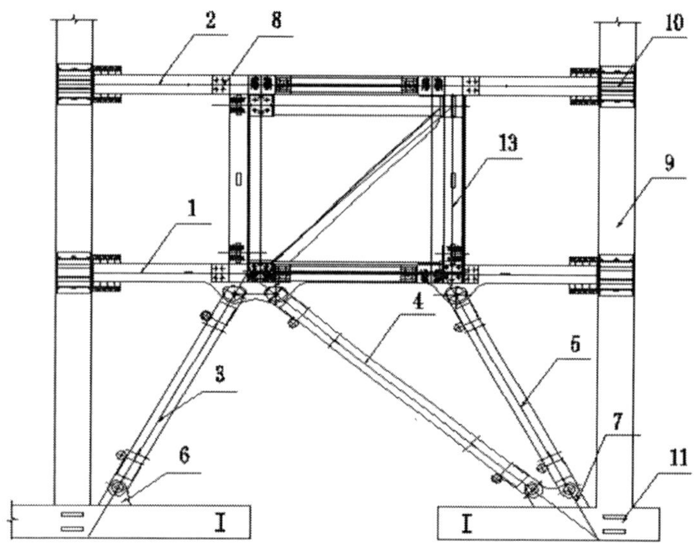

Figure 1. The supporting system of inner.

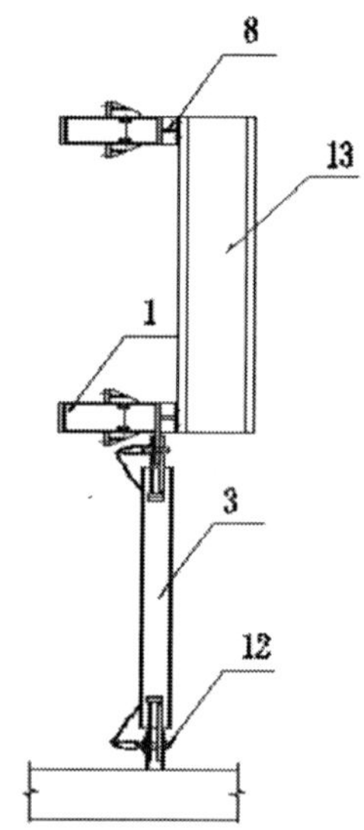

Figure 2. The side view of climbing tower crane.

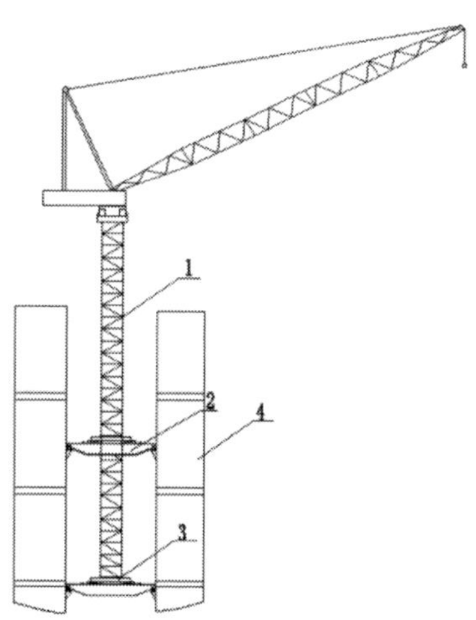

Figure 3. The installation diagram of the supporting system of ZSL2700 tower crane.

After using in independent working conditions, the tower crane needs to be converted into the inner climbing condition. Firstly, the location of the embedded parts shall be welded in a pre-designed connection support which buried in the wall 9 of the building to complete the support welding. Secondly, the main beam 1 is connected with the first, second and third inclined supporting rods by pin shafts, and then the main beam 1 is lifted to form a combination with the supporting rods. After the corresponding position is reached, the main beam 1 is adjusted in a horizontal state, and the main beam 1 is fixed with the welded joint support on the wall 9 by the high strength bolt. Then the first ear plate 6 and the second ear plate 7 are welded. The main beam 2 is hoisted according to the same installation mode, and is less connected to the ear plate than the main beam 1. Finally, the connect frame 13 is installed, which completes the installation of a set of support frames.

2.2. Structural forms of supporting beams
The box section has good structural performance and has been widely used in water conservancy, hydro power, civil building and bridge engineering. Large torsional rigidity and good stability in construction and use is the main advantages of box section. It can effectively resist negative bending moment and adapt to the structure with positive and negative bending moment, such as continuous beam, etc. The combination of load-bearing structure and transmission structure makes the components be forced together and the cross section be efficiency. What's more it adapts to the requirements of modern construction methods.

The supporting beam of this paper is a box-shaped beam of variable cross-section, which is part of the end beam [17, 18]. Its height at the end is smaller than the main beam and shows a linear change from small to large, the height also within a certain range, as shown in figure 4, the thickness of the upper and lower cover plates and the web of the girder is unchanged. In this paper, the equivalent length calculation method of the equivalent variable cross-section beam in the small height beam at the end part is studied under the condition of the angular displacement is produced with the action of the unit bending moment.

Figure 4. Schematic diagram of beam structure change of supporting.

3. Calculation methods and principles

3.1. Calculation of equivalent length
We often encounter the calculation problem of variable cross-section beams deformation in daily engineering design [19 - 21]. For the deformation calculation of the equal cross-section beam, various design manuals and mechanics books can be examined.

However, there is no readily available formula for the deformation calculation of variable cross-section beams. Therefore, it is necessary to convert the variable cross-sections into the equivalent sections of the beam and transformed the main beam into a ladder deformation cross-section beam which can be calculated by using the diagrammatic multiplication method [22], and the desired unknown values can be obtained.

The method used in this paper is to calculate the stiffness of the cross-section beam by using the angle difference between the variable cross-section beam and the equal cross-section beam as the minimum value under the unit torque.

So that the beam of the variable cross-section is transformed into the equal cross-section beam. It provides a simple and practical new method for calculating that deformation of the variable cross-section beam, the principle of which is simple and easy to be mastered by engineering technician and convenient to be used in engineering design.

Figure 5 represents a variable cross-section cantilever beam whose cross-section is continuously changed, the moment of inertia of any cross-section is I(x). In order to transform the variable cross-section beam shown in figure 5 into the same equal cross-section beam with the same stiffness EI as shown in figure 6, under the action of bending moment, a certain angular displacement value θ will be generated under the action of bending moment M.

It can be assumed that there is an imaginary beam whose length value is the same as the end beam, and when it produces an angular displacement under the same bending moment, the length value of which is certain, and this length value is the equivalent calculation length of the variable cross-section beam.

The length of this equivalent end beam is solved below and the moment of inertia of any section on the variable cross-section end beam is set as I(x). The variable cross-section beam shown in figure 5. It produces angular displacement θ under the action of the unit bending moment M = 1. Figure 6 shows a hypothetical end beam with a height h_0 as the height of the end, and it can be obtained a fixed length L_1 produces a rotational angle displacement θ under the end bending moment of M = 1, then L_1 is the equivalent calculation length of the variable cross-section beam portion.

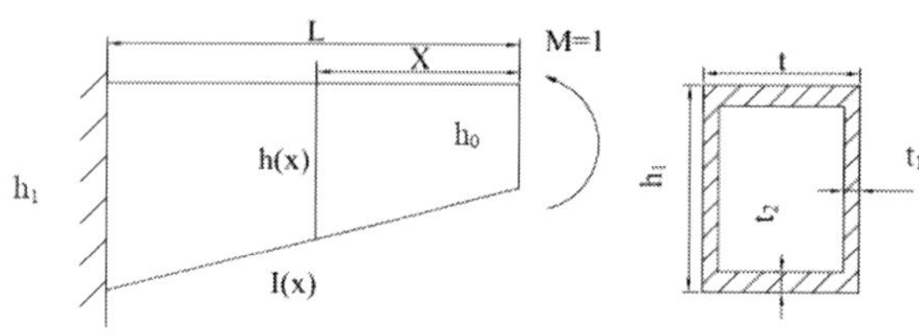

Figure 5. Variable cross-section cantilever.

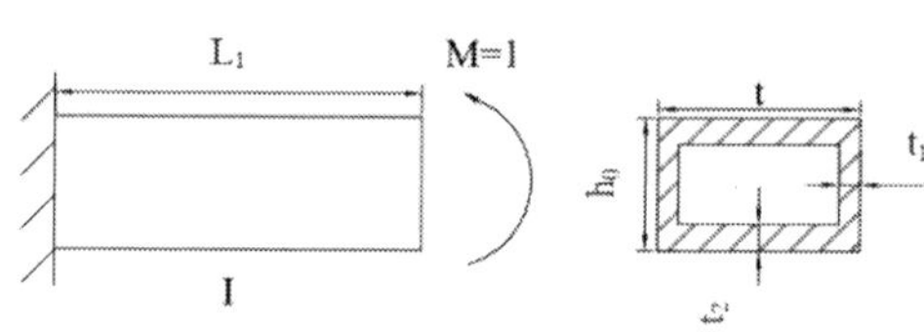

Figure 6. Another cross-sectional cantilever.

The rotation angle of the two cantilever beams:

$$\theta_a = \int_0^L \frac{M}{EI(x)} x \qquad \theta_b = \int_0^{L_1} \frac{M}{EI} dx = \frac{L_1}{EI} \tag{1}$$

Expression of parameter of variable cross-section beam:
Height of cross-section beam:

$$h(x) = h_0 + \frac{x}{L}(h_1 - h_0) \tag{2}$$

Moment of inertia of cross-section beam:

$$I(x) = \frac{th^3(x) - (t - 2t_1)(h(x) - 2t_2)^3}{12}$$

$$= \frac{t[h_0 + \frac{x}{L}(h_1 - h_0)]^3 - (t - 2t_1)[h_0 + \frac{x}{L}(h_1 - h_0) - 2t_2)^3}{12} \tag{3}$$

$$I = \frac{th_0^3 - (t - 2t_1)(h_0 - 2t_2)^3}{12} \tag{4}$$

Expression of rotation angle of variable cross-section fixed beam:

$$\theta_a = \int_0^L \frac{M}{EI(x)}dx = \frac{12}{E}\int_0^L \frac{1}{t[h_0 + \frac{x}{L}(h_1 - h_0)]^3 - (t - 2t_1)[h_0 + \frac{x}{L}(h_1 - h_0) - 2t_2)^3}dx \tag{5}$$

It can be seen from the above formula that the solution to the rotation angle of the variable cross-section fixed beam is a integrate rational function. The direct solution itself is very troublesome. In this way, we can change our minds and simplify θ_α firstly, and it can be simplified into the form as $\int_0^L \frac{1}{ax^3 + bx^2 + cx + d}dx$, among them, *a, b, c, d* are coefficients. Then:

$$a = 2t_1(\frac{h_1 - h_0}{L})^3$$

$$b = 6(\frac{h_1 - h_0}{L})^2[h_0 t_1 + tt_2 - 2t_1 t_2]$$

$$c = 6(\frac{h_1 - h_0}{L})[2tt_2(h_0 - t_2) + t_1(h_0 - 2t_2)^2] \tag{6}$$

$$d = h_0^3 t - (h_0 - 2t_2)^3 (t - 2t_1)$$

The solution of θ_a can be converted to find the roots of a cubic equation.

The root process of the cubic equation.

Find the three roots of equation $ax^3+bx^2+cx+d=0$ (*a, b, c, d* and $a \neq 0$). It's solved by the Shengjin formula [23], a derivation of the Kardan formula. The heavy root discriminant:

$$\begin{aligned} A &= b^2 - 3ac \\ B &= bc - 9ad \\ C &= c^2 - 3bd \end{aligned} \tag{7}$$

General discriminant: $\Delta = B^2 - 4AC$

The root of the equation can be obtained according to the Shengjin formula. Then the general formula of θ_b can be solved by the deformation of the identity equation.

Identity distortion:

$$\frac{1}{ax^3 + bx^2 + cx + d} = \frac{1}{a(x^3 + \frac{b}{a}x^2 + \frac{c}{a}x + \frac{d}{a})} = \frac{1}{a}(\frac{c_1}{x - x_1} + \frac{c_2}{x - x_2} + \frac{c_3}{x - x_3}) \tag{8}$$

Among them:

$$\begin{aligned} &c_1 + c_2 + c_3 = 0 \\ &c_1(x_2 x_3) + c_2(x_1 x_3) + c_3(x_1 x_2) = 1 \\ &c_1(x_2 + x_3) + c_2(x_1 + x_3) + c_3(x_1 + x_2) = 0 \end{aligned} \tag{9}$$

c_1, c_2, c_3 are the setting coefficient.

Integrate:

$$\int_0^L \frac{1}{ax^3 + bx^2 + cx + d}dx = \frac{1}{a}[c_1 \ln(x - x_1) + c_2 \ln(x - x_2) + c_3 \ln(x - x_3)]\Big|_0^L \tag{10}$$

θ_a can be obtain by integrating rational function. The equivalent length L_1 of the cantilever beam can be solved by $\theta_a = \theta_b$.

3.2. Solution of fixed beam bending moment at both ends

Through the solution of the equivalent length above, the equivalent length L_l of the cross section of the variable cross-section beam (as shown in figure 7) can be gained. Now, the trapezoidal cross section at both ends of the variable beam is equivalent to a fixed cross section. (as shown in figure 8). The length of the beam with small height is L_1, and the bending stiffness is EI_1.

The total length of the fixed beam at both ends is L, the bending stiffness of the high beam is, EI_2 a concentration force F is applied perpendicular to the axis of the beam, the position of concentrated force is shown in figure 4, 5. The supporting main beam is simplified into the following form to analyze the end bending moment.

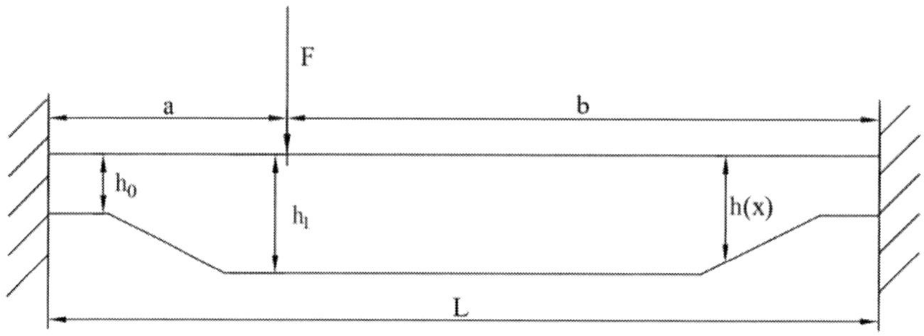

Figure 7. Variable cross-section beam fixed at both ends of the sketch.

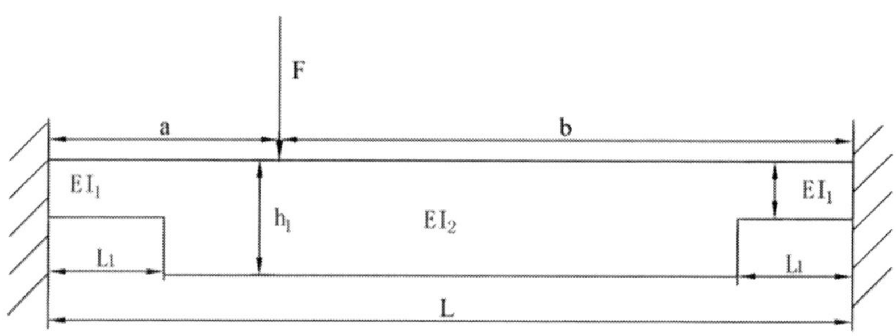

Figure 8. Calculation diagram of stepped cross-section beam fixed at both ends.

As shown in figure 8, this beam is a statically indeterminate structures, and the force method is the first method to be developed in various statically indeterminate calculation methods, which is the most basic method of calculation and being widely used.

To solve the statically indeterminate structure, the redundant connection of the statically indeterminate structure should be removed to obtain the static structure, and the redundant unknown force is used as the basic unknown quantity. The redundant unknown force can be obtained by displacement condition which following the base system should deform in the same way as the original. Then the balance condition can be calculated for the remaining counter force and internal forces.

The simply supported beam is taken as the basic structure, the basic system is shown in figure 9, the redundant unknown force are the beam end bending moment X_1, X_2, and the horizontal counter force X_3.

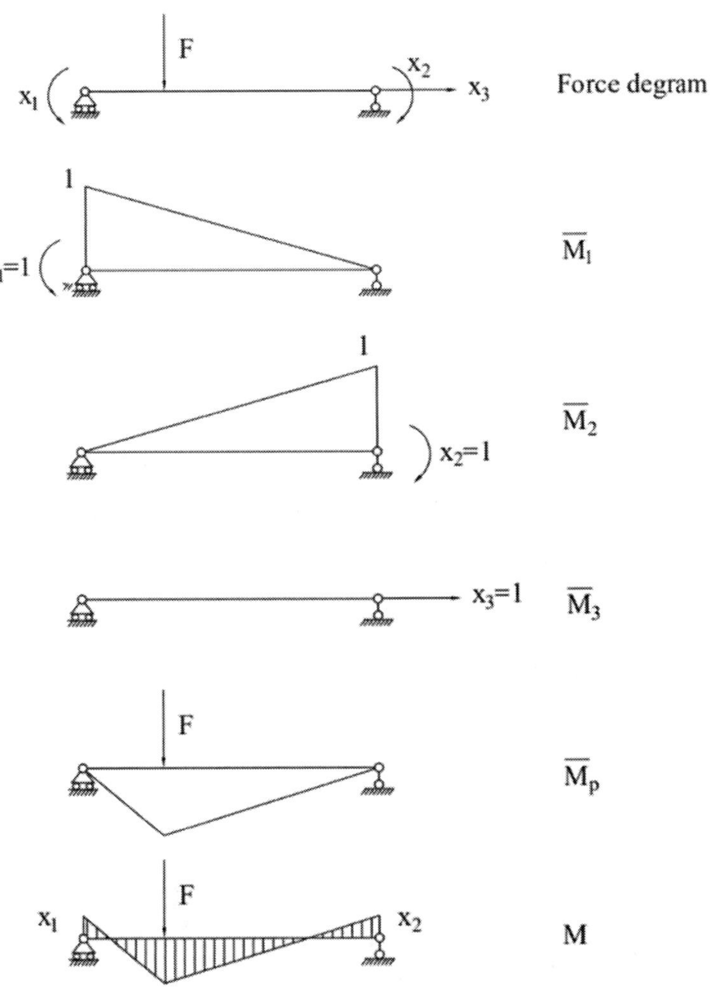

Figure 9. The simplify system of beam fixed at both ends.

It can be assumed that when the unit redundant unknown force $\overline{X}_1=1$, $\overline{X}_2=1$, $\overline{X}_3=1$, and load F are acting on the basic system respectively, the displacement of A point along the direction of X_1 are δ_{11}, δ_{12}, δ_{13}, and Δ_{1p}, the displacement in the direction of X_2, are δ_{21}, δ_{22}, δ_{23} and Δ_{2p}, the displacement along the direction of X_1, are δ_{31}, δ_{32}, δ_{33} and Δ_{3p}, respectively, according to the principle of superposition, the displacement conditions of above can be written as:

$$\begin{aligned} \Delta_1 &= \delta_{11}X_1 + \delta_{12}X_2 + \delta_{13}X_3 + \Delta_{1p} = 0 \\ \Delta_2 &= \delta_{21}X_1 + \delta_{22}X_2 + \delta_{23}X_3 + \Delta_{2p} = 0 \\ \Delta_3 &= \delta_{31}X_1 + \delta_{32}X_2 + \delta_{33}X_3 + \Delta_{3p} = 0 \end{aligned} \tag{11}$$

δ_{11}, δ_{12}, δ_{13} in the system of equations are called principal coefficients or principal displacements, whose values are always positive and not equal to zero.

The rest δ are called the auxiliary coefficient or the auxiliary displacement, and the last term in the formula Δ_p is called the free term. The values of the deputy coefficient and the free term may be positive, negative or zero. According to the reciprocal theorem of displacement, the two deputy coefficients of the symmetric position on both sides of the main slant line δ_{12} and δ_{21} are equal to each other, namely, $\delta_{ij}=\delta_{ji}$. The coefficients and the free items in the system are the displacement under the action of the already force, so it can be solved completely. For the plane structure, the calculation of these displacements can be written as:

$$\delta_{ii}=\sum\int\frac{\overline{M_i^2}}{EI}ds+\sum\int\frac{\overline{F_{Ni}^2}}{EA}ds+\sum\int\frac{\overline{kF_{si}^2}}{GA}ds$$

$$\delta_{ij}=\delta_{ji}=\sum\int\frac{\overline{M_i}\,\overline{M_j}}{EI}ds+\sum\int\frac{\overline{F_{Ni}}\,\overline{F_{Nj}}}{EA}ds+\sum\int\frac{k\overline{F_{si}}\,\overline{F_{sj}}}{GA}ds$$

$$\Delta_{ip}=\sum\int\frac{\overline{M_i}M_p}{EI}ds+\sum\int\frac{\overline{F_{Ni}}F_{Np}}{EA}ds+\sum\int\frac{k\overline{F_{si}}F_{sp}}{GA}ds \quad (12)$$

This is the calculation formula for displacement of the planar rod structure under load. The three items on the right of the upper equation are the calculation formula of bending deformation, axial deformation and shear deformation of the structure for the solved displacement.

The $\overline{M}$ diagrams and M_P diagrams of the basic structure are shown in the figure above. Due to $\overline{M}_3=0$, $\overline{F}_{S3}=0$, and $\overline{F}_{N1}=\overline{F}_{N2}=\overline{F}_{NP}=0$, it can be obtained from the calculation formula of displacement that $\overline{F}_{N1}=\overline{F}_{N2}=\overline{F}_{NP}=0$, $\delta_{23}=\delta_{32}=0$, $\Delta_{3p}=0$. So, the third equation of the above equation becomes

$$\delta_{33}X_3=0 \quad (13)$$

In the calculation of δ_{33}, the influence of bending moment and axial force should be considered at the same time

$$\delta_{33}=\sum\int\frac{\overline{M_3^2}}{EI}ds+\sum\int\frac{\overline{F_{N3}^2}}{EA}ds=0+\frac{L}{EI}=\frac{L}{EI}\neq 0 \quad (14)$$

Then,

$$X_3=0 \quad (15)$$

This indicates that the beams fixed at both ends do not produce a horizontal counter force with a load perpendicular to the beam axis. Therefore, it can be simplified to solve the problem of two redundant unknown forces, and the typical equation becomes:

$$\delta_{11}X_1+\delta_{12}X_2+\Delta_{1p}=0$$

$$\delta_{21}X_1+\delta_{22}X_2+\Delta_{2p}=0 \quad (16)$$

For the beam and the rigid frame, the displacement is mainly caused by the bending moment. The influence of axial force and shear force is too small, so it can be omitted (only the effect bending moment is considered). The coefficients and the free items can be obtained by the diagrammatic multiplication method.

$$\delta_{11}=\sum\int\frac{\overline{M_1^2}}{EI}ds$$

$$\delta_{22}=\sum\int\frac{\overline{M_2^2}}{EI}ds$$

$$\delta_{12} = \delta_{21} = \sum \int \frac{\overline{M_1}\,\overline{M_2}}{EI} ds$$

$$\Delta_{1p} = \sum \int \frac{\overline{M_1} M_p}{EI} ds$$

$$\Delta_{2p} = \sum \int \frac{\overline{M_2} M_p}{EI} ds \qquad (17)$$

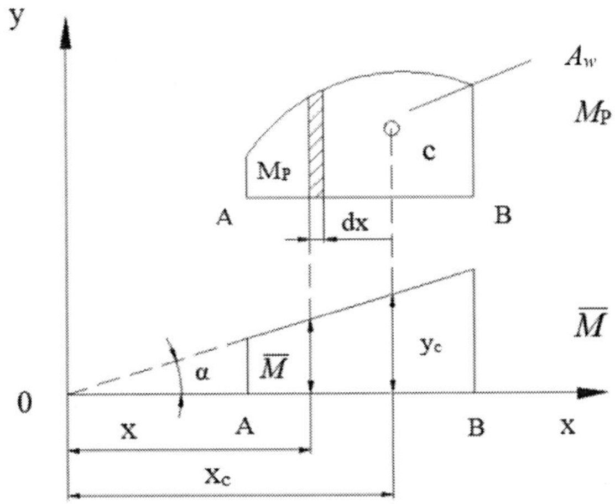

Figure 10. Diagrammatic multiplication method.

As shown in figure 10, it can be assumed that in the two bending moments graph of the AB segment of the equal cross-section rod, the diagram of $\overline{M}$ is a straight line, while the diagram of M_p is in any shape. The rod axis is taken as the X axis, and the intersection of the extension of $\overline{M}$ diagram $\int \overline{M}M_P/EIds$ and the X axis is regarded as the origin, the y axis is set, then the integral form is written as, the dx in the formula is available to be substituted with dx.

The image of $\overline{M}$ varies according to the line, $\overline{M}$ =xtanx, and tanα is constant, so the integral form becomes:

$$\int \frac{\overline{M}M_p}{EI} ds = \frac{\tan\alpha}{EI} \int x M_p dx = \frac{\tan\alpha}{EI} \int x dA_w \qquad (18)$$

In the formula $dA_w = M_P dx$, dA_w is the differential area of the shadow line in the graph of M_P, so $x dA_w$ is the static moment of the differential area to the axis, and $\int x dA_w$ is the static moment of total area of the M_P graph to the Y-axis, according to the theorem on moment of resultant force, it should be equal to the value which the area A_w of the graph M_P times the distance x_c from the C to the Y-axis, substitute $\int x dA_w = A_w x_c$ into the upper formula, it can be obtained :

$$\int \frac{\overline{M}M_p}{EI} ds = \frac{\tan\alpha}{EI} A_w x_c = \frac{A_w y_c}{EI} \qquad (19)$$

In the formula y_c is the vertical coordinate of the $\overline{M}$ graph corresponding to the center C of M_P graph. Since the use of the diagrammatic multiplication method is conditional, the division integration method is performed in order to find the coefficients in the equation. It can be gained:

IOP Conf. Series: Materials Science and Engineering **789** (2020) 012070 doi:10.1088/1757-899X/789/1/012070

$$\delta_{11}=\frac{1}{EI_1}\left[\frac{6L_1L^2-6LL_1^2+4L_1^3}{6L^2}\right]+\frac{1}{EI_2}\left[\frac{2L^3-4L_1^3+6LL_1^2-6L^2L_1}{6L^2}\right]$$

$$\delta_{12}=\frac{1}{EI_1}\left[\frac{6LL_1^2-4L_1^3}{6L^2}\right]+\frac{1}{EI_2}\left[\frac{L^3+4L_1^3-6LL_1^2}{6L^2}\right]$$

$$\Delta_{1p}=-\frac{1}{EI_1}\frac{F}{6L^2}\left[2aL_1^3+3bLL_1^2-2bL_1^3\right]$$

$$-\frac{1}{EI_2}\frac{F}{6L^2}\begin{bmatrix}b(a-L_1)(2LL_1-2L_1^2+bL_1+aL-aL_1+2ab)\\+a(b-L_1)(2b^2+2bL_1+2L_1^2)\end{bmatrix} \tag{20}$$

$$\Delta_{2p}=-\frac{1}{EI_1}\frac{F}{6L^2}\left[2bL_1^3+3aLL_1^2-2aL_1^3\right]$$

$$-\frac{1}{EI_2}\frac{F}{6L^2}\begin{bmatrix}a(b-L_1)(2LL_1-2L_1^2+aL_1+bL-bL_1+2ab)\\+b(a-L_1)(2a^2+2aL_1+2L_1^2)\end{bmatrix}$$

Here $\delta_{11}=\delta_{22}$, $\delta_{12}=\delta_{21}$.

In order to simplify the formula, the stiffness ratio of the step variable cross-section and the ratio of length of the variable cross-section are set as

$$\alpha=\frac{I_1}{I_2}\qquad \beta=\frac{L_1}{L} \tag{21}$$

The range of values of α and β are: $\alpha\in(0,1]$ $\beta\in(0,0.5]$,

The expressions of the coefficients α and β are obtained:

$$\delta_{11}=\delta_{22}=\frac{L}{EI}\left[\frac{1}{\alpha}(\beta-\beta^2+\frac{2}{3}\beta^3)+\frac{1}{3}-\frac{2}{3}\beta^3+\beta^2-\beta\right] \tag{22}$$

$$\delta_{12}=\delta_{21}=\frac{L}{EI}\left[\frac{1}{\alpha}(\beta^2-\frac{2}{3}\beta^3)+\frac{1}{6}+\frac{2}{3}\beta^3-\beta^2\right]$$

In addition, let $\gamma=a/L$

$$\Delta_{1p}=-\frac{L^2}{EI}\frac{F}{6}\begin{bmatrix}\frac{1}{\alpha}\left(4\gamma\beta^3-2\beta^3+3\beta^2-3\gamma\beta^2\right)\\+\left(\gamma-\beta-\gamma^2+\beta\gamma\right)\left(3\beta-2\beta^2-2\beta\gamma+3\gamma-2\gamma^2\right)\\+\left(\gamma-\gamma^2-\beta\gamma\right)\left(2+2\gamma^2-4\gamma+2\beta-2\beta\gamma+2\beta^2\right)\end{bmatrix} \tag{23}$$

$$\Delta_{2p}=-\frac{L^2}{EI}\frac{F}{6}\begin{bmatrix}\frac{1}{\alpha}\left(2\beta^3-4\gamma\beta^3+3\gamma\beta^2\right)\\+\left(\gamma-\gamma^2-\beta\gamma\right)\left(\beta-2\beta^2+2\beta\gamma+\gamma-2\gamma^2+1\right)\\+\left(1-\gamma\right)\left(\gamma-\beta\right)\left(2\gamma^2+2\beta\gamma+2\beta^2\right)\end{bmatrix} \tag{24}$$

Through the above process, the coefficients and the free items in the typical equation of the force method are solved by the diagrammatic multiplication method and represented by two parameters of design. In this way the end bending moment value of the fixed end beam be gained.

3.3. Simulation analysis

It appears that above expressions are complex, so the two-element equation system can be solved by MATLAB software and the expression for the end bending moment can be gotten immediately. Here are end bending moment of the main beam of a few concentrated forces in different points. A few special points of concentrated force are given in the following figure 11, in which the relationship between the end bending moment x and the moment of inertia ratio α and the variable cross-section length ratio β can be seen.

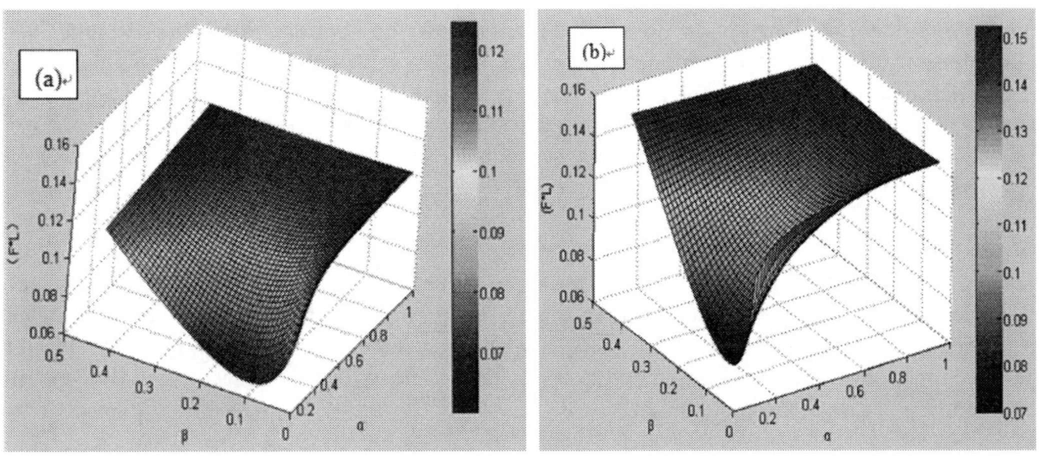

Figure 11. Fixed beam end bending moment diagram.

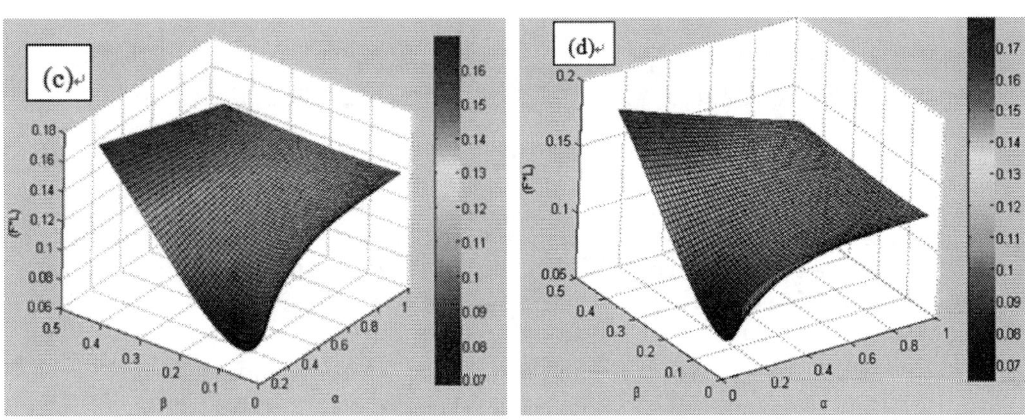

Figure 12. Fixed beam end bending moment diagram.

The bending moment value in the graph refers to the bending moment that is close to the central force F. The results given above will be discussed with a few special positions of concentrated force.

The concentrated force is applied at 1/2L, when the ratio of the length of the variable cross-section is $\beta = 0$ or $\beta = 0.5$, at this time, the stepped variable cross-section beam which our research is the equivalent cross-section beam.

At which moment, the bending moment of the fixed ends achieve the maximum and have the equal value $1/8FL$ when the moment of inertia α is closer to 1, that is, the bending stiffness of the two step beams is not quite different, so that it can be regarded as the equal cross-section beam, and the bending moment value is the most significant 1/8FL. It can be seen from the structure mechanics book, when a=b=1/2L, the bending moment at both ends of AB are 1/8FL. From the figure 12(a) above, it can be seen that the maximum value of the bending moment is also 1/8FL. When $\alpha \in (0, 0.2)$, $\beta \in (0.05, 0.2)$, the end bending moment is minimized.

The concentrated force is applied at 1/4L and 1/5L, as shown in figure 12(b), under the same condition of equal cross-section beam, the influence of the variable cross-section length ratio β on the fixed end bending moment is greater than that of the moment of inertia ratio α. In the area of $\alpha \in (0, 0.2)$and $\beta \in (0.4, 0.5)$, the maximum bending moment occurred in the inner end of the region. The design of the main beam should avoid the selection of parameters in this area.

From this we can conclude that the dark blue area of the bending moment diagram is the main beam design zone, the light blue area is the optional zone, the green and yellow area is the reference selection zone, and the red area is the design rejection zone. In this way, we can obviously know that magnitude of the end bending moment when the main beam is designed by the bending moment diagram [24], and it is important to design the wall body with special requirement for some bearing capacity.

4. Optimization analysis of main beam structure

4.1. Structure of ZSL2700 main beam

Many scholars have done research on the simulation of real tower cranes [25 - 26]. The structure of ZSL2700 supporting main beam designed in the above chapter is analyzed as an example. The main beam is a rectangular box type structure welded with steel plate. The structure is shown in figure 13 below: we apply the bending moment weakening technique which is described above to give end bending moment value of the main beam.

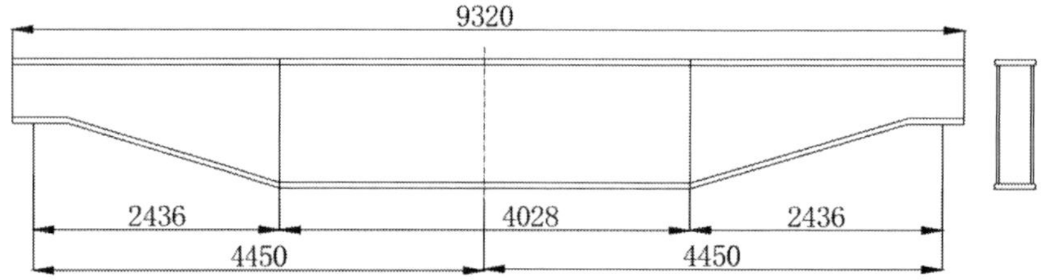

Figure 13. Schematic diagram of ZSL2700 main beam structure.

According to the design dimension parameters of the main beam, and the equivalent method of the variation cross-section of the upper section, the equal cross-section length of the oblique cross-section of ZSL2700 main beam can be obtained: L1=550/2+738=1013since the main beam is designed as L=9320mm, the moment of inertia of main beam in small cross-section height and that in large cross-section height can obtained.

$$I_1 = \frac{1}{12} \times (400 \times 1200^3 - 350 \times 1100^3) = 18779166666 mm^2$$

$$I_2 = \frac{1}{12} \times (400 \times 600^3 - 350 \times 500^3) = 3554166667 mm^2 \tag{25}$$

The moment of inertia ratio and the variable cross section ratio of the two trapezoidal beams:

$$\alpha = \frac{I_1}{I_2} = 0.189 \quad \beta = \frac{L_1}{L} = 0.109 \tag{26}$$

IOP Conf. Series: Materials Science and Engineering **789** (2020) 012070 doi:10.1088/1757-899X/789/1/012070

In the case of the main beam, the concentration force is applied at 1/6L, and the relationship between the bending moment X of the main beam and the ratio of the moment of inertia α and the length ratio of the variable cross-section β is shown in figure 14.

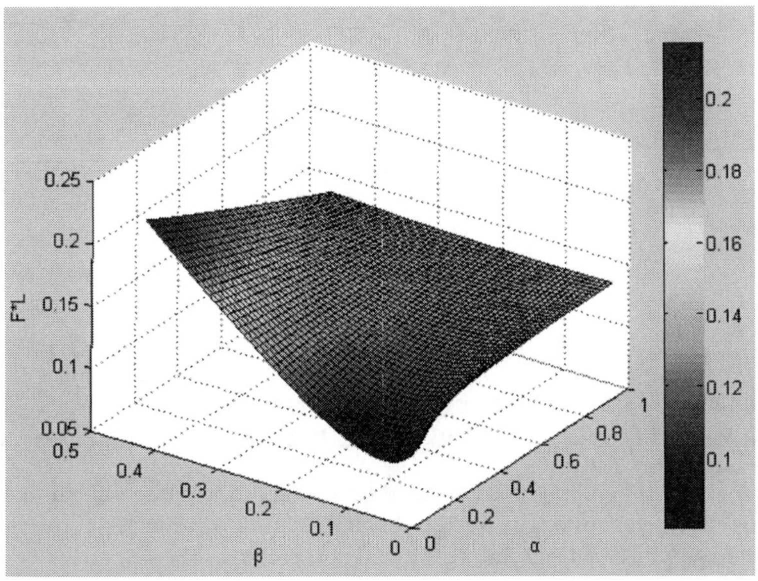

Figure 14. Schematic diagram of ZSL2700 beam end bending moment.

Value of special point, if α=1, β=0, the beam is equal cross-section beam at this time. It can be seen from structural mechanics:

Bending moment of both ends:

$$X_1 = \frac{Fab^2}{L^2} \quad X_2 = \frac{Fa^2b}{L^2} \tag{27}$$

If a=1/6L, b=5/6L, substitute them into the formula:

$$X_1 = \frac{F\frac{1}{6}L(\frac{5}{6}L)^2}{L^2} = 0.12FL$$

$$X_2 = \frac{F\frac{5}{6}L(\frac{1}{6}L)^2}{L^2} = 0.02FL \tag{28}$$

If α=0.189, β=0.109, a=1/6L substitute them into the equation system:

Bending moment of both ends of variable cross section beam:

$$X_1 = 0.0769FL \quad X_2 = 0.0213FL \tag{29}$$

When β=0.109, the relationship between bending moment X_1 and α is given below: at the time α=0.189, the diagram of the relationship between bending moment X_1 and β is shown as follows:

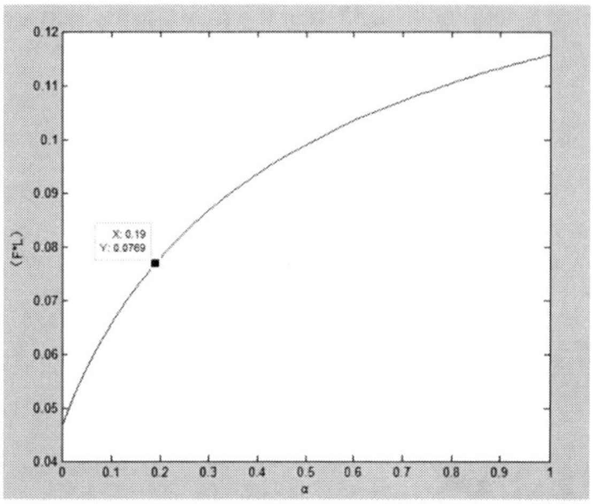

Figure 15. Diagram of ZSL2700 main beam end bending moment and section length ratio.

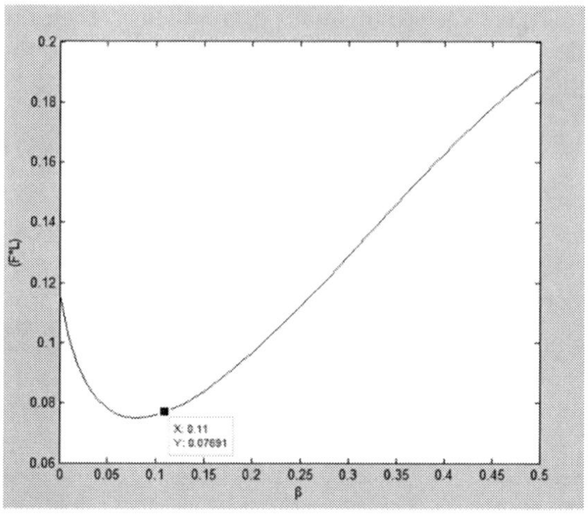

Figure 16. Diagram of ZSL2700 main beam bending moment and inertia moment ratio.

We can see the value of bending moment transmitted by the main beam end to the wall during the solving process as described above. The design of the structural parameters of the main beam is within the allowable range, which is basically the optimized parameter.

5. Conclusions

Based on the study of the supporting structure of ZSL2700 tower, in order to reduce the force transformed from the structure of the main beam to the building, the innovation technology of end bending moment weakening is used in the main beam of the structure, which is proposed for the first time to reduce the bending moment effect of the main beam end to the wall.

There is a certain function relation between the end bending moment and the main beam parameter α, β:$M=f(\alpha, \beta)$, that is, given suitable α, β, the bending moment of the wall can be reduced to a certain extent, which is of great significance to the thin wall support, among them, α is the ratio between the small height of the main beam end and the height of the large end, β is the ratio of the small height of the main beam end to the overall length of the main beam.

The proposition of this technique can ensure the safety of the tower in the thin walls of the building and the research method adopted in the project provides the design idea and theoretical support for the support of tower crane in similar building construction.

References

[1] Zhai R K 2017 *J. Engineering Technology* (China: Southwest Information Center of the Ministry of Science and Technology) **12** 00018-00018

[2] Vulcu C, Stratan A, Ciutina A and Dubina D 2017 *J. Journal of Structural Engineering* (Reston of United States: American Society of Civil Engineers) **143**(5) 04017002 p1-19

[3] El-Zohairy A, Salim H and Saucier A 2019 *J. Journal of Structural Engineering* (Reston of United States: American Society of Civil Engineers) **145**(5) 04019029 p1-11

[4] Behnam H and Kuang J S 2018 *J. Journal of Structural Engineering* (Reston of United States: American Society of Civil Engineers) **144**(4) 14018013 p1-13

[5] Amr S, Hothifa R and Khaled E S 2018 *J. International Journal of Structural Integrity* (Australia: Monash University of Australia) **9(1)** p625-645

[6] Liu X and Lam F 2017 *J. Journal of Structural Engineering* (Reston of United States: American Society of Civil Engineers) **143**(12) 04017171 p1-10

[7] Yu Q Q and Wu Y F 2016 *J. Journal of Composites for Construction* (Reston of United States:

American Society of Civil Engineers) **21**(2) 04016093 p1-12
[8] Mohammad M, Kashani J M and Ebrahim A D 2019 *J. Journal of Bridge Engineering* (Reston of United States: American Society of Civil Engineers) **24**(7) 03119001 p1-16
[9] Salem O 2018 *J. Journal of Structural Fire Engineering* (United States: Multi-Science) **9**(4) p342-360
[10] Kaya C, Kütan S, Sheikh M and İlki A 2017 *J. Journal of Composites for Construction* (Reston of United States: American Society of Civil Engineers) **21**(1) 04016072 p1-16
[11] Xie S, Gao M M and C D D 2017 *J. Journal of China Coal Society* (Beijing of China: China Coal Society) **42**(6) p1420-1428
[12] Sabahattin A, Rohullah J and Eray O 2019 *J. Journal of Structural Engineering* (Reston of United States: American Society of Civil Engineers) **145**(1) 04018223 p1-12
[13] Nestor I 2004 *J. Practice Periodical on Structural Design and Construction* (Reston of United States: American Society of Civil Engineers) **9(2)** p87-92
[14] Mohajeri N F, Abbasnia R and Rashidian O 2016 *J. Journal of Performance of Constructed Facilities* (United States: American Society of Civil Engineers) **30**(5) 04016025 p1-12
[15] Fei Peng S M, Weichen X and Yuan T 2018 *J. Journal of Structural Engineering* (Reston of United States: American Society of Civil Engineers) **44**(12) 14018215 p1-11
[16] 2002 *S. Crane safety manual for operating and maintenance personnel* (Milwaukee WI: Association of Equipment Manufacturers AEM)
[17] Du H 2017 *J.* (Taiyuan of China: Journal of Taiyuan University of Science and Technology) p49-64
[18] Du Y B and Zheng B 2017 *J. Jiangsu Architecture* (Jiangsu of China: Jiangsu Provincial Civil and Architectural Association) **4** p80-82
[19] Bian X W, Yang D D and Zhang F 2017 *J. Construction Machinery* (Beijing of China: Construction Machinery Magazine) **(10)**
[20] Zhang Y G 2011 *J. Construction Machinery Technology & Management* (Beijing: National Construction Urban Construction Machinery Quality Supervision and Inspection Center) **5** p101-104
[21] Yuan W B and Wang L D 2011 *J. Advanced Materials Research* (IET: Institution of Engineering Technology) **v243-249** p419-425
[22] Yang Y F and Zhao Y J 2017 *J. Analysis of continuous collapse resistance of mountain falling layer RC frame structure* (Chongqing University) p1-125
[23] Fan S J 1989 *J. Natural Science* (Hainan of China: Journal of Hainan Normal University Natural Science) **12** **2**(2) p91-98
[24] Blajer W and Kolodziejczyk K 2011 *J. Multibody System Dynamics* (Netherandsn: Springer Netherlands) **25**(2) 131-143
[25] Yang W H, Li Y R, Fang Z F and He K D 2011 *J. Mechanic Automation and Control Engineering* (Beijing: Institute of Electrical and Electronics Engineers) p1660-1663
[26] Meng X Y 2010 *J. China mechanical engineering* (China: ASME International Mechanical Engineering Congress and Exposition) **v13** 105-108

IOP Conf. Series: Materials Science and Engineering **789** (2020) 012071 doi:10.1088/1757-899X/789/1/012071

Research status and development trend of concrete spiral distributor

D F Zou[1,2], W D Yu[1,2*], L T Fan[1,2], D Li[1,2], S Y Zhang[1,2] and L Wang[3]

[1] Mechanical Engineering College of Shenyang Jianzhu University, Shenyang Liaoning, China
[2] National-Local Joint Engineering Laboratory of NC Machining Equipment and Technology of High-Grade Stone, Shenyang Jianzhu University, Shenyang, China
[3] Jinyu Jidong Concrete Co. Ltd. Shenyang Liaoning, China

[*] yuwd@sjzu.edu.cn

Abstract. Concrete distributor is one of the key equipment for the industrial production of concrete product parts. Its performance directly affects the product quality and production efficiency of precast concrete components, which in turn affects the level of architecture industrialization of China. At present, the domestic concrete distributor has a low degree of automation and a simple mechanical structure. There is too much manual intervention in the production. It is impossible to achieve a real automatic distribution. Also, BIM data cannot be shared with the control system, resulting in low production efficiency. This paper summarized these problems which are in the context of big data, internet of things and intelligence, and proposed the upgrade ideas for precast concrete components production equipment. In order to break through the bottleneck that restricts the stability, refinement and intelligent development of precast concrete components distribution. The research and exploration in BIM data analysis mechanical structure optimization and control system upgrade were conducted. Then the overall development of mechanical equipment that meets construction standards of China will be promoted.

1. Introduction

With the proposal and implementation of Made-in-China 2025 [1], the concept of assembly building [2] has become a hot topic in the construction industry. The Guidelines on Developing Assembly Buildings are issued by the State Council, that pointed out that in the future. We will vigorously promote Assembly Buildings which is making up 30% of the new building area. As a modern building technology of energy saving, high efficiency and environmental protection.

Assembly building is gradually accepted by major countries all over the world which puts forward higher requirements for building industrial equipment [3-5].

Concrete distributor is the core equipment of precast concrete components production line in assembly building (see figure 1). Its performance will directly affect the precision and production efficiency of precast concrete components. The new generation of artificial intelligence is represented by big data and in-depth learning, will have a subversive impact on manufacturing industry which will be an inevitable trend for factories to realize information and intelligent production in the future.

The research of concrete distributor in developed countries is early. Users, manufacturers and designers are represented by the famous international enterprises such as German Ebawe Company, Avermann Company, Vollert Company and Soma Company. They are mainly studied the modern technology of scientific use, optimum design and intelligent control of concrete distributor. Ebawe's concrete distributor has a disposable width of 3000 mm. It can extract and transform the drawing information of concrete product parts, transmit it to the distribution system before production, also guide the distribution path planning and realize automatic and accurate distribution. The concrete distributor of Vollert Company has two ways: screw forced discharge and star-shaped axle-slip valve discharge. They can adapt to different concrete consistency and realize manual or automatic control functions. Soma has studied the seamless connection between the production information of precast concrete components and Building Information Modeling (BIM). It can optimize the data processing ability of the main control computer and developed the equipment interface for NC data exchange and feedback.

In China, the designers and the researchers of concrete distributor mainly study the mechanical structure, automatic control and operation efficiency of the distributor. At present, the distributor can be controlled manually according to the size of the components only through manual intervention.

Most domestic concrete distributors can be realized:

- automatic weighing and judging whether feeding is needed;
- the hopper of the distributor can be vertical lifting, adjusting the height of the material according to the thickness of the product. Also adjusting the distance between the material receiving and the feeding trolley to prevent concrete splashing.
- Manual and wireless control mode are adopted to operate the distributor. High degree of automation can also achieve interlocking control with feeding trolley, die table walking and vibration equipment etc, that can ensure that the production process of the distributing station is reasonable and reliable. However, there are many phenomena such as inaccurate blanking quantity of concrete distributor and excessive manual participation, which have not really realized automatic and accurate distribution.

It is obviously that there is still a big gap between domestic concrete distributors and developed countries' automation and intellectualization. While foreign equipment can achieve refined distribution, but it cannot meet the domestic production status (such as the problem of steel bar). At present, the technologies of data exchange between the distributor system and BIM system which are path planning before concrete product parts manufacturing, precise control of distributor and intelligent production of concrete product parts, have become the key technologies that restrict the development of concrete distributor towards modernization, automation, precision and intellectualization.

Figure 1. Concrete distributor.

2. Structure and operation principle of concrete distributor

2.1. Mechanical structure and operation principle of concrete distributor

The concrete distributor includes the distributing system and the walking system. The distributing system includes hopper, scatter device, spiral conveying appliance, gate assembly and bottom plate assembly.

The walking system includes steel structure bracket, trolley walking device and cart walking device. The walking device of the cart is a frame structure which can move forward and backward in the Y direction. It is driven by dual variable frequency motor. Also, its parallel track acts as the wheel tread of the walking device of the trolley.

The walking device of the trolley can move along the X direction on the walking device of the cart. It is generally driven by a single variable-frequency moto. The hydraulic cylinder is vertically mounted lifting. It can realize Z direction movement of the hopper. A weighing sensor is installed on the frame of the hopper and trolley walking so that it can device to monitor the concrete weight in the hopper.

The hopper is a component which is used to undertake concrete and complete the distribution. The dispersing device mixes the concrete in the hopper forcibly. In this way it can ensure the good fluidity of the concrete.

The spiral conveying appliance uses screw to discharge material forcibly. It arranges multiple screw rods in the width of material distribution. Each screw is driven by a separate variable frequency motor. It can be driven and controlled speed independently so that meet the requirements of fine distribution. The structure of the concrete distributor shows in figure 2.

The working process are as follows: concrete is dumped into hopper of distributor through torpedo tank; starting of scatter device; opening of gate corresponding to distributor area according to component shape; starting of screw drive motor to distribute material.

Through the movement of the walking device of the cart and trolley, the concrete is evenly laid on the die table. The switch of the gate and the start and stop of the screw motor completed during the traveling process. After the distributing finished. In order to prevent the residual concrete from solidifying inside the hopper, the bottom plate assembly is opened. Also, the hopper is quickly washed.

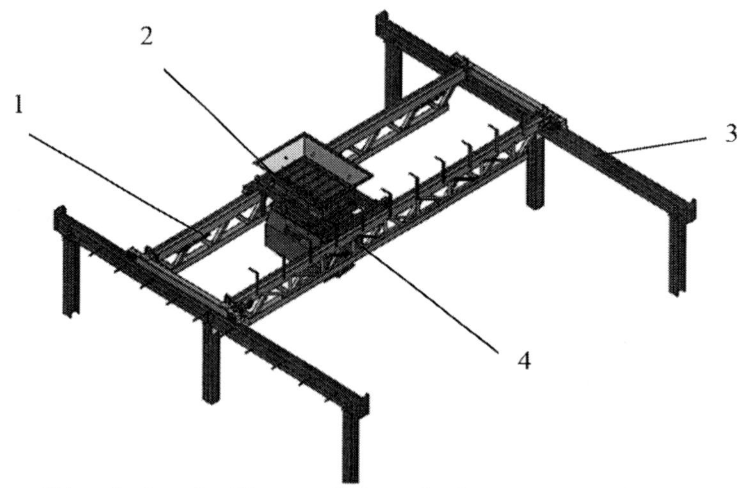

1 - Cart walking device; 2 – Hopper; 3 - Bracket beam; 4 - Trolley walking device

Figure 2. Structure of concrete distributor.

2.2. Control system and operation principle of concrete distributor

The computer is connected with the basic automation controller. The instructions are downloaded to the controller through PLC programming. The controller sends instructions to the variable frequency motors that can drive the cart and the trolley to the starting point of the material distribution.

IOP Conf. Series: Materials Science and Engineering **789** (2020) 012071 doi:10.1088/1757-899X/789/1/012071

Then the controller generates instructions to start the scatter motor and sends instructions to open the gate of the corresponding material distribution area. Finally it drives the variable frequency motor of the corresponding spiral conveying appliance and also distribute concrete material. A weighing sensor is installed at the joint of hopper and trolley that can detect the concrete weight in hopper. The weighing sensor transmits data with the PLC controller. When the concrete content in the distribution equipment is insufficient, the PLC controller will send out alarm instructions to remind the operator.

Distribution control system belongs to primary control (see figure 3). In the production, it is impossible to realize the operation of automatic control equipment. It needs to be controlled by human through external wireless remote control for the distribution. According to the size information of the component, the operator can determine the distribution path and the numbers of openings by himself.

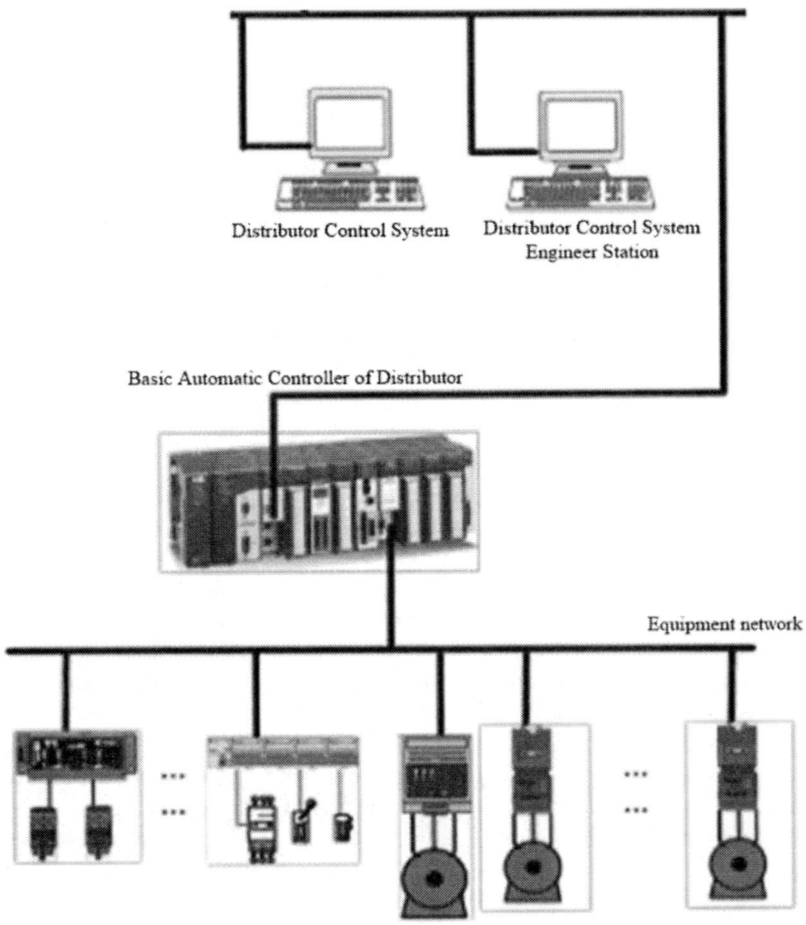

Figure 3. Distribution control system.

3. Research direction and development trend of spiral distributor

As a big construction country, China is moving towards strongly. The extensive construction mode needs to be upgraded and transformed [5]. It needs to be supported by industrialization and informatization as the core. The full life cycle of the product, the digitization of the entire manufacturing process and the modular integration of ICTs will result in a production model of highly flexible, personalized products and services. Under this new production mode, the assembly-type construction equipment will be promoted for information fusion and intelligent production. The spiral distributor is the core equipment in the production line of precast components. The importance of product optimization and upgrading is self-evident.

3.1. Optimization and upgrade of the mechanical structure of the spiral distributor

The wheel-rail structure is adopted in the walking device of the distributor traditionally. The walking wheel driven by the variable frequency motor that travels along the track in the direction of X and Y. Slipping and creeping will inevitably occur. The precise positioning cannot be achieved, resulting in inaccurate blanking point and uneven blanking volume. Adding a rack and pinion mechanism to the cart and the trolley. Using the servo motor to drive. Changing the gear to the active part. Driving the walking wheel to walk on the track. Walking wheels play only support and guide roles, gear and rack drive which can effectively avoid skidding, creeping and other problems. The rotor speed of the servo motor is controlled by the input signal which can respond quickly and has high positional accuracy. It can also ensure the smooth running of the low speed and heavy load for the cart and trolley.

The spiral conveying appliance is driven by a variable frequency motor. The discharge amount is adjusted by frequency conversion speed regulation. But there is a gap between the actual speed and the theoretical speed of the screw, resulting in inaccurate discharge. Especially in the later stage of screw wear. The difference between the theoretical discharge amount and the actual discharge amount is larger, it cannot achieve fine distribution. The relationship between screw blade diameter, screw shaft diameter, pitch, screw speed, power consumption and wear are vague. It needs further study and optimization. Wear-resistant material is needed for the helical blade in the design. It can effectively increase the service life of the screw. The encoder is added to each screw variable-frequency motor so that the motor can achieve closed-loop control in speed regulation. It can keep the actual speed consistent with the theoretical speed. The screw is optimized by discrete element method, computational fluid dynamics, lightweight structure design and experimental method. The power consumption and wear are minimal when the materials output is satisfied then. In the later stage of screw wear, the screw speed is compensated by predicted life model that ensure the blanking amount stable.

3.2. Optimal upgrade of control system of spiral distributor

Usually the control strategy of the distributor control system is simple. There is even no control strategy. It is impossible to achieve intelligent and accurate distribution. Component BIM information cannot be transmitted to the control system through the interface, which requires manual conversion and low efficiency. Lacking of positioning sensors, the distributor cannot locate quickly and accurately. The control system has no prediction model of the materials output and cannot accurately control the materials output; in the later stage of screw wear. There is no rotational speed compensation control model to ensure uniform materials output. Heavy sensors become detection components which do not play a role in the control system. Lacking of production process monitoring and prediction result in the distributor cannot operate reliably and accurately.

On the basis of the lower computer programming controller, the upper computer industrial control system is added to the distribution control system. The programming controller mainly realizes the multi-direction motion control of the distributor, the control of the spiral conveying motor and the control of the gate cylinder. The function of the upper computer reads the graphical data of the components parsed by BIM, divide, process the distributor area in the upper computer software and transmit the distributor area data to the PLC controller. The controller controls the work of the executing mechanism [6-8]. The encoder added to the screw motor and the weighing sensor provided by the distributor are used to collect the signals in the process of distribution. The collected data are fed back to the control system to form a closed-loop PID regulation. The programmable controller is implemented in the form of main program calling subroutine, which is convenient for system debugging and upgrading. A control algorithm is added to the program to ensure the stability of the motor starting and stopping stages.

BIM is an important technology to promote the development of new industrialization and construction industry informatization [9]. Based on BIM technology, the BIM-compatible information parsing interface is designed by using data compatibility extraction, symbol recognition of building components and topological relationship tree building method. The component is identified by vector graph symbols so that the component has its own "ID number". The component data parameters are transmitted to the distribution control system through the corresponding interface. The optimization theory is adopted in the system in order to the generation and calculation of the distribution path. It can realize the automatic distribution path planning and guide the automatic distribution of the distributor. The visual control interface of the distributor is shown in figure 4.

For the automated intelligent distribution system, the detection module is indispensable. The component thickness information and positioning information of the distributor need to be extracted and fed back through the detection technology. In the process of distributing material, the pre-calibrated position of the distributor can be measured by laser displacement sensor. The thickness information of the component can be measured by ranging method. The information can be fed back to the control system. Through the feedback information, the position compensation and quality compensation can be carried out by the control system. It ensures the distributor can deliver material accurately.

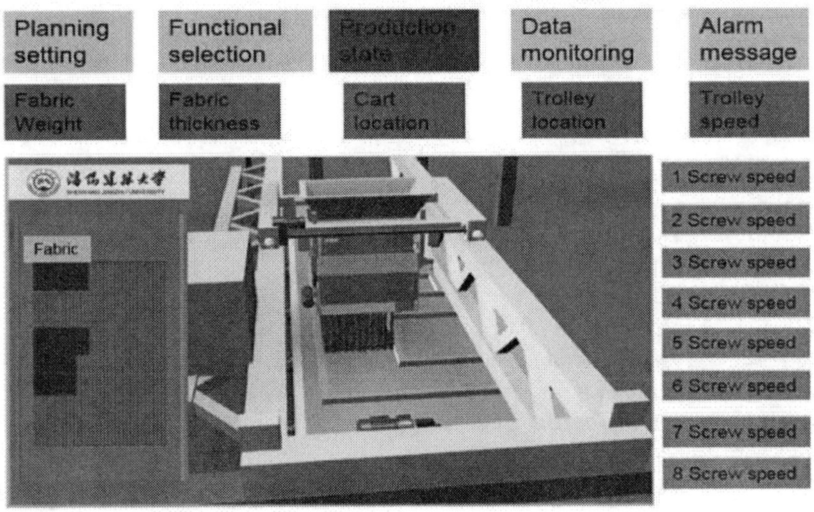

Figure 4. Visual control interface of the distributor.

Before the distribution production, the concrete distribution machine is driven by the distribution cart and the distribution trolley. It locates the pre-production starting such as the end point of the edge touching. According to the BIM analysis of the concrete component information to calculate the required concrete quality, distribution path, spiral speed and other production process information for distribution production. The distribution trolley is equipped with the distribution machine. According to the calculated distribution path, it moves in the corresponding position in the side touch on the distribution crane beam. At the same time, the screw of each distribution port rotates at a certain speed. It pushes out the concrete in the hopper. Finally pour it into the distribution area surrounded by the side formwork of the bottom formwork tray. When the distribution trolley moves to the side mold of the door and window hole, the corresponding screw stops rotating. The prediction quality of concrete materials transported by screw. The quality is detected by the weighing sensor of the distributor. The quality of materials is analyzed by BIM information are applied to the distribution control system that achieve closed-loop accurate discharge.

4. Conclusion

With the rapid development of computer technology and artificial intelligence technology, driven by the national manufacturing 2025 policy. The era of intelligent production is on the way. Digital, automated, refined and intelligent distribution technology can liberate the labor force to the greatest extent, realize automatic, accurate and stable operation of the distribution machine. Also, it improves production efficiency comprehensively.

Based on the optimization and upgrading of the mechanical structure of the distributor. To meet the requirements of the automation control system, improve the accuracy and stability of the distribution and upgrade the distribution control system. The control system can plan the distribution path according to the BIM analytical data and carry out automated production. The application of detection and positioning technology can guide the automation production by real-time feedback signals. It also ensures the accuracy of the distribution.

Acknowledgements
This research is supported by National Key Research and Development Projects (2017YFC0704003). The natural science foundation project supported by Liaoning province (2018011512-301). Science and technology project supported by The Education Department of Liaoning (LJZ2017031). Research and Innovation Cultivation Project of Shenyang Jianzhu University (CXPY2017005). General Project of Shenyang Jianzhu University (2017016).

References

[1] Gao Q S and Li T 2018 *J. Industrial Technol and Economy* **10** pp 59-66
[2] Huang X K, Tian C Y et al 2018 *J. Architectural Science* **9** pp 50-55
[3] Qi B K and Zhang Y 2015 *J. Journal of Shenyang Jianzhu Univ* (Social Science Edition) **17** pp 156-159
[4] Zhang H 2017 *J. Concrete World* **04** pp 24-30
[5] Li B 2014 *J. China Science and Technol Information* **07** pp 114-115
[6] Cao Y G 2016 *J. Exploration Sci* **07** pp 35-39
[7] Zhang Z P 2017 *J. Internal combustion engine and accessories* **04** pp 123
[8] Dong Y G Xie W et al 2016 *J. Industrial technol* **11** pp 125
[9] Yu L F Zhang J C 2015 *J. Journal of Civil Eng and Management* **32** pp 73-78

AUTHOR INDEX

Institute of Physics
Dirac House, Temple Back
Bristol BS1 6BE UK

ISSN: 1757-8981
ISBN 978-1-7138-3504-2

9 781713 835042

CURRAN ASSOCIATES INC.
proceedings
.com